T0180948

Lecture Notes in Computer Science 11507

Commenced Publication in 1973
Founding and Former Series Editors:
Gerhard Goos, Juris Hartmanis, and Jan van Leeuwen

More information about this series at http://www.springer.com/series/7407

Ignacio Rojas · Gonzalo Joya ·
Andreu Catala (Eds.)

Advances in Computational Intelligence

15th International Work-Conference
on Artificial Neural Networks, IWANN 2019
Gran Canaria, Spain, June 12–14, 2019
Proceedings, Part II

 Springer

Editors
Ignacio Rojas 🆔
University of Granada
Granada, Spain

Gonzalo Joya
University of Malaga
Malaga, Spain

Andreu Catala
Polytechnic University of Catalonia
Barcelona, Spain

ISSN 0302-9743   ISSN 1611-3349 (electronic)
Lecture Notes in Computer Science
ISBN 978-3-030-20517-1   ISBN 978-3-030-20518-8 (eBook)
https://doi.org/10.1007/978-3-030-20518-8

LNCS Sublibrary: SL1 – Theoretical Computer Science and General Issues

This Springer imprint is published by the registered company Springer Nature Switzerland AG
The registered company address is: Gewerbestrasse 11, 6330 Cham, Switzerland

Preface

We are proud to present the set of final accepted papers for the 13th edition of the IWANN conference—the International Work-Conference on Artificial Neural Networks—held in Gran Canaria, (Spain) during June 12–14, 2019.

IWANN is a biennial conference that seeks to provide a discussion forum for scientists, engineers, educators, and students about the latest ideas and realizations in the foundations, theory, models, and applications of hybrid systems inspired by nature (neural networks, fuzzy logic and evolutionary systems) as well as in emerging areas related to these topics. As in previous editions of IWANN, it also aims to create a friendly environment that could lead to the establishment of scientific collaborations and exchanges among attendees. The proceedings will include all the communications presented at the conference. A publication of an extended version of selected papers in a special issue of several specialized journals (such as *Neural Computing and Applications, PLOS One,* and *Neural Proccesing Letters*) is also foreseen.

Since the first edition in Granada (LNCS 540, 1991), the conference has evolved and matured. The list of topics in the successive Call for Papers has also evolved, resulting in the following list for the present edition:

1. **Mathematical and theoretical methods in computational intelligence**. Mathematics for neural networks. RBF structures. Self-organizing networks and methods. Support vector machines and kernel methods. Fuzzy logic. Evolutionary and genetic algorithms.
2. **Neurocomputational formulations**. Single-neuron modeling. Perceptual modeling. System-level neural modeling. Spiking neurons. Models of biological learning.
3. **Learning and adaptation**. Adaptive systems. Imitation learning. Reconfigurable systems. Supervised, non-supervised, reinforcement, and statistical algorithms.
4. **Emulation of cognitive functions**. Decision-making. Multi-agent systems. Sensor mesh. Natural language. Pattern recognition. Perceptual and motor functions (visual, auditory, tactile, virtual reality, etc.). Robotics. Planning motor control.
5. **Bio-inspired systems and neuro-engineering**. Embedded intelligent systems. Evolvable computing. Evolving hardware. Microelectronics for neural, fuzzy and bioinspired systems. Neural prostheses. Retinomorphic systems. Brain–computer interfaces (BCI) Nanosystems. Nanocognitive systems.
6. **Advanced topics in computational intelligence**. Intelligent networks. Knowledge-intensive problem-solving techniques. Multi-sensor data fusion using computational intelligence. Search and meta-heuristics. Soft computing. Neuro-fuzzy systems. Neuro-evolutionary systems. Neuro-swarm. Hybridization with novel computing paradigms.
7. **Applications**. Expert systems. Image and signal processing. Ambient intelligence. Biomimetic applications. System identification, process control, and manufacturing. Computational biology and bioinformatics. Parallel and distributed computing. Human-computer interaction, Internet modeling, communication and networking.

Intelligent systems in education. Human–robot interaction. Multi-agent systems. Time series analysis and prediction. Data mining and knowledge discovery.

At the end of the submission process, and after a careful peer review and evaluation process (each submission was reviewed by at least 2, and on average 2.9, Program Committee members or additional reviewers), 150 papers were accepted for oral or poster presentation, according to the reviewers' recommendations and the authors' preferences.

In this edition of IWANN 2019, a workshop entitled "Artificial Intelligence in Nanophotonics" was presented, organized by Dr. Nikolay Zheludev, University of Southampton, UK, and NTU Singapore and Dr. Cesare Soci, NTU, Singapore.

During IWANN 2019, several special sessions were held. Special sessions are a very useful tool for complementing the regular program with new and emerging topics of particular interest for the participating community. Special sessions that emphasize multi-disciplinary and transversal aspects, as well as cutting-edge topics, are especially encouraged and welcome, and in this edition of IWANN 2019 comprised the following:

- **SS01: Artificial Neural Network for Biomedical Image Processing**
  Organized by: Dr. Yu-Dong Zhan
- **SS02: Deep Learning Models in Health Care and Biomedicine**
  Organized by: Dr. Leonardo Franco, Dr. Ruxandra Stoean and Dr. Francisco Veredas
- **SS03: Deep Learning Beyond Convolution**
  Organized by: Dr. Miguel Atencia
- **SS04: Machine Learning in Vision and Robotics**
  Organized by: Dr. José García-Rodríguez, Dr. Enrique Domínguez and Dr. Ramón Moreno
- **SS05: Data-Driven Intelligent Transportation Systems**
  Organized by: Dr. Ignacio J. Turías Domínguez, Dr. David Elizondo and Dr. Francisco Ortega Zamorano
- **SS06: Software Testing and Intelligent Systems**
  Organized by: Dr. Juan Boubeta, Dr. Pablo C. Cañizares and Dr. Gregorio Díaz
- **SS07: Deep Learning and Natural Language Processing**
  Organized by: Dr. Leonor Becerra-Bonache, Dr. M. Dolores Jiménez-López and Dr. Benoit Favre
- **SS08: Random-Weights Neural Networks**
  Organized by: Dr. Claudio Gallicchio
- **SS09: New and Future Tendencies in Brain–Computer Interface Systems**
  Organized by: Dr. Ricardo Ron and Dr. Ivan Volosyak
- **SS10: Human Activity Recognition**
  Organized by: Dr.-Ing. habil. Matthias Pätzold
- **SS11: Computational Intelligence Methods for Time Series**
  Organized by: Dr. Héctor Pomares
- **SS12: Advanced Methods for Personalized/Precision Medicine**
  Organized by: Dr. Luis Javier Herrera and Dr. Fernando Rojas

- **SS13: Exploring Document Information to Improve Neural Summarization Models**
  Organized by: Dr. Luigi Di Caro
- **SS15: Machine Learning in Weather Observation and Forecasting**
  Organized by: Dr. Juan Luis Navarro-Mesa, Dr. Antonio Ravelo-García and Dr. Carmen Paz Suárez Araujo

In this edition of IWANN, we were honored to have the presence of the following invited speakers:

1. Dr. Nuria Oliver, Director of Research in Data Science, Vodafone Chief Data Scientist, Data-Pop Alliance
2. Dr. Aureli Soria-Frisch, Director of Neuroscience, Starlab Consulting Division
3. Dr. Jose C. Principe, Distinguished Professor ECE, Eckis Professor of ECE, Director Computational NeuroEngineering Lab, University of Florida
4. Dr. Marin Soljacic, Professor of Physics at MIT

It is important to note that for the sake of consistency and readability of the book the presented papers are not organized as they were presented in the IWANN 2019 sessions, but classified under 22 chapters. The papers are organized in two volumes arranged basically following the topics list included in the call for papers. The first volume (LNCS 11506), entitled *Advances in Computational Intelligence. IWANN 2019. Part I*, is divided into ten main parts and includes contributions on:

1. Machine learning in weather observation and forecasting
2. Computational intelligence methods for time series
3. Human activity recognition
4. New and future tendencies in brain–computer interface systems
5. Random-weights neural networks
6. Pattern recognition
7. Deep learning and natural language processing
8. Software testing and intelligent systems
9. Data-driven intelligent transportation systems
10. Deep learning models in health care and biomedicine

In the second volume (LNCS 11507), entitled *Advances in Computational Intelligence. IWANN 2019. Part II*, is divided into 12 main parts and includes contributions on:

1. Deep learning beyond convolution
2. Artificial neural network for biomedical image processing
3. Machine learning in vision and robotics
4. System identification, process control, and manufacturing
5. Image and signal processing
6. Soft computing
7. Mathematics for neural networks
8. Internet modeling, communication, and networking
9. Expert systems

10. Evolutionary and genetic algorithms
11. Advances in computational intelligence
12. Computational biology and bioinformatics

The 14th edition of the IWANN conference was organized by the University of Granada, University of Malaga, and Polytechnical University of Catalonia. We wish to thank to the University of Gran Canaria for their support and grants.

We would also like to express our gratitude to the members of the different committees for their support, collaboration, and good work. We especially thank our honorary chairs (Prof. Joan Cabestany, Prof. Alberto Prieto and Prof. Francisco Sandoval), the technical program chairs (Prof. Miguel Atencia, Prof. Francisco García-Lagos, Prof. Luis Javier Herrera and Prof. Fernando Rojas), the local Organizing Committee (Prof. Domingo J. Benítez Díaz, Prof. Carmen Paz Suárez Araujo and Prof. Juan Luis Navarro Mesa), the Program Committee, the reviewers, invited speaker, and special session organizers. Finally, we want to thank Springer and especially Alfred Hofmann and Anna Kramer for their continuous support and cooperation.

June 2019   Ignacio Rojas
                                             Gonzalo Joya
                                             Andreu Catala

Organization

Program Committee

Kouzou Abdellah	Djelfa University, Algeria
Vanessa Aguiar-Pulido	Weill Cornell Medicine, Cornell University, USA
Arnulfo Alanis Garza	Instituto Tecnologico de Tijuana, Mexico
Ali Alkaya	Marmara University, Turkey
Amparo Alonso-Betanzos	University of A Coruña, Spain
Jhon Edgar Amaya	University of Tachira, Venezuela
Gabriela Andrejkova	Slovakia
Davide Anguita	University of Genoa, Italy
Javier Antich Tobaruela	University of the Balearic Islands, Spain
Alfonso Ariza	University of Málaga, Spain
Angelo Arleo	CNRS - University Pierre and Marie Curie Paris VI, France
Corneliu Arsene	SC IPA SA, Romania
Miguel Atencia	University of Málaga, Spain
Jorge Azorín-López	University of Alicante, Spain
Antonio Bahamonde	University of Oviedo at Gijón, Asturias, Spain
Halima Bahi	University of Annaba, Algeria
Javier Bajo	Polytechnic University of Madrid, Spain
Juan Pedro Bandera Rubio	ISIS Group, University of Malaga, Spain
Oresti Banos	University of Granada, Spain
Bruno Baruque	University of Burgos, Spain
Leonor Becerra Bonache	Laboratoire Hubert Curien, France
Lluís Belanche	Universitat Politècnica de Catalunya, Spain
Sergio Bermejo	Universitat Politècnica de Catalunya, Spain
Francisco Bonin-Font	University of the Balearic Islands, Spain
Juan Boubeta-Puig	University of Cádiz, Spain
Antoni Burguera	Universitat de les Illes Balears, Spain
Pablo C. Cañizares	Complutense University of Madrid, Spain
Tomasa Calvo	University of Alcala, Spain
Azahara Camacho	Carbures Defense, Spain
David Camacho	Autonomous University of Madrid, Spain
Francesco Camastra	University of Naples Parthenope, Italy
Hoang-Long Cao	Vrije Universiteit Brussel, Belgium
Carlos Carrascosa	GTI-IA DSIC University Politecnica de Valencia, Spain
Pedro Castillo	University of Granada, Spain
Andreu Catala	Universitat Politècnica de Catalunya, Spain
Ana Cavalli	Institut Mines-Telecom/Telecom SudParis, France

Miguel Cazorla	University of Alicante, Spain
Wei Chen	Fudan University, China
Valentina Colla	Scuola Superiore S. Anna, Italy
Francesco Corona	Aalto University, Finland
Ulises Cortés	Universitat Politècnica de Catalunya, Spain
Marie Cottrell	SAMM Université Paris 1 Panthéon-Sorbonne, France
Raúl Cruz-Barbosa	University Tecnológica de la Mixteca, Mexico
Erzsébet Csuhaj-Varjú	Eötvös Loránd University, Hungary
Daniela Danciu	University of Craiova, Romania
Angel Pascual Del Pobil	University of Jaume I, Spain
Enrique Dominguez	University of Malaga, Spain
Richard Duro	Universidade da Coruna, Spain
Gregorio Díaz	University of Castilla - La Mancha, Spain
David Elizondo	Centre for Computational Intelligence, UK
Enrique Fernandez-Blanco	University of A Coruña, Spain
Carlos Fernandez-Lozano	University of A Coruña, Spain
Jose Manuel Ferrandez	P. University of Cartagena, Spain
Oscar Fontenla-Romero	University of A Coruña, Spain
Leonardo Franco	University of Málaga, Spain
Claudio Gallicchio	University of Pisa, Italy
Esther Garcia Garaluz	Eneso Tecnología de Adaptación SL, Spain
Francisco Garcia-Lagos	University of Malaga, Spain
Jose Garcia-Rodriguez	University of Alicante, Spain
Pablo García Sánchez	University of Granada, Spain
Rodolfo García-Bermúdez	University Técnica de Manabí, Ecuador
Angelo Genovese	University of Milan, Italy
Peter Gloesekoetter	Münster University of Applied Sciences, Germany
Juan Gomez Romero	University of Granada, Spain
Karl Goser	Technical University Dortmund, Germany
Manuel Graña	UPV/EHU, Spain
Jose Guerrero	Universitat de les Illes Balears, Spain
Bertha Guijarro-Berdiñas	University of A Coruña, Spain
Nicolás Guil Mata	University of Málaga, Spain
Alberto Guillen	University of Granada, Spain
Pedro Antonio Gutierrez	University of Cordoba, Spain
F. Luis Gutiérrez Vela	University of Granada, Spain
Marco A. Gómez-Martín	Complutense University of Madrid, Spain
Luis Herrera	University of Granada, Spain
Cesar Hervas	University of Cordoba, Spain
Mercedes Hidalgo-Herrero	Complutense University of Madrid, Spain
Wei-Chiang Hong	Jiangsu Normal University, China
Petr Hurtik	IRAFM, Czechia
Jose M. Jerez	University of Málaga, Spain
M. Dolores Jimenez-Lopez	Rovira i Virgili University, Spain
Juan Luis Jiménez Laredo	Université du Havre Normandie, France
Gonzalo Joya	University of Málaga, Spain

Vladimir Rasvan	University of Craiova, Romania
Antonio Ravelo-García	University of Las Palmas de Gran Canaria, Spain
Ismael Rodriguez	Complutense University of Madrid, Spain
Fernando Rojas	University of Granada, Spain
Ignacio Rojas	University of Granada, Spain
Ricardo Ron-Angevin	University of Málaga, Spain
Francesc Rossello	University of the Balearic Islands, Spain
Fabrice Rossi	SAMM - Université Paris 1, France
Peter M. Roth	Graz University of Technology, Austria
Fernando Rubio	Complutense University of Madrid, Spain
Ulrich Rueckert	Bielefeld University, Germany
Addisson Salazar	Universitat Politècnica de València, Spain
Francisco Sandoval	University of Málaga, Spain
Jorge Santos	ISEP, Portugal
Jose Santos	University of A Coruña, Spain
Jose A. Seoane	Stanford Cancer Institute, Stanford University, USA
Cesare Soci	Nanyang Technological University, Singapore
Jordi Solé-Casals	University of Vic - Central University of Catalonia, Spain
Catalin Stoean	University of Craiova, Romania
Ruxandra Stoean	University of Craiova, Romania
Carmen Paz Suárez-Araujo	University Las Palmas de Gran Canaria, Spain
Peter Szolgay	Pazmany Peter Catholic University, Hungary
Claude Touzet	Aix-Marseille University, France
Ignacio Turias	University of Cádiz, Spain
Daniel Urda	University of Cádiz, Spain
Olga Valenzuela	University of Granada, Spain
Oscar Valero	University of las Islas Baleares, Spain
Francisco Velasco-Alvarez	University of Málaga, Spain
Marley Vellasco	Pontifical Catholic University of Rio de Janeiro, Brazil
Alfredo Vellido	Universitat Politècnica de Catalunya, Spain
Francisco J. Veredas	University of Málaga, Spain
Ivan Volosyak	Rhine-Waal University of Applied Sciences, Germany
Yudong Zhang	Nanjing Normal University, China
Nikolay I. Zheludev	University of Southampton, UK
Igor Zubrycki	Lodz University of Technology, Poland
Juan Antonio Álvarez García	University of Seville, Spain

Additional Reviewers

Abdelgawwad, Ahmed	Benito-Picazo, Jesus
Almendros-Jimenez, Jesus M.	Bermejo, Sergio
Azorín-López, Jorge	Borhani, Alireza
Basterrech, Sebastian	Brazalez-Segovia, Enrique

Cazorla, Miguel
Cuartero, Fernando
Dapena, Adriana
Delecraz, Sebastien
Duro, Richard
Escalona, Félix
Fuster-Guillo, Andres
Garcia-Garcia, Alberto
García-González, Jorge
Gomez-Donoso, Francisco
Gorostegui, Eider
Graña, Manuel
Hicheri, Rym
Hinaut, Xavier
Hoermann, Timm
Korthals, Timo
Kouzou, Abdellah
Lachmair, Jan
Luque-Baena, Rafael M.
López-García, Guillermo
López-Rubio, Ezequiel
Macià Soler, Hermenegilda

Mattos, César Lincoln
McCabe, Philippa Grace
Medina-Bulo, Inmaculada
Molina-Cabello, Miguel A.
Muaaz, Muhammad
Muniategui, Ander
Nguyen, Huu Nghia
Oneto, Luca
Oprea, Sergiu
Ortiz-De-Lazcano-Lobato, Juan Miguel
Orts-Escolano, Sergio
Palomo, Esteban José
Pedrelli, Luca
Riaza Valverde, José Antonio
Riley, Patrick
Rincon, Jaime A.
Ruiz Delgado, M. Carmen
Safont, Gonzalo
Saval-Calvo, Marcelo
Scardapane, Simone
Segovia, Mariana
Thurnhofer-Hemsi, Karl

Contents – Part II

Machine Learning in Vision and Robotics

System Identification, Process Control, and Manufacturing

Image and Signal Processing

Soft Computing

Mathematics for Neural Networks

Internet Modeling, Communication and Networking

Expert Systems

Evolutionary and Genetic Algorithms

Advances in Computational Intelligence

Computational Biology and Bioinformatics

Contents – Part I

Human Activity Recognition

New and Future Tendencies in Brain-Computer Interface Systems

Random-Weights Neural Networks

Pattern Recognition

Deep Learning and Natural Language Processing

Software Testing and Intelligent Systems

Data-Driven Intelligent Transportation Systems

Deep Learning Models in Healthcare and Biomedicine

Deep Learning Beyond Convolution

Fuzzy Preprocessing for Semi-supervised Image Classification in Modern Industry

Petr Hurtik$^{(\boxtimes)}$ and Vojtěch Molek

Institute for Research and Applications of Fuzzy Modeling, Centre of Excellence
IT4Innovations, University of Ostrava, Ostrava, Czech Republic
{petr.hurtik,vojtech.molek}@osu.cz

Abstract. We are focusing on image classification in industrial processing taking into account the most problematic issue of the processing: the lack of labeled data. Here, we are considering three datasets: the first one is an unsorted collection of all types of manufactured products and includes 100 images per class. The second one consists of products sorted into particular classes by a specialized employee and includes only ten images per class. The last one includes a massive volume of labeled images, but it is used only for the proposal validation. As the configuration is challenging for neural networks, we propose to use Image Represented by a Fuzzy Function in order to enrich original image information. We solve the task using various autoencoder architectures and prove that such the proposal increases the autoencoders success rate.

Keywords: Unsupervised learning · Image classification ·
Image Represented by a Fuzzy Function · IRFF · Autoencoder

1 Paper Context, Motivation, and Aim

Working with industrial partners, we observed that one trend is always present: it is necessary to make a manufacturing process autonomous. Therefore, we are witnessing, the replacement of workers with machines, e.g., by autonomous manipulating or welding robots. Such robots are managed using controlling units which are usually coded in advance. As computer vision and artificial intelligence areas become more reliable and able to solve tasks handled by humankind, so-called smart cameras become replacing workers even in an optical quality control area. The benefits of smart cameras are more than evident: processing speed can be increased, human subjectivity can be ignored, and the non-stop manufacturing regime can be easily adopted. On the other hand, taking into account that smart cameras utilize algorithms based on neural networks (NN), there are still several issues to be solved.

In this paper, we are going to focus on two issues complicating the deployment of NN-based applications in industrial processing. The first one lies in the necessity of getting a reasonably big training dataset [1]. The second one is

© Springer Nature Switzerland AG 2019
I. Rojas et al. (Eds.): IWANN 2019, LNCS 11507, pp. 3–13, 2019.
https://doi.org/10.1007/978-3-030-20518-8_1

that the training dataset should be labeled. In the literature, there exist several approaches, how the two issues can be suppressed – the first issue by, e.g., transfer learning [15], so that a pre-trained model is taken and fine-tuned for a specific task. The solution of the second problem lies in a usage a non-supervised architecture, called autoencoder [2].

To outline the issues, we will recall the basics of autoencoder [3] and discuss a way, how its accuracy can be improved. We will address a unique form of pre-processing given by a transformation of standard images into a fuzzy representation of images, which enriches the information serving as an input into an autoencoder. Such enriched information could be useful for datasets which suffer from a small number of samples problem, which is precisely the case of industrial processing [5,10]. Solving this issue makes the deployment of NN-based algorithms more feasible.

2 Necessarily Short Necessary Preliminaries

Here, we are going to recall two bases which our paper uses: autoencoder, i.e., type of neural network architecture and IRFF, i.e., an alternative representation of an image.

2.1 Autoencoders

The general idea behind autoencoders is to create a way, how a network can learn a meaningful data representation without their labels. That is done by learning a transformation of input data into low-dimensional space together with an inverse transformation back into the original space. As an evaluation process, the utilization of the distance function between the input and the output [4] is used. The general schema of an autoencoder is shown in Fig. 1. As it is obvious from the schema, the first part of a network, called encoder, forces the network to represent an input data with the use of only a few neurons. To ensure that the neurons – a coded representation – have learned meaningful representation, the second part, called decoder, is used. Such a schema can be viewed also as a process of feature extraction [9], compression/decompression [12], or dimensionality reduction [14]. Because the autoencoder learns to select only the most descriptive information, and because the less descriptive one is suppressed, it achieves great results in image denoising task [13].

The described process can be performed fully in an unsupervised way. In order to use encoder as a classifier, the classification part is added in the following way: the first part of autoencoder (encoder) is kept, but a classification network is added instead of the decoder, and the new architecture is fine-tuned. In the opposite of autoencoder, the training of classifier is supervised, i.e., it requires labeled data. So the whole process of training encoder, decoder and classifier can be considered as semi-supervised learning.

It is entirely natural that in order to train autoencoder, i.e., to learn important dependencies between data, a training dataset should be as big as possible

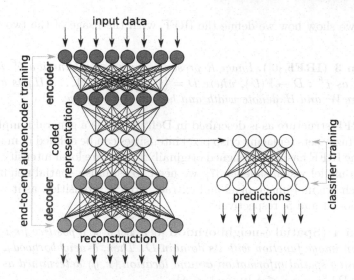

input data

end-to-end autoencoder training

encoder

coded representation

decoder

reconstruction

predictions

classifier training

Fig. 1. The illustrative schema shows an autoencoder consisting of encoder and decoder parts with added classifier part. The encoder and decoder realize unsupervised part; the classifier is supervised so the whole process can be viewed as semi-supervised.

to capture all the essential dependencies and reveal inner data structures. In this paper, **we consider real industry conditions, so the number of samples in training dataset is highly limited**, which is opposite to the requirement. To help the network, we propose to enrich the dataset with the usage of our proposed representation, Image Represented by a Fuzzy Function.

2.2 Image Represented by a Fuzzy Function (IRFF)

The IRFF representation has been originally proposed by Madrid and Hurtik [8] in order to improve processing speed and robustness of the Hough transform algorithm. Later, in [6] it has been used as a universal pre-processing for neural networks with the aim of decreasing their classification error. Here, we are going to recall the basis of IRFF and demonstrate its practical impact on the said assumptions.

In this work, we are working with images in their raster form according to the following definition.

Definition 1 (2D grayscale image). *Let $f : D \to L$, be a grayscale image function where $D = \{1, \ldots, W\} \times \{1, \ldots, H\}$ is a discrete space, W and H denote width and height of the image and $L \subset \mathbb{R}$.*

As we are dealing with fuzzy sets, we are going to define universum of a fuzzy set at first.

Definition 2 (Universum of fuzzy sets [8]). *Let $\mathbb{F}(U)$ be a universum of fuzzy sets F over unit interval given as $\mathbb{F}(U) = \{F \mid F : U \to [0, 1]\}$.*

Then, we show how we define the IRFF with the usage of the two previous definitions.

Definition 3 (IRFF [6]). *Image Represented by a Fuzzy Function f^F is a mapping given as $f^F : D \to \mathbb{F}(U)$, where $D = \{1, \ldots, W\} \times \{1, \ldots, H\}$ is a discrete space where W and H denote width and height of the image.*

The IRFF structure as is described in Definition 3 is a general template, i.e., particular fuzzy sets replacing crisp pixel intensities can be defined in many ways. Because the IRFF has been designed originally to capture local intensity contrast in order to model visual saliency [7], we need to establish spatial neighborhood around each original image pixel and extract from it intensities, as it was given by Definitions 4 and 5, respectively.

Definition 4 (Spatial δ-neighborhood [6]). *Let δ be a positive finite integer, and f be an image function with its domain D. Then, δ-neighborhood $\omega_{f,x,y}$ is a set containing spatial information around location (x, y) determined as $\omega_{f,x,y} = \{(x_i, y_i) \in D \mid |x - x_i| \leq \delta, |y - y_i| \leq \delta\}$.*

Definition 5 (Intensity δ-neighborhood [6]). *Let $\omega_{f,x,y}$ be a spatial δ-neighborhood of a point (x, y) of an image function f with its domain D. Then, intensity δ-neighborhood $\Omega_{f,x,y}$ of such the point is set $\Omega_{f,x,y} = \{f(x_i, y_i) \mid (x_i, y_i) \in \omega_{f,x,y}\}$.*

The last remaining piece to be done is encoding $\Omega_{f,x,y}$ into $f^F(x, y)$, i.e., an intensity δ-neighborhood into a fuzzy set given by its membership function. Let us recall that our membership function determines what is membership degree $\gamma \in [0, 1]$ of element $\beta \in [0, 255]$ into fuzzy set $f^F(x, y)$. In order to have a reasonable membership function lead by motivation, we define three conditions.

1. $f^F(x, y)(f(x, y)) = 1$,
2. $\forall \beta \in \Omega_{f,x,y} : f^F(x, y)(\beta) > 0$,
3. $f^F(x, y)(\beta) = 0$ iff $\beta \notin [\min(\Omega_{f,x,y}), \max(\Omega_{f,x,y})]$.

According to the three conditions, we can use the various shape of a membership function (sinusoidal, triangular, trapezoidal, ...) and design it using various values from intensity δ-neighborhood (minimum/maximum, quantiles, means, ...). To achieve as simple as possible processing and because all possible local intensity variance is in $\Omega_{f,x,y}$, we will use values of $\max(\Omega_{f,x,y})$ and $\min(\Omega_{f,x,y})$ in order to define membership functions as follows:

$$f^F(x, y)(\beta) = \begin{cases} \frac{\beta - f_L^F(x,y)}{f_C^F(x,y) - f_L^F(x,y) + \varepsilon} & \text{if } f_L^F(x, y) \leq \beta \leq f_C^F(x, y) \\ \frac{f_R^F(x,y) - \beta}{f_R^F(x,y) - f_C^F(x,y) + \varepsilon} & \text{if } f_C^F(x, y) < \beta \leq f_R^F(x, y) \, , \\ 0 & \text{otherwise} \end{cases} \quad (1)$$

where ε is an extremely small positive constant and we are using the following substitutions in order to shorten the formula:

$$f_L^F(x, y) = \min(\Omega_{f,x,y}),$$
$$f_C^F(x, y) = f(x, y),$$
$$f_R^F(x, y) = \max(\Omega_{f,x,y}).$$
(2)

The benefits of the triangular-based membership function lie in its easy manipulation. We can use known apparatus, namely interval approach [11], i.e., we can identify our membership function only with the triplet $(f_L^F(x, y),$ $f_C^F(x, y),$ $f_R^F(x, y))$. Computations with only three points are extremely fast – see, e.g., [8]. Also, the usage of min and max values seems to be reasonable when we are working with neural networks, because of these non-linearly extracted values cannot be obtained by a network in another way than with the usage of min, or max-pooling.

3 Combination of IRFF with NN

Let us recall; the IRFF preprocessing replaces each original pixel intensity by a fuzzy set identified by its membership function. Such representation – a set of membership functions – is not suitable as a direct input to an arbitrary neural network, which is our aim; therefore, another representation has to be found. As we mentioned in the previous section, the advantage of the triangular membership function is that we can work with it on the basis of the interval approach, i.e., we can process each of the three value describing a particular triangular membership function separately. So each of the three values can be placed into its matrix representing an image. It results in three matrices, one for the minimum, one for the center and one for the maximum values. That coincides with multi-channel images, such as three-channel image representing IRFF for a grayscale input image or nine-channel image representing IRFF for an RGB image.

Such a new composed image can be processed channel-by-channel, i.e., each channel will be processed by a separate neural network and the networks will be assembled by defuzzification. The disadvantages of such processing are evident: the significant increase of the original network complexity and the necessity of choosing suitable defuzzification function. Therefore, we propose to use the multi-channel image as a standard input into a neural network and leave the neural network to select automatically such information, which is suitable for it. An illustration of the integration is shown in Fig. 2.

4 Benchmark

This section is divided into three parts: in Sect. 4.1, we present the dataset. Further, we discuss used architectures and their settings in Sect. 4.2 and finally, we give the numerical results in Sect. 4.3.

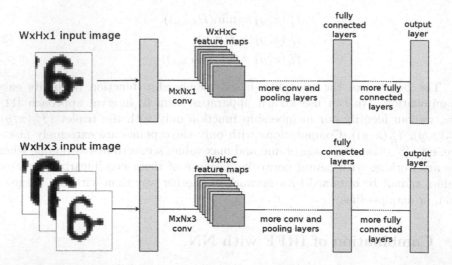

Fig. 2. The illustrative schema shows the difference between a standard NN (top) and an NN with the proposed preprocessing. The only variation is in the first layer, which is three times bigger for the NN with IRFF input.

4.1 Used Dataset

Based on our experience, a producer usually has three types of datasets. The first one, reasonably big, is an unsorted collection of all types of manufactured products. The second one, a tiny one, consists of products sorted into particular classes due to their labels assigned by a specialized employee. The last one, almost infinite one, is available after an optical recognition system is installed and includes all product types taken by the system. So the third dataset is not available when a system is designed and therefore cannot be used for its training.

To mimic the conditions, we propose the following datasets. The first one consists of 100 images per class, without their labels. The second one includes ten images per class together with their labels. The last one should be as big as possible and serves only for testing (the overall system validation) purposes. Let us note, the usage of only ten labeled images per class is challenging for neural networks and it makes the problem almost unsolvable.

As the task, we artificially design a dataset consisting of images with the resolution of 26×28px. The subject of the dataset is to classify 0–9 numbers, i.e., to classify them into ten classes. Therefore, we have 1000 images in our first, unsorted database; 100 images with their labels in the second one and for the testing dataset, it lefts us 63350 images. In contrast to the well known MNIST dataset, the numbers vary in size, intensity, are damaged, and are not centered. Moreover, an image sometimes captures a part(s) of another number. Visualization of several numbers is shown in Fig. 3. With such dataset, one may meet in all areas of industrial processing where objects are marked by their IDs,

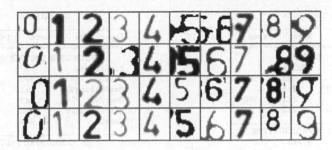

Fig. 3. The illustration shows images which are being taken from the dataset used in the benchmark. Notice that the images are not centered and includes distortions.

e.g., production of steel ingots, car lights, medical equipment, etc., and where is the requirement to read such markings.

4.2 Methodology and Used Methods

To benchmark the results, we propose to use two convolutional autoencoder models according to Figs. 4 and 5 respectively. The first one is a stacked convolution autoencoder, and the second one doubles the convolution layers of the former one. For both autoencoder and classifier networks, we use hyperbolic tangent as the activation function in all but the last layer, because it is – based on our experience – more stable than ReLU activation function. Also, each network is trained for 150 epochs for autoencoder and 150 epochs for classifier with an automatic decrease of learning rate on a plateau.

In order to suppress the problem of the initialization with random weights, we executed training/testing parts 30× and measure maximum, average and standard deviation of the network success rate.

Standardly, when an engineer is facing a lack of data, it is natural to use data augmentation. According to Fig. 3, it may be observed that the numbers are partially shifted, and their slope is not equal. Therefore, we realize the benchmark for two training datasets: the original one and a new, augmented one, with the setting of rotation_range = 20, width_shift_range = 0.2, and height_shift_range = 0.2.

4.3 Results

Firstly, let us comment on the size of the used architectures. We want to recall that the integration of IRFF means that only the first layer (and the last one in the case of autoencoder) is tripled, the size of the remaining ones is the same. The advantage is evident: the number of parameters to be learned is increased only negligibly, namely by 17.94% and 6.19% in the case of the two autoencoder models. For the classifier, the increase even smaller: the IRFF version is bigger only by 0.14% for Model 1 and by 0.14% for Model 2. In other words, the training time is not significantly changed when IRFF preprocessing is used.

```
         AUTOENCODER                              CLASSIFIER
------------------------------       ------------------------------------
Layer type    Note    Param #        Layer type      Note    Param #
==============================       ====================================
         ENCODER                       AUTOENCODER ENCODER PART
InputLayer   28x26x1/3  0             InputLayer                0
Conv2D       16 3x3   160/448         Conv2D         16 3x3   160/448
MaxPooling2D  2x1      0              MaxPooling2D    2x1      0
Conv2D        8 3x3   1160            Conv2D          8 3x3   1160
MaxPooling2D  2x2      0              MaxPooling2D    2x2      0
                                      Conv2D          4 3x3    292
----------------------------
     LATENT REPRESENTATION            ------------------------------------
Conv2D        4 3x3    292                    CLASSIFIER PART
                                      Flatten
----------------------------         Dense          364    132860
         DECODER                      Dropout        0.5
UpSampling2D  2x2      0              Dense          182     66430
Conv2D        8 3x3    296            Dropout        0.5
UpSampling2D  2x1      0              Dense          10       1830
Conv2D       16 3x3   1168           ====================================
Conv2D      1/3 3x3   145/435        Trainable params: 202,732/203,020
==============================

Trainable params:  3,221/3,799
```

Fig. 4. The simpler architecture marked as Model 1.

```
         AUTOENCODER                              CLASSIFIER
----------------------------------   ------------------------------------
Layer type    Note    Param #        Layer type      Note    Param #
==================================   ====================================
         ENCODER                       AUTOENCODER ENCODER PART
InputLayer   28x26x1/3  0             InputLayer                0
Conv2D       16 3x3   160/448         Conv2D         16 3x3   160/448
Conv2D       16 3x3   2320            Conv2D         16 3x3   2320
MaxPooling2D  2x1      0              MaxPooling2D    2x1      0
Conv2D        8 3x3   1160            Conv2D          8 3x3   1160
Conv2D        8 3x3    584            Conv2D          8 3x3    584
MaxPooling2D  2x2      0              MaxPooling2D    2x2      0
Conv2D        4 3x3    292            Conv2D          4 3x3    292
                                      Conv2D          1 3x3    148
----------------------------------
     LATENT REPRESENTATION            ------------------------------------
Conv2D        4 3x3    148                    CLASSIFIER PART
                                      Flatten
----------------------------------   Dense          364    132860
         DECODER                      Dropout        0.5
Conv2D        4 3x3    148            Dense          182     66430
UpSampling2D  2x2      0              Dropout        0.5
Conv2D        8 3x3    296            Dense          10       1830
Conv2D        8 3x3    584            ====================================
UpSampling2D  2x1      0              Trainable params: 205,784/206,072
Conv2D       16 3x3   1168
Conv2D       16 3x3   2320
Conv2D      1/3 3x3   145/435
==================================

Trainable params:  9,325/9,903
```

Fig. 5. The architecture marked as Model 2, where the convolution layers are doubled.

Secondly, a visualization of an autoencoder coding/decoding process can be illustrated – see Fig. 6. By the first look, both autoencoders for standard and IRFF input seems to work valid. If we focus on details, it can be shown that IRFF input can repair damaged numbers, as it is visible on the case of the number "8". The reason is that IRFF is constructed with the usage of δ-neighborhood, so gaps with the size smaller than 2δ can be filled. According to the visualization, the neural network uses the information and therefore the output of autoencoder is repaired too.

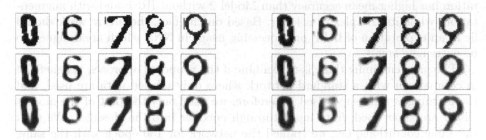

Fig. 6. Visualization of autoencoder processing. Top row: input into autoencoder. Middle row: output of Model 1 after coding/decoding. Bottom row: output of Model 2 after coding/decoding. The left side presents autoencoders with the standard input; the right side presents autoencoders with the IRFF input.

Table 1. Classification accuracy

	Baseline			IRFF		
	Mean [%]	Std [±]	Max [%]	Mean [%]	Std [±]	Max [%]
Model 1	66.27	3.38	72.87	69.69	3.94	74.43
Model 1 + Augment	81.41	3.55	86.55	83.40	2.73	88.99
Model 2	78.51	0.71	79.50	78.95	1.16	81.34
Model 2 + Augment	82.28	2.45	88.46	83.46	2.57	88.94

In Table 1, we present the measured metrics for all the test cases. Note, the classification accuracy is in percentage. According to the table, several comments have to be made.

1. When the issue of lack of data occurs, it is highly beneficial to use data augmentation.
2. IRFF preprocessing reached higher mean accuracy for both standard and augmentation datasets, on average by 1.76% points.
3. IRFF preprocessing reached higher maximum accuracy for both standard and augmentation datasets, on average by 1.57% points.

In other words, the usage of the preprocessing, which means to add only several lines of source code, classify correctly additional 1108 images on the tested dataset consisting of 63350 images.

The expectable fact is that the impact of IRFF preprocessing is higher when we have less data or when the model is simpler. The following numbers can support the claim. For Model 1 (both with/without augmentation), we reached average accuracy improvement by 2.70%, for Model 2 only by 0.81%. If we focus only on Model 1, the improvement is 3.42% with no augmentation and 1.99% with augmentation. The interesting fact is that Model 1 with IRFF and augmentation has higher mean accuracy than Model 2 without IRFF and with augmentation, while its complexity is lower. Based on the fact, the future work should lie in an exploration of IRFF preprocessing usage in NNs, which are designed to be extremely small.

The last remaining thing is to examine if the proposed approach is beneficial in comparison with a standard network where only labeled data are used, and the learning is fully supervised. Therefore, we used only the part of the models which are supervised: from input through encoder into the classifier. We used the same conditions, i.e., we trained the network for 150 epoch with the same data augmentations, run multiple-time and computed the same measures (max, min, st dev). The supervised Model 1 reached the mean accuracy of 78.32%, maximum 81.55%, and st dev ±2.32. The supervised Model 2 reached the mean accuracy of 81.08%, maximum 85.09%, and st dev ±3.77. Based on these results, we proved the two-stage learning consisting of non labeled and labeled data is beneficial in the task of image classification.

5   Summary

In the paper, we deal with the problem of image classification under conditions which we met during working with industrial partners. Therefore, we proposed the dataset of 100 images per class, but without labels and the dataset of 10 images per class, with labels. To solve the task, we used a convolutional autoencoder trained in an unsupervised way over the first dataset followed by a classifier trained over the labeled dataset.

Further, we described a universal preprocessing for neural networks, which replaces each standard image by its fuzzy representation, which serves as a new input into the same network. Based on the presented benchmark, we demonstrated that the preprocessing does not increase the number of parameters to be learned significantly, but can improve the neural network accuracy, on average by 1.75 percent points.

Acknowledgment. The work was supported from ERDF/ESF "Centre for the development of Artificial Intelligence Methods for the Automotive Industry of the region" (No. CZ.02.1.01/0.0/0.0/17_049/0008414).

For more supplementary materials and overview of our lab work see http://graphicwg. irafm.osu.cz/storage/pr/links.html.

References

1. Alwosheel, A., van Cranenburgh, S., Chorus, C.G.: Is your dataset big enough? sample size requirements when using artificial neural networks for discrete choice analysis. J. Choice Model. **28**, 167–182 (2018)
2. Baldi, P.: Autoencoders, unsupervised learning, and deep architectures. In: Proceedings of ICML Workshop on Unsupervised and Transfer Learning, pp. 37–49 (2012)
3. Geng, J., Fan, J., Wang, H., Ma, X., Li, B., Chen, F.: High-resolution SAR image classification via deep convolutional autoencoders. IEEE Geosci. Remote Sens. Lett. **12**(11), 2351–2355 (2015)
4. Hinton, G.E., Salakhutdinov, R.R.: Reducing the dimensionality of data with neural networks. Science **313**(5786), 504–507 (2006)
5. Hurtik, P., Burda, M., Perfilieva, I.: An image recognition approach to classification of jewelry stone defects. In: 2013 Joint IFSA World Congress and NAFIPS Annual Meeting (IFSA/NAFIPS), pp. 727–732. IEEE (2013)
6. Hurtik, P., Molek, V., Hula, J.: Data preprocessing technique for neural networks based on image represented by a fuzzy function. IEEE Trans. Fuzzy Syst. 1–10 (2019). submitted
7. Itti, L., Koch, C.: A saliency-based search mechanism for overt and covert shifts of visual attention. Vis. Res. **40**(10), 1489–1506 (2000)
8. Madrid, N., Hurtik, P.: Lane departure warning for mobile devices based on a fuzzy representation of images. Fuzzy Sets Syst. **291**, 144–159 (2016)
9. Masci, J., Meier, U., Cireşan, D., Schmidhuber, J.: Stacked convolutional autoencoders for hierarchical feature extraction. In: Honkela, T., Duch, W., Girolami, M., Kaski, S. (eds.) ICANN 2011. LNCS, vol. 6791, pp. 52–59. Springer, Heidelberg (2011). https://doi.org/10.1007/978-3-642-21735-7_7
10. Novák, V., Hurtík, P., Habiballa, H.: Recognition of distorted characters printed on metal using fuzzy logic methods. In: IFSA World Congress and NAFIPS Annual Meeting (IFSA/NAFIPS), 2013 Joint, pp. 733–738. IEEE (2013)
11. Novák, V., Perfilieva, I., Močkoř, J.: Mathematical Principles of Fuzzy Logic, vol. 517. Springer, Heidelberg (2012)
12. Theis, L., Shi, W., Cunningham, A., Huszár, F.: Lossy image compression with compressive autoencoders. arXiv preprint arXiv:1703.00395 (2017)
13. Vincent, P., Larochelle, H., Lajoie, I., Bengio, Y., Manzagol, P.A.: Stacked denoising autoencoders: learning useful representations in a deep network with a local denoising criterion. J. Mach. Learn. Res. **11**(Dec), 3371–3408 (2010)
14. Wang, W., Huang, Y., Wang, Y., Wang, L.: Generalized autoencoder: a neural network framework for dimensionality reduction. In: Proceedings of the IEEE Conference on Computer Vision and Pattern Recognition Workshops, pp. 490–497 (2014)
15. Weiss, K., Khoshgoftaar, T.M., Wang, D.: A survey of transfer learning. J. Big Data **3**(1), 9 (2016)

Interpretability of Recurrent Neural Networks Trained on Regular Languages

Christian Oliva[✉] and Luis F. Lago-Fernández

Escuela Politécnica Superior, Universidad Autónoma de Madrid,
28049 Madrid, Spain
christian.oliva@estudiante.uam.es, luis.lago@uam.es

Abstract. We study the ability of recurrent neural networks to model and recognize simple regular languages. Training the networks under different levels of noise and regularization, we analyze their response in terms of accuracy and interpretability using a complete set of validation data. Our results show that a small noise level improves the generalization of the networks, while regularization provides a higher interpretability. Under proper levels of noise and regularization, the networks are able to obtain a high accuracy, and the hidden units display activation patterns that could be related to discrete states in a deterministic finite automaton.

1 Introduction

Recurrent Neural Networks (RNNs) [1–3] are a kind of neural networks specially designed to process and model sequences. They introduce feedback connections in their architecture, in such a way that the state of the network at any given time depends both on the current input and on the previous network state. RNNs have been successfully applied in many domains where it is necessary to model the temporal structure of a sequence. These include speech recognition [4], natural language modeling [5], automatic translation [6,7], music composition [8] or robot control [9], among others.

Almost since their introduction, many authors have recognized the parallelism between RNNs and finite automata, and the ability of RNNs to model formal languages has been widely studied [10–12]. RNNs have been shown to be Turing equivalent [13], and many works have studied the ability of RNNs to model regular and context independent languages [14,15]. Nevertheless, and despite their capacity to model a formal language, whether or not RNNs are in fact internally implementing automata is not completely clear. Since the pioneer work of Giles et al. in the 90's [16], many authors have tried to analyze the complex activation space of RNNs trained on regular languages, searching for activation clusters that could be related to the states of the corresponding finite automaton [17]. The main idea of these approaches is based on a quantization of the activation space into a set of discrete

I. Rojas et al. (Eds.): IWANN 2019, LNCS 11507, pp. 14–25, 2019.
https://doi.org/10.1007/978-3-030-20518-8_2

states, followed by a determination of the transitions between states in response to each of the input symbols. For a recent review of the field, sometimes known as rule extraction, we refer the reader to [18]. Although these approaches have been frequently criticized [19] due to the difficulty of relating a continuous dynamical system to a finite state machine, in practice the extracted automata show a good performance, often increasing the generalization capability of the networks they are extracted from [18].

Rule extraction can be seen as an attempt to add interpretability to RNNs. In fact one of the main drawbacks of the use of RNNs is the lack of methods that allow to interpret and visualize what the networks are learning. Hence RNNs are usually treated as black boxes that are able to solve some problems with a high accuracy but are difficult to understand. In this context, the recent work of Karpathy et al. [20] provides some empirical evidence that in RNNs trained on real-world data there are neurons that respond to specific and meaningful patterns. In particular, using texts as input, they observed neurons responding to patterns such as quotation, indentation or line length.

In this article we analyze the behavior and interpretability of RNNs when they are trained to recognize simple regular languages under different levels of noise and L1 regularization. Using the color activation plots introduced by [20], we study how the trained networks respond to different input strings. Our results show that small levels of noise and L1 regularization are able to improve the network's interpretability without affecting its overall accuracy. Furthermore, under some circumstances the network activation patterns admit an interpretation that resembles an automaton. This suggests that automatic extraction of the transition rules could in some cases be feasible.

The article is organized as follows. In Sects. 2 and 3 we review the concepts of recurrent neural networks and deterministic finite automata, respectively. Section 4 describes the experiments performed. In Sect. 5 we analyze the main results of our study, and finally Sect. 6 presents the conclusions and discusses future lines of research.

2   Recurrent Neural Networks

In its simplest form, the activity of a RNN can be described by the following equations [1]:

$$h_t = \sigma(W_{xh}x_t + W_{hh}h_{t-1} + b_h) \tag{1}$$

$$y_t = \sigma(W_{hy}h_t + b_y) \tag{2}$$

where x_t, h_t and y_t represent the activation vectors for the input, hidden and output layers, respectively, at time t. The rest of symbols include the weight matrices (W_{xh}, W_{hh} and W_{hy}), the bias vector for each layer (b_h and b_y), and the sigmoid activation function (σ). Note the dependence of the hidden activation h_t on its value for the previous time step h_{t-1}, modulated by the weight matrix W_{hh}. This basic model is known as Elman RNN [1], and it is sometimes referred to as vanilla RNN. More complex architectures, such as LSTMs [2] or GRUs [7], develop this idea to improve the results on a variety of problems. Throughout this article we focus on the Elman model.

3   Regular Languages and Deterministic Finite Automata

Regular languages are the simplest set of formal languages according to the Chomsky hierarchy. They can be described by type-3 grammars (regular grammars), and are formally equivalent to Kleene's regular expressions. The kind of automata associated to regular languages are called Deterministic Finite Automata (DFA). A DFA is an abstract computational model consisting of a finite number of states, and a transition function that determines changes of state in response to input symbols [21].

Formally, a DFA is defined by the quintuple $M = (Q, \Sigma, \delta, q_0, F)$, where Q is the set of internal states, Σ is the set of input symbols, $\delta : Q \times \Sigma \to Q$ is the transition function, $q_0 \in Q$ is the initial state and $F \subseteq Q$ is the set of final states. Starting from the initial state q_0, a DFA processes an input string by reading one symbol at a time and performing state transitions according to the transition function δ. A given string w is said to be accepted by the DFA M if, after processing the whole string w the automaton M ends in a final state $q_f \in F$. Otherwise the string w is rejected by M. The set of all strings accepted by M, denoted $L(M)$, is a regular language.

4   Experiments

In this section we describe the experiments carried out to analyze the network interpretability in terms of the number of hidden units, the noise level and the regularization parameter. As a proof of concept, we focus on two simple regular languages whose automata are very easy to visualize. In the following sections we describe the RNN that we use and the problems we address.

4.1   Network Parameters

In all the experiments we use the simple Elman RNN architecture described in Sect. 2, with the activation function in the hidden layer set to *tanh*. Additionally, we introduce Gaussian noise into the recurrent connection, as follows:

$$h_t = \tanh(W_{xh}x_t + (W_{hh} + X_\nu \mathbb{I})h_{t-1} + b_h) \tag{3}$$

where X_ν is a normally distributed random variable centered at 0 and with amplitude ν, and $\mathbb{I}$ is the $n_h \times n_h$ identity matrix. Note that this form of noise does not affect hidden units whose activity is close to 0. We adopt this form so that it does not counterbalance the regularization (see Sect. 5.4). We train the network to minimize a cross-entropy cost function with L1 regularization:

$$L = -\sum_{t=1}^{n}[\hat{y}_t \log(y_t) + (1-\hat{y}_t)\log(1-y_t)] + \gamma(||W_{xh}||_1 + ||W_{hh}||_1 + ||W_{hy}||_1) \quad (4)$$

where $\hat{y}_t$ is the expected output, γ is the regularization parameter and the expression $|| \cdot ||_1$ denotes the L1 matrix norm. The parameter n is the number of steps in the unrolled version of the network (25 in all our experiments). The size of the input layer depends on the number of input symbols (3 for all the addressed problems, see below), and we consider one single output unit whose interpretation is to accept $(y = 1)$ or reject $(y = 0)$ a given input string. The number n_h of units in the hidden layer is varied in our experiments in the range $[2, 10]$.

4.2 Data

We consider two simple regular languages on the alphabet of symbols $\{a, b\}$: (i) the set of strings with an even number of a's; and (ii) the set of strings that start with a b and end with an a. They will be referred to as the *parity* problem and the *bxa* problem, respectively. The problems are defined by the regular expressions $(b^*ab^*a)^*b^*$ (*parity*) and $b(a+b)^*a$ (*bxa*), where $*$ means zero or more repetitions of the preceding symbol, and $+$ may be interpreted as a boolean or.

In order to train the network, we generate training and test data consisting of long random sequences containing the characters a, b and \$. The role of the \$ symbol is to separate the different strings contained within the data. It can be interpreted as the end of the actual string and the beginning of the next one, or it can be thought of as an *empty* string. The use of this additional symbol avoids the need to reset the network state whenever a new string is received. For each symbol in an input sequence we compute the expected output of the network as a 1 (accept) or 0 (reject). Table 1 shows some input strings together with the expected output for the *parity* and *bxa* problems.

Table 1. Some examples of input strings and the expected output for the *parity* and *bxa* problems.

Input	\$aaaaabbb	\$baaa	\$aabaababbbbab	\$baaaaabbbbbaaaaa
parity	101010000	11010	10110110000011	1101010000010101
bxa	000000000	00111	00000000000000	0011111000011111

We have generated 5 different datasets to be used in our experiments (see summary in Table 2). The *train* dataset contains 50000 symbols, with equal probabilities for symbols a and b, and a probability of appearance for the \$ symbol of 0.1. The *big* dataset is generated under the same conditions, but it contains 100000 symbols. The *long* dataset contains 20000 characters, but much longer strings. It has been generated by reducing the \$ probability to 0.01. Finally, the *all as* (*all bs*) dataset contains 15000 characters and strings composed mainly of *as* (*bs*). All the networks are trained using the *train* dataset, while the remaining datasets are used for validation.

Table 2. Description of the datasets used to train and validate the neural networks. The table shows, for each dataset, the number of input characters, the probability of each input symbol (a, b, \$), and the average, minimum and maximum string lengths.

Data	# chars	a prob.	b prob.	\$ prob.	avg len	min len	max len
Train	50000	0.45	0.45	0.1	8.9	0	95
Big	100000	0.45	0.45	0.1	9.0	0	81
Long	20000	0.495	0.495	0.01	88.3	0	477
All as	15000	0.98	0.01	0.01	113.5	0	566
All bs	15000	0.01	0.98	0.01	95.8	0	475

4.3   Network Training

For each experimental condition we train 20 different networks starting from random weight initialization. The networks are trained for a maximum of 50 epochs (100000 iterations). The cost function is minimized using a standard gradient descent algorithm with a learning rate of 0.01. All the results reported in next section are averages over the 20 executions for a given set of parameters.

5   Results and Analysis

We measure the quality of our results both in terms of accuracy and interpretability. The accuracy of a network is evaluated on 4 validation sets with different properties (see Table 2). Special attention is payed to extreme string instances: long strings composed mainly of one single symbol that have not been presented to the network during the training phase. The network interpretability is analyzed by plotting color activation maps of the hidden units in response to a sequence of input symbols.

5.1   Initial Settings

In the first set of experiments we train the networks without noise or regularization ($\nu = 0$, $\gamma = 0$). We try network configurations with different number of hidden units n_h between 2 and 10, and compute the average accuracy over 20 different executions. The results on the training and the four validation sets are shown in Fig. 1. We observe that when the networks are complex enough

($n_h \geq 4$) the *parity* problem is solved with no generalization error. However, networks trained to solve the *bxa* problem fail consistently for long strings containing only the *b* symbol.

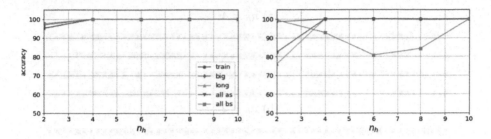

Fig. 1. Average accuracy versus number of hidden neurons for the *parity* (left) and *bxa* (right) problems. Both noise (ν) and regularization (γ) parameters are set to 0.

To analyze the interpretability of the networks, we show in Fig. 2 (top) a color activation map for the hidden units in a network with $n_h = 4$ that obtains 0 validation error on the *parity* problem. The main observation is that, in spite of achieving a perfect generalization, this network is using all the available resources to solve the problem, coding the solution across all the hidden neurons. As a result the network interpretability is reduced and it is difficult to understand the concepts being coded by each unit. This problem has also been observed for higher n_h.

On the other hand, if we force the network with $n_h = 2$, we obtain the color map in Fig. 2 (bottom) for one of the networks that achieves a good generalization. In this case the role of each hidden unit seems quite clear. The first unit is responding to all even *a*s, and also to the $ symbol whenever it appears after a rejected string. The second unit responds to all *b*s that follow after an even number of *a*s, and also to the $ symbol when it appears after an accepted string. Taking altogether, we may interpret the network solution as a logical OR of the activation of the hidden units. The activation patterns in the hidden layer might also be interpreted as discrete states in a DFA, as frequently reported. In conclusion, by reducing the number of hidden neurons we are gaining interpretability at the cost of not always being able to solve the problem. One natural question is whether the interpretability of more complex networks could be improved by introducing regularization. This is explored in Sect. 5.2.

In the *bxa* case there is the additional problem of generalization when long strings containing only the *b* symbol are presented to the network. The fact that RNNs fail for long strings has been reported before [18]. Looking at the color activation maps we can gain some insight into the origin of this problem. Figure 3 plots the color activation map for a network with $n_h = 4$ trained to solve the *bxa* problem, when a long string of *b*s is used as input. As far as the input symbol does not change, the network state should remain constant and the string should be rejected. Nevertheless what we observe is that some neurons experiment a smooth change of state, which leads to the observed generalization error. In Sect. 5.3 we explore if the addition of noise can help to avoid this problem.

(a) $n_h = 4$

(b) $n_h = 2$

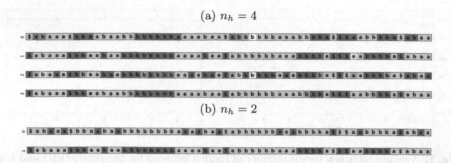

Fig. 2. Color activation plots for two networks with $n_h = 4$ (top) and $n_h = 2$ (bottom) trained on the *parity* problem with $\nu = 0$ and $\gamma = 0$. Red and cyan mean positive and negative activation, respectively. (Color figure online)

Fig. 3. Color activation plot for a network with $n_h = 4$ trained on the *bxa* problem with $\nu = 0$ and $\gamma = 0$. (Color figure online)

5.2 L1 Regularization

In this experiment we test whether the introduction of L1 regularization can improve the network's interpretability. We have trained networks with $n_h = 4$ hidden neurons and a regularization parameter γ varying in the range $[10^{-5}, 10^{-3}]$. Figure 4 shows the accuracy on the training and the 4 validation sets for networks trained on the *parity* (left) and the *bxa* problems (right). As before, the points are averages over 20 executions. We observe a decrease of the accuracy as the regularization parameter is increased. The right end of the plots corresponds to a situation where the regularization is so high that no neurons become active, with the networks giving a constant output and rejecting all the input strings. That is the reason for the apparent improve in the accuracy for the *all as* and *all bs* validation sets in the *bxa* problem. Note that, for this problem, all strings containing one single symbol must be rejected.

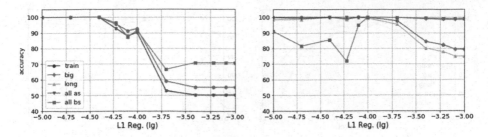

Fig. 4. Average accuracy versus regularization ($\log(\gamma)$) for the *parity* (left) and *bxa* (right) problems. The number of hidden neurons is $n_h = 4$ and the noise parameter is $\nu = 0$.

For small regularization (left end of the plots) the situation is equivalent to the non regularized case (Sect. 5.1). Finally, the intermediate region shows a progressive decrease of the accuracy. For γ in the range of 10^{-4} not all the networks are able to obtain a good generalization. Nevertheless the networks that perform well on the validation sets are now more interpretable, as shown in Fig. 5 for a typical case with $\gamma = 10^{-4}$ in the *parity* problem. Note that only 2 out of the 4 available units are being used to code the solution, and that they can be given the same interpretation as those in Fig. 2b. In conclusion, we observe that L1 regularization is able to improve the network's interpretability by forcing the network to use less hidden units.

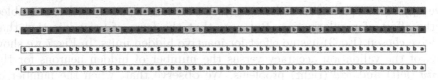

Fig. 5. Color activation plot for a network with $n_h = 4$ trained on the *parity* problem with $\nu = 0$ and $\gamma = 10^{-4}$. (Color figure online)

5.3 Noise

In this section we explore whether the injection of controlled amounts of noise is able to improve the generalization capability of the trained networks. The underlying idea is that noise will force the neurons to be more saturated in order to provide a stable response. Hence we expect more stability when constant strings are presented to the network. We train networks with $n_h = 4$ hidden units, setting the regularization parameter to $\gamma = 0$. The noise level is varied between 0 and 1, and we perform 20 different executions for each case to compute the average accuracy. The results are shown in Fig. 6 for both the *parity* and the *bxa* problems. As expected, the average accuracy experiments a gradual decrease as the noise level raises from 0 to 1, but it seems that the *bxa* problem is less affected.

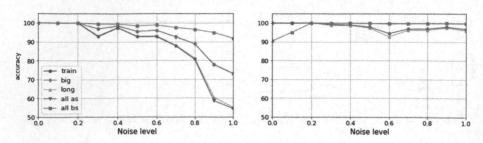

Fig. 6. Average accuracy versus noise level for the *parity* (left) and *bxa* (right) problems. The number of hidden neurons is $n_h = 4$ and the regularization parameter is $\gamma = 0$.

More interestingly, we observe that when the training accuracy converges to 1 in the presence of moderate noise, the final network obtains a very high validation score. In such a situation the problem with long strings is no longer present. In conclusion, noise introduction makes the network training phase more difficult, but when training converges the resulting networks generalize better to new, unobserved strings.

5.4   Putting All Together: L1 Regularization and Noise

In the last experiment we combine L1 regularization and noise injection in order to obtain networks that generalize well to new data and at the same time can be easily interpreted. In view of our previous results, we consider the values $\gamma = 10^{-4}$ and $\nu = 0.4$. In principle one could think that noise and regularization have opposite effects, since noise can activate a neuron that has been disabled by the effect of regularization. But note that the form of noise injection that we have chosen (Eq. 3) has no effect for inactive hidden units. In Fig. 7 we show plots of the validation accuracy versus the number of hidden neurons for the *parity* (left) and *bxa* (right) problems. We observe that, when the number of hidden units is high enough, the networks are able to do almost perfectly for both problems.

In order to illustrate the interpretability of the networks, we show in Figs. 8 and 9 the color activation maps for two networks with $n_h = 6$ trained to solve the *parity* and the *bxa* problems respectively. Note that, despite the availability of 6 hidden neurons, the effect of regularization is forcing the networks to use only 2 neurons for the *parity* problem and 3 neurons for the *bxa* problem. For the *parity* problem, the two active neurons have the same interpretation of those in Figs. 2 and 5, although the colors are inverted. For the *bxa* problem there are 3 active neurons. The first one is marking (in cyan, negative activation) the position of *a*s in strings that started with a *b*. The second one is signaling (in red, positive activation) *b*s in strings that started with a *b*. Finally, the third neuron is responding (in red, positive activation) to all the $ symbols, so marking the beginning of a new string. A similar behavior is observed in networks with a different number of hidden neurons. These results show that by tuning the noise level and the regularization parameter we can obtain highly interpretable networks that generalize well to new data.

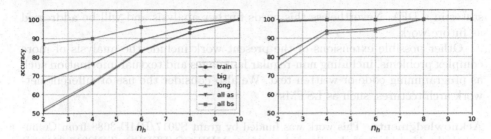

Fig. 7. Average accuracy versus number of hidden neurons for the *parity* (left) and *bxa* (right) problems. The noise level is $\nu = 0.4$ and the regularization parameter is $\gamma = 10^{-4}$.

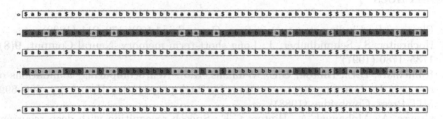

Fig. 8. Color activation plot for a network with $n_h = 6$ trained on the *parity* problem with $\nu = 0.4$ and $\gamma = 10^{-4}$. (Color figure online)

Fig. 9. Color activation plot for a network with $n_h = 6$ trained on the *bxa* problem with $\nu = 0.4$ and $\gamma = 10^{-4}$. (Color figure online)

6 Conclusions and Future Work

In this article we have studied the behavior of RNNs when trained to recognize simple regular languages. We tested the networks under different levels of noise and L1 regularization, and analyzed their response in terms of accuracy and interpretability. Our results show that networks that are trained under moderate levels of noise and regularization are in general more interpretable and obtain higher accuracy on validation sets. In these conditions, the network's hidden units seem to respond to specific and meaningful input patterns. Although we have not performed a complete analysis, some preliminary tests show that it may be possible to relate these activation patterns observed in the hidden layer to states in a DFA. Nevertheless this needs further analysis and will be addressed as future work.

Other possible extensions of the present work include the analysis of more complex problems, including non-regular languages and textual information such as programming code or written text. We also consider the use of different network architectures, such as LSTMs.

Acknowledgments. This work was funded by grant S2017/BMD-3688 from Comunidad de Madrid, and by Spanish project MINECO/FEDER TIN2017-84452-R (http://www.mineco.gob.es/).

References

1. Elman, J.L.: Finding structure in time. Cogn. Sci. **14**(2), 179–211 (1990)
2. Hochreiter, S., Schmidhuber, J.: Long short-term memory. Neural Comput. **9**(8), 1735–1780 (1997)
3. Rumelhart, D.E., Hinton, G.E., Williams, R.J.: Neurocomputing: foundations of research. In: Learning Representations by Back-propagating Errors, pp. 696–699. MIT Press, Cambridge (1988)
4. Graves, A., Mohamed, A., Hinton, G.E.: Speech recognition with deep recurrent neural networks. In: IEEE International Conference on Acoustics, Speech and Signal Processing, ICASSP 2013, Vancouver, BC, Canada, 26–31 May 2013, pp. 6645–6649 (2013)
5. Mikolov, T., Karafiát, M., Burget, L., Cernocký, J., Khudanpur, S.: Recurrent neural network based language model. In: 11th Annual Conference of the International Speech Communication Association, INTERSPEECH 2010, Makuhari, Chiba, Japan, 26–30 September 2010, pp. 1045–1048 (2010)
6. Sutskever, I., Vinyals, O., Le, Q.V.: Sequence to sequence learning with neural networks. In: Advances in Neural Information Processing Systems 27: Annual Conference on Neural Information Processing Systems 2014, 8–13 December 2014, Montreal, Quebec, Canada, pp. 3104–3112 (2014)
7. Cho, K., et al.: Learning phrase representations using RNN encoder-decoder for statistical machine translation. In: Proceedings of the 2014 Conference on Empirical Methods in Natural Language Processing, EMNLP 2014, 25–29 October 2014, Doha, Qatar. A meeting of SIGDAT, a Special Interest Group of the ACL, pp. 1724–1734 (2014)
8. Boulanger-Lewandowski, N., Bengio, Y., Vincent, P.: Modeling temporal dependencies in high-dimensional sequences: application to polyphonic music generation and transcription. In: Proceedings of the 29th International Conference on Machine Learning, ICML 2012, Edinburgh, Scotland, UK, 26 June–1 July 2012 (2012)
9. Mayer, H., Gomez, F., Wierstra, D., Nagy, I., Knoll, A., Schmidhuber, J.: A system for robotic heart surgery that learns to tie knots using recurrent neural networks. In: 2006 IEEE/RSJ International Conference on Intelligent Robots and Systems, pp. 543–548, October 2006
10. Zeng, Z., Goodman, R.M., Smyth, P.: Learning finite state machines with self-clustering recurrent networks. Neural Comput. **5**(6), 976–990 (1993)
11. Omlin, C.W., Giles, C.L.: Extraction of rules from discrete-time recurrent neural networks. Neural Netw. **9**(1), 41–52 (1996)

12. Casey, M.: The dynamics of discrete-time computation, with application to recurrent neural networks and finite state machine extraction. Neural Comput. **8**(6), 1135–1178 (1996)
13. Siegelmann, H.T., Sontag, E.D.: On the computational power of neural nets. J. Comput. Syst. Sci. **50**(1), 132–150 (1995)
14. Cohen, M., Caciularu, A., Rejwan, I., Berant, J.: Inducing regular grammars using recurrent neural networks. CoRR, abs/1710.10453 (2017)
15. Gers, F.A., Schmidhuber, E.: LSTM recurrent networks learn simple context-free and context-sensitive languages. Trans. Neural Netw. **12**(6), 1333–1340 (2001)
16. Giles, C.L., Miller, C.B., Chen, D., Sun, G., Chen, H., Lee, Y.: Extracting and learning an unknown grammar with recurrent neural networks. In: Advances in Neural Information Processing Systems 4, [NIPS Conference, Denver, Colorado, USA, 2–5 December 1991], pp. 317–324 (1991)
17. Jacobsson, H.: Rule extraction from recurrent neural networks: a taxonomy and review. Neural Comput. **17**(6), 1223–1263 (2005)
18. Wang, Q., Zhang, K., Ororbia II, A.G., Xing, X., Liu, X., Giles, C.L.: An empirical evaluation of rule extraction from recurrent neural networks. Neural Comput. **30**(9), 2568–2591 (2018)
19. Kolen, J.F.: Fool's gold: extracting finite state machines from recurrent network dynamics. In: Advances in Neural Information Processing Systems 6, [7th NIPS Conference, Denver, Colorado, USA, 1993], pp. 501–508 (1993)
20. Karpathy, A., Johnson, J., Fei-Fei, L.: Visualizing and understanding recurrent networks. CoRR, abs/1506.02078 (2015)
21. Linz, P.: An Introduction to Formal Languages and Automata, 4th edn. Jones and Bartlett Publishers, Burlington (2006)

Unsupervised Learning as a Complement to Convolutional Neural Network Classification in the Analysis of Saccadic Eye Movement in Spino-Cerebellar Ataxia Type 2

Catalin Stoean[1]([✉]), Ruxandra Stoean[1], Roberto Antonio Becerra-García[2],
Rodolfo García-Bermúdez[3], Miguel Atencia[2], Francisco García-Lagos[2],
Luis Velázquez-Pérez[4], and Gonzalo Joya[2]

[1] University of Craiova, Craiova, Romania
{catalin.stoean,ruxandra.stoean}@inf.ucv.ro
[2] Universidad de Málaga, Málaga, Spain
idertator@gmail.com, matencia@ctima.uma.es, lagos@dte.uma.es,
gjoya@uma.es
[3] Universidad Técnica de Manabí, Portoviejo, Ecuador
rodgarberm@gmail.com
[4] Centro para la Investigación y Rehabilitación de las Ataxias Hereditarias,
Holguin, Cuba
velazq63@gmail.com

Abstract. This paper aims at assessing spino-cerebellar type 2 ataxia by classifying electrooculography records into registers corresponding to healthy, presymptomatic and ill individuals. The primary used technique is the convolutional neural network applied to the time series of eye movements, called saccades. The problem is exceptionally hard, though, because the recorded saccadic movements for presymptomatic cases often do not substantially differ from those of healthy individuals. Precisely this distinction is of the utmost clinical importance, since early intervention on presymptomatic patients can ameliorate symptoms or at least slow their progression. Yet, each register contains a number of saccades that, although not consistent with the current label, have not been considered indicative of another class by the examining physicians. As a consequence, an unsupervised learning mechanism may be more suitable to handle this form of misclassification. Thus, our proposal introduces the k-means approach and the SOM method, as complementary techniques to analyse the time series. The three techniques operating in tandem lead to a well performing solution to this diagnosis problem.

Keywords: Classification · Convolutional neural networks ·
Unsupervised learning · k-means · Self-organizing maps ·
Saccadic eye movement

© Springer Nature Switzerland AG 2019
I. Rojas et al. (Eds.): IWANN 2019, LNCS 11507, pp. 26–37, 2019.
https://doi.org/10.1007/978-3-030-20518-8_3

1   Introduction

Spino-cerebellar ataxia of type 2 (SCA2) is an incurable neurodegenerative disorder that progressively and, at first imperceptibly, affects the nervous system. It can be diagnosed by very expensive means such as genetic analysis, and its course is visible in the impairment of certain body movements. But currently, the easiest, cheapest and most widely available procedure is based on electrooculography, by recording and examining the weak electrical potentials generated by the eye movement of a person tracking the trajectory of an object. This movements, induced by an abrupt displacement of the object, are called saccades. They have proved to be a valuable marker in common neurological disorders and their form can be used to perform a diagnosis at a pre-clinical stage of the disease. Consequently, the computational classification of these saccades can not only support a correct distinction between healthy, presymptomatic and ill people, but more importantly an early detection of the presymptomatic cases, thus triggering a timely medical assessment and intervention.

In the last couple of decades, medicine has become the playground of traditional machine learning techniques, such as random forests, support vector machines and neural networks [1,3,5,8,10] and more recently of deep learning methods [4,7,11]. The current problem has been tackled by shallow learning methods, however only to grasp the complexity of the classification task: healthy and presymptomatic registers are easily mistaken for each other [2]. This paper therefore goes further in exploring the potential of saccade classification through deep convolutional neural networks (CNN).

An electrooculographic test consists in a point alternatively appearing at each side of a screen, thus inducing a particular angular deviation in the patient's eyes. The evaluation proceeds by repeating the same object's trajectory several times, and the corresponding saccadic samples form a register, which is labelled by the expert either from its subjective analysis or from additional extra knowledge. The vast majority of saccades in the register of an ill person are clearly distinct from those of a healthy person. However, most saccades in the register of a presymptomatic person are practically indistinguishable from the standard healthy one, whereas a small number present a form slightly (but appreciably) different. Hence, supervised classification of registers comprising both positive and negative examples might get confused in learning the correspondence between a saccade and its label from the whole register. This is the case of CNN, thus our study also appoints two unsupervised approaches to perform a label-free analysis of the saccadic patterns, namely k-means (KM) and self-organizing maps (SOM). Additionally, the two methods provide an informative visualization of the form and disposition of saccades that provides more insight into the problem, unlike black box learning of the CNN. The creation of an ensemble out of the three methods is finally considered, leading to a 93.75% test accuracy on registers.

The paper is organized as follows. Section 2 describes the data and the preprocessing steps before feeding it to the learners. Section 3 outlines the three chosen techniques, with their architecture and supplementary mechanisms for the current problem. The experimental setup and results are presented in Sect. 4. Section 5 draws the conclusions and suggests directions for future improvement.

2 Data

Electrooculograms are elecrophysiologic signals obtained by electrooculography, which measures the electrical potential between the cornea and Bruch's membrane. This potential varies with the angular position of the eye, allowing us to track different eye movements such as saccades. The data used in this work was extracted from electrooculograms recorded with a sampling frequency of 200 Hz.

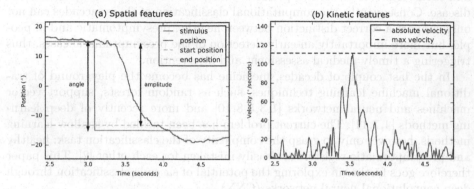

Fig. 1. Useful saccadic spatial and kinetic features

We have registers of 88 subjects at different health status: 29 of them are healthy people (control subjects), 29 are subjects with initial symptoms (presymptomatic individuals), and 30 are already suffering from active SCA2. For each person we have registers corresponding to electrooculograms recorded using protocols with different stimuli angles (10°, 20°, 30°, 60°). After calibration, the electrical potential samples are converted to angular position time series, and the velocity profiles are computed together with some important features, shown in Fig. 1. The saccade extraction process is described in the following procedure:

1. Position signals are filtered using a median filter of 9 points.
2. Velocity profiles are calculated using a Lanczos method with 11 points, and the output filtered again with another median filter of 9 points.
3. For detecting the saccades, KM with 2 clusters (having only a preprocessing role here) is applied to samples of the velocity profile. The samples in the cluster with high velocities are considered belonging to a saccade and the other ones to a fixation. Saccadic samples grouped together are selected as individual saccades. The sample with the minimum abscissa value (timestamp) in the group is considered the onset point, and the sample with the maximum abscissa value, the offset point.

4. For each pair of onset and offset points:
   (a) If the amplitude of the saccade is too low or too high the saccade is considered noisy and discarded. Skip the remaining steps.
   (b) Extract the saccade window using the following procedure:
      – Find the sample index with the maximal velocity between the onset and offset points of the saccade, and define it as mvi.
      – Define the saccade window as the set of 192 samples in the interval $[mvi - 96, mvi + 95]$.
   (c) The central value of the saccade, defined as the average of the minimal and maximal values, is subtracted from saccade windows, thus resulting centred saccades with null central value.
   (d) Left and right saccades can be distinguished where in the former the onset has a lower value than the offset. Right saccades are flipped using a symmetrical transformation with respect to the vertical axis, thus resulting in all the saccades with the same direction.

Finally, 6124 saccade windows of 192 samples each are obtained and vertically stacked resulting in a 6124×192 matrix X, which is then normalized component-wise into the interval $[-0.5, 0.5]$ to enhance the performance of the optimization algorithms involved in the learning process. The normalized matrix X_N results from the following formula: $X_N = \frac{X - \min(X)}{\max(X) - \min(X)} - 0.5$, where the operations $\min, \max$ act over all the matrix components.

The three classes of the problem are denoted as follows: C (control, healthy subjects), P (presymptomatic cases) and S (sick patients). The number of saccades per register in each class ranges in the interval $[49, 169]$ for C, $[38, 172]$ for P, and $[6, 169]$ for S.

3   Methodology

The architectural design of the CNN and the tuning of the KM and SOM for the task at hand are presented in the following subsections.

3.1   Convolutional Neural Networks

Since saccades are time series data, a 1D CNN (with temporal convolutions) is considered. Its architecture is chosen after manual testing. According to extensive preexperimentation, the CNN models applied to this task tend to overfit. Hence several dropout layers are interposed in between the feature extraction and the classification steps. The flow of layers is established as follows:

– A pair of convolutional 1D layers with size 3, a number of filters equal to 128 and a ReLU activation function.
– A max pooling 1D layer with a window size of 2 and a stride of 2.
– A dropout layer with rate 0.4.
– A second pair of convolutional layers with 256 filters of size 3.

- Another dropout layer with rate 0.4.
- A global average pooling 1D layer to further limit overfitting [6].
- A last dropout layer with rate 0.8.
- The final dense layer with the three outputs, corresponding to the given labels. Experiments showed that a sigmoid activation function works better than the usual softmax one, due to the overlapping nature of the outcomes.

The number of epochs is set to 100, the batch size to 25 and the optimizer is Adam. Data are split into training, test, and validation sets by considering whole registers, although the model is trained on individual saccades, whose class is given by the register to which it belongs. The model with the best accuracy on the validation saccadic samples is the selected one.

After training, the saccades in the test set are then taken one by one and labeled with the class predicted by the CNN model. The majority label in one register establishes its class. A register contains several saccades and the medical decision is made on the observed behavior of all these examples. As such, there are several samples that exhibit features distinct from the expected shape for that class. As a consequence, supervised learning will wrongly attribute some samples to classes that correspond to the shape of those series. Therefore, an unsupervised treatment might discover a more accurate grouping of saccades, according to shapes and not to labels given to a whole register. Accordingly, two conceptually different representatives of this type of learning are used for an unsupervised analysis of the problem in the next section.

3.2   K-Means

The simplest form of unsupervised learning is represented by the KM algorithm. The KM approach used herein follows the standard procedure for training: k cluster centroids are established starting from random positions and moved as samples are assigned to them based on proximal Euclidean distances. From pre-experimentation, it is observed that the value for k must be higher than 7, which gives a first confirmation of the multimodality of the saccadic forms present in the data set. Once the different shapes of the given saccades are discovered, the connection between clusters and the three classes of the problem has to be established.

Algorithm 1 outlines the entire KM procedure, from the generation of centroids (line 1) to the labeling of test cases. The determination of the profiles for each of the three categories is performed in the validation phase. Saccades in every validation register are attributed to a generated cluster centroid by Euclidean distance (line 5). Then, for each register, the percentage of saccades that are assigned to each cluster is determined (line 7). At this point, the three labels of the problem are also taken into account and the average percentage is now computed over all registers for each category (line 9). A profile for every label, regarding the percentage of saccades corresponding to each cluster, is therefore obtained. However, these profiles have many intersection points. It is therefore of interest to grasp what is the difference between them, i.e. which clusters are more prominent for each label (those that have more saccades assigned or, on the contrary, less examples as opposed to the amount for the other classes). Consequently, the difference between the cluster values of one category versus the average of the other two is next calculated by Manhattan distance (line 13). An average of this vector of differences is computed (line 14) and those positions that have values above it denote the clusters that are discriminating for that class (line 15). A ranking of the importance of each of these prominent clusters still has to be quantified and therefore weights proportional to the corresponding value above the average of discerning differences give the measure of this degree (line 16).

In the test phase, the form for each register (in percentage of its saccades attributed to each of the k clusters) is acquired (line 22). Finally, the weighted Euclidean distance (on the base of the weights calculated in line 16) between this form and the profile of each class (obtained in the validation phase—line 9) is calculated (line 23), naturally taking into account only the discriminative clusters (line 15). The distances corresponding to the three classes are then divided by the number of positions found important for each label (line 24). The minimum distance points to the label that will be predicted by the model for the current test register (line 25).

3.3   Self Organizing Maps

Although simple and direct, KM may be a too general algorithm for clustering the temporal saccadic samples. SOM, on the other hand, introduces a more sophisticated dimensionality reduction that is more appropriate for time series analysis, while at the same time provides a different visualization angle of the learning. The SOM generates a two-dimensional map from the initial high-dimensional input, based on the competition between neurons in response to the training saccades and a neighborhood function that preserves the topology of the initial space.

Once the map is unsupervisedly produced from unlabeled saccades in the training phase, the mapping of the test registers proceeds in the following manner. The winning neuron is determined for each saccade in a test register. For that position, each training saccadic sample mapped within the same place is collected and the class to which it belongs gets one vote. The test saccade thus gets a triplet of votes, corresponding to votes for each label of the training samples found in that position. The register is assigned the sum of the triplets of each of its saccades and the class with the maximum value gives its final label. In case of equality, the leftmost (less severe) condition is taken as the label.

4   Experimental Results

The data is split into training-validation-test with the percentages of 40%-40%-20% of randomly taken registers from each category: 12-12-5 registers for C, 12-12-5 for P and 12-12-6 for S.

Algorithm 1. KM detection of saccadic centroids, class profiles and test predictions

Require: Data set of saccadic registers and the three outcomes
Ensure: k centroids and predicted labels for test observations
 1: Determine the k centroids from the *training* saccades
 2: **for** each label **do**
 3:     **for** each *validation* register **do**
 4:         **for** each saccade **do**
 5:             Attribute to a cluster centroid
 6:         **end for**
 7:         Compute the percentage of saccades attributed to each cluster
 8:     **end for**
 9:     Calculate the average percentage of saccades for every class
10: **end for**
11: **for** each label **do**
12:     Compute the average of the validation profiles of the complementary two labels

13:     Get the vector of differences between the profile of the current label and the computed average profile of the other categories
14:     Average the obtained vector of differences
15:     Take positions whose values are above average in the vector as class defining clusters
16:     Calculate weights for each position proportional to the distance to the average
17: **end for**
18: **for** each *test* register **do**
19:     **for** each saccade **do**
20:         Attribute to a cluster centroid
21:     **end for**
22:     Compute the percentage of saccades attributed to each cluster
23:     Compute the weighted distances between the percentage form of the register and the validation profile of each label, taking into account only the class defining clusters
24:     Divide the distances by the number of positions found for each class
25:     Label the register with the class of the validation profile that led to the minimum distance
26: **end for**

The following comparative results come from a split where the number of saccades for training is 901 C - 1039 P - 491 S (2431 in total), for validation 977 C - 845 P - 771 S (2593 in total) and for test 360 C - 421 P - 319 S (1100 in total). There are accordingly 2238 C saccades, 2305 P and 1581 S ones.

Given the random nature for the initial weighting in CNN and SOM, and for the KM initialization of centroids, 30 runs are conducted for each method in order to be conclusive of the obtained test results. The three methods are implemented in Python, using the following libraries: Keras with TensorFlow as back end (CNN), Scikit-learn (KM) and MiniSom (SOM). The architecture and corresponding parametrization of the CNN is appointed after manual selection.

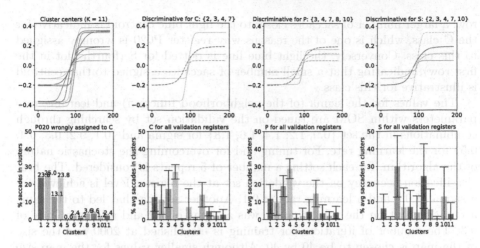

Fig. 2. The plots in the first row indicate the cluster centers as found by the KM in the first picture, followed by the centroids that are discovered as important for each class in turn. The first plot on the second row indicates the profile of the saccades in a presymptomatic register wrongly classified and the subsequent plots show the mean percentages of saccades from the validation set that are assigned to each class in turn. The numbers and colors shown in the plots are consistent among all of them. The centroids in dotted line indicate that the amount of saccades associated with that cluster should be smaller than the corresponding average number from the other two classes; continuously drawn ones mean the opposite. (Color figure online)

The value for k in KM is also the result of manual tuning. Figure 2 shows in the first plot the 11 centroids discovered by the KM on the training data. Each saccade from the validation and from the test set is then compared to these cluster centers. For every validation register the amount of saccades that are attributed to each cluster (the assignment is set by the minimum Euclidean distance) is computed and a percentage of the saccades distribution is further obtained (line 7 in Algorithm 1). Then, averages are computed over the registers with the same class for validation (line 9): the results are illustrated in the last 3 plots from the second row. The next step regards the identification of the discriminating centroids for every label in turn, based on these computed validation class averages. The discriminative clusters are selected for each class against the other two in such a way that they are either much smaller or much larger than for the complementary labels (line 15). The distinctive centroids for every class in turn are illustrated in the plots 2, 3 and 4 from the first row: dotted line means that the amount of saccades in that register should be smaller for that cluster, while continuous ones indicate the reverse. It should be underlined that the colors for the clusters are kept similar for all plots, be that they contain lines or bars.

The shapes for the C and P classes are very similar, as observed from the profiles for each label found in the validation phase. However, the discriminative clusters are not identical for the two, as indicated in the second plot from the first

row: a larger number of saccades closer to the light blue centroid is indicative for
the C class, which is one of the reasons why register P020 is wrongly assigned
to this class. Conversely, the light blue line is dotted for S (fourth plot in the
first row), indicating that a small number of saccades assigned to that centroid
is illustrative for this class.

The values for the sigma (of the neighborhood function) and learning rate
parameters within SOM are tuned on the validation set by searching through
combinations in the set {0.9, 1.2, 1.4, 1.6, 1.8} for sigma and {0.1, 0.2, 0.3, 0.4,
0.5} for the learning rate. For tuning and for overcoming the stochastic nature
of the algorithm, for each setting a number of 5 repeats is considered. The best
classification accuracy on the validation set at the register level is achieved for
a sigma of 0.9 and a learning rate of 0.1. Another setting that led to the same
accuracy result on the validation set was a sigma of 1.2 and a learning rate of
0.3. The number of iterations for training is established at 2000 and the size
of the map is chosen to be 40 by 40. Although smaller values for the map size
lead to faster running time, the accuracy of the results decreases in these cases,
while larger values for this parameter (it was tried only up to 50) did not lead to
significant improvement. Using these settings, the training phase conducts to the
map illustrated in Fig. 3. Although the problem has only three classes, there are
many more clusters obtained by SOM. This comes in line with the KM results
which showed that better results were obtained when the number of clusters was
high, concretely 11. Moreover, there are many cells in which winning neurons
are chosen from different classes—especially the C class is frequently mixed with
the P category.

Figure 4 shows the confusion matrices for the three methodologies. The misclassified test registers in the 30 repeated runs for each of the three methods
are illustrated in Fig. 5. Although for all three matrices the number of confused
registers of types C and P are identical, the actual mistaken registers are not
the same: as observed in Fig. 5, P020 is the only one that is confused by both
KM and SOM. Therefore if an ensemble is created out of the three methods by
majority vote, there is only one error out of 16 registers, leading to a 93.75%
test accuracy.

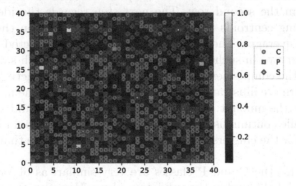

Fig. 3. SOM representation with the winning neurons as obtained from the training
phase.

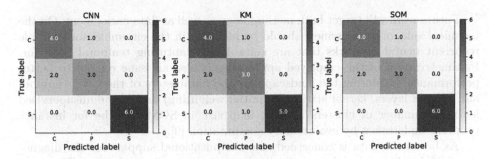

Fig. 4. The confusion matrices for CNN, KM and SOM. The mistaken registers by each model in turn are illustrated in Fig. 5.

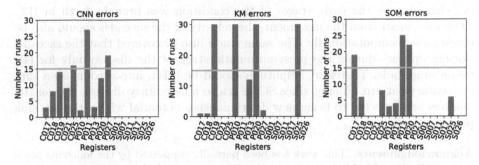

Fig. 5. The number of times the class of each test register is mistaken out of 30 different runs. When a register is mistaken in more than half of the 30 times, it is considered wrongly classified.

5   Conclusions and Future Work

The paper aims to discover the insights of SCA 2 diagnosis from electrooculography data and does so through representatives of both supervised and unsupervised learning. Three techniques are employed for saccade pattern recognition in connection to the three classes (healthy, presymptomatic and ill) of the problem: a convolutional neural network, a k-means and a self-organizing map.

KM performs the worst of the three as a single decision maker (with four wrongly attributed registers from all three categories), which was expected, as it is the simplest form of learning. The CNN and the SOM mistake three registers, two of which are presymptomatic and one healthy. Moreover, it was interesting to see that the three algorithms have each a very different perspective upon the saccades-class correspondence, which demonstrates both the multiple facets of the problem and the efficiency of the ensemble, with only one final wrongly labeled test register and a corresponding accuracy of 93.75%. Although each of the chosen techniques misclassifies more presymptomatic cases than samples from the complementary classes, the overall ensemble mistakes only one, which makes it a suitable solution for the task of an early recognition of the disease.

Future work will target both methodology as well as data enrichment. On the technical side, other classifiers should be included in the ensemble, for example recurrent neural networks that are suitable for capturing temporal behavior. Parametrization of the employed approaches is also an issue related to better performance. Exploring the landscape for the parameters of the CNN (number and type of layers, kernel attributes, initial weighting) and of the unsupervised learners (number of clusters in KM, map size in SOM) can be for instance properly undertaken by evolutionary computation [9].

As far as the data is concerned, the computational support for the diagnosis of SCA 2 has only taken so far into account saccadic movement. There are however some other preclinical indicators of the disease that should be further included as input for the learning methodologies. Evidence that saccadic slowing characterizes the early stages of the condition was brought forth in [12]. Therefore, apart from the movement, the velocity of the saccades should also be investigated computationally. The same study also discovered that the saccadic slowing appears during the presymptomatic stage of the disease only for 60° target amplitude. Therefore, amplitude should be taken into account as a supplementary pattern. Finally, since SCA2 is also a hereditary disease, the genetic markers are expected to bring new discriminating potential within the learning process.

Acknowledgements. This work has been partially supported by the following institutions: the Spanish Ministry of Science, Innovation and Universities, through the Plan Estatal de Investigación Científica y Técnica y de Innovación, Project TIN2017-88728-C2-1-R, and the University of Málaga-Andalucía-Tech through the Plan Propio de Investigación y Transferencia, Project DIATAX: Integración de nuevas tecnologías para el diagnóstico temprano de las Ataxias Hereditarias.

References

1. Atencia, M.A., García-Garaluz, E., de Arazoza, H., Joya, G.: Estimation of parameters based on artificial neural networks and threshold of HIV/AIDS epidemic system in cuba. Math. Comput. ModelL. **57**(11–12), 2971–2983 (2013)
2. Becerra-García, R.A., et al.: Data mining process for identification of nonspontaneous saccadic movements in clinical electrooculography. Neurocomputing **250**, 28–36 (2017)
3. Cleophas, T.J., Zwinderman, A.H.: Machine Learning in Medicine - A Complete Overview. Springer, Cham (2015). https://doi.org/10.1007/978-3-319-15195-3
4. Esteva, A., et al.: A guide to deep learning in healthcare. Nat. Med. **25**, 24–29 (2019)
5. Lichtblau, D., Stoean, C.: Cancer diagnosis through a tandem of classifiers for digitized histopathological slides. PLoS ONE **14**(1), 1–20 (2019)
6. Lin, M., Chen, Q., Yan, S.: Network in network. CoRR abs/1312.4400 (2014)
7. Litjens, G., et al.: A survey on deep learning in medical image analysis. Med. Image Anal. **42**, 60–88 (2017)
8. Obermeyer, Z., Emanuel, E.J.: Predicting the future—big data, machine learning, and clinical medicine. N. Engl. J. Med. **375**(13), 1216–1219 (2016)

9. Preuss, M., Stoean, C., Stoean, R.: Niching foundations: basin identification on fixed-property generated landscapes. In: Krasnogor, N., Lanzi, P.L. (eds.) 13th Annual Conference on Genetic and Evolutionary Computation (GECCO-2011), pp. 837–844. ACM (2011)

10. Stoean, C., Stoean, R., Sandita, A., Mesina, C., Gruia, C.L., Ciobanu, D.: How much and where to use manual guidance in the computational detection of contours for histopathological images? Soft Comput. (2018). https://doi.org/10.1007/s00500-018-3029-9

11. Stoean, R.: Analysis on the potential of an EA-surrogate modelling tandem for deep learning parametrization: an example for cancer classification from medical images. Neural Comput. Appl. (2018). https://doi.org/10.1007/s00521-018-3709-5

12. Velázquez-Pérez, L., et al.: Saccade velocity is reduced in presymptomatic spinocerebellar ataxia type 2. Clin. Neurophysiol. **120**, 632–5 (2009)

Scale-Space Theory, F-transform Kernels and CNN Realization

Vojtech Molek[✉] [iD] and Irina Perfilieva [iD]

Institute for Research and Applications of Fuzzy Modeling,
Centre of Excellence IT4Innovations, University of Ostrava,
30. dubna 22, Ostrava, Czech Republic
{vojtech.molek,irina.perfilieva}@osu.cz

Abstract. We present scale-space and F-transform inspired modification to convolutional neural networks. The proposed modification improves network classification accuracy using multi-scale image representation and F-transform kernels pre-training. We evaluate our model on two databases and show better performance than networks without F-transform pre-training.

Keywords: F-transform · Initialization · Scale-space ·
Convolutional neural network · Classification

1 Introduction

Convolutional Neural Networks (CNNs) are a special type of neural networks (NNs) exploiting the spatial data structures. NNs success comes from the ability to learn [10], that in general means to approximate function mapping input variables to output variables. The approximation fitness depends on network parameters. To learn is essentially to solve an optimization problem [20].

The whole learning process is highly dependent on the initial network parameters configuration [8]. There are general strategies on how to meaningfully initialize networks parameters such as random initialization from some distribution or Xavier initialization [6], but perhaps the most used one is to use parameters from other networks that have been already trained on similar or the same data [2,3,5,13,16]. The efficiency of initialization depends on a distance between initialized and initializing networks training datasets [27]. At the same time, [27] authors postulate that it is always better to initialize with pre-trained parameters than initialize parameters randomly. Due to the hierarchical nature of feedforward neural networks, parameters in lower layers capture general non-domain specific knowledge [28], and therefore those parameters initialization can be the same or similar through the different domains.

In this paper, we use scale-space theory to create and evaluate CNN with a convolutional layer with multi-scale kernels [26] based on F-transform [18]. We

Supported by Ostrava University.

draw inspiration from the scale-space theory that defines a pyramid scheme to process data in different scales. Specifically, we exploit the F-transform kernels ability to be scaled in different sizes and by this, to emulate the pyramid scheme. We justify our selection, referring to the fact that the F-transform kernels extract meaningful features sufficient for reconstruction of an original object [19] and are, therefore, suitable candidates for NN initialization. Similary, the scale-spaces inspiration was used in Inception [25], with reference to [1, 22]. However, Inception does not focus on scalable weight initialization, but instead on classification and overall performance in the ILSVRC 2014[1].

Our contribution consists in setting up the initialization of a **convolutional layer with multi-scale F-transform kernels**. We expand FTNet from our previous work [14] and evaluate FTNet performance on two datasets using hyperparameters comparative search. Further, we analyze, how kernels are changing during learning. We analyzed kernels from the first convolutional layer of several known CNNs trained on the ImageNet and analyzed similarities among kernels across the networks.

2 Scale-Space CNN

The functioning of a CNN depends on its ability to extract relevant features from a dataset, and then use them for the CNN training and testing. From the literature, we know that a CNN architecture is determined by two main factors: a problem to be solved and a training dataset. Both parts are immanent to the process of a CNN creation.

Several parameters should be selected on the stage of a CNN design. Mostly, there are results of the trial and error approach to problem-solving. Among fundamental parameters, the window/kernel sizes of convolutional layers are of principal importance. It turned out that tuning of these parameters is not thoroughly discussed in connection with the NN design. The pre-described values are given by designers and are taken as a default setting. However, the question of their proper selection has not been discussed yet. In this contribution, we will be focused on one possible answer to this question.

We stem from the so-called scale-space theory [11] that is focused on a multi-scale representation of images. This theory is widely used in the area of computer vision and in particular, in applications connecting with noise removing, inpainting, segmentation, and scale invariant feature transform (SIFT). The last application is of our particular interest due to its close relationship to the problem of feature extraction. Below, we remind relevant details [12].

2.1 Scale-Space Theory in Computer Vision

Typical tasks in the field of machine vision are object recognition, object manipulation, and visually guided navigation. For all these tasks, the problem of image

---

representation is of the first importance since it makes relevant aspects of the information content immediately accessible without additional processing. All-purpose features can be extracted by uncommitted vision systems whose operations are convolutions with Gaussian-like kernels and their derivatives at different scales. This conclusion was shared by various authors and motivated the elaboration of the SIFT, successfully used for image matching and recognition. SIFT extracts distinctive invariant features that can be used to perform reliable matching between different views of an object or scene. The SIFT features can also be used for object recognition. Without going into details, we say that SIFT features are 2D keypoints on the image each associated to a vector of low-level descriptors. The latter are invariant by rescaling, in-plane rotating, noise addition and in some cases by changes of illumination.

If we analyze how the mentioned invariance is achieved, then we see that the principal trick consists in looking through all involved scales where a particular keypoint is still detected. This refers to the scale-space methodology and to the characterization of kernels that are used for scaling. In the SIFT realization, the Gaussian function was used as a scale-space kernel; moreover, it was claimed that no other function is suitable for this purpose [12].

2.2 The Higher Degree F-transform Kernels for the Scale-Space Methodology

In this contribution, we wish to show that together with the deep learning strategy, the scale-space methodology can be efficiently applied for the object recognition task. We also want to break the restriction on Gaussian functions as only suitable for this purpose. Instead, we propose to apply the higher degree F-transform kernels as an alternative for creating a scale-space representation.

We rely on our previous successful experiments with up to the second degree F-transform (F^2-transform) kernels combined with the LeNet-5-type architecture [14]. That time, we proposed a new CNN FTNet where we made use of F^2-transform kernels parametrized by kernel sizes, learnability, the choice of strides or pooling, etc. The FTNet was trained on the database MNIST and tested on handwritten inputs. The obtained results demonstrated better recognition accuracy than the automatically trained LeNet-5.

Another argument in favor of F^2-transform kernels was obtained after we analyzed experiments with several CNNs. The purpose was to find analogies between the F-transform-based and the convolution kernels in the first layer of the considered CNNs. In more detail, we selected six networks: VGG16, VGG19, InceptionV3, MobileNet, ResNet, and AlexNet, that were trained on ImageNet, using the same training dataset, consisting of ≈ 1.2M color images with various resolution (usually downsampled to 256×256). We analyzed kernels from the first convolutional layer of each considered network and discussed whether there are similarities among kernels across the networks. To reduce the space of kernels, we first apply the hierarchical clustering on every network kernel set separately, and then, look for the similarities among clusters. We observed that the extracted clusters contain similar elements (kernels) across the different networks that fall into one of the following categories:

- *Gaussian-like*;
- *Edge detectors (with various angle specifications)*;
- *Texture detectors*;
- *Color extraction blobs*.

We compared the semantic meaning of the extracted clusters (in terms of the above-given characteristics) with that of the F-transform kernels in the FTNet and observed the coincidence in the first two items from the above-given list. To be more precise, the F^0-transform kernels are the Gaussian-like, and the F^1-transform kernels are (horizontal or vertical) edge detectors. This again supports our motivation regarding the suitability of the F-transform kernels in the first layers of CNNs.

2.3 CNN with the Scale-Space Methodology and Higher Degree F-transform Kernels

We analyze the efficiency of the combined methodologies: the scale-space and the deep convolution-based learning, where scaling is performed by the higher degree F-transform kernels. The estimation and comparison are considered on the problem of image recognition. The whole process is realized by a CNN whose architecture differs almost coincides with the FTNet [14] and from it in the first layer, where we use the FT kernels with three different sizes.

We remind that the higher degree F-transform extracts the features that characterize the following universal properties of any functional object. They are average values of a function range, its first and higher derivatives within a selected area. If a functional object is an image, then these universal properties/features are average values of brightness, edginess, convexity and so on. The degree of averageness depends on the size of the corresponding area (kernel).

With the scale-space methodology, we expect to increase the universality of the FTNet in the sense of its applicability to non-specified datasets of images. To support this conjecture, we did experiments with the datasets MNIST and CIFAR-10.

3 Experiments Setup and Results

In this experimental section, we confirm the above discussed theoretical conjectures, attempting the similar tests as those in our previous research [14]. In [14], we have studied the impact of the F-transform kernels and other learning options on FTNet. Here, we introduce a new searching option for the first convolutional layer - a variation of kernel sizes (VACL) and estimate its impact on the FTNet accuracy.

In Table 1, we remind the FTNet architecture. The FTNet is parametrized by a set of *hyperparameters*. The hyperparameters define, among other things,

Table 1. FTNet architecture [14]. C_i – a convolution layer, S_i – a subsampling layer, FC_i – a fully connected layer.

Hyperparameters	C_1	S_2	C_3	S_4	FC_5	FC_6
Kernel size	5×5	-	5×5	-	-	-
# feature maps	8	-	64	-	-	-
Stride	1×1	2×2	1×1	2×2	-	-
Pooling size	-	2×2	-	2×2	-	-
# fully connected units	-	-	-	-	500	10

an architecture of a network and details of the learning procedure. One of the key hyperparameters is the on/off learning of layers weights. The "on" state of a network allows the network to reach the optimal weights configuration.

In new experiments we look for optimal combinations of the four hyperparameters: **Initialization** $\mathcal{I}$, **kernels size** $\mathcal{D}$, **Trainability** $\mathcal{T}$ and **Subsampling** $\mathcal{S}$. In order to analyze the FTNet accuracy with respect to the four hyperparameters we perform a complete search through the space. This space is determined by all possible combinations of hyperparameters values, displayed in Table 2.

Table 2. The hyperparameters values for C_1 and C_3. Note that VACL is exclusive for the first convolutional layer due to model size restriction.

$\mathcal{D}$	$\mathcal{S}$	$\mathcal{T}$	$\mathcal{I}$
3×3	None	Trainable	F-transform kernels
5×5	Max-pool	Non-trainable	Kernels $\sim \mathcal{N}(0,1)$
7×7	Stride	-	-
VACL	-	-	-

Let us characterize the above-mentioned hyperparameters:

- $\mathcal{I}$ species whether layer C_1 and C_3 are initialized with F-transform kernels or random ones with $\sim \mathcal{N}(0,1)$.
- $\mathcal{D}$ species whether size of C_1 and C_3 convolution kernels.
- $\mathcal{T}$ species whether C_1 and C_3 are trained or not.
- $\mathcal{S}$ species whether C_1 and C_3 are followed by *maxpooling layer*[2] or *strided convolution* [23] or neither of them.

Scale-space inspired VACL value of $\mathcal{D}$ specifies C_1 convolution sizes such that all three sizes (3×3, 5×5, 7×7), are present. With the value VACL of $\mathcal{D}$, we force the network to process the input image through multiple scales (resolutions). VACL schema is in Fig. 1.

---

[2] Subsampling operation originates from Hubel and Wiesel [7]; comparison of pooling can be found in [21].

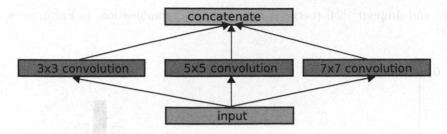

Fig. 1. Scheme of the convolutional layer with variable kernels sizes, realized as multiple convolution layers with their outputs concatenated.

Let us notify that other hyperparameters: optimizer (*Stochastic Gradient Descent* with *Nesterov momentum* [15]), *learning rate* (0.01), regularization, and *cross-entropy* loss function are not subject to grid search.

We evaluate all combinations of hyperparameters' values on two datasets: **MNIST** [4] and **CIFAR-10** [9]. We have chosen MNIST and CIFAR10 due to their relative simplicity, which allows us to conduct experiments without significant added variability from data. Because the F-transform kernels extract features with functional meaning and insensitive to colors, we convert CIFAR-10 images to the grayscale. Additionally we normalize and center [17] both datasets. To minimize randomness in our experiment we use pre-sampled weights, whenever $\mathcal{I}$ is initialized with random kernel weights. Further, we run ten iterations of every combination. All combinations are trained for three epochs, due to the time restrictions. Nevertheless, the achieved test accuracy is more than 97% on MNIST and 52% on CIFAR-10. In total, we have 1728 hyperparameters combinations to evaluate. Lastly, we train one of the best combinations for 100 epochs to see whether the F-transform kernels in C_1 change.

The results of our searching are sorted with respect to the accuracy on testing datasets. Figures 3 and 2 contain relative frequencies of hyperparameters of the 500 best combinations. In the case of MNIST (Fig. 3), the most prevalent kernel size for C_1 is VACL; this confirms our hypothesis regarding the scale-space methodology beneficence with respect to the learning process. Among the optimal initialization of C_1 and C_3 we see significantly more combinations with **trainable** F-transform kernels. We conclude that the F-transform kernels initialization has serious, positive impact on the accuracy of FTNet. The results on CIFAR-10 database have the same statistics and additionally prove the usefulness of VACL and F-transform initialization.

An unexpected result is high relative frequency of the "no subsampling" in both cases while stride being worst out of the three.

As an additional argument in favor of our technology, we visualize the F-transform kernels in C_1 before and after 100 epochs of training with the following setting: $\mathcal{I}$ = *F-transform kernels*, $\mathcal{D}$ = {VACL, 3×3}, $\mathcal{T}$ = *trainable*, $\mathcal{S}$ = *max pool* and dropout[3] [24] (between FC_5 and FC_6) combination. In Fig. 4, we see that

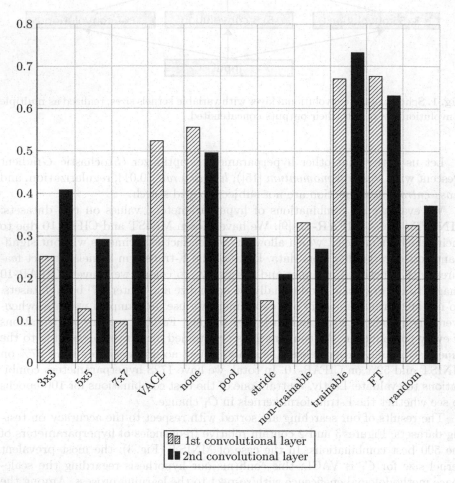

Fig. 2. Relative frequencies of the hyperparameters values for the first and second convolutional layers within the first 500 best combinations in terms of accuracy after 3 epochs of learning on CIFAR-10.

1. Up to the small contrast changes, 3×3 kernels are kept.
2. The shapes of 5×5 kernels are kept in general; however, some variational details were added after training.
3. The shapes of 7×7 are kept for F^1 and F^2 kernels; however the F^0 kernels became similar to the rotated F^1.

---

[3] We have employed dropout to reduce network overfitting.

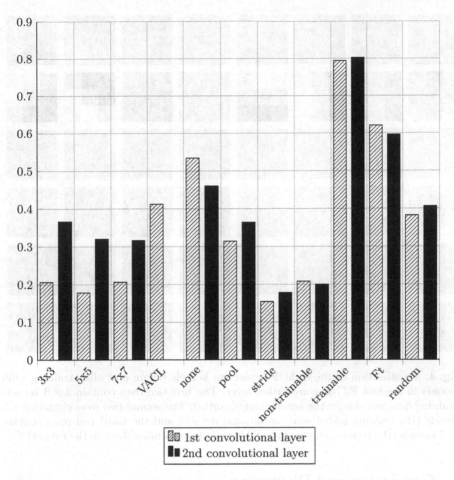

Fig. 3. Relative frequencies of the hyperparameters values for the first and second convolutional layers within the first 500 best combinations in terms of accuracy after 3 epochs of learning on MNIST.

Thus we draw the following conclusions:

1. The scale-space inspired VACL is the most frequent *"size"* of the convolution kernels in the first convolutional layer of the considered CNN, trained on both datasets.
2. The initialization of C_1 and C_3 convolutional layers with the F-transform kernels leads to the higher network accuracy.
3. Excluding subsampling from CNN's architecture increases network accuracy; however, it contributes to an undesirable effect of overfitting.
4. Including subsampling into CNN's architecture leads to a quicker decrease of a loss function.
5. The F-transform kernels in the first layer C_1, do not significantly change their shapes during training. Therefore they are an ideal choice for features extraction.

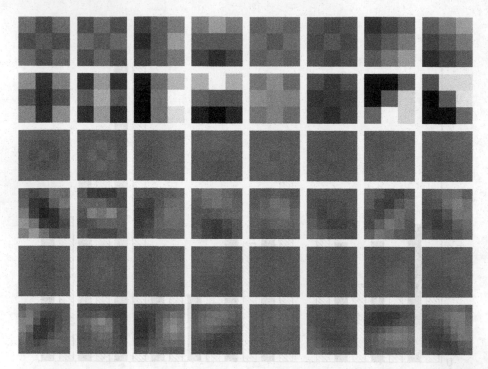

Fig. 4. Visualization of the eight F-transform kernels before and after training (100 epochs) in the first FTNet convolution layer. The first two rows contain 3×3 kernels (training does not change the kernels significantly). The second two rows contain 5×5 kernels (the training added some variational details) and the third two rows contain 7×7 kernels (the training changed the F^0 kernels transforming them to the rotated F^1).

4 Conclusion and Discussion

In this paper, we introduced a novel approach combing the F-transform technique with the scale-space theory. We described the FTNet initialization with F-transform kernels in their multiple scale representation. We showed that FTNet with the proposed modification achieved higher accuracy in comparison with the randomly initialized FTNet with single sized convolution kernels. Because the F-transform kernels extract general features with functional semantics, we confirmed that they serve as a good choice for deep CNNs initialization, being independent of the target domain. We supported our FTNet with well-developed theory of the F-transform that can serve as a bridge between Fuzzy Logic and Neural networks. Further research could answer the question of whether the proposed methodology can be efficiently propagated to higher layers and applied to colorful images as well as expand current architecture.

Acknowledgment. The work was supported by ERDF/ESF "Centre for the development of Artificial Intelligence Methods for the Automotive Industry of the region" (No. CZ.02.1.01/0.0/0.0/17_049/0008414).

References

1. Arora, S., Bhaskara, A., Ge, R., Ma, T.: Provable bounds for learning some deep representations. In: International Conference on Machine Learning, pp. 584–592 (2014)
2. Chen, T., Goodfellow, I., Shlens, J.: Net2Net: accelerating learning via knowledge transfer. arXiv preprint arXiv:1511.05641 (2015)
3. Cruz Jr, G.V., Du, Y., Taylor, M.E.: Pre-training neural networks with human demonstrations for deep reinforcement learning. arXiv preprint arXiv:1709.04083 (2017)
4. Deng, L.: The mnist database of handwritten digit images for machine learning research (best of the web). IEEE Signal Process. Mag. **29**(6), 141–142 (2012)
5. Erhan, D., Bengio, Y., Courville, A., Manzagol, P.A., Vincent, P., Bengio, S.: Why does unsupervised pre-training help deep learning? J. Mach. Learn. Res. **11**(Feb), 625–660 (2010)
6. Glorot, X., Bengio, Y.: Understanding the difficulty of training deep feedforward neural networks. In: Proceedings of the Thirteenth International Conference on Artificial Intelligence and Statistics, pp. 249–256 (2010)
7. Hubel, D.H., Wiesel, T.N.: Receptive fields, binocular interaction and functional architecture in the cat's visual cortex. J. Physiol. **160**(1), 106–154 (1962)
8. Krähenbühl, P., Doersch, C., Donahue, J., Darrell, T.: Data-dependent initializations of convolutional neural networks. arXiv preprint arXiv:1511.06856 (2015)
9. Krizhevsky, A., Nair, V., Hinton, G.: The cifar-10 dataset. http://www.cs.toronto.edu/kriz/cifar.html (2014)
10. LeCun, Y., Bengio, Y., Hinton, G.: Deep learning. Nature **521**(7553), 436 (2015)
11. Lindeberg, T.: Scale-space theory: a basic tool for analyzing structures at different scales. J. Appl. Stat. **21**(1–2), 225–270 (1994)
12. Lowe, D.G.: Distinctive image features from scale-invariant keypoints. Int. J. Comput. Vis. **60**(2), 91–110 (2004)
13. Mishkin, D., Matas, J.: All you need is a good init. arXiv preprint arXiv:1511.06422 (2015)
14. Molek, V., Perfilieva, I.: Convolutional neural networks with the F-transform kernels. In: Rojas, I., Joya, G., Catala, A. (eds.) IWANN 2017. LNCS, vol. 10305, pp. 396–407. Springer, Cham (2017). https://doi.org/10.1007/978-3-319-59153-7_35
15. Nesterov, Y.E.: A method for solving the convex programming problem with convergence rate $\mathcal{O}(1/k^2)$. Dokl. Akad. Nauk SSSR **269**, 543–547 (1983). https://ci.nii.ac.jp/naid/10029946121/en/
16. Oquab, M., Bottou, L., Laptev, I., Sivic, J.: Learning and transferring mid-level image representations using convolutional neural networks. In: Proceedings of the IEEE Conference on Computer Vision and Pattern Recognition, pp. 1717–1724 (2014)
17. Pal, K.K., Sudeep, K.: Preprocessing for image classification by convolutional neural networks. In: IEEE International Conference on Recent Trends in Electronics, Information & Communication Technology (RTEICT), pp. 1778–1781. IEEE (2016)

18. Perfilieva, I., Haldeeva, E.: Fuzzy transformation. In: Proceedings Joint 9th IFSA World Congress and 20th NAFIPS International Conference (Cat. No. 01TH8569), vol. 4, pp. 1946–1948. IEEE (2001)

19. Perfilieva, I., Holčapek, M., Kreinovich, V.: A new reconstruction from the F-transform components. Fuzzy Sets Syst. **288**, 3–25 (2016)

20. Ruder, S.: An overview of gradient descent optimization algorithms. arXiv preprint arXiv:1609.04747 (2016)

21. Scherer, D., Müller, A., Behnke, S.: Evaluation of pooling operations in convolutional architectures for object recognition. In: Diamantaras, K., Duch, W., Iliadis, L.S. (eds.) ICANN 2010. LNCS, vol. 6354, pp. 92–101. Springer, Heidelberg (2010). https://doi.org/10.1007/978-3-642-15825-4_10

22. Serre, T., Wolf, L., Bileschi, S., Riesenhuber, M., Poggio, T.: Robust object recognition with cortex-like mechanisms. IEEE Trans. Pattern Anal. Mach. Intell. **3**, 411–426 (2007)

23. Springenberg, J.T., Dosovitskiy, A., Brox, T., Riedmiller, M.: Striving for simplicity: the all convolutional net. arXiv preprint arXiv:1412.6806 (2014)

24. Srivastava, N., Hinton, G., Krizhevsky, A., Sutskever, I., Salakhutdinov, R.: Dropout: a simple way to prevent neural networks from overfitting. J. Mach. Learn. Res. **15**(1), 1929–1958 (2014)

25. Szegedy, C., et al.: Going deeper with convolutions. In: Proceedings of the IEEE Conference on Computer Vision and Pattern Recognition, pp. 1–9 (2015)

26. Vlašánek, P., Perfilieva, I.: The F-transform in terms of image processing tools. J. Fuzzy Set Valued Anal. **2016**(1), 54–62 (2016)

27. Yosinski, J., Clune, J., Bengio, Y., Lipson, H.: How transferable are features in deep neural networks? In: Advances in Neural Information Processing Systems, pp. 3320–3328 (2014)

28. Zeiler, M.D., Fergus, R.: Visualizing and understanding convolutional networks. In: Fleet, D., Pajdla, T., Schiele, B., Tuytelaars, T. (eds.) ECCV 2014. LNCS, vol. 8689, pp. 818–833. Springer, Cham (2014). https://doi.org/10.1007/978-3-319-10590-1_53

Numerosity Representation in InfoGAN:
An Empirical Study

Andrea Zanetti[1,2]([✉]), Alberto Testolin[3], Marco Zorzi[3,4],
and Pawel Wawrzynski[1]

[1] Warsaw University of Technology, Warsaw, Poland
pawel.wawrzynski@pw.edu.pl
[2] Intel Technology Poland, Gdansk, Poland
andrea.zanetti@intel.com, a.zanetti@ii.pw.edu.pl
[3] Department of General Psychology and Padova Neuroscience Center,
University of Padova, Padua, Italy
{alberto.testolin,marco.zorzi}@unipd.it
[4] IRCCS San Camillo Hospital, Venice, Italy

Abstract. It has been shown that *"visual numerosity emerges as a statistical property of images in 'deep networks' that learn a hierarchical generative model of the sensory input"*, through unsupervised deep learning [1]. The original deep generative model was based on stochastic neurons and, more importantly, on input (image) reconstruction. Statistical analysis highlighted a correlation between the numerosity present in the input and the population activity of some neurons in the second hidden layer of the network, whereas population activity of neurons in the first hidden layer correlated with total area (i.e., number of pixels) of the objects in the image. Here we further investigate whether numerosity information can be isolated as a disentangled factor of variation of the visual input. We train in unsupervised and semi-supervised fashion a latent-space generative model that has been shown capable of disentangling relevant semantic features in a variety of complex datasets, and we test its generative performance under different conditions. We then propose an approach to the problem based on the assumption that, in order to let numerosity emerge as disentangled factor of variation, we need to cancel out the sources of variation at graphical level.

1 Introduction

There is general consensus that humans's ability to perceive numerosity in visual stimuli relies on two core neuro-cognitive systems [2]: the Approximate Number System (ANS) enables to roughly estimate numerosity when there are many items in the visual display, whereas a second system processes small numerosities (in the "subitizing" range, typically up to four items) and it is tied to tracking objects in time and space. In order to represent numerical quantity at a semantic level, a cognitive system would need to abstract it away from the many low-level (e.g., graphical) features present in the sensory input, thereby extracting numerosity as a common factor of variation between the images.

© Springer Nature Switzerland AG 2019
I. Rojas et al. (Eds.): IWANN 2019, LNCS 11507, pp. 49–60, 2019.
https://doi.org/10.1007/978-3-030-20518-8_5

Recent simulation work [1,5,6] has shown that deep learning models can reproduce human performance in numerosity discrimination tasks that tap the ANS, suggesting that our numerical abilities might emerge from domain general learning mechanisms [3]. In particular, Stoianov and Zorzi [1] argued that in their model numerosity was *"computed through the combination of local computations and a simple global image statistic (cumulative area), without explicit individuation and size normalization of visual objects"*. However, despite these initial findings it is still unknown whether deep networks could learn to encode numerosity as a single explicit dimension, which would allow to control the generative process in an efficient and interpretable way, rather than having numerosity encoded as distributed pattern of activation of some neurons that can be "decoded" via the use of a trained linear classifier, with no direct access and control over it.

In the present work we address this question by focusing on InfoGAN [4], a powerful deep learning model that has been shown capable of disentangling the most relevant factors of variations in many different datasets, ranging from handwritten digits to faces [16]. We study if and how this type of generative model could learn to map numerosity into one or possibly more latent variables. Although it has been proved that it is theoretically impossible for an arbitrary generative model to learn disentangled representations of the input data in unsupervised settings [7], this impossibility holds *a priori* only for models without any inductive biases suitable for the task at hand (that is, the set of solutions the unsupervised model is able to produce and their probability under the model). Inductive biases can be expressed in many ways (model architecture, training algorithm, initialization scheme, etc.). In our study, we explore the role of different biases in the InfoGAN by adding cost components and varying the model architecture, latent space dimensionality and other hyperparameters.

Our main contributions can be summarized as follows. Three different models are analyzed with the aim of investigating the emergence of single elements of the latent code that would represent numerosity; the models are based on the following assumptions:

- with the first InfoGAN model, we implicitly make the assumption that no particular strategies must be considered to abstract numerosity from other graphical features in the input data in an unsupervised fashion;
- with the second model, we move to a semi-supervised setting to overcome the challenges resulting from a completely unsupervised learning regimen;
- with the third model, we tackle the problem of mapping numerosity as a disentangled dimension in the latent space assuming that this might emerge if we cancel out the sources of statistical variation at the graphical level.

Overall, one of the models appears to have the greater potential for disentangling numerosity. Our experiments also confirm the difficulties indicated in [5]. The outline of the paper is as follows: Sect. 2 overviews related literature. Section 3 formulates the problem. Section 4 presents experimental results, which are discussed in Sect. 5. The last section concludes the paper and outlines further research.

2   Related Work

An important assumption in representation learning [7,9] is that real-world data (like images or videos) can be thought as generated by a generative process that has two phases: first a latent random variable is sampled from a (possibly multivariate) prior distribution $P(z)$, where z can be thought of as the "cause" of the semantic factor of variations in the input data, and then the real-word data x would be generated by sampling a conditional probability $P(x|z)$. A classic unsupervised learning task is to find the "best" representation of the data, meaning a representation that embodies as much information about the input data as possible, being at the same time constrained to meet some conditions that are application-specific. In the present work, we have taken the position for which "best" corresponds to "disentangled", meaning a data representation that attempts to disentangle the sources of variation underlying the data distribution such that the dimensions of the representations are statistically independent [9].

The InfoGAN [4] is a variation of Generative Adversarial Networks [10] that extends the basic adversarial setup with a regularization based on the maximization of the Mutual Information between the Generator output and part of the latent code fed into the Generator itself. The InfoGAN model considers a minmax game that starts from the fundamental GAN minmax game:

$$\min_G \max_D V(D, G) \tag{1}$$

$$V(D, G) = \mathbb{E}_{x \sim P_{data}}[\log(D(x))] + \mathbb{E}_{(c,z) \sim P(c,z)}[\log(1 - D(G(c, z)))] \tag{2}$$

and adds a regularizer that represents the Mutual Information between the generated output $G(c, z)$ and the coding part c of the noise fed into the Generator to obtain a modified minmax game:

$$I(c; G(c, z)) = H(c) - H(c|G(c, z)) = H(G(c, z)) - H(G(c, z)|c) \tag{3}$$

$$\min_G \max_D V_I(D, G) \tag{4}$$

$$V_I(D, G) = V(D, G) - \lambda I(c; G(c, z)) \tag{5}$$

Also the related CatGAN model [11] is of interest here, as the author introduced the extension for the Discriminator to classify the output of the Generator either in a number of classes known a priori, when it is possible to access labels for the dataset at hand, or simply using an "estimated" number of classes. When the labels are accessible, it is possible to add a cost component corresponding to the cross-entropy between the sought distribution and the one obtained at the classifier output. When the labels are not available, by assuming a specific number of classes (but ignoring the labels for each entry) it is possible to add a cost component maximizing a ratio between the entropy of class y assigned to $G(c, z)$ by the classifier (which is supposed to be high) and the entropy of the assigned label conditioned to $G(c, z)$, so $y|G(c, z)$, (which is supposed to be low, for the choice to be sure); this is basically the same idea used in the Inception Score [12]. It can also be shown [13] that maximizing the Inception Score corresponds to maximizing the Mutual Information between the input being classified, in this case $G(c, z)$ and the class y being outputted by the additional classifier.

3 Problem Formulation and Methods

3.1 Problem Formulation

The main problem considered here is to study **if** and **how** it is possible for the InfoGAN model to learn to represent numerosity as independent factor of variation in its latent space. To this aim, we extended the Info-GAN to make it develop the same disentangled representations seen on the MNIST dataset, but extracted from a dataset composed of images containing a differ-ent number of items [1] (see samples in Fig. 1). Though

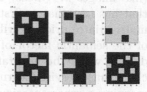

Fig. 1. Dataset samples.

the visual structure of these images might look simpler than MNIST images, the most relevant *semantic* direction of variation is not strictly graphical but more abstract in nature, being indeed numerosity.

To test the hypothesis behind each model we consider, and assess the quality of the related learning process, we explicitly look for a "minimum degree" of disentanglement in the representation learned by each model, investigating what kind of latent space representation is induced by the learning process for each model, and we *visualize it* by changing one latent variable at a time and generating new data (images) with the trained Generator of the model under test. The visual inspection of the Generator output obtained in this way is a first qualitative indicator of whether any disentanglement has been reached in the process. Though this approach may seem fuzzy, there is no formal definition (yet) of disentanglement which is widely accepted, and there is no unified protocol to quantify it [7]; therefore, in this paper we accept the visual inspection as first qualitative evaluation of disentanglement of the latent space, as it has been done in the original InfoGAN paper [4].

3.2 Methods

We start out with the InfoGAN model as used in [4] (experiment 1 in that paper), but using a synthetic dataset obtained from the one used in [1] after applying the following basic transformations ad data augmentation techniques:

1. We reduce the numerosity range from 1 to 8, with most the experiments actually focusing on the range from 1 to 4. Though these intervals are somewhat limited, they still leave the possibility to build instructive parallels between the simulations and the ANS/Subitizing distinction.
2. We apply simple data augmentation procedures based on image reflection along orthogonal and diagonal axes.
3. We invert the background with the foreground, therefore doubling the total amount of images.

In so doing, we obtain a full dataset, for numerosity 1 to 8, composed of 128.000 images, half of in black over white, and half in white over black. In particular, the third transformation is motivated by the fact that numerosity should be

minimally linked to any graphical representation, being a concept connected to
"areas of coherence or correlation" in the perceptual input domain, whatever this
could be (visual, audio or else). Our investigation is carried out incrementally,
in terms of complexity of the models considered:

1. Infogan: Fully unsupervised with InfoGAN
2. Label-aware: InfoGAN with unsupervised G but D aware of labels
3. R-oriented: 'Relational' oriented learning

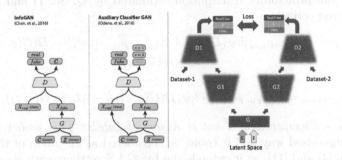

Fig. 2. Pictorial representations of InfoGAN, CatGAN, and R-oriented model (Info-
GAN and CatGAN pictures from [17]).

We conduct many simulations with each model, with different latent code
dimensionalities (that is c in the equations; the uncompressible noise z is always
set to 32), for both the Categorical variable (number of possible choices) and the
number of continuous Uniform$[-1,1]$ variables in c. We also experiment changing
the numerosity of the dataset from 1 to a maximum (2,3,4,...8), with the aim to
study in each case the latent space with the methodology clarified at the beginning of this section, and compare qualitatively the degree of disentanglement
reached in each case. For brevity, in this report, we provide illustrations only of
specific but (we believe) representative cases of the results.

Infogan - Fully Unsupervised with InfoGAN: the model is the same as the one
described in [4] with the exception of the latent space configuration, which for
us is a field of exploration, and a change in the activation function in the first
convolutional layer of the Discriminator, using the Absolute Value Rectification
which has been shown [9] to be well suited for features that are invariant under a
polarity reversal, like in our dataset. Few other modifications are also attempted
(reducing the number of feature maps in the convolutional layers) aimed at simplifying the Generator and Discriminator networks, as the visual patterns in our
synthetic data are simpler than the ones in the MNIST dataset. However, no significant differences were found in the overall results, both in terms of graphical
reconstruction and, more importantly, with respect to numerosity disentanglement. In [4] it is shown how to derive a variational lower bound to the Mutual
Information between the coding part of the latent "noise" and the output of the
Generator $G(c, z)$. The minmax game is then re-defined as follows:

$$\min_{G,Q} \max_{D} V_{InfoGAN}(D,G,Q) \tag{6}$$

$$V_{InfoGAN}(D,G,Q) = V(D,G) - \lambda L_I(G,Q) \tag{7}$$

where D can intervene on L_I as Q in practice is often implemented using some layers of D. The L_I lower bound is shown in the same paper to be:

$$L_I(G,Q) = \mathbb{E}_{c' \sim P(c'),(c,z) \sim P(c,z)} log\hat{P}(c|G(c,z)) + H(c) \tag{8}$$

where $\hat{P}$ is the probability distribution estimated by Q. So, D and G aim at minimizing respectively:

$$D_{tot.loss} = -[\mathbb{E}_{x \sim P_{data}}[\log(D(x))] + \mathbb{E}_{(c,z) \sim P(c,z)}[\log(1 - D(G(c,z)))]] - \lambda L_I(c;G(c,z)) \tag{9}$$

$$G_{tot.loss} = \mathbb{E}_{(c,z) \sim P(c,z)}[-\log(D(G(c,z)))] - \lambda L_I(c;G(c,z)) \tag{10}$$

Label-Aware - Unsupervised G but D Aware of Labels: This model is trained in a semi-supervised way and it could be considered as an union of the models presented in [4] and [11], as it extends the InfoGAN settings with an additional classifier whose output corresponds to the numerosity of the image at its input, and it introduces a regularization component to the total cost function in the following way: during the Discriminator training we expect the output of the classifier to be the correct class of the real data being submitted, while the output of the classifier is ignored when the generated data is passed to the Discriminator. During the Generator training, conversely, the output of the Generator -in every single instance in the batch- is expected to be classified with very low entropy (that is $P(y|G(z))$ expected to have very low entropy), and at batch level we expect the entropy of the classifier output to be high (that is, $P(y)$ expected to have very high entropy). Based on these considerations, it is possible to add a cost component (IS^{-1} in the following equations) to the total loss function for the Generator. This approach is similar to the reasoning behind the "Inception Score" metric for generative models, proposed in [12], and the whole setup used in this case has also similarities to the one proposed in [11]. In mathematical terms this translates in the following cost functions for D and G.

$$D_{tot.loss} = -[\mathbb{E}_{x \sim P_{data}}[\log(D(x))] + \mathbb{E}_{(c,z) \sim P(c,z)}[\log(1 - D(G(c,z)))]] - \lambda L_I(c;G(c,z)) + CE(labels, y) \tag{11}$$

$$G_{tot.loss} = \mathbb{E}_{(c,z) \sim P(c,z)}[-\log(D(G(c,z)))] - \lambda L_I(c;G(c,z)) + IS^{-1}(P(y|x), P(y)) \tag{12}$$

where CE is the Cross Entropy and IS^{-1} is as just described[1].

---

[1] It is worth clarifying that for *each* component of the cost functions shown in all the equations, for all the three models considered, we apply a weighting hyper-parameter (thus, not only for the Information based reguliarized of the InfoGAN model), and we investigate empirically the effect of changing them.

R-oriented - Relational Oriented Learning: This model is trained in a semi-supervised way and it is based on the idea that in order to learn a disentangled representation of numerosity, we need to abstract it from the specific graphical appearance of an image. We thus force the model to represent features that are shared between two datasets, but expressed through different graphical representations, and whose informative content is what we want the model to learn to represent. This can be thought of as learning to represent a relation between sets rather than the features that one object (image) has or a set of objects (image dataset) exposes in statistical terms. This approach is inspired by the fact that a natural number can be defined as an equivalence class of finite sets under the equivalence relation of equinumerosity. We thus force the Generator to be multi-output over *the same latent space*. This in practice means that the whole setup described in Label-aware model is duplicated, sharing a bottom layer for the Generators which use *the same single latent space*. In this new duplicated setup we also add a component to the Generator loss function, the Jensen-Shannon divergence, which is symmetric with respect to its arguments, and we calculate it between the probability distributions over the output class predicted by the two classifiers. This has the goal to force to learn a probability distribution in the bottom layers of the Generator which must contain the necessary information to make both the reconstructions possible, while representing the same numerosity[2]. A pictorial representation of the model is shown in Fig. 2. In this setup the total loss functions used are slightly more complex than in the previous cases, as we have split the Generator in two lines of generation, rooted on the same latent space. With JS being the Jensen-Shannon divergence as per above, and with all the remaining quantities averaged between the two lines of the model, D and G aim at minimize respectively:

$$D_{tot.loss} = -[\mathbb{E}_{x \sim P_{data}}[\log(D(x))] + \mathbb{E}_{(c,z) \sim P(c,z)}[\log(1 - D(G(c,z)))]]$$
$$- \lambda L_I(c; G(c,z)) + CE(labels, y) \tag{13}$$

$$G_{tot.loss} = \mathbb{E}_{(c,z) \sim P(c,z)}[-\log(D(G(c,z)))] - \lambda L_I(c; G(c,z))$$
$$+ IS^{-1}(P(y|x), P(y)) + JS(P_1(y), P_2(y)) \tag{14}$$

4 Experiments and Results

4.1 InfoGAN

With this model, we do not observe a clear "departure" from graphical features, in the sense that the study of the latent space always shows a strong connection with the graphical appearance of the images being generated, and very weak

---

[2] In our first setup to investigate this model we used, as second dataset, the labels themselves, feeding one line of the model with the labels and the other line with images. It must be noted however that this approach can be extended to a setup that does not use labels at all, however we leave this for future developments.

control over numerosity results from changes of the latent code dimensions. We note that whenever the dimensionality of the Categorical space is set to 2 (so with possible values (1,0) or (0,1)), the model appears to be capable of associating the Categorical part of the code to the kind of background/foreground of the image being generated (white over black versus black over white), but still not in a totally independent way, as changing from (1,0) to (0,1) it usually changes also the numerosity being represented (columns in the upper half, left in Fig. 3). Apart from that, no evidence of a correlation with numerosity, rather than with graphical features, is detected in any of the latent variables in all the tested options.

We provide latent space visualizations for two cases that we believe are representative of what we observe with this model. The first case (upper half of Fig. 3) has the latent code c configured as 2D Categorical variable and 1D continuous Uniform variable, with dataset numerosity equal to 2, abbreviated to (2D-1C-1,2); in this case the learning process cause the Categorical latent to represent the relation background/foreground of the image (columns on the upper left of Fig. 3), while the continuous variable shows a weak tendency to represent the numerosity contained in the dataset (1 or 2), only occasionally changing the numerosity displayed when it moves from negative values (left side of the rows in the upper part of Fig. 3) to positive values (right side of the rows).

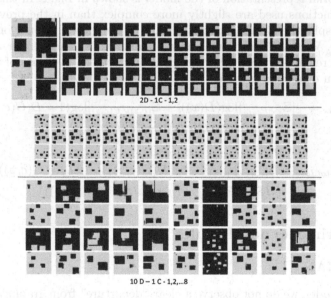

Fig. 3. Example of latent space exploration with the InfoGAN model, (2D-1C-1,2) upper part, (10D-1C-1,2,...8) lower part.

The second case (lower half Fig. 3), abbreviated as (10D-1C-1,2,...8) shows no evident signs of numerosity control in the continuous variable (rows in the center of Fig. 3) and the 10-dimensional Categorical variable does not seem to have picked any particular role (columns at the bottom of Fig. 3). We notice an increasing difficulty of the model to deal with the task as we increase the numerosity of the dataset, but we still obtain acceptable graphical quality of the generated samples. More interesting, we observe that a greater dimensionality of the Categorical part of the latent code never corresponds[3] to numerosity as learned descriptive (disentangled) dimension of the data.

4.2 Label-Aware

Similar results are obtained with the Label-aware model; we provide illustration for the same two cases considered in the previous section. We note a deterioration in the quality of the reconstructions contrasted by a slightly increase in the consistency with numerosity, as being represented in the Categorical part of the code (Fig. 5). As found for the previous model, in the case of binary Categorical latent variable, (2D-1C-1,2) in Fig. 5, we can see again that the Categorical variable picks the type of background while the continuous part of the latent space models some aspects that are fully connected to graphical features of the dataset. Generally, we notice that as we

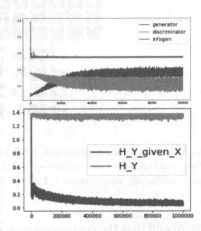

Fig. 4. Typical Losses and Entropies in Label-aware and R-oriented.

increase the numerosity of the input also this model fails to show any possible correlation between any latent code variable and numerosity.

4.3 R-oriented

This model is designed to take as input two datasets, both with numerosity as either a strong or weak factor of variation. Here, we take a first exploratory step and we choose the inputs to be the dataset of images for one generative line, whereas the other generative line takes in, at the corresponding generator, the set of labels represented in one-hot encoding. This means that the two Generation lines are now expected to generate credible images on one side and valid coding for labels on the other, with *a priori* no relation since the Generators are never exposed to the association between labels and images from the dataset. For this model too, we provide latent space visualizations for the (2D-1C-1,2) case, as well as for (3D-1C-1,2,3), (4D-1C,1,2,3,4) and (6D-1C-1,2,...6), and we show that with this model, in all cases, numerosity is mapped to the Categorical variable

---

[3] Even when the Categorical dimensionality is somehow compatible with the numerosity being analized, for example with numerosity 5 and Categorical dimension 5, or Categorical dimension 10 to account for 8 quantities and 2 possible graphical expressions, w/b or b/w.

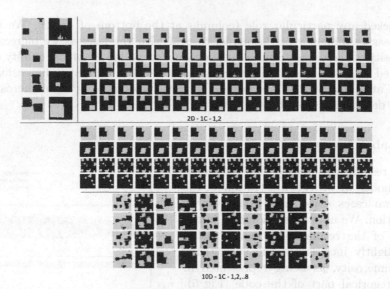

Fig. 5. Latent space exploration with Label-aware model with 2-dim. Categorical and 1-dim. continuous code; numerosity 1,2 (2D-1C-1,2); and 10-dim. Categorical and 1-dim. continuous, numerosity 1,2,...8 (10D-1C-1,2...8).

(Fig. 6) and that the continuous code only changes graphical aspects of the image being generated. Each picture shows the generated output image from the image generative line, while each digit on top each picture is the output of the classifier on the label generative line.

When setting the Categorical variable to a fixed value and changing randomly the remaining part the latent code, image numerosity (number of coherent colored areas in the image) stays the same in most of the cases, while the pattern of pixel activation of the image changes, giving rise to different ways of expressing the same numerosity. We note however that the graphical quality of the reconstruction is deteriorated compared to previous models; this might be connected to the setup used, in which the full latent code (c, z) is shared between the generative lines, even the uncompressible noise z, which seems neither necessary nor helpful. An appropriate calibration of the weights of the various cost functions components might also help improving the graphical reconstruction. However, our main goal was to investigate disentanglement, thus we accept a deterioration in graphical appearance leaving to future work taking care of this improvements.

5   Discussion

From our experiments it turned out, maybe not surprisingly, that it is not easy to map numerosity to any of the latent codes, either discrete or continuous, regardless of the dimensionality used in the various attempts. Being numerosity a concept of discrete nature, the first naive approach would be to model the

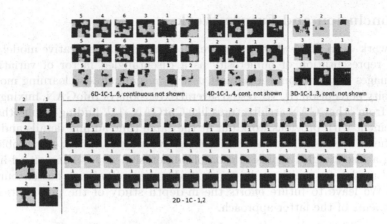

Fig. 6. Latent space study for the R-oriented model, for a (2D-1C-1,2), lower part of the image, for which we report a latent space study for both discrete and continuous variable, and, in the upper part, the latent space exploration with the Categorical variable for (6D-1C-1,...,6), (4D-1C-1,...,4), (3D-1C-1,...,3); the number above each single picture is the corresponding output of a classifier on the image generative line

InfoGAN code space as a Categorical code whose dimensionality is equal to the maximum number of objects present in the input images. This appeared to be effective with the first two models only with numerosity of 2 (see Sect. 4). From a semantic perspective, numerosity can be understood as an *inclusion* concept, for an image with 2 objects includes an image with 1 and so on; in this view one may think that the optimal representation for the concept might be discrete, but perhaps not simply Categorical. Similarly, representing numerosity using a continuous variable seems counter-intuitive, as the latent space should learn a distribution that is peaked around some values (that is, multimodal) to provide a "Categorical-like" representation of numerosity, along one continuous direction only. In this scenario, we might end up with parts of this continuous axis where the mapped images would be "morphing" from one numerosity to another one, leaving the result in this "transitory part of the latent space" undefined. The results we obtained confirm that learning a high level concept like numerosity in a generative model, in an unsupervised or semi-supervised fashion, and map it to an independent dimension of the latent space is not a straightforward task even for the InfoGAN model. This is in harmony with the findings in [8], and it may interpreted thinking about the semantically meaningful variations of "higher level of abstraction" as being overwhelmed by other statistical variations in the data, connected to much lower level features, like purely graphical for example. These variations must be somehow ignored by the learning process in order to let emerge the variations that carry the relevant semantic information. In this, the way the learning process is driven seems to be a way to leave the higher level semantic features to emerge, and we proposed here a possible approach, the R-oriented model, which is also inspired by the findings in [15].

6 Conclusion and Future Work

In this work we verified whether a powerful latent space generative model could learn a representation of numerosity as a disentangled factor of variation of input images derived from the dataset used in a seminal deep learning model of numerosity perception [1]. We tested several extensions of InfoGAN, mixing ideas coming from other GAN architectures like CatGAN [11], finding results that are in agreement with previous work but at the same time adding details and ideas to the field. We also introduced an alternative way to attack the problem of learning a disentangled representation of numerosity, introducing an ad-hoc *R-oriented* model, for which we have reported some preliminary but encouraging results. We leave to future efforts the in-depth study of the possibilities and developments of the latter approach.

References

1. Stoianov, I., Zorzi, M.: Emergence of a 'visual number sense' in hierarchical generative models. Nat. Neurosci. **15**(2), 194–196 (2012)
2. Feigenson, L., Dehaene, S., Spelke, E.: Core systems of number. Trends Cogn. Sci. **8**(7), 307–314 (2004)
3. Zorzi, M., Testolin, A.: An emergentist perspective on the origin of number sense. Philos. Trans. Royal Soc. B Biol. Sci. **373**(1740) (2018)
4. Chen, X., Duan, Y., Houthooft, R., Schulman, J., Sutskever, I., Abbeel, P.: Info-GAN: Interpretable Representation Learning by Information Maximizing Generative Adversarial Nets (2016), arXiv:1606.03657
5. Wu, X., Zhang, X., Shu,X.: Cognitive Deficit of Deep Learning in Numerosity (2018), arXiv:1802.05160
6. Chen, S.Y., Zhou, Z., Fang, M., McClelland, J.L.: Can Generic Neural Networks Estimate Numerosity Like Humans? (2014)
7. Locatello, F., Bauer, S., Lucic, M., Gelly, S., Schölkopf, B., Bachem, O.: Challenging Common Assumptions in the Unsupervised Learning of Disentangled Representations (2018), arXiv:1811.12359
8. Zhao, S., Ren, H., Yuan, A., Song, J., Goodman, N., Ermon, S.: Bias and Generalization in Deep Generative Models: An Empirical Study arXiv:1811.03259v1 (2018)
9. Goodfellow, I., Bengio, Y., Courville, A.: Deep Learning. MIT Press, Cambridge (2016)
10. Goodfellow, I., et al.: Generative Adversarial Networks (2014), arXiv:1406.2661
11. Springenberg,J.: Unsupervised and Semi-supervised Learning with Categorical Generative Adversarial Networks (2015), arXiv:1511.06390
12. Salimans, T., Goodfellow, I., Zaremba, W., Cheung, V., Radford, A., Chen, X.: Improved Techniques for Training GANs (2016), arXiv:1606.03498
13. Barratt, S., Sharma, R.: A Note on the Inception Score (2018), arXiv:1801.01973
14. Katrina E., Drozdov, A.: Understanding Mutual Information and its Use in Info-GAN (2016)
15. Hill, F., Santoro, A., Barrett, D., Morcos, A., Lillicrap,T.: Learning to make analogies by contrasting abstract relational structure (2019), arXiv:1902.00120v1
16. Liu, Z., Luo, P., Wang, X., Tang, X.: Deep learning face attributes in the wild. In: ICCV (2015)
17. https://github.com/lukedeo/keras-acgan/blob/master/acgan-analysis.ipynb

Deep Residual Learning for Human Identification Based on Facial Landmarks

Abdelgader Abdelwhab Abdelgader[1] and Serestina Viriri[2(✉)]

[1] College of Computer Science and Information Technology,
Sudan University of Science and Technology, Khartoum, Sudan
gadoradatabase@gmail.com
[2] School of Mathematics, Statistics and Computer Science,
University of KwaZulu-Natal, Durban, South Africa
viriris@ukzn.ac.za

Abstract. The face detection and recognition are still challenging research areas, since up to date there is no accurate integral model that works in every situation. As a result, the focus has been shifted to Convolutional Neural Networks (CNNs) and fusion techniques with the hope of better solution. The CNNs have enhanced the state-of-the-art of the human facial identification. However, the CNNs are not easy to train due to degradation problem called (gradient vanishing) when the network depth increased, so there is a need of residual network to solve this problem by going deeper without losing the gradient. In this paper, a pre-trained deep residual network for features extraction and ensembles of classifiers are implemented and facial landmarks are extracted and passed to the pre-trained model. Support Vector Machine (SVM) and random forest classifiers are fused at decision level using weighted and majority voting techniques. The experimental results conducted on ORL database show an excellent mean accuracy rate of about 100%. The accuracy rate of about 100% was achieved on LFW dataset with a minimum 70 facial images per person, and 99% with a minimum 10 facial images per person.

Keywords: Deep residual network · Convolutional Neural Networks · Facial landmarks · Feature extraction · SVM · Random forest · Fusion

1 Introduction

The advancement of the technology has resulted in high demands of security in terms of personal data, personal authentication, identification, etc. For instance, a question can be asked; is this a citizen, or the right person to do such work. Biometrics is one of the solutions because it relies on the human physiological, behavioral features, which can not be shared nor forgotten called ownership factor [1,2]. Passwords, keys and ID card can be shared or stolen.

© Springer Nature Switzerland AG 2019
I. Rojas et al. (Eds.): IWANN 2019, LNCS 11507, pp. 61–72, 2019.
https://doi.org/10.1007/978-3-030-20518-8_6

1.1 Biometrics

Biometrics can be defined as a research area focused on measuring and analyzing a person's unique characteristics [3]. Alphonse Bertillon was the first person to introduce the idea of personal identification system based on biometrics, using mathematical morphology operations, and traits as eyes, hair and skin color in 1896. Biometrics can be divided into hard or primary traits and soft traits [4,5].

Face is one of the most promising biometrics traits and widely used in research because it has high permanence over time. The soft biometric features are ancillary information that provide a partial description which is not fully distinctive and permanent [6]. Moreover, facial soft biometrics traits are user-friendly, can be acquired at distance and at the same time when acquiring face [7]. Fusing a hard biometrics with multiple biometrics, sensors, or algorithm is called hybrid biometric system [8]. Biometrics data may change over time or affected by environmental condition so by fusing more than one trait or same trait from more than one source can overcome the unimodal limitations and reduce the rejection, acceptance errors rate based on the system requirements [9].

1.2 Deep Residual Network

Neural networks allow the computer to learn through observation while deep learning is the way to train neural networks [10]. Deep residual networks (ResNets) is a deep neural networks (DNNs) using residual representation with identity as shortcut connections [11]. It achieved first place in ILSVRC 2015 and COCO 2015, image classification competitions with 3.57% error rate. The main difference between ResNets and plain DNNs is a shortcut (identity) connection for the input as it makes the layers learning gradually [11]. Where, (X) is residual function, x is identity input of the layer, H(X) is the output so residual is:

$$F(X) = H(x) - x \tag{1}$$

Then desired mapping is:

$$H(x) = F(X) - X \tag{2}$$

Deep metric learning is used as key technique to minimize the difference between same person and maximize it otherwise. However, it returns feature vector instead of one result label [12].

2 Methods and Techniques

Our proposed method is using pre-trained model for feature extraction which is included in DLIB library created by King et al. [13]. This model is based on deep residual network (ResNet-34) introduced by Zhang et al. [11] with some modification.

2.1   Dataset

Two dataset are used for this experiment, one is classified as easy dataset and the other is difficult. They are commonly used in literature.

2.1.1   ORL Dataset

ORL dataset is used for evaluation purpose [14]. It consists of 400 images for 40 person with 10 images per person. These images were taken at different time between 1992 and 1994 with frontal view, and were presented with different expressions such as (open and closed eyes, smiling) and with glass.

2.1.2   LFW Dataset

Labeled faces in the wild (LFW) dataset shows images in natural environment with different lighting, pose, make-up, facial expression, age and background [15]. It has 5570 person faces with total images of 13244 and some person has up to 500 images. People with minimum 10 and 70 faces per person are selected and used for evaluation.

2.2   Proposed Methods Implementation

At enrollment data is collected, split into train and test set, and threshold tunned. The prediction and accuracy calculation done at identification phase as shown in Fig. 1.

2.2.1   Face Detection

Dlib face detector is used, which is combination of hog features and linear support vector machine classifier. HOG is widely used as a feature descriptor [16]. HOG face detection steps:

1. The input image is divided into blocks.
2. Each block is divided into small region called cell.
3. Vertical and horizontal gradients are calculated for each pixel in the cell and 1-D Sobel vertical and horizontal operators is used ($[1,0, 1]$ and $[-1, 0, 1]$) as denoted in Eqs. (3) and (4):
4. HOG is created for each cell with Q bins number as shown in Fig. 2.
5. SVM classifier trained with this HOG feature and return whether this face or none.

$$Gx(x,y) = H(x+1,y) - H(x-1,y) \qquad (3)$$

$$Gy(x,y) = H(x,y+1) - H(x,y-1) \qquad (4)$$

Where Gx(x,y) is the horizontal gradient, and Gy(x,y) is the vertical gradient and H(x,y) provides pixel intensity at coordinates x and y. Then the magnitude and direction are obtained using the equation:

$$G(x,y) = \sqrt{G_x^2(x,y) + G_y^2(x,y)}\,\delta(x,y) = \arctan(\frac{Gy(x,y)}{Gx(x,y)}) \qquad (5)$$

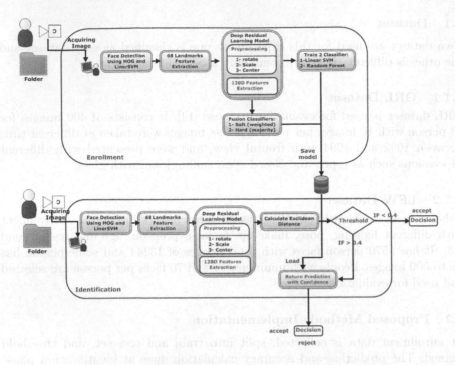

Fig. 1. Proposed deep residual learning model overview

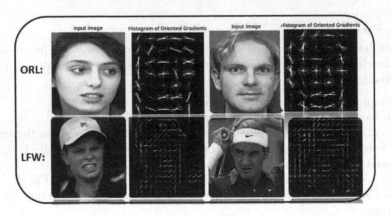

Fig. 2. Histogram of oriented gradients

2.2.2 Facial Landmarks Detector

Dlib face landmarks detector is used to estimate the location of 68 (x, y)-coordinates of the face feature location, such as Eyes, mouth, eyebrows, chin,top_lip and nose [17] as shown in Fig. 3. After applying HOG and Linear SVM for face detection, we apply facial landmark detector to collect the x, y coordinates of the landmarks. Then face ROI and coordinates are passed to Deep learning network.

Fig. 3. The 68 facial landmarks from the ORL and LFW datasets

2.2.3 Deep Learned Features Extraction Network

At this stage preprocessing and features extraction are done. The model takes preprocessed detected face image and 68 landmarks location as input then 128D measurable and distinctive features vector is extracted which representing each individual face. The model used to minimize the distance between same faces and maximize the distance between different faces.

2.2.4 Euclidean Distance

Euclidean Distance is most used to measures the distance between any two points in space corresponds to the a straight line drawn between them [18]. For similarity detection between two faces we calculate the distance between. The distance between two vectors is defined as follows:

$$d(X, Y) = \sqrt{\sum (X_1 i - X_2 i)^2} \qquad (6)$$

Euclidean distance takes the 128D features returned by CNN network and calculate the distance between them and compared with threshold. The given threshold is tuned on training set till we get the optimum result. If the result is less than threshold faces are considered for the same person otherwise we go to next step by calling support vector machine and random forest classifiers.

2.2.5 Support Vector Machine Classifier

A Support Vector Machine (SVM) is supervised machine learning classifier separating classes [19]. SVM one-vs-the-rest technique is implemented for multiclass classification, the probability for each class is returned for classification.

2.2.6   Random Forest Classifier

Random Forest Classifier is ensemble algorithm, which combines more than one algorithms [20]. It randomly created decision trees and it predicts based on the majority voting. For each tree in forest classifier data set is divided randomly.

2.2.7   Decision Level Fusion

Decision level fusion implementation is considered easy and straightforward We have applied ensemble classifiers at decision level fusion where support vector machine and random forest classifiers are fused using soft (weighted) and hard (majority) voting rules. Majority voting algorithm predict the final class label as the class label that predicted most repeatedly and frequently by the classification models as shown in followed equation:

$$y = mode\{C1(x), C2(x), ..., Cm(x)\} \tag{7}$$

where y = class label and C1, C2, Cm are classifiers. It's totally based on the class labels [21]. Where in soft voting, class predicted based on the averaging classifier probability as shown followed:

$$y = \arg\max \sum_{j=1}^{m} W_j P_{ij} \tag{8}$$

where wj is the weight that can be assigned to the jth classifier.

3   Experimental Results and Discussion

Experiment conducted on ORL database [14] and LFW datasets [15]. Scikit-learn library is used for evaluation procedures and metrics. SVM, Random Forest and Fusion (majority, weighted) classifiers are used [22].

3.1   Evaluation Procedures

This model is evaluated under three different test as shown in Table 1.

3.1.1   Training Accuracy

The entire dataset is used for training and testing at the same time to see the model complexity.

3.1.2   Train-Test Split

Dataset is split into tow parts randomly according to the percentage of training and testing, first part is used for training and the other used for testing. 80% is used for training and 20% for testing. this step is repeated 5 times and average accuracy calculated.

3.1.3 K-Fold Cross-Validation

Dataset is divided into K number of folders, the last folder left for testing and the rest is used for training. This procedure repeated K times. K = 5 is used. Finally the average accuracy is calculated against each folders.

3.2 Evaluation Metrics

Different performance measurements are used for evaluation and Tables 1, 2 and 3 give a summary of the proposed system evaluation results.:

1. Accuracy: is the simplest performance evaluation metric with the equation but does not shows the classifier lack against false accepted or rejected classes [23]

$$Accuracy = \frac{\text{Truly predicted}}{\text{All the predicted}} = \frac{TP + TN}{TP + TN + FP + FN} \tag{9}$$

   Table 1 shows all the classifiers got up to 100% on training accuracy while on test accuracy, SVM and soft weighted fusion achieved the highest accuracy. Table 2 shows SVM and soft voting achieved 100% on mean accuracy and f1 score but 99% on k-folder split. As shown in Table 3 when using person with minimum 10 faces, the highest mean accuracy achieved by SVM on both train test split and K-folder.
2. Receiver operating characteristic (ROC): plots true positive rate against false positive rate as shown in following equation:

$$\text{True Positive Rate} = \frac{\text{true positives}}{\text{actual positives}} = \frac{TP}{TP + FN} \tag{10}$$

$$\text{False Positive Rate} = \frac{\text{false positives}}{\text{actual negatives}} = \frac{FP}{FP + TN} \tag{11}$$

   ROC is better than accuracy when there is imbalanced data for each class. Left upper corner is the optimal point with zero false positive rate. However closer point to the corner maximizes the area under curve and gives high accuracy [24]. According to Fig. 4, SVM classifier achieves high micro average score overall 40 classes and 67% for class no 37. Figure 5 shows SVM classifer achieves 100% true positive rate for all the 7 person with minimum 70 faces per person.
3. Recall and precision: They are backward related when recall increases precision decreases. Recall knowns as sensitivity to predict correctly all the positive values and the ability to predicts true negative not as true positive is called precision or positive predicted value [25] as shown in following equations:

$$Recall = \text{True Positive Rate} = \frac{TP}{TP + FN} \tag{12}$$

$$Precision = PositivePredictedValue = \frac{TP}{TP + FP} \tag{13}$$

Precision-recall curve can show the relation between the precision and recall, the ideal point for perfect classifier model is on the top right corner where no false rejection or acceptance and only true positive but higher area under curve means good classifier. Moreover, Precision-recall works better than ROC when there is imbalanced data because it does not count true negative [26].

4. Learning curve: Fig. 6 shows increasing the dataset size can increase the performance of SVM classifier since both training and validation score getting high gradually where no need for more training data on LFW datset to increase the performance since training and cross-validation scores are going as same from the point 350 training set as shown in Fig. 7

5. F1-Score: is a well tuned average of the recall and precision since, they cannot tell if a classifier has higher precision and lower recall and misses some unclassified instances but with f1-score can do [27] and Tables 1, 2 and 3 show 100% accuracy for SVM and soft weighted fusion.

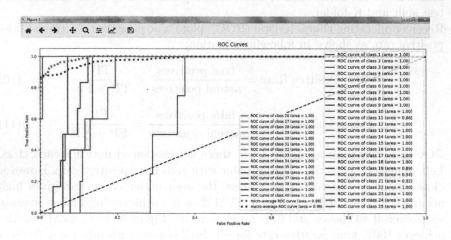

Fig. 4. ROC curve for ORL dataset 40 classes.

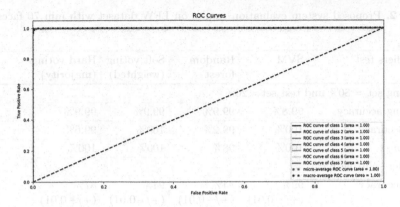

Fig. 5. ROC curve for LFW dataset 7 classes.

Table 1. Proposed system evaluation results on ORL

Classifiers test	SVM	Random forest	Soft voting (Weighted)	Hard voting (Majority)
Training set = 80% and Test set= 20%				
Training accuracy	100%	100%	100%	100%
Test accuracy	100%	98.75%	100%	98.75%
F1-score	100%	98%	100%	99%
K-folder splitting K=5				
Test accuracy	100%	96% (+/−0.01)	99% (+/−0.02)	99% (+/−0.01)
Euclidean distance threshold = 0.4 or 0.5				
Training accuracy Test accuracy on 20% test set, 80% training set	100%			

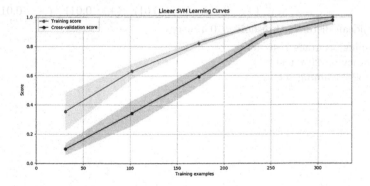

Fig. 6. Learning curve for SVM with linear kernel on ORL dataset

Table 2. Proposed system evaluation results on LFW dataset with min 70 faces per person

Classifiers test	SVM	Random forest	Soft voting (weighted)	Hard voting (majority)
Training set = 80% and test set= 20%				
Training accuracy	99.8%	99.9%	99.9%	99.9%
Test accuracy	100%	99.2%	100%	99.6%
F1-score	100%	98%	100%	100%
K-folder splitting K=5				
Test accuracy	99% (+/−0.01)	84% (+/−0.01)	94% (+/−0.01)	87% (+/−0.01)
Euclidean distance threshold = 0.4 or 0.5				
Training accuracy test accuracy on 20% test set, 80% training set	99%			

Table 3. Proposed system evaluation results on LFW dataset with min 10 faces per person

Classifiers test	SVM	Random forest	Soft voting (weighted)	Hard voting (majority)
Training set = 80% and Test set = 20%				
Training accuracy	99.5%	99.9%	99.9%	99.7%
Test accuracy	96.4%	80.9 %	92.5%	84.4%
F1-score	95%	79.5 %	91.7%	82.7%
K-folder splitting K = 5				
Test accuracy	99% (+/−0.00)	84% (+/−0.01)	94% (+/−0.01)	87% (+/−0.01)
Euclidean distance threshold = 0.4 or 0.5				
Training accuracy test accuracy on 20% test set, 80% training set	99%			

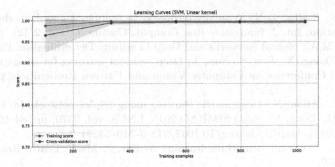

Fig. 7. Learning curve for SVM with linear kernel on LFW dataset

4 Conclusion

Human identification proposed model is developed based on deep residual network (ResNets pre-trained model) using facial landmarks and fusing soft (weighted) voting and hard (majority) classifier at decision level. Multi-algorithm as liner support vector machine and random forest classifiers are used. According to the performance measurement metrics and procedures SVM and soft (weighted) fusion have highest accuracy on ORL and LFW datasets with balanced and imbalanced classes since dataset is divided randomly into train and test sets. The results show that Dlib-model works perfect for 128 dimensional feature vector extraction and fusion at decision level still can improves the week classifiers. In future work another dataset can be used for evaluation, since deep learning mode is perfect as the training is.

References

1. Clarke, R.: Human identification in information systems: management challenges and public policy issues. Inf. Technol. People **7**(4), 6–37 (1994)
2. Srinivasa, K.G., Gosukonda, S.: Continuous multimodal user authentication: coupling hard and soft biometrics with support vector machines to attenuate noise. CSI Trans. ICT **2**(2), 129–40 (2014)
3. Lu, X., Jain, A.K.: Ethnicity identification from face images. SPIE **2**(2), 114–123 (2004)
4. Jain, A.K., Ross, A., Prabhakar, S.: An introduction to biometric recognition. IEEE Trans. Circuits Syst. video Technol. **14**(1), 4–20 (2004)
5. Lafta, H.A., Abbas, S.S.: Effectiveness of extended invariant moments in fingerprint analysis. Asian J. Comput. Inf. Syst. **1**(3), 2321–5658 (2013)
6. Dantcheva, A., Elia, P., Ross, A.: What else does your biometric data reveal? a survey on soft biometrics. IEEE Trans. Inf. Forensics and Secur. **11**(3), 441–67 (2016)
7. Lin, D., Tang, X.: Recognize high resolution faces: from macrocosm to microcosm. In: IEEE Computer Vision and Pattern Recognition, pp. 1355–1362 (2006)
8. Chang, K.I., Bowyer, K.W., Flynn, P.J.: An evaluation of multimodal 2D+3D face biometrics. IEEE Trans. Pattern Anal. Mach. Intell. **27**(4), 619–24 (2005)

9. Sree, S.S., Radha, D.N.: A survey on fusion techniques for multimodal biometric identification. Int. J. Innovative Res. Comput. Commun. Eng. **2**(12), 1–5 (2014)
10. Nielsen, M.A.: Neural Networks and Deep Learning. Determination Press (2015)
11. He, K., Zhang, X., Ren, S., Sun, J.: Deep residual learning for image recognition. In: IEEE Conference on Computer Vision and Pattern Recognition, pp. 770–778 (2016)
12. Hoffer, E., Ailon, N.: Deep metric learning using triplet network. In: Feragen, A., Pelillo, M., Loog, M. (eds.) SIMBAD 2015. LNCS, vol. 9370, pp. 84–92. Springer, Cham (2015). https://doi.org/10.1007/978-3-319-24261-3_7
13. King, D.E.: Dlib-ml: a machine learning toolkit. J. Mach. Learn. Res. **10**, 1755–1758 (2009)
14. Samaria, F.S., Harter, A.C.: Parameterisation of a stochastic model for human face identification. In: IEEE Workshop on Applications of Computer Vision, pp. 138–142 (1994)
15. Huang, G.B., Jain, V., Learned-Millerm, E.: Unsupervised joint alignment of complex images. In: IEEE 11th International Conference on Computer Vision, pp. 1–8 (2007)
16. Dalal, N., Triggs, B.: Histograms of oriented gradients for human detection. In: Computer Vision and Pattern Recognition, 2005. CVPR 2005, pp. 886–893 (2005)
17. Kazemi, V., Sullivan, J.: One millisecond face alignment with an ensemble of regression trees. In: IEEE Conference on Computer Vision and Pattern Recognition, pp. 1867–1874 (2014)
18. Technical white papers. http://www.pbarrett.net/techpapers/euclid.pdf. Accessed 24 June 2018
19. Hearst, M.A., Dumais, S.T., Osuna, E., Platt, J., Scholkopf, B.: IEEE Intell. Syst. Appl. **13**(4), 18–28 (1998)
20. Breiman, L.: Random forests. Mach. Learn. **45**(1), 5–32 (2005)
21. Boyer, R.S.: Automated Reasoning: Essays in Honor of Woody Bledsoe, 2nd edn. Springer, Heidelberg (2012). https://doi.org/10.1007/978-94-011-3488-0
22. Pedregosa, F., et al.: Scikit-learn: machine learning in Python. J. Mach. Learn. Res. **12**(10), 2825–2830 (2011)
23. Sokolova, M., Lapalme, G.: A systematic analysis of performance measures for classification tasks. Inf. Process. Manag. **45**(4), 427–437 (2009)
24. Fawcett, T.: An introduction to ROC analysis. Pattern Recogn. Lett. **27**(8), 861–874 (2006)
25. Saito, T., Rehmsmeier, M.: The precision-recall plot is more informative than the ROC plot when evaluating binary classifiers on imbalanced datasets. PloS One **10**(3), e0118432 (2015)
26. Davis, J., Goadrich, M.: The relationship between precision-recall and ROC curves. In: Proceedings of the 23rd International Conference on Machine Learning, pp. 233–240, June 2006
27. Sasaki, Y.: The truth of the F-measure. Teach Tutor Mater **1**(5), 1–5 (2007)

Dynamic Clustering of Time Series with Echo State Networks

Miguel Atencia[1(✉)], Catalin Stoean[2], Ruxandra Stoean[2],
Roberto Rodríguez-Labrada[3], and Gonzalo Joya[1]

[1] Universidad de Málaga, Málaga, Spain
{matencia,gjoya}@uma.es
[2] University of Craiova, Craiova, Romania
{catalin.stoean,ruxandra.stoean}@inf.ucv.ro
[3] Centro para la Investigación y Rehabilitación de las Ataxias Hereditarias,
Holguin, Cuba
roberto@ataxia.hlg.sld.cu

Abstract. In this paper we introduce a novel methodology for unsupervised analysis of time series, based upon the iterative implementation of a clustering algorithm embedded into the evolution of a recurrent Echo State Network. The main features of the temporal data are captured by the dynamical evolution of the network states, which are then subject to a clustering procedure. We apply the proposed algorithm to time series coming from records of eye movements, called saccades, which are recorded for diagnosis of a neurodegenerative form of ataxia. This is a hard classification problem, since saccades from patients at an early stage of the disease are practically indistinguishable from those coming from healthy subjects. The unsupervised clustering algorithm implanted within the recurrent network produces more compact clusters, compared to conventional clustering of static data, and provides a source of information that could aid diagnosis and assessment of the disease.

Keywords: Echo State Networks · Clustering · k-means · Saccadic eye movement · Time series

1 Introduction

In this work we propose a novel methodology for unsupervised clustering of time series, by inputting the sequences to an Echo State Network (ESN), which is a recurrent neural network. Cluster analysis refers to the *discovery* of classes or categories within data that were previously unknown, thus it is included in the broad category of unsupervised learning tasks. Cluster analysis has a long history in the fields of statistics and machine learning (see e.g. the recent review

Partially supported by the Spanish Ministry of Science, Innovation and Universities, through the Plan Estatal de Investigación Científica y Técnica y de Innovación, Project TIN2017-88728-C2-1-R, as well as the Universidad de Málaga.

© Springer Nature Switzerland AG 2019
I. Rojas et al. (Eds.): IWANN 2019, LNCS 11507, pp. 73–83, 2019.
https://doi.org/10.1007/978-3-030-20518-8_7

[11] and references therein), and probably the best known clustering algorithm is k-means [6]. However, handling data with temporal features introduces a number of complications. First of all, the length of the sequence may be not known in advance, may be different for each sequence, or may even be infinite, which in all cases breaks the usual assumption that data is arranged as a rectangular (finite) matrix. Besides, time series can contain long term correlations, where very large chunks of data must be taken into account in order to fully capture the qualitative dynamical behaviour of the series. Finally, even if fixed-length data are available, considering time series as a vector neglects the temporal information, e.g. the correlation of data values at distant time points may be more significant that the find of similar values in successive components. Specific cluster analysis of temporal data has been tackled from the complementary viewpoints of time series [1] and data streams [3]. Many clustering algorithms for time series are applied on a *window* of data, which must then be carefully chosen: if it is too small, long-term relationships will be missed, whereas a large data window will introduce a signifcant computational cost.

This work is prompted by the need to analyse data coming from electrooocu-lographic records, in order to implement a tool to aid in the diagnosis and assessment of spino-cerebellar ataxia of type 2 [12]. Data coming from ataxia patients in Cuba, as well as control, healthy subjects, have been analysed and labelled by physicians. Previously, data mining techniques have been applied to these data for the extraction of significant clinical events [2], but much work is needed in order to improve the accuracy of classification. In particular, the records from healthy individuals and patients at an early stage of the disease are almost indistinguishable, even for human experts.

Echo state networks are recurrent neural networks that can be classified into the paradigm of reservoir computing (see [8] and references therein). The most striking feature of ESNs is the absence of learning, in the conventional sense of modification of weights or connections between units. Instead, a set of units is fully connected by recurrent connections, whereas the connection values are constant. The feedback loops induce a dynamical behaviour that is intended to capture the important features of the time series that is presented as input. Echo State Networks have been applied to time series coming from medical applications [4], but there is much margin for improving our knowledge of both the fundamental analysis of these models, and their practical applicability. In particular, to the best of our knowledge, our proposal that a clustering algorithm is embedded inside the temporal evolution of an ESN is original.

In Sect. 2 the formulation of Echo State Networks is briefly reviewed, emphasizing the need to choose the network hyperparameters in order to produce a rich dynamical behaviour. The characteristics of the dataset of electrooculographic records are gathered in Sect. 3, stressing the difficulty of detecting the disease when eye movements are not yet significantly impaired. The proposed algorithm is described in Sect. 4, contrasting its principled definition to usual methods for clustering of time series. Experimental results are discussed in Sect. 5, whereas

Sect. 6 puts an end to the paper with some final remarks and directions for further research.

2   Echo State Networks

In their original formulation, Echo State Networks are supervised classification models that are applied to data stemming from time series. Given an input signal $x(t)$, the objective is to predict at every instant t a target output $o(t)$, where $t = 0 \ldots T \ldots$ is the discrete time whose final time t_f is not necessarily finite or known in advance. It is often the case that the final aim is time series prediction, and then the target output is simply the one-step ahead input $o(t) = x(t+1)$. The architecture of the network consists of a fixed number d of units that store a value at every time step, so the *state* of the network is a d-dimensional vector $y(t) \in \mathbb{R}^d$, which is time varying. The output is computed from the reservoir state and the input by a feedforward connectionist layer, which is usually linear:

$$\hat{o}(t) = W_{\text{io}}\, x(t) + W_{\text{out}}\, y(t) + b \tag{1}$$

where b is a bias vector. Therefore, learning proceeds by solving a regression problem through minimization of the error $E(t) = o(t) - \hat{o}(t)$ thus computing the (time-varying) parameters W_{io}, W_{out}, b. Note that, the dependence of the error on parameters being linear, this regression problem is particularly simple.

In an ESN, the reservoir states $y(t)$ are dynamically updated by the following recursive rule:

$$y(t) = (1 - l)\, y(t-1) + l\, \tanh\left(W_{\text{in}}\, x(t) + W\, y(t-1)\right) \tag{2}$$

where $l \in [0, 1]$ is a *leaking* hyperparameter. The distinguishing feature of ESNs, compared to e.g. backpropagation through time learning [5], is that the weight matrices W_{in}, W are constant, and fixed once and for all at the beginning of the reservoir time evolution. The rationale behind the construction of ESNs is that the dynamical evolution of the reservoir is able to grasp the important features of the input, which is thus *echoed* by the states, so the influence of the particular chosen values of the weight matrices is negligible. When this objective is achieved, the network is said to possess the *echo state* property.

In order to construct an ESN, choices must be made regarding the value of several hyperparameters. First of all, the reservoir size d must attain a trade-off between capacity and computational cost, and it is important that the number of units is related to the length and dimensionality of input data, since a reservoir too large fed by insufficient data may lead to overfitting. The already mentioned leaking rate l adjusts the balance between long and short term memory, and it can be regarded as a measure of the velocity of the dynamical evolution of the network. Finally, the strategy to define the weight matrix W is the choice that most influences the qualitative dynamical behaviour of the network. There are several—to some extent, equivalent—criteria to characterize the network dynamics with a single hyperparameter that rules the construction of the weight matrix. One of these concepts is the maximum Lyapunov exponent [13], which, if positive, signals that the dynamics is chaotic. Another useful measure of the complexity of the network dynamics is the spectral radius R, which is the maximum absolute value of the eigenvalues of the weight matrix W. A stability analysis by linearization shows that if $R < 1$ then the network state tends to an equilibrium, in the absence of input. This behaviour is considered undesirable, since a single fixed point would not carry the whole information of the input. On the contrary, a very large value of R would cause the network states to grow without bound. The rule of thumb for fixing the spectral radius suggests that its value should be slightly larger than 1, to produce a rich dynamics while the states remain bounded [8]. Anyway, these results can hardly be rigorously applied in practice since, on the one hand, the influence of inputs cannot be neglected and, on the other hand, the presence of the hyperbolic tangent introduces a non-linearity.

Since the objective of this paper is unsupervised learning, we omit the output layer, together with the regression computation to fix the parameters W_{io}, W_{out}, b. Our focus is on the network dynamics being able to capture the most significant features of the time series, which will allow for clustering.

3   Dataset of Saccadic Movements

The motivating problem for the current work is to aid to the diagnosis of spino-cerebellar ataxia of type 2 (SCA2), which is a degenerative disorder that causes uncoordinated movement, among other neurological symptoms. In particular, weakening eye muscles produce a slowing of fast eye movements (saccades). The onset of the disease is progressive and its duration usually ranges between 10 and 15 years [9]. At initial stages, patients can undergo relatively mild symptoms, or even none at all, thus diagnosis may be delayed until clinical manifestations are severely disabling. Contrarily, early diagnosis can help establish a supportive program that includes physical therapy and life style adaptations. Since clinical manifestations are often inconclusive, the definitive diagnosis can only be established by genetic testing. However, specific genetic analysis is costly and cannot be used as a general screening method in areas where the disease is prevalent, such as Cuba. Instead, the examination of the degree of alteration of saccadic movements provides a cheap and accessible diagnostic tool.

Saccades are measured by means of an electrooculograpy device, which samples the weak electrical potentials due to eye movements, when the subject is trying to track the trajectory of an object on a screen. Typically, the test is repeated with several amplitudes that lead to different angular displacements of the subject's eyes. In this work we have used a database that contains 88 registers, which have been labelled as C (control, healthy individuals), P (presymptomatic subjects, including those with very mild degeneration), and D (diseased patients with severe clinical manifestations). The electrooculograpic records are analysed to isolate individual saccades, which are then preprocessed leading to a database of 6124 saccades, each containing 192 samples. The details of this preprocessing are included in a companion paper. As an example, three saccades each corresponding to one class are shown in Fig. 1.

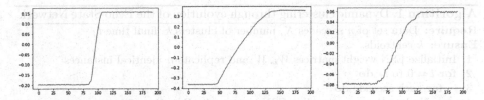

Fig. 1. Sample saccades (at different vertical scales) corresponding to a healthy individual, a presymptomatic subject, and an ill patient, from left to right. Observe the different amplitude, speed, and steadiness.

It is important to emphasize that labelling has been performed by medical experts from their subjective analysis, knowledge of the subject's family history, genetic tests, or other means, because observation of saccades alone does not allow to establish a diagnosis. In particular, individual saccades from presymptomatic patients are often indistinguishable from those recorded in healthy subjects. Thus from the point of view of classification as a supervised learning task, the problem at hand can be considered as *ill-posed*, since some instances from presymptomatic patients possess almost exactly the same data features as those from healthy subjects. Therefore a supervised classification method is prone to low accuracy, since data are mislabelled. This suggests the use of unsupervised techniques to provide a finer clustering than the one obtained from the standard classification procedure.

4   Dynamic Clustering

The proposed procedure can be summarised as the application of a clustering method inside every step of the ESN. Starting from a dataset with n saccades, n identically initialized ESNs are evolved, each accepting as input the series of the corresponding saccade. Then, at every evolution step t, the reservoir states of all ESNs are stacked to form a new database, which is subject to a clustering method, thus obtaining k clusters, each represented by its centroid. At the next time step, the clustering process is repeated, but now the initial clusters are set to those coming from the previous step $t - 1$. The final produced clustering is the one that results from the last time step in the series. A formal algorithmic description of the proposed method is shown in Algorithm 1.

The rationale behind the described algorithm is that reservoir states should capture time series dynamics, without the need to explicitly compute frequency-domain features, such as Fourier transforms. In contrast, methods based upon sliding windows on the series suffer from a limited bandwidth and cannot achieve long-term memory. Therefore our novel proposal would constitute a sort of hybrid between time-domain and frequency-domain techniques. In the description above, we have not specified which clustering method is used. In principle, any unsupervised clustering technique would work, but the minimal adaptations would be needed for iterative, partitioning methods, where the cluster

---

Algorithm 1. Dynamic clustering through evolution of the Echo State Network.

Require: Data set of n saccades X, number of clusters k, final time t_f
Ensure: k centroids
 1: Initialize ESN weight matrices W_i, W and replicate n identical instances
 2: **for** $t = 0$ to t_f **do**
 3:     **for** each saccade **do**
 4:         Update the corresponding ESN instance by Equation (2)
 5:     **end for**
 6:     **if** $t = 0$ **then**
 7:         Initialize centroids
 8:     **else**
 9:         Set initial centroids to centroids resulting from step $t - 1$
10:     **end if**
11:     Build the dataset Y of n reservoir states
12:     Compute centroids at step t from clustering of dataset Y
13: **end for**

---

centroids obtained at the previous step can be used for the next initialization. We have implemented the experiments with the well-known method k-means, which is computationally rather efficient. In contrast, hierarchical methods not only would require some tweaking in the algorithm, but also they usually lead to unaffordable computational cost.

A significant advantage of our proposal is the ability to deal with variable-length time series. Since the evolution of all ESNs, each corresponding to one series, is simultaneous, shorter input sequences should be set to a null value. However, the corresponding reservoir states would continue their evolution autonomously, hopefully having memorised the dynamical behaviour acquired from the series. This feature is particularly advantageous in the example described in Sect. 3 since, although apparently all saccades have the same length, this is the result of the preprocessing, which includes a somewhat arbitrary clipping to the established length, whereas saccades themselves are intrinsically variable-length time series.

5   Experimental Results

In order to provide a proof of concept for the proposed approach, we have implemented the dynamical clustering method embedded into the ESNs evolution. As an illustrative example of the key issues of the method, we use as input the saccades time series. Four kinds of experiments have been performed to assess the performance gain resulting from the introduction of the ESN, compared to conventional forms of unsupervised clustering:

- First of all, a baseline has been established by performing clustering on the vectors comprising the whole saccades, disregarding the fact that components are temporally related.

- Also, for the sake of comparison, clustering has been carried out on an instantaneous snapshot of the reservoir states, which is obtained by averaging of the states along the whole evolution.
- Another fixed view of the states results from considering the instantaneous state at the final time, once the input presentation ends.
- Finally, our main proposal is the repeated implementation of clustering iteratively at every step along the ESNs evolution.

For all experiments, the k-means algorithm has been chosen for clustering, due to its simplicity and computational efficiency.

After preliminary experimentation, the number of clusters is set to $k = 10$. The hyperparameters of the ESNs have been set as follows: the reservoir size is $1/5$ of the length of the saccades, i.e. $d = 38$, which is considered enough to memorize the input dynamics while keeping a limited computational cost; the leaking rate is $l = 0.3$; finally, several runs have been performed with different values of the weight matrix spectral radius R, which is known to be a critical hyperparameter in the dynamical behaviour of the ESN. For the sake of brevity, only experiments with $R = 2$ and $R = 3$ are here reported.

A critical concern in unsupervised learning is the assessment of the quality of the result since, unlike in supervised classification, there is no obvious accuracy measure. Among the large numbers of clustering quality measures, we have chosen the silhouette coefficient [10]. For each data instance i assigned to cluster A, the silhouette coefficient s_i is defined by

$$a_i = \frac{1}{|A| - 1} \sum_{j \in A - \{i\}} d(i, j)$$

$$b_i = \min_{B \neq A} \frac{1}{|B|} \sum_{j \in B} d(i, j) \qquad (3)$$

$$s_i = \frac{b_i - a_i}{\max(a_i, b_i)}$$

It is easy to see that $-1 \leq s_i \leq 1$ and values of s_i intuitively represent the certainty that the instance i really belongs to cluster A. For the clusterings obtained in the four procedures described above, we have computed the average of the silhouette coefficients, and results are shown in Table 1. It can be observed that the clustering obtained from the reservoir states is significantly more meaningful that the one directly resulting from saccades data. This reinforces the notion that the Echo State Network, with its recurrent dynamics, is able to extract and memorise temporal features of data, which are not obvious when saccades are simply considered as a vector.

A second set of experiments aims at obtaining information from the unsupervised clustering about the health status of subjects, rather than individual saccades, thus providing supplementary information to aid a diagnosis. Therefore, we now incorporate the knowledge of the class labels by computing a *severity index* from the clustering results, by means of an averaging. Specifically, the following procedure is performed:

Table 1. Values of the silhouette coefficient for clustering with the original saccade time series, the average of the reservoir states of the corresponding ESNs, the final reservoir state after the complete presentation of the saccade, and the repeated process at every step of the ESNs evolution.

Clustering data	Spectral radius	
	$R = 2$	$R = 3$
Original time series	0.42	
Reservoir state average	0.43	0.51
Final reservoir state	0.52	0.49
Dynamical reservoir states	**0.54**	0.53

1. For each saccade, the severity value 0, 1, or 2 is assigned according to whether the saccade is labelled as C, P, or D, respectively.
2. For each cluster, these values are averaged for all the saccades that are assigned to the same cluster. The assignment results from minimum distance to the cluster centroid, among all clusters. The obtained value $SI(c) \in [0, 2]$ is considered the *severity index* of the cluster c.

The rationale behind this computation is that the classification in three disease stages C, P, D is too coarse and there is significant overlap in the characteristics of individual saccades, especially between healthy subjects and presymptomatic ones. The distribution of labels and the severity indexes of the ten clusters are shown in Fig. 2 and Table 2, respectively. It is obvious that classes are not identically distributed among clusters, which suggests that unsupervised clustering has captured some information of the disease stage that could be used to inform a diagnosis. In order to rigorously confirm this postulate, we have performed a χ^2 independence test. The corresponding contingency table is shown in Table 3. The hypothesis test results in a p-value of $2.39\,10^{-56}$, i.e. negligible, thus the hypothesis that clustering is statistically independent from class labels can be rejected. A deeper analysis of the form of this dependence, and the validation of the severity index methodology by medical experts is left for further research, so we have not performed here a systematic experimentation that results in a competitive classification method.

Table 2. Values of the severity index for the ten clusters, shown in the same order as in Fig. 2.

Cluster #	0	1	2	3	4	5	6	7	8	9
S.I		1.051	0.696	1.179	0.950	1.073	0.761	1.080	1.089	0.872	0.720

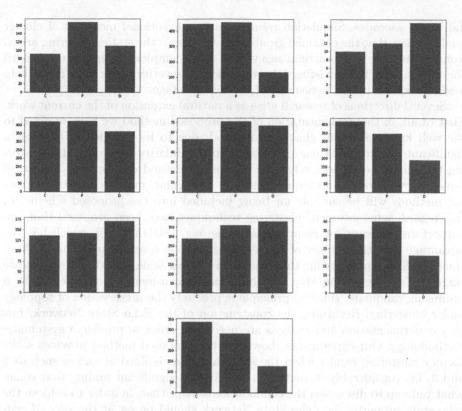

Fig. 2. Distribution of labels in the ten clusters obtained from dynamic clustering with $R = 3$. Clusters are shown in the same order as in Table 2.

Table 3. Contingency table between health status classes and unsupervised clusters.

Class/Cluster #	0	1	2	3	4	5	6	7	8	9
C	92	447	10	422	53	418	136	289	33	338
P	168	458	12	424	71	345	143	361	40	283
D	111	132	17	362	67	190	172	381	21	128

6   Conclusions

In this work we have presented a novel method for unsupervised clustering of time series, based upon the introduction of an Echo State Network. This recurrent model possesses a natural dynamics, due to the presence of feedback connections, which captures the relevant features of the presented sequences. The state values then form a data table that goes through a conventional clustering algorithm *at every time step*. The method is applied to the task of analysing eye movement records in order to aid in the diagnosis of ataxia, where usual classification methods achieve limited results, due to the mixed up labelling of individual saccades. Simulation results, using conventional measures of cluster quality, show that the obtained grouping outperforms the static clustering arisen from considering temporal data as a vector. Also, complementing this result with the information from existing labels provides a severity index that could help medical experts in the assessment of the disease stage.

Several directions of research arise as a natural extension of the current work. First of all, in the implementation of the proposed method we have resorted to the well known k-means clustering method, due to its simplicity. However, a significant advantage of our proposal is its modularity regarding the clustering technique, i.e. a more advanced clustering method can be used instead of k-means, with some mild requirements. In particular, most iterative partitioning methods will be suitable for being included into the proposed scheme. In this regard, some advanced clustering techniques have been proposed that can outperform k-means for many applications, e.g. ISODATA [6], which has the advantage that the number of clusters must not be specified beforehand. Also, since the final aim is placing the patients in a linear scale regarding their disease stage, a Self Organizing Map [7] with a one-dimensional topology could be a promising candidate, since its principle is precisely the preservation of topology under clustering. Regarding the construction of the Echo State Network, further experimentation and analysis are needed in order to provide a systematic methodology. Our experiments show that the proposed method provides satisfactory clustering results when the spectral radius is fixed at values such as 2 and 3, i.e. considerably distant from 1. This is a significant finding that somewhat puts up to discussion the common knowledge that, in order to achieve the echo state property, the Echo State Network should be set at the *edge of criticality* through values of the spectral radius only slightly larger than one. Our ongoing work aims at a theoretical explanation of these results.

References

1. Aghabozorgi, S., Seyed Shirkhorshidi, A., Ying Wah, T.: Time-series clustering - a decade review. Inf. Syst. **53**, 16–38 (2015)
2. Becerra-García, R.A., et al.: Data mining process for identification of non-spontaneous saccadic movements in clinical electrooculography. Neurocomputing **250**, 28–36 (2017)
3. Ding, S., Wu, F., Qian, J., Jia, H., Jin, F.: Research on data stream clustering algorithms. Artif. Intell. Rev. **43**(4), 593–600 (2015)
4. Gallicchio, C., Micheli, A., Pedrelli, L.: Deep echo state networks for diagnosis of Parkinson's Disease. In: Verleysen, M. (ed.) Proceedings of the European Symposium on Artificial Neural Networks, Computational Intelligence and Machine Learning (ESANN 2018). pp. 397–402. i6doc, Bruges (2018)
5. Goodfellow, I., Bengio, Y., Courville, A.: Deep Learning. The MIT Press, Cambridge (2016)
6. Jain, A.K.: Data clustering: 50 years beyond K-means. Pattern Recogn. Lett. **31**(8), 651–666 (2010)
7. Kohonen, T.: Essentials of the self-organizing map. Neural Netw. **37**, 52–65 (2013)

8. Lukoševičius, M.: A practical guide to applying echo state networks. In: Montavon, G., Orr, G.B., Müller, K.-R. (eds.) Neural Networks: Tricks of the Trade. LNCS, vol. 7700, pp. 659–686. Springer, Heidelberg (2012). https://doi.org/10.1007/978-3-642-35289-8_36

9. Pulst, S.M.: Spinocerebellar ataxia type 2. In: GeneReviews®. University of Washington, Seattle (1993). http://www.ncbi.nlm.nih.gov/pubmed/20301452

10. Rousseeuw, P.J.: Silhouettes: a graphical aid to the interpretation and validation of cluster analysis. J. Comput. Appl. Math. **20**(C), 53–65 (1987)

11. Saxena, A., et al.: A review of clustering techniques and developments. Neurocomputing **267**, 664–681 (2017)

12. Velázquez-Mariño, M., Atencia, M., García-Bermúdez, R., Sandoval, F., Pupo-Ricardo, D.: Architecture for neurological coordination tests implementation. In: Rojas, I., Joya, G., Catala, A. (eds.) IWANN 2017. LNCS, vol. 10306, pp. 26–37. Springer, Cham (2017). https://doi.org/10.1007/978-3-319-59147-6_3

13. Verstraeten, D., Schrauwen, B., D'Haene, M., Stroobandt, D.: An experimental unification of reservoir computing methods. Neural Netw. **20**(3), 391–403 (2007)

Artificial Neural Network for Biomedical Image Processing

Multiple Sclerosis Detection via Wavelet Entropy and Feedforward Neural Network Trained by Adaptive Genetic Algorithm

Ji Han and Shou-Ming Hou[✉]

School of Computer Science and Technology, Henan Polytechnic University,
Jiaozuo, Henan 454000, People's Republic of China
housm@163.com

Abstract. Multiple sclerosis is a disease that damages the central nervous system. Current medical treatments can only prevent or relieve symptoms. The target of this study is to improve the detection efficiency and classification accuracy. We propose a method based on wavelet entropy and feedforward neural network trained by adaptive genetic algorithm that is implemented over 10 runs of 10-fold cross validation. In which the wavelet entropy serves as a feature extractor and the feedforward neural network is employed as a classifier. Adaptive genetic algorithm work as a training algorithm. We also use the three-level decomposition of db2 wavelet to make a frequency analysis. According to the experimental results, the global optimization capability of adaptive genetic algorithm is more powerful than ordinary genetic algorithm. Comparing to the HWT-LR method, the accuracy of our method detection is higher.

Keywords: Multiple sclerosis · Wavelet entropy ·
Feedforward neural network · Adaptive genetic algorithm

1 Introduction

Multiple sclerosis is a common chronic central nervous system autoimmune nervous system disease [1–4]. The age of onset of the disease is concentrated between 20 and 40 years old, with more women than men. Clinical manifestations include delirium, numbness, painful paralysis, aphasia, visual impairment, ataxia, mental symptoms or mental disorders. If multiple sclerosis can be detected early, it can be of great help to the patient's treatment.

The cause of multiple sclerosis is usually associated with genetic factors, environmental factors such as viral infections and geographic autoimmune responses [5]. There is actually no effective treatment for multiple sclerosis. Current treatments include acute phase treatment and palliative care. Acute treatment mainly relies on relieving symptoms, shortening the course of disease, reducing nerve damage and preventing complications [6]. The main purpose of treatment during remission is to control the disease. Therefore, researchers focus on the detection and identification of multiple sclerosis.

© Springer Nature Switzerland AG 2019
I. Rojas et al. (Eds.): IWANN 2019, LNCS 11507, pp. 87–97, 2019.
https://doi.org/10.1007/978-3-030-20518-8_8

Recently, researchers have tended to use computer vision and image processing techniques to perform multiple sclerosis detection tasks. Guo, Qin [7] proposed an improved multimodal local steering nucleus method to detect multiple sclerosis in the brain. The method uses multimodal brain MR images and a priori knowledge of the brain's approximate axisymmetry to detect changes in brain conditions. Lopez [8] employed a Haar wavelet transform (HWT) with logistic regression (LR) method.

This paper proposes a wavelet entropy and adaptive genetic algorithm to train the feedforward neural network to detect cerebral sclerosis. The experimental results show that the detection effect is greatly improved in terms of speed and stability. The rest of this paper is organized as follows. Section 2 shows the dataset and experiments. Section 3 will summary an overview of the methods including wavelet entropy, and the adaptive genetic algorithm and genetic algorithm and feedforward neural network. Section 4 is devoted to analyzing the experimental results. In Sect. 5, we make a conclusion.

2　Dataset

The dataset used in this experiment were obtained from Ref. [9], the healthy controls were used from 681 slices of 26 healthy controls provided in Ref. [9]. MS images were obtained from the eHealth laboratory [10], the healthy controls were used from 676 slices of 38 healthy controls provided in the eHealth laboratory. All cases of brain lesions were identified and delineated by experienced MS neurologists and confirmed by radiologists. Table 1 shows the demographic characteristics of two datasets.

Table 1. Demographic characteristics of two datasets

Dataset	Source	# Subjects	Number of slice	Age	Gender (m/f)
Multiple sclerosis [10]	eHealth laboratory	38	676	34.1 ± 10.5	17/21
Healthy control [9]	Dr. Pan et al.	26	681	33.5 ± 8.3	12/14

This study chose a histogram stretching method to normalize these two datasets. HS aims to enhance the contrast by stretching the range of intensity values of two sources of images to the same range, providing the effect of inter-scan normalization. We do contrast normalization for both two data of different sources, and finally combine them together, forming a $676 + 681 = 1,357$-image dataset.

3　Methodology

The whole implementation of our method is based on three success components: wavelet entropy, feedforward neural network, and adaptive genetic algorithm (AGA). The wavelet entropy serves as a feature extractor. The feedforward neural network is employed as a classifier. Finally, the AGA work as a training algorithm due to its powerful global optimization capability. Deep learning methods [11–15] were not used, since our dataset is small.

3.1 Wavelet Entropy

Wavelet entropy [16] is a nonlinear dynamic analysis method proposed by combining wavelet transform and information entropy theory. In information theory, entropy is used to indicate that all symbols in a source contain an average of information and uncertainty, which represents useful information about the dynamic process of the signal [17–21]. A multi-scale wavelet transform is applied to the signal, and the obtained wavelet coefficients of each scale are converted into a probability distribution sequence. The magnitude of the entropy value calculated by the sequence reflects the sparsity of the wavelet coefficient matrix. The calculation process of the wavelet entropy algorithm can be described as follows.

Let the signal x(t) be discrete wavelet transform, the detail coefficient at time k at the j-th decomposition scale is $d_{j,k}$, the approximation coefficient is $a_{j,k}$, then the detail energy of the scale j (j = 0, 1, …, M) is defined as in Eq. (1)

$$E_j = \sum_k |c_j(k)|^2 \tag{1}$$

where $c_j(k)$ is the corresponding wavelet coefficient, $c_j(k) = [x(t), y_{j,k}(t)]$. The total energy of the signal is defined as in Eq. (2)

$$E = ||x(t)||^2 = \sum_j \sum_k |c_j(k)|^2 = \sum_j E_j \tag{2}$$

Relative wavelet energy can be expressed as $P_j = E_j/E$. Divide the detail coefficient $c_j(k)$ of the j-th layer decomposition into n sub-intervals, defined as in Eq. (3).

$$E_j(k) = \sum_k^{m/n} |c_j(k)|^2 \tag{3}$$

where m is the number of sampling points. The relative wavelet energy $P_{j,i}$ of each subinterval can be obtained by dividing the wavelet energy $E_{j,i}$ of the corresponding i-th subinterval by the total energy of the layer wavelet coefficients, defined as in Eqs. (4) and (5)

$$P_{j,i} = E_{j,i}/E_j \tag{4}$$

$$E_j = \sum_{j=1}^n E_{j,i} \tag{5}$$

The wavelet entropy W_i corresponding to the i-th subinterval can be defined as in Eq. (6).

$$W_i = -\sum_j P_{j,i} \ln(P_{j,i}) \qquad (6)$$

3.2 Feedforward Neural Network

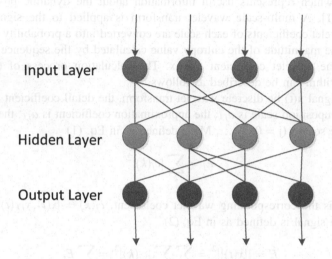

Fig. 1. Structure of feedforward neural network

The wavelet entropy values from Sect. 3.1 are the input to ANN. Figure 1 shows the structure of feedforward neural network. The feedforward neural network includes an input layer, a hidden layer, and an output layer. Each neuron is arranged in layers, every neuron is only connected to the neurons of the previous layer, and receives the output of the previous layer, and outputs it to the next layer [22]. There is no feedback between the layers. At the same time, the feedforward neural network has a simple structure and is widely used, and can approximate arbitrary continuous functions and square integrable functions with arbitrary precision [23–27]. In addition, any limited set of training samples can be accurately implemented.

3.3 Genetic Algorithm

Genetic algorithm is an adaptive global optimization search algorithm that simulates the genetic and evolutionary processes of living things in the natural environment [28, 29]. The principle of genetic algorithms is to use the "survival of the fittest" principle to generate approximate optimal solutions in a potential solution population [30].

The genetic algorithm consists of selection operator, crossover operator and mutation operator, which has strong advantages for solving the optimal solution in the global scope [31]. It can perform multi-point search at the same time, and directly uses the fitness as the search information, and the operation object is the code compiled by a certain method [32, 33]. The basic idea of genetic algorithm is to gradually optimize the problem through the combination of gene crossover and mutation, so as to find the optimal solution. Genetic algorithm flow chart shown in Fig. 2.

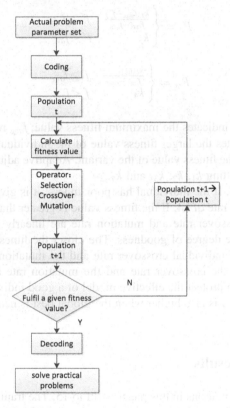

Fig. 2. Genetic algorithm flow chart

Genetic algorithm is an efficient and stable optimization method with good fault tolerance and robustness [34]. But genetic algorithms also have premature convergence problems. Because some individuals may have high fitness values earlier, their individual gene coding rapidly spreads in the population, making the population genes lack diversity, which leads to iterative solution falling into local optimum and it is difficult to find the full space optimal solution. In order to ensure the validity of the research, this study proposes an adaptive genetic algorithm based on genetic algorithm.

3.4 Adaptive Genetic Algorithm

Adaptive genetic algorithm is an improvement of the basic genetic algorithm, which maintains the diversity of the population while ensuring the convergence of the genetic algorithm. The adaptive genetic algorithm is proposed by Srinvivas et al. [35], its idea is to increase the crossover probability P_c and the mutation probability P_m when the population fitness is more concentrated [36]. When the population fitness is more dispersed, reducing the P_c and P_m. P_c and P_m are defined in Eqs. (7) and (8) respectively.

$$P_c = \begin{cases} \frac{k_1(f_{max}-f')}{f_{max}-f_{avg}} & f' \geq f_{avg} \\ k_3 & f' < f_{avg} \end{cases} \tag{7}$$

$$P_m = \begin{cases} \frac{k_2(f_{max}-f)}{f_{max}-f_{avg}} & f \geq f_{avg} \\ k_4 & f < f_{avg} \end{cases} \tag{8}$$

Among them: f_{max} indicates the maximum fitness value; f_{avg} represents the average fitness value; f' indicates the larger fitness value of the individual doing the crossover operation; f indicates the fitness value of the variant. Adaptive adjustment of Pc and Pm can be achieved by setting k_1, k_2, k_3, and k_4.

In Eqs. (7) and (8), If the individual has poor fitness, it is given a larger crossover rate k3 and a mutation rate of k4. If the fitness value is greater than the average fitness, the corresponding crossover rate and mutation rate are linearly assigned to the individual according to the degree of goodness. The closer the fitness is to the maximum fitness, the smaller the individual crossover rate and the mutation rate. When equal to the maximum fitness, the crossover rate and the mutation rate are zero. This cross-adjustment method can protect the effective model of a good individual, and the search efficiency of the method is also higher when the global best is near the local extremum point.

4 Experiment Results

The number of hidden neurons in this paper is set to 15. The training algorithm is back propagation method. The maximum training iteration is 1,000. The parameters of AGA is the same as Ref. [35]. All trials were run 10 times, and the average and standard deviation were offered in this Section.

4.1   Statistical Analysis

This proposed MS detection method "WE-FNN-AGA" is implemented over 10 runs of 10-fold cross validation. In this study, Three-level decomposition of db2 wavelet was utilized. The results are shown in Table 2. Here we report the sensitivity, specificity, precision, and accuracy, in total four measures. From Table 2, we can observe our method procured a sensitivity of 91.91 ± 1.24%, a specificity of 91.98 ± 1.36%, a precision of 91.97 ± 1.32%, and an accuracy of 91.95 ± 1.19%.

Table 2. 10 runs of 10-fold cross validation of our method

Run	Sensitivity	Specificity	Precision	Accuracy
1	91.28	90.16	90.27	90.71
2	90.83	92.50	92.41	91.67
3	90.97	89.58	89.67	90.27
4	92.32	92.22	92.19	92.26
5	91.27	92.21	92.12	91.75
6	91.56	91.19	91.20	91.38
7	90.54	92.51	92.35	91.53
8	93.49	93.39	93.37	93.44
9	94.68	94.42	94.47	94.55
10	92.16	91.63	91.63	91.90
Average	91.91 ± 1.24	91.98 ± 1.36	91.97 ± 1.32	91.95 ± 1.19

4.2   Optimal Decomposition Level

Table 3 shows the performance of different decomposition levels of using WEs. We can observe that the performance achieves the greatest when decomposition level of 3. For ease of comparison, the comparison results were shown in Fig. 3.

Table 3. Comparison of wavelet decomposition level

Decomposition level	Sensitivity	Specificity	Precision	Accuracy
1	86.32 ± 1.09	86.36 ± 1.06	86.32 ± 0.86	86.34 ± 1.12
2	88.28 ± 1.71	88.31 ± 1.26	88.27 ± 1.24	88.30 ± 1.29
3	91.91 ± 1.24	91.98 ± 1.36	91.97 ± 1.32	91.95 ± 1.19
4	87.91 ± 1.32	88.03 ± 1.59	87.99 ± 1.49	87.97 ± 1.25

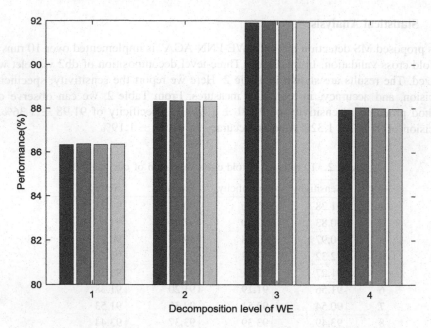

Fig. 3. Comparison of decomposition levels

4.3 Comparison with GA Method

AGA is an improved version of ordinary GA. In this experiment, we compared AGA with ordinary GA approach. Other parameters are the same as our method. The results of 10 runs of 10-fold cross validation using GA is listed in Table 4. Comparing with the results obtained by proposed AGA method, we can find GA approach reduces the performance.

Table 4. 10 runs of 10-fold cross validation using ordinary GA method

Run	Sensitivity	Specificity	Precision	Accuracy
1	88.02	89.87	89.63	88.95
2	90.54	90.31	90.30	90.42
3	88.46	88.11	88.11	88.28
4	88.46	89.87	89.68	89.17
5	90.54	89.28	89.36	89.90
6	89.50	89.28	89.25	89.39
7	89.34	88.11	88.20	88.72
8	86.25	87.66	87.43	86.96
9	90.53	89.86	89.91	90.20
10	88.46	88.11	88.08	88.28
Average	89.01 ± 1.30	89.05 ± 0.91	89.00 ± 0.91	89.03 ± 0.99

4.4   Comparison to HWT-LR Method

Table 5 compares our method with HWT-LR approach [8]. Within their paper, different parameter settings were tested. They proposed a novel MS slice identification system, based on Haar wavelet transform, principal component analysis, and logistic regression. Then they used 2-level, 3-level, and 4-level decomposition to test the accuracy respectively. Finally, they achieve the highest accuracy of 89.72 ± 1.18%. In Table 5, we can observe that our WE-FNN-AGA method is better than HWT-LR method [8]. Here we only select accuracy for ease of comparison.

Table 5. Comparison with HWT-LR method

Method	Accuracy
HWT-LR [8]	89.72 ± 1.18
WE-FNN-AGA (Ours)	**91.95 ± 1.19**

(Bold means the best)

5   Conclusion

This paper concentrates on the problem of multiple sclerosis detection. A detection system via wavelet entropy and feedforward neural network trained by adaptive genetic algorithm was proposed for detecting Multiple sclerosis. The main contributions are list as follows:

(1) In this paper, we first applied wavelet entropy and feedforward neural network for the multiple sclerosis detection whose early diagnosis helps patients with subsequent treatment.
(2) In order to improve the global optimization capability, we propose adaptive genetic algorithm based on genetic algorithm.
(3) The proposed method has relatively good performance in terms of sensitivity, specificity, precision and accuracy.

The results show the precision and accuracy of our method are 91.97 ± 1.32% and 91.95 ± 1.19%, it is superior to the HWT-LR method. In later studies, we will continue to look for better ways to improve detection accuracy.

Acknowledgement. This paper was supported by National Natural Science Foundation of China (No. 61503124), key scientific and technological project of Henan province (No. 17210-2210273, 182102210086).

References

1. Zhou, X.-X.: Comparison of machine learning methods for stationary wavelet entropy-based multiple sclerosis detection: decision tree, k-nearest neighbors, and support vector machine. Simulation **92**(9), 861–871 (2016). https://doi.org/10.1177/0037549716666962

2. Dong, Z.: Synthetic minority oversampling technique and fractal dimension for identifying multiple sclerosis. Fractals **25**(4) (2017). Article ID: 1740010

3. Cheng, H.: Multiple sclerosis identification based on fractional Fourier entropy and a modified Jaya algorithm. Entropy **20**(4) (2018). Article ID: 254

4. Huang, C.: Multiple sclerosis identification by 14-layer convolutional neural network with batch normalization, dropout, and stochastic pooling. Front. Neurosci. **12** (2018). Article ID: 818. https://doi.org/10.3389/fnins.2018.00818

5. Zhan, T.M., Chen, Y.: Multiple sclerosis detection based on biorthogonal wavelet transform, RBF kernel principal component analysis, and logistic regression. IEEE Access **4**, 7567–7576 (2016). https://doi.org/10.1109/ACCESS.2016.2620996

6. Azadmehr, A., et al.: Immunomodulatory and anti-inflammatory effects of Scrophularia megalantha Ethanol extract on an experimental model of multiple sclerosis. Res. J. Pharmacognosy **6**(1), 43–50 (2019). https://doi.org/10.22127/rjp.2018.80370

7. Guo, Y., Qin, P.L.: Research on detection algorithm of multiple sclerosis of brain based on multimode local steering nucleus. Comput. Sci. **45**(3), 243–248 (2018)

8. Lopez, M.: Multiple sclerosis slice identification by Haar wavelet transform and logistic regression. Adv. Eng. Res. **114**, 50–55 (2017)

9. Pan, C.: Multiple sclerosis identification by convolutional neural network with dropout and parametric ReLU. J. Comput. Sci. **28**, 1–10 (2018). https://doi.org/10.1016/j.jocs.2018.07.003

10. MRI Lesion Segmentation in Multiple Sclerosis Database (2018). eHealth laboratory, University of Cyprus

11. Wang, S., Chen, Y.: Fruit category classification via an eight-layer convolutional neural network with parametric rectified linear unit and dropout technique. Multimedia Tools Appl. (2018). https://doi.org/10.1007/s11042-018-6661-6

12. Pan, C.: Abnormal breast identification by nine-layer convolutional neural network with parametric rectified linear unit and rank-based stochastic pooling. J. Comput. Sci. **27**, 57–68 (2018). https://doi.org/10.1016/j.jocs.2018.05.005

13. Zhao, G.: Polarimetric synthetic aperture radar image segmentation by convolutional neural network using graphical processing units. J. Real-Time Image Proc. **15**(3), 631–642 (2018). https://doi.org/10.1007/s11554-017-0717-0

14. Sangaiah, A.K.: Alcoholism identification via convolutional neural network based on parametric ReLU, dropout, and batch normalization. Neural Comput. Appl. (2019). https://doi.org/10.1007/s00521-018-3924-0

15. Xie, S.: Alcoholism identification based on an AlexNet transfer learning model. Front. Psychiatry (2019). https://doi.org/10.3389/fpsyt.2019.00205

16. Sik, H.H., Gao, J.L., Fan, J.C., Wu, B.W.Y., Leung, H.K., Hung, Y.S.: Using wavelet entropy to demonstrate how mindfulness practice increases coordination between irregular cerebral and cardiac activities. Jove-J. Visualized Exp. (123), 10 (2017). Article ID: e55455. https://doi.org/10.3791/55455

17. Gorriz, J.M.: Multivariate approach for Alzheimer's disease detection using stationary wavelet entropy and predator-prey particle swarm optimization. J. Alzheimers Dis. **65**(3), 855–869 (2018). https://doi.org/10.3233/JAD-170069

18. Li, Y.-J.: Single slice based detection for Alzheimer's disease via wavelet entropy and multilayer perceptron trained by biogeography-based optimization. Multimedia Tools Appl. **77**(9), 10393–10417 (2018). https://doi.org/10.1007/s11042-016-4222-4

19. Han, L.: Identification of Alcoholism based on wavelet Renyi entropy and three-segment encoded Jaya algorithm. Complexity (2018). Article ID: 3198184
20. Phillips, P.: Intelligent facial emotion recognition based on stationary wavelet entropy and Jaya algorithm. Neurocomputing **272**, 668–676 (2018). https://doi.org/10.1016/j.neucom.2017.08.015
21. Li, P., Liu, G.: Pathological brain detection via wavelet packet Tsallis entropy and real-coded biogeography-based optimization. Fundamenta Informaticae **151**(1–4), 275–291 (2017)
22. Guliyev, N.J., Ismailov, V.E.: Approximation capability of two hidden layer feedforward neural networks with fixed weights. Neurocomputing **316**, 262–269 (2018). https://doi.org/10.1016/j.neucom.2018.07.075
23. Naggaz, N.: Remote-sensing image classification based on an improved probabilistic neural network. Sensors **9**(9), 7516–7539 (2009)
24. Zhang, Y.: Stock market prediction of S&P 500 via combination of improved BCO approach and BP neural network. Expert Syst. Appl. **36**(5), 8849–8854 (2009)
25. Wu, L.: Weights optimization of neural network via improved BCO approach. Prog. Electromagnet. Res. **83**, 185–198 (2008). https://doi.org/10.2528/PIER08051403
26. Wu, L.: Crop classification by forward neural network with adaptive chaotic particle swarm optimization. Sensors **11**(5), 4721–4743 (2011)
27. Ji, G.: Fruit classification using computer vision and feedforward neural network. J. Food Eng. **143**, 167–177 (2014). https://doi.org/10.1016/j.jfoodeng.2014.07.001
28. Ji, G.: Genetic pattern search and its application to brain image classification. Math. Probl. Eng. (2013). Article ID: 580876. https://doi.org/10.1155/2013/580876
29. Ji, G.L.: A rule-based model for bankruptcy prediction based on an improved genetic ant colony algorithm. Math. Prob. Eng. (2013). Article ID: 753251. https://doi.org/10.1155/2013/753251
30. Nurcahyo, S., Nhita, F., Adiwijaya, K.: Rainfall prediction in Kemayoran Jakarta using hybrid Genetic Algorithm (GA) and Partially Connected Feedforward Neural Network (PCFNN). In: 2nd International Conference on Information and Communication Technology (ICOICT), Bandung, Indonesia, pp. 166–171. IEEE (2014)
31. Gagnon, R., Gosselin, L., Park, S., Stratbucker, S., Decker, S.: Comparison between two genetic algorithms minimizing carbon footprint of energy and materials in a residential building. J. Build. Perform. Simul. **12**(2), 224–242 (2019). https://doi.org/10.1080/19401493.2018.1501095
32. Wang, S., Wu, L., Huo, Y., Wu, X., Wang, H., Zhang, Y.: Predict two-dimensional protein folding based on hydrophobic-polar lattice model and chaotic clonal genetic algorithm. In: Yin, H., et al. (eds.) IDEAL 2016. LNCS, vol. 9937, pp. 10–17. Springer, Cham (2016). https://doi.org/10.1007/978-3-319-46257-8_2
33. Wei, L., Yang, J.: Fitness-scaling adaptive genetic algorithm with local search for solving the Multiple Depot Vehicle Routing Problem. Simulation **92**(7), 601–616 (2016). https://doi.org/10.1177/0037549715603481
34. Kerr, A., Mullen, K.: A comparison of genetic algorithms and simulated annealing in maximizing the thermal conductance of harmonic lattices. Comput. Mater. Sci. **157**, 31–36 (2019). https://doi.org/10.1016/j.commatsci.2018.10.007
35. Srinivas, M., Patnaik, L.M.: Adaptive probabilities of crossover and mutation in genetic algorithms. IEEE Trans. Syst. Man Cybern. **24**(4), 656–667 (1994)
36. Li, J.: Texture analysis method based on fractional Fourier entropy and fitness-scaling adaptive genetic algorithm for detecting left-sided and right-sided sensorineural hearing loss. Fundamenta Informaticae **151**(1–4), 505–521 (2017)

Multi-mother Wavelet Neural Network Training Using Genetic Algorithm-Based Approach to Optimize and Improves the Robustness of Gradient-Descent Algorithms: 3D Mesh Deformation Application

Naziha Dhibi[(✉)] and Chokri Ben Amar

REGIM-Lab: Research Group on Intelligent Machines Engineering,
National School of Sfax (ENIS), 3038 Sfax, Tunisia
dhibi.naziha@gmail.com, chokri.benamar@ieee.org

Abstract. This paper presents the implementation of genetic algorithm which aims at searching for an optimal or near optimal solution to the deformation 3D objects problem based on multi-mother wavelet neural network training. First, we introduce the problem of 3D high mesh deformation using Multi-Mother Wavelet Neural Network architecture (MMWNN). Furthermore, gradient training limits of wavelet networks are characterized by their inability to evade local optima. The idea is to integrate genetic algorithms into the wavelet network to avoid both insufficiency and local minima in the 3D mesh deformation technique. Simulation results validate the generalization ability and efficiency of the proposed network based on genetic algorithms (MMWNN-GA). Thus the significant improvement of the performances in terms of quality of 3D meshes deformation.

Keywords: 3D mesh deformation · Multi-mother wavelet neural network · Genetic algorithm · Gradient descent

1 Introduction

Geometric mesh parameters are motivated habitually via an optimization technique. The ability to automatically update a current mesh to conform to a deformed geometry is necessary to allow rapid prototyping of many other geometric designs. Indeed, deformation of geometric meshes plays a fundamental role in computer graphics, essentially in areas such as computer aided design and computer animation. It is also an important element in the analysis of moving bodies and shape optimization. Deformation can be also defined as the process of interactively transforming the surface of a mesh in response to some control mechanism. Typically, mesh deformation techniques require global knowledge of the object structure (such as a skeleton) which results fairly time consuming. In fact, features alignment process is necessary in order to guarantee a successful deformation process. The idea is comes to define a set features of interest on both source and target objects and apply a warping/deformation of the

© Springer Nature Switzerland AG 2019
I. Rojas et al. (Eds.): IWANN 2019, LNCS 11507, pp. 98–108, 2019.
https://doi.org/10.1007/978-3-030-20518-8_9

parametric domain using wavelet network in order to guarantee that the parametric position of the corresponding features are as closed as possible for both models. The initial feature alignment process reduces the distances between features without modifying the local vertex positions. Since an essential goal to achieve when considering mesh deformation concerns the detail preservation, differential representations can capture information about the local shape properties of a mesh, such as curvature, scale or orientation. One of the most popular differential representations of a 3D geometry concerns the so-called Laplacian coordinates (differential coordinates). Laplacian mesh editing allows deforming 3D objects while their surface details are preserved because it allows the simulation of realistic deformations. This representation ensures a direct detail-preserving reconstruction of the modified mesh by solving a linear least squares system. We propose a new mesh deformation framework with an intuitive interface and efficient reconstruction algorithm based on multi-mother wavelet neural network to optimize, align features for mesh and minimize distortion with fixed features in which we used a linear differentia coordinates as means to preserve the high frequency details of the surface in order to approximate features alignment [7].

The proposed wavelet neural network structure provides a link between the neural network and the wavelet decomposition. The role of the wavelet network is to construct a discretized wavelet family by adapting the network parameters to the displacement data; the new version constructs the network by the implementation of several mother wavelets in the hidden layer. The objective is to maximize the potentiality of wavelet selection that deformed better the signal; on the other hand deep learning neural network can be expensive and require massive datasets to train itself on. But the resolutions of wavelet networks training problems by gradient are characterized by their noticed inability to escape of local optima, to overcome this problem we propose to use genetic algorithm for the design of wavelet network in order to determine the optimal network, which aims at searching for an optimal or near optimal solution to the coverage holes problem.

2 Related Work

Neural networks are used in process modeling, data mining, artificial intelligence, machine learning, compression and many other applications. Artificial Neural Networks (ANNs) have a limited ability to characterize local features, such as discontinuities in curvature, jumps in value or other edges. These local features, which are located in time and/or frequency, typically embody important process-critical information such as aberrant process modes or faults [1]. Zhang and Benveniste [2] and Bakshi et al. [3] improved upon this weakness of ANNs by constructing Wavelet Neural Networks (WNNs).

However, since objective functions used in guiding the network construction in ANNs and WNNs is based on global Mean Squared Error (MSE), the modeling effectiveness of such key local features is not emphasized. More importantly, as addressed in Martell [4], existing wavelet-based model selection methods [5] focus on data de-noising and use an excessive number of wavelet coefficients/bases in their approximation models. In many recent applications, this limits wavelet's applicability to potentially large size data encountered. The wavelet network structure proposed by Zhang and Benveniste [2, 6], provides a link between the neural network and the wavelet decomposition. Among wavelet network applications we can talk about 3D mesh deformation, it has been proposed in [7] to use multi mother wavelet network as an approximation tool to align features of mesh and minimize distortion with a fixed feature. This structure has the advantage of constructing the network by several mother wavelets and gives important results in 3D meshes modeling proposed by [8] and [16].

To estimate the network parameters, we need an optimization algorithm. It then implements iterative methods such as "backpropagation". Besides the choice of selection method, we must choose an optimization algorithm to optimize the parameters. From the Levenberg method [17], originally scheduled for solving linear systems by least squares. The method combines two algorithms into one. The simple gradient method (the steepest descent method) which is merely a less accurate estimate but has a slow convergence and the Gauss-Newton which has a quadratic convergence but requires an initial vector near to the solution. Thus, leading the gradient method to a local optimum makes the Gauss-Newton ineffective since the Jacobian becomes singular. The weakness of the Levenberg-Marquardt method is the convergence.

Therefore, the resolutions of wavelet networks training problems by gradient are characterized by their noticed inability to escape of local optima mostly when using high-dimensional meshes. For these reasons we propose the implementation of genetic algorithm which aims at searching for an optimal or near optimal solution to the deformation objects problem based on multi-mother wavelet neural network training and to optimize and improves the robustness of gradient-descent algorithms, this architecture gives a good result for the approximation of the 1D and 2D functions in [15].

3 MMWN Based Genetic Algorithm and Gradient Descent for 3D Mesh Deformation

In this section, we present the suggested approach. We present the assumptions of the network for 3D mesh deformation model, and we discuss the approach based on the genetic algorithm.

We proposed to use genetic algorithm for the design of wavelet network. The problem was to find the optimal network structure and parameters. In order to determine the optimal network, the proposed algorithms modify the number of wavelets in the library. The performance of the algorithm is achieved by evolving the initial population and by using operators that alter the structure of the wavelets library.

3.1 Network Assumptions

Alternative algorithms for wavelet network construction were proposed in [6] to handle better problems of large dimension. The wavelet number was considerably reduced by eliminating the wavelets whose supports do not contain any data points. Some regression techniques, such as stepwise selection by orthogonalization and backward elimination, were then used to further reduce the wavelet number. The wavelet networks with radial wavelet functions can be considered as an RBF network, since both of them have the localized basis functions. The difference is that wavelet function is localized both in the input and frequency domains [9]. Besides retaining the advantage of faster training, wavelet networks have a guaranteed upper bound on the accuracy of approximation [10] with a multi-scale structure. Wavelet networks have been used in non-parametric regression estimation [6] and were trained based on noisy observation data to avoid the problem of undesirable local minima [10]. In [11], wavelet networks and stepwise selection by orthogonalization regression technique were used to build a neural network model for a 2.9 GHz microwave power amplifier. In this technique, a library of wavelets was built according to the training data and the wavelet that best fits the training data was selected. Later in an iterative manner, wavelets in the remainder of the library that best fits the data in combination with the previously selected wavelets were selected. For computational efficiency, later selected wavelets were orthonormalized to previously selected ones.

The proposed network structure is similar to the classic network, but it has some differences; the classic network uses dilation and translation versions of only one mother wavelet, besides the new version constructs the network by the implementation of several mother wavelets in the hidden layer. The objective is to maximize the potentiality of wavelet selection [12] that approximates better the signal. In our work we used this network for 3D mesh deformation to align features of mesh and minimize distortion with a fixed feature but the resolutions of wavelet networks training problems by gradient are characterized by their noticed inability to escape of local optima for 3D mesh high resolution, for this we propose to use genetic algorithm as an optimization method to validate the generalization ability and efficiency of the proposed Multi Mother Wavelet Neural Network (Fig. 1).

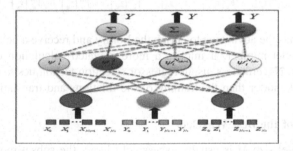

Fig. 1. Graphical representation of the new network architecture

3.2   Creation of the Multi-wavelet Family Library

There are five main parameters to adjust in an MMWNN: the basis function type (mother wavelet) for each wavelet, the wavelets number in the hidden layer, the two parameters (translation, dilation) of the mother wavelet and the connections weights between the hidden layer of wavelet and the output layer. Changing one of these parameters causes a change in the network behavior.

The creation of the multi wavelet family's library consists in constructing a several mother wavelet family for the network construction. The library elements, generated by distributing the parameters on a dyadic grid, are grouped in levels on the basis of the dilation and translation parameters. This choice presents the advantage not only to enrich the library, but also to get a better performance for a given wavelet number. The drawback of this choice is the size of the library. A wavelet library having several wavelet families N_l is more voluminous than the one that possesses the same wavelet mother. Each wavelet ψ_{ji} associates a vector whose components are the values of these wavelets corresponding to the training examples.

We create a matrix consisting of V_{MW} vector blocks representing the wavelets of each mother wavelet we constitute a matrix which is the constituent of the blocks of the regressor represents the wavelets of each mother wavelet or the expression is defined by:

$$V_{MW} = \left\{V_i^j\right\}_{i=[1,...,N], j=[1,...,M]} \tag{1}$$

The matrix V_{MW} is defined by:

$$V_{Mw} = \begin{vmatrix} V_1^1(x_1) & ... & V_N^1(x_1) & ... & V_1^j(x_1) & ... & V_{N_M}^j(x_1) & ... & V_1^M(x_1) & ... & V_{N_M}^M(x_1) \\ V_1^1(x_2) & ... & V_N^1(x_2) & ... & V_1^j(x_2) & ... & V_{N_M}^j(x_2) & ... & V_1^M(x_2) & ... & V_{N_M}^M(x_2) \\ \vdots & ... & \vdots & ... & \vdots & ... & \vdots & ... & \vdots & ... & \vdots \\ \vdots & ... & \vdots & & \vdots & ... & \vdots & & \vdots & ... & \vdots \\ \vdots & ... & \vdots & ... & \vdots & ... & \vdots & ... & \vdots & ... & \vdots \\ V_1^1(x_N) & ... & V_N^1(x_{N_1}) & ... & V_1^j(x_N) & ... & V_{N_M}^j(x_{N_1}) & ... & V_1^M(x_{N_1}) & ... & V_{N_M}^M(x_{N_1}) \end{vmatrix} \tag{2}$$

This choice has the advantage to enrich the library and receive a better performance for a given wavelet number. It implies a more elevated calculation cost during the selection process. Nevertheless, using regressor selection techniques to choose the best wavelets is often shorter than the training of the dilations and translations.

3.3   Selection of the Best Wavelets

Because the model size M is usually excessively large, the subset model selection is required. The optimal subset selection techniques are computationally prohibitive and impractical. The practical method is the forward selection, and the OLS (Orthogonal Least Squares) procedure [13, 14] which are an efficient implementation of this subset selection procedure. For the best selection, we propose an improved version of OLS procedure for the subset model selection.

The improved version of OLS procedure selects columns from the input library sequentially. The column that provides the best combination with the output f to model the signal y will be picked to form the new output f. Then we apply a summary of optimization of the parameters of this one in order to bring it closer to the signal.

3.4   Change of the Library Dimension

Choosing the number of wavelets to be selected to construct the wavelet network is an important and difficult problem that is related to the standard model order in the determination of the problem [6]. For this we used genetic algorithm to change the library dimension this algorithm used two crossover operators:

- **The Crossover1 Operator**

  One of them changes the number of columns of chromosome so it changes the number of mother wavelets and establish in the library a new version of wavelets issued of the new mother wavelet. After the selection of the two chromosomes to which we will affect this operator, we choose an arbitrary position a in the first chromosome and a position b in the second according to a. After that, we exchange the second parts of the two chromosomes.

- **The Crossover2 Operator**

  For the second operator, we select an arbitrary position c in the first chromosome and a position d in the second chromosome according to c.

  $$That \quad Min_{point} = Min(c, d) \tag{3}$$

  First, we change the values of c and d to Min_{point}. Then, we exchange the second parts of the two chromosomes. In this case, we have necessarily first children having the same length as the second chromosome and the second children having the same length as the first chromosome.

  The second operator does not modify the number of columns of every chromosome.

- **Mutation Operator**

  In the main, the initial population does not have all the information that is indispensable to the solution. The objective of applying the mutation operator is to introduce some new information (wavelets) into the population. Mutation consists in changing one or more gene(s) in chromosome chosen randomly according to a mutation probability pm. Nevertheless, the muted gene may be the optimal one therefore, the new gene will not replace the old but it will be added to this chromosome.

3.5   Network Size Determination

For the first iteration, once the regressor is selected and optimized, we are going to construct the network with only one wavelet. In this step, we have a uniform crossover operator and a mutation operator applied to structural parameters of WN (translations and dilatations) and we increment the number of wavelets to the following iterations.

After crossing, we will apply the mutation operator. We propose a uniform mutation that involves random mutation of the T and D vectors. For the next iteration, we reset the wavelet array using this new population which will replace the old one and the algorithm will be continued.

Now the genetic algorithm is applied in the wavelet network Its goal is to solve both problems by learning the existing approaches named the local minima generated by the gradient descent and the cost of selecting wavelets generated by the large size of the library network of multi-mothers wavelet. We can use this network for 3D mesh deformation (Fig. 2).

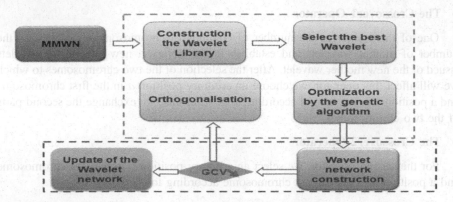

Fig. 2. MMWNN based genetic algorithm

Our 3D mesh deformation algorithm is shown in Fig. 3, in 3D mesh deformation the interpolation between objects can be done simply by determining the trajectory of the corresponding vertices on the representation obtained in the previous step. The simplest way to interpolate between these points is a linear interpolation. The interpolation in the wavelet domain makes it possible to control interpolation starting time and speed at various resolutions.

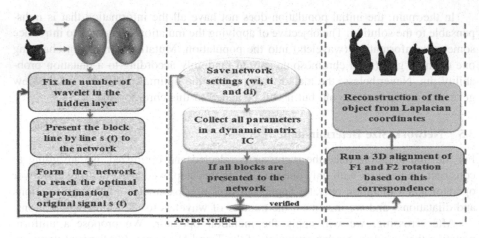

Fig. 3. 3D mesh deformation based MMWN-GA

4 Experimental Results

In this section, we present some experimental results of the proposed Multi Mother Wavelet Neural Networks relies on genetic algorithm (MMWNN-GA) for 3D mesh deformation we conclude by a comparison in terms of MSE, rate deformation, and the running time of the two approaches when we used 15 wavelet in the hidden layer.

To evaluate the quality of the reconstructed object we use the Mean-Squared Error (MSE) and the Standardized Square Root of the least square error in Eqs. (4) and (5).

$$MSE = \frac{1}{N_i} \sum_{k=1}^{N_i} \left(O_N(x_{Nk}, y_{Nk}, z_{Nk}) - O(x_k, y_k, z_k) \right)^2 \tag{4}$$

$$NSRMSE = \sqrt{\frac{\sum_{k=1}^{N} \left(O_N(x_{Nk}, y_{Nk}, z_{Nk}) - O(x_k, y_k, z_k) \right)^2}{\sum_{k=1}^{N} \left(\bar{O} - O(x_k, y_k, z_k) \right)^2}} \tag{5}$$

M is the mesh to be deformed; K is the number of observations. $\bar{O} = \sum_{k=1}^{N} O(x_k, y_k, z_k)$ is the results of the 3D representation, resulting from the multi-mother wavelet library are based on an improved version of the OLS.

The ratio $r\left(\vec{x} \right)$ is defined as the distance of the neighbour vertex $\vec{x}$ to the next handle vertex $\vec{x}_h$ divided by the minimal distance of a fixed vertex to a handle one.

$$r(\vec{x}) = \frac{|\vec{x} - \vec{x}_h|}{min\left(|\vec{x}_f - \vec{x}_h| \right)} \tag{6}$$

For this deformation ratio, it is valid that

$$r(\vec{x}) \in [0, 1] \, \forall \vec{x} \in N$$

The approximation result and object deformation of «horse», «Scape», «Face» and «Flamingo» using 15 wavelets for a single iteration are presented in the following Table 1:

Table 1. Comparative study in terms of MSE and deformation rate by MMWNN and MMWNN-GA

	MMWNN			MMWNN-GA		
Objects	MSE	Deformation rate	Time (s)	MSE	Deformation rate	Time (s)
Scape	4.829e–7	27	17964.431	1.874e–2	20	9655.258
Horse	5.649e–4	14	12001.305	1.470e–2	10	5412.001
Face	9.655e–5	23	15878.235	4.540e–2	15	7856.524
Flamingo	8.956e–5	32	18210.784	5.365e–3	22	1055.825

Considering the results mentioned above, we can note that the MSE reaches very weak values for example of 1.470e–2 for the Horse object, using multi-mother wavelet network based genetic algorithm constructed with 15 wavelet, against an MSE equal to 5.649e–4 with the multi-mother wavelet network based on the best mother wavelet.

In 3D mesh deformation compared to MMWN our approach using MMWNN-GA minimizes the deformation rate between all corresponding vertices, and minimizes distortion with fixed features to have a good reconstructed object in the deformation process, as able to greatly reduce the MSE and the execution time. We can deduce the superiority of the MMWNN-GA algorithm over classical MMWNN in term of 3D mesh deformation and approximation as they much reduce the number of gradient iterations since initialization step has already achieved acceptable results and are always much better than the ones obtained with MMWNN.

The Figs. 4, 5, 6 and 7 represent our training results using 15 wavelets in the hidden layer for the 3D deformation of the four objects (Scape, Horse, Face and Flamingo), obtained by the MMWNN-GA.

Fig. 4. Example of our 3D mesh deformation technique based MMWNN-GA on the Scape model

Fig. 5. Example of our 3D mesh deformation technique based MMWNN-GA on the Flamingo model

Fig. 6. Example of our 3D mesh deformation technique based MMWNN-GA on the Horse model

Fig. 7. Example of our 3D mesh deformation technique based MMWNN-GA on the Face model

Table 2 gives the NSRMSE error after 100 learning iterations, indicating the wavelet number used in the hidden layer.

Table 2. NSRMSE and the number of selected wavelets for the Flamingo object using MMWNN-GA

Nbr wavelet	M-hat	Polywog1	Slog1	Beta1	Beta2	Beta3	NSRMSE
15	5	1	2	4	1	2	3.584e–3
40	10	5	12	8	4	1	2.451e–3
150	25	40	55	10	8	12	1.205e–2
200	30	30	55	20	52	13	1.024e–1

To solve the problem of best wavelet selection that represent well the object, and to enhance the deformation quality, we used several wavelets (Beta, Polywog1, Mexican Hat, Slog1 and Polywog1) to construct the wavelet network. So, when comparing the results given by Table 2, we can say that the performances obtained in terms of NSRMSE the MMWNN-GA structure are always much better than the ones obtained with MMWNN. In addition, we see that increasing the wavelet number improves the modeling quality. This shows that the proposed procedure brings effectively a better capacity of deformation using the wavelet network based genetic algorithm.

5 Conclusion

The goal of this paper is to present a robust and fast geometric mesh deformation algorithm using MMWNN based genetic algorithm. Founded on these experimental results, we can say that the proposed procedure brings effectively a better deformation capacity using the several mother wavelets and the genetic algorithm to align features of mesh and minimize distortion with a fixed feature. The performance of the algorithm is achieved by evolving the initial population and by using operators that alter the structure of the wavelets library. Comparing to MMWNN algorithms, results show significant improvement in the resulting performance and topology.

References

1. Jin, J., Shi, J.: Feature-preserving data compression of stamping tonnage information using wavelets. Technometrics **41**(4), 327–339 (1999)
2. Zhang, Q., Benveniste, A.: Wavelet networks. IEEE Trans. Neural Networks **3**(6), 889–898 (1992)
3. Bakshi, B., Stephanopoulos, G.: Wave-net: a multiresolution, hierarchial neural network with localized learning. AIChE J. **39**(1), 57–81 (1993)
4. Martell, L.: Wavelet-based Data Reduction and De-Noising Procedures, Ph.D. thesis, North Carolina State University (2000)
5. Yang, S.H., Wang, X.Z., Chen, B.H., McGreavy, C.: Application of wavelets and neural networks to diagnostic system development, 2, an integrated framework and its application. Comput. Chem. Eng. **23**, 945–954 (1999)
6. Zhang, Q.: Using wavelet network in nonparametric estimation. IEEE Trans. Neural Networks **8**(2), 227–236 (1997)
7. Naziha, D., Akram, E., Wajdi, B., Chokri, B.: 3D high resolution mesh deformation based on multi library wavelet neural network architecture. 3D Res. J. **7**(31), 1–14 (2016)
8. Othmani, M., Bellil, W., Ben Amar, C., Alimi, M.A.: A novel approach for high dimension 3D object representation using multi-mother wavelet network. Int. J. Multimedia Tools Appl. MTAP **59**, 7–24 (2012)
9. Bakshi, B.R., Koulouris, A., Stephanopoulos, G.: Wave-nets: novel learning techniques, and the induction of physically interpretable models. Wavelet Appl. **2242**, 637–648 (1994)
10. Delyon, B., Juditsky, A., Benveniste, A.: Accuracy analysis for wavelet approximations. IEEE Trans. Neural Networks **6**, 332–348 (1995)
11. Harkouss, Y., Rousset, J., Chehade, H., Barataud, D., Teyssier, J.P.: Modeling microwave devices and circuits for telecommunications system design. In: Proceedings of IEEE International Conference on Neural Networks, Anchorage, Alaska, pp. 128–133, May 1998
12. Ruqiang, Y., Gao Robert, X.: Base wavelet selection for bearing vibration signal analysis. Int. J. Wavelets Multiresolut. Inf. Process. (IJWMIP) **7**(4), 411–426 (2009)
13. Titsias, M.K., Likas, A.C.: Shared kernel model for class conditional density estimation. IEEE Trans. Neural Network **12**(5), 987–996 (2001)
14. Colla, V., Reyneri, L.M., Sgarbi, M.: Orthogonal least squares algorithm applied to the initialization of multi–layer perceptrons. In: ESANN 1999 Proceedings – European Symposium on Artificial Neural Networks Bruges (Belgium), 21–23 April 1999, D–Facto public., pp. 363–369 (1999). ISBN 2–600049–9–X
15. Chihaoui, M., Bellil, W., Amar, C.: Multimother wavelet neural network based on genetic algorithm for 1D and 2D functions approximation. In: Proceedings of the International Conference on Fuzzy Computation and 2nd International Conference on Neural Computation, pp. 429–434 (2010)
16. Dhibi, N., Bellil, W., Ben Amar, Ch.: Study implementation of a new training algorithm for wavelet network based on genetic algorithm and multiresolution analysis for 3D objects modeling. In: IEEE Mediterranean Electrotechnical Conference (2012)
17. Levy, B.: Laplace–Beltrami eigenfunctions: towards an algorithm that understands geometry. In: IEEE International Conference on Shape Modeling and Applications (2006)

A Clinical Decision Support System to Help the Interpretation of Laboratory Results and to Elaborate a Clinical Diagnosis in Blood Coagulation Domain

Francois Lasson[2]([✉]), Alban Delamarre[1,2], Pascal Redou[3], and Cedric Buche[2]

[1] VISAGE Laboratory, FIU, Miami, USA
adela177@fiu.edu
[2] LAB-STICC, ENIB, Brest, France
{lasson,redou,buche}@enib.fr
[3] LATIM, Brest, France

Abstract. Hemophilia is a rare hemorrhagic disorder caused by clotting factor deficiencies that leads to a less efficient coagulation system. Treatments of this pathology rely on a patient's subjective assessment which reflects a need for a laboratory assay able to predict the clinical patient phenotype. According to the literature, global assays such as thrombin generation (TG), are good predictors of bleeding episodes and therefore seem to be good candidates to fit this need. Nevertheless, the result of the TG assay, known as thrombogram, is difficult to interpret for non-expert clinicians. In this paper, we present a machine learning-based clinical decision support system which goal is to help clinical decision making. In doing so, we have adopted several approaches in order to evaluate well-known machine learning algorithms, in terms of accuracy and robustness, on a thrombogram database generated using numerical simulations. Obtained results, 95.57% of accuracy using a cascade of a SVM and MLPs to classify all categories and 98.10% of accuracy for the binary case hemophilia A/B, prove that our proposal can efficiently diagnose hemophilia.

Keywords: Machine learning-based clinical decision support system · Thrombin generation assay · Hemophilia diagnosis · Scikit-Learn

1 Introduction

Blood coagulation is a biological phenomenon leading to clot formation that prevents and stops bleeding after vascular damage. This complex system is regulated by a judicious equilibrium between the procoagulant pathways that are responsible for clot formation and the anticoagulant pathways that regulate and inhibit this process. An imbalance in this equilibrium may cause two kinds of pathologies: thrombotic disorders and hemorrhagic disorders. Hemophilia is a

© Springer Nature Switzerland AG 2019
I. Rojas et al. (Eds.): IWANN 2019, LNCS 11507, pp. 109–122, 2019.
https://doi.org/10.1007/978-3-030-20518-8_10

hemorrhagic disorder caused by deficient clotting factors that leads to a less efficient coagulation system. The main treatment for this pathology is replacement therapy that consists of clotting factor concentrate administrations. There are well-known laboratory assays that quantify the concentrations of these proteins in plasma but most of them are bad predictors of a bleeding episode. Hence, current treatments of hemophilia rely on a patient's subjective assessment using physiological parameters such as a persistence of pain or a decrease of joint mobility. According to the literature, in particular [24], a global assay called Thrombin Generation (TG) seems to be a good candidate to predict patient phenotype. The result of this test, the thrombogram, is a curve that plots thrombin (the key enzyme of the blood clotting system) concentration over time. Although biologists have identified several discriminating features from these curves (e.g. lag time, time to peak, peak or endogenous thrombin potential), this result is difficult to interpret for non-expert clinicians. To deal with this issue, Clinical Decision Supports (CDS) can be used. The goal of these systems is to help clinical decision making in order to increase the quality of care, the health outcomes but also to improve the cost-benefit by avoiding, in the hemophilia case, any overdose of clotting factor concentrates. A clinical decision can be diagnostic elaboration and also therapeutic orientation. In the context of hemophilia, a model able to individualize therapy (choice of the drug and optimization of its dose) will be really helpful. We present in this paper a machine learning based CDS system, which goal is to diagnose hemophilia from thrombin generation curves. Although there is no added clinical value of hemophilia diagnosis through TG, this work can nevertheless be seen as a preliminary study. Its aim is to demonstrate the interest of such a system to assist clinicians in interpreting thrombograms, a first step toward a therapeutic orientation.

This model has to determine the type of hemophilia (A or B) and its severity (Mild, Moderate, or Severe). Given the machine learning context, efficient learning algorithm strategies need large data sets. In the clinical domain, obtaining a huge data set is a long, complex and expensive process. As a consequence, a numerical thrombin generation model has been used in this study to generate data [10].

The next section presents the context of this study. Section 3 presents the state of the art of classification techniques and a summary of the different types of CDS systems. Section 4 details the approach used in our context. In Sect. 5, evaluation criteria are identified and described according to the context; classification results are then presented. Section 6 contains results analysis. We finally discuss the issues of this work and the directions for future work in Sect. 7.

2   Context

2.1   Problem Specification

Since the analysis of coagulation curves is a very complicated task and can be source of serious consequences if misinterpreted, our objective is to create a CDS. Given a TG curve as an input, this system should be able to provide a complete diagnosis to clinicians, i.e. determining whether the patient is hemophiliac or not and if so, the type of hemophilia and its severity. Given a space $\mathcal{V}$ of unlabeled data and $\mathcal{Y}$ a finite set of labels, we have $\mathcal{X} = \mathcal{V} \times \mathcal{Y}$ the space of labeled samples. Let $\mathcal{D} = \{x_1, x_2, ..., x_n\}$ be a dataset composed of n labeled instances, where $x_i = \langle v_i \in \mathcal{V}, y_i \in \mathcal{Y} \rangle$ and v_i a vector representing a time series of length m such as $v_i = \{t_1, t_2, ..., t_m\}$. The objective is to find the best classifier $\mathcal{C}$ which for a given time series v associates a label y such as $\mathcal{C}(v) = y$ with $\langle v, y \rangle \in \mathcal{X}$.

2.2 Thrombogram Dataset

In the clinical context, the construction of a large database is a complex, long and expensive process. This is particularly true for the hemophilia context owing to the prevalence of this disorder. To deal with this difficulty, we have implement a numerical thrombin generation model in order to generate thrombograms using numerical simulations. As defined in [10], 41 biochemical reactions between 35 proteins are taken into account to construct a system of ordinary differential equations. Its resolution provides the thrombin concentration overtime which is a thrombogram. The simulation of a hemophilia patient simply consists of lowering factor VIII (hemophilia A) or factor IX (hemophilia B) initial concentration. Using artificially generated data provides two advantages in the clinical domain. As noted before, data acquisition is complicated and very expensive as it requires finding patients with the corresponding pathologies who are not undergoing treatment. Furthermore, a numerical model allows us to generate large amounts of data covering a wide range of thrombogram types.

Our dataset $\mathcal{D}$ is made of 7 categories labeled as $\mathcal{Y} = \{Healthy, Hemophiliac A mild, Hemophiliac A moderate, Hemophiliac A severe, Hemophiliac B mild, Hemophiliac B moderate, Hemophiliac B severe\}$ and it is composed of 14000 thrombograms with the following proportions $Quantity = \{5000, 1500, 1500, 1500, 1500, 1500, 1500\}$ where each thrombogram contains 181 points. The integration step used to generate these data is equal to 5 mHz. This dataset provides the ability to perform different types of classification: healthy or hemophiliac, hemophiliac A/B, hemophiliac severity and all these categories at once.

2.3 Overview of the Approach

The performance of this CDS is based on its ability to successfully classify thrombograms using machine learning techniques. Therefore, 6 well-known classification models have been evaluated in this study. These are presented in the next section. The different steps of our approach are briefly described below:

1. Firstly, we created a dataset that includes the different categories of hemophilia. A large amount of thrombograms were generated in order to realize an efficient training set for each classification method.
2. Secondly, in order to optimize classification performances of each model, we tuned their hyper-parameters. Because of their interdependence, search algorithms such as grid search and random search have been applied. In an

attempt to improve results, we also reduce the dimensionality of each thrombogram using feature extraction techniques.

3. Next, we compared the different classification techniques based on established medical criteria: accuracy, precision, recall and False positive rate.
4. Since some methods perform better on a specific classification set, we developed a cascade classification technique, using a combination of binary classifiers to separate the different types of hemophilia step by step.

3   State of the Art

3.1   Clinical Decision Support

In the clinical domain, diagnosis mistakes can have disastrous consequences. A CDS is a system which goal is to advise clinicians during the process of decision making in order to reduce diagnostic errors. In this domain, two approaches are conceivable: Knowledge-based CDS and non knowledge-based CDS.

A knowledge-based CDS is composed of inference and association rules established by experts. Patients' data is fed into the system to produce a diagnosis suggestion. MYCIN [21] for example, is a system composed of about 600 rules able to detect severe infections like bacteremia or meningitis. Systems like Arden Syntax [19] GLIF3 [4] PROforma [22] were developed to allow health professionals to directly build their own CDS systems. According to the literature, there is a plurality of application scope of this kind of CDS. However, in the domain of coagulation, even though it exists features regarded as the most discriminant by experts, the interpretation of thrombogram and therefore the detection of hemophilia and its characteristics remains a challenging task.

Non knowledge-based CDS systems, for their part, use artificial intelligence techniques to associate patterns in the data with pathologies. For instance, Shin et al. created a system able to detect cancer based on mass spectrometry and machine learning algorithms [20]. Due to the complexity of the thrombogram analysis, application of artificial intelligence techniques on TG curves seems to be a good option [24]. Therefore, we directed our choice towards non knowledge-based CDS.

3.2   Classification Methods

We want to classify time series using a static approach since all of the TG curves of our database have the same length. Therefore, an assortment of supervised parametric and non-parametric techniques can be used for comparison, such as well-known Support Vector Machines, SVM [8,18,26] or K-Nearest Neighbors, KNN [6] which use distance functions to discriminate categories. Moreover, using a large dataset we can also correctly perform neural network training using a MultiLayer Perceptron MLP [13,25]. Although this model is not recent, it forms the basis of the very popular deep learning techniques. In addition, for the purposes of completeness, we also performed a linear discriminant analysis LDA and applied the Adaboost [2] and the Decision tree [16] algorithms. All the classification results presented in this paper were obtained using the Scikit-Learn framework.

3.3 Extraction Methods

Extracting features from thrombograms and thus, reduce their dimensionality, could improve classification results. Therefore, we identify several feature extraction techniques. Piecewise Linear Approximation PLA [9], for example, applies the last square method on segments of the data. The Symbolic Aggregate approXimation algorithm combined with a Piecewise Trend Approximation, SAX/PTA [7] translates time series into strings, the SAX method is based on values whereas the PTA is based on slope. Thrombograms can also be represented by coefficients, using Discrete Fourier Transform, DFT [1, 23] and Discrete Wavelet Transform, DWT [5, 15]. The obtained coefficients contain information in both the temporal and the frequency domain. To perform these feature extractions, we use the Pyts Python package.

4 Workflow

This section describes in detail the steps of classification performance optimization.

4.1 Hyper-parameter Search

In order to optimize classification performance for each model, we have tuned their hyper-parameters using search algorithms. In other words, for each type of classification, we have optimized the penalty parameter, the kernel function and its associated coefficients of the SVM classifier. We also searched for the optimal number of layers, the number of neurons in each hidden layers, the learning rate, the random seed, the activation function and the solver algorithm for a shallow MLP. In the case of the Decision Tree, we have tuned its maximum depth, its maximal number of features and the minimum number of samples needed to split an internal node or to create a new leaf node. Moreover, we have taken into consideration the number of nearest neighbors and the distance metric of the KNN, the number of estimators and the learning rate of the Adaboost classifier and the solver of the LDA. Given the limited number of these meta-variables, classical search algorithms are sufficient [12]. A grid search requires fixing all hyper-parameters to a given value except one, which varies across a finite set of values. A random search consists in setting all hyper-parameters to random values chosen in an established range. We initially used a grid search with a logarithmic scale, in order to query a wider range of values for each hyper-parameter, and to determine a subset of values. Next, we aimed to reduce this subset by using a second grid search with a linear scale. Then, we finally tested a large number of random hyper-parameter values within this subset. This process is used to gradually reduce the subset of possible hyper-parameter values and allows the system to find one of the best configurations.

4.2 Classification

Thrombograms are fed into the classification algorithms which, in turn, output a class label (healthy, hemophiliac etc.). As is common in practice, we carry out a k-fold cross-validation (CV) to train, optimize and evaluate these supervised models. First of all, the dataset is divided into two parts, taking into consideration the proportions of the label values. The first one is used for the training phase and the second one for testing. In order to tune each hyper-parameter, the training sub-dataset is then randomly split into k folds of approximately equal size. A rotation principle in which, $k - 1$ of these folds are used to train the classifier while the remaining fold is used to validate its performance, is repeated k times until each fold has been used as a validation dataset [11]. At the end of this process, when its set of optimal hyper-parameter values has been found, the classifier is trained on the whole training sub-dataset and evaluated on the remaining part of the database. In the clinical context, due to the lack of data induced by the prevalence of hemophilia disorder, performing a k-fold cross-validation with a high k value could improve the classifier performances. However, the database used in this study is generated by our numerical model. Large amounts of data covering a wide range of thrombogram types can be simulated. Therefore, we hypothesized that carrying-out a complex cross-validation would be unnecessarily computationally expensive. Considering that point, we use 80% of our numerical database for the training phase and 20% for testing. Then, the cross-validation is performed with two different numbers of folds: $k \in \{3,10\}$. The obtained results are presented and discussed in Sect. 5.

4.3 Features Extraction

In order to improve classification, we realized a feature extraction as a preprocessing step using techniques identified during the state of the art. Each technique was applied on the thrombogram dataset to reduce their dimensionality and, as a consequence, facilitate the learning process. For the purposes of comparison, we also used 4 features regarded as the most discriminant by experts: Time to Peak, Peak, Lag-Time and Endogenous Thrombin Potential.

4.4 Cascade

A method can be used to identify all 7 categories in a single classification process. However, another approach consists in using a cascade of classification models. Some categories can be pulled out of the dataset by a specific classifier. For example, thrombograms of healthy patients can be extracted using a classifier trained on discriminating healthy patient from hemophiliac. On the remaining hemophiliac sub-dataset we can then separate hemophiliac A from B using another classifier and so on until all categories are isolated. With this principle, we can use the best method at each step of the cascade.

5   Evaluation

To measure the efficiency and the robustness of our system, given the clinical context, we need to take into account specific criteria, they are presented in this first subsection. Results obtained are shown in the second one.

5.1   Evaluation Criteria

The main objective of a CDS is to help clinicians in the decisions, and thus to reduce medical errors. The worst possible case is the prediction of an absence of illness for an infected patient. No measures would therefore be taken to insure the safety of the patient. Our goal is thus to minimize these cases, they can be measured using the False Positive Rate (FPR). A second objective is to reduce the quantity of assays used for pathology detection. This goal is reached by detecting a majority of healthy patients: Recall. Finally, as we are working with artificially generated thrombograms, data are by definition clean of experimental noise. One last point is to study the noise robustness of the selected model.

5.2   Results

First of all, we have hypothesized in Sect. 4, that carrying-out a complex cross-validation would be unnecessarily computationally expensive in this study. In order to test this assumption, we compared results obtained using cross-validation with two different numbers of folds: $k \in \{3,10\}$. To that end, an SVM classifier has been trained to discriminate the Hemophilia A/B case and tuned using a cross-validated grid search. On the same hardware, the 10-fold cross-validation took approximately 5 times longer than the 3-fold cross validation. Regarding the CV accuracies, $k = 10$ leads to $98.0 \pm 0.4\%$ and $k = 3$ to $97.8 \pm 0.3\%$. Beyond the fact that these performances are very similar, the same optimal hyper-parameters values have been found by these two CV-grid search: polynomial kernel of degree 6 and a low penalty parameter equal to 5. Therefore, the last step which consists in training the classifier on the whole training sub-dataset leads to the same accuracy scores on the remaining part of the database: 98,1%. Hence, our assumption holds true.

Table 1 shows the result obtained with the different classification techniques using a 3-fold CV without feature extraction techniques. We can observe that SVM performs the best, above 94.49% for each type of classification. Moreover, it appears that the discrimination between hemophilia A and B is the most challenging one. Even thought *SVM* and *MLP* obtained high accuracies on this classification, all other methods failed to correctly discriminate these 2 categories. Regarding these MLPs, the hyper-parameter tuning phase results in shallow architectures composed of 3 hidden layers at most, optimal learning rates equal to $1e - 3$, hyperbolic tangent activation function and 'LBFGS' solver.

Table 1. Averaged performance of each method without feature extraction techniques. Highest classification rate, Recall, Precision and F-measure and lowest FPR are shown in bold.

Dataset	Method	Accuracy	Recall	Precision	F-Measure	FPR
All Categories	Decision Tree	75.00	68.16	67.94	67.97	32.06
	Adaboost	56.60	43.14	41.16	42.13	56.86
	KNN	78.95	72.68	72.53	72.51	27.47
	LDA	90.29	48.67	74.52	58.88	2.77
	SVM	**94.49**	**93.17**	**93.17**	**93.14**	**0.91**
	MLP	90.00	87.14	87.12	87.08	1.63
Healthy/ Hemophiliac	Decision Tree	97.88	97.78	97.60	97.69	2.40
	Adaboost	88.07	87.42	86.95	87.17	13.05
	KNN	98.21	97.98	98.13	98.05	1.87
	LDA	96.73	97.09	94.04	95.54	3.45
	SVM	99.04	98.83	99.06	98.95	1.16
	MLP	**99.18**	**99.01**	**99.22**	**99.11**	**0.99**
Hemophilia A/B	Decision Tree	69.63	69.62	69.64	69.62	30.36
	Adaboost	52.85	52.86	52.87	52.81	47.13
	KNN	74.37	74.41	74.45	74.37	25.55
	LDA	57.53	57.31	57.69	57.50	42.24
	SVM	**98.1**	**98.1**	**98.1**	**98.1**	**1.9**
	MLP	90.83	90.83	90.83	90.83	9.17
Hemophilia Severity	Decision Tree	91.66	91.65	91.67	91.66	8.33
	Adaboost	84.04	83.87	84.14	84.01	16.13
	KNN	93.22	93.21	93.27	93.17	6.73
	LDA	92.95	89.48	89.38	89.43	5.31
	SVM	96.03	96.03	96.1	96.03	1.98
	MLP	**96.23**	**96.23**	**96.29**	**96.23**	**1.88**

6 Discussion

In this section we compare the performance of each classification method. We also analyse the impact of using feature extraction and cascade classification. Finally, according to the identified criteria we present the most accurate method.

6.1 Performance of Hemophilia Detection

The main objective of this study is to detect hemophilia using thrombograms. Table 1 shows results obtained for the healthy and hemophiliac classification. Decision Tree, KNN, LDA, SVM and MLP can accurately discriminate thrombograms, their accuracies are above 96%. We can notice that SVM and MLP are slightly better for this classification. On the other hand, Adaboost performs relatively worst. In the performance evaluation section we pointed out that classifying a hemophiliac patient as healthy can have disastrous consequences. MLP outperforms other techniques regarding the *False Positive rate*, 0.99%. Moreover, in order to save hemophilia detection tests, we want to identify a majority of healthy patients, the best *Recall* is also obtained by MLP, 99.01%. Thus, a CDS based on a MLP is able to fulfil the two main criteria i.e., avoiding clinical errors and reducing costs.

6.2   Classification Comparison After Feature Extraction

All classification techniques were also tested on extracted features in order to improve their performances. However, classification results using these features are less accurate than the ones obtained using whole thrombograms. Regarding the thrombogram database reduced to the 4 features identified by experts, a SVM obtains 85.81% of accuracy to classify All Categories, 98.83% of accuracy for the binary case Healthy/Hemophiliac, 83.62% for the Hemophilia A/B case and 95.33% for the Severity. In other words, this kind of dimensionality reduction results in a 8.7% decrease of accuracy for the All Categories classifier, 14.5% for the A/B classifier but only around 1% for the Healthy/Hemophiliac and the Severity classifiers. Results obtained by other techniques such as PLA, SAX/PTA, DFT and DWT are very similar. These lower results can be explained by the loss information induce by feature extraction techniques, particularly in the most challenging case A/B where the whole TG curve seems to be relevant.

6.3   Classification Comparison Using Cascade Classification

The cascade classification discriminates thrombograms categories by categories, healthy and hemophiliac first, hemophiliac A/B next, and finally the hemophilia severity. This technique allows to use the best classifiers for each classification tasks. We can notice a major contribution brought by the cascade technique. First of all, it allows to divide the problem and to identify where classification performances were not satisfying. Secondly, we incremented the number of classifier required by the system by decreasing the classification complexity, i.e. the cascade uses binary classifiers, except for the severity which could also be reduced to a combination of binary classifiers. However, this kind of approach also have a drawback. In fact, sets of instances received by classifiers in the lower levels of the cascade already contains some misclassified thrombograms, thus errors made early in the process increase error rates of the following steps. In comparison to a single multiclass-SVM, the global performance achieved by a cascade composed of the best classifiers: a SVM for the A/B case and two MLPs for the others, is equal to 95.57%. In other words, this cascade principle results in a 1.08% increase of accuracy.

6.4   Complete Diagnosis Performance

In order to provide a complete diagnosis to clinicians specifying the type and the severity of the hemophilia, we consider results obtained for the All Categories classification with a cascade composed of a SVM and two MLPs. It achieves 95.57% of accuracy on this classification. We can therefore create a CDS able to provide a complete accurate diagnosis to clinicians.

6.5   Robustness

As mentioned in the previous sections, the database used in this study is composed of artificially generated thrombograms. Hence, data are clean of noise and are bound to be different than experimental ones. In order to approach an actual application, we added a Gaussian noise to our data with the variance $var(X) = k$ in which X is a centered Gaussian distribution and $k \in \{0, 5, 10, 15, 20\}$. It should be noted that the variance of this kind of experimental data usually seems to be in the range [5, 10]. To determine influence of this external parameter, some metrics were computed using noised datasets for the training, tuning and testing phases of these models. We focused on the two most challenging cases: Hemophilia A/B and All Categories and therefore, we used SVM classifiers. Given the fact that hyper-parameters are closely linked to the type of data used, we applied the process of hyper-parameter search mentioned in the previous section for each level of noise. Figure 1 shows that the addition of noise strongly degrades the entire model performance. Obviously, noise addition using $var = 20$ resulted in a 23.8% decrease of accuracy for the All Categories classifier, 34.7% for the A/B classifier. The poor performances induced by the noise addition can be explained using the Table 1. All used algorithms accurately classify Healthy/Hemophiliac and severity types because thrombograms are really different for these categories. Yet, it's not the case for hemophiliac A/B which is more complex to discriminate. In fact, noise addition increases similarities between thrombograms, particularly for thrombograms which have weak amplitude (severe hemophilia).

These results point out the well-known difficulty of SVM to work on raw data, without preprocessing step [14]. To go beyond this limit, we decided to smooth and fit noised data using a median filter and a Savitzky-Golay filter with a constant window size. We transposed these results in the Table 2. Yet, due to the constant window size, this process alters the shape of the peak and thus, no classification improvement has been noted (Fig. 2).

In a second experiment, we also computed learning curves in which the quantity of noised data used for the training phase varies from 1–13999. As we can see in the Fig. 1, noise addition does not affect the quantity of needed data for training since only the amplitude is impacted and not the convergence speed.

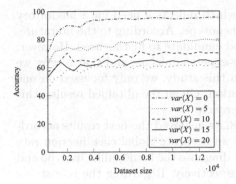

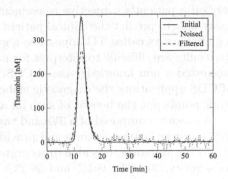

Fig. 1. The influence of noise on the hemophiliac A/B classifier performance

Fig. 2. Instance of a hemophiliac A mild illustrating the difference between raw, noised and filtered thrombograms

Table 2. Noise robustness and application of a Savitzky-Golay and a median filter whose window sizes are $size = 9$ (P: Hemophiliac B, N: Hemophiliac A)

Smooth & fit process	Variance	Accuracy	No. hemophiliac B classified as hemophiliac A	FPR	No. hemophiliac A classified as hemophiliac A	Recall
Without	0	98.1	17/900	1.9	883/900	98.1
	5	79.74	194/900	21.63	706/900	81.11
	10	72.15	244/900	27.19	656/900	71.48
	15	67.44	268/900	29.78	632/900	64.67
	20	63.44	333/900	37.04	567/900	63.93
With	0	95.22	40/900	4.52	860/900	94.96
	5	81.74	164/900	18.30	736/900	81.78
	10	71.22	266/900	29.56	634/900	72
	15	66.78	294/900	32.67	606/900	66.22
	20	61.78	325/900	36.22	575/900	59.78

7 Conclusion

This final section highlights issues of a CDS to detect hemophilia and suggests ways for further research.

Hemophilia is a rare bleeding disorder that leads to a less efficient coagulation system. The main treatment for this pathology is a replacement therapy that consists of clotting factor concentrate administrations. Even though well-known biological assays that quantify the concentrations of these factors in plasma exist, there are not good predictors of bleeding episodes. Therefore, treatments rely on a patient's subjective assessment which reflects a need for a laboratory assay able to predict the clinical patient phenotype. According to the literature, a global assays called TG appears as a good candidate to fit this need. However, its results are difficult to interpret for non-expert clinicians and that is why we suggested a non knowledge-based CDS. In this study, we only focussed on one of CDS application: the diagnostic elaboration. Given the obtained results, this work points out the benefit of such an approach.

A cascade composed of a SVM and two MLPs achieved the best results according to our evaluation criteria and provides an accurate global classification rate (95.57%). Moreover, it is able to accurately diagnose the hemophilia, its type and its severity, 99.18%, 98,1% and 96.23% respectively. Regarding the robustness of our model, adding a Gaussian noise strongly degrades the performance of the hemophiliac A/B classifier which could create an issue within an experimental context application. Moreover, as seen in the previous section, despite the theoretical interest of our noise filter, obtained results are disappointing.

For a first approach using machine learning techniques in the field of blood coagulation, few methods were used. Plenty of other well-known techniques could be tested on thrombograms. HMM [27] and DTW [17] generally perform well on time series and could appear as a great contribution to this work. Nevertheless, this kind of techniques have limitations like the SVM to deal with raw data. So, two options can be considered: (1) First of all, we could go more in depth during the smoothing process using a dynamic size window rather than a static one. This could potentially reduce the issue of peak alteration, and therefore increase the noise robustness of our model. (2) We could use deep learning methods. The main advantage of deep architectures, owing to their large number of hidden layers and the no-linearity associated to each ones, is their ability to extract highly abstract features from data. In addition, they can deal with very different types of data and can be applied to supervised but also unsupervised problems [3].

This study showed the ability of machine learning techniques to diagnose hemophilia. Obtained results open doors for other clinical application in the domain of blood illness, such as thrombophilia diagnosis and therapeutic orientation.

References

1. Agrawal, R., Faloutsos, C., Swami, A.: Efficient similarity search in sequence databases. In: Lomet, D.B. (ed.) FODO 1993. LNCS, vol. 730, pp. 69–84. Springer, Heidelberg (1993). https://doi.org/10.1007/3-540-57301-1_5
2. Alonso, C., Rodriguez, J.: Time series classification by boosting interval based literals. Inteligencia Artif. 4(11), 2–11 (2000)
3. Bengio, Y., Lamblin, P., Popovici, D., Larochelle, H.: Greedy layer-wise training of deep networks. In Proceedings of the 19th International Conference on Neural Information Processing Systems, NIPS 2006, pp. 153–160. MIT Press, Cambridge (2006)
4. Boxwala, A.A., et al.: GLIF3: a representation format for sharable computer-interpretable clinical practice guidelines (2004)

5. Chaovalit, P., Gangopadhyay, A., Karabatis, G., Chen, Z.: Discrete wavelet transform-based time series analysis and mining. ACM Comput. Surv. **43**(2), 1–37 (2011)
6. Chaovalitwongse, W.A., Fan, Y.-J., Sachdeo, R.C.: On the time series K-Nearest Neighbor classification of abnormal brain activity. IEEE Trans. Syst. Man Cybern. Part A Syst. Hum. **37**(6), 1005–1016 (2007)
7. Esmael, B., Arnaout, A., Fruhwirth, R.K., Thonhauser, G.: Multivariate time series classification by combining trend-based and value-based approximations. In: Murgante, B., et al. (eds.) ICCSA 2012, Part IV. LNCS, vol. 7336, pp. 392–403. Springer, Heidelberg (2012). https://doi.org/10.1007/978-3-642-31128-4_29
8. Gudmundsson, S., Runarsson, T.P., Sigurdsson, S.: Support vector machines and dynamic time warping for time series. In: 2008 IEEE International Joint Conference on Neural Networks (IEEE World Congress on Computational Intelligence), pp. 2772–2776 (2008)
9. Keogh, E., Chu, S., Hart, D., Pazzani, M.: Segmenting time series: a survey and novel approach. In: Data Mining in Time Series Databases, pp. 1–21 (2003)
10. Kerdelo: Méthodes informatiques pour l'expérimentation in virtuo de la cinétique biochimique. Application à la coagulation du sang. Modélisation et simulation. Thèse, Université Rennes 1 (2006)
11. Kohavi, R.: A study of cross-validation and bootstrap for accuracy estimation and model selection. In: International Joint Conference on Artificial Intelligence, vol. 14, no. 12, pp. 1137–1143 (1995)
12. Mantovani, R., Rossi, A., Vanschoren, J., Bischl, B., De Carvalho, A.: Effectiveness of random search in SVM hyper-parameter tuning. In: International Joint Conference on Neural Networks, pp. 1–8. IEEE (2015)
13. Nanopoulos, A., Alcock, R., Manolopoulos, Y.: Feature-based classification of time-series data. Inf. Process. Manage. **0056**, 49–61 (2001)
14. Nguyen, M.H., de la Torre, F.: Optimal feature selection for support vector machines. Pattern Recogn. **43**(3), 584–591 (2010)
15. Popivanov, I., Miller, R.: Similarity search over time-series data using wavelets. In: Proceedings 18th International Conference on Data Engineering, pp. 212–221. IEEE Computer Society (2002)
16. Quinlan, J.R.: C4.5: Programs for Machine Learning, vol. 1 (1992)
17. Ratanamahatana, C., Keogh, E.: Everything you know about dynamic time warping is wrong. In: Third Workshop on Mining Temporal and Sequential Data, pp. 22–25 (2004)
18. Rüping, S.: SVM kernels for time series analysis. Time, p. 8 (2001)
19. Samwald, M., Fehre, K., de Bruin, J., Adlassnig, K.P.: The Arden Syntax standard for clinical decision support: experiences and directions. J. Biomed. Inform. **45**(4), 711–718 (2012)
20. Shin, H., Markey, M.K.: A machine learning perspective on the development of clinical decision support systems utilizing mass spectra of blood samples (2006)
21. Shortliffe, E.H.: Computer-Based Medical Consultations: Mycin. Elsevier, New York (1976)
22. Sutton, D.R., Fox, J.: The syntax and semantics of the PROforma guideline modeling language. J. Am. Med. Inform. Assoc. **10**(5), 433–443 (2003)
23. Wu, Y.-L., Agrawal, D., El Abbadi, A.: A comparison of DFT and DWT based similarity search in time-series databases. In: Proceedings of the 9th International Conference on Information and Knowledge Management (CIKM), vol. 35(2000-08), pp. 488–495 (2000)

24. Young, G., Sorensen, B., Dargaud, Y., Negrier, C., Brummel-Ziedins, K., Key, N.: Thrombin generation and whole blood viscoelastic assays in the management of hemophilia: current state of art and future perspectives. Blood **121**(11), 1944–1950 (2013)

25. Yu, C.H., Bhatnagar, M., Hogen, R., Mao, D., Farzindar, A., Dhanireddy, K.: Anemic status prediction using multilayer perceptron neural network model, vol. 50, pp. 213–220 (2017)

26. Zhang, D., Zuo, W., Zhang, D., Zhang, H.: Time series classification using support vector machine with Gaussian elastic metric kernel. In: Proceedings - International Conference on Pattern Recognition, pp. 29–32 (2010)

27. Zhang, Y.: Prediction of financial time series with hidden Markov models (2004). https://core.ac.uk/download/pdf/56371948.pdf

Machine Learning in Vision and Robotics

Real-Time Logo Detection in Brand-Related Social Media Images

Oscar Orti[1,3], Ruben Tous[2,3(✉)], Mauro Gomez[1,3], Jonatan Poveda[1,3],
Leonel Cruz[2,3], and Otto Wust[1,3]

[1] Adsmurai, Barcelona, Spain
[2] Universitat Politècnica de Catalunya (UPC), Barcelona, Spain
rtous@ac.upc.edu
[3] Barcelona Supercomputing Center (BSC), Barcelona, Spain

Abstract. This paper presents a work consisting in using deep convolutional neural networks (CNNs) for real-time logo detection in brand-related social media images. The final goal is to facilitate searching and discovering user-generated content (UGC) with potential value for digital marketing tasks. The images are captured in real time and automatically annotated with two CNNs designed for object detection, SSD InceptionV2 and Faster Atrous InceptionV4 (that provides better performance on small objects). We report experiments with 2 real brands, Estrella Damm and Futbol Club Barcelona. We examine the impact of different configurations and derive conclusions aiming to pave the way towards systematic and optimized methodologies for automatic logo detection in UGC.

Keywords: Social media · Instagram · User generated content ·
Deep learning · Marketing · Object detection

1 Introduction

Nowadays, there is a growing interest by profit and non-profit organizations in deriving benefit from the photos that users share on social networks such as Instagram or Twitter [17], a part of the so-called user-generated content (UGC). A significant part of these images has potential value for organizational intelligence and digital marketing tasks. On the one hand, users' photos can be analyzed to obtain knowledge about users behavior and opinions in general, or with respect to a certain products, services or brands. On the other hand, some users' photos can be of value themselves, as original and authentic content that can be used, upon users' permission, in the different organizations' communication channels.

Platforms for photo-centric UGC are proliferating rapidly nowadays (e.g. Adsmurai [15], Olapic [11], Chute [1] and Curalate [2]), but analyzing images on social media streams is challenging. The potential bandwidth to analyze is huge and, while they help, user defined tags are scarce and noisy. Any automated processing component needs to be extremely efficient and scalable. A large part

© Springer Nature Switzerland AG 2019
I. Rojas et al. (Eds.): IWANN 2019, LNCS 11507, pp. 125–136, 2019.
https://doi.org/10.1007/978-3-030-20518-8_11

of current solutions relies on costly manual curation tasks over random samples. This way many contents are not even processed, and many valuable photos go unnoticed. Adoption of image recognition techniques in commercial UGC systems is currently very limited (Fig. 1).

Fig. 1. Example images containing one or more occurrences of the Estrella Damm logo.

In this work, we address the logo detection problem, one of most challenging tasks that an automated UGC system needs to face. We have developed a system that automatically processes, in real-time, an incoming stream of social media images and detects and localizes all the occurrences of any of a set of supported logotypes. The system makes use of two state-of-the-art deep convolutional neural networks (CNNs) designed for object detection, SSD InceptionV2 and Faster Atrous InceptionV4 (that provides better performance on small objects). The resulting system is currently being integrated within a real commercial service, the Adsmurai's Visual Commerce Platform [16].

In this paper we describe the technical design of the system and the results of the performance evaluation experiments in which real images related with two commercial brands, Estrella Damm and Futbol Club Barcelona, have been used. We examine the impact of different configurations and derive conclusions aiming to pave the way towards systematic and optimized methodologies for automatic logo detection in UGC.

2   Related Work

The work presented in this paper is related to recent works attempting to facilitate the classification and search of images in social networks such as Instagram and Twitter. Some works, such as [10,15] or [12], also apply scene-based and object-based image recognition techniques to enrich the metadata originally present in the images in order to facilitate their processing. All latest works rely on CNNs as an underlying technique. In our case, the applied image recognition techniques, while also relying in CNNs, are tuned for content curation for digital marketing tasks. This implies new problems, such as the need to deal with small datasets (e.g. brand-based image datasets). Previous works such as [5,18] and [4] also classify social media data paying special attention to brands and products.

Regarding the detection of logos, one of the current state-of-art architectures is the Single Shot MultiBox Detector (SSD) [9]. SSD works similar than YOLO [13] but includes a RPN (Region Proposal Network, technique popularized by [3]) to improve the diversity of prior boxes by running it on multiple convolutional layers and different depth levels. This CNN is based on VGG feature extractor and performs faster than the Faster-RCNN [14]. Since SSD appeared some variations of that object detection architectures, but the main approach changed to object segmentation and instance segmentation with new architectures like MaskRCNN [7] or RetinaNet [8]. Nowadays it is possible to instantiate objects in almost real time video, but not in average computers, although it does not require as much computation capacity as ResNet because of there is an awesome work of memory and calculus optimization.

3   Methodology

3.1   Outline of a UGC Curation Platform

Logo detection is just one of the tasks of a UGC curation platform. A UGC curation platform is a software system that processes a continuous stream (usually with a huge and volatile bandwidth) of social media images and prepares them (e.g. by indexing them, annotating them with the proper metadata, etc.) to be exploited for organizational intelligence or digital marketing tasks. The functionality of these platforms is usually divided into two different stages, data acquisition and data consumption. Both stages interact through a common database containing metadata about the images, and they can occur concurrently (once images start feeding the database users or analytics tools can start using them). During data acquisition new images are captured in real-time, as they are published on the underlying sources (Instagram in our case). Descriptors of the images (including the URL pointing to the image content) are acquired using the APIs provided by these underlying sources. Once a new image is captured, it is processed by different components (e.g. scene recognition, face detection, logo detection) that automatically enrich the image's metadata with annotations that describe their visual content. During the data consumption stage users (or analytics tools) can navigate, search and select images from the generated database.

3.2   Logo Detection

The logo detection system described in this paper has two main functionalities, training and inference. Both functionalities are provided as APIs to facilitate their integration within the UGC curation platform. On the one hand, given dataset of training images depicting a logotype and labeled in PASCALVOC style, the system is able to train and store a new model for detecting that logotype. The system uses one model for each logo, to facilitate the inclusion of new logotypes dynamically. Figure 2 outlines the workflow of the train API. The API is fed with training images for a given logotype. The checkpoints generated during training store the model's weights at a given training step, and are used by the inference API to generate a model's inference graph. The saved checkpoints are also used as pre-loaded weights to re-train the model. The model validation is conducted in parallel to the training (each 10 min).

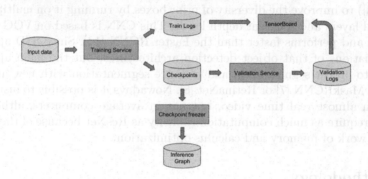

Fig. 2. Train API workflow

On the other hand, a scalable inference API is provided. The inference API downloads the checkpoints of the available models (the ones generated during training) from cloud storage (AWS S3). The API receives JSON requests with a list of new images and the identifier of the model to be used. The requested images are downloaded to the filesystem. If still not loaded, the API loads the inference graph. The inference graph is applied to detect all the occurrences of the requested logotype in each image. The bounding boxes of all the detections of each input image are finally returned in PASCALVOC style.

All the explained functionalities are integrated following the human-in-the-loop (HITL) pattern (see Fig. 3). When the inferred logo detections for a given image are validated during the consumption stage of the UGC curation platform (e.g. if the image is selected by the user), the image is added to the model's training dataset, thus enabling to continuously improve the model's performance.

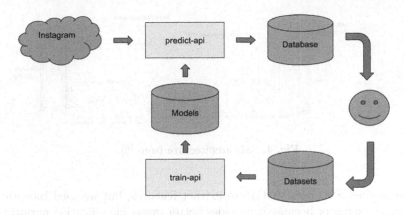

Fig. 3. Object detector HITL

3.3 Networks

Internally, the developed logo detection APIs rely on one or more object detection algorithms. With the help of the TensorFlow Object Detection API, the system has been designed to enable the implementation of any state-of-the-art logo detection CNN-based method. It can be configured with different feature extractors such as InceptionV2, InceptionV3, InceptionV4, ResNet50, ResNet101 or MobileNet. Each one could be highly customized by configuring the activation functions or batch normalization among others. It is also possible to define which kind of region proposal to use among SSD, FasterRCNN or RFCN and customize its hyper-parameters. To define the training it is possible to choose the batch size and the learning-rate optimizer, which could be RMSProp, momentum or scheduled. Different kind of data augmentations are available too, like random horizontal flip or random crops, among others.

The implementation described in this paper uses two CNNs, SSD InceptionV2 and Faster Atrous InceptionV4 (to improve the performance on small objects). On the one hand, Faster R-CNN uses a region proposal network to create boundary boxes and utilizes those boxes to classify objects. While it is considered the start-of-the-art in accuracy, the whole process runs at 7 frames per second. Far below what a real-time processing needs. SSD speeds up the process by eliminating the need of the region proposal network. To recover the drop in accuracy, SSD applies a few improvements including multi-scale features and default boxes. These improvements allow SSD to match the Faster R-CNN's accuracy using lower resolution images, which further pushes the speed higher (Fig. 4).

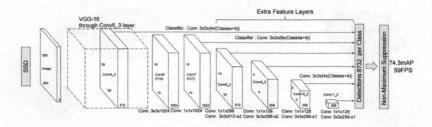

Fig. 4. SSD architecture from [9].

The original SSD uses VGG16 to extract features, but we used InceptionV2 as a feature extractor because it provides better image classification performance in our context. The SSD uses multi-scale feature maps to detect objects, so it uses higher resolution feature maps to detect small objects and lower resolution for bigger objects. SSD uses hard negative example mining because we make far more predictions than the number of objects presence. So there are much more negative matches than positive matches. This creates a class imbalance which hurts training. We are training the model to learn background space rather than detecting objects. However, SSD still requires negative sampling so it can learn what constitutes a bad prediction. So, instead of using all the negatives, we sort those negatives by their calculated confidence loss. SSD picks the negatives with the top loss and makes sure the ratio between the picked negatives and positives is at most 3:1. This leads to a faster and more stable training.

On the other hand, Faster Atrous InceptionV4 is designed to achieve the best performance without taking into account the training or inference time. This CNN uses InceptionV4 (also called InceptionResNetV2) as a feature extractor, which achieves the state-of-the-art in image classification accuracy. The atrous version of the Faster RCNN [6], uses dilated convolutions to achieve better results. Before the Faster RCNN, the slowest part in Fast RCNN was Selective Search or Edge boxes. Faster RCNN replaces selective search with a very small convolutional network called Region Proposal Network to generate regions of interests.

4   Experiments

We annotated two datasets related to two real brands, Estrella Damm and Futbol Club Barcelona. The manual annotation was performed using the LabelImg Tool. Each dataset is composed with 650 images. We have used transfer learning to train the networks because of data scarcity. So the 650 images (70% training and 30% validation) were used to fine-tune the CNN architectures that had been pre-trained with the COCO dataset. The validation set is not presented to the neural network along all the training phase. The datasets were converted to TensorFlow's TFRecord format. During evaluation we measured the precision, recall and the mean average precision (mAP), which requires to define an intersection over union (IoU) condition (the ratio between the intersection and the union of the predicted boxes and the ground truth boxes). We used two IoU thresholds, 0.5 and 0.75, to determine if a detection is a true positive. The whole project is developed in Python 3.6. The CNNs have been implemented in TensorFlow 1.4. All data (datasets, models and inference graphs) are stored in Amazon AWS S3. APIs endpoints are implemented in Flask. Docker containers are used for the different system functionalities. One container is used for the predict API. The train API is divided into four containers, one for each service: train, validation, TensorBoard and checkpoint freezer. Docker Swarm is used for container orchestration.

4.1   Single Shot Detection InceptionV2 Configuration

The SSD requires a fixed input shape, so all image are resized in the preprocessing step into (512, 512, 3). The SSD uses an anchor generator with the configured aspect ratios (1, 2, 3, 0.5, 0.33) to propose regions of interest. Then InceptionV2 configured with batch normalization is used as a feature extractor, with batch normalization. The classification loss is defined with the sigmoid, whereas the localization loss is the smooth L1. It is also configured a hard example minner to get between 0 to 3 negatives examples for each positive example. This allows the network to learn negatives and improve its performance. The training batch size is 24 (because of hardware limitations) and the learning rate is optimized using RMSprop. So, The SSD is trained using Mini-Batch Gradient Descent.

4.2   Faster Atrous RCNN InceptionV4 Configuration

The Faster RCNN does not require a fixed input shape. So, in this case, the image size is bounded by a minimum (600) and a maximum (1024) dimension size. If the dimension of an image is lower or higher than a threshold, then the image is resized, maintaining its aspect ratio, to respect the bounds. The feature extractor is implemented using InceptionV4 (or Inception ResNet V2). The loss functions used are softmax. The training batch size is 1 because it's not possible to generate batch sizes greater than 1 with images of different sizes. So, this CNN is trained with stochastic gradient descent.

5   Results

Figures 5 and 7 show the mAP and recall results for both, the FC Barcelona and the Estrella Damm, datasets respectively.

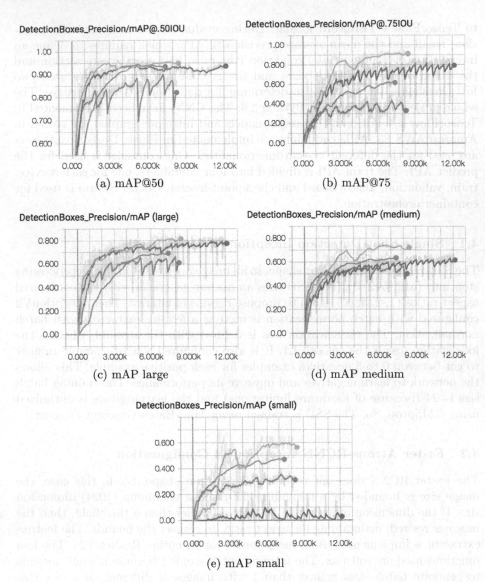

Fig. 5. Evaluation mAP results for both datasets (with different IoU thresholds, (a) 0.5 and (b) 0.75, and different logotype sizes, (c) large, (d) medium, (e) small). Dark blue: Estrella Damm SSD, Light blue: Estrella Damm Faster R-CNN, Green: FCB SSD, Orange: FCB Faster R-CNN. (Color figure online)

The results show that the SSD provides good results, around 0.9 mAP with a 0.5 IoU threshold, for both datasets. The results for FC Barcelona are inferior to the ones for Estrella Damm mainly because the FC Barcelona logotype is frequently found over deformable objects (e.g. clothes). Considering that SSD is faster and also easier to train than Faster R-CNN, we conclude that it is well suited for real-time logo detection because of its good trade-off between inference time and accuracy.

However, the results obtained with SSD for small objects are not satisfactory. This is caused mainly because small logos may not appear in all feature maps (the more an object appears on a feature map, the more likely that the MultiBox algorithm can detect it). One way to address this limitation is to increase the size of the input image, but that reduces the speed at which SSD can run and does not completely alleviate the problem of detecting small objects. Faster R-CNN obtains better results with small logotypes for both datasets.

Fig. 6. Validation images along the training

Figures 6 and 8 show some example detections for both, the FC Barcelona and the Estrella Damm, datasets respectively.

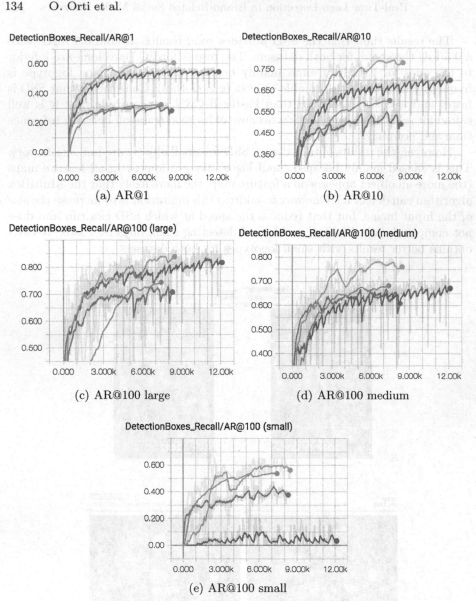

Fig. 7. Evaluation recall results for both datasets (with different IoU thresholds, (a) 0.5 and (b) 0.75, and different logotype sizes, (c) large, (d) medium, (e) small). Dark blue: Estrella Damm SSD, Light blue: Estrella Damm Faster R-CNN, Green: FCB SSD, Orange: FCB Faster R-CNN. (Color figure online)

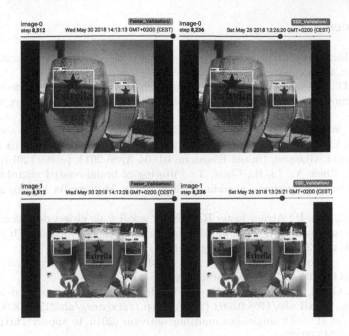

Fig. 8. Validation images along the training

6   Conclusions

This paper describes the design and implementation of a software architecture for real-time logo detection in brand-related social media images. The final goal is to facilitate searching and discovering user-generated content (UGC) with potential value for digital marketing tasks. The images are captured in real time and automatically annotated with two CNNs designed for object detection, SSD InceptionV2 and Faster Atrous InceptionV4. We report experiments with 2 real brands, Estrella Damm and Futbol Club Barcelona. Unlike Faster R-CNNs, which contain multiple moving parts and components, SSDs are unified, encapsulated in a single end-to-end network, making SSDs easier to train and well suited for real-time object detection because of their good trade-of between inference time and accuracy. However, our experiments show that SSDs provide poor performance for small objects, mainly because small objects may not appear on all feature maps. For this reason our proposed solution includes both, an SSD network and a Faster R-CNN network that can be selected depending on the goal.

Acknowledgements. This work is partially supported by the Spanish Ministry of Economy and Competitivity under contract TIN2015-65316-P and by the SGR programme (2014-SGR-1051 and 2017-SGR-962) of the Catalan Government.

References

1. Chute. Enterprise UGC. http://www.getchute.com/. Accessed 6 June 2017
2. Curalate. https://www.curalate.com/. Accessed 6 June 2017
3. Erhan, D., Szegedy, C., Toshev, A., Anguelov, D.: Scalable object detection using deep neural networks. CoRR abs/1312.2249 (2013). http://arxiv.org/abs/1312.2249
4. Gao, Y., Wang, F., Luan, H., Chua, T.: Brand data gathering from live social media streams. In: Proceedings of the International Conference on Multimedia Retrieval, ICMR 2014, Glasgow, United Kingdom, 01–04 April 2014, p. 169 (2014)
5. Gao, Y., Zhen, Y., Li, H., Chua, T.: Filtering of brand-related microblogs using social-smooth multiview embedding. IEEE Trans. Multimedia **18**(10), 2115–2126 (2016)
6. Guan, T., Zhu, H.: Atrous faster R-CNN for small scale object detection. In: 2017 2nd International Conference on Multimedia and Image Processing (ICMIP), pp. 16–21, March 2017. https://doi.org/10.1109/ICMIP.2017.37
7. He, K., Gkioxari, G., Dollár, P., Girshick, R.B.: Mask R-CNN. CoRR abs/1703.06870 (2017). http://arxiv.org/abs/1703.06870
8. Lin, T., Goyal, P., Girshick, R.B., He, K., Dollár, P.: Focal loss for dense object detection. CoRR abs/1708.02002 (2017). http://arxiv.org/abs/1708.02002
9. Liu, W., et al.: SSD: single shot multibox detector (2016, to appear). http://arxiv.org/abs/1512.02325
10. Nguyen, D.T., Alam, F., Ofli, F., Imran, M.: Automatic image filtering on social networks using deep learning and perceptual hashing during crises. CoRR abs/1704.02602 (2017). http://arxiv.org/abs/1704.02602
11. Olapic. Earned content platform. http://www.olapic.com/. Accessed 6 June 2017
12. Park, M., Li, H., Kim, J.: HARRISON: a benchmark on hashtag recommendation for real-world images in social networks. CoRR abs/1605.05054 (2016)
13. Redmon, J., Divvala, S.K., Girshick, R.B., Farhadi, A.: You only look once: unified, real-time object detection. CoRR abs/1506.02640 (2015). http://arxiv.org/abs/1506.02640
14. Ren, S., He, K., Girshick, R.B., Sun, J.: Faster R-CNN: towards real-time object detection with region proposal networks. CoRR abs/1506.01497 (2015). http://arxiv.org/abs/1506.01497
15. Tous, R., et al.: Automated curation of brand-related social media images with deep learning. Multimedia Tools Appl. **77**(20), 27123–27142 (2018). https://doi.org/10.1007/s11042-018-5910-z
16. Tous, R., et al.: User-generated content curation with deep convolutional neural networks. In: Proceedings of the 2016 IEEE International Conference on Big Data, BigData 2016, Washington DC, USA, 5–8 December 2016, pp. 2535–2540 (2016)
17. Tous, R., Torres, J., Ayguadé, E.: Multimedia big data computing for in-depth event analysis. In: Proceedings of the 2015 IEEE International Conference on Multimedia Big Data (BigMM), Beijing, China, 20–22 April 2015, pp. 144–147. IEEE (2015)
18. Zhao, S., Yao, H., Zhao, S., Jiang, X., Jiang, X.: Multi-modal microblog classification via multi-task learning. Multimedia Tools Appl. **75**(15), 8921–8938 (2016)

A Novel Framework for Fine Grained Action Recognition in Soccer

Yaparla Ganesh[✉], Allaparthi Sri Teja, Sai Krishna Munnangi,
and Garimella Rama Murthy

International Institute of Information Technology, Hyderabad, Hyderabad, India
{ganesh.yaparla,sriteja.allaparthi,krishna.munnangi}@research.iiit.ac.in,
rammurthy@iiit.ac.in

Abstract. Sports analytics have become a topic of interest in the field of
Artificial intelligence. With the availability of huge volumes of high level
data, significant progress has been made in the domain of action recognition in the past. Though video based action recognition has progressed
well using state of the art deep learning techniques, its applications are
limited to some higher level actions like throwing, jumping, running etc.
There has been some work in fine-grained action recognition technique,
such as, identification of type of throws in Basketball, and the type of a
player's shots in Tennis. However with larger play field and with many
players on field, multi player sports such as Soccer, Rugby, Hockey and
etc. pose bigger challenges and remain unexplored. These games in general are live fed through field view cameras or skycams which aren't stationary. For these reasons, we chose to recognize player's actions in the
game of Soccer and thereby, explore the capabilities of existing architectures and deep neural networks for these kind of games. Our main
contributions are the proposed framework that can automatically recognize actions of players in live football game which will be helpful for text
query based video search, for extracting stats in a football game and to
generate textual commentary and the Soccer-8k dataset which consists
of different action clips in the soccer play.

1 Introduction

The problem of Action Recognition requires a system to identify the subject's
action from a series of observations on the subject's movements. Action Recognition can be further classified into Sensor-based [1] and Vision-based [2] Action
Recognition problems.

Vision-based Action Recognition problems are the most researched problems,
with major share taken by Sports Analytics [1,3–5]. With volumes of rich data
available, many video based action recognition techniques are on the rise, but
to the best of our knowledge there is no vision based application that delivers
some stats related to the sport being played in real time. This might be due
to lack of availability of event or player centric datasets for fine grained action
recognition in multi player sports. This might be the reason why sensor based

© Springer Nature Switzerland AG 2019
I. Rojas et al. (Eds.): IWANN 2019, LNCS 11507, pp. 137–150, 2019.
https://doi.org/10.1007/978-3-030-20518-8_12

activity recognition predominantly dominates the area of sport analytics. In multi player sports these sensors are used to track people and objects involved so as to extract player specific statistics (heat-maps, traits) and team level statistics (coordination, formations).

We took a step forward to recognize fine grained actions in soccer. Understanding and recognizing complex events in a soccer match like passing, shooting, heading from videos is a challenging task. This is because of the coverage from different angles by multiple cameras and the existence of multiple players in the region of interest. In this paper, we introduce a framework exploiting the recent developments in the deep learning models for action recognition.

The attempts of earlier works in these directions to classify the player action in futsal matches aren't player centric or event centric [3]. However, such frameworks need in depth labeling of the actions at each player in every frame. Due to lack of such big datasets and the huge manual work for creating the required labels made the application of the well defined neural network architectures limited in sports.

Many models for fine-grained action recognition in sports have been proposed. These fine-gained action recognition models have been based on Two stream convolutional neural networks [6], Optical flow analysis and spatio-temporal analysis based on 3D CNN [2] and RNN's [7]. In this paper, we are particularly interested in the recognizing different actions like passing, shooting, heading, dribbling etc. in the game of soccer, based on spatio-temporal analysis. This requires a special dataset that is soccer specific. With lack of such a standard dataset, we created the Soccer-8k dataset. The samples in the Soccer-8k dataset are RGB, monocular video clips generated from Soccer matches, with dynamic background and contains occlusions. These videos include different moving backgrounds with multiple persons involved in the scene of interest making the probability of finding the common features for the same labeled videos very less and potentially covering all the features.

We propose a novel framework for event based action recognition in soccer. The framework takes the video as input and outputs the same video with each individual action tagged and bounded within a box, as represented in Fig. 1. The framework consists of primarily two modules, Event Detector and Action Classifier. The Event Detector module identifies the desired events and generates a video clip containing the actions surrounding the event. This clip is inputted to Action Classifier module. We introduce GAWAC (GAussian Weighted event based Action Classifier) architecture, an integral part of Action Classifier module, which classifies the input clip. Other subsequent modules portrays each classified clip onto the original video by tagging the action and confining it within a bounded box in the input. The framework, and the Soccer-8k dataset are the main contributions of this paper.

The rest of the paper is organized as follows. Section 2 refers to the related work done and Sect. 3 discusses how the Soccer-8k dataset has been created from the Event Detector module of our framework. Section 4 discusses the architecture and implementation details of GAWAC. Section 5 deals with the experiments done along with their results on the Soccer-8k dataset respectively. Section 6 discusses the results. Section 7 concludes the paper along with details about the future work.

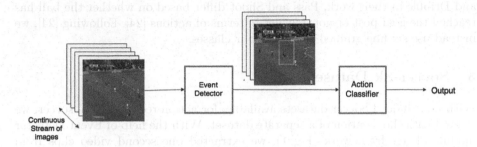

Fig. 1. Block diagram of the framework

2   Related Work

State of the art techniques developed in sport analytics so far are limited to few specific settings. The reason for this is, the problem of Action Recognition incorporates a broad range of scenarios or settings and many other subproblems, each of which can have significant affect. For example, in the case of video based action recognition, an action captured through a dynamic camera needs to be handled differently from the one captured through static cameras. And within the same video itself, monocular view and multi-view may need different processing. With such influential factors and lack of proper standard datasets, the progress made so far is comparatively less.

Many of the previous works using deep learning models in the action recognition deals with the interaction of multiple objects with each other in the scene of interest in different sports. These interactions are used for tracking the ball, finding the possession, person tracking etc. [8–11]. In addition to all of the above models, many have explored the two stream deep learning frameworks for action recognition which deals with temporal and optical flow [6,7,12–17]. Along with these deep learning models, some of the existing methods have been exploring the hand-made features like HOG, HOF, MBH extracted from trajectory-based representations computed using optical flow [18–21].

Deep learning models based on 3D CNN and RNN's have shown promising results on fine grained video based action recognition. The reason being 3D CNN and CNN plus LSTM based architectures have capability to extract spatio-temporal information present in the video dataset. [4] have presented a 3 layered LSTM model for fine grained action recognition in the game of Tennis, trained on their custom made dataset, THETIS [22]. First, the authors have extracted the features through the Inception model [23] which were then used to train the model. [5] provided three LSTM based frameworks for scoring athletic events, within each of which, clip level features are aggregated for video level description differently providing different expressions on quality of the action, that helped in scoring the actions. [3] presents a hierarchical model of CNN and LSTMs for team activity recognition in Soccer. The authors have used 3 classes, Shoot, Pass and Dribble in their work. Pass and Shoot differ based on whether the ball has reached the goal post or not, but not in terms of actions [24]. Following [24], we instead use six fine grained actions as our classes.

3   Soccer-8k Dataset

With no standard Soccer datasets available, for action recognition in soccer, we started with the creation of a separate dataset. With the help of Event Detector module of our framework (Fig. 1), we extracted one second video clips from input videos and created the Soccer-8k dataset. We took full match Laliga and few Champions League full HD videos available on YouTube for the creation of our soccer action recognition dataset. We extract event information in the video in order to localize the area we present to the classifier. Event is said to happen when a ball is released or gathered in open play. Figure 2 shows examples of correct event detections. Identification and tracking of both the ball and player in an input video are two sub tasks involved in solving the problem of event detection. Given we have only 2D information, occlusion and the mix up of ball with the players along with blur in video can lead to wrong event identification. Figure 3 shows examples of incorrect event detections, which are discarded.

It is required of us to first identify the region of video which relates to player's action i.e. who is in contact with the ball and then labeling each of those before passing them as an input to the model. Since it is not easy to manually identify the clips relating a player's action with the ball, we automated the process using ball tracking and player tracking in soccer. However, changes in the camera angles, occlusion of the ball, camera motion and the mix up of ball with the players posed serious challenges.

Handling all those issues, we built the Event Detector module on top of these steps to generate clips that contain only the players' actions with the ball. Eight thousand of such generated clips from special input videos are saved which formed the Soccer-8k dataset.

1. We begin with identifying the relevance and usefulness of each frame. Generally broadcasters broadcast clips that contain closeup view of players or replays of a particular incident from different views for better user experience. Such frames are deemed irrelevant and discarded. Based on predefined norms such as the number of players in the frame, their sizes and play area shown, the frame is analyzed on the visual information contained. Depending on relevance of the frame, ball tracking or ball identification is performed.
2. We define a frame as Pivot frame if the ball is in contact with a player in the current frame, but not in the preceding or succeeding frames. A Pivot Frame is labeled as Type-1 if an action takes place in succeeding frames. Else it is labeled as a Type-2 Pivot frame. In general, Type-1 frames occur when the

Fig. 2. Examples of correct event detection. The inner red box represents the area where ball has been detected to be in contact with the player. Yellow box represents the area (300 × 300) around the event detection point that is to be cropped while creating localized video clip from clip generated by stacking frames around the pivot frame. (Color figure online)

Fig. 3. Examples of incorrect event detection. In few cases, because of incorrect depth perception of the image, the system wrongly detects the occurrence of an event. All such incorrect samples are filtered out during the labeling process.

ball has just been received by a player and Type-2 occur when the ball has been released by a player. Figure 4 shows examples of Pivot Frames.

3. Each relevant frame is checked for Pivot Frame. If P_i is pivot frame, then all the frames in between P_{i-n} and P_{i+m} capture the action that took place in between. It has been observed that most of the actions in soccer games have a span of no more than 24 frames. So we choose n, m as 9, 14 respectively, if the pivot frame is identified as Type-1 and 14, 9, if it is a Type-2 frame.

4. The frames in between P_{i-n} and P_{i+m} are then extracted and a clip is created using them.

5. Off ball player's movements which have no impact on the action can affect the classifier. Hence, we crop $p \times p$ area in video clip centering around the point of event detection forming a new localized video clip that has to be classified. To find the most suitable value for p, we began with low values of p (for better localization of a specific action). Point of event detection is shifted in consecutive frames because of factors like camera movement. This results in the failure to capture the whole action. Hence, we had to leverage

Fig. 4. Examples of randomly sampled frames from our dataset

on localization for camera movement and based on the experiments carried out, value of 300 for p seemed to work well for all the clips generated.

6. The saved clips for dataset are outsourced to soccer enthusiasts who have labeled the actions in each clip. For every sample, the label agreed by at least 50% of the labelers is assigned. Else, the sample is discarded.

The Soccer-8k dataset is created using the above labeled clips. The dataset contains a total of 7942 action clips. We train our model on this dataset, splitting the dataset into training and test sets. To support future work in this area, we made Soccer-8k dataset publicly available.

The table below details the classes and the total number of samples in each class.

Class	Number of samples
Short pass	2622
Long pass	670
Header	87
Trapping	1946
Dribbling	1991
Turning	626

Passing implies keeping possession of the ball by maneuvering it on the ground between different players with the objective of advancing it up the playing field. There are 2 types of passes, long and short. A long pass is an attempt to move the ball a long distance down the field via a cross, without the intention to pass it to the feet of the receiving player. A short pass involves keeping the ball low and making it easier for a team mate to control. Header is technique that is used to control the ball using the head to pass, shoot or clear. Trapping refers to intercepting the ball and controlling it. Dribbling is maneuvering of the ball by a single player while moving in a given direction, avoiding defender's attempt to intercept ball. Turning refers to instantaneous shift in player's direction along with the ball to move away from defense [25]. These classes are chosen based on the individual techniques a player should posses in Soccer [24].

4   Classifier and Implementation Details

GAWAC is a Convolutional Neural Network where in which instead of using a 2D kernel for convolution, 3D kernel is used. This helps in capturing the temporal dependencies between the frames of a video along with spatial dependencies.

GAWAC (Fig. 5) has 5 convolution, 4 max-pooling, and 2 fully connected layers, followed by a softmax output layer. After the fourth convolution layer, we use a Gaussian distributed weight filter to emphasize the information at the center. We do this to make the model learn that the action is concentrated around the ball, centering which the clip is extracted. This helps us in removing irrelevant information about other players in the 300×300 image.

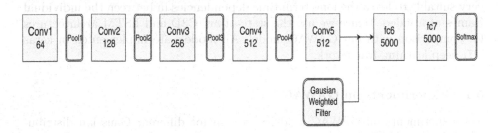

Fig. 5. The GAWAC architecture

All 3D convolution kernels are $3 \times 3 \times 3$ with stride 1 in both spatial and temporal dimensions. Number of filters in convolution layers are 64, 128, 256, 256 and 512 respectively. Except for the first maxpool layer having dimensions $1 \times 2 \times 2$, all other maxpool layers have dimensions $2 \times 2 \times 2$. Each fully connected layer has 5000 output units. The input dimensions are $3 \times 12 \times 100 \times 100$. Figure 1 presents the work flow of the model.

5   Experiments on Soccer-8k Dataset

The experiments are conducted on our Soccer-8k dataset. It contains 7942 videos of six different soccer actions performed by different players in different matches at different venues. The six actions are:

- Short Pass
- Long Pass
- Heading
- Trapping
- Turning
- Dribbling

All of these videos are RGB and have moving backgrounds which is one of the challenges. All these video clips have 24 frames. Figures 11, 12, 13, 14, 15 and 16 shows sample frames from few video clips.

For each experiment, 70% of Soccer-8k dataset is reserved as training set and rest 30% is divided equally as test and validation datasets.

Imbalance in number of samples of each class may lead to biased classification. To handle this we up-sample the classes with low number of samples ensuring equal number in all classes. To avoid repetition the same example to network we augment the data which ensures better learning, reducing the chances of over-fitting.

C3D model is the best performer on UCF101 dataset for video-based sport identification, which is rich in visual content including video collections of Baseball Pitch, Basketball Shooting, Diving and etc. [26] has proved that LSTMs are very suitable to learn the long-term time dependencies in between the individual frames of a video. Hence we use the state-of-art C3D and LSTM architectures for comparison with our GAWAC model. All models are implemented in keras [27] deep learning framework.

5.1   Experiments on GAWAC

The experiments on GAWAC model were run for different Gaussian distributions. We tried different Gaussian distributions, with Mean (μ) and Variance (σ^2) as (0, 1.0), (0, 0.5) and (0, 0.2). These experiments were run on Nvidia GeForce GTX 1080 Ti GPU, with 11 GB of VRAM. With batch size as 32, the experiments are run on the augmented dataset with approximate size of 15,000 samples. The model is trained on new augmented dataset which gets generated on the fly in each epoch. This ensures that the model never gets the same example in training. Techniques such as shifting, rotation, centering, flip etc., are used to generate augmented video samples. The input clip is resized from $12 \times 3 \times 300 \times 300$ to $12 \times 3 \times 100 \times 100$ and inputted to the model. Here 12, 3, 100 and 100 denote number of frames in the clip, number of channels in each frame, height and width of video respectively.

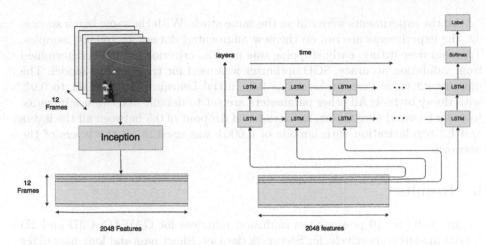

Fig. 6. Block diagram of the Inception model

Fig. 7. The 2 layered LSTM architecture, which works on the features obtained from the Inception model

To avoid over-fitting, early stopping was used as criterion, which is determined from validation accuracy. Stochastic Gradient Descent (SGD) optimizer with Nestrov Momentum parameter set to 0.9 was used to train the model. Initial Learning rate was set to 0.005, with a decay of 1e-4. All other parameters are set to the default values of SGD optimizer in keras. To avoid over-fitting, dropout of 0.8 was introduced between fully connected layers of network, L2 weight regularization was used with lambda of 0.0001 in all the convolution layers and 0.00005 in fully connected layers of network.

5.2 Experiments on C3D

All the experiments on C3D were run on same stack with similar settings and hyper parameters.

5.3 Experiments on LSTM Architecture

For this model, with data augmentation, a total of 100,000 samples are generated with equal number of samples in each class. We use shifting, rotation, centering, flip etc. as techniques for augmentation. This dataset is fed as input to Inception model (Fig. 6) and features were extracted from the penultimate layer. These features are fed to two layered LSTM (Fig. 7) that has 400 LSTM cells in each layer.

All the experiments were run on the same stack. With the same batch size i.e. 32, the experiments are run on the new augmented dataset of 100,000 samples. To avoid over-fitting, early stopping was used as criterion, which is determined from validation accuracy. SGD optimizer was used for training this model. The momentum value here is also set as 0.9. Initial Learning rate was set to 0.01, with decay of 1e-4. All other parameters are set to default SGD values in keras. In order to avoid over-fitting, we introduced dropout of 0.5 between all the layers and L2 regularization with lambda of 0.0001 was used in all the layers of the network.

6   Results

Figures 8, 9 and 10 presents the confusion matrices for GAWAC, C3D and 2D LSTM models respectively, for Soccer-8k dataset. Short pass and long pass differ slightly with respect to the action of the player. The differentiation between the two actions becomes difficult if the motion of the ball is not noticed. This results in confusion in the models. Though the models are expected to perform well in Header class, the very few samples in the dataset resulted in improper learning for this class. Similar to Long and Short passes, the act of Turning is much similar to that of Trapping and Dribbling, and hence higher confusion in case of Turning samples.

Moreover, the common actions that are possible in every sample, such as running, jumping can impact the decision to an extent. This can often result in wrong classification of the sample. Cropping the video to a size of 300 × 300 around the point of event with an aim to capture the entire action has however resulted in the capturing of unnecessary information in the samples. This can have significant effect on the features learnt. Addition of Gaussian filter emphasized on the learning features around the ball and hence accuracy is higher in GAWAC.

The table below conveys the F1 scores of each model.

Model	F1 Score
C3D	46.02
Inception + LSTM	41.88
GAWAC with $\sigma^2 = 1.0$	55.72
GAWAC with $\sigma^2 = 0.5$	62.75
GAWAC with $\sigma^2 = 0.2$	52.8

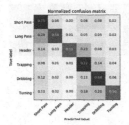
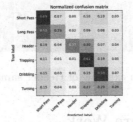
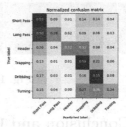

Fig. 8. Confusion matrix of GAWAC trained on Soccer-8k, $\sigma^2 = 0.5$

Fig. 9. Confusion matrix of C3D trained on Soccer-8k

Fig. 10. Confusion matrix of 2 layered LSTM trained on Soccer-8k

Fig. 11. An example of short pass sample

Fig. 12. An example of long pass sample

Fig. 13. An example of Turning sample

Fig. 14. An example of Trapping sample

Fig. 15. An example of Dribbling sample

Fig. 16. An example of Header sample

7 Conclusion and Future Work

We have presented a novel event based framework for fine grained action recognition in the sport of Soccer, *which is first of its kind*. We compared our model with the state-of-art techniques C3D and LSTM. The models are trained on the newly created custom dataset, Soccer-8k, that contains approximately 8000 samples and a total of 6 classes. In addition, our framework can recognize actions of players from live feed and can be used for stats generation. With slight modification, our framework can be generalized to other multi-player sports too. Moreover, the errors made by the models are because of the similarity in classes implying that the features they have learned are significant and are semantically meaningful.

This work can be extended for much finer action recognition, such as classifying an action into subtypes. For example a pass can be classified as either left or right footed pass or either outside or inside foot pass. This work opens new dimensions in analysis of soccer games without any need of additional infrastructure like electronic chips and can be afforded even by low budget clubs. This work with addition of multi player tracking and identification can be used for generation of textual commentary on fly and for text query based video search. Player statistics, team statistics and coordination among the team members are few things that can be extracted and analyzed.

References

1. Chen, L., Hoey, J., Nugent, C.D., Cook, D.J., Yu, Z.: Sensor-based activity recognition. IEEE Trans. Syst. Man Cybern. Part C (Appl. Rev.) **42**, 790–808 (2012)
2. Ji, S., Xu, W., Yang, M., Yu, K.: 3D convolutional neural networks for human action recognition. IEEE Trans. Pattern Anal. Mach. Intell. **35**, 221–231 (2013)
3. Tsunoda, T., Komori, Y., Matsugu, M., Harada, T.: Football action recognition using hierarchical LSTM. In: 2017 IEEE Conference on Computer Vision and Pattern Recognition Workshops (CVPRW), pp. 155–163. IEEE (2017)
4. Mora, S.V., Knottenbelt, W.J.: Deep learning for domain-specific action recognition in tennis. In: 2017 IEEE Conference on Computer Vision and Pattern Recognition Workshops (CVPRW), pp. 170–178. IEEE (2017)
5. Parmar, P., Morris, B.T.: Learning to score olympic events. arXiv preprint arXiv:1611.05125 (2016)
6. Simonyan, K., Zisserman, A.: Two-stream convolutional networks for action recognition in videos. In: Advances in Neural Information Processing Systems, pp. 568–576 (2014)

7. Singh, B., Marks, T.K., Jones, M., Tuzel, O., Shao, M.: A multi-stream bi-directional recurrent neural network for fine-grained action detection. In: Proceedings of the IEEE Conference on Computer Vision and Pattern Recognition, pp. 1961–1970 (2016)
8. Bagautdinov, T., Alahi, A., Fleuret, F., Fua, P., Savarese, S.: Social scene understanding: end-to-end multi-person action localization and collective activity recognition. arXiv preprint arXiv:1611.09078 (2016)
9. Chen, S., et al.: Play type recognition in real-world football video. In: 2014 IEEE Winter Conference on Applications of Computer Vision (WACV), pp. 652–659. IEEE (2014)
10. Ibrahim, M.S., Muralidharan, S., Deng, Z., Vahdat, A., Mori, G.: A hierarchical deep temporal model for group activity recognition. In: Proceedings of the IEEE Conference on Computer Vision and Pattern Recognition, pp. 1971–1980 (2016)
11. Maksai, A., Wang, X., Fua, P.: What players do with the ball: a physically constrained interaction modeling. In: Proceedings of the IEEE Conference on Computer Vision and Pattern Recognition, pp. 972–981 (2016)
12. Wang, H., Schmid, C.: Action recognition with improved trajectories. In: Proceedings of the IEEE International Conference on Computer Vision, pp. 3551–3558 (2013)
13. Yue-Hei Ng, J., Hausknecht, M., Vijayanarasimhan, S., Vinyals, O., Monga, R., Toderici, G.: Beyond short snippets: deep networks for video classification. In: Proceedings of the IEEE Conference on Computer Vision and Pattern Recognition, pp. 4694–4702 (2015)
14. Feichtenhofer, C., Pinz, A., Zisserman, A.: Convolutional two-stream network fusion for video action recognition. In: Proceedings of the IEEE Conference on Computer Vision and Pattern Recognition, pp. 1933–1941 (2016)
15. He, K., Zhang, X., Ren, S., Sun, J.: Deep residual learning for image recognition. In: Proceedings of the IEEE Conference on Computer Vision and Pattern Recognition, pp. 770–778 (2016)
16. Zhang, B., Wang, L., Wang, Z., Qiao, Y., Wang, H.: Real-time action recognition with enhanced motion vector CNNs. In: Proceedings of the IEEE Conference on Computer Vision and Pattern Recognition, pp. 2718–2726 (2016)
17. Feichtenhofer, C., Pinz, A., Wildes, R.: Spatiotemporal residual networks for video action recognition. In: Advances in Neural Information Processing Systems, pp. 3468–3476 (2016)
18. Dalal, N., Triggs, B.: Histograms of oriented gradients for human detection. In: IEEE Computer Society Conference on Computer Vision and Pattern Recognition, CVPR 2005, vol. 1, pp. 886–893. IEEE (2005)
19. Laptev, I., Marszalek, M., Schmid, C., Rozenfeld, B.: Learning realistic human actions from movies. In: IEEE Conference on Computer Vision and Pattern Recognition, CVPR 2008, pp. 1–8. IEEE (2008)
20. Wang, H., Kläser, A., Schmid, C., Liu, C.L.: Dense trajectories and motion boundary descriptors for action recognition. Int. J. Comput. Vision 103, 60–79 (2013)
21. Perronnin, F., Sánchez, J., Mensink, T.: Improving the Fisher kernel for large-scale image classification. In: Daniilidis, K., Maragos, P., Paragios, N. (eds.) ECCV 2010. LNCS, vol. 6314, pp. 143–156. Springer, Heidelberg (2010). https://doi.org/10.1007/978-3-642-15561-1_11
22. Gourgari, S., Goudelis, G., Karpouzis, K., Kollias, S.: THETIS: three dimensional tennis shots a human action dataset. In: Proceedings of the IEEE Conference on Computer Vision and Pattern Recognition Workshops, pp. 676–681 (2013)

23. Szegedy, C., Ioffe, S., Vanhoucke, V., Alemi, A.A.: Inception-v4, inception-ResNet and the impact of residual connections on learning. In: AAAI, pp. 4278–4284 (2017)
24. Secrets to Sports AS: 50 selected soccer skills and drills (2003)
25. Wikipedia: Association football – Wikipedia, the free encyclopedia (2017). Accessed 1 Dec 2017
26. Hochreiter, S., Schmidhuber, J.: Long short-term memory. Neural Comput. **9**, 1735–1780 (1997)
27. Chollet, F., et al.: Keras (2015)

Towards Automatic Crack Detection by Deep Learning and Active Thermography

Ramón Moreno[✉][iD], Eider Gorostegui-Colinas[iD], Pablo López de Uralde[iD], and Ander Muniategui[iD]

LORTEK, Arranomendia Kalea 4A, 20240 Ordizia, Spain
rmoreno@lortek.es
http://www.lortek.es

Abstract. Metal joining processes are crucial in current technological devices. To grant the quality of the weldings is the key to ensure a long life cycle of a component. This work faces crack detection in Electron-Bean Welding (EBW) and Tungsten Inert Gas (TIG) weldings using Inductive Thermography with the aim to substitute traditional Non-Destructive Testing (NDT) inspection techniques. The novel method presented in this work can be divided up into two main phases. The first one corresponds to the thermographic inspection, where the thermographic recordings are reconstructed and processed, whereas the second one deals with cracks detection. Last phase is a Convolutional Neural Network inspired in the well-known VGG model which segments the thermographic information, detecting accurately where the cracks are. The thermographic inspection has been complemented with measurements in an optical microscope, showing a good correlation between the experimental and the prediction of this novel solution.

Keywords: Deep learning · Cracks detection · Image segmentation · Thermography · Induction · EBW · TIG · NDT

1   Introduction

Cracks detection is key in many materials, having special relevance metals. Many structures on buildings and security devices are made with metals. To detect the good quality of welds on manufactured devices and building structures is a very important task. Cracks can grow and lead to complete fracture of the component posing significant threats to component life and may lead to serious injuries or loss of life. Fractures in metals occurs with little or no visible warning. To discover of any kind of crack is key for immediate reparation in order to avoid the propagation of the fracture.

Several Non Destructive Testing techniques have been developed in the last decades: dye penetrant testing, magnetic particles, etc. As the name indicates, these techniques are characterized by inspecting components or samples without destroying them. However, they are usually are time consuming and user

© Springer Nature Switzerland AG 2019
I. Rojas et al. (Eds.): IWANN 2019, LNCS 11507, pp. 151–162, 2019.
https://doi.org/10.1007/978-3-030-20518-8_13

dependant. Among the wide group of NDT techniques, Active Infrared Thermography (IRT) is getting relevance in the last years because of the advances both in the infrared camera technology and in the computational world, which has lead to new automatic inspection methods that can be employed in industrial environments.

Active Thermography [1] is characterized by the evaluation of dynamic temperature changes. These changes can be produced in a controlled way by several excition sources like laser, lamps or induction heating. It intends to overcome some of the drawbacks of other traditional methods not only reducing the inspection time, but also providing clean and contactless inspections, something that for many industries is invaluable. Another important advantage of thermography is that this technique is robust enough to provide repeatability in the measurements and in the posterior defect detection, something that traditional inspection techniques will always lack [2–4].

This work aims to go a step forward on the automatic crack detection on Inconel 718 Electron Beam Welded (EBW) and Tungsten Inert Gas (TIG) Welded components, where EBW crack sizes can go down to 180 μm, which is currently at the limit of the resolution of the employed thermographic equipment. It is a continuation of our previous work [5], where it was shown that directional induction enlightens all existing cracks in a surface and a detection algorithm was proposed in the frequency domain. In this work a different method is used without the need of turning into this frequency domain. Instead, now we have added automatic segmentation by Convolutional Neural Networks (CNN) in the thermograms themselves.

Deep convolutional networks [6,7] have outperformed the state of the art in many research areas. It is well-known that one of the requirements for training these deep networks is the need of thousand annotated training examples. Unfortunately in many cases it is quite difficult to obtain a proper database of labeled examples so as to undertake a suited training. Nonetheless, there exist a collection of tricks which allows to overcome this lack. The most used is over-sampling [8] (also known as data augmentation), which consists into multiply the number of samples by synthetic techniques. Other technique known as fine tunning [9], that uses previously trained models and initialize the weights of the new network with the weights of previous one. Usually these networks requires less training because it is supposed that the old model is close to the solution. A last technique that we use in this work is transfer learning, which consist in to train a CNN with a database of images with long similarity to our problem.

This work exploits over-sampling, fine-tunning and transfer-learning so as to train CNN able to detect the cracks properly. At this moment and as far as we know, there not exist a public database of metal cracks samples with the ground true labeled, nonetheless there exist many data bases in the Internet with data of retinal vessels together with it annotation. There is a lot of research around Diabetic retinopathy because it affects to a large part of the world population. By chance, the retinal vessels shapes are very similar to the breaks covered in this study. That is why this work uses a data base of retinal images for training, and the results have been validated on true thermograms.

The introduced method on this paper can be summarized in few steps. The first one corresponds to the experimental measurement which consists of four thermographic recordings per zone. In the second step an unique image is generated from the four recordings, unifying all crack information in a single 2D image. Finally, the third step corresponds to the defect detection itself using deep convolutional networks. Figure 1 illustrates the method in a blocks chain diagram.

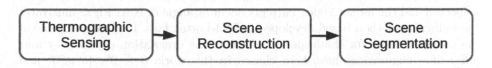

Fig. 1. Scheme of the cracks detection system. It starts with a thermographical sensing, afterwards a phase of scene reconstruction, an finally image segmentation for cracks detection.

Ongoing this paper is outlined as follow: Sect. 2 gives an introduction to the induction on thermography followed in this work. Section 3 explains the eye vessels similarity with the cracks and gives a description of the databases used for training. Section 4 talks about the fine-tuning carried out. Section 5 gives the experimental results and finally Sect. 6 ends with the conclusions and further works.

2   Induction Thermography

Induction thermography [10–12] or pulsed eddy current thermography uses electromagnetic pulses to excite eddy currents in electrically conductive materials. These currents give heat off by resistive losses and release heat. The heat or irradiance energy can be detected on the surface by a thermographical camera. Surface or underneath cracks cause local changes of the current electrical densities which are easily detectable on infrared captured images.

The tests consist on a short duration pulse (less than 300 ms) applied to the inductor, which is located on top of the region of interest The thermal camera is placed perpendicular to the specimen, at a distance about 110 mm, focalized on the analysis zone. The measurements are developed with a synchronization between the camera and the induction equipment, in order to obtain a thermal sequence coherent and relevant to the post analysis planned. The total length of the sequence is about one second because recording the cooling of the specimen is interesting for frequency domain analysis.

2.1   Inductor Orientation Importance on the Measurements

The inductor considered in the present work for the thermal excitation of the samples is orientation dependent. This means that the heating of a crack will be different depending on the relative orientation of the crack (α) considered and the inductor (β) (see Fig. 2(a)(b)). This affects the thermogram and moreover the irradiance behavior on cracks. Has been observed on cracks a clear difference with respect to the rest of the scene by using the fast Fourier transformation [5]. The optimal configuration occurs when the angle β of the inductor coincides with the perpendicular of the crack. In other words, the best configuration corresponds to the condition when the magnetic field is parallel to the crack to be inspected and the induced eddy currents occur at angle β. Next Fig. 2 illustrates the induction method based developed in IK4-Lortek labs. It shows a scheme of the prepared setup for the inspection. Due to the orientation dependency four different configurations have been chosen in this work: $\alpha = 0°$, $45°$, $90°$ and $135°$.

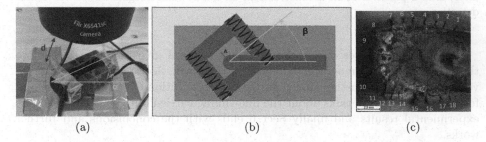

(a)   (b)   (c)

Fig. 2. Setup Overview. On the left, image (a) shows the TIG Sample with inductor at $\beta \neq 0°$. On the middle, sketch (b) illustrates a scheme showing the relative orientation of the inductor, β, and the crack, α. On the right, image (c) shows an example of macrography with cracks.

2.2   Scene Reconstruction

Four thermographic recordings are necessary to build a scene reconstruction. One for each considered induction direction. A *summary image* (frame) is selected containing all relevant information with respect to the cracks.

Starting from the original thermal sequence a background subtraction has been applied (first frame subtracted) in order to evidence the temperature changes on the cracks. In addition to this, the thermogram corresponding to the end of the induction pulse has been selected. Combining this two methods, the resulting thermogram has the maximum contrast between cracks and background sound zones. To prove the importance of the inductor orientation related to the cracks, an FFT has been applied to the thermal sequences [5]. By doing this, we can check that the "butterfly" signals appear on cracks parallel to the field direction (perpendicular to the Eddy currents generated) and not on the others.

The overall scene reconstruction is given by the average of all directional *image summaries*.

Next Fig. 3 illustrates a scene reconstruction. Given the four image summary obtained from each directional thermography, the sum of all of them is shown in (e). Notice that on (a) to (d) *summary images*, not all cracks are shown in every direction. The sum image (d) is the only one that contains all cracks existing in the observed material.

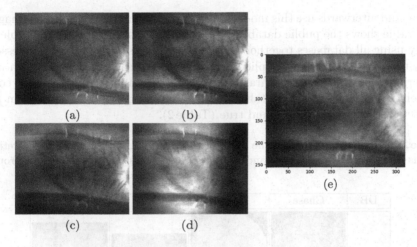

Fig. 3. Scene reconstruction. From the (a) to (d) the four captures corresponding to $\alpha = 0°, 45°, 90°$ and $135°$ orientations. Image (e) shows the sum of (a) to (d) captures.

3   Eye Vessels Similarity

At the moment it does not exist any public database with weld cracks images plus the annotations. To develop a database is rather difficult for a main reason: it is necessary to use a powerful microscopy together with an specific software so as to mark the cracks accurately. Given this lack, we have seek for public databases useful for our goal; cracks detection.

Medical image research is very active and many times are braking edge in a lot of fields. Nowadays private companies and public researchers are putting efforts facing the diabetic retinopathy disease [13]. One of the key steps to analyze retinal fundus images is to detect accurately the vessels. The literature is plenty of these works. Fortunately for our goals, there exist public databases of retinal images with the vessels marked.

In spite there are clear differences between cracks and vessels shapes, there are also many similarities as are the orientations and curves shape. Our goal here, is to train a segmentation model using the retinal images plus the vessels

Table 1. Public databases with retinal images. First column shows the number of images available, the second one the name of the database ant the third one a link to the sources.

#Images	db_name	Link
28	Chase	https://blogs.kingston.ac.uk/retinal/chasedb1/
45	HRF	https://www5.cs.fau.de/research/data/fundus-images/
40	Drive	https://www.isi.uu.nl/Research/Databases/DRIVE/
20	Stare	http://cecas.clemson.edu/~ahoover/stare/

marks, and afterwards use this model for cracks segmentation on welding images. Next table shows the public databases that we have used in this work (Table 1).

By using all databases together we have 133 images with the marked vessels. From these images we have applied a process of oversampling by rotation and mirroring which gives a final dataset up to one thousand images with the complementary ground true. Next table shows some examples of the images in the datasets together with the ground true (Table 2).

Table 2. Examples. First row is the name of the database, second row shows a retinal fundus image of every database and the third one shows the corresponding ground true.

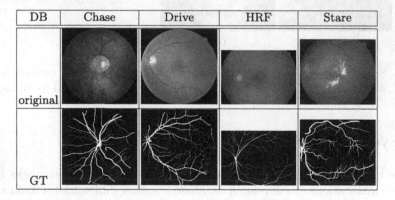

4 Transfer Learning from VGG Model

The third phase of this work relies on convolutional neural networks (CNN). Literature is plenty of works demonstrating the great capabilities of this computational approach for machine learning. It is also well-known that this kind of networks need huge amount of data in order to accomplish the training process properly. This requirement is far to be covered by our small image dataset. In order to overcome this shortcoming, we hold our CNN upon the VGG model [14] public available on the Internet[1]. The VGG model (with 16 layers) has been

---

[1] http://www.robots.ox.ac.uk/~vgg/research/very_deep/.

trained using up to 1.5 million images, the network has more the 1.3 million parameters and a weight of 528 MB. It is useful, in fact the literature is plenty of works based of these models [15–18]

Our proposal uses the same first layers: from conv1_1 to conv3_3. It allows us to use the VGG trained model weights, that is known in the literature as transfer learning [9]. Starting from this knowledge, we have added some convolutional layers to the VGG network, addressing then the network towards our needs. The structure is as follows: First nine layers (from conv1_1 to conv3_3) are exactly the same as VGG. The rest extend from some layers of the initial structure, adding an extra convolution plus a deconvolution and a crop in two extensions. The extensions hang on relu_1_2, relu_2_2 and relu_3_3. The three branches finally cover into a single branch though concatenation. This ending branch has convolution and a softmax function.

5   Experimental Results

This section gives the details of the experiments. First we show detail on training the CNN. Secondly, it is described the technicalities of thermographical sensing accomplished in this work. Afterwards we explain the particularities on model training, and finally we show the results. It follows the scheme shown in Fig. 1 explained at the introduction.

5.1   Training the CNN

The learning has been carried out with 30 epoch, a learning rate of 0.01 with optimization algorithm of stochastic gradient descent (SGD). Next Fig. 4 shows the training convergence for 30 epoch.

As for the training, it has been executed in a server with processor Intel Xeon with 12 Cores, 32G RAM and a NVIDIA QUADRO card. It took 37 min for 30 epochs and 19 h for 600 epochs. The best result is obtained after 600 epochs, where an accuracy of **0.96** and a loss of **0.08** are reached.

5.2   Thermography Sensing; Setup and Parameters

The employed inductor generator is a mid-frequency 3 kW generator. As for the inductors, the inductor designed at the research center itself has been employed, which despite not being optimized, has provided good experimental results.

A Flir Infrared Cooled Thermal Camera has been used to perform the recordings, specifically a Flir X6541sc with the standard 50mm optics and an extension ring, reaching a resolution of $45 - 50 \, m/pixel$. The frame rate has been set to 300 Hz for all measurements. Finally, integration time has been set to 400–700 μs.

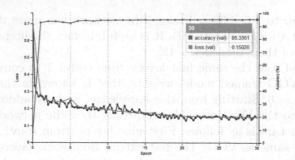

Fig. 4. Network training. The graph illustrates the training with 30 epoch. Orange: accuracy, green: loos in validation, and blue: loos in training. The main convergence is reached before the 10th epoch. After 30 epochs it got an accuracy of 95.33 (in percent) and a loss 0.15. (Color figure online)

During the inspection, the sample is scanned step by step, being the inspection window 20 mm × 20 mm approximately. Overlapping is always allowed in between the scans to avoid skipping any possible defects at the edges. Pulses from 100 ms to 200 ms have been employed. This interval has provided the best results together with a frequency that ranges from 10 to 30 KHz, depending on the type of sample considered. The employed setup is described schematically in Fig. 5.

For EBW samples best thermographic results have been obtained for pulses ranging between 100 ms and 200 ms. As for the induction frequency, the heating produced at 20 kHz, which is the highest, has been concluded to be the optimal. Integration time has been set to 700 µs in all measurements. In the case of TIG samples, the considered optimal pulses also range from 100 ms to 200 ms. However, the optimal induction frequency range is wider: 10 kHz to 35 kHz.

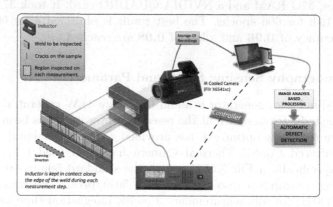

Fig. 5. Schematic figure showing the employed setup in order to inspect EBW and TIG Inconel samples using Induction Thermography.

5.3   Scene Segmentation

The proposed CNN network has been trained using a Intel(R) Xeon(R) CPU E5-2620 v4 @ 2.10 GHz with a Quadro M4000. We have trained the network over the vessels image database giving out an accuracy on training of 0.93 that is not quite far from the state of the art [19]. Notice that this measurement is pixel-wise and the background is the predominant class.

The experiment has been carried out following the method introduced in this paper:

1. Thermographic sensing: in this phase. We get a collection of 4 thermographic captures corresponding to $\alpha = 0°, 45°, 90°$ and $135°$ orientations.
2. Scene reconstruction: Given the *summary image* obtained from each thermography, the sum image is computed highlighting the cracks detected on every capture. At this point, we have computed the negative form (255 - *summary image*). We make this trick in order give an output similar to the vessels in fundus retinal images.
3. The model obtained from training the CNN over the retinal image database is tested on the negative version of the scene reconstruction.

Next Fig. 6 shows the experimental results on some scenes. On the first column an example of the four captures is shown in order to demonstrate that not

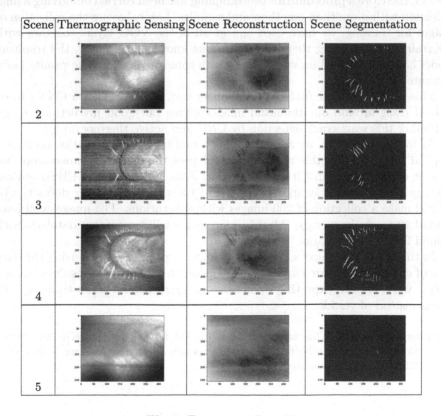

Fig. 6. Experimental results

all cracks appears in (cracks are corresponds to bright regions). In the middle columns, it is shown the negative version of the *summary image* (cracks are darker than background). Notice that this image contains all cracks existing in the observed metal. On the right is the scene segmentation.

The results are accurate, the CNN is detecting all craks existing in the observed surface of the experimental samples. On scenes 2, 3 and 4 the cracks are detected. 5th scene has no cracks, however the CNN has detected small regions as possible cracks, which is a clear case of false positives. These results have been analyzed in IK4-Lortek laboratory using an optical micorscope.

6   Conclusions and Further work

This work has proved that CNN are useful for automatic crack detection on inspections based on active thermography. The experimental outcomes give accurate crack detection even though it has been trained with a vessel database, instead of a crack database.

The main problem to accomplish this work emerge from the lack of labeled samples with welds and cracks. This drawback has been overcomed by using some retinal image databases available in the Internet. The sum of the images are 133, therefore a procedure for oversampling has been carried out giving a final collection with more than one thousand samples. Later, having the collection of images for training, we have used almost all of the VGG layers together with the trained model using therein the previous knowledge. Finally, the resulting model has been applied on weld images with some cracks, and the results looks accurate.

This work is just the first step in a longer way. It proofs that a CNN can be trained using around one thousand images giving out good predictions, so we will follow this way: crack detection by CNN and active thermography.

As for ongoing work, we are developing a tool for make annotation on microscopical images. With this annotation tool plus the electronic microscope we have at our facilities (an Ultra Plus de Carl Zeiss with EDS and EBSD sensors (Energy Dispersive X-ray spectroscopy, and Electron backscatter diffraction)), we will make a collection of 200 images with annotations. The images will have around 10 cracks in average. All this information will turn into a database with around 2000 labeled cracks.

In this way, in the next work we will train the proposed CNN using the data base of cracks that we are currently developing. Given the good results obtained in this work, we are expecting even more accurate results when training with cracks instead of vessels.

Akcnowledgements. This work has been funded by the project KK-2018/00104 (Departamento de Desarrollo Económico e Infraestructuras del Gobierno Vasco, Programa ELKARTEK-Convovatoria 2018).

References

1. Stankovičová, Z., Dekýš, V., Nový, F., Novák, P.: Nondestructive testing of metal parts by using infrared camera. Procedia Eng. **177**, 562–567 (2017). XXI Polish-Slovak Scientific Conference Machine Modeling and Simulations MMS 2016. September 6–8, 2016, Hucisko, Poland
2. Srajbr, C., Dillenz, A., Bräutigam, K.: Crack detection at aluminum fuselages by induction excited thermography. In: 4th International Symposium on NDT in Aerospace 2012 - Th.3.A.3 (2012)
3. Srajbr, C.: Induction excited thermography in industrial applications. In: 19th World Conference on Non-Destructive Testing 2016 (2016)
4. Netzelmann, U., Walle, G., Lugin, S., Ehlen, A., Bessert, S., Valeske, B.: Induction thermography: principle, applications and first steps towards standardisation. Quant. InfraRed Thermography J. **13**(2), 170–181 (2016)
5. Gorostegui-Colinas, E., Muniategui, A., de Uralde, P.L., Gorosmendi, I., Hériz, B., Sabalza, X.: A novel automatic defect detection method for electron beam welded inconel 718 components using inductive thermography. In: 14th Quantitative InfraRed Thermography Conference (2018)
6. Garcia-Garcia, A., Orts-Escolano, S., Oprea, S., Villena-Martinez, V., Martinez-Gonzalez, P., Garcia-Rodriguez, J.: A survey on deep learning techniques for image and video semantic segmentation. Appl. Soft Comput. **70**, 41–65 (2018)
7. Gu, J., et al.: Recent advances in convolutional neural networks. Pattern Recogn. **77**, 354–377 (2018)
8. Simpson, A.J.R.: Over-sampling in a deep neural network. CoRR, vol. abs/1502.03648 (2015)
9. Yosinski, J., Clune, J., Bengio, Y., Lipson, H.: How transferable are features in deep neural networks? CoRR, vol. abs/1411.1792 (2014)
10. Oswald-Tranta, B.: Thermoinductive investigations of magnetic materials for surface cracks. Quant. InfraRed Thermography J. **1**(1), 33–46 (2004)
11. Oswald-Tranta, B., Wally, G.: Thermo-Inductive Surface Crack Detection in Metallic Parts (2006)
12. Oswald-Tranta, B., Sorger, M.: Localizing surface cracks with inductive thermographical inspection: from measurement to image processing. Quant. InfraRed Thermography J. **8**(2), 149–164 (2011)
13. JM, T., TJ, G., RM, G.: Retinopathy. JAMA **298**(8), 944 (2007)
14. Simonyan, K., Zisserman, A.: Very deep convolutional networks for large-scale image recognition. CoRR, vol. abs/1409.1556 (2014)
15. Park, B., Matsushita, M.: Estimating comic content from the book cover information using fine-tuned VGG model for comic search. In: Kompatsiaris, I., Huet, B., Mezaris, V., Gurrin, C., Cheng, W.-H., Vrochidis, S. (eds.) MMM 2019. LNCS, vol. 11296, pp. 650–661. Springer, Cham (2019). https://doi.org/10.1007/978-3-030-05716-9_58
16. Sengupta, A., Ye, Y., Wang, R., Liu, C., Roy, K.: Going deeper in spiking neural networks: VGG and residual architectures. Front. Neurosci. **13**, 95–95 (2019)

17. Wu, S., Liu, C., Wang, Z., Wu, S., Xiao, K.: Regression with support vector machines and VGG neural networks. In: Hassanien, A.E., Azar, A.T., Gaber, T., Bhatnagar, R., F. Tolba, M. (eds.) AMLTA 2019. AISC, vol. 921, pp. 301–311. Springer, Cham (2020). https://doi.org/10.1007/978-3-030-14118-9_30
18. Mateen, M., Wen, J., Nasrullah, Song, S., Huang, Z.: Fundus image classification using VGG-19 architecture with PCA and SVD. Symmetry **11**, 1 (2019)
19. Liskowski, P., Krawiec, K.: Segmenting retinal blood vessels with deep neural networks. IEEE Trans. Med. Imaging **35**, 2369–2380 (2016)

Optimization of Convolutional Neural Network Ensemble Classifiers by Genetic Algorithms

Miguel A. Molina-Cabello[✉], Cristian Accino, Ezequiel López-Rubio, and Karl Thurnhofer-Hemsi

Department of Computer Languages and Computer Science, University of Málaga, Bulevar Louis Pasteur, 35, 29071 Málaga, Spain
{miguelangel,ezeqlr,karlkhader}@lcc.uma.es, cristian.accino@gmail.com
http://www.lcc.uma.es/~ezeqlr/index-en.html

Abstract. Breast cancer exhibits a high mortality rate and it is the most invasive cancer in women. An analysis from histopathological images could predict this disease. In this way, computational image processing might support this task. In this work a proposal which employs deep learning convolutional neural networks is presented. Then, an ensemble of networks is considered in order to obtain an enhanced recognition performance of the system by the consensus of the networks of the ensemble. Finally, a genetic algorithm is also considered to choose the networks that belong to the ensemble. The proposal has been tested by carrying out several experiments with a set of benchmark images.

Keywords: Breast cancer classification · Medical image processing · Convolutional neural networks

1 Introduction

Medicine fields are being enhanced by employing digital image processing. These images are obtained in medical test such as X-ray image, ultrasound image and resonance imaging, among others. According to this information, digital image processing facilitates the analysis of the medical images due to an improvement of them by emphasizing the parts where medical staff focus on. In addition, image processing can be used in order to predict a disease. In this way, a system like this kind could detect an illness by processing an image input. In fact, image processing is essential for pathology detection [3, 8, 12, 16].

An example of the application of image processing could be found in blood sample images obtained in laboratory by microscopy. The hematocrit is the percentage occupied by red blood cells in relation to the total blood. Early detection of several diseases, like anemia, can be indicated by a decrease or growth of the hematocrit value. The image processing supports the analysis of blood images by counting the red blood cells. Several model kinds can be used to this purpose [2, 7, 9].

© Springer Nature Switzerland AG 2019
I. Rojas et al. (Eds.): IWANN 2019, LNCS 11507, pp. 163–173, 2019.
https://doi.org/10.1007/978-3-030-20518-8_14

Among all types of cancer, breast cancer is the most invasive cancer in women and presents a high mortality rate. Histopathological analysis is currently the most widely used method for breast cancer diagnosis. Thus, automatic classification of histopathological images can help health professionals to diagnose breast cancer more quickly and effectively. A significant breakthrough in this field was the collection of over 7,900 histopathological image samples, which formed the BreakHis dataset [14], as the previous automated histopathology image recognition systems had the limitation of working with small datasets. One of the popular kinds of image recognition is based on the visual feature descriptors to identify patterns on the image [14]. On the other hand, the recent deep learning schema can also be applied to detect and classify the desired parts of the image. In this way, convolutional neuronal networks can be applied to this purpose, for example [13]. In that work, a special kind of Convolutional Neural Network (CNN) has been used, namely AlexNet [5]. It has been previously used in a wide range of fields like vehicle classification in traffic videos [10] as well as blood classification [9].

In this work we outperform the reference model from [13]. In addition, in order to enhance the performance of a network, an ensemble might be considered to this task. In this way, several networks provide their output and an improved output can be obtained by the consensus of the networks [10]. Finally, a genetic algorithm is also considered to choose the networks that belong to the ensemble.

The paper is structured as follows. Section 2 presents the methodology of the proposal, differentiating between the considered ensemble types and the genetic algorithm to choose the best possible option for the set of networks which comprise the ensemble. The experiments have been carried out performed in Sect. 3, where an optimisation parameter values process and the performance of the ensembles are reported. Finally, the conclusions are provided in Sect. 4.

2   Methodology

In this section we propose our ensemble methodology in order to improve the performance of Convolutional Neural Network (CNN) classifiers. Given M classes and N CNNs, i.e. N classifiers to be combined, let $\mathbf{y}_i$ be the output vector of the i-th CNN, with $i \in \{1, ..., N\}$. That is, y_{ij} is the predicted score for the j-th class by the i-th CNN, for $j \in \{1, ..., M\}$. Then a subset $\mathcal{S} \subseteq \{1, ...N\}$ of the CNNs can be chosen to form an ensemble of classifiers. Four possible ensemble types are considered:

– Maximum ensemble. The final score for a class is given by the maxima of the scores associated to that class:

$$\mathbf{z}_{Max} = \max \{\mathbf{y}_i \mid i \in \mathcal{S}\} \tag{1}$$

– Mean ensemble. The final score for a class is given by the arithmetic mean of the scores associated to that class:

$$\mathbf{z}_{Mean} = \mathrm{mean} \{\mathbf{y}_i \mid i \in \mathcal{S}\} \tag{2}$$

– Median ensemble. The final score for a class is given by the median of the scores associated to that class:

$$\mathbf{z}_{Median} = \text{median}\left\{\mathbf{y}_i \mid i \in \mathcal{S}\right\} \tag{3}$$

– Voting ensemble. The final score for a class is given by the number of times that it ranks the highest among the scores yielded by a CNN:

$$\mathbf{z}_{Voting} = \left(\left|\left\{i \in \mathcal{S} \mid j = \arg\max_{k \in \{1,\dots,M\}} \{y_{ik}\}\right\}\right|\right)_{j \in \{1,\dots,M\}} \tag{4}$$

where $|\cdot|$ stands for the cardinal of a set.

After the ensemble scores are computed, the predicted class is the one which attains the highest score. In order to choose the best possible option for the set of classifiers $\mathcal{S}$ which comprise the ensemble, we propose to use a genetic algorithm. Each individual has a chromosome made of N binary variables, which indicate whether a specific CNN belongs to the ensemble. The fitness function is the accuracy of the resulting ensemble, measured over a suitable validation set.

3 Experiments

The carried out experiments apply the optimization of the architecture of the neural network proposed in [13] and the fine-tuning of the parameter configuration of the trained model openly provided by the authors on [1]. Furthermore, the proposed ensemble methodology is applied as well as the genetic algorithm.

The structure of this section is as follows. First of all, Subsect. 3.1 shows the software and hardware that have been used. Then, the tested image dataset is specified in Subsect. 3.2. After that, the obtained results from the parameter configuration optimisation process are described in Subsect. 3.3. And finally, 3.4 exhibits the ensemble process results.

3.1 Methods

Caffe (Convolutional Architecture for Fast Feature Embedding) [4] is the open source deep learning framework chosen for carrying out the experiments. Written in C++, it is developed by Berkeley AI Research (BAIR)[1], as well as community contributors. Caffe provides GPU (Graphical Processing Unit) acceleration support with CUDA [11]. This is what really makes Caffe fast, as it takes advantage of the fact that images are floating-point matrices that can usually be processed across several computational nodes. In this way, a powerful enough GPU can dramatically accelerate the training of deep neural networks and even becomes vital in order to complete it in a reasonable time.

BAIR and its open community provide a repository of trained models, called Model Zoo, where there are models for a wide variety of purposes. Some of

---

[1] http://bair.berkeley.edu/.

the most well known networks can be found there, such as AlexNet [6] and GoogleNet [15]. As training a network from scratch is always a time-consuming process, sometimes these pre-trained models are taken as starting point. In cases when the purpose of the pre-trained network have something in common, it may be enough to just train some additional layer to achieve a good performance (sometimes even better than by starting from scratch). This technique is known as transfer learning and it is the reason why this kind of repository is very active and supported.

The architecture of the neural network to be optimised, comprises thirteen layers:

- A data layer that expects a LMDB containing 64 × 64 images. The batch size for training is set to 100.
- Three successive sequences of a convolutional layer, a pooling layer and a ReLU layer. This is a distinctive characteristic of AlexNet.
- Two fully-connected layers. The last of them, which outputs the class scores, is connected to the last layer, a loss layer with *softmax* activation.

All the experiments have been carried out on a 64-bit Personal Computer with an Intel Core i3 Processor (2 × 2.0 GHz), 4 GB RAM, NVIDIA GeForce GT 710 as GPU and standard hardware.

3.2   Dataset

In order to evaluate the proposed methodology, a subset from the BreakHist image dataset has been considered to test the approach. It can be downloaded from its website[2] upon request.

The dataset is composed of the images that were acquired at 40X magnification, which consists of 652 benign and 1,370 malignant tumor samples. In addition, we generate 1,000 random 64 × 64 patches for each image in both sets, as

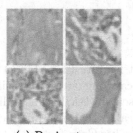
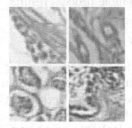

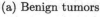

(a) Benign tumors   (b) Malignant tumors

Fig. 1. 64 × 64 pixel sample patch images used for training the neural networks. They have been generated from random images which belong to the BreakHis dataset. The random patch extraction is already applied on each displayed region. (a) exhibits benign tumors while (b) shows malignant tumors.

---

[2] https://web.inf.ufpr.br/vri/databases/breast-cancer-histopathological-database-breakhis/.

strategy #4 defined in [13] indicates. The resulting dataset is then split into training and test set, which account for 65% and 35%, respectively. Figure 1 exhibits several images from the dataset which show benign and malignant tumors.

3.3 Parameter Configuration Optimisation Process Results

In order to compare the performance of the system from a quantitative point of view a well-known measure has been selected: the accuracy (Acc). This measure provides values in the interval $[0, 1]$, where higher is better, and represents the percentage of hits of the system, i.e. recognition rate. The definition of this measure can be described as follow:

$$Acc = \frac{TP + TN}{TP + FP + FN + TN} \tag{5}$$

where TP is the True positives (number of hits), TN is the True negatives (correct rejections), FN is the False negatives (false alarms) and FP is the False positives (misses).

To quantify our goal, we test the reference model ten times on the test dataset generated. The mean accuracy shown by this model is 0.872 ± 0.015.

The fine-tuning of the parameter configuration is divided into two phases, as it is shown in Fig. 2. Each one of this two phases tests the performance of a network by training a network considering different parameters: the first phase tunes the base learning rate and the weight decay, while the second phase tunes the solver type and performs net surgery by either adding or removing a fully-connected layer, so the number of fully-connected layers is also considered. Table 1 exhibits the considered parameter values in this process. The first phase performs the parameter tuning on the reference model and the second one on the model achieving the highest mean accuracy in the first phase, called the first phase candidate model. Finally, the second phase produces the final candidate model, which is the model showing the best performance within such phase.

As training a single network is a quite time-consuming task, the number of iterations is limited to 10,000. Using a batch size of 100 images, there is enough information to extract a tendency of the performance and identify potential candidates.

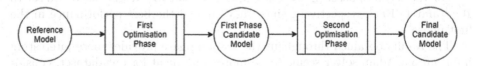

Fig. 2. CNN optimisation process diagram. Given the reference CNN model, we have tuned several parameter configurations in order to improve the performance of the approach. This process has two different steps: the first one involves the base learning rate and the weight decay, while the second step studies the solver type and the number of fully-connected layers.

Table 1. Considered parameter values in the CNN optimisation process.

Parameter	Values
Base learning rate	$\{10^{-4}, 10^{-3}\}$
Weight decay	$\{4 \cdot 10^{-5}, 4 \cdot 10^{-4}, 4 \cdot 10^{-3}\}$
Solver type	$\{\text{SGD, Adam}\}$
Number of fully-connected layers	$\{1, 2, 3\}$

Besides, 10 models are generated for each configuration so as to reliably measure the performance, since training is not deterministic. Therefore, 60 models are generated at the end of each phase. They will also be useful to apply ensemble learning later on.

Table 2 gathers all of the combinations tested within the first phase and the performance obtained by each of them in terms of mean accuracy. As it illustrates, a weight decay of $4 \cdot 10^{-3}$ clearly outperforms the rest, while the base learning rates tested do not present any relevant pattern. As a possible explanation of this, we understand that a too lower weight decay does not allow the model to converge fast, since weights can grow too large.

Table 2. Mean accuracy obtained by each parameter configuration tested within the first phase. Best result is highlighted in **bold**.

Base learning rate	Weight decay	Mean accuracy	Standard deviation
10^{-3}	$4 \cdot 10^{-3}$	0.846	0.024
10^{-3}	$4 \cdot 10^{-4}$	0.832	0.032
10^{-3}	$4 \cdot 10^{-5}$	0.840	0.020
10^{-4}	$4 \cdot 10^{-3}$	**0.847**	0.014
10^{-4}	$4 \cdot 10^{-4}$	0.842	0.020
10^{-4}	$4 \cdot 10^{-5}$	0.836	0.177

For the second phase, the base learning rate and the weight decay are set to 10^{-4} and $4 \cdot 10^{-3}$, respectively, since they showed the best performance in the first phase.

The results obtained throughout the second phase provides more interesting information. Adam solver seems to be more convenient for our dataset, considering that it surpasses SGD for each case. Furthermore, a single fully-connected is not enough to extract rich information, whereas incrementing the number of them does not necessarily means better class scores. Table 3 gathers these results.

Given the commented results, the final candidate configuration reaches an accuracy of 0.852 ± 0.011 at the 10,000th iteration and consists the parameter values which are shown in Table 4.

Table 3. Mean accuracy obtained by each parameter configuration tested within the second phase. Best result is highlighted in **bold**.

Fully-connected layers	Solver type	Mean accuracy	Standard deviation
1	SGD	0.840	0.017
1	Adam	0.843	0.019
2	SGD	0.847	0.139
2	Adam	0.851	0.120
3	SGD	0.843	0.021
3	Adam	**0.852**	0.011

Table 4. Considered parameter values after the CNN optimisation process.

Parameter	Value
Base learning rate	10^{-3}
Weight decay	$4 \cdot 10^{-3}$
Solver type	Adam
Number of fully-connected layers	3

Then, we resumed the training for the model with this configuration until the 50,000th iteration. The mean accuracy achieved at the 50,000th is 0.891 ± 0.020, which surpasses the 0.872 ± 0.015 obtained by the reference model on our test dataset.

3.4 Ensemble Process Results

Ensemble learning is a well-known technique used in machine learning to improve robustness over a single estimator. To do that, ensemble methods combine the predictions of several estimators trained with a given algorithm, as illustrated in Fig. 3. In this case, we have considered 30 networks.

Firstly, we built an voting ensemble comprising the generated neural networks with the final candidate configuration.

In order to measure its performance, we undertook 10 test iterations with a batch size of 100 using the test dataset. Here, each iteration performs a step forward on a model, which computes the class scores for each picture in the batch, and outputs the accuracy achieved on the batch.

The ensembles, with a mean accuracy of 0.863 ± 0.037, outperforms the mean accuracy obtained by a single model, 0.851 ± 0.033. This single model is the one within the ensemble that achieved the best performance in the second phase of the hyperparameter optimisation carried out. Besides, the ensemble shows higher performance at the vast majority of test iterations. However, these accuracy values are lower than those obtained in the second phase. This is basically due to the test images used this time, since the reference model reaches a lower

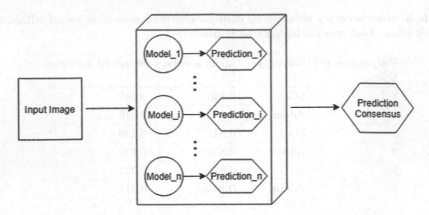

Fig. 3. Diagram of an ensemble. It receives an image as input and its output is a consensus on all the predictions generated by each model composing it.

mean accuracy value as well, 0.824 ± 0.047. Therefore, the ensemble represents a performance improvement of 4% over the reference model.

As it seems to be worthwile to apply ensemble learning, we now apply the genetic algorithm proposed in our methodology. In order to do so, we run the evolution of a population of 300 individuals over 1000 generations. Each individual is an array of 30 boolean elements. A value of 0 means that the neural network represented by the element is left out of the ensemble, whereas a value of 1 includes the neural network within the ensemble. The fitness function is the testing process previously mentioned, which measures the performance of the ensemble that the individual represents. Therefore, the fitness value of each element is the mean accuracy achieved by the ensemble of neural networks being represented. The aim of the evolution is to maximise the fitness value, whose maximum value is 1. If an individual reached such performance, the evolution would stop.

The first generation starts with 300 individuals generated randomly and each generation undertakes the following actions:

- At the beginning, each pair of individuals within the population may be crossed over with a probability of 0.5.
- Within a generation, each individual may mutate by flipping some of its elements with a probability of 0.2.
- At the end of each a generation, each evolved individual is evaluated by the fitness function. The next generation starts with the resulting individuals.

The performance of all of the ensemble types considered are shown in Table 5. As it can be observed, Mean Ensemble obtains the best performance of the considered ensembles. However, the performance is very similar among them.

The most significant result is the number of network that the best of each ensemble types uses. The Maximum and Mean ensembles have into account only 18 and 20 networks from the 30 available ones. Nevertheless, the Median

Table 5. Best individual after the evolution for each type of ensemble and its mean accuracy achieved. The ones mean the presence of the i-th network in the ensemble whereas the zeros mean the abscence (the number of used networks in the ensemble is shown within the brackets). Best result is highlighted in **bold**.

Ensemble type	Best individual	Mean accuracy
Maximum ensemble	1, 1, 1, 0, 1, 1, 1, 0, 1, 0, 0, 0, 1, 1, 1, 1, 0, 0, 1, 1, 0, 0, 0, 1, 1, 1, 1, 0, 1, 0 (18)	0.8610
Mean ensemble	1, 1, 1, 1, 0, 1, 1, 1, 0, 1, 1, 1, 1, 1, 1, 0, 0, 1, 0, 1, 1, 0, 0, 0, 1, 0, 0, 1, 1, 1 (20)	**0.8650**
Median ensemble	1, 1, 0, 1, 1, 1, 1, 1, 1, 0, 1, 1, 0, 1, 1, 1, 1, 1, 1, 1, 1, 1, 0, 1, 0, 0, 1, 1, 1, 1 (24)	0.8640
Voting ensemble	1, 1, 1, 1, 1, 1, 1, 1, 0, 1, 0, 1, 1, 1, 1, 0, 1, 1, 1, 1, 1, 1, 1, 1, 1, 1, 1, 0, 1, 1 (26)	0.8640

and Voting ensembles employ a higher number of the available networks. Thus, a lower computational and memory costs is achieved with the two first approaches.

4   Conclusion

An approach for the detection of breast cancer by using histopathological images has been presented in this work. This proposal is based on the use of the deep convolutional neural networks, in particular the Alexnet model. Given a model as a base, a two-phase process to optimise the parameter values has been proposed. According to the exhibited results, a not very low weight decay outperforms other higher considered options, while the tested base learning rates do not present any relevant pattern. In addition, Adam solver type works better than SGD type. Furthermore, the more fully-connected layers the architecture employs a higher performance is yielded.

Moreover, several ensemble types have been considered in order to enhance the performance of the system. A genetic algorithm is also applied to improve it. Although the different tested ensemble types provide a similar performance, Maximum and Mean ensembles employ a lower number of networks, so that, they need a lower memory and computational requirements. The obtained results indicate that the presented proposal is suitable to classify breast cancer images correctly with a high efficiency.

Acknowledgments. This work is partially supported by the Ministry of Economy and Competitiveness of Spain under grant TIN2014-53465-R, project name Video surveillance by active search of anomalous events and TIN2016-75097-P. It is also partially supported by the Autonomous Government of Andalusia (Spain) under projects TIC-6213, project name Development of Self-Organizing Neural Networks for Information Technologies; and TIC-657, project name Self-organizing systems and robust estimators for video surveillance. All of them include funds from the European Regional Development Fund (ERDF). The authors thankfully acknowledge the computer resources, technical expertise and assistance provided by the SCBI (Supercomputing and Bioinformatics) center of the University of Málaga. They also gratefully acknowledge the support of NVIDIA Corporation with the donation of two Titan X GPUs used for this research. The authors would like to thank the grant of the Universidad de Málaga.

References

1. Caffe models trained on the images of breakhist acquired with 40x magnification factor.   https://web.inf.ufpr.br/vri/databases/breast-cancer-histopathological-database-breakhis/. Accessed 25 May 2017
2. Bagui, O.K., Zoueu, J.T.: Red blood cells counting by circular hough transform using multispectral images. J. Appl. Sci. **14**(24), 3591–3594 (2014)
3. Davis, R., Boyers, S.: The role of digital image analysis in reproductive biology and medicine. Arch. Pathol. Lab. Med. **116**(4), 351–363 (1992)
4. Jia, Y., et al.: Caffe: convolutional architecture for fast feature embedding. In: MM 2014 - Proceedings of the 2014 ACM Conference on Multimedia, pp. 675–678 (2014)
5. Krizhevsky, A., Sutskever, I., Hinton, G.: ImageNet classification with deep convolutional neural networks. Adv. Neural Inf. Process. Syst. **25**, 1097–1105 (2012)
6. Krizhevsky, A., Sutskever, I., Hinton, G.E.: Imagenet classification with deep convolutional neural networks. Adv. Neural Inf. Process. Syst. **2**, 1097–1105 (2012)
7. Mazalan, S.M., Mahmood, N.H., Razak, M.A.A.: Automated red blood cells counting in peripheral blood smear image using circular hough transform. In: 2013 1st International Conference on Artificial Intelligence, Modelling and Simulation (AIMS), pp. 320–324. IEEE (2013)
8. McAuliffe, M.J., Lalonde, F.M., McGarry, D., Gandler, W., Csaky, K., Trus, B.L.: Medical image processing, analysis and visualization in clinical research. In: 2001 Proceedings of 14th IEEE Symposium on Computer-Based Medical Systems, CBMS 2001, pp. 381–386. IEEE (2001)
9. Molina-Cabello, M.A., López-Rubio, E., Luque-Baena, R.M., Rodríguez-Espinosa, M.J., Thurnhofer-Hemsi, K.: Blood cell classification using the hough transform and convolutional neural networks. In: Rocha, Á., Adeli, H., Reis, L.P., Costanzo, S. (eds.) WorldCIST'18 2018. AISC, vol. 746, pp. 669–678. Springer, Cham (2018). https://doi.org/10.1007/978-3-319-77712-2_62
10. Molina-Cabello, M.A., Luque-Baena, R.M., López-Rubio, E., Thurnhofer-Hemsi, K.: Vehicle type detection by ensembles of convolutional neural networks operating on super resolved images. Integr. Comput. Aided Eng. **25**(4), 321–333 (2018). https://doi.org/10.3233/ICA-180577
11. Nickolls, J., Buck, I., Garland, M., Skadron, K.: Scalable parallel programming with CUDA. Queue **6**(2), 40–53 (2008). https://doi.org/10.1145/1365490.1365500
12. Nogueira, P.A., Teófilo, L.F.: A multi-layered segmentation method for nucleus detection in highly clustered microscopy imaging: a practical application and validation using human u2os cytoplasm-nucleus translocation images. Artif. Intell. Rev. **42**(3), 331–346 (2014)
13. Spanhol, F.A., Oliveira, L.S., Petitjean, C., Heutte, L.: Breast cancer histopathological image classification using convolutional neural network. In: International Joint Conference on Neural Networks (IJCNN 2016), Vancouver, Canada, pp. 2560–2567 (2016)

14. Spanhol, F.A., Oliveira, L.S., Petitjean, C., Heutte, L.: A dataset for breast cancer histopathological image classification. IEEE Trans. Biomed. Eng. **63**(7), 1455–1462 (2016)
15. Szegedy, C., et al.: Going deeper with convolutions. In: Computer Vision and Pattern Recognition (CVPR) (2015). http://arxiv.org/abs/1409.4842
16. Vishnuvarthanan, A., Rajasekaran, M.P., Govindaraj, V., Zhang, Y., Thiyagarajan, A.: Development of a combinational framework to concurrently perform tissue segmentation and tumor identification in t1-w, t2-w, flair and mpr type magnetic resonance brain images. Expert Syst. Appl. **95**, 280–311 (2018)

One Dimensional Fourier Transform on Deep Learning for Industrial Welding Quality Control

Ander Muniategui(✉)⬤, Jon Ander del Barrio⬤, Xabier Angulo Vinuesa⬤,
Manuel Masenlle⬤, Aitor García de la Yedra⬤, and Ramón Moreno⬤

LORTEK, Arranomendia Kalea 4A, 20240 Ordizia, Spain
amuniategui@lortek.es
http://www.lortek.es

Abstract. This paper presents a method for industrial welding quality control. It focuses on the detection of Lack of Fusions (LoF) in joined parts produced in a rotational welding process. The solutions are based on the LeNet and AlexNet networks that are extended with previous convolutional layers based on 1D-pDFT (1D Polar Discrete Fourier Transform) and Gabor filters. The new layers add to the network the ability to deal with the images by means of knowledge arising from the physical process. In this paper a detailed description of the optical setup and the procedure to obtain defectives samples is also given.

Keywords: 1D Polar Discrete Fourier Transform · Deep learning · AlexNet · Gabor filter · Industrial application · Quality control

1 Introduction

In industrial production lines, quality control of produced parts is an important final step [1]. Advances in optical systems (cameras, lenses, illumination), processors and computational techniques have fostered visual inspection of produced parts in industry, even in processes with low defective rates (<1 s). In this respect, Convolutional Neural Networks (CNN)[2–4] have emerged as a powerful tool to ensure very high accuracy rates in image classification, hence being incorporated rapidly in industry for many different objectives such as defect detection [5], classification [6], or fault detection [7].

CNNs are designed to automatically learn spatial hierarchies of features by backpropagation, using multiple building blocks such as convolutional, pooling, Relu and fully connected layers. This convolutional approach for machine learning has been widely applied in computer vision, where the main applications are related with image segmentation [8] and image classification [9].

It is well-known that one of the requirements of Deep Learning algorithms is the need of large amounts of annotated data to train the models correctly [10]. Even though over-sampling methods [11,12] have shown to be a real alternative

© Springer Nature Switzerland AG 2019
I. Rojas et al. (Eds.): IWANN 2019, LNCS 11507, pp. 174–185, 2019.
https://doi.org/10.1007/978-3-030-20518-8_15

to create synthetic samples from real data. Anyway, human-based annotation is required for building datsets of labeled data. In many highly adjusted industrial processes with defective rates lower than 1% or even in the order of ppm (parts per million), it is difficult to generate a database with enough defective parts to train the models.

In this paper, a first approach for the quality control of parts produced within an industrial welding process with a defective rate lower than 1% is shown. The presented approach has been divided into three main steps. First, the optical system was developed and implanted in the productive line. Second, a few reference images of defects were obtained. These are used to develop a preliminary image analysis software for the detection of Lack of Fusion (LoF) type of defects that ensured no false negatives (false positive rate was sacrificed) serving to populate a database of defective parts. And third, once few thousands of defective parts are captured and manually curated, a balanced database was created and used to train deep learning models. Here, LeNet and AlexNet nets [13] have been extended with a pre-processing step that incorporates the developed image analysis method. The aim has been to decipher whether the incorporation of the pre-processing steps add relevant information concerning LoF type defects to the model.

The presented image analysis method is based on 1D-polar Discrete Fourier Transform [14] (1D-pDFT) filters and Gabor filters tuned for this application. 1D-pDFT filter banks produce wave-like patterns within LoF borders and Gabor filter banks allow their identification. As a result, LoF regions on the image are highlighted while the rest is blurred.

The paper is outlined as follows: Sect. 2 explains the methods, where Subsect. 2.1 describes the optical sensing system. Later Subsect. 2.2 introduces the 1D-pDFT and Gabor filters used, and Subsect. 2.3 indicates the modified LeNet and AlexNet networks. Afterwards, Sect. 3 shows the experimental results, and finally Sect. 4 ends with the conclusions.

2   Methods

2.1   System Description

A common feature of industrial manufacturing processes is the low rate of produced defectives parts, what makes complex the creation of a proper database. In the use case of this work, the defective rate is lower than 1%.

Produced parts are obtained by inserting two metallic pieces one inside the other and welding their circular contacting surface, i.e. the parts and their contact borders are rotationally symmetric. Correctly welded parts show non uniform regions, while defective parts are characterized for having circular shaped non welded regions (LoF). The inspection system and examples of correct and defectives parts are indicated in Fig. 1.

The inspection system for defective parts identification has been developed based on an optical system (lens, lighting and cameras), where the image analysis software makes a semi-supervised annotation discriminating between correct and

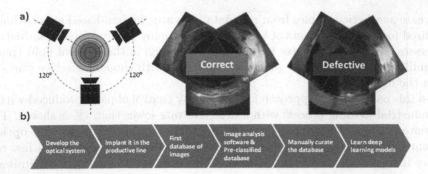

Fig. 1. Inspection system implanted in the production line to capture the whole welded region to inspect. The parts have rotational symmetry. Defects are characterized by Lack of Fusions (LoF) with circular borders.

defective pieces. The inspection system was developed in several consecutive steps, as shown in Fig. 1b.

1. First, the optical system has been implanted in the productive line. Since parts are rotationally symmetric, the system consists in three cameras at 120° at the same distance from the rotational symmetry edge of the produced parts. This system allows to capture three images of the whole welded region simultaneously every 2 s. The aim of this first step consisted of obtaining sufficient images of defectives parts so as to carry out a preliminary image analysis software.

2. Second, an image analysis software has been developed from a small set of reference images and implanted in the productive line. The objective was to carry out a preliminary classification of parts to considerably reduce the time required in the creation of the manually curated database to be used in deep learning model learning. The software was developed to retrieve no false negatives, result that was obtained by sacrificing the false positive rate.

3. Third, as a result of six months of data collection, more than one million parts were preliminary classified from which few thousands were pre-classified as potential real defectives. The obtained database of defective parts was then manually curated to obtain a reduced and balanced database of images.

4. Finally, deep learning models have been trained with the aim of considerably reducing the false positive rate. Here, the knowledge gained in the image analysis software was used to extend LeNet and AlexNet nets with an image pre-processing step.

2.2   Periodical Patterns and LoF Detection

The image analysis software has been developed with the aim of carrying out a preliminary classification of produced parts directly in the productive line. This classification serves to populate the database with the images of produced defectives and correct parts. Input images consist of three gray-scaled images per each produced part (one per camera). In order to develop the image analysis software, the circular shape of the LoF was exploited (see examples of LoF region in Fig. 2 (a) and (b) images labeled as "IN").

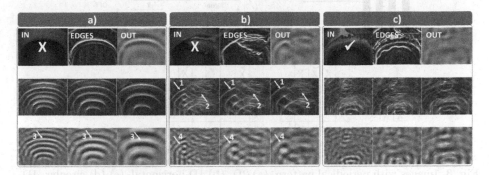

Fig. 2. Example of the detection of LoF regions. (a) not welded part, (b) partially welded part and (c) correctly welded part.

Having into consideration the rotational symmetry of the produced parts, the position of the cameras and the welding process, the sets of images captured by any of the cameras are very similar each other. Owed to this, a unique image (from the three captured) has been used for classification goals. The final classification of the produced part into correct or defective is obtained combining the three results.

The core of the application consists on the detection of the characteristic circular region of a LoF. For this aim two filter banks are generated: 1D-polar Discrete Fourier Transform-based filters and Gabor filters with specific parameter values (see Fig. 5).

1D Polar Discrete Fourier Transform: Fourier Transform (FT), with its discrete and fast variants, is extensively used in signal and image analysis. It allows extracting relevant signal features related with periodicity. In image analysis, 2D Fourier Transform is generally applied in texture detection, periodical 2D pattern recognition, image compression, etc. (see Fig. 3a) [15,16]. One dimensional Fourier Transform is also extensively used in 1D signal processing.

Beacuse of an image is a set of pixels with a bi-dimensional structure (discrete 2D), it can be analyzed as a 1D signal for pixels in a specific 1D paths, i.e. a line, a curve, a circle, a rectangle. Let consider the image shown in Fig. 3(b). Its vertical pattern can be analyzed by considering the intensity values of the pixels that lay within the dotted red line. The signal obtained this way corresponds to a square wave whose DFT (Discrete Fourier Transform) has well known properties (see the amplitude of the DFT a zero mean square wave in Fig. 3(d)).

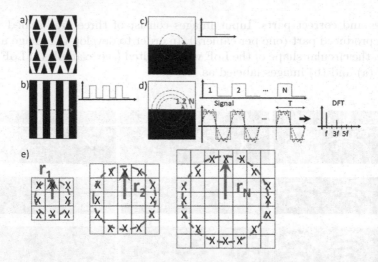

Fig. 3. Images with periodical pattern. (a) 2D, (b) 1D-horizontal, (c)-(d) circular. 1D-DFT can be used to analyze the periodicity of the signals obtained by traversing the images in (b) from the left to the right within a horizontal path, and (d) from the intersection of white and black paths within a circular path of any radii. Periodical signal obtained from (b) and (d) are shown in the right. A representation of the coefficients of the 1D-Discrete Fourier Transform of the signal of (e) is shown in the right.(Color figure online)

Similarly, square waves will also arise from boundary regions if circular paths centered in the boundary pixels are considered (see the red dotted line of Fig. 3(c)). Note, however that in this case, the signal consists of a unique period/cycle of the previous square wave. In order to obtain a square wave pattern, 1D signals derived from the circular path must be concatenated. Notice that concatenated signals must have the same number of points, i.e. the same number of pixels must be interrogated in case circular paths of different radii are considered.

As mentioned above, the characteristic feature of LoF is that they have circular borders (see "IN" image of Fig. 4). Owed to this, square wave signals are obtained if circular paths centered in the boundary region are considered to extract the 1D signal. These signals have a periodicity of 360° and hence their DFT will show a high amplitude at its equivalent frequency ($f_{360°}$). Note that if the value of the amplitude at $f_{360°}$ of the DFT for the 1D signals extracted for circles centered at every pixel of the image is determined, only those circles centered at border regions will show high values, i.e. any other 1D signal obtained this way will not be periodic. This operation can be easily implemented by convolutional operations.

Let us consider the circular filters for radii r_1, r_2 and r_N shown in Fig. 3(e). These are constructed as follows:

– For radius r a set number of points of its perimeter are determined

- The equivalent discrete circle is determined, i.e. a 2D discrete matrix is obtained
- The non zero pixels of the discrete circle is traversed anti-close-wise and the coefficients of the DFT for $f_{360°}$ are consecutively assigned

Note that the obtained matrix corresponds to a filter of size $(2 \times r + 1, 2 \times r + 1)$ with non zero values at pixels that correspond to the discrete circle. Non zero values are those of the coefficients of the DFT for $f_{360°}$. Note also that the above steps allow creating filters for combinations of different radii, i.e. required to obtain a periodical signal with more than one cycle within boundary regions, see Fig. 3(d).

In the following, these filters will be referred as 1D-pFFT filters. Here, these filters are used to find edges and wave-like patterns associated to LoF regions.

Some examples of the edges found using 1D-pFFT filters is shown in Fig. 2. Image is labeled as "EDGES".

2.3   Network Description

A preliminary image analysis software was developed for a first classification of produced parts. Afterwards, this method was incorporated into LeNet and AlexNet nets as pre-processing steps.

Preliminary Image Analysis Method for LoF Detection. A scheme of the developed image analysis software for LoF detection is shown in Fig. 4.

The first step consists of determining the edges of the input image using a 1D-pFFT filter of radius 4. On the 2nd, 3rd and 4th steps a wave-like pattern is forced in LoF regions using a set of 1D-pFFT filters and afterwards, a Gabor Filter bank is applied in order to search this pattern. Finally, in the 5th step, the output of all Gabor Filters are summarized into a unique image determining the mean image. As a result, images that show LoF patterns show wave-like patterns while the rest do not. See the output images obtained for these steps for two input images with LoF and an image of a correct part shown in Fig. 2. In the first row of (a), (b) and (c) the input image, the edges found using 1D-pFFT filters and the output result are indicated. Note that in the output image only those borders associated to LoF are highlighted. In the second row of (a), (b) and (c) the results of applying three different 1D-pFFT filters with different radius values to the detected borders shown in the first row are indicated. Observed that within these three images wave-like patterns are only obtained nearby the LoF regions (see white arrows), the rest is blurred. Finally, the third row shows the output obtained when Gabor filters are applied to the output images of the second row. Note that wave-like patterns are the only regions that are kept highlighted. Image annotated as "OUTPUT" is obtained by just getting the mean image of these last three images. In these images the regions that are highlighted in the three image of the third row are shown also as highlighted, the rest regions are blurred, i.e. LoF regions are identified. The complete cascade of filters is better described in Fig. 4.

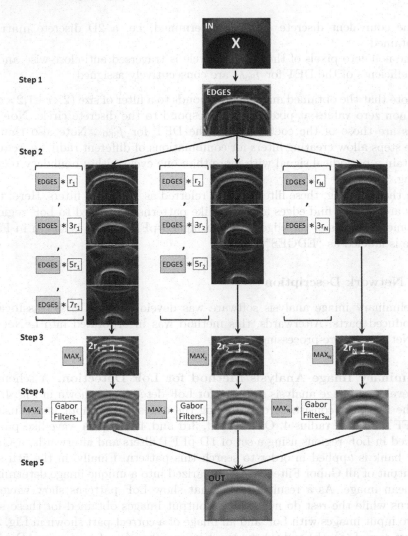

Fig. 4. Scheme of the image analysis method developed to detect LoF regions. Step 1: border detection using 1D-pFFT or radius 4. Step 2: cascade of sets of 1D-pFFT filters of sizes r_i, $3r_i$, $5r_i$ and $7r_i$. Step 3: each output is normalized and the maximum per pixel is determined to obtain the wave-like pattern. Step 4: Gabor filter bank is applied and Step 5: common "active" regions are identified by determining the mean per pixel of every output of the previous step.

1D-pDFT Filter. In order to get the wave-pattern a specific set of 1D-pDFT filters must be applied. Let consider the image of a non welded part shown in Fig. 4. The resulting image obtained when a 1D-pFFT filter of radius r_1 is applied to the image with the detected edges is shown in the upper left most image of Fig. 4-step 2. Note that two parallel curves are obtained. The distance between these two patterns is $2r_1$ (a distance of r_1 above and below the edge).

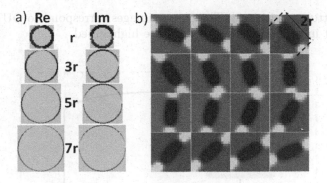

Fig. 5. 1D-pFFT (a) and Gabor (b) filters used for LoF type of defects detection.

Let now assume that a filter of radius $3r_1$ is used, the parallel curves appear at a distance of $6r_1$ (a distance of $3r_1$ above and below the edge). Considering that both results are summarized using the maximum per pixel into a unique image, then, the resulting image will have four parallel curves at a distance of $2r_1$ between consecutive ones. If other filters of radius $5r_1$ and $7r_1$ are also considered, the summarized image obtained will be just the one shown on Fig. 4-step 3, i.e. wave-like patterns are observed. This procedure is carried out for different values of r. Hence, a set of 1D-pDFT filters is applied: $\{F(r_i), F(3r_i), F(5r_i), F(7r_i)\}$ for $i = 1, .., L$, being $F(r_i)$ the 1D-pDFT filter of radius r_i and L the number of sets. The real and imaginary values 1D-pDFT filters are shown in Fig. 5.

Gabor Filter Banks. As indicated, within the wave-like patterns the distance between consecutive curves is of $2r_i$. This step aim is to detect these patterns. The detection is performed using a set of Gabor filters. Each set looks like the matrix of filters shown in Fig. 5 and it is adjusted for a specific r_i of the 1D-pFFT filters used in the previous step. In fact, Gabor Filters have been adjusted to have their two peaks at a distance of $2r_i$ and 16 different orientations between $[0, \pi]$. If Gabor Filters with peak distance of $2r_1$ are applied to the output image of 1D-pDFT set $\{F(r_1), F(3r_1), F(5r_1), F(7r_1)\}$ then peaks in the middle of two consecutive curves will be obtained, i.e. consecutive curves are at a distance of $2r_1$. This procedure is applied for the output of every set of 1D-pDFT filters of radius r_i. The obtained outputs are shown in Fig. 5-step 4. In all of them there will be a common highlighted region: the region of the border that generates the wave-pattern. Hence, if all the outputs are summarized, this common region will be highlighted and not common ones will be blurred (see last image in Fig. 5).

Next Fig. 6 depicts the CNN architecture. The outputs of the image analysis steps are shown. Wave-like patterns are only observed in regions with LoF (see (a) and (b) on the figure). In (b) arrow 1 indicates the LoF region with wave-like patterns and arrow 2 points to a region without LoF and hence without a wave-like pattern. Finally, the part of figure (c) has no LoF and as a consequence no wave-like pattern is observed. The output images correspond to OUT in Fig. 2. Observe that in (a) and (b) LoF regions are highlighted.

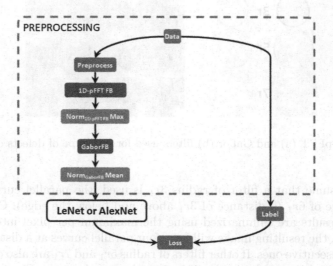

Fig. 6. Extended AlexNet net to incorporate 1D-pDFT and Gabor filters used in the image analysis method developed for LoF detection. Convolutional layers have been added to AlexNet net as a pre-processing step. 1D-pDFT and Gabor filters have been incorporated as convolutional layers with fixed weights and normalization and pixelwise maximum/mean operations as layers. Red patches correspond to convolution operations, orange to normalization results and gray to intermediate layers.

Incorporation of the LoF Image Analysis into LeNet and AlexNet Nets. In order to determine whether the image analysis scheme of Fig. 4 could serve as a relevant information for Deep Learning models, the following steps are incorporated into LeNet and AlexNet nets as a pre-processing scheme. (1) 1D-pDFT and Gabor Filters with fixed weights implemented as consecutive convolutional layers. (2) Normalization and sum steps ($\sqrt{real^2 + imaginary^2}$, $[0,1]$ interval normalization, maximum per pixel and mean per pixel) are added as layers. The complete net is shown in Fig. 6.

3   Experimental Results

LeNet, modified LeNet, AlexNet and modified AlexNet nets have been trained with a balanced database of correct and defective 3000 gray-scaled images captured from one of the cameras (C1-camera 1). Due to the configuration of the inspection system, the images captured by the three cameras are very similar. Although owed to the set up of the inspection system the sets of images obtained by the three cameras are supposed to be very similar between them, there are minimum differences related to the centering of the parts in the image and the lighting. Hence, a possible method to test whether pre-processing steps added to deep learning nets incorporate relevant information associated with LoFs is to train them for one the images of one of the cameras and test them with the images of the other two cameras. With this aim, the networks have been learned using a balanced database of images from camera 1 (C1) and afterwards trained models have been used for prediction on the images obtained by the cameras C2 and C3. Results are shown in Table 1. In this table the accuracy and loss values of the learned models are shown and the percentages of correctly predicted classes in C2 and C3 cameras are indicated (Defective (%) = percentage of correctly identified defective parts and Correct (%) = percentage of correctly identified non defective parts)

Table 1. Results for LeNet, modified LeNet, AlexNet and modified AlexNet. These have been learned with images of camera C1 and afterwards the trained models have been used for prediction for images of cameras C2 and C3. Defective (%) = percentage of defective welds (LoF) classified as defectives. Good (%) = percentage of correct welds classified as good.

Net	Learning		Prediction			
	C1		C2		C3	
	Accuracy	Loss	Defective (%)	Good (%)	Defective (%)	Good (%)
LeNet	99.5	0.023	**95.25**	4.88	72.95	42.01
LeNet+1D-pDFT	97.04	0.105	90.51	**79.07**	**94.96**	**61.72**
AlexNet	99.37	0.018	**98.09**	74.68	**98.31**	62.10
AlexNet+1D-pDFT	97.37	0.078	97.71	**80.22**	97.77	**67.37**

Results show that in both cases, the original LeNet and AlexNet nets have higher accuracy and lower loss values than modified LeNet and AlexNet nets (99% vs 97% and 0.02 vs 0.1 respectively). However, prediction results are worse for original LeNet and AlexNet than for their modified counterparts. In fact, in most of the cases the percentages of correctly predicted classes are higher for the modified nets. In particular, the percentages of correctly classified non defective parts is much higher if 1D-pDFT layers are added to the model, i.e. 5% vs 79% and 42% vs 62% in LeNet and LeNet-1D-pDFT for C2 and C3 and 74% vs 80% and 62% vs 67% in AlexNet and AlextNet-1D-pDFT for C2 and C3 respectively.

These results show that adding pre-processing layers reduce False Positive rates with a minimal impact in False Negative rates. Results are preliminary and require further work, however, they seem encouraging.

4   Conclusions and Further Work

In this paper an inspection system for the quality control of parts produced in an industrial welding process with a defective rate less than 1% has been presented.

Results showed that adding pre-processing steps - 1D-pDFT and tuned Gabor filters - allow reducing False Positive rates in both nets (better results are obtained for LeNet than for AlexNet). Although results are preliminary, they are encouraging since pre-processing steps seem to improve the transference capabilities, reduces false positive rate, of the model to other similar scenarios while keeping the accuracy and true positive rates.

As further work we plan to reduce the sizes of 1D-pDFT filters by separating the geometric information and DFT coefficients into different convolutional layers. This should reduce the amount of required memory considerably.

Acknowledgments. This work has been funded by the project KK-2018/00104 (Departamento de Desarrollo Económico e Infraestructuras del Gobierno Vasco, Programa ELKARTEK-Convovatoria 2018).

Contributions. AM, JADB and RM developed the image analysis software and contributed mainly in the writing of the paper. XAV, MM and AGDY developed the inspection system (optics, lighting and automation). AM and JADB made the manual curation for classification by visual inspection of captured images.

References

1. Ferguson, M., Ak, R., Tina Lee, Y.-T., Law, K.H.: Detection and Segmentation of Manufacturing Defects With Convolutional Neural Networks and Transfer Learning, vol. 2, September 2018
2. Sharma, N., Jain, V., Mishra, A.: An analysis of convolutional neural networks for image classification. Procedia Comput. Sci. **132**, 377–384 (2018). International conference on computational intelligence and data science
3. Garcia-Garcia, A., Orts-Escolano, S., Oprea, S., Villena-Martinez, V., Martinez-Gonzalez, P., Garcia-Rodriguez, J.: A survey on deep learning techniques for image and video semantic segmentation. Appl. Soft Comput. **70**, 41–65 (2018)
4. Gu, J., et al.: Recent advances in convolutional neural networks. Pattern Recogn. **77**, 354–377 (2018)
5. Jung, S.Y., Tsai, Y.H., Chiu, W.Y., Hu, J.S., Sun, C.T.: Defect detection on randomly textured surfaces by convolutional neural networks. In: 2018 IEEE/ASME International Conference on Advanced Intelligent Mechatronics (AIM), pp. 1456–1461, July 2018
6. Xuan, Q., Fang, B., Liu, Y., Wang, J., Zhang, J., Zheng, Y., Bao, G.: Automatic pearl classification machine based on a multistream convolutional neural network. IEEE Trans. Ind. Electron. **65**, 6538–6547 (2018)
7. Kiranyaz, S., Gastli, A., Ben-Brahim, L., Alemadi, N., Gabbouj, M.: Real-time fault detection and identification for MMC using 1D convolutional neural networks. IEEE Trans. Ind. Electron. 1–1 (2018)
8. Liu, F., Lin, G., Shen, C.: CRF learning with CNN features for image segmentation. Pattern Recogn. **48**(10), 2983–2992 (2015). Discriminative feature learning from big data for visual recognition
9. Wang, J., Yang, Y., Mao, J., Huang, Z., Huang, C., Xu, W.: CNN-RNN: a unified framework for multi-label image classification. In: The IEEE Conference on Computer Vision and Pattern Recognition (CVPR), June 2016

10. Lake, B.M., Salakhutdinov, R., Tenenbaum, J.B.: Human-level concept learning through probabilistic program induction. Science **350**(6266), 1332–1338 (2015)
11. Simpson, A.J.R.: Over-sampling in a deep neural network. CoRR, vol. abs/1502.03648 (2015)
12. Ando, S., Huang, C.: Deep over-sampling framework for classifying imbalanced data. CoRR, vol. abs/1704.07515 (2017)
13. Krizhevsky, A., Sutskever, I., Hinton, G.E.: Imagenet classification with deep convolutional neural networks. In: Proceedings of the 25th International Conference on Neural Information Processing Systems, NIPS 2012 (USA), vol. 1, pp. 1097–1105. Curran Associates Inc. (2012)
14. Averbuch, A., Coifman, R., Donoho, D., Elad, M., Israeli, M.: Fast and accurate polar fourier transform. Appl. Computat. Harmonic Anal. **21**(2), 145–167 (2006)
15. Watkins, P., Kao, J., Kanold, P.: Spatial pattern of intra-laminar connectivity in supragranular mouse auditory cortex. Front. Neural Circuits **8**, 15 (2014)
16. Stephant, N., Rondeau, B., Gauthier, J.-P., Cody, J., Fritsch, E.: Investigation of hidden periodic structures on SEM images of opal-like materials using FFT and IFFT. Scanning **2014**(36), 487–499 (2014)

A Serious Game to Build a Database
for ErrP Signal Recognition

Adam Pinto[1]($\boxtimes$), Guilherme Nardari[1], Marco Mijam[2], Edgard Morya[3],
and Roseli Romero[1]

[1] Institute of Mathematics and Computer Sciences, USP, São Carlos, Brazil
{adam,guinardari,rafrance}@icmc.usp.br
[2] São Carlos School of Engineering, USP, São Carlos, Brazil
marco.mijam@usp.br
[3] Edmond and Lily Safra International Neuroscience Institute (IIN-ELS),
Santos Dumont Institute, Macaiba, Brazil
edgard.morya@isd.org

Abstract. Brain wave signals allow communication between user and computer in a system called Brain-Computer Interface. Signal processing can detect attention, engagement, and errors in a task. Error-Related Potentials (ErrP) can be extracted from brain signals with noise, however, it is quite complicated to be recognized and accurate. This paper presents a new database, using gaming and a humanoid robot to induce the occurrence of user errors and methods to extract signal features. A Haar wavelet was used to feature extraction, and a MultiLayer Perceptron (MLP) and a Convolutional Neural Network (CNN) to signal classification. Several experiments are presented demonstrating that wavelet extraction outperformed Fourier transform to extract the error and MLP performed a consistent accuracy.

Keywords: Brain Computer Interface · Biomedical signals · Robotics

1 Introduction

The ability to focus on a task depends on several factors, both psychological and physical. Added to these factors is the mental load a task requires, which differs from more mechanical tasks, *i.e.* of repetitive effort, and more mental tasks such as study and research. The greater the mental workload of a task, greater is the possibility, over time, of a person making some mistakes.

Brain-Computer Interfaces (BCI) allow the signal from the user's brain to be interpreted by a machine. Brain signal can be acquired invasive (electrode implant) or non-invasively (surface electrode), and the signal processing can recognize brain states, diseases and some everyday actions. In education, it can detect student's mistakes and inform a teacher or even an educational device that a strategy has to change to improve student's learning and experience. This paper aims to detect participants error from brain signals. Error-Related

© Springer Nature Switzerland AG 2019
I. Rojas et al. (Eds.): IWANN 2019, LNCS 11507, pp. 186–197, 2019.
https://doi.org/10.1007/978-3-030-20518-8_16

Potential (ErrP) is extracted from electroencephalogram (EEG) and is related to error awareness. A game developed with PyGame induces errors over time, and after the game, the user was invited to talk to a humanoid robot that also induces errors. A database containing the information about user behavior during the game has been created to test the results obtained in previous works, and also to build a base with higher relation to education. From our knowledge, this is not found in the literature yet.

This preliminary study for ErrP recognition compared two techniques for features extraction and a neural network, Multilayer Perceptrons (MLP), classified the signal. The hypothesis is that the database creates behaviors in the same way as previously available databases on the Internet and the results of a recognition methodology already applied is close to each other.

This paper is divided as follows. In Sect. 2 concepts, such as the form of obtaining the signal and the devices are presented. In Sect. 3 the literature review related to this research. In Sect. 4, it is presented our proposal database and in Sect. 5, are shown the experiments setup. In Sect. 6, are shown the results and discussion. Finally, in Sect. 7, the conclusion and future works are presented.

2 Obtain EEG Information

Electrophysiological activities of neuronal groups can be recorded through electrodes, either invasive or noninvasive, and processed on a computer. Electroencephalogram uses noninvasive electrodes positioned on the scalp. This technique lacks the need for surgery, but its signal gets noisier as it records information from populations of neurons. When well calibrated it can records populations of neurons to investigate brain diseases, seizures, motor imagery, engagement, and even error awareness [12].

The first challenge of a BCI system is the automatic interpretation of the EEG signal and the non-localized occurrence of the components. ERD (Event-Related Desynchronization) and ERS (Event-Related Synchronization) are well-located components related to sensorimotor rhythms [10], and the ErrP, associated with personal errors.

Patterns of brain activity may be associated with specific activities and events, but users must undergo a process of training, perform mental exercises, imagine specific bodily functions so that the system can understand the particular pattern of each person. Factors such as age and gender influence the EEG signal, and some patterns may be shared at one age but be pathological in another.

Another factor that influences readings is the different positioning of the electrodes on the scalp, which has already generated comparative studies. The most common system used in the literature is the 10–20 system, proposed by [5], positioning to allow a more uniform coverage of the scalp. In Fig. 1, it is shown the positioning of the electrodes in this system, in which the letters indicate the lobe (F for Frontal, P for Parietal, C for Central, T for Temporal and O for Occipital). Odd numbers refer to left hemisphere, even to right hemisphere, and z relates to midline sagittal plane. The name 10–20 is derived from the total distance between the pairs of electrodes, ranging from 10% to 20%.

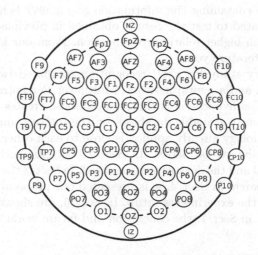

Fig. 1. International 10–20 position system [8].

3 Related Works

Learning has a cognitive load to understand and retain the subject. Understanding is best while the student remains focused, but various events, such as fatigue and difficulty, cause the student to make mistakes. Using a robot, researchers attempted to verify the cognitive load in children in collaborative storytelling context [15]. The experiments varied in two ways: insertion of a content related to the existing story and the insertion of a new story not related to the existing story. It was found that although the first strategy creates a heavier cognitive load, it was as enjoyable as the second. So, the workload was not related to the enjoyable the activity was.

Traditional teaching methods and new technologies available as tools for learning are discussed in a national context in the work [3]. Despite initial mistrust, students demonstrated a better focus on more complex content after insertion of educational robotics and had significant improvements in the tests. In a summary of 30 years of operation of the NIED (Nucleus of Informatics Applied to Education), an institute that specifically researches in education, d'Abreu and Loureno [4], reported a good students' acceptance of robotics and an improvement in learning in elementary schools. Using well-known robotic kits, with the main block replaced by an easy-to-program electronic board, the children and adolescents should apply the theoretical knowledge learned in the classroom, programming the Arduino to move the plastic parts of the robot. Experiments showed that even students who had little interaction with the robot improved their performance in school subjects.

The use of biomedical signs was used to test improvements in the safety of drivers, especially on roads, where driving time is higher. Since fatigue is strongly related to car accidents, in [9] it was created a wearable platform using EEG and Pulse sensors, microcontrollers, smartphones and Gaze tracking, named BioTracker©. To record the EEG, they used a *Neurosky Mindwave Mobile* and an Arduino UNO as microcontroller. In this research was made a combination of several sensors that, alone, were not able to predict accidents. Together, while providing more information to the system, improving the sensitivity to errors. The combination of techniques is important for ErrP research linked to education. Once the errors are detected, a system for the teacher's help also needs to understand the conditions of the student, trying to understand the reasons for the errors.

For an automatic detection of ErrP, in [17] three volunteers subjects should move a virtual robot to one side of a room (left or right) by delivering repetitive commands until the robot reaches the target. At each step, there was a 20% probability that the robot made a wrong direction movement. A Portable Biosemic ActiveTwo headset was in charge of capturing the signal of 32 electrodes scattered following the 10–20 system. The data was recorded at 512 Hz, then a bandpass filter was applied between 1–10 Hz and the signal was decreased to 128 Hz. Using a Gaussian classifier and verifying only the Fz and Cz channels, considered the best channels to find the ErrP record, an accuracy of 72% was obtained in the recognition of the errors during the experiment.

In a survey on BCI applied to gaming [6], some of the biggest challenges of BCI applications were presented. According to [6], gaming for BCI research has been neglected, because the data are sensitive to noise, depends on user mood and most of the EEG devices are expensive or generate data with many artifacts and noise. Most of papers in BCI to games focuses on well-know Event Related Potentials (like the P-300) and Evoked Potentials (as Steady-state Visually Evoked Potential - SSVEP). Medical areas, neurofeedback and motor imagery are more commun research subjects, being difficult to find research with educational applications.

4   Method and Material

The creation of the database involved the creation of a game to provoke the ErrP event and the interaction with a robot. For the game, two phases were created: one of calibration, when no error occurred, and another one of test, when the errors happened randomly. The following subsections decrypt each step of this process.

4.1   Proposed Game

A Pacman style game was developed based on PyGame to create the database. The player must move a zombie to eat each brain every 30 s. The goal is to eat most brains before time runs out. Since the proposed game does not have any other obstacle, a priori it is a fairly simple game and the player should not have problems to play for a long time. Thus, after 15 s, the game is programmed to randomly generate three types of errors:

- The zombie simply stops for 1 s, without executing any user-given commands;
- The zombie starts walking more slowly for 2 s, respecting the directions imposed by the user;
- The zombie moves randomly to a side other than the one indicated by the user, for 2 s. Each time the bug happens, the zombie can behave differently.

Since there is no statistical difference in the types of errors that the user experiences [13], the different types of errors were created only to have slightly different ErrP waves for neural network training.

4.2   Database Setup

To create the database for this work, 8 healthy individuals, being five men and three women, were called to play the proposed game[1]. All individuals were volunteers to participate, and they were not informed of the experiment's objectives. As the behavior of the brain is related to age [16], this variation was the only controlled variable, since all individuals had the same level of education and relatively the same knowledge of robotics and biomedical signs. All of them were masters or undergraduate students in neuroengineering.

The students were instructed not to talk to each other about the experiment until everyone finished its participation. The participants were taken to an isolated room, without noise or any other disturbance. Inside the room was only a computer, the robot, the subject and the researcher, who gave all the instructions. The subject sat and wore the EEG cap, washed and sterilized with alcohol before each experiment. It was used the BrainProducts V-AMP, with 16 channels, placing the ground in the left ear and the reference electrode in the right ear. Only passive electrodes was used. The process is exemplified in Fig. 2. Once the lower impedance phase was finalized, the user performed some tests, such as blinking the eyes and wiggling the fingers, to verify the signal. When the tests were all positive, the second step of the experiment was started, the calibration.

4.3   Calibration

In the calibration step, the user was only informed of how the game controls work. The 15-second count was started, and the user should be able to pick up the brains present in the game, but the clock was not increased at this step. On the game screen there was only a timer, a scratched scoreboard and brain-counting information. Since that was the first time playing, users scored low, but could understand how the mechanics of the game worked.

---

[1] Ethical Committee approval CAAE: 79649717.0.0000.5292.

Fig. 2. Researcher applying gel to improve signal impedance prior to start of EEG recording.

A second round of calibration was started with the same characteristics, but now the scoreboard showed the user's name and the number of brains collected in the previous round. The researcher then instigated the participant to score higher to be able to move on to the next phase. At the end of the calibration phase, whether the user reached the goal or not, a congratulation screen was displayed, with the words "Go bite the world" and a creep zombie laugh sound, indicating that the user would have been able to move to the second level. In this second phase, it was possible to do tests on the user's attention, since they should concentrate to be better than the previous round.

4.4 Playing the Game

For the real game phase, the user was warned of the rules change (30 s in the start time and increase of time for each brain eaten). At this stage, the game screen was basically the same with only the difference of the scoreboard, which indicated a name and a high score. The game interface can be seen in Fig. 3. The researcher warned that the goal of the player was to exceed that score to continue to a supposed phase three. The player was not warned of the bugs in the game and the score, a priori fairly easy to achieve, was unreachable due to these forced errors. The researcher had no further interaction with the player, even if he complained about constant bugs after a while. After the first flaw, the player was invited to try again, but this time it was explained that the bugs presented were programmed to annoy the player. This second round was set up to check if the ErrP would be different, since now that the user knew that errors could happen, but did not know when nor what error would exactly happen.

Fig. 3. Game interface on real game phase

In both scenarios a background music was smooth, and the bugs had distinctive sounds: a bug-like sound of operating systems when the zombie stopped and a growl of the zombie whenever it changed direction or slowed down. The countdown was counted in a macabre voice, while the clock turned red in the final five seconds. The fault screen was a graveyard with the words "You're not good enough" with the sound of the zombie being shot down. All elements of the game were created to instigate the player and imitate situations that can happen to students in times of study and tests: desperation, the need to act quickly and frustration. This information is important to see the behavior of brain waves at all times of the game.

4.5   Robot Interaction

Finished the game, the user was invited to participate in the third part of the experiment: interaction with a robot. For this step, was chosen the well-known robot NAO, of SoftBank Robotics, for being a humanoid robot with several resources for the interaction. For a better acceptance by users [1], the robot acted as naturally as possible, introducing itself and complaining about the weather.

Then, the robot called himself a freshman student of neuroengineering, and who would ask some questions to enhance its knowledge about the human brain and computer programming. In Fig. 4, are shown the phases of this experiment.

Fig. 4. Individuals participating in the experiment with the game and the robot.

The questions were divided into easy and difficult questions, considering the participants' knowledge. The easy questions were about brain information such as what are the brain lobes or whether the interaction with EEG was considered invasive or not. Because they were health care professionals, and because they also had already completed anatomy and biomedical signal processing classes, everyone was able to easily answer those questions. The same did not happen to the difficult questions that were about programming.

After asking the questions the robot gave alternative answers, which should be chosen by the participant. The robot had 3 behaviors:

- **Behavior 1:** the robot hears the answer and acts as it was expected: says that it is right if the users answer is correct and says that it is wrong if the users answer is wrong;
- **Behavior 2:** the robot hears the response and acts contrary to expectations: it says that it is right if the users answer is wrong and says that it is wrong if the users answer is right;
- **Behavior 3:** the robot hears the answer but it gets confused: if the user answers "letter b" the robot understands "letter d". In the group of easy questions, the robot always understood the wrong answer and in the group of difficult questions, the robot always understood the right answer.

Questions and behaviors were chosen randomly, ensuring only that all behaviors would happen and that the group of easy questions was completed before the difficult questions were asked. All questions should be answered, even if they did not know the correct answer.

5 Experiments

The 16 electrodes of the V-AMP were placed in the (Fpz, Fz, F8, FC3, FCz, FC4, C3, Cz, C4, CP3, P3, P4, P8, POz, O1, Oz) positions, but for ErrP recognition, the most important were FCz and Cz. These signals were bandpass filtered for eliminating artifacts and cleaning the signal. The zero-phase Butterworth IIR filter has been set to reject signals above 0.5 Hz and below 100 Hz.

One of the problems of the database created was the noise, like the blink of the eyes, that was manually removed. An important point is that in this dataset it is very well defined the moments in which any kind of error started and finished, making it easier to cut out the epochs of interest for the creation of the training set (Fig. 5).

To extract the features, 2 methods were considered: 8 layers Haar Wavelet [14] and Fast Fourier Transform (FFT) [2] from the raw input data. This methodology was the same used in [11]. It is important to emphasize that the database was created already extracting a 3 Haar Wavelet layers. However, to compare the results, this information was discarded using the raw signal.

In these experiments, we aim at demonstrating the viability of the proposed dataset for ErrP wave event detection tasks. For this reason, the models are trained and evaluated using two sessions from a single subject, considering 80 windows, being half of each class. The reported results correspond to the average of running a 10-fold cross validation three separate times with different random shuffling seeds. For the classification, a MultiLayer Perceptrons network (MLP) was used with the algorithm Resilient Backpropagation (R-PROP), using the hyperbolic tangent as activation function, fixing the number of hidden layers at 2 and changing the number of neurons in each layer. A second neural network, composed by convolutional and long-short-term memory layers, as described in Table 1 trained via binary cross entropy loss and Adam optimizer [7] was also trained in this dataset. The second model has the advantage of encoding the temporal information and both channels at each time while the multilayer perceptron receives the entire concatenated window as input.

6   Results

In Table 2 is shown the accuracy average of the 2 methods described in Sect. 5. For the comparison to the work in [11], the 8 layers of Wavelet (Haar complete) and only the final layer generated (Haar Final) were used. For the Fourier transform only one side (FFT) was used and both sides (RFFT). The dataset was normalized utilizing data from the training set only, subtracting the mean and scaling to unit variance. The reported accuracy corresponds to an average of 3 rounds of 10-fold cross-validation in errors obtained using the game session.

The automatic selection was not yet accurate, requiring a manual revision to create the test and training sets, but using the knowledge generated in previous works has helped the algorithm to choose the correct windows. The accuracy obtained was very close to the results of previous studies. However, this time a much larger set of data was used, which shows that full knowledge of the database helped to find the most important points.

As it was expected, the wavelet transform did a better features extraction, obtaining better accuracy results. The variation of MLP topologies was statistically insignificant, since a larger number of different networks were used and

Table 1. Architecture used in the experiments

Operation	Kernel size	Output	Activation
Input	-	2048,2	-
Conv	100	2948,32	ReLu
MaxPooling	5	409,32	-
Conv	5	409,64	ReLu
MaxPooling	5	81,64	-
LSTM	-	100	Tanh
Dense	-	1	Sigmoid

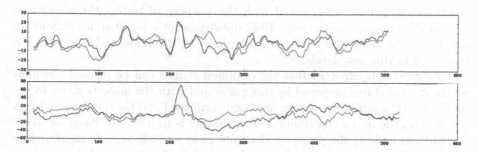

Fig. 5. Two examples of the windows used in this work. At the top of the image is showed the occurrence of an ErrP (between 190 and 210 ms), and the bottom, a normal moment of the experiment. The channel FCz being represented by blue line and the channel Cz by the green. (Color figure online)

Table 2. Average of accuracy using all features extraction methods

Topology	Haar Final	Haar Complete	FFT	RFFT
MLP 3000/3000	79.16%	77.60%	63.33%	73.33%
MLP 1000/1000	73.33%	79.16%	67.16%	**75.00%**
MLP 500/500	77.60%	79.16%	**70.00%**	69.16%
MLP 100/100	69.16%	82.50%	63.33%	60.00%
CNN + LSTM	**94.16%**	**92.50%**	NA	NA
Average	78.68%	82.18%	65.95%	69.37%

the result remained very close within those feature extraction methods. It is noted that while the accuracy variation varies by an average of 3.36% between the topologies, the average variation between the two methods (wavelet and Fourier) varies in the average of 8.74%. Therefore, the variation of the features extraction technique is significantly important

CNN's results were much higher than those of MLP. This clearly demonstrates that temporal information is important for signal analysis, and the two channels being considered together, and not concatenated like the one made in

MLP, had many relevant features extracted. There is still a probability of modifying the kernels and increasing the number of layers, but the network can do very well, when compared to literature, for an individual.

7   Conclusion

In this paper, a new database for ErrP detection was presented. For its creation, a serious game was also developed that allowed this database construction. The ErrP classification result was compared with other known databases. The recognition of ErrP was compared with those existing in the literature and the results showed that it is consistent and competitive. Further, albeit the proposed game is non-educational, it was created with the purpose of testing the algorithms for educational applications, once EEG databases for education are difficult to find. It is even more with concerns in the focus of attention and frustration as considered in this new database.

Therefore, it is expected that the obtained signals can be studied considering the mental states proposed by this game and with the answers given to the robot. Once the initial hypothesis has been confirmed and the database behaves as expected, the major contribution of this paper is to make the database available for use, making the proposed elements clear and all possible mental state situations. It is necessary to increase the number of volunteers for the experiment, adding more variables (such as different knowledge of technology) and other types of errors.

As future work, it is intended to better explore the basis created, and understand the changes of the brain according to the elements created by the game. Thus, one of the applications is to use the algorithm trained in the humanoid robot for real-time recognition of the signals. The robot must adapt its behavior according to the measures of engagement, frustration and errors committed by the student. Also, it is important to study better filtering of the signal, especially with respect to the more complicated artifacts to be removed. A smart search can be proposed, improving the automatic selection of the moment of error, decreasing the manual service, using previous information.

References

1. Breazeal, C.L.: Designing Sociable Robots. MIT press, Cambridge (2004)
2. Brigham, E.O., Brigham, E.O.: The Fast Fourier Transform and its Applications, vol. 448. Prentice Hall, Englewood Cliffs (1988)
3. dAbreu, J.V.V., Bastos, B.L.: Robtica pedaggica: Uma reflexo sobre a apropriao de professores da escola elza maria pellegrini de aguiar. Anais dos Workshops do Congresso Brasileiro de Informtica na Educao (2013)
4. dAbreu, J.V.V., Lourenço, M.: Núcleo de informática aplicada à educação (nied)-30 anos de atuação (1983–2013). In: Anais dos Workshops do Congresso Brasileiro de Informática na Educação, vol. 1 (2013)
5. Jasper, H.: Ten-twenty electrode system of the international federation. Electroenceph. Clin. Neurophysiol. **10**, 371–375 (1958)

6. Kerous, B., Skola, F., Liarokapis, F.: EEG-based BCI and video games: a progress report. Virtual Reality **22**(2), 119–135 (2018). https://doi.org/10.1007/s10055-017-0328-x
7. Kingma, D.P., Ba, J.: Adam: a method for stochastic optimization. arXiv preprint arXiv:1412.6980 (2014)
8. Klem, G.H., Lüders, H.O., Jasper, H.H., Elger, C.: The ten-twenty electrode system of the international federation. The international federation of clinical neurophysiology. Electroenceph. Clin. Neurophysiol. **52**, 3–6 (1999)
9. Morales, J.M., Di Stasi, L.L., Díaz-Piedra, C., Morillas, C., Romero, S.: Real-time monitoring of biomedical signals to improve road safety. In: Rojas, I., Joya, G., Catala, A. (eds.) IWANN 2015. LNCS, vol. 9094, pp. 89–97. Springer, Cham (2015). https://doi.org/10.1007/978-3-319-19258-1_8
10. Pfurtscheller, G., Neuper, C., Flotzinger, D., Pregenzer, M.: EEG-based discrimination between imagination of right and left hand movement. Electroenceph. Clin. Neurophysiol. **103**(6), 642–651 (1997)
11. Pinto, A.H.M., Mijan, M.A.M., Nardari, G.V., Romero, R.A.F.: Comparing features extraction and classification methods to recognize ERRP signals. In: 2018 Latin American Robotic Symposium, 2018 Brazilian Symposium on Robotics (SBR) and 2018 Workshop on Robotics in Education (WRE), pp. 1–7, November 2018
12. Sanei, S., Chambers, J.A.: EEG Signal Processing. Wiley, Chichester (2007)
13. Spüler, M., Niethammer, C.: Error-related potentials during continuous feedback: using EEG to detect errors of different type and severity. Front. Hum. Neurosci. **9**, 155 (2015)
14. Stanković, R.S., Falkowski, B.J.: The Haar wavelet transform: its status and achievements. Comput. Electr. Eng. **29**(1), 25–44 (2003)
15. Sun, M., Leite, I., Lehman, J.F., Li, B.: Collaborative Storytelling Between Robot and Child: A Feasibility Study (2017)
16. Tatum, W.O.: Ellen R. Grass lecture: extraordinary EEG. Neurodiagnostic J. **54**(1), 3–21 (2014). https://doi.org/10.1080/21646821.2014.11079932. https://www.tandfonline.com/doi/abs/10.1080/21646821.2014.11079932
17. W. Ferrez, P., Millan, J.d.R.: You are Wrong!-automatic Detection of Interaction Errors from Brain Waves, pp. 1413–1418, January 2005

Using Inferred Gestures from sEMG Signal to Teleoperate a Domestic Robot for the Disabled

Nadia Nasri[✉], Francisco Gomez-Donoso, Sergio Orts-Escolano, and Miguel Cazorla

University Institute for Computer Research, University of Alicante, Alicante, Spain
{nnasri,fgomez,sorts,miguel}@dccia.ua.es

Abstract. With the lightning speed of technological evolution, several methods have been proposed with the aim of controlling robots and using them to serve humanity. In this work, we present and evaluate a novel learning-based system to control Pepper, the humanoid robot. We leveraged an existing low-cost surface electromyography (sEMG) sensor, that is in the consumer market, Myo armband. To achieve our goal, we created a dataset including 6 hand gestures recorded from 35 intact people by the usage of the Myo Armband device, which has 8 non-intrusive sEMG sensors. Using raw signals extracted from Myo armband, we have been able to train a gated recurrent unit-based network to perform gesture classification. Afterwards, we integrated our system with a live hand gesture recognition application, transmitting the commands to the robot for implementing a live teleoperation method. In this way, we are able to evaluate in real-time the capabilities of our system. According to the experiments, the teleoperation of a Pepper robot achieved an average of 77.5% accuracy during test.

Keywords: Surface electromyography sensor · Dataset · Gated recurrent units · Gesture recognition · Pepper

1 Introduction

Every year by the reason of having an accident or a chronic illness, a considerable number of people become disabled, losing parts of their body or even becoming paralyzed. Amputees and disabled people are the widest group, whose comforts during lifetime depend on technology improvements. They have difficulties in self-care and face various issues in their daily routines; hence, some of them need a person to take care of their daily living due to their disabilities. The important aspects of science and technology improvement in human health and the simplification of their lifestyle have been studied in many works [1,2].

The methods for controlling robots and teleoperation are a topic of interest for the scientific community and robotic field [3–5]. As a matter of fact, surface electromyography (sEMG) sensors have been used widely for this purpose.

© Springer Nature Switzerland AG 2019
I. Rojas et al. (Eds.): IWANN 2019, LNCS 11507, pp. 198–207, 2019.
https://doi.org/10.1007/978-3-030-20518-8_17

These sensors are useful non-invasive methods to record the electrical activity produced by muscles and they can be applied for controlling robotic arms [6], robotic wheelchairs [7] and mobile robots [8], among others. In order to keep pace with the latest technology of the day, the most recent methods for controlling robots through gesture recognition systems take advantage of Machine Learning techniques [6,9]. These classifiers learn specific features of each pose from EMG sensors labeled data, classify those gestures and send the corresponding command for each of them to the robot. Also, there are other studies conducted with diverse methods and different neural networks with aim of to develop controlling methods for robots [10–12].

With a view to create the maximum benefit at the minimum cost, our proposed framework utilizes the Myo Gesture Control Armband, which is made of low-cost non-intrusive sEMG sensors to record muscle activities and capture surface EMG signals. A gated recurrent unit (GRU) architecture is also used to classify the gestures and control the robot pepper.

The main objective of this works is optimization the lives of people with disabilities and assisting them in their daily activities at home. Our proposed method can also be used as a teleoperation method in high-risk jobs to replace manual labors in many areas. An operator can by using Myo armband, remote and control a robot through 6 static hand gestures. Several teleoperation methods were proposed [13,14] and there also exist some works which benefit from Myo armband to control robots with different methods [6,9,15].

Hand gesture recognition via surface (sEMG) sensors and high-density surface electromyography (HD-sEMG) sensors placed on the arm have been used to control muscle-computer interfaces (MCIs) in several researches [16–18]. Moreover, there are several studies focused on deep learning techniques, especially using ConvNet as a feature extractor in hand gesture classification [6,19,20].

Low-cost Myo armband was used in some recent research and it has been proven that it is proper for being used in gesture classification works and controlling a robot [6,9]. One of the most recent studies done with a Myo armband (and its default gesture recognition pattern), combined the EMG signal with a Virtual Reality (VR) headset output to control a robot through eye movements and facial gestures [21]. However, the Myo armband default system in real time has a considerable error rate and contains only 5 hand gestures.

In our previous work, we explained about our experiments for recording data and creating a new dataset. We also indicated various reasons for choosing GRU neural network to train raw sEMG signals captured by Myo armband, performing for the final step a test examination for 5 new subjects where we proved a GRU architecture is accurate enough to be used in hand gesture recognition systems [22].

The rest of the paper is organized as follows: Sect. 2, the details of the robot, the capture device and relevant details of the process of acquiring dataset are explained. Section 3 describes the system and architecture of the neural network used to train the hand gestures recognition and describes the communication with the robot. Section 4 shows the obtained results and describes the experimental process. Section 5 presents the main conclusions of the present work.

2 Robotic Teleoperation System and Hardware

The development of humanoid robots and assistive robots has increased research interest in teleoperation. Teleoperation (or remote operation) signifies the technique of controlling a robot or a machine from a distance. From a remote location an operator can send commands to the robot or machine, over a communication channel. Our immediate aim is to allow a disabled person to teleoperate a humanoid robot using raw signals captured by a sEMG sensor. Therefore, we used a humanoid robot, Pepper and the Myo armband device.

2.1 Robot Pepper

Pepper is a humanoid robot introduced by Aldebaran and SoftBank Robotics[1] in June 2014 and marketed in February 2015. It is designed to be a social robot that recognizes faces and analyzes expressions and tones of voice. The robot has a height of 1.21 m, a width of 480 mm and a depth of 425 mm. The Robot has motors in its joints similar to human body, except its fingers, which can not be moved independently and neither includes sensors.

Pepper is intended for human interaction and several works were done on various aspects of using it [2,23,24]. In this work we implemented a teleoperation method on Pepper through Myo armband (see Fig. 1). We developed a deep learning method that is able to distinguish among 6 different gestures. Each gesture corresponds to a robot action (move forward, move backward, move to the right, move to the left, raise arms, move elbows and hands to grab an object). As it can be noticed, the actions are intended to move the robot and gran objects, so the disabled people can use it for reaching distant items.

2.2 Recording Hardware

In recent years, Thalmic Labs released the Myo[2] armband, a gesture control device composed of a low-consumption ARM Cortex-M4 120 MHZ microprocessor, 8 dry sEMG for gestural data and an inertial measurement unit (IMU) for spatial data(9-axis) with low-sampling rates (200 Hz) (see Fig. 2). In this work, we focus on gestural data that gives the information about the electrical activity produced by skeletal muscles. The use of the Myo armband device has been studied in multiple works using deep learning architectures [25–28].

---

[1] https://www.softbankrobotics.com/.
[2] https://www.myo.com/.

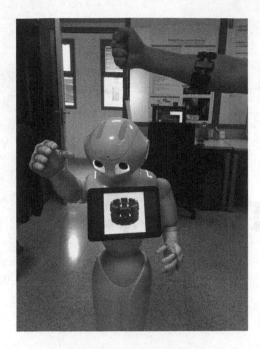

Fig. 1. Myo armband connected to robot Pepper

Fig. 2. Myo armband

2.3 Recording Data

Our classifier, which is proposed and exhaustively tested in [22], is able to distinguish 6 dissimilar static hand gestures (open hand, closed hand, wrist extension, wrist flexion, tap and victory sign) recorded from 35 intact subjects (see Fig. 3). Myo armband seemed stable to the external factors [28], but as there are many factors that can affect the sEMG signals, while using sensors [29,30], we determined some conditions during our experiments .

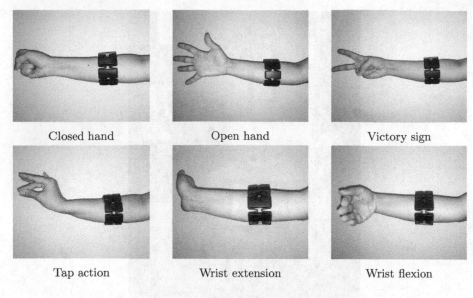

Closed hand   Open hand   Victory sign

Tap action   Wrist extension   Wrist flexion

Fig. 3. Hand gestures

Before each live test, we calibrated the Myo armband for being used by the right arm of each subject. Myo data was transmitted via Bluetooth at slightly less than 200 Hz (8 channels).

3   System Description

In this paper, we used the same neural network with the same characteristics as our previous work [22]. It featured three GRU layers created from 150 units with a hyperbolic tangent (tanh) activation function and followed by a fully connected layer. Every GRU had connection to a dropout to avoid overfitting with a 0.5 probability of neuron activation. The last layer in our proposed neural network was a fully connected (FC) layer with 6 neurons according to the number of gestures in our work. The GRU layer's output, which carried important features of the sEMG signals, had complete linkage with fully connected neurons and fully connected classify the input signals according these features. We applied a softmax activation function in the last layer to set up the network for classification.

Moreover, we used Pynaoqi library to introduce our desired command to Pepper for each gesture. In order to connect our system with the robot we used a local area network (LAN) and sent commands through WIFI to the robot (see Fig. 4).

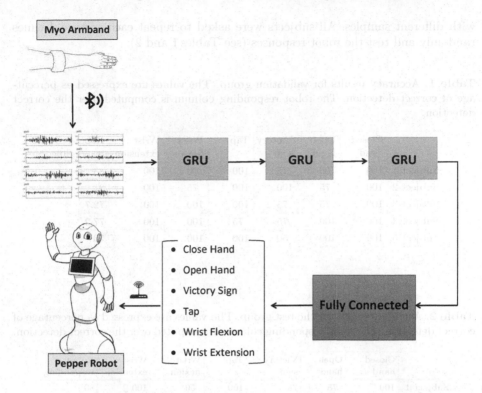

Fig. 4. System and neural network components

4   Experiments and Results

In order to test our system, we trained the proposed neural network with a whole dataset of 35 subjects and implemented a live system to give commands to Pepper through Myo armband. Our system, which is depicted in Fig. 4, takes as input the raw signal of the Myo armband. Then, a stream of sEMG signals are forwarded to the recurrent neural network (GRU-based) which states the hand gesture the user is performing. Finally, the corresponding action is sent to the Pepper robot. Regarding the run-time, the system can predict the gestures in 42 ms, however for teleoperation tasks Pepper needs more time to terminate with requested commands. Therefore, to avoid conflicts between commands, we considered a minimum (5 s) delay for sending new commands to the robot. Pepper was programmed to move in 4 directions for 4 hand gestures (close hand, open hand, wrist flexion and wrist extension) and for the rest of gestures we considered *rise arms, move elbows and hands to grab an object.*

With the view to having more realistic results, we chose 10 subjects to test our live remote system. We divided them into two groups: Test group included 5 new subjects whose data was not seen by the system during the training process. Validation group included 5 subjects who participated in the training process with different samples. All subjects were asked to repeat each gesture 4 times randomly and test the robot responses (see Tables 1 and 2).

Table 1. Accuracy results for validation group. The values are expressed as percentage of correct detection. The robot responding column is computed over the correct detection.

	Closed hand	Open hand	Victory sign	Tap	Wrist flexion	Wrist extension	Robot responding
Subject 1	100	100	75	100	100	100	83
Subject 2	100	75	100	100	75	100	81
Subject 3	100	75	75	100	100	100	72.7
Subject 4	100	100	75	75	100	100	77.2
Subject 5	100	100	50	100	100	100	70

Table 2. Accuracy results for the test group. The values are expressed as percentage of correct detection. The robot responding column is computed over the correct detection.

	Closed hand	Open hand	Victory sign	Tap	Wrist flexion	Wrist extension	Robot responding
Subject 1	100	75	75	100	50	100	80
Subject 2	100	75	75	100	75	75	75
Subject 3	75	75	25	75	100	100	72.2
Subject 4	100	100	75	75	50	75	78.9
Subject 5	75	100	50	100	75	100	85

Following the results of the live remote system, the victory-sign gesture had the highest error and for the rest of gestures achieved appropriate results in both groups. However, Pepper had some issues with responding to correct predictions and failed several times. Some of the reasons for errors in Pepper response were due to issues in network connection or problems with the robot's motors (motor synchronization and temperature matters). As Pepper hands do not contain motors and tactile sensors for each finger and only have the options *close hand* and *open hand*, this type of robot has problems in the task of grabbing objects. We choose objects made of flexible materials so as not to be broken or push robot fingers in the opposite direction of fingers movements (see Fig. 5).

The proposed system could recognize approximately 92.4% of gestures (validation group) in live prediction gestures examination and obtained an accuracy of 76.7% for the robot responding to the gestures. For the test group with new subjects, our system achieved an average accuracy of 80.8% for the prediction gestures examination and an accuracy of 78.2% for the robot responding to the correct prediction of gestures.

Fig. 5. Pepper holding an object

5   Conclusions

In this paper, we proposed a novel method for live gesture classification and for teleoperating a humanoid robot by using sEMG raw signals received from the Myo armband. The proven system performance studied for the two different test groups achieved an average accuracy of 77.5% for the teleoperation task. Regarding our experiment results, we can conclude that the Pepper robot can be controlled by a low-cost non-intrusive Myo armband. Therefore, it can be used to improve the lives of paralyzed people, so they can use the robot for helping them in their daily tasks and live more independently. To demonstrate our results we have published a video in Youtube[3].

Acknowledgements. This work was supported by the Spanish Government TIN2016-76515R grant, supported with Feder funds. It has also been funded by the University of Alicante project GRE16-19, by the Valencian Government project GV/2018/022, and by a Spanish grant for PhD studies ACIF/2017/243. The authors would like to thank all the subjects for their participation in our experiments. We would also like to thank NVIDIA (Santa Clara, California, USA) for the generous donation of a Titan Xp and a Quadro P6000.

---

[3] https://youtu.be/b0AoS3aE7Mk.

References

1. Cook, A.M., Polgar, J.M.: Essentials of Assistive Technologies. ELSEVIER Mosby (2012)
2. Costa, A., Martinez-Martin, E., Cazorla, M., Julian, V.: Pharos-physical assistant robot system. Sensors **18**(8), 2633 (2018)
3. Kowalczuk, Z., Czubenko, M.: Model of human psychology for controlling autonomous robots. In: 2010 15th International Conference on Methods and Models in Automation and Robotics, pp. 31–36, August 2010
4. Li, M., Li, W., Zhao, J., Meng, Q., Sun, F., Chen, G.: An adaptive P300 model for controlling a humanoid robot with mind. In: 2013 IEEE International Conference on Robotics and Biomimetics (ROBIO), pp. 1390–1395, December 2013
5. Lamiroy, B., Espiau, B., Andreff, N., Horaud, R.: Controlling robots with two cameras: how to do it properly. In: IEEE International Conference on Robotics and Automation (ICRA 2000), San Francisco, USA, pp. 2100–2105. IEEE Computer Society, April 2000
6. Allard, U.C., et al.: A convolutional neural network for robotic arm guidance using sEMG based frequency-features. In: Intelligent Robots and Systems (IROS). IEEE 2016, pp. 2464–2470 (2016)
7. Kucukyildiz, G., Ocak, H., Karakaya, S., Sayli, O.: Design and implementation of a multi sensor based brain computer interface for a robotic wheelchair. J. Intell. Robot. Syst. **87**, 247–263 (2017)
8. Shin, S., Kim, D., Seo, Y.: Controlling mobile robot using IMU and EMG sensor-based gesture recognition. In: 2014 Ninth International Conference on Broadband and Wireless Computing, Communication and Applications (BWCCA), pp. 554–557, November 2015
9. Bisi, S., De Luca, L., Shrestha, B., Yang, Z., Gandhi, V.: Development of an EMG-controlled mobile robot. Robotics **7**, 36 (2018)
10. Wang, H.-B., Liu, M.: Design of robotic visual servo control based on neural network and genetic algorithm. Int. J. Autom. Comput. **9**, 24–29 (2012)
11. Stanton, C., Bogdanovych, A., Ratanasena, E.: Teleoperation of a humanoid robot using full-body motion capture, example movements, and machine learning, December 2012
12. Morris, A.S., Mansor, A.: Finding the inverse kinematics of manipulator arm using artificial neural network with lookup table. Robotica **15**, 617–625 (1997)
13. Yang, C., Chang, S., Liang, P., Li, Z., Su, C.-Y.: Teleoperated robot writing using EMG signals. In: 2015 IEEE International Conference on Information and Automation, pp. 2264–2269 (2015)
14. Reddivari, H., Yang, C., Ju, Z., Liang, P., Li, Z., Xu, B.: Teleoperation control of Baxter robot using body motion tracking. In: 2014 International Conference on Multisensor Fusion and Information Integration for Intelligent Systems (MFI), pp. 1–6, September 2014
15. Xu, Y., Yang, C., Liang, P., Zhao, L., Li, Z.: Development of a hybrid motion capture method using MYO armband with application to teleoperation. In: 2016 IEEE International Conference on Mechatronics and Automation, pp. 1179–1184, August 2016
16. Saponas, T.S., Tan, D.S., Morris, D., Balakrishnan, R.: Demonstrating the feasibility of using forearm electromyography for muscle-computer interfaces. In: Proceedings of the SIGCHI Conference on Human Factors in Computing Systems, CHI 2008, pp. 515–524. ACM, New York (2008)

17. Rojas-Martinez, M., Manyanas, M., Alonso, J., Merletti, R.: Identification of iso-metric contractions based on high density EMG maps. Electromyogr. Kinesiol. **23**, 33–42 (2013)
18. Zhang, X., Zhou, P.: High-density myoelectric pattern recognition toward improved stroke rehabilitation. IEEE Trans. Biomed. Eng. **59**, 1649–1657 (2012)
19. Atzori, M., Cognolato, M., Müller, H.: Deep learning with convolutional neural networks applied to electromyography data: a resource for the classification of movements for prosthetic hands. Front. Neurorobot. **10**, 9 (2016)
20. Geng, W., Du, Y., Jin, W., Wei, W., Hu, Y., Li, J.: Gesture recognition by instan-taneous surface EMG images. Sci. Rep. **6**, 36571 (2016)
21. Wang, K.-J., Tung, H.-W., Huang, Z., Thakur, P., Mao, Z.-H., You, M.-X.: EXG-buds: universal wearable assistive device for disabled people to interact with the environment seamlessly, pp. 369–370, March 2018
22. Nasri, N., Orts-Escolano, S., Gomez-Donoso, F., Cazorla, M.: Inferring static hand poses from a low-cost non-intrusive sEMG sensor. Sensors **19**(2), 371 (2019)
23. Bauer, Z., Escalona, F., Cruz, E., Cazorla, M., Gomez-Donoso, F.: Improving 3D estimation for the pepper robot using monocular depth prediction. In: Workshop de Agentes Físicos (WAF) (2018)
24. Cruz, E., et al.: Geoffrey: an automated schedule system on a social robot for the intellectually challenged. Comput. Intell. Neurosci. **2018**, 17 (2018)
25. Pomboza-Junez, G., Terriza, J.H.: Hand gesture recognition based on sEMG signals using support vector machines. In: Consumer Electronics-Berlin (2016)
26. Allard, U.C., et al.: Deep learning for electromyographic hand gesture signal clas-sification by leveraging transfer learning. CoRR, vol. abs/1801.07756 (2018)
27. Cote-Allard, U., Fall, C.L., Campeau-Lecours, A., Gosselin, C., Laviolette, F., Gos-selin, B.: Transfer learning for sEMG hand gestures recognition using convolutional neural networks. In: IEEE International Conference on Systems (2017)
28. Pizzolato, S., Tagliapietra, L., Cognolato, M., Reggiani, M., Muller, H., Atzori, M.: Comparison of six electromyography acquisition setups on hand movement classification tasks. PloS One **12**, e0186132 (2017)
29. Farina, D., Cescon, C., Merletti, R.: Influence of anatomical, physical, and detection-system parameters on surface emg. Biol. Cybern. **86**, 445–456 (2002)
30. Kuiken, T.A., Lowery, M.M., Stoykov, N.S.: The effect of subcutaneous fat on myoelectric signal amplitude and cross-talk. Prosthet. Orthot. Int. **27**(1), 48–54 (2003). PMID: 12812327

3D Orientation Estimation
of Pharmaceutical Minitablets
with Convolutional Neural Network

Gregor Podrekar[1]([✉]), Domen Kitak[1], Andraž Mehle[1], Domen Rački[1],
Rok Dreu[2], and Dejan Tomaževič[1,3]

[1] Sensum d.o.o., Computer Vision Systems, Ljubljana, Slovenia
gregor.podrekar@sensum.eu
[2] Faculty of Pharmacy, University of Ljubljana, Ljubljana, Slovenia
[3] Faculty of Electrical Engineering, University of Ljubljana, Ljubljana, Slovenia

Abstract. We present a Convolutional Neural Network for 3D orientation estimation of pharmaceutical minitablets, i.e., round tablets with diameter less than 3 mm. The network inputs a single grayscale image with the minitablet positioned approximately in the center and predicts a 3D unit orientation vector that fully describes the 3D orientation of the imaged minitablet. We trained the network on synthetic images, generated by rendering CAD models of minitablets at realistic conditions by varying the orientation, scale, camera distance, position within the imaging plane, and surface properties. No manual 3D orientation labeling of training images was therefore required. We evaluated the accuracy of the approach on both synthetic and real images. The real images were acquired during pharmaceutical film coating processes. Accuracies of 1.388° and 2.657° were achieved on synthetic and real image datasets, respectively. We tested two different minitablet shapes. Obtained results indicate that good performance can be obtained on a real image datasets despite training the network on synthetic data only. The estimated 3D orientations provide means for further automated analysis of the images, which we demonstrated by measuring an important coating process parameter (coating thickness) during the minitablet coating process. Although tested only for minitablets, the 3D orientation estimation approach should perform well also for other symmetrical shapes.

Keywords: CNN · 3D orientation estimation · Minitablets · Film coating

1 Introduction

3D orientation estimation plays a crucial role in automated visual inspection systems, when products that are to be inspected cannot be precisely positioned by appropriate mechanical manipulation systems. An example of such a scenario in the pharmaceutical industry is a film coating process of pharmaceutical

© Springer Nature Switzerland AG 2019
I. Rojas et al. (Eds.): IWANN 2019, LNCS 11507, pp. 208–219, 2019.
https://doi.org/10.1007/978-3-030-20518-8_18

minitablets (tablets with diameter less than 3 mm). Same as tablets, minitablets are often film coated. The applied coating can either serve cosmetic purposes, mask unpleasant taste or smell, increase physical and chemical stability or modify the release of the drug. Coating of minitablets is mostly performed in Wurster fluid-bed equipment where a circular motion of the minitablets is achieved by distributing a stream of air into the equipment chamber, that pneumatically transports them repeatedly through the spray zone where the coating is being applied [7]. A predetermined, fixed amount of the coating material is normally sprayed onto the minitablet cores while new approaches, encouraged by the Food and Drug Administration through a PAT (Process Analytical Technology) initiative in 2004, tend to rely on in-line process measurements in order to ensure a predefined final coating thickness. This is especially important in case of modified release coatings where the coating thickness directly affects the safety and efficacy of the drug.

One way of measuring the coating thickness during the coating process is by using a machine vision system. The size of minitablets is constantly measured from the acquired images and the mean coating thickness is estimated from the mean size increase over the coating time [22]. Since the minitablets cannot be mechanically manipulated inside the coating equipment while they are being imaged, their orientations alongside the positions have to be determined from the acquired images in order to appropriately measure their size in different dimensions. Although 3D orientation estimation can be avoided by using appropriate data interpretation heuristics, as demonstrated in [22], estimating the 3D orientations can improve the size measurements and is a more general solution.

In the past, various approaches for pose (orientation and location) estimation from 2D images have been proposed. Early attempts aimed to estimate the pose of the objects by iteratively establishing the correspondences between the distinctive lines on the model and the image [9,18]. These methods require an initial pose estimation that is close to the actual pose because they tend to get stuck in local minima and are therefore not directly applicable. Descriptor-based methods, on the other hand, try to circumvent these limitations. Here, the representative description of the local regions on both the query and the reference image are being matched, where the reference images are generated for all relevant views [2,19,20]. Descriptor-based methods perform best with textured objects while different features have also been proposed for textureless objects [5,8,10,23]. Template matching presents the third group of methods, where exhaustive search is usually employed using a priori edge information for matching. Different variations have been proposed over the decades, such as Chamfer matching and its improved variations [1,3,17]. In addition, methods replacing the exhaustive search and chamfer matching with other techniques [4,6,12,15] were also introduced, either reducing computational loads, increasing accuracy or both.

Recently, convolutional neural networks (CNNs) have been shown to be a promising tool for both 2D and 3D orientation estimation from single 2D images [11,16]. The idea here is that given an adequate number of object views at different orientations during the training, the network learns to predict the orientation of the object from a single image. CNN approaches, in comparison to other state of the art approaches, do not require handcrafted image preprocessing or feature extraction. The network learns the required image transformations by itself from the training data. On the other hand, the preparation of an appropriate training dataset, consisting of vast amount of object views at known orientations is needed. This requires manually labeling the orientations of the objects for each training image. A tedious and time-consuming task, not always practical for industrial applications.

In this work, we present a CNN based approach for 3D orientation estimation of minitablets from single grayscale images. The approach in general follows the one presented in [16] but is specialized for symmetrical, minitablet like shapes. We overcome the problem of tedious manual training dataset creation by training the network solely on generated synthetic images while evaluating its performance on real image data sets, created as part of this work.

2   Method

We assume that the object is positioned approximately in the center of the image. This image is fed to the network that outputs the estimated orientation of the object. In contrast to 2D orientation estimation, where classifying the 2D orientation performs better then regressing it [11], the same is not feasible for 3D orientation estimation due to a much larger amount of possible views. We therefore formulate the 3D orientation estimation as a regression problem, following the approaches presented in [13,16]. We start this section by defining the object's orientation as a 3D unit vector. Next, we describe the network's architecture and training procedure.

2.1   3D Orientation

Minitablets are of round, symmetrical shape. For such shapes, we can describe their 3D orientation with only two Euler angles (for general objects three angles are needed). Although these angles can be directly used as a predicted output of the network, such approach is not optimal, since these values lie in a non-Euclidean space. This makes the use of a typical mean square error (MSE) loss function for training the network less efficient [11]. In [13,16], authors overcame this issue by using the quaternions, while in this work, we define the 3D orientation as a unit vector in a 3D space: $v = (v_x, v_y, v_z)$. This is possible due to the symmetrical shape of the minitablets, where rotating the minitablet around its vertical axis does not change its appearance on the image. The 3D orientation unit vector can be interpreted as a vector, normal to the minitablet's cap area (Fig. 1).

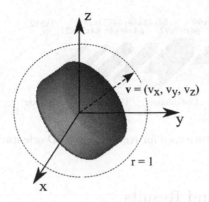

Fig. 1. 3D orientation unit vector ($\boldsymbol{v}$) representing the 3D orientation of the minitablet.

2.2 Architecture and Training

Combination of four convolutional and max pooling layers, followed by four fully connected layers was empirically chosen (Fig. 2). Relu activation functions were used for all layers except for the last one, where linear activation function was used. Grayscale images of 200×200 pixels presented an input to the network, while the output came in a form of a 3D unit orientation vector. Mean squared error (MSE) was used as a training cost function, measuring the distance between the ground truth ($\boldsymbol{v_g}$) and the predicted 3D vector ($\boldsymbol{v_p}$):

$$MSE(\boldsymbol{v_g}, \boldsymbol{v_p}) = (v_{gx} - v_{px})^2 + (v_{gy} - v_{py})^2 + (v_{gz} - v_{pz})^2 \qquad (1)$$

while absolute angular error (AAE), derived from the dot product equation:

$$AAE(\boldsymbol{v_g}, \boldsymbol{v_p}) = \arccos(|\boldsymbol{v_g} * \boldsymbol{v_p}|) \qquad (2)$$

served for validation and testing purposes. The absolute value of a dot product was used due to symmetrical shapes of the minitablets. AAE shows the smallest rotation needed to align the prediction with the ground truth. It is a more intuitive distance metric that provides more insight into actual angular error between the ground truth and the predicted 3D orientations in comparison to the MSE. Use of AAE as a training cost function on the other hand in our case caused the network not to converge during the training and was therefore not used for this purpose.

The output tensor was normalized to unit length, both during the training and the prediction. Adam optimizer [14] was used for training the network although little difference was observed in comparison to other optimization algorithms. Input images were normalized by intensity mean subtraction and standard deviation division. During the training, the input images were additionally augmented in order to prevent the network from overfitting. The images were randomly smoothed, augmented with random noise and the position of the minitablet was randomly shifted from the image center.

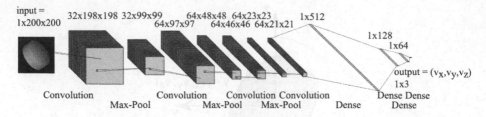

Fig. 2. Network architecture used for regressing the 3D orientation of minitablets from a single grayscale image.

3   Experiment and Results

We performed the experiments with two different tablet shapes (Shape A and B). For each shape, a synthetic and a real image datasets were created, where the synthetic datasets served the training purposes, while the orientation estimation performance was evaluated on both synthetic and real image datasets. All images were prepared in a way so that the center of mass of each minitablet was coincident with the center of the image. This was a trivial task in case of synthetic images but required preprocessing in case of real images, described in more detail below.

3.1   Synthetic Image Datasets

CAD models of both shapes were rendered at different random orientations, scales, distances from the camera (depth of field effect), positions within the imaging plane (different shading), and with varying random surface properties such as roughness, edge smoothness and reflection (Fig. 3).

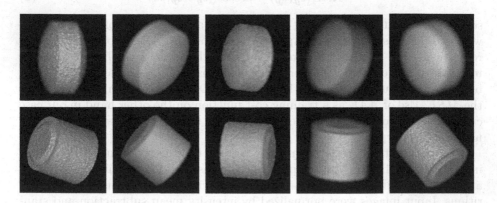

Fig. 3. Synthetic images of minitablets rendered at different orientations, scales, distances from the camera, positions within the imaging plane and with random surface properties (roughness, edge smoothness and reflection). Shape A (upper row) and shape B (lower row).

Blender's Cycles render v2.79b (Blender Foundation, Netherlands) was used for rendering. The rendering configuration was set in a way to mimic the actual imaging setup used for acquiring the real images (circular lighting, telecentric camera projection). Each synthetic image dataset consisted of 100,000 images, rendering of which took approximately 24 h on a GPU (NVIDIA, GeForce GTX 1050 Ti) supported computer. We split the datasets up into 80,000, 10,000 and 10,000 images, used for training, validation and testing, respectively.

3.2 Real Image Datasets

The image datasets (one per minitablet shape) were recorded with a process analytical technology visual inspection system for monitoring, understanding and optimization of fluid-bed pharmaceutical production processes (PATVIS APA, Sensum, Slovenia). PATVIS APA consists of a digital grayscale camera, a telecentric lens and an illumination unit. It provides images of moving particles from the process to a dedicated image processing software, running on attached processing unit. PATVIS APA was mounted on a laboratory scale fluid bed equipment (Fig. 4) and recorded images of minitablets being coated through an observation window at 100 frames per second with an exposure time of 7µs. The acquired images had a resolution of 848 × 848 pixels, which covered a 15.96 × 15.96 mm image region at 18.82 pixel size.

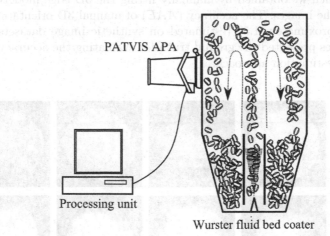

Fig. 4. Visual inspection system (PATVIS APA) mounted on a Wurster fluid bed coater was used for acquiring the images of minitablets during the coating processes.

Both image datasets were preprocessed in order to extract the individual minitablet views. The procedure described in [22] was followed, placing each segmented minitablet in the center of a 200 × 200 pixels sized image. The preprocessing included center detection by consecutive binary thresholding, morphological closing and labeling operations (Fig. 5, middle). Additionally, the focus of each minitablet was estimated based on average border gradient values (Fig. 5, right). Minitablet views exceeding a predetermined focus threshold were omitted because such out-of-focus images were not appropriate for further analysis. Example extracted images for both minitablet shapes are shown in Fig. 6.

Fig. 5. The preprocessing of acquired images (left) included binary thresholding, morphological closing and labeling (middle). Focus was also estimated for every detected minitablet based on its average border gradient values (right). Each detected minitablet was placed in the center of 200 × 200 pixels sized image that was later fed to the CNN.

The resulting datasets included 29,125 and 35,369 images for shape A and B, respectively. We annotated 1000 images per dataset with 3D orientation information, which we obtained by manually fitting the 3D edge models of the minitablets to the images. The accuracy (AAE) of manual 3D orientation estimation was approximately 1.5° (estimated on synthetic image datasets). The annotated images presented the ground truth for evaluating the accuracy of the 3D orientation estimation approach.

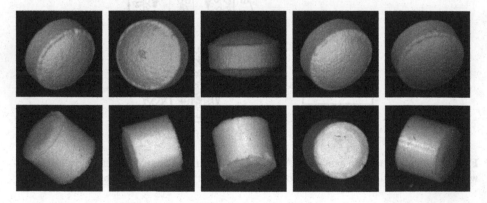

Fig. 6. Images of minitablets extracted from the images acquired during the coating processes. Shape A (upper row) and shape B (lower row).

3.3   Implementation Details

The CNNs were implemented in Keras with Tensorflow backend and were trained on a Workstation with an Intel(R) Core(TM) i7-3770 CPU, 16 GB RAM and a NVIDIA GeForce GTX 1050 Ti graphic card. The training took approximately 8 h and was terminated when the validation loss stopped improving significantly (100 epochs). Minibatches of 64 images were used. Training parameters (learning rate, momentum ...) were set to default values as no noticeable improvement was observed while adjusting them.

The parameters used for additionally augmenting the training images were the following. Each minitablet's position shift was randomly chosen from the $[-5, 5]$ interval, both in vertical and horizontal direction. Gaussian noise was randomly chosen from normal distribution (mean $= 0$, standard deviation $=$ 10% of max image intensity value) while the image smoothing was performed with Gaussian kernel (size 3×3) with its standard deviation chosen randomly from the $[0, 5]$ interval.

3.4   Evaluation

We evaluated the accuracy of the 3D orientation estimation approach on both synthetic (10,000 images per dataset) and real (1000 images per dataset) image datasets by comparing the predicted orientations to the ground truth orientations. The ground truth orientations were known in advance in case of synthetic images and were manually determined in case of real images. AAE was used as a distance metric (2) because of its more intuitive nature in comparison to MSE (1) that was used during the training of the network. Table 1 shows the obtained errors in a form of mean and median AAEs. Additionally, minimum and maximum AAEs are presented alongside with 99th percentile value. Distributions of AAEs are also plotted in boxplot for both the synthetic and the real image datasets (Fig. 7).

Table 1. 3D orientation estimation errors (AAE).

Image dataset	Size	Median AAE (°)	Mean AAE (°)	Min AAE (°)	Max AAE (°)	Q99 AAE (°)
Shape A (synthetic)	10,000	1.283	1.663	0.025	80.900	7.654
Shape A (real)	1000	2.531	2.986	0.091	86.622	9.407
Shape B (synthetic)	10,000	1.492	1.750	0.018	88.430	5.097
Shape B (real)	1000	2.783	3.335	0.0472	75.756	11.646

Mean median AAE for both synthetic image datasets (Shape A and B) accounted to 1.388°, while in case of real image datasets increased to 2.657°. An approximately two times larger error can be observed when estimating the orientations on real images. This is much better than expected and reported by other authors [16], making further automated coating analysis based on estimated orientations feasible. Presence of outliers can be observed from the boxplot, where approximately 1% of all orientation estimations exceed AAE of 10.5° in case of both real image datasets. Good overall performance can by our opinion be accounted to simple, symmetrical shapes, typical for minitablets. It also proves that the synthetic image datasets were appropriately generated and managed to generalize well for time changing appearance variations of minitablets during the coating process.

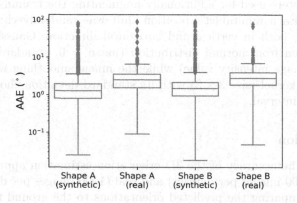

Fig. 7. AAEs for both tablet shapes (A and B) and for both image dataset types (synthetic and real). Logarithmic scale is used due to the presence of outliers.

Slightly better accuracies were obtained for Shape A minitablets, both in case of synthetic and real image datasets. This is by our opinion mostly a consequence of random network weights initialization and random input data augmentation. For that reason, training the same network architecture multiple times with the same training data does not result in the network performing exactly the same.

3.5   Example Application: Mean Coating Thickness Estimation

Knowing the orientations of the imaged minitablets enables contactless and non-invasive in-line measurements of various properties of the coating. For example, the presence of coating defects [21] and the mean coating thickness [22]. Here, we demonstrate the mean coating thickness measurements.

During the coating process, the mean size of minitablets increases due to the increasing coating thickness. By measuring this size increase, we can therefore at any coating time (t) estimate the mean coating thickness ($\overline{CT}(t)$) as:

$$\overline{CT}(t) = \frac{\overline{diameter}(t) - \overline{diameter}(t = 0)}{2}. \tag{3}$$

The diameters of imaged minitablets were measured as a distance from one side of each minitablet to the opposite one (Fig. 8), where the estimated 3D

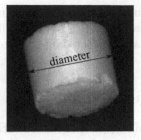

Fig. 8. Diameters of minitablets were measured as a distance between the two opposite border points detected in the direction defined by estimated 3D orientations. Shape A (left) and shape B (right).

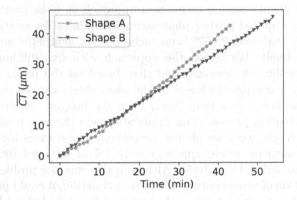

Fig. 9. Mean coating thickness time trends for both real image datasets measured in-line. The mean coating thicknesses were estimated as mean size increase of the minitablets during the coating processes, which was measured from the acquired images based on the estimated 3D orientations with the proposed approach.

orientations defined the direction. The border of the minitablets was detected as the maximum gradient point, located on transition from the minitablet's region to the background. It could be robustly detected due to a good contrast between the two regions.

The diameters of all extracted minitablet images (29,125 and 35,369 for shape A and B, respectively) were measured as described and aggregated over a one-minute time intervals. Mean coating thicknesses for all time intervals were calculated, following (3). The obtained mean coating thickness trends in Fig. 9 show the measured mean coating thicknesses for both real image datasets (shape A and B). Both trends are monotonically increasing during coating in a linear manner as expected. Different inclination of the trends can be observed which is a consequence of different coating process parameters (spraying rate ...). The final measured mean coating thicknesses were verified with an optical microscope, where good agreement was observed between the final points of in-line measurements based on image analysis and the reference microscope measurements performed on fully coated samples of minitablets. The absolute error between the two was less than 2μm, which indicates an accuracy of the mean coating thickness estimation using the proposed CNN orientation estimation approach of 5% or better. This is exceptional for an in-line measurement approach.

4   Conclusion

We presented an approach for 3D orientation estimation of pharmaceutical minitablets with convolutional neural network. The input to the network was a single grayscale image while its output was a 3D unit orientation vector. We trained the network solely with synthetic images, created by rendering the CAD models of minitablets at realistic conditions. On the other hand, we evaluated the accuracy of the 3D orientation estimation approach on both synthetic and real image datasets, acquired during pharmaceutical minitablet coating processes. Accuracies of 1.388° and 2.657° were achieved on synthetic and real image datasets, respectively. We tested the approach with two different minitablet shapes. Additionally, we demonstrated that based on the predicted 3D orientations, the mean coating thickness of the minitablets can be estimated with a satisfactory accuracy (less than 2μm) from the images, acquired during the minitablet film coating process. This application was the final goal of this work. As part of our future work we plan to investigate the reasons for the observed occurrences of outliers, where approximately 1% of all (real image datasets) estimations exceeded AAE of 10.5°. Although this was not problematic for our specific application of mean coating thickness estimation, it could present a problem in some other cases (detection of coating defects) and should be therefore carefully inspected.

References

1. Barrow, H.G., Tenenbaum, J.M., Bolles, R.C., Wolf, H.C.: Parametric correspondence and chamfer matching: two new techniques for image matching. In: Proceedings of the 5th International Joint Conference on Artificial Intelligence - Volume 2, IJCAI 1977, Cambridge, USA, pp. 659–663. Morgan Kaufmann Publishers Inc. http://dl.acm.org/citation.cfm?id=1622943.1622971
2. Belongie, S., Malik, J., Puzicha, J.: Shape matching and object recognition using shape contexts 24(4), 509–522. https://doi.org/10.1109/34.993558
3. Borgefors, G.: Hierarchical chamfer matching: a parametric edge matching algorithm 10(6), 849–865. https://doi.org/10.1109/34.9107
4. Bratanič, B., Pernuš, F., Likar, B., Tomaževič, D.: Real-time pose estimation of rigid objects in heavily cluttered environments. https://doi.org/10.1016/j.cviu.2015.09.002, http://www.sciencedirect.com/science/article/pii/S1077314215001976
5. Carmichael, O.T., Hebert, M.: Object recognition by a cascade of edge probes. In: BMVC. https://doi.org/10.5244/C.16.8
6. Choi, C., Christensen, H.I.: 3D textureless object detection and tracking: an edge-based approach. In: 2012 IEEE/RSJ International Conference on Intelligent Robots and Systems, pp. 3877–3884. https://doi.org/10.1109/IROS.2012.6386065

7. Cole, G., Hogan, J., Aulton, M.: Pharmaceutical Coating Technology Cole. Taylor & Francis Ltd., London (1995)
8. Damen, D., Bunnun, P., Calway, A., Mayol-Cuevas, W.W.: Real-time learning and detection of 3D texture-less objects: a scalable approach. In: BMVC. https://doi.org/10.5244/C.26.23
9. Dementhon, D.F., Davis, L.S.: Model-based object pose in 25 lines of code 15(1), 123–141. https://doi.org/10.1007/BF01450852, http://www.springerlink.com/content/k3670132h2815858/
10. Ferrari, V., Jurie, F., Schmid, C.: From images to shape models for object detection 87(3), 284–303. https://doi.org/10.1007/s11263-009-0270-9
11. Hara, K., Vemulapalli, R., Chellappa, R.: Designing deep convolutional neural networks for continuous object orientation estimation. http://arxiv.org/abs/1702.01499
12. Hinterstoisser, S., et al.: Gradient response maps for real-time detection of textureless objects 34(5), 876–888. https://doi.org/10.1109/TPAMI.2011.206
13. Kendall, A., Grimes, M., Cipolla, R.: PoseNet: a convolutional network for real-time 6-DOF camera relocalization. http://arxiv.org/abs/1505.07427
14. Kingma, D.P., Ba, J.: Adam: a method for stochastic optimization. http://arxiv.org/abs/1412.6980
15. Lampert, C.H., Blaschko, M.B., Hofmann, T.: Beyond sliding windows: object localization by efficient subwindow search. In: 2008 IEEE Conference on Computer Vision and Pattern Recognition, pp. 1–8. https://doi.org/10.1109/CVPR.2008.4587586
16. Langlois, J., Mouchère, H., Normand, N., Viard-Gaudin, C.: 3D orientation estimation of industrial parts from 2D images using neural networks, vol. 2, pp. 409–416. SCITEPRESS. https://doi.org/10.5220/0006597604090416
17. Liu, M., Tuzel, O., Veeraraghavan, A., Chellappa, R., Agrawal, A., Okuda, H.: Pose estimation in heavy clutter using a multi-flash camera. In: 2010 IEEE International Conference on Robotics and Automation, pp. 2028–2035. https://doi.org/10.1109/ROBOT.2010.5509897
18. Lowe, D.G.: Three-dimensional object recognition from single two-dimensional images 31(3), 355–395. https://doi.org/10.1016/0004-3702(87)90070-1
19. Lowe, D.G.: Distinctive image features from scale-invariant keypoints 60(2), 91–110. https://doi.org/10.1023/B:VISI.0000029664.99615.94
20. Mikolajczyk, K., Schmid, C.: A performance evaluation of local descriptors 27(10), 1615–1630. https://doi.org/10.1109/TPAMI.2005.188
21. Podrekar, G., Bratanic, B., Likar, B., Pernus, F., Tomazevic, D.: Automated visual inspection of pharmaceutical tablets in heavily cluttered dynamic environments. In: 2015 14th IAPR International Conference on Machine Vision Applications (MVA), pp. 206–209. https://doi.org/10.1109/MVA.2015.7153168
22. Podrekar, G., et al.: In-line film coating thickness estimation of minitablets in a fluid-bed coating equipment 19(8), 3440–3453. https://doi.org/10.1208/s12249-018-1186-x
23. Tombari, F., Franchi, A., Di, L.: BOLD features to detect texture-less objects. In: 2013 IEEE International Conference on Computer Vision, pp. 1265–1272. https://doi.org/10.1109/ICCV.2013.160

Flatness Defect Detection and Classification in Hot Rolled Steel Strips Using Convolutional Neural Networks

Marco Vannocci, Antonio Ritacco, Angelo Castellano, Filippo Galli[✉],
Marco Vannucci, Vincenzo Iannino, and Valentina Colla

TeCIP Institute, ICT-COISP Center, Scuola Superiore Sant'Anna, Via G. Moruzzi 1,
56124 Pisa, Italy
{m.vannocci,a.ritacco,a.castellano,f.galli,m.vannucci,v.iannino,
v.colla}@santannapisa.it
https://ict-coisp.it

Abstract. This paper addresses the improvement of flatness defect detection and classification in the steel industry. Localization and classification of the defects is respectively taken care of by a detector and a classifier. The pipeline can start with either CSV or image files coming straight from the plant sensors. To probe the performance of the system, it was used to detect and classify flatness defects in hot steel strips. A total of about 513 strips produced in a real steelworks were used for this purpose for a total of about 4806 defect images. A comparison between different traditional machine learning and deep learning models was carried out showing better performances with the latter approach.

Keywords: Flatness defects · Detection · Classification ·
Deep learning · Convolutional Neural Networks · Hot-rolling mill ·
Steel strip

1 Introduction

The field of computer vision has benefited in the last two decades of major advancements due to Deep Neural Networks (DNNs), particularly so in object detection and classification, thanks to the introduction of Convolutional Neural Networks (CNNs) [23]. This is demonstrated by widespread benchmark competitions such as the ImageNet ILSVRC [29], where around 14 million labelled images over 1000 classes are provided to researchers to evaluate the performance of their algorithms. The scale of the challenge is such that a reduction in the error rate comes with architectural improvements in the design of the networks that generalize to different datasets and provide greater insights as to how gradient-based learning can be optimized. Among others, the following architectures have contributed significantly in such advancements; VGG [31] shows how deeper architectures could improve the performance of the model using smaller convolutional filters. Inception [33] consists of multiple inception modules which build

© Springer Nature Switzerland AG 2019
I. Rojas et al. (Eds.): IWANN 2019, LNCS 11507, pp. 220–234, 2019.
https://doi.org/10.1007/978-3-030-20518-8_19

on top of previous architectures as [24] where 1×1 convolutional kernels are used to reduce the total number of parameters in the network, effectively reducing training time and allowing for deeper networks. ResNet [19] tackles the challenge of training ever deeper networks by adopting a residual learning perspective, so that each layer learns the residual functions with respect to its input instead of unreferenced signals. This is achieved by adding identity skip connections to the stacked layers. Similarly to [32], information from one layer reaches the next via different paths, and with different levels of processing, speeding up training. DenseNet [21] takes on results produced by [19] and extends skip connections from one layer to the following one, by effectively shortening the paths between inputs and outputs while jointly increasing the overall network depth.

Training networks from scratch for image classification problems can be extremely demanding in terms of computational time, memory and resources in general. This is due to the high learning capacity of the models which translates in a huge number of learnable parameters and multiply-add operations, as shown in [8]. Nevertheless, learning to recognize objects belonging to a set of classes requires knowledge that can be transferred to perform similar tasks, like recognizing objects belonging to a different set of classes, by thus exploiting a-priori knowledge [26,28]. In the context of CNNs, the assumption is that features learned at different levels of depth in the network might be similar across different training sets. This means that the weights of networks trained on a dataset can be reused to provide a good weight initialization for a network that needs to be trained on a new dataset. Alternatively, the first layers of the trained network can be plugged, with their weights fixed, to a trainable fully connected network.

Here an application of CNN to the classification of images coming from a real industrial application is proposed and compared with classical Machine Learning (ML) approaches, by proving their effectiveness and general validity, which allow a fast deployment in the industrial context.

The paper is organised as follows: Sect. 2 briefly depicts the industrial application; Sect. 3 describes the generation of the dataset which are fed to the different ML approaches starting from real world data coming form the industrial plant; Sect. 4 provides an overview of the explored learning architectures relying on traditional ML and Deep Learning (DL) algorithm. Validation results and that obtained through the best performing method on the test set coming from the generated dataset are presented in Sect. 5. Finally, some concluding remarks are provided in Sect. 6.

2   The Industrial Context

In steel strip production, the hot rolling process plays a fundamental role in the quality of the final product. For flat products, the starting point of such process is the slab, a long steel ingot with rectangular section, which is initially heated in a reheating furnace in order to make it workable and allow thickness reduction in the Hot Rolling Mill (HRM). The most significant thickness reduction is obtained in the initial stage of the HRM, namely the roughing mill, where the heated slab enters after a descaling phase, which removes the oxidized carbon from the slab. Then the semi-manufactured product enters the finishing mill, which provides the strips with the final thickness. The strip, whose length can be up to several hundreds meter, undergoes a final cooling phase before being coiled.

Hot rolled products surface and shape defects can strongly affect the final quality of the final flat steel products. In particular, strip flatness defects are very relevant, as they are indicators of lack of uniformity in the hot rolling process but they can be detected only at the end of the finishing stage. Therefore they are not recoverable and their presence leads to quality downgrading with consequent economic losses. Flatness defects are typically due to different elongation in the strip fibre caused by uneven stresses acting across the width, by the high rolling speed, that lead to fluttering strip or by an uneven thermal gradient across the strip which can generate waviness. Uneven heating/cooling is the main cause of this last type of defect, due to the appearance of internal stresses that can locally overcome the yield stress of the material leading to plastic deformation of the strip [15,16]. These types of defects are divided into two sub-classes depending on whether the edge of the strip is affected or not. In the former case it is a *wave* defect while in the latter it is a *buckle* defect. Detection and classification of defects are performed in a non-automatic way via human operators dedicated to this scope with a serious concern related to the operators' turn-over affecting also the efficiency of the overall process. Thus, an automatic system with detection and classification capabilities can really improve such efficiency helping to mitigate the lack of knowledge left by the retired personnel.

Several successful applications of ML methods in the detection of surface defects on flat steel products can be found in the literature [1–3,10,38], although they mostly deal with the cold rolling stage. In particular, very recently Deep CNNs have been applied to classify surface defects in cold rolled strips [25].

Concerning HRM surface defects of steel strips, different methods and techniques have been used for their detection and classification. Classification starts after defect localization through Bounding Box (BB) extraction via image segmentation. Detection can be done using co-occurrence matrix [7], morphological operation [37], spatial domain filtering [6,12,35,39], frequency domain analysis [36] and joint spatial/spatial-frequency analysis [14,37]. Classification typically exploits supervised classifiers: NNs with Back Propagation [7], Support Vector Machines (SVM) [14,35] or unsupervised classifier via Self-Organizing Maps (SOM) [7] or Learning Vector Quantiser (LVQ) [36].

3   Available Dataset and Preprocessing Stages

3.1   The Raw Data

Strip planarity is usually measured by considering the strip as formed by elements of longitudinal fibre: if all have the same length, the strip is perfectly flat, otherwise the strip is affected by flatness defects. Thus, as the fibre elements do not stretch independently, when they show different lengths, flatness defects appear as waves on the strip.

The transversal flatness profiles of each strip are usually concatenated and represented as a bi-dimensional map of the strip flatness, such as the image reported in Fig. 1, which is produced directly from a measuring system installed at the end of the finishing mill: the green colour indicates good flatness, while the red one stands for bad flatness. A camera provides a line scan along the transversal direction of the strip, with a step of a few meters per line and each line is composed of about 20 readings. Hence, the initial dataset is made up by a list of sensor readings each containing the coordinates of a strip point and a number denoting the flatness level: the higher the value, the worse the flatness.

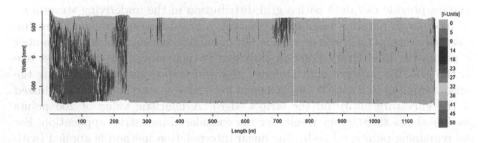

Fig. 1. Example of coloured flatness map. (Color figure online)

3.2 Data Pre-processing

In surface inspection systems commonly applied in the steel industry, especially on hot products, raw data often need to be prepared and pre-processed before the subsequent elaboration stages, in order to remove unreliable data [9] and reduce them in a form suitable to ML systems [30].

In the present application, raw data plots reveal the presence of incorrect sensor readings for a limited number of lines. These errors negatively affect the subsequent processing stages and they need to be removed. Misread sensor lines are represented by a poor alignment with the neighbouring lines, such as reported in Fig. 2a. A subset of faulty lines can be spotted scanning along the strip borders for misaligned points. DBSCAN [13] has been exploited for the purpose, due to its evaluation of point spatial density. In particular, the algorithm has been set up to run on the top and bottom edges of the strip, thus clustering the fringe points into two different classes: the first containing points above a certain density limit, the second holding points far enough from the others to be considered isolated (see Fig. 2b). Upon detection of isolated points, the whole vertical line is discarded. The result of DBSCAN enforcement is a cleaner dataset in both border regularity and planar uniformity.

The following stage consists of planar interpolation of numerical data and can be correctly done when input data are aligned, i.e. every strip point is anchored

(a)   (b)

Fig. 2. (a) Unreliable sensor reading, where some columns are misaligned; (b) DBSCAN clustering: brighter points are outliers

onto a grid. This requirement is additionally enforced by the need to make the whole system capable to work on several input data format, as some monitoring systems provide raw data with a grid distribution in the underlying structure.

In order to obtain a regular distribution, a fitted-for-the-problem algorithm was used: a sampling grid is determined by manually choosing the desired number of points per line. Every point in each line is then moved to the nearest one on the grid. To preserve data integrity, the grid step is chosen considering the sensor sampling step and the displacement between line starting detection (linked to the curvature entity on the strip's edge). A heuristic value of 200 points per line shows satisfactory results for the considered industrial application. For the remaining points, a line-by-line linear interpolation method is applied (with extrapolation part performed using zero values).

3.3   Image Building and Segmentation

The image builder module transforms the input tabular data into a complete image of the strip. Final dimensions of the image are extracted from the different values of coordinates: every different longitudinal coordinate (X) contributes to increase the image width, while transversal coordinates (Y) contribute to the image height, being the strip represented as a long horizontal slice. To preserve as much as possible the aspect ratio of the strip appearance (despite the huge difference between its length – hundreds of meters – and width – around 1 m –) a width expansion by a factor 10 is performed.

A matrix is built through two-dimensional linear interpolation exploiting input data as sample points. A grey-scale image is obtained in which the pixel intensity is proportional to the flatness value, pixel-per-pixel (Fig. 3a). An upper bound was determined by the original dataset and used to cap the flatness for all the processed strips.

To find defects and define their BBs, a segmentation algorithm is required. The selected one is the thresholding, applied to the grey-scale image with a manual threshold, followed by a connected-components regions definition according to [18]. The BBs are computed using border coordinates of every region (see Fig. 3b): a set of features is computed on each BB and, on the basis of such features, a split is evaluated. The split operates on a region separating its upper half part from its lower one by recalculating all the features on the obtained pieces. No recursive splits are performed: after the first, the region is accepted and the process goes on with the other BBs.

In order to distinguish different types of defects, the following set of features have been selected for each region:

- weighted centroid on the strip;
- bounding box information;
- vertical distance from nearest edge;
- horizontal distance from head or tail (whichever is nearer);
- coverage percentage of total BB area;
- maximum and average flatness level detected;
- coverage percentage of central BB band;
- coverage similarity between upper and lower bands.

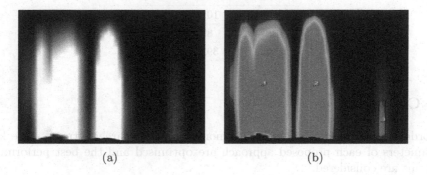

(a)   (b)

Fig. 3. (a) Final greyscale image of the initial portion of a strip. (b) BBs detection stage: the detected BBs are numbered.

3.4 Labelling

The training stage of supervised machine learning-based classifiers requires labelled data, namely, in the present case, a (possibly large) set of sample shape defects which are already correctly classified [34].

In the present application, strip defects recognized by the BB algorithms are manually classified in 4 categories - Wave, Buckle, Multiwave and Multi-buckle - (see Fig. 4) by expert personnel devoted to this purpose within the project. Dataset images come from all available steel strip images each containing a number of defects which is not constant. Of these, 80% are devoted to the training and validation sets while the remaining 20% are test images. This results in a dataset composed of 4806 images: 3938 images were used for training and validation in a 70–30% split, 868 images were used for testing. The classes distribution is shown in Table 1.

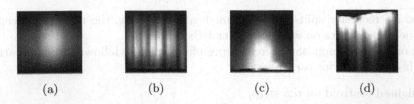

| (a) | (b) | (c) | (d) |

Fig. 4. Exemplar defects extracted from the grey-scale strip image: (a) buckle, (b) multibuckle, (c) wave, (d) multiwave. BBs are resized to squares for better visualization.

Table 1. Classes distribution in the final labelled images dataset

Class	TR&VD	TS
Wave	1435	225
Buckle	1463	261
Multiwave	1019	179
Multibuckle	891	203
Total	3938	868

4   Classification

In order to make a fair comparison among the tested architectures, some key parameters of each proposed approach are optimised and the best performing systems are considered.

4.1   Classical Machine Learning Approaches

Among the classical ML approaches to defect classification, based on the features of the available images and on previous literature results, we have tested Multi-Layer Perceptron (MLP) trained through Back Propagation, SVM, Decision Tree (DT), K-nearest-neighbors (KNN) and Ensemble Methods (EMs). For all the tested classical ML approaches, the best configuration was selected through a grid search over some key hyperparameters. In particular, a 10-fold cross validation (10FCVD) grid search is applied on a dataset obtained collecting together training (TR) and validation (VD) sets.

As far as MLP is concerned, the grid search involves the following hyperparameters:

- Hidden neurons: 5, 20, 50, 100;
- Hidden layer activation functions: Sigmoid, ReLU, Tanh;
- Optimization techniques: Adam [22], Stochastic Gradient Descent (SGD)[1];
- Batch sizes: 32, 128, 512, 1024;
- Number of Epochs: 200, 500, 1000;

---

[1] In this work it is used with Mini-batches and momentum.

- Learning Rates: 0.1, 0.01, 0.001;
- Momentum (SGD only): 0.9, 0.7;
- Nesterov momentum (SGD only): False, True;
- Regularization:
  - L2 (SGD only): 0.001, 0.005, None;
  - Dropout rate: 0.2, 0.4, None;
  - Weight Constrains: 1, 2, 3, None.

SVM is a binary classification algorithm, thus the classification task has been handled in 3 different ways: (i) by embedding the classifiers into a One-vs-One (OVO) scheme (using libsvm implementation); (ii) a Crammer&Singer (CS) scheme (only for linear SVM with liblinear implementation); (iii) a One-vs-Rest (OVR) scheme (only for linear SVM using liblinear implementation).

Concerning the Decision Trees, 3 parameters were considered for the selection of the best configuration: the Maximum Depth *Max Depth*, the minimum number of samples required to split internal nodes *Min sample split* and the minimum number of samples to be a leaf node *Min sample leaf*. The search grid contains the following values:

$$\text{Max Depth (MD): } 3, 5, 8, 10, 15, 20; \tag{1}$$

$$\text{Min sample leaf (MSL): } 5, 10, 15, 20, \ldots, 50; \tag{2}$$

$$\text{Min sample split (MSS): } 2, 12, 22, 32, \ldots, 200. \tag{3}$$

For KNN the considered hyperparameters are the weight function (uniform and weighted) and the number of neighbours (from 2 to 200 with a step of 5).

As far as EMs are concerned, both Bagging and Random Forest have been tested. Bagging is a common EM that exploits bootstrap sampling [4], while Random Forest represents an enhancement of Bagging that can improve variable selection [5]. The considered hyperparameter was the number of different estimators used in the training procedure (10, 40, 100, 200, 500) using DT with the same parameter listed in (1)–(3) as weak learners.

4.2   Convolutional Neural Networks

Given the ever increasing interest in image classification in academic research and industry, a huge number of different architectures is available in the literature. In order to narrow down the many different available options, we used a number of models that proved in the past to be successful in similar challenges and have marked milestones in the topic. In particular, the models described in Sect. 1 provided a valuable starting point for building a CNN suitable for the classification task at hand. Such models are open-sourced and weights are available to download for the particular instances pre-trained on the ImageNet database. Keras [11] allows for a practical implementation of said models and for their weights initialization with transfer learning, such as described in Sect. 1, which makes it the go-to framework for this problem. In order to mitigate the potential overfitting due to the high number of parameters, 4 different instances of each architecture have been trained after gradually pruning the graph, by practically cutting the depth of the network at specific levels. In other words, for each architecture we train 4 different instances with increasing depth. The rationale behind this approach is two-fold:

- One way of trying to avoid overfitting is limiting the actual number of learnable parameters contained within a network;
- Pre-trained CNNs have feature maps that tend to specialize in identifying patterns that are more specific to the dataset the deeper they are in the network. By contrast, we might expect feature maps in earlier layers to generalize better to other datasets, such as ours.

Each pruned model is then plugged to a Global Average Pooling Layer introduced by [24], its output fed to a 100 ReLU units hidden layer which connects to the 4 units output Softmax layer.

5   Results

Table 2 shows the results obtained with the MLP-based classifiers. Such table highlights three remarkable results: (i) (Adam100) shows the best performance; (ii) (SGD5) represents the best compromise between number of learnable parameters and 10FCVD accuracy; (iii) (SGD20) provides the best compromise between number of parameter updates and 10FCVD accuracy. On one hand, resorting to a 10-fold cross validation increases the confidence on the performance of the models potentially reducing their variance; on the other hand ML models can be trained with less computational effort with respect to that required by DL models, making this approach technically feasible.

MLP-based architectures with gradient descent appear more accurate with bigger batch sizes requiring less parameter updates, while the best Adam metaparameters configuration was the same as in [22]. Concerning SGD, our search showed that Nesterov Momentum always leads to better performances compared to standard Momentum. All the tested models improve their performance using regularization: no model trained without regularization is in the top 20 cross validation ranking. Moreover, in our top 20 ranking no improvements are obtained in configurations with Dropout with max-norm. In the output layer a Softmax activation function is used, and the loss function used on training is the Categorical Cross-Entropy. Xavier initialization [17] is used for Sigmoid and Tanh activation functions while He [20] is used with ReLUs.

The performed experiments show that small networks (most of the time with SGD optimization) with Nesterov Momentum and L2 regularization represent the best compromise in terms of accuracy and training time, while in order to achieve the highest accuracy Adam with Dropout is preferable (in accordance with the results obtained in [27]). Noticeably, even with a small number of hidden neurons (5) a good average accuracy can be achieved.

Table 3 shows the performances of the best SVM-based classifiers according to the 10FCVD accuracy for the 3 multiclass encoding schemes. Noticeably Adam100 outperforms all the SVM-based classifiers.

Table 2. MLP grid search results.

Optim	Hid	Batch	Ep	Act	Init	Lr	Mom[a]	Reg	10FCVD	N[b]
Adam	100	512	500	ReLU	He	0.001	None	Dr(0.4)	0.8451	2004
SGD	5	1024	1000	Tanh	Xa	0.10	0.9	L2(0.001)	0.8359	104
SGD	20	1024	200	ReLU	He	0.10	0.9	L2(0.001)	0.8370	404

[a]Non-zero values refer to Nesterov Momentum
[b]Trainable Parameters

Table 3. SVM: best grid search results.

Kernel	c_0	γ	Scheme	10FCVD
Rbf	1000	0.01	OVO	0.8387
Linear	10	None	CS	0.8334
Linear	10	None	OVR	0.8095

The best DT-based classifier on 10FCVD resulted with MD = 8, MSL = 12, MSS = 42 achieving a mean validation accuracy of 0.7912, which is far worse than the one provided by both MLPs and SVMs.

The best KNN-based classifier on 10FCVD resulted with 17 neighbours and a weighted distance, achieving a mean validation accuracy of 0.762, which represents the worst performance among classical ML approaches.

For each of the two considered EMs a 10FCVD involving a total of 12000 different configurations was carried out, obtaining the results reported in Table 4, where NWL stands for Number of Weak Learners.

Table 4. Ensemble methods: best grid search results.

Method	MD	MSL	MSS	NWL	10FCVD
Random	20	2	2	100	0.8225
Bagging	15	2	2	100	0.8245

As far as CNNs are concerned, the particular incarnations of the original architectures, their depth (i.e. the number of hidden layers), and their learnable parameters are summarized in Table 5. Each instance is fully trained for up to 100 epochs to minimize the Categorical Cross Entropy over batches of 128 images via Adam [22] optimizer. Training is stopped when the training loss keeps decreasing but the validation loss starts increasing which happens when overfitting data (early stopping method). During training the number of images is artificially augmented by applying random affine transformations to the inputs: shear, zoom, vertical and horizontal flip, width and height shift. Validation occurs at every epoch for every architecture and for every instance, with the 30% of the images not included in the test set. Input scaling in the $[0,1]$ range is performed throughout every set, as well as a resizing to a 100×100 squared image.

Table 5. CNNs depth and learnable parameters

Architecture	VGG				Inception			
Depth	6	10	14	18	64	133	229	311
Learnable Params (10^6)	0.273	1.761	7.686	14.766	0.733	3.515	9.033	21.973
Validation Categorical Acc	0.917	0.317	0.925	0.924	0.918	0.925	0.923	0.927
Architecture	ResNet				DenseNet			
Depth	50	92	132	174	49	137	365	593
Learnable Params (10^6)	0.659	3.067	7.542	23.739	0.370	1.349	5.874	12.648
Validation Categorical Acc	0.893	0.910	0.902	0.914	0.898	0.915	0.919	0.921

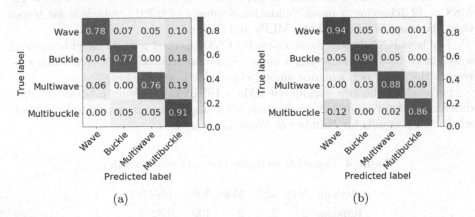

(a)   (b)

Fig. 5. Normalized confusion matrix for the Adam100 (a) and Inception311 (b) models on the test set.

To sum up, the best classical ML classifier (Adam100) shows a validation accuracy of 0.8451; applied to the test set this value goes to 0.782 and the corresponding confusion matrix is shown in Fig. 5.

Additionally, based on the results on the validation set in Table 5, the CNN-based Inception311 classifier shows the best performances with a categorical accuracy of 0.927. Applied to the test set, this value goes to 0.892, and the corresponding confusion matrix is shown in Fig. 5 where the only misclassification reaching more than 10% of the specific class images is when telling Multi-buckles and Waves apart. Note that this model is the second biggest in terms of trainable parameters but has about half the layers of the deepest network,

DenseNet593. In turn DenseNet593 has a validation accuracy of 0.921 but with way less parameters, close to half of those in Inception311. In this context we are just interested in the categorical accuracy, but in memory-critical applications a trade off between number of parameters and performance would be necessary and show a different winner. Curiously, VGG10's drop in accuracy is not easily explainable, and persists while varying train-validation splits, optimizer and a reduction to 9 layers. Further investigation may be carried out in future work.

6   Conclusions and Future Work

Flatness defect detection and classification on hot-rolling steel strip has been faced via image segmentation coupled with classical ML and DL techniques. CNNs proved to outperform a number of classical ML approaches (MLP, SVM, Decision Trees, KNN) and the best model is Inception311, with a categorical accuracy of about 90% on the test set. The worst performing classes are Multibuckle and Multiwave. This might be explained by the fact that these classes typically show a longer elongation on the longitudinal direction which is lost during the scaling of the input to a squared image. If we inspect the confusion matrix of the best performing ML model (see Fig. 5a) to make a comparison, this tendency is mitigated by exploiting other features that are lost when plugging the BBs straight to the CNN, such as the BB location with respect to the image coordinates. In fact, Waves and Multiwaves tend to distribute over the edge while Buckles and Multibuckles emerge in the middle of the strip. Moreover, the proposed solution consists in a novel complete and flexible framework which is fed with real world data coming from the plant and automatically detects and classifies flatness defects. Currently 4 different defect classes are distinguished, but the proposed approach can be easily extended by exploiting a wider dataset. The proposed method also represents a complete pipeline to support plant personnel in detecting and recognizing defective strips. Possible improvements include developing a single network architecture to perform simultaneous localization and classification.

Acknowledgments. The work described in the present paper was developed within the project entitled *Integration of complex measurement information of thick products to optimise the through process geometry of hot rolled material for direct application* INFOMAP (Contract No. RFSR-CT-2015-00008) that has received funding from the Research Fund for Coal and Steel of the European Union. The sole responsibility of the issues treated in the present paper lies with the authors; the Commission is not responsible for any use that may be made of the information contained therein.

References

1. Borselli, A., Colla, V., Vannucci, M.: Surface defects classification in steel products: a comparison between different artificial intelligence-based approaches. In: Proceedings of the 11th IASTED International Conference on Artificial Intelligence and Applications, AIA 2011, pp. 129–134 (2011)
2. Borselli, A., Colla, V., Vannucci, M., Veroli, M.: A fuzzy inference system applied to defect detection in flat steel production. In: IEEE World Congress on Computational Intelligence, WCCI 2010 (2010)
3. Brandemburger, J., Colla, V., Nastasi, G., Ferro, F., Schirm, C., Melcher, J.: Big data solution for quality monitoring and improvement on flat steel production. IFAC-PapersOnLine **49**(10), 55–60 (2016)
4. Breiman, L.: Bagging predictors. Mach. Learn. **24**(2), 123–140 (1996)
5. Breiman, L.: Random forests. Mach. Learn. **45**(1), 5–32 (2001)
6. Bulnes, F., Usamentiaga, R., Garcia, D., Molleda, J.: Vision-based sensor for early detection of periodical defects in web materials. Sensors **12**(8), 10788–10809 (2012)
7. Caleb, P., Steuer, M.: Classification of surface defects on hot rolled steel using adaptive learning methods. In: Proceedings of the Fourth International Conference on Knowledge-Based Intelligent Engineering Systems and Allied Technologies, KES 2000 (Cat. No. 00TH8516), vol. 1, pp. 103–108, August 2000
8. Canziani, A., Paszke, A., Culurciello, E.: An analysis of deep neural network models for practical applications. CoRR abs/1605.07678 (2016)
9. Cateni, S., Colla, V., Nastasi, G.: A multivariate fuzzy system applied for outliers detection. J. Intell. Fuzzy Syst. **24**(4), 889–903 (2013)
10. Cateni, S., Colla, V., Vignali, A., Brandenburger, J.: Cause and effect analysis in a real industrial context: study of a particular application devoted to quality improvement. In: Esposito, A., Faundez-Zanuy, M., Morabito, F.C., Pasero, E. (eds.) WIRN 2017. SIST, vol. 102, pp. 219–228. Springer, Cham (2019). https://doi.org/10.1007/978-3-319-95098-3_20
11. Chollet, F.: Keras documentation (2015). https://keras.io
12. Djukic, D., Spuzic, S.: Statistical discriminator of surface defects on hot rolled steel. In: Image and Vision Computing, New Zealand, pp. 158–163, December 2007
13. Ester, M., Kriegel, H.P., Sander, J., Xu, X.: A density-based algorithm for discovering clusters a density-based algorithm for discovering clusters in large spatial databases with noise. In: Proceedings of the Second International Conference on Knowledge Discovery and Data Mining, KDD 1996, pp. 226–231. AAAI Press (1996)
14. Ghorai, S., Mukherjee, A., Gangadaran, M., Dutta, P.K.: Automatic defect detection on hot-rolled flat steel products. IEEE Trans. Instrum. Meas. **62**(3), 612–621 (2013)
15. Ginzburg, V.B.: High-Quality Steel Rolling: Theory and Practice. Marcel Dekker, New York (1993)
16. Ginzburg, V.B., Ballas, R.: Flat Rolling Fundamentals. Marcel Dekker, New York (2000)
17. Glorot, X., Bengio, Y.: Understanding the difficulty of training deep feedforward neural networks. In: Teh, Y.W., Titterington, M. (eds.) Proceedings of the Thirteenth International Conference on Artificial Intelligence and Statistics. Proceedings of Machine Learning Research, vol. 9, pp. 249–256. PMLR, Chia Laguna Resort, Sardinia, Italy, 13–15 May 2010

18. Haralick, R.M., Shapiro, L.G.: Computer and Robot Vision, 1st edn. Addison-Wesley Longman Publishing Co. Inc., Boston (1992)
19. He, K., Zhang, X., Ren, S., Sun, J.: Deep residual learning for image recognition. CoRR abs/1512.03385 (2015)
20. He, K., Zhang, X., Ren, S., Sun, J.: Delving deep into rectifiers: Surpassing human-level performance on imagenet classification. In: Proceedings of the 2015 IEEE International Conference on Computer Vision (ICCV), ICCV 2015, pp. 1026–1034. IEEE Computer Society, Washington, DC, USA (2015)
21. Huang, G., Liu, Z., Weinberger, K.Q.: Densely connected convolutional networks. CoRR abs/1608.06993 (2016)
22. Kingma, D.P., Ba, J.: Adam: a method for stochastic optimization. CoRR abs/1412.6980 (2014)
23. Lecun, Y., Bottou, L., Bengio, Y., Haffner, P.: Gradient-based learning applied to document recognition. Proc. IEEE **86**(11), 2278–2324 (1998)
24. Lin, M., Chen, Q., Yan, S.: Network in network. CoRR abs/1312.4400 (2013)
25. Liu, Y., Geng, J., Su, Z., Zhang, W., Li, J.: Real-time classification of steel strip surface defects based on deep CNNs. In: Lecture Notes in Electrical Engineering, pp. 257–266. No. 529 (2019)
26. Long, M., Wang, J., Ding, G., Sun, J., Yu, P.S.: Transfer feature learning with joint distribution adaptation. In: IEEE International Conference on Computer Vision, pp. 2200–2207, December 2013
27. Phaisangittisagul, E.: An analysis of the regularization between L2 and dropout in single hidden layer neural network. In: 7th International Conference on Intelligent Systems, Modelling and Simulation (ISMS), pp. 174–179, January 2016
28. Razavian, A.S., Azizpour, H., Sullivan, J., Carlsson, S.: CNN features off-the-shelf: an astounding baseline for recognition. CoRR abs/1403.6382 (2014)
29. Russakovsky, O., et al.: Imagenet large scale visual recognition challenge. Int. J. Comput. Vision (IJCV) **115**(3), 211–252 (2015)
30. Sgarbi, M., Colla, V., Cateni, S., Higson, S.: Pre-processing of data coming from a laser-EMAT system for non-destructive testing of steel slabs. ISA Trans. **51**(1), 181–188 (2012)
31. Simonyan, K., Zisserman, A.: Very deep convolutional networks for large-scale image recognition. CoRR abs/1409.1556 (2014)
32. Srivastava, R.K., Greff, K., Schmidhuber, J.: Highway networks. CoRR abs/1505.00387 (2015)
33. Szegedy, C., et al.: Going deeper with convolutions. In: IEEE Conference on Computer Vision and Pattern Recognition (CVPR), pp. 1–9 (2015)
34. Vannucci, M., Colla, V., Sgarbi, M., Toscanelli, O.: Thresholded neural networks for sensitive industrial classification tasks. In: Cabestany, J., Sandoval, F., Prieto, A., Corchado, J.M. (eds.) IWANN 2009, Part I. LNCS, vol. 5517, pp. 1320–1327. Springer, Heidelberg (2009). https://doi.org/10.1007/978-3-642-02478-8_165
35. Wu, G., Kwak, H., Jang, S., Xu, K., Xu, J.: Design of online surface inspection system of hot rolled strips. In: Proceedings of the IEEE International Conference on Automation and Logistics (ICAL), pp. 2291–2295 (2008)
36. Wu, G., Zhang, H., Sun, X., Xu, J., Xu, K.: A bran-new feature extraction method and its application to surface defect recognition of hot rolled strips. In: IEEE International Conference on Automation and Logistics, pp. 2069–2074, August 2007
37. Wu, X., Xu, K., Xu, J.: Application of undecimated wavelet transform to surface defect detection of hot rolled steel plates. In: Congress on Image and Signal Processing, vol. 4, pp. 528–532, May 2008

234   M. Vannocci et al.

38. Xu, K., Xu, Y., Zhou, P., Wang, L.: Application of rnamlet to surface defect identification of steels. Opt. Lasers Eng. **105**, 110–117 (2018)
39. Yang, S., He, Y., Wang, Z., Zhao, W.: A method of steel strip image segmentation based on local gray information. In: IEEE International Conference on Industrial Technology, pp. 1–4, April 2008

Image Completion with Filtered
Low-Rank Tensor Train Approximations

Rafał Zdunek$^{(\boxtimes)}$, Krzysztof Fonał, and Tomasz Sadowski

Faculty of Electronics, Wroclaw University of Science and Technology,
Wybrzeze Wyspianskiego 27, 50-370 Wroclaw, Poland
{rafal.zdunek,krzysztof.fonal,tomasz.sadowski}@pwr.edu.pl

Abstract. The topic of image completion has received increasing attention in recent years, motivated by many important applications in computer vision, data mining and image processing. In this study, we consider
the problem of recovering missing values of pixels in highly incomplete
images with a random or irregular structure. The analyzed gray-scale or
colour images are transformed to multi-way arrays which are then recursively approximated by low-rank tensor decomposition models. In our
approach, the multi-way array is represented by the tensor train model,
and in each iterative step, the low-rank approximation is filtered with
the Gaussian low-pass filter. As a result, the proposed algorithms considerably outperform the state-of-the art methods for matrix and tensor
completion problems, especially when an incompleteness degree is very
high, e.g. with 90% of missing pixels.

Keywords: Image completion · Tensor train decomposition ·
Low-rank approximations

1 Introduction

Image completion aims to automatically fill-in target regions in incomplete
images with alternative contents that are estimated from accessible and unper-
turbed regions. The target regions can be regarded as unwanted objects, occlusion, or "dead" pixels that are not observed or considerably perturbed. In the
literature, the topic of automatic image completion or image inpainting has been
studied for many years and several computational approaches exist to address
this problem. Bertalmio *et al.* [1] formalized the image inpainting problem and
proposed a milestone algorithm that is based on a diffusion-based technique
to fill-in missing regions based on the neighboring information. To fill-in large
regions, the exemplar-based image synthesis algorithms seem to be more appropriate [2–4]. Various neural network architectures, e.g. convolutional neural network (CNN) [5,6], have been used recently to tackle this problem.

When the target regions are small and irregularly dissipated, the best
alternative seems to be the methods that approximate an observed incomplete image with a low-rank model. This approach is based on the assumption that the original image has a low-rank structure usually represented by

© Springer Nature Switzerland AG 2019
I. Rojas et al. (Eds.): IWANN 2019, LNCS 11507, pp. 235–245, 2019.
https://doi.org/10.1007/978-3-030-20518-8_20

clusters of similar patches. Many computational strategies for matrix completion problems involve the nuclear norm minimization as a convex relaxation to a rank-minimization problem which is a NP-hard. The examples include the soft-thresholding algorithm (SVT) [7], truncated nuclear norm minimization (TNNM) [8], and weighted nuclear norm regularization (WNNR) [9]

The nuclear norm minimization problems are usually solved iteratively, and the singular value decomposition (SVD) of the approximated matrix is computed in each iteration, which leads to a high computational cost. To tackle this problem, low-rank approximations can be performed using other low-rank factorization models (unnecessary restricted to SVD). For example, Wen *et al.* [10] use the nonlinear successive over-relaxation (SOR) scheme to iteratively update low-rank factors in matrix completion problems. Wang and Zhang [11] used nonnegative matrix factorization (NMF) with sparsity constraints to sequential recover a low-rank matrix based on the idea of patch propagation inpainting. NMF has also been used in other approaches, e.g. together with smoothness constraints [12].

NMF, despite its intrinsic nonnegativity constraints and its flexibility to other constraints (sparsity, smoothness, orthogonality, etc.), has one important drawback, which is its susceptibility to initialization due to non-convex alternating optimization. The uniqueness conditions are more relaxed in tensor decomposition models, e.g. in the CP model [13]. Moreover, multi-linear decomposition models are more robust to modelling cross-modal interactions, which is particularly useful for image processing. The idea of tensor inpainting with the CP model has been initialized by Acar *et al.* [14], and then developed in other works, e.g. including the Bayesian CP with a hierarchical probabilistic model [15] or the smooth CP [16], where the smoothness constraints are imposed through regularization terms.

Tensor decomposition models are useful not only in color image completion, i.e. 3D arrays, but they can also be applied to gray-scale images by tensorisation operations. A matrix can be transformed to a multi-way array by its reshaping or by using more advanced scanning operations. One of them is the ket augmentation (KA) that was introduced in [17]. Using the KA or other reshaping operations, any N-way tensor to be completed can be transformed to a multi-way array with a large number of modes. A low-rank approximation of such a multi-way observation array can be robustly performed with many tensor network models, e.g. such as the tensor train (TT) decomposition [18]. The recent works [19–21] show that TT models work also very efficiently for image completion problems.

Motivated by the efficiency of TT models in solving image completion problems, we further develop this research area by proposing new models and related numerical algorithms for TT-based image completion. In our study, we consider two versions of a tensor completion model. In the first approach, an observed incomplete RGB tensor is given to mode permutation in such a way that the mode representing the colors is the middle mode. This approach allows us to set higher ranks for decomposition. In the other version, we use the KA to formulate a multi-modal incomplete tensor which is then approximated with the SVD-based TT model. In both versions, we propose to use the Gaussian low-pass filtration in each iterative step to smooth a recovered image additionally (besides a low-rank approximation). The numerical experiments demonstrate high efficiently of the proposed algorithms with respect to the state-of-the art image completion algorithms.

The remainder of this paper is organized as follows. The tensor completion problem and its relaxation with the TT model are presented in Sect. 2. The proposed algorithms are described in Sect. 3. The numerical experiments performed on image completion problems with various degrees of incompleteness are given in Sect. 4. The last section contains the conclusions.

2   Model

Let $\mathcal{M} = [m_{i_1,\ldots,i_M}] \in \mathbb{R}_+^{I_1 \times \ldots \times I_M}$ be an original incomplete M-order tensor, and Ω be the set of indices of known entries in $\mathcal{M}$. The aim of the tensor completion problem is to find the minimum-rank tensor $\mathcal{X} = [x_{i_1,\ldots,i_M}] \in \mathbb{R}_+^{I_1 \times \ldots \times I_M}$ that has the same entries as the tensor $\mathcal{M}$ in the items indicated by the set Ω. Such a problem can be formulated as follows:

$$\min_{\mathcal{X}} \operatorname{rank}(\mathcal{X}), \quad \text{s.t.} \quad \mathcal{X}(\Omega) = \mathcal{M}(\Omega). \tag{1}$$

The problem (1) is NP-hard and its solution can be approximated by its convex surrogate: $\min_{\mathcal{X}} ||\mathcal{X}||_*$, s.t. $\mathcal{X}(\Omega) = \mathcal{M}(\Omega)$, where $|| \cdot ||_*$ stands for the nuclear norm.

Since the number of modes in an input tensor $\mathcal{M}$ is usually small, i.e. $M = 2$ for a gray-scale image (matrix), and $M = 3$ for a color image, $\mathcal{M}$ can be transformed to a higher-order tensor $\bar{\mathcal{M}} \in \mathbb{R}_+^{I_1 \times \ldots \times I_N}$, where $N > M$. Let $\bar{\mathcal{M}} = \Psi(\mathcal{M}, [I_1, \ldots, I_N])$, where $\Psi(\cdot)$ denotes the reordering operation that can be expressed by a simple reshaping or the KA operation.

In this study, we assume that $\bar{\mathcal{X}} = \Psi(\mathcal{X}, [I_1, \ldots, I_N]) \in \mathbb{R}_+^{I_1 \times \ldots \times I_N}$ is approximated by the tensor $\bar{\mathcal{Z}}$ that is given in the TT format:

$$\bar{\mathcal{Z}} = \sum_{r_1=1}^{R_1} \sum_{r_2=1}^{R_2} \ldots \sum_{r_{N-1}=1}^{R_{N-1}} \bar{\mathcal{Z}}_1(1, :, r_1) \circ \bar{\mathcal{Z}}_2(r_1, :, r_2) \circ \ldots \circ \bar{\mathcal{Z}}_N(r_{N-1}, :, 1)$$

$$= \bar{\mathcal{Z}}_1 \bullet \bar{\mathcal{Z}}_2 \bullet \ldots \bullet \bar{\mathcal{Z}}_N, \tag{2}$$

where $\bar{\mathcal{Z}}_n \in \mathbb{R}^{R_{n-1} \times I_n \times R_n}$ is the core tensor for the n-th mode $(n = 1, \ldots, N)$, and the symbols $\circ$ and $\bullet$ denote the outer product and the tensor train contraction operator[1], respectively. The TT-ranks in (2) are determined by the set $\{R_0, R_1, \ldots, R_{N-1}, R_N\}$, where $R_0 = R_N = 1$. Following the concept proposed

---

[1] The tensor train contraction between the tensors $\mathcal{A} \in \mathbb{R}_+^{I_1 \times \ldots \times I_N}$ and $\mathcal{B} \in \mathbb{R}_+^{J_1 \times \ldots \times J_M}$ with $I_N = J_1$ performs a tensor contraction between the last mode of $\mathcal{A}$ and the first mode of $\mathcal{B}$, which results in $\mathcal{C} = \mathcal{A} \bullet \mathcal{B} \in \mathbb{R}_+^{I_1 \times \ldots \times I_{N-1} \times J_2 \times \ldots \times J_M}$.

in [16] for the tensor completion with the CP model, we formulate the following optimization problem:

$$\min_{\bar{\mathcal{X}}, \bar{\mathcal{Z}}_1, \ldots, \bar{\mathcal{Z}}_N} \frac{1}{2} \|\bar{\mathcal{X}} - \bar{\mathcal{Z}}\|_F^2 + \Phi(\bar{\mathcal{Z}}),$$

$$\text{s.t.} \, \bar{\mathcal{Z}} = \bar{\mathcal{Z}}_1 \bullet \ldots \bullet \bar{\mathcal{Z}}_N, \quad \mathcal{X}(\Omega) = \mathcal{M}(\Omega), \quad \text{and} \quad \mathcal{X}(\Omega) \geq \mathbf{0}, \tag{3}$$

where $\Phi(\bar{\mathcal{Z}})$ is the function enforcing a certain property of $\bar{\mathcal{Z}}$, e.g. smoothness.

3   Image Completion Algorithm

To estimate the core tensors $\{\bar{\mathcal{Z}}_n\}$ in (2), given $\bar{\mathcal{Z}}$ and the TT-ranks, we use the SVD-based computational scheme that was introduced in [18], and then developed in many papers [19–22]. The concept assumes a sequential computation of the core tensors with SVD which is applied to reshaped tensor sub-trains. This procedure is illustrated by Algorithm 1.

Algorithm 1. SVD-TT

    Input : $\bar{\mathcal{Z}} \in \mathbb{R}^{I_1 \times \ldots \times I_N}$ – input tensor, $\{R_1, \ldots, R_{N-1}\}$ – TT-ranks
    Output: $\{\bar{\mathcal{Z}}_n\}$ – estimated core tensors

1 $R_0 = 1$
2 $\bar{Z} = \bar{\mathcal{Z}}_{(1)} \in \mathbb{R}_+^{I_1 \times \prod_{p>1}^N I_p}$ % Mode-1 unfolding
3 **for** $n = 1, \ldots, N-1$ **do**
4   Compute: $[\tilde{U}, \tilde{S}, \tilde{V}] = \text{svd}(\bar{Z}, R_n)$ % truncated SVD given rank R_n
5      $\bar{\mathcal{Z}}_n = \text{reshape}(\tilde{U}, [R_{n-1}, I_n, R_n])$
6      $\bar{Z} = \text{reshape}(\tilde{S}\tilde{V}^T, [R_n I_{n+1}, l_n])$, where $l_n = \max(1, \prod_{p=n+2}^N I_p)$,
7 $\bar{\mathcal{Z}}_N = \text{reshape}(\bar{Z}, [R_{N-1}, I_N, 1])$

In this study, we do not formulate the penalty function Φ in (3) explicitly but we assume an expected result of using such penalty, and this effect is obtained by the usage of a certain filtration in each iterative step. We expect local smoothness in the estimated image, and hence, a Q-order low-pass (LP) Gaussian filter $\mathcal{F}$ is applied to iterative low-rank approximations, including the initialization. To satisfy the nonnegativity constraint in (3), a low-rank approximation obtained with the SVD-TT (Algorithm 1) is projected onto the nonnegative orthant by the projection $\mathcal{P}_+[\cdot]$. The final version of the proposed algorithm is presented by Algorithm 2.

Motivated by [16], we introduced in Step 8 a possibility of changing the rank of low-rank updates versus iterations. In general, the TT-ranks can be determined by the mapping $\mathcal{F}_R([R_0, \ldots, R_N], k)$ in the k-th iteration. This operation can be imposed onto all or selected ranks, e.g. only the rank R_2.

Algorithm 2 can be run with various options, mostly determined by the operation Ψ in Steps 1, 2, and 14. We consider the following approaches:

Algorithm 2. IC-TT

Input : $\mathcal{M} \in \mathbb{R}^{I_1 \times \cdots \times I_M}$ – incomplete M-th order tensor, $\Omega \in \mathbb{R}^{I_1 \times \cdots \times I_M}$ –
binary tensor of indices of known entries in $\mathcal{M}$, $\Psi(\cdot)$ – reordering
operator, $R = [R_0, \ldots, R_N]$ – TT-ranks, Q – order of LP filter, τ -
threshold for stopping iterations, k_{max} - maximum number of
iterations

Output: $\mathcal{X}$ - completed tensor

1 $\bar{\mathcal{M}} = \Psi(\mathcal{M}, [I_1, \ldots, I_N]) \in \mathbb{R}_+^{I_1 \times \cdots \times I_N}$;   // Reordering of incomplete image
2 $\bar{\Omega} = \Psi(\Omega, [I_1, \ldots, I_N]) \in \mathbb{R}_+^{I_1 \times \cdots \times I_N}$;   // Reordering of Ω
3 **Initialization:** $\bar{\mathcal{X}} = \mathcal{F}(\bar{\mathcal{M}}, Q)$;   // Low-pass filtration
4 $\epsilon_0 = \frac{\|\bar{\mathcal{X}}(\Omega) - \bar{\mathcal{M}}(\Omega)\|_F}{\|\mathcal{M}(\Omega)\|_F}$;   // Initial residual error
5 $\bar{\mathcal{X}}(\Omega) = \bar{\mathcal{M}}(\Omega)$, $k = 1$;

6 **while** $\epsilon_{k-1} - \epsilon_k > \tau$ and $2 < k \leq k_{max}$ **do**
7      $k \leftarrow k + 1$;
8   Update TT-ranks: $[\tilde{R}_0, \ldots, \tilde{R}_N] \leftarrow \mathcal{F}_R([R_0, \ldots, R_N], k)$;
9      $[\bar{\mathcal{X}}_1, \ldots, \bar{\mathcal{X}}_N] = \text{SVD} - \text{TT}(\bar{\mathcal{X}}, [\tilde{R}_0, \ldots, \tilde{R}_N])$;   // Algorithm 1
10     $\bar{\mathcal{X}} = \mathcal{P}_+ [\bar{\mathcal{X}}_1 \bullet \ldots \bullet \bar{\mathcal{X}}_N]$;   // Projection
11     $\bar{\mathcal{X}} \leftarrow \mathcal{F}(\bar{\mathcal{X}}, Q)$;   // Low-pass filtration
12     $\epsilon_k = \frac{\|\bar{\mathcal{X}}(\Omega) - \bar{\mathcal{M}}(\Omega)\|_F}{\|\mathcal{M}(\Omega)\|_F}$;   // Residual error
13     $\bar{\mathcal{X}}(\Omega) = \bar{\mathcal{M}}(\Omega)$;   // Correction

14 $\mathcal{X} = \Psi^{-1}(\bar{\mathcal{X}}, [I_1, \ldots, I_M]) \in \mathbb{R}_+^{I_1 \times \cdots \times I_M}$;   // Reordering inverse operation

---

- **P-TT** (Mode-permutation): Let $\mathcal{M} \in \mathbb{R}_+^{I_1 \times I_2 \times 3}$ be a RGB image of the resolution $I_1 \times I_2$. Then: $\bar{\mathcal{M}} = \Psi(\mathcal{M}, [I_1, I_3, I_2]) \in \mathbb{R}_+^{I_1 \times 3 \times I_2}$ (the last mode is permuted with the middle mode).
- **KA-TT** (Ket augmentation): $\bar{\mathcal{M}} = \Psi(\mathcal{M}, [I_1, \ldots, I_N]) \in \mathbb{R}_+^{I_1 \times \cdots \times I_N}$, where N results from the size of the patch used for scanning in the KA procedure [17].

The P-TT option allows us to avoid scanning of an image, such as in the KA, which reduces on overall computational complexity. Due to the mode permutation, the ranks R_1 and R_2 can be set to higher values. Without this operation, R_2 could not be set a higher value than three because $\bar{Z} \in \mathbb{R}^{R_1 I_2 \times I_3}$, where $I_3 = 3$, and the SVD could not provide more than three right singular vectors. After the mode-permutation, we have $I_3 >> 3$ and if R_1 is large, R_2 can be set to the maximum value $R_2^{(max)} = \min\{R_1 I_2, I_3\}$, which is usually larger than three. The KA-TT gives us better flexibility in selection of TT-ranks, however, the scanning operation in the KA may be time-consuming, especially for a large image. Moreover, the input image must be reshaped to the resolution $[2^{d_1} \times 2^{d_2}]$, where d_1 and d_2 are natural numbers.

4   Experiments

4.1   Setup

The proposed algorithm is evaluated for various image completion problems using the natural image that is illustrated in Fig. 1.

Fig. 1. Original image ($512 \times 512 \times 3$, RGB colors) (Color figure online)

The experiments are carried out on the workstation with the following parameters: Windows 10, CPU Intel i7-7820k 3.60 GHz with Intel Turbo Boost Max Technology 3.0 enabled, 32 GB RAM, using Matlab 2016a and its parallelization pool.

The tests are performed on the benchmark image `barbara` ($512 \times 512 \times 3$ pixels, RGB color). The incomplete images are obtained by removing from the original one: (a) randomly selected 50%, 70% and 90% of pixels, (b) single lines of pixels forming a regular grid of 10 pixels wide. In this experiment, the proposed methods (P-TT and KA-TT) are compared with following algorithms: the HALS for image completion [23], SVT [7], LMaFit [10], SmPC-QC [16], TMAC-inc and TMAC-dec [24].

Each algorithm is re-run 100 times for various random initializations, with a maximum number of iterations limited to 100, the rank is set to 50 for all but P-TT and KA-TT, and the threshold $\tau = 10^{-8}$. For P-TT, $R_1 = R_2 = 50$, and we set $[R_1, \ldots, R_4] = [20, 30, 20, 3]$ for KA-TT with the window of 16×16 pixels in the KA.

4.2   Results

The recovered images are validated quantitatively using the Signal-to-Interference Ratio (SIR) measure [13], defined as $\mathrm{SIR} = 20 \log_{10} \frac{\|\mathcal{M}\|_F}{\|\mathcal{M} - \mathcal{X}\|_F}$. The SIR-values are averaged over the colormaps. The boxplots of SIR values are presented in Fig. 2. The incomplete and randomly selected completed images are shown in Figs. 3 and 4. The averaged runtime [in seconds] of each algorithm is listed in Table 1.

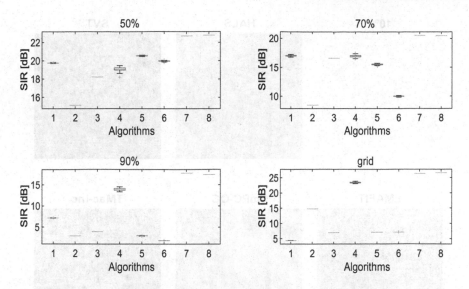

Fig. 2. SIR statistics obtained with the algorithms: 1 – HALS, 2 – SVT, 3 – LMaFit, 4 – SmPC-QC, 5 – TMAC-inc, 6 – TMAC-dec, 7 – P-TT, 8 – KA-TT.

Table 1. Averaged runtime [sec.] and the standard deviation obtained with the tested algorithms for various image completion problems.

	50%	70%	90%	grid
HALS	120.51 ± 4.6	104.06 ± 3.59	103.8 ± 4.76	130.68 ± 12.12
SVT	56.66 ± 7.34	55.48 ± 6.51	54.55 ± 6.59	48.71 ± 6.76
LMaFit	3.87 ± 0.67	2.68 ± 0.39	1.93 ± 0.41	4.75 ± 0.89
SmPC-QC	81.62 ± 190.34	57.48 ± 6.6	43.83 ± 9.84	48.72 ± 7.59
TMac-inc	17.57 ± 2.87	15.83 ± 2.21	12.27 ± 2.26	19.59 ± 3.27
TMac-dec	17.34 ± 2.87	16.86 ± 2.37	13.6 ± 2.43	17.95 ± 3.23
P-TT	51.56 ± 7.59	49.82 ± 6.84	47.52 ± 6.66	47.85 ± 7.47
KA-TT	97.1 ± 3.94	94.33 ± 4.13	91.78 ± 5.16	87.22 ± 8.26

4.3   Analysis of Experimental Results

In this study, we proposed a new algorithmic approach to an image completion problem. The image to be completed is iteratively approximated by a low-rank model expressed by the TT decomposition. We considered two options. First, an incomplete image is assumed to be represented by a 3D tensor which is approximated with a short TT model (only three core tensors) after the mode permutation. In the other approach, an incomplete image (could be a gray-scale) is scanned by the KA operation to a multi-way array with many modes, and such a tensor is approximated by a long TT model. The numerical experiments performed for various scenarios of image degradation (randomly missing pixels

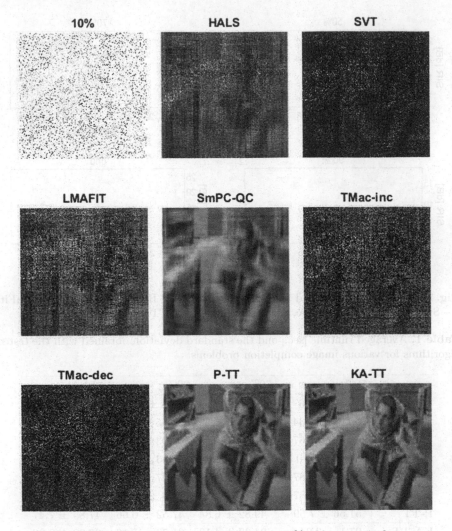

Fig. 3. Results of image completion from 10% of known pixels.

or a regular grid) showed that there is no significant difference in the quality of recovered images between both options (see Fig. 2) but the P-TT algorithm is noticeably faster than KA-TT (see Table 1). This is an interesting observation which suggests that a simple 3-core TT model is sufficient to perform the image completion of good quality.

Another important contribution of this study is the usage of the low-pass Gaussian filter to smooth approximations in each iterative step. It is particularly important when the incompleteness of an image is very high (e.g. 90% of randomly missing entries). In this case, the smoothing prevents from spiky disturbances which occur despite a low-rank approximation is used, and additionally it allows us to increase the TT-ranks. Obviously, lower ranks diminish

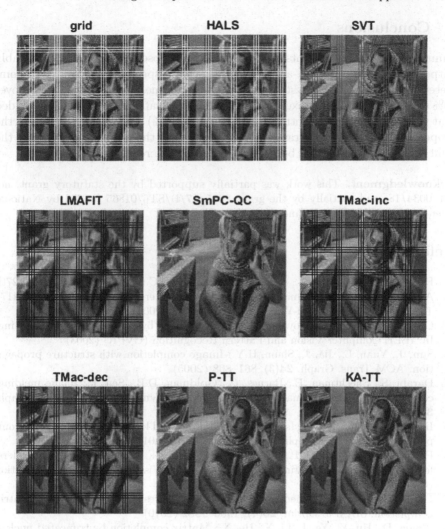

Fig. 4. Results of image completion from grid-disturbed data.

spiky disturbances but also restrict scene details considerably. The experiments demonstrated that the usage of 2D low-pass filtration and higher ranks is more efficient than lower ranks without filtration.

The results illustrated in Figs. 2 and 3 show that the proposed algorithms (P-TT and KA-TT) substantially outperform all the tested algorithms for the image completion problem when only 10% of pixels is available. For the grid-disturbed image (Fig. 4), the proposed algorithms offer similar but still better quality of reconstruction as SmPC-QC. The other algorithms give much worse results. When the incompleteness is not very strong (app. 50%), the other algorithms yield the reconstructed images of acceptable quality but still much lower than we obtained with P-TT and KA-TT (see Fig. 2).

5 Conclusions

Summing up, the experiments showed that the proposed algorithms considerably outperform the state-of-the-art algorithms with respect to the quality of completed images for various tests as well as the runtime. They are able to recover 90% of randomly missing pixels in an image with a satisfactory quality, provided that the indices of known entries (the remaining 10%) are known. Obviously, the proposed algorithms are parameter-dependent, and the problem of selecting the right order of the filter will be studied in our future research.

Acknowledgment. This work was partially supported by the statutory grant, no. 401/0034/18, and partially by the grant 2015/17/B/ST6/01865 funded by National Science Center (NCN) in Poland.

References

1. Bertalmio, M., Sapiro, G., Caselles, V., Ballester, C.: Image inpainting. In: 27th Annual Conference on Computer Graphics and Interactive Techniques, pp. 417–424. ACM Press/Addison-Wesley Publishing Co. (2000)
2. Criminisi, A., Perez, P., Toyama, K.: Object removal by exemplar-based inpainting. In: IEEE Computer Vision and Pattern Recognition (CVPR) (2003)
3. Sun, J., Yuan, L., Jia, J., Shum, H.Y.: Image completion with structure propagation. ACM Trans. Graph. **24**(3), 861–868 (2005)
4. Darabi, S., Shechtman, E., Barnes, C., Goldman, D.B., Sen, P.: Image melding: combining inconsistent images using patch-based synthesis. ACM Trans. Graph. **31**(4), 82:1–82:10 (2012)
5. Iizuka, S., Simo-Serra, E., Ishikawa, H.: Globally and locally consistent image completion. ACM Trans. Graph. **36**(4), 107:1–107:14 (2017)
6. Pathak, D., Krähenbühl, P., Donahue, J., Darrell, T., Efros, A.: Context encoders: feature learning by inpainting. In: IEEE Computer Vision and Pattern Recognition (CVPR) (2016)
7. Cai, J.F., Candès, E.J., Shen, Z.: A singular value thresholding algorithm for matrix completion. SIAM J. Optim. **20**(4), 1956–1982 (2010)
8. Zhang, D., Hu, Y., Ye, J., Li, X., He, X.: Matrix completion by truncated nuclear norm regularization. In: IEEE Conference on Computer Vision and Pattern Recognition, pp. 2192–2199 (2012)
9. Zhang, M., Desrosiers, C.: Image completion with global structure and weighted nuclear norm regularization. In: 2017 International Joint Conference on Neural Networks (IJCNN), pp. 4187–4193 (2017)
10. Wen, Z., Yin, W., Zhang, Y.: Solving a low-rank factorization model for matrix completion by a nonlinear successive over-relaxation algorithm. Math. Program. Comput. **4**(4), 333–361 (2012)
11. Wang, Y., Zhang, Y.: Image inpainting via weighted sparse non-negative matrix factorization. In: 18th IEEE International Conference on Image Processing, pp. 3409–3412 (2011)
12. Sadowski, T., Zdunek, R.: Image completion with smooth nonnegative matrix factorization. In: Rutkowski, L., Scherer, R., Korytkowski, M., Pedrycz, W., Tadeusiewicz, R., Zurada, J.M. (eds.) ICAISC 2018. LNCS (LNAI), vol. 10842, pp. 62–72. Springer, Cham (2018). https://doi.org/10.1007/978-3-319-91262-2_6

13. Cichocki, A., Zdunek, R., Phan, A.H., Amari, S.I.: Nonnegative Matrix and Tensor Factorizations: Applications to Exploratory Multi-way Data Analysis and Blind Source Separation. Wiley, Hoboken (2009)
14. Acar, E., Dunlavy, D.M., Kolda, T.G., Mørup, M.: Scalable tensor factorizations with missing data. In: 2010 SIAM International Conference on Data Mining, pp. 701–712 (2010)
15. Zhao, Q., Zhang, L., Cichocki, A.: Bayesian CP factorization of incomplete tensors with automatic rank determination. IEEE Trans. Pattern Anal. Mach. Intell. **37**(9), 1751–1763 (2015)
16. Yokota, T., Zhao, Q., Cichocki, A.: Smooth PARAFAC decomposition for tensor completion. IEEE Trans. Signal Process. **64**(20), 5423–5436 (2016)
17. Latorre, J.I.: Image compression and entanglement. arXiv:quant-ph/0510031, Quantum Physics (2005)
18. Oseledets, I.: Tensor-train decomposition. SIAM J. Sci. Comput. **33**(5), 2295–2317 (2011)
19. Bengua, J.A., Phien, H.N., Tuan, H.D., Do, M.N.: Efficient tensor completion for color image and video recovery: low-rank tensor train. IEEE Trans. Image Process. **26**(5), 2466–2479 (2017)
20. Ko, C.Y., Batselier, K., Yu, W., Wong, N.: Fast and accurate tensor completion with tensor trains: a system identification approach. CoRR abs/1804.06128 (2018)
21. Wang, W., Aggarwal, V., Aeron, S.: Efficient low rank tensor ring completion. In: IEEE International Conference on Computer Vision (ICCV), pp. 5698–5706 (2018)
22. Cichocki, A., Lee, N., Oseledets, I.V., Phan, A.H., Zhao, Q., Mandic, D.P.: Tensor networks for dimensionality reduction and large-scale optimization: Part 1 low-rank tensor decompositions. Found. Trends Mach. Learn. **9**(4–5), 249–429 (2016)
23. Sadowski, T., Zdunek, R.: Modified HALS algorithm for image completion and recommendation system. In: Świątek, J., Borzemski, L., Wilimowska, Z. (eds.) ISAT 2017. AISC, vol. 656, pp. 17–27. Springer, Cham (2018). https://doi.org/10.1007/978-3-319-67229-8_2
24. Xu, Y., Hao, R., Yin, W., Su, Z.: Parallel matrix factorization for low-rank tensor completion. Inverse Probl. Imaging **9**(2), 601–624 (2015)

Knowledge Construction Through Semantic Interpretation of Visual Information

Cristiano Russo[1]([📧])[iD], Kurosh Madani[1], and Antonio Maria Rinaldi[2,3][iD]

[1] Université Paris-Est, LISSI laboratory (EA 3956), Creteil, France
cristiano.russo@univ-paris-est.fr, madani@u-pec.fr
[2] Department of Electrical Engineering and Information Technologies,
University of Naples Federico II, Naples, Italy
antoniomaria.rinaldi@unina.it
[3] IKNOS-LAB Intelligent and Knowledge Systems (LUPT), Naples, Italy

Abstract. The skills required by machines in the last decade have grown exponentially. Recent efforts made by the scientific community have shown amazing results in the field of research related to artificial intelligence and robotics. Recent studies show that machines may be superior to humans in carrying out certain tasks. However, in many approaches they still fail to achieve high-level skills required to support humans and interact with them. Furthermore, the "intelligence" exhibited is hardly ever the result of a real autonomous decision. In this article we propose a novel approach, with the aim of providing a machine with the ability to evolve and build its own knowledge by combining both semantic and visual information. The proposed concept, its implementation and experimental results are shown and discussed.

Keywords: Knowledge construction · Ontologies · Neural networks

1 Introduction

Astonishing progress is being witnessed today in the field of research related to artificial intelligence, smart robotics and cognitive computing.

Among the various approaches pursued over the years, one of the most attractive is the one which takes inspiration from the biological processes related to humans. That is what we call the bio-inspired approach. Since the design of the first models of neural networks, we have tried to emulate the biological processes used in the brains of the most advanced animal species. Following these principles, numerous achievements have been obtained over the years for individual tasks, such as object manipulation, object detection, object recognition and so on that have allowed an increasingly pervasive use of machines, especially in the industrial field. Recently, we started to design and develop humanoid robots that could coexist with human beings, and that were able to make decisions in complete autonomy to assist and support humans in their daily activities. While

© Springer Nature Switzerland AG 2019
I. Rojas et al. (Eds.): IWANN 2019, LNCS 11507, pp. 246–257, 2019.
https://doi.org/10.1007/978-3-030-20518-8_21

the objective of designing intelligent machines was somehow achieved, albeit with many limitations, we are still very far from obtaining machines able to flawlessly cohabit and interact with humans in the real world as autonomously-thinking machines. Cognitive architectures in artificial intelligence aims to allow artificial systems to exhibit intelligent behaviors in a general context through a detailed analogy with the constitutive and evolutionary functioning and mechanisms underlying human cognition [5]. However, to be considered autonomous, an intelligent system requires also the ability to evolve its knowledge, to make decisions and to select among alternatives like humans do in their everyday life. For instance, a marketing director must choose which of the new proposals is the best for his company, a judge must decide whether to convict or not a defendant whose case has no precedent, and a striker in front of the opposing goalkeeper must choose whether to shoot on goal or pass the ball to his team mate. Such decisions are not associated with the recognition of a situation or pattern, but rather require individual reasoning based on a combination of intuition and one's own knowledge, while most cognitive architectures are based on the recognition of situations already seen as a recognize-act cycle that underlies all cognitive behavior. Providing the ability of "autonomy" is crucial, and at the same time challenging, in order to build systems that can take the form of general virtual humans for companionship, versatile service or personal robots, software agents for interactive tutoring or personal assistant. Clearly, it is not the goal of this work to find a definitive solution to the above-cited question, as it is a very ambitious goal, which will likely take years to be achieved. We are rather interested in defining a possible approach toward that goal, looking for a kind of autonomous knowledge acquisition and construction method driven by ontologies and classical machine learning techniques.

In this work we will refer to machines of any kind, or more generally to artificial intelligent systems as *agents*. Agents are intended to live in the same environment, and they often need to transfer knowledge from one to the other. So, in order to make this possible it is desirable to have a common understanding on how knowledge is represented. Whatever the modality through which this occurs, a communicating agent must represent the knowledge that it aims to convey or that it believes another agent intends for it. Two or more agents can communicate about categories recognized and decisions made, about perceptions and actions, etc., or one agent can asks the others for general knowledge regarding a particular domain of interest, or how to accomplish a more specific manipulation task and so forth.

Before introducing the general idea of the work, we first assume true the hypothesis that it is possible to associate a semantically coherent set of labels for an indefinite set of unlabeled observations. It doesn't matter which mechanism is used. The question which we try to answer in this work is the following. Once we have such labels, can we define a process of acquisition and construction of knowledge based on the semantic labels obtained? We define a general approach for knowledge construction, then we present the methodologies and techniques used to accomplish this task. In our vision, ontologies and linked open data can be used as a powerful tool by which we can guide the process of knowledge construction. Moreover, such a design, could be easily extended, by combining the knowledge learned through one's own experience with the knowledge shared by others, enabling a collaborative knowledge acquisition environment.

The remainder of this paper is structured as follows. In Sect. 2 we start with an overview over related work; in Sect. 3 we present our approach for knowledge construction and continue with a a prototype of implementation in Sect. 4. Section 5 is devoted to conclusions and future works.

2   Related Work

Given the multiplicity of the fields involved, in this section we limit ourselves to provide a brief overview, far from being complete, of works with intentions similar to ours from the point of view of the general idea, which is that of the construction of knowledge and the autonomous or semi-autonomous learning. Then, we focus on some strategies presented in the literature which used hierarchies of semantic concepts for classification purposes.

In [10] an intelligent machine vision system able to learn autonomously individual objects based on salient object detection is presented. This system extracts salient objects which can be efficiently used for training the machine learning-based object detection and recognition units.

The same authors present in [8,9] a cognitive system based on "artificial curiosity" for high-level knowledge acquisition from visual patterns. The curiosity (perceptual curiosity and epistemic curiosity) is realized through combining perceptual saliency detection and Machine-Learning based approaches. The learning is accomplished by autonomous observation of visual patterns and by interaction with an expert (a human tutor) detaining semantic knowledge about the detected visual patterns. The use of perception and biologically inspired neural architectures as primary sources for building smart robotics applications is also witnessed and widely discussed in [1].

In [3] an approach for knowledge acquisition about the environment is proposed using introspective mental processes to infer if a domain concept is well-known or not. The purpose is to endow a robot with the ability to model its knowledge about the environment and to acquire new knowledge when it occurs in a kind of self-consciousness process. In [16] a method for spatial Concept Acquisition is presented. The authors use an unsupervised learning method for the lexical acquisition of words related to places visited by robots, from human continuous speech signals. Another work which aims to give robots the ability to adaptively understand their operational world is [17]. The problem of learning dialogue policies to support semantic attribute acquisition is discussed, so that the effort required by humans in providing knowledge to the robot through dialogue is minimized. The use of ontologies as an effective tool for knowledge integration and construction is witnessed in [2,13]. A multimedia semantic knowledge base is also presented in [14,15]. Multimedia ontologies could be effectively used also for the purpose of associating semantic labels to images [11, 12]. The methodology is based on multimedia ontologies organised following a formal model. Multimedia data and linguistic properties are used to bridge the gap between the target semantic classes and the available low-level multimedia descriptors.

By integrating concept hierarchy for semantic image concept organization, a hierarchical mixture model is proposed in [4]. The goal is to enable multi-level image concept modeling and hierarchical classifier training. A Multi-level Semantic Classification Trees is proposed in [7]. It combines different information sources for predicting speech events (e.g. word chains, phrases, etc.). Our approach is based on an evolutionary process, combining established methodologies, such as ontologies and machine learning techniques that guide the process of acquisition and construction of knowledge.

3   The Proposed Approach

As we pointed out in Sect. 1, acquiring new knowledge in an autonomous or semi-autonomous way is one of the main challenges of next future decade. Our purpose is to set up a framework for local knowledge construction exploiting well-known techniques and methodologies. To clarify the context in which we place ourselves, we could imagine the following scenario, also depicted in Fig. 1. We have a set of observations, i.e. unlabeled images, made by an agent. At this stage we suppose by hypothesis that it is possible (and indeed it is) to associate semantic labels related to what is actually represented in these observations. Once we have obtained the set of labels, we suppose we are able to organize and link these labels defining a network of interconnections between them, i.e. a graph representing the acquired knowledge. The generated structure is then exploited for the development of a multilevel semantic classifier to independently recognize the concepts represented by future observations made by the agent. As new observations and possibly new labels are discovered, it is possible to reorganize the knowledge graph automatically adding new knowledge, according to the previously defined strategy. This approach allows an agent to evolve its local knowledge and to make its own inferences and make its own decisions autonomously. For example, a domestic robot may be able to recognize new classes, which at first it is not able to recognize. This is in contrast to the classical learning approach used by the vast majority of the instruments available today, which, although very complex and evolved, are to be considered as black boxes.

Various methodologies can be adopted to guide the two phases of the process just described, that is the mapping between the set of observations made and the set of semantic labels and the transformation of the latter according to a well-defined semantic structure. These choices significantly influence the quality of the final result. Regarding the first step, we decided to automate it, using a convolutional neural network, already trained on ImageNet, given its popularity and high performances. Nothing forbids using different tools, or combinations of them, in order to extend the number of classes to be added to the local knowledge base that we want to evolve. The use of a network, initially trained on ImageNet

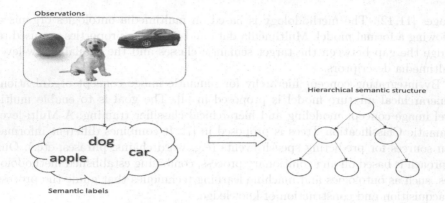

Fig. 1. The general idea

is also justified by the fact that a direct mapping exist between ImageNet and the lexical-semantic dictionary WordNet [6], whose relationships can be exploited for the second step of this first phase, that is the construction of the graph containing the concepts acquired locally. Also in this case, the mechanism that drives the creation of the graph can be different. In our case study, for demonstration purposes, we used an ontology-driven approach, using the semantic relations of hyponymy and hyperonymy contained in WordNet to have a meaningful hierarchy. In general, the ultimate goal is to have a comprehensive knowledge base, possibly integrating multiple sources of knowledge related to various application domains. Heterogeneity plays a primary role in this context. It can arise at different levels of abstraction from the linguistic one to the conceptual one, therefore, in our vision, the use of ontology matching and merging mechanisms, and in general the use of linked open data could be very helpful at this stage, giving the whole process a higher extensibility level and a more robust setting, avoiding ambiguities and breakdowns. The only requirement to be met in this first phase is that the output of the first step for obtaining the semantic labels is correctly processed in the second step, so an intermediate step of preprocessing of labels may be necessary at hand. The second phase of our approach consists in exploiting the graph of acquired knowledge, training a multilevel semantic classifier. The purpose is to allow an agent to autonomously recognize the observations it makes based on its own knowledge, and automatically extend its capabilities when new knowledge is added, in a continuous learning process. It is worth noting that for this particular case, the graph can be seen as a taxonomy, given the use of the only relations of hyponymy and hyperonymy of WordNet, which generate a hierarchical structure. Adding other relationships could lead to new solutions, having a more complex graph-based structure. For example such relationships could be actively exploited during the learning process, or to define new classifiers. A generic architecture of such a classifier is shown in Fig. 2.

We decided to use a double approach based on the hierarchy obtained in order to have a more robust prediction. The idea is to make a first pass, vertically along

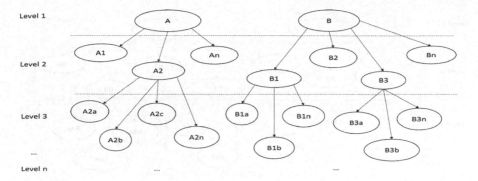

Fig. 2. The multilevel semantic classifier

our structure. For each level, we define a number of classifiers equal to the number of classes of the previous level. Based on the classification resulting at the n-th level, the observation is further processed by the corresponding classifier of the next level. In this way, the final result will be a chain of semantic predictions, from the more general and abstract to the more specific. Since the errors made at a certain level are inevitably propagated to the successive levels, we make a second passage, this time in a horizontal sense, defining a single classifier for each level of the hierarchy. Figure 3, shows how classifiers were defined at each level. The red dashed lines identify the classes on which classifiers are trained in the "vertical way", the green lines group all the classes of a level, since we have one classifier per level in the *horizontal way*. Finally, the results are combined and compared. This approach does not guarantee that all errors that occur in the first phase are corrected. In fact, both classification chains may provide the same chain of predictions, but incorrect. However, it has some advantages. For example, the comparison of the two classification chains improves the robustness of the system, since it can be used to have a confidence measure of the prediction obtained or to appropriately define corrective weights, giving more importance to one or the other depending on the case. Furthermore in a context of interaction between a human and a robot, the disparity of the two classification could lead the robot to ask the human for solving such cases, actively involving him/her in its knowledge acquisition process.

Having a semantic structure of the classes available has many advantages. In the first place, a more abstract level of knowledge, of concepts not yet known, is already possible at an early stage of the evolutionary process. In other words, the layered hierarchical structure allows us to still have a correct, more abstract classification, even for never seen before objects. The Fig. 4 exemplifies such a scenario. On the left side there is an example of a local knowledge base with only few concepts already acquired. Imagine that the agent observes for the first time a new type of object, not present in the knowledge base, we say an *orange*. Even if a mistake is made in the last level, given the absence of this concept, the agent will still be able to recognize at the previous level that it is a type of *fruit*

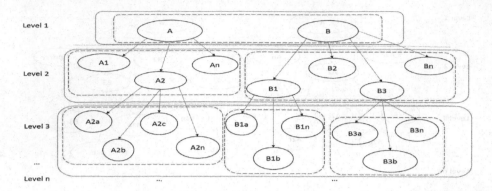

Fig. 3. Classifiers chain

(assuming that the classification at this level is correct). When the knowledge related to the misclassified concept is added, the local knowledge is reorganized as previously described, and the levels which changes are retrained accordingly.

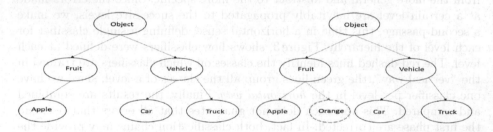

Fig. 4. Categorizing unseen concepts

4   Experimental Study

Since we start from the point where our agent has already acquired certain knowledge, which will contain the experiences and observations already made by itself, we first retrieve a number of observations representing well-known concepts. We pair them with semantic labels by using the deep neural network VGG-16 pretrained on ImageNet. Given the not excessive number of observations used in this demonstration, we also decided to perform a manual check of the labels associated with the sample observations and to correct the errors occurred. Then, we defined the semantic hierarchy along the lines of that provided by hyponymy and hyperonymy in WordNet. An automatic approach could in this case be to follow the exact structure of concepts as defined in WordNet, however, given the excessive depth of the latter, and the lack of utility with respect to our purpose, which is to show the new knowledge construction approach, in the initialization phase, we have specifically defined our most abstract concepts to be used in the highest levels of the hierarchy. In other words, some specializations have been grouped into a single class as shown in Fig. 5. For each dashed line we go down a level in the hierarchy. For this set of experiments the number of levels in the hierarchy was 3.

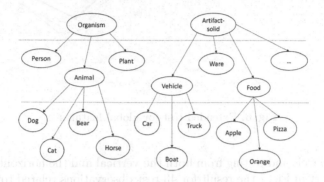

Fig. 5. An example of the defined taxonomy

For example, the chain of hyponymies from the animal concept to the dog concept has a length equal to 8 (animal - chordate - vertebrate - mammal - placental mammal - carnivore - canid - dog), so we eliminate these intermediate concepts, considering as direct the relationships between the concept animal and the concept dog. Again, we remark that different strategies could be adopted at this stage, according to how the knowledge acquisition process is guided. Hence, one could use his/her self-defined ontology, or other well-known ontologies if using an ontology-driven approach, or other strategies according to the needs, the kind of knowledge to be acquired and the individual purposes.

Then, we developed the multilevel semantic classifier based on the generated structure. The first experiment was to test independently all the classifiers at different levels of the hierarchy. We restricted our analysis using the multi-layer perceptron and global descriptors as features. A first test was conducted for the analysis using 8 different global descriptors (Joint composite Descriptor (JCD), Pyramid of Histograms of oriented gradients (PHOG), Edge Histogram (EH), Scalable color (SC), Fuzzy color and texture histogram (FCTH), Color and Edge directivity descriptor (CEDD), Color Layout (CL), Auto color correlation (ACC)). We reported the result of test accuracy for each descriptor and each level (vertical) in Fig. 6. It is worth noting two specific results from the chart. First, a trend emerges related to the behavior of all the features, which are very similar, and secondly, the noticeable decline of performance in the lower levels. This effect was predictable, considering both the small number of samples and the greater number of classes at the lower levels.

After the classifiers are trained on the local knowledge at our disposal, it is possible, at this point, to proceed to test the approach described in the Sect. 3,

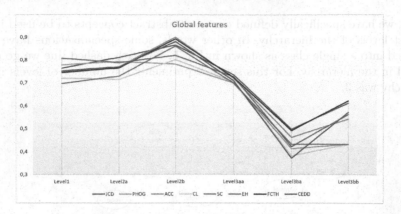

Fig. 6. Extended test for global features

chaining the levels and testing from both the vertical and the horizontal perspective. We report in Fig. 7 the result for 43 test observations related to the classes of the domain food. Each pair in the figure represent the number of correct classification (over the total of 43) and the percentage. The first 3 columns are related to the final classification result for combination (both vertical and horizontal are correct), vertical and horizontal respectively. The second 3 columns are related to the classification result until the second level, which is the second last level in our case study.

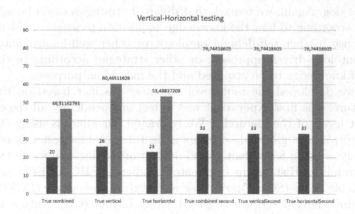

Fig. 7. Combined test for vertical and horizontal classification. On the left side there are the first 3 columns pairs (related to last level (final) results). On the right side the same results at the second last level.

As we expected the results obtained shows a higher performance in the second level, i.e. most of the observations are correctly classified as food. It is interesting to note also that the combined use of the two methods of classification does not penalize in particular the vertical classification (from the most abstract to the most specific). However, this point requires a more in-depth analysis in order to verify the behavior in more complex structures, with a greater number of classes or, again, with different features. Hence, this justifies its use for further learning by the agent in case of doubt to improve the performance. This strategy can be useful for different tasks, since it provides fast intermediate results with a trade-off on abstraction of concepts retrieved. Moreover, its peculiarity is that of being easily extensible, which is ultimately what we want to achieve.

4.1   Case Study

We show in this subsection an interactive use case of our framework in a real office environment using the semi-humanoid robot *Pepper* as agent. Pepper is designed to be fully interactive, as it can hold full conversations, and it is endowed with a number of sensors, allowing it to speak, to detect obstacles, to take photos/videos, to move and so on. For visual sensors, in particular, it is equipped with two 2D cameras and an ASUS Xtion 3D sensor. The scenario is that of a human asking the robot to accomplish certain tasks and we will see how the robot meets the requests and how it processes the knowledge acquired. The robot is equipped with a speech recognition engine which provides the capabilities of translating to text what a human says. For this demonstration, we want to test if, after a first training phase, the robot is able to correctly recognize the upper level of an object never seen before, in this case an apple. When the robot receives the command, that is searching for objects, it raises its head down and takes a photo of what is in front of him, as shown in Fig. 8.

(a) The Pepper robot searching (b) Image acquired by Pepper
for objects

Fig. 8. The robot in action

The robot has no awareness of the *apple* concept, hence it misclassifies it with the *pear* concept in the vertical way and with the *banana* concept in the horizontal way, see top of Fig. 9, but it is able to correctly recognize that the object in front of it belongs to the food domain.

When the robot is undergoing a new training phase containing knowledge related to the *apple* concept, the class is added to the graph, and the robot is able to correctly recognize the observation as shown in the lower part of Fig. 9.

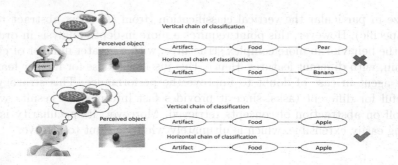

Fig. 9. Results obtained from classification. On top, the robot misclassifies the object, in the lower side it correctly classify the object

5 Conclusion and Future Work

In this paper we have presented an extensible approach for knowledge construction combining both ontologies, which are used for a standardized representation of knowledge in an automatic way, and machine learning techniques which helps in the object recognition task. This double process allows us to combine low-level features with other high-level (semantic) interconnected in a graph-based structure. The use of an upper ontology also facilitates the integration of the conceptualization from other ontologies, and achieves a high degree of reusability and interoperability among heterogeneous sources of knowledge. Then, we have provided a use case in a real indoor environment with the Pepper robotic platform. The future work aims to extend current activities on the ontological model, including the recognition of actions to do when a certain object, or set of object, is seen. The semantic multilevel classifier performances will be further tested extending the number of classes. In addition we plan to test it with different features, e.g. using local descriptors and features extracted from activation layers of deep neural networks. Extensive experiments will be carried out to better evaluate the overall performances of our approach and searching for ameliorating techniques of knowledge construction which exploit the combined classification involving a human in the learning loop.

References

1. Bülthoff, H.H., Wallraven, C., Giese, M.A.: Perceptual robotics. In: Siciliano, B., Khatib, O. (eds.) Springer Handbook of Robotics, pp. 1481–1498. Springer, Heidelberg (2008). https://doi.org/10.1007/978-3-540-30301-5_64
2. Caldarola, E.G., Rinaldi, A.M.: An approach to ontology integration for ontology reuse. In: 2016 IEEE 17th International Conference on Information Reuse and Integration (IRI), pp. 384–393. IEEE (2016)
3. Chella, A., Lanza, F., Pipitone, A., Seidita, V.: Knowledge acquisition through introspection in human-robot cooperation. Biol. Inspired Cogn. Arch. **25**, 1–7 (2018)

4. Fan, J., Luo, H., Gao, Y.: Learning the semantics of images by using unlabeled samples. In: IEEE Computer Society Conference on Computer Vision and Pattern Recognition, CVPR 2005, vol. 2, pp. 704–710. IEEE (2005)
5. Lieto, A., Bhatt, M., Oltramari, A., Vernon, D.: The role of cognitive architectures in general artificial intelligence (2018)
6. Miller, G.A.: WordNet: a lexical database for English. Commun. ACM **38**(11), 39–41 (1995)
7. Nöth, E., et al.: An integrated model of acoustics and language using semantic classification trees (1996)
8. Ramík, D.M., Madani, K., Sabourin, C.: From visual patterns to semantic description: a cognitive approach using artificial curiosity as the foundation. Pattern Recognit. Lett. **34**(14), 1577–1588 (2013)
9. Ramík, D.M., Sabourin, C., Madani, K.: Autonomous knowledge acquisition based on artificial curiosity: application to mobile robots in an indoor environment. Robot. Auton. Syst. **61**(12), 1680–1695 (2013)
10. Ramík, D.M., Sabourin, C., Moreno, R., Madani, K.: A machine learning based intelligent vision system for autonomous object detection and recognition. Appl. Intell. **40**(2), 358–375 (2014)
11. Rinaldi, A.M.: A multimedia ontology model based on linguistic properties and audio-visual features. Inf. Sci. **277**, 234–246 (2014)
12. Rinaldi, A.M.: A complete framework to manage multimedia ontologies in digital ecosystems. Int. J. Bus. Process. Integr. Manag. **7**(4), 274–288 (2015)
13. Rinaldi, A.M., Russo, C.: A matching framework for multimedia data integration using semantics and ontologies. In: 2018 IEEE 12th International Conference on Semantic Computing (ICSC), pp. 363–368. IEEE (2018)
14. Rinaldi, A.M., Russo, C.: A semantic-based model to represent multimedia big data. In: Proceedings of the 10th International Conference on Management of Digital EcoSystems, pp. 31–38. ACM (2018)
15. Rinaldi, A.M., Russo, C.: User-centered information retrieval using semantic multimedia big data. In: 2018 IEEE International Conference on Big Data (Big Data), pp. 2304–2313. IEEE (2018)
16. Taniguchi, A., Taniguchi, T., Inamura, T.: Spatial concept acquisition for a mobile robot that integrates self-localization and unsupervised word discovery from spoken sentences. IEEE Trans. Cogn. Dev. Syst. **8**(4), 285–297 (2016)
17. Vanzo, A., Part, J.L., Yu, Y., Nardi, D., Lemon, O.: Incrementally learning semantic attributes through dialogue interaction. In: Proceedings of the 17th International Conference on Autonomous Agents and MultiAgent Systems, pp. 865–873. International Foundation for Autonomous Agents and Multiagent Systems (2018)

Ensemble Transfer Learning Framework for Vessel Size Estimation from 2D Images

Mario Milićević[(✉)] ⓘ, Krunoslav Žubrinić ⓘ, Ivan Grbavac ⓘ, and Ana Kešelj ⓘ

University of Dubrovnik, Dubrovnik, Croatia
mario.milicevic@unidu.hr

Abstract. The term gross tonnage refers to the internal volume of a vessel and it has several legal, administrative and safety uses. Therefore, there is significant value in developing a mechanism for the automatic estimation of vessel size based on 2D images taken in uncontrolled conditions. However, this is a demanding task as vessels can be photographed from various angles and distances, a part of a vessel can be obstructed, or a vessel can blend with the background. We proposed an ensemble of fine-tuned transfer learning models, which we trained on 20,000 images in a training dataset consisting of randomly downloaded images from the Shipspotting website. Multiple deep learning methods were applied and modified for regression problems, together with two classical machine learning algorithms. A detailed analysis of model performances was given, based on which it can be concluded that such an approach results in a vessel size evaluation of the same quality as with the best human experts from the corresponding field.

Keywords: Deep learning · Convolutional neural networks · Transfer learning · Regression · Ensemble methods · Computer vision · Vessel size estimation

1 Introduction

The estimation of physical object properties (such as the shape, volume or mass) is a challenge even for a human expert in a corresponding field of expertise. The shape of an object is defined by a certain arrangement of edges. Humans perceive the shape of an object by employing both bottom-up and top-down processes [1]. At the beginning of a visual experience, data sensed and received is analyzed and shapes are recognized within this data (bottom-up processing). The next step is conceptually driven or top-down processing, where the focus is placed on concepts known to the observer and on how the observer's (conceptual) world seems to be organized. This is then followed by perceptual interpolation [2], during which the mind does its best to fill in the gaps of what is not well visible, obstructed or is blending in with the background. Even with all this, human

© Springer Nature Switzerland AG 2019
I. Rojas et al. (Eds.): IWANN 2019, LNCS 11507, pp. 258–269, 2019.
https://doi.org/10.1007/978-3-030-20518-8_22

perception of size still depends on how we perceive depth, an action defined by binocular disparity, that is, by an almost imperceptible but important difference between how images are made on either of the retinae [3]. The fact that human perception is influenced by prior knowledge, expectations and intentions explains how complex the procedure is, for example, to estimate the size of a vessel expressed in gross tonnage, all based on an image. Gross tonnage forms the basis for safety regulations, registration fees and port dues. The gross tonnage is a function of the moulded volume of all enclosed spaces of the ship. It is defined in accordance with the 1969 International Convention on Tonnage Measurement of Ships [4], and it is calculated by multiplying the interior volume V of the ship in cubic meters by a variable K, which is a multiplier based on ship volume, where K is larger for larger ships.

Fig. 1. Examples of vessels photographed in various conditions. The gross tonnage of these vessels is 38,844 GT, 11,153 GT, 317 GT and 95,128 GT respectively.

From the examples in Fig. 1, it can be seen how complex the process of expert vessel size estimation based just on photographs is, as the ships were photographed from various angles and distances from the observer. An additional issue is the fact that on some pictures, a part of the vessel is obstructed, or the vessel silhouette is blending with the background, so that it becomes increasingly difficult to correctly assess the proportions. This is why the reliability of assessment results is significantly influenced by previous experience and intuition. Due to all the named issues, it can be assumed that the estimation of vessel size based on just images, especially in the context of a limited learning dataset, will be a great challenge even for machine learning algorithms.

2   Related Works

To our knowledge, there are few papers dealing with the assessment of physical object properties such as mass or volume in an uncontrolled environment. Most papers focus on observing objects in a known and controlled environment, or even have reference measures within the images. For example, in [5], authors obtained product information about 3 million Amazon.com products, as well as a dataset containing household objects which were manually photographed and weighed. They presented a model that takes into account an estimate of the

3D shape of the object. That model compared favorably to the performance of humans. Similarly, there are several papers on the assessment of weight, that is, mass of a narrowly-defined class of objects in relatively controlled conditions. In [6,7], the authors focus on the weight of cattle, and in [8,9] the weight of pigs. In [10], weight estimation of bread wheat and durum wheat in different amounts was performed by using image processing techniques and counting wheat kernels. In [11], the authors tried to measure the width and length of eggs by real-time image processing and optimizing the model to find the best relation between image processing outputs and the weight of the egg.

A completely different approach is used by authors in [12], where the possibility of learning physical object properties from video material is discussed. They proposed an unsupervised representation learning model which explicitly encodes basic laws of physics into the structure and uses them to understand and analyze automatically discovered observations from the provided videos. Multiple authors deal with the analysis of volume or mass of food. For example, in [13], the authors described image analysis tools used to determine the regions where a particular food is located, identify the food type and estimate the weight of the food item. Chae et al. in [14] analyzed the possibility of automatic estimation of food volume with the use of food-specific shape templates corresponding to each segmented food. Their system reconstructs the three-dimensional properties of the food shape from a single image by extracting feature points in order to size the food shape template, and at the end of the process finally evaluates food portion size. Similarly, authors of a number of papers address the problem of 3D reconstruction from one or more images [15,16].

The authors of the already mentioned MARVEL dataset [17] state the main objective of their research to be vessel type classification, which is crucial for maritime surveillance. They also state other research possibilities - e.g. specific attribute prediction and classification, where the objective is to estimate draught, length, gross tonnage or summer deadweight. The authors have conducted initial research for these attributes as well, but the results have shown that this issue was not delved deeper into. For example, the Pearson correlation coefficient for Convolutional Neural Network (CNN) models for gross tonnage is only 0.2699.

3   Dataset

The dataset used in this study is extracted from the MARVEL dataset [18] which was made from 2 million marine vessel images collected from the Shipspotting website [19], where the following attributes are available for most images: beam, year built, draught, flag, gross tonnage, IMO number, name, length, category, summer deadweight, MMSI, vessel type. Gundogdu et al. detect 1,607,190 images with valid annotated vessel categories, and 1,583,882 images with valid annotated labels for gross tonnage. We randomly downloaded 25,000 images from the Shipspotting website with valid gross tonnage labels. All images were resized to 256×256 pixels, and then various outliers were detected and eliminated. The decision to use a relatively limited dataset gives an additional dimension to the project.

Fig. 2. Examples of outliers eliminated from the original dataset.

Among the manually eliminated outliers (Fig. 2) were duplicate images of the same vessel, monochromatic images, as well as various potentially confusing situations, such as images displaying multiple ships or where a ship is not in its expected environment. Finally, through random selection, training datasets with 20,000 images, validation and test datasets with 2,000 images respectively were formed.

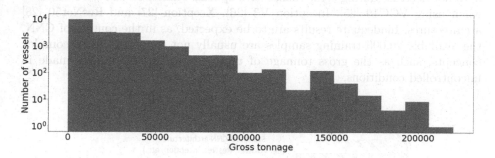

Fig. 3. Distribution of gross tonnage values for learning dataset examples.

The distribution of values (Fig. 3) in the learning dataset is between 17 GT and 211,450 GT, the mean value is 24,361 GT, and the median is 14,358 GT. The Shapiro-Wilk test of normality [20] shows that the data is not normally distributed (p-value = 0.018, alpha = 0.05).

4   Experimental Study

CNN training is implemented with the Keras [21] and TensorFlow [22] deep learning frameworks, using a NVidia GeForce GTX 1080 Ti GPU with 11 GB memory on Ubuntu 16.04 Linux OS. In the previous research [23], a similar dataset, also extracted from the MARVEL dataset, was used to study the problem of fine-grained ship type classification in 26 different classes. This type of multiclass classification is an area in which CNNs achieve the best results. There are fewer available examples of similar architectures being used for regression issues, that is, for numeric estimations which include the assessment of gross tonnage. For example, in [24], the authors performed deep regression experiments on three vision problems, and they showed that correctly fine-tuned deep regression networks can compete even with problem-specific methods that were designed specifically for solving a single task.

We conducted the first experiment using conventional deep learning architectures in which the last layers were replaced with new layers, which allow regression instead of classification. First, a modified version of the VGG16 architecture was used in which multiple combinations of the top layers were tested. For example, one of the variants of the top layers was as follows: *FL-DN(1024)-BN-ACT(relu)-DN(1)-ACT(linear)* - where FL denotes a flatten layer that transforms a 2D feature map into a vector, DN() denotes a dense layer, BN denotes a batch normalization layer and ACT() denotes an activation function. This combination achieved a correlation coefficient of 0.66 and mean absolute error (MAE) of 12,703 GT for the test dataset. We used early stopping with a patience equal to eight epochs (complete cycles through the training set). The results showed that the learning process stops after only 25 epochs, because validation loss does not decrease during eight consecutive epochs. Similar results were achieved when using VGG19 [25], Inception V3 [26], Xception [27] and ResNet50 [28] architectures. Inadequate results are to be expected, as in the context of CNN, the available 20,000 training samples are usually not enough to learn complex concepts, such as the gross tonnage of vessels based on 2D images made in uncontrolled conditions.

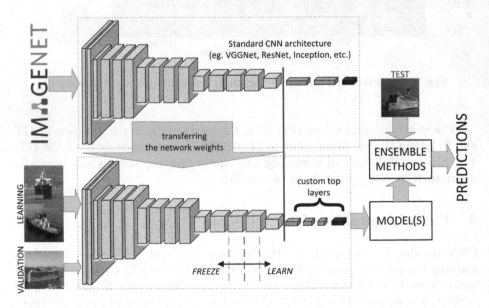

Fig. 4. Ensemble transfer learning using pretrained CNN model initialized with weights trained on ImageNet.

One possible solution is using transfer learning [29]. Donahue et al. [30] showed that features extracted from the activation of a deep convolutional network trained on a large dataset can be reused for novel generic tasks (Fig. 4). New tasks may differ significantly from the originally trained tasks and there may be insufficient labelled data to conventionally train deep network architecture to the new tasks.

ImageNet [31] trained features are the most popular starting point for fine-tuning on transfer tasks. In [32], the authors concluded that there is still no definitive answer to the question "What makes ImageNet good for transfer learning?", but it is obvious that traditional CNN architectures can extract high quality generic low/middle level features from an ImageNet dataset. Therefore, it is necessary to build a custom network which will act as a source network for the lower and middle layers from which pretrained ImageNet weights will be extracted. Top layers (fully connected and softmax layers) will be replaced with a configuration of layers which best fit the target concept. Since we are dealing with regression problem, the last activation function must be "linear" instead of "softmax".

Fine-tuned transfer learning has been performed using different pretrained CNN architectures, such as VGG16, VGG19, Xception, InceptionV3 and ResNet50. Also, we used the mentioned models as feature extractors for our learning dataset, where the deep features which were returned as an output of different custom top layers were processed with machine learning algorithms, such as Support Vector Regression (SVR) [33] and Random Forest (RF) [34]. Furthermore, we performed a simple grid search for all the models across the important network's hyperparameters by training multiple models and choosing the one with the best performances.

4.1 Results and Discussion

The results are summarized in Table 1.

Table 1. Transfer learning/feature extraction models' initial results for test dataset.

Model	MAE (GT)	RMSE (GT)	Corr. coef.
VGG16 (transfer learning)	5,143	8,451	0.936
VGG19 (transfer learning)	5,465	8,942	0.921
Xception (transfer learning)	9,542	14,887	0.827
InceptionV3 (transfer learning)	6,751	11,047	0.893
ResNet50 (transfer learning)	7,158	11,713	0.866
SVM (feature extraction)	7,220	10,547	0.872
RF (feature extraction)	5,557	9,093	0.919

Root Mean Square Error (RMSE) is the standard deviation of the prediction errors and it is significantly affected by large errors and outliers. The reason behind the stark difference between MAE and RMSE will become clearer further in the paper where the results are explained in detail. As can be seen from the table, the best results for the test dataset were achieved using the VGG16 architecture. The configuration for custom top layers, which replace the original top layers, is as follows: $CO(512)$-$ACT(relu)$-BN-FL-$DN(512)$-BN-$ACT(relu)$-$DN(1)$-$ACT(linear)$ - where $CO(n)$ denotes the 2D convolution layer with n filters, $ACT()$ denotes the activation function, BN denotes the batch normalization layer, FL denotes a flatten layer and $DN(n)$ denotes a dense layer. Other pretrained layers from the original network are frozen, except for the last four layers, which are retrained using the training data.

A significant advancement was achieved by employing Batch Normalization. A well-known problem of CNNs is covariate shift [35] which degrades the efficiency of their training. A covariate shift is a change in the distributions of internal nodes of a deep network, during the training. Ioffe et al. proposed a new technique, Batch Normalization, that reduces internal covariate shift and accelerates the training of deep networks. This is achieved via a normalization step that fixes the means and variances of each mini-batch, and backpropagates the gradients through the normalization parameters. Further research confirmed that deep neural networks trained with Batch Normalization converge faster and generalize better, reducing the need for Dropout [36].

It has also been shown that the size of each mini-batch is an important hyperparameter in cases when CNNs are used either for classification or for regression. Authors in [24] tested the influence of mini-batch sizes on the results and speed of learning. They came to the conclusion that some deep learning algorithms are more flexible to the influence of batch size, no matter the batch size employed. Also, they concluded that using a larger batch size is a acceptable heuristics towards good optimization. In [37], authors showed that increasing the mini-batch size progressively reduces the range of learning rates which provide convergence of learning process and acceptable test performance. On the other hand, small mini-batch sizes provide more accurate gradient calculations, which results in more stable and trustworthy training. They also found that the best results were achieved for mini-batch sizes between $m = 2$ and $m = 32$, which contrasts with recent works proposing the use of mini-batch sizes in the hundreds or thousands. Our experiences have confirmed the findings that the effect of mini-batch sizes varies for different algorithms. However, we reached the conclusion that the nature of the data also effects the choice of the optimal mini-batch size. Therefore, in our case, the best results were achieved with $m = 16$.

The best model is described in Table 2, Figs. 5 and 6, where Fig. 6 also shows the reasons why there is a gap between MAE and RMSE for all the models: (a) there is a small number of examples for which the estimation has significant deviations, and (b) for the biggest vessels the model tends to underestimate, likely due to the small number of training examples.

Further improvements can be achieved by using ensemble methods, a well-known technique in machine learning [38], where combining the outputs of several predictors improves the performance of a single generic one. Authors in [39]

Table 2. Descriptive statistics for the test dataset and predictions with the best model.

	Test dataset	Predictions
Mean	23,601	22,882
Mode	13,312	14,420
Standard deviation	27,845	24,508
Kurtosis	5.6	4.3
Skewness	2.1	1.8

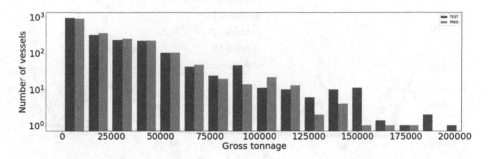

Fig. 5. Histogram comparison between the test dataset and predictions with the best fine-tuned VGG16 model.

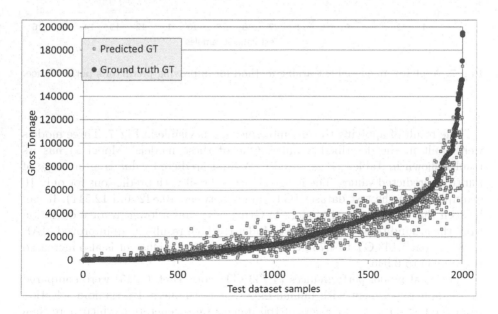

Fig. 6. Prediction results versus ground truth for test dataset and the best model.

compared these methods in the context of deep networks across multiple architectures and datasets, and they showed that some standard techniques may not be suitable for deep ensembles, since the new approaches improve performance.

Our research has shown that there is a very simple mechanism at hand which further improves predictor performances. For the best architecture (modified VGG16), multiple models were developed by changing the data which comprise a mini-batch. This was achieved by modifying the values of random seed - a number used to initialize a pseudorandom number generator which chooses the data for a mini-batch.

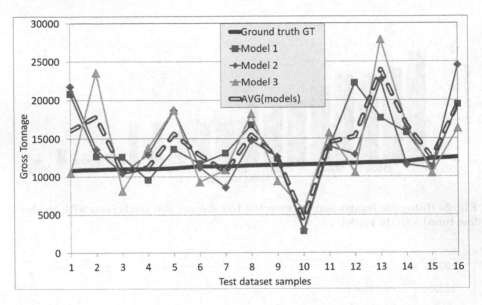

Fig. 7. Applying an ensemble learning method for improving predictor's performance.

The result of applying the ensemble method is visible in Fig. 7. Three models were built in the described manner. One of these models (Model 1) has its results shown in Table 1, Figs. 5 and 6. A final predicted value is an average of multiple predicted values. The image shows a detail with predictions for only 16 examples from the test dataset (GT values between 10,842 and 12,511). It can be seen that the combined model (AVG(models)) has a lower average deviation from the ground truth than the best Model 1, which results in an improved MAE by 8% (from 5,143 GT to 4,733 GT). The correlation coefficient is also increased from 0.936 to 0.945.

The final model performances (4,733 GT, corr. coef. 0.945) were compared in a simple experiment with human expert performances. By random selection from the test dataset, three sets of 100 images were generated, which were then given to three experts to assess the size of the vessels. It was shown that humans can reach a correlation coefficient between 0.90 and 0.95 as they can recognize relative relations between the sizes of two ships quite well. However, MAE is more important in practice, that is, the absolute level of correctness in assessing the size of a vessel, where the average result of 7850 GT is poorer than the one achieved by the model.

5 Conclusions

We have demonstrated that the CNNs can also be used successfully for regression tasks. Due to the limited amount of data in the learning dataset, it was necessary to apply transfer learning. We also proposed a simple ensemble-based approach to boost performance. By analyzing a small number of test cases in which a larger error has been observed, it can be concluded that there is a room for further improvements, which will be the goal and direction for future research. Therefore, by using the optimal architecture, it is possible to achieve good results in the field of automatic vessel size estimation based on 2D images.

References

1. Foley, H., Matlin, M.: Sensation and Perception. Psychology Press, London (2015)
2. Yantis, S.: Sensation and Perception. Macmillan International Higher Education (2013)
3. DeAngelis, G.C., Cumming, B.G., Newsome, W.T.: Cortical area MT and the perception of stereoscopic depth. Nature **394**(6694), 677 (1998). https://doi.org/10.1038/29299
4. International Convention on Tonnage Measurement of Ships. http://www.imo.org/en/about/conventions/listofconventions/pages/international-convention-on-tonnage-measurement-of-ships.aspx. Accessed 4 Feb 2019
5. Standley, T., Sener, O., Chen, D., Savarese, S.: image2mass: estimating the mass of an object from its image. In: Conference on Robot Learning, pp. 324–333 (2017)
6. Tasdemir, S., Urkmez, A., Inal, S.: Determination of body measurements on the Holstein cows using digital image analysis and estimation of live weight with regression analysis. Comput. Electron. Agric. **76**(2), 189–197 (2011). https://doi.org/10.1016/j.compag.2011.02.001
7. Bozkurt, Y., Aktan, S., Ozkaya, S.: Body weight prediction using digital image analysis for slaughtered beef cattle. J. Appl. Anim. Res. **32**(2), 195–198 (2007). https://doi.org/10.1080/09712119.2007.9706877
8. Yang, Y., Teng, G.: Estimating pig weight from 2D images. In: Li, D. (ed.) CCTA 2007. TIFIP, vol. 259, pp. 1471–1474. Springer, Boston (2008). https://doi.org/10.1007/978-0-387-77253-0_100
9. Pezzuolo, A., Guarino, M., Sartori, L., González, L.A., Marinello, F.: On-barn pig weight estimation based on body measurements by a Kinect v1 depth camera. Comput. Electron. Agric. **148**, 29–36 (2018). https://doi.org/10.1016/j.compag.2018.03.003
10. Sabanci, K., Ekinci, S., Karahan, A.M., Aydin, C.: Weight estimation of wheat by using image processing techniques. J. Image Graph. **4**(1), 51–54 (2016). https://doi.org/10.18178/joig.4.1.51-54

11. Javadikia, P., Dehrouyeh, M.H., Naderloo, L., Rabbani, H., Lorestani, A.N.: Measuring the weight of egg with image processing and ANFIS model. In: Panigrahi, B.K., Suganthan, P.N., Das, S., Satapathy, S.C. (eds.) SEMCCO 2011. LNCS, vol. 7076, pp. 407–416. Springer, Heidelberg (2011). https://doi.org/10.1007/978-3-642-27172-4_50

12. Wu, J., Lim, J.J., Zhang, H., Tenenbaum, J.B., Freeman, W.T.: Physics 101: learning physical object properties from unlabeled videos. In: BMVC, vol. 2 (2016). https://doi.org/10.5244/c.30.39

13. He, Y., Xu, C., Khanna, N., Boushey, C.J., Delp, E.J.: Food image analysis: segmentation, identification and weight estimation. In: IEEE International Conference on Multimedia and Expo (ICME), pp. 1–6 (2013). https://doi.org/10.1109/icme.2013.6607548

14. Chae, J., et al.: Volume estimation using food specific shape templates in mobile image-based dietary assessment. In: Computational Imaging IX, vol. 7873. International Society for Optics and Photonics (2011). https://doi.org/10.1117/12.876669

15. Choy, C.B., Xu, D., Gwak, J.Y., Chen, K., Savarese, S.: 3D-R2N2: a unified approach for single and multi-view 3D object reconstruction. In: Leibe, B., Matas, J., Sebe, N., Welling, M. (eds.) ECCV 2016. LNCS, vol. 9912, pp. 628–644. Springer, Cham (2016). https://doi.org/10.1007/978-3-319-46484-8_38

16. Fan, H., Su, H., Guibas, L.J.: A point set generation network for 3D object reconstruction from a single image. In: Proceedings of the IEEE Conference on Computer Vision and Pattern Recognition, pp. 605–613 (2017). https://doi.org/10.1109/cvpr.2017.264

17. Solmaz, B., Gundogdu, E., Yucesoy, V., Koc, A.: Generic and attribute-specific deep representations for maritime vessels. IPSJ T. Comput. Vis. Appl. 9, 1–18 (2017). https://doi.org/10.1186/s41074-017-0033-4

18. Gundogdu, E., Solmaz, B., Yücesoy, V., Koç, A.: MARVEL: a large-scale image dataset for maritime vessels. In: Lai, S.-H., Lepetit, V., Nishino, K., Sato, Y. (eds.) ACCV 2016. LNCS, vol. 10115, pp. 165–180. Springer, Cham (2017). https://doi.org/10.1007/978-3-319-54193-8_11

19. Ship Photos and Ship Tracker. http://www.shipspotting.com. Accessed 10 Oct 2018

20. Shapiro, S.S., Wilk, M.B.: An analysis of variance test for normality (complete samples). Biometrika 52(3/4), 591–611 (1965). https://doi.org/10.2307/2333709

21. Chollet, F.: Deep Learning with Python, 1st edn. Manning Publications Co., Greenwich (2017)

22. Abadi, M. et al.: TensorFlow: large-scale machine learning on heterogeneous systems. http://www.tensorflow.org (2015)

23. Miličević, M., Žubrinić, K., Obradović, I., Sjekavica, T.: Data augmentation and transfer learning for limited dataset ship classification. WSEAS Trans. Syst. Control 13, 460–465 (2018)

24. Lathuilière, S., Mesejo, P., Alameda-Pineda, X., Horaud, R.: A comprehensive analysis of deep regression. arXiv preprint arXiv:1803.08450 (2018)

25. Simonyan, K., Zisserman, A.: Very deep convolutional networks for large-scale image recognition. arXiv preprint arXiv:1409.1556 (2014)

26. Szegedy, C., Vanhoucke, V., Ioffe, S., Shlens, J., Wojna, Z.: Rethinking the inception architecture for computer vision. In: Proceedings of the IEEE Conference on Computer Vision and Pattern Recognition, pp. 2818–2826 (2016). https://doi.org/10.1109/cvpr.2016.308

27. Chollet, F.: Xception: Deep learning with depthwise separable convolutions. In: Proceedings of the IEEE Conference on Computer Vision and Pattern Recognition, pp. 1251–1258 (2017). https://doi.org/10.1109/cvpr.2017.195

28. He, K., Zhang, X., Ren, S., Sun, J.: Deep residual learning for image recognition. In: Proceedings of the IEEE Conference on Computer Vision and Pattern Recognition, pp. 770–778 (2016). https://doi.org/10.1109/cvpr.2016.90

29. Pan, S.J., Yang, Q.: A survey on transfer learning. IEEE Trans. Knowl. Data Eng. **22**(10), 1345–1359 (2010). https://doi.org/10.1109/tkde.2009.191

30. Donahue, J., et al.: DeCAF: a deep convolutional activation feature for generic visual recognition. In: International Conference on Machine Learning, pp. 647–655 (2014)

31. Krizhevsky, A., Sutskever, I., Hinton, G.E.: Imagenet classification with deep convolutional neural networks. In: Advances in Neural Information Processing Systems, pp. 1097–1105 (2012)

32. Huh, M., Agrawal, P., Efros, A.A.: What makes ImageNet good for transfer learning? arXiv preprint arXiv:1608.08614 (2016)

33. Drucker, H., Burges, C.J., Kaufman, L., Smola, A.J., Vapnik, V.: Support vector regression machines. In: Advances in Neural Information Processing Systems, pp. 155–161 (1997)

34. Breiman, L.: Random forests. Mach. Learn. **45**(1), 5–32 (2001)

35. Ioffe, S., Szegedy, C.: Batch normalization: Accelerating deep network training by reducing internal covariate shift. arXiv preprint arXiv:1502.03167 (2015)

36. Srivastava, N., Hinton, G., Krizhevsky, A., Sutskever, I., Salakhutdinov, R.: Dropout: a simple way to prevent neural networks from overfitting. J. Mach. Learn. Res. **15**(1), 1929–1958 (2014)

37. Masters, D., Luschi, C.: Revisiting small batch training for deep neural networks. arXiv preprint arXiv:1804.07612 (2018)

38. Dietterich, T.G.: Ensemble methods in machine learning. In: Kittler, J., Roli, F. (eds.) MCS 2000. LNCS, vol. 1857, pp. 1–15. Springer, Heidelberg (2000). https://doi.org/10.1007/3-540-45014-9_1

39. Lee, S., Purushwalkam, S., Cogswell, M., Crandall, D., Batra, D.: Why M heads are better than one: training a diverse ensemble of deep networks. arXiv preprint arXiv:1511.06314 (2015)

Analyzing Digital Image by Deep Learning for Melanoma Diagnosis

Karl Thurnhofer-Hemsi[✉] and Enrique Domínguez

Department of Computer Languages and Computer Sciences, University of Málaga,
Boulevar Louis Pasteur, 35, 29071 Málaga, Spain
{karlkhader,enriqued}@lcc.uma.es

Abstract. Image classification is an important task in many medical applications, in order to achieve an adequate diagnostic of different lesions. Melanoma is a frequent kind of skin cancer, which most of them can be detected by visual exploration. Heterogeneity and database size are the most important difficulties to overcome in order to obtain a good classification performance. In this work, a deep learning based method for accurate classification of wound regions is proposed. Raw images are fed into a Convolutional Neural Network (CNN) producing a probability of being a melanoma or a non-melanoma. Alexnet and GoogLeNet were used due to their well-known effectiveness. Moreover, data augmentation was used to increase the number of input images. Experiments show that the compared models can achieve high performance in terms of mean accuracy with very few data and without any preprocessing.

Keywords: Image processing · Deep learning · Melanoma

1 Introduction

Skin cancer was the most commonly diagnosed cancer in the United States in 2016. Melanoma accounts for only 1% of all skin cancer cases, but the vast majority of skin cancer deaths. This is a type of skin cancer caused by abnormal multiplication of pigment producing cells that give color to the skin: melanocytes.

Melanoma is highly curable when detected in its earliest stages, it is more likely than other skin cancer to spread to other parts of the body [1]. Melanoma, in their initial growth phases, and other benign moles are similarities in their characteristic, which makes the diagnosis difficult between what is malignant and what is benign for experienced dermatologists [8].

The best way to detect skin cancer early is to recognize new or changing skin growths, particularly those that look different from other moles. The ABCDE rule (Fig. 1) outlines warning signs of the most common type of melanoma: A is for asymmetry, B is for border irregularity, C is for color, D is for diameter greater than 6 millimeters and E is for evolution [11].

Many previous techniques in dermatological computer aided classification has lacked the generalization capability of medical practitioners due to insufficient data and require extensive preprocessing [2], lesion segmentation [6,7] and

© Springer Nature Switzerland AG 2019
I. Rojas et al. (Eds.): IWANN 2019, LNCS 11507, pp. 270–279, 2019.
https://doi.org/10.1007/978-3-030-20518-8_23

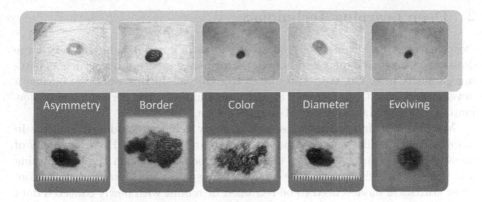

Fig. 1. Traditional clinical analysis followed by dermatologists (ABCDE rule).

extraction of specific features before classification [4,16]. Several learning-based methods were applied form melanoma classification [3,13,15] using the special features of this kind of lesion. By contrast, our proposal requires no human intervention, since it is trained directly from the labelled images with a single neural network.

Convolutional Neural Networks (CNNs) is one of the most popular deep learning techniques for image analysis. CNNs were inspired by the animal visual cortex and they are one of the first truly successful deep learning architectures, which have shown outstanding performance in processing images and videos. Nowadays, with the help of GPU-accelerated computing techniques, CNNs have been successfully applied to object recognition (e.g. handwriting, face, behavior...), recommender systems or image classification. Recent works show that deep networks are being a powerful tools for medical image analysis [10], and therefore for melanoma classification [12,17].

In this work, two different deep learning based methods have been implemented on a computer for detection of melanoma lesions, which could assist a dermatologist in early diagnosis of this cancer. Both methods were evaluated on the provided images from the DermIS and DermQuest databases (now integrated in Derm101[1]). The whole dataset is composed by 533 images, where 329 are melanomas and 204 nevi. The provided experimental results are really promising, even more if we think about the few images that have been used. Deep learning often requires hundreds of thousands or millions of images for the best results.

The rest of the paper is organized as follows: Sect. 2 describes the deep learning techniques used in this work. Experiments and discussed results are reflected in Sect. 3. Finally, conclusions and future works are summarized in Sect. 4.

---

[1] Comprehensive digital resource for healthcare professionals including information from world-renowned experts in the field. https://www.derm101.com/.

2   Deep Learning Techniques

Deep learning algorithms, powered by advances in computation and very large datasets, have shown to exceed human performance in visual tasks such as playing games (Chess, Go, etc.) and object recognition. In this work, we outline a development of a CNN in order to assist and improve the performance of dermatologists in melanoma image based detection.

Machine learning technologies have already achieved significant success in many areas including classification, regression and clustering. However, many of these techniques work well only under a common assumption the the training and test data are drawn from the same feature space and the same distribution. Most statistical models need to be retrained or rebuilt with newly collected data when the distribution changes. Unfortunately, in many real world applications this process could be expensive or simply impossible. In such cases, transfer learning can truly be suitable and beneficial.

The need for transfer learning may arise when the data is easily outdated or when the distribution of data is very different among the distint classes. In such situations, it is easier to adapt a classification model, which has been trained on a similar problem, to help learn classification models for equal nature data. There exists a large amount of works on transfer learning for reinforcement in the literature. Nevertheless, in this paper, we only focus on transfer learning for classification problems by using two widely known CNNs, such as AlexNet and GoogLeNet, that are related more closely to the addressed problem.

AlexNet [9] consists of five convolutional layers and three fully connected layers, that won the ImageNet Large Scale Visual Recognition Challenge (ILSVRC) in 2012. A schematic draw of the net's architecture is shown in Fig. 2a. The input is an RGB image of size 256×256, and it corresponds to one of 1000 different classes. The output is a vector of probabilities with one value for each class. Random crops of size 227×227 are generated from inside the 256×256 images to feed the first layer of AlexNet. The first two convolutional layers are followed by the max pooling layers. The third, fourth and fifth convolutional layers are connected directly. The last is followed by a maxpool, and connected to a series of two fully connected layers, to end into a softmax classifier with 1000 class labels. ReLU activation is applied after all the convolution and fully connected layers.

GoogLeNet [14] is a deep neural network architecture for computer vision codenamed Inception. It was presented at the ILSVRC 2014 classification and detection challenges, being one of the best the state-of-the-art deep networks. The philosophy of this model is the deeper the better. This consists on both increasing the depth of the network and the number of neurons at each layer. It is composed by a total of 22 layers, most of them based on Inception modules as depicted in Fig. 2b. All the convolutions use ReLU activation. The input of the network is 224×224 taking RGB color channels with mean subtraction. The output, as Alexnet, is a vector of 1000 components to determine between 1000 classes.

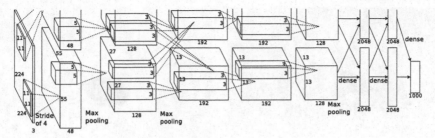

(a) Alexnet (image extracted from [9])

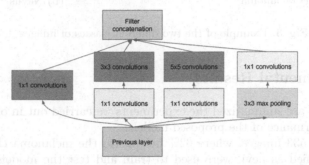

(b) Inception of GoogLeNet (image extracted from [14])

Fig. 2. Architecture of the networks.

Both networks were trained on 1.2 million images from the ImageNet database[2]. We used these pre-trained models and trained them on our dataset using transfer learning in order to transfer the knowledge of these models to our target problem.

A typical problem of these deep learning techniques are the requirement of large datasets to perform a good model training. Small datasets provoke deep networks do not generalize well data from the test set. Thus, these networks suffer from the problem of over-fitting. Data augmentation is a simply way to reduce over-fitting on models. The amount of training data is increased using only its information. Data augmentation techniques have commonly been applied to medical classification problems during last years with success [5]. Depending on the problem we are dealing with, different augmentation methods can be applied. The main techniques are based on data warping, which is an approach which augment the input data in data space. A commonly practice for augmenting image data is to perform geometric deformations, such as reflections, crops and translations of the image. Also, changing the color and luminosity of the image is useful in some cases. All of them are performed as an affine transformation of the original image.

---

[2] http://www.image-net.org.

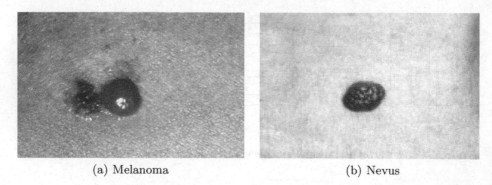

(a) Melanoma　　　　　　　　　　　　(b) Nevus

Fig. 3. Example of the two different classes of images.

3　Experimental Results

In this section are summarized the experiments we carried out in order to evaluate the performance of the proposed models.

A total of 533 images, where 62% belongs to the melanoma class and the rest are classified as nevi, were used to train and test the models (Fig. 3). It is a small dataset for a deep learning model, but one of our objectives is to show that a high accuracy can be achieved with few samples. For this purpose, data augmentation techniques are employed for precision improvement. These are methods for constructing iterative sampling algorithms via the introduction of unobserved data. In these experiments, we carried out reflections, crops and rotations to the data.

There are several parameters that can be tuned in a deep neural network. However, the main ones we are focusing on are:

- Batch Size (BS): indicates the amount of images processed in one iteration.
- Initial learning rate (LR): establishes with which rate is going to start the learning procedure.
- Validation frequencies (Val. Freq.): is the number of iterations between evaluations of the validation metrics.

The parameter values selected for analysis are summarized in Table 1.

From a quantitative point of view, we have chosen a well-known measure in order to compare the performance of the detection of the different studied networks: the Accuracy (ACC). It measures the proportion of true detections among the total number of tested images.

$$ACC = \frac{TP + TN}{TP + TN + FN + FP} \tag{1}$$

where TP, TN, FP, and FN denotes the true positives, true negatives, false positives and false negatives, respectively.

Table 1. Parameters for neural network tuning

Parameter	Tested values
BS	$\{16, 32, 64, 128\}$
LR	$\{0.0001, 0.00001, 0.000001\}$
Val. Freq.	$\{5, 10, 15, 20\}$

All the possible combinations of the values mentioned in Table 1 were used for training and the outcomes of the test. In Table 2 are summarized the outcomes of the proposed models, using both raw dataset and data augmentation.

With the non use of data augmentation, GoogLeNet performs slightly better than Alexnet in average. Using batch sizes 16 or 32, there exist learning rates with which the mean accuracy of Alexnet overcome GoogLeNet, specially using validations frequencies 15 and 20. However, for higher batch sizes and independently from the learning rate used, GoogLeNet obtains the best results with a best accuracy of almost 88%.

The best configurations are summarized in Table 3. When we use the original dataset without applying data augmentation, smaller learning rates perform better, whereas that when the data is augmented more stability is achieved with a high learning rate for both networks. The optimal validation frequency for any model is almost the same, which would say that this parameter in not too relevant. On the other hand, the batch size values are discrepant and symmetric with respect the data augmentation technique: Alexnet generates the best accuracy with a batch size of 128 with more data and GoogLeNet needs 16, but when the original dataset is used, GoogLeNet carried on a better optimization with 128 and Alexnet with 32, reflecting their dependencies on the dataset size.

Figure 4 shows the training loss and accuracy of the best configuration of the Alexnet network using data augmentation. The smoothness of the plots indicates that the model learned properly, reducing the fluctuations while the number of epochs increased. The loss stabilization indicates that no over-fitting has occurred and the absence of significant drops in the accuracy curve suggest that the model is robust. This panorama changes when data augmentation is employed during the training phase. Despite the fact that both methods improve substantially their results, now Alexnet is the best model for almost any configuration of the parameters, depicting an accuracy of 93,4% of detections in front of the best 92% of detections of the GoogLeNet.

Finally, average confusion matrices for the best configurations are presented in Table 4. The outcomes of the raw dataset (Tables 4a and c) are worse than the augmented one. The amount of false negatives is around 8–9% for both Alexnet and GoogLeNet, which is still quite high. Using Alexnet, only 52% of the dataset is detected as melanoma (in front of the 62% that are present on the dataset), and results reach 54% of true positives for GoogLeNet. The best rate of false negatives (lower is better) is achieved using data augmentation (Tables 4b and d), with 3% of FN for Alexnet. Moreover, 58% of the melanomas

Table 2. Average accuracy and standard deviation of the proposed models. Best results are boldfaced.

BS	LR	Val. Freq.	Non-augmented		Augmented	
			Alexnet	GoogleNet	Alexnet	GoogleNet
16	1E-06	5	0,839 ± 0,06	0,845 ± 0,02	0,927 ± 0,02	0,886 ± 0,05
		10	0,834 ± 0,03	0,845 ± 0,06	0,927 ± 0,01	0,906 ± 0,03
		15	0,843 ± 0,05	0,831 ± 0,05	0,928 ± 0,01	0,897 ± 0,07
		20	0,835 ± 0,05	0,837 ± 0,04	0,931 ± 0,01	0,896 ± 0,02
	1E-05	5	0,835 ± 0,04	0,859 ± 0,04	0,930 ± 0,01	0,897 ± 0,06
		10	0,826 ± 0,05	0,826 ± 0,04	0,923 ± 0,01	0,914 ± 0,03
		15	0,826 ± 0,04	0,841 ± 0,03	0,930 ± 0,01	0,886 ± 0,04
		20	0,811 ± 0,03	0,857 ± 0,06	0,926 ± 0,01	0,912 ± 0,02
	1E-04	5	0,842 ± 0,06	0,847 ± 0,04	0,923 ± 0,01	0,912 ± 0,03
		10	0,831 ± 0,04	0,837 ± 0,06	0,927 ± 0,01	0,909 ± 0,02
		15	0,816 ± 0,09	0,828 ± 0,03	0,928 ± 0,01	**0,920 ± 0,02**
		20	0,836 ± 0,04	0,864 ± 0,03	0,925 ± 0,02	0,895 ± 0,03
32	1E-06	5	0,841 ± 0,04	0,857 ± 0,03	0,924 ± 0,01	0,889 ± 0,03
		10	0,820 ± 0,04	0,802 ± 0,07	0,930 ± 0,01	0,896 ± 0,03
		15	0,839 ± 0,03	0,819 ± 0,05	0,930 ± 0,01	0,903 ± 0,02
		20	0,833 ± 0,03	0,833 ± 0,04	0,929 ± 0,01	0,888 ± 0,03
	1E-05	5	0,833 ± 0,03	0,839 ± 0,05	0,925 ± 0,01	0,901 ± 0,03
		10	0,842 ± 0,05	0,868 ± 0,05	0,925 ± 0,03	0,907 ± 0,02
		15	0,834 ± 0,04	0,856 ± 0,07	0,927 ± 0,01	0,894 ± 0,02
		20	**0,851 ± 0,03**	0,842 ± 0,05	0,926 ± 0,01	0,911 ± 0,02
	1E-04	5	0,809 ± 0,06	0,830 ± 0,03	0,929 ± 0,01	0,905 ± 0,02
		10	0,841 ± 0,06	0,837 ± 0,05	0,926 ± 0,01	0,914 ± 0,01
		15	0,832 ± 0,04	0,837 ± 0,07	0,922 ± 0,01	0,904 ± 0,02
		20	0,848 ± 0,02	0,856 ± 0,07	0,928 ± 0,01	0,910 ± 0,03
64	1E-06	5	0,827 ± 0,05	0,834 ± 0,04	0,930 ± 0,01	0,879 ± 0,05
		10	0,835 ± 0,08	0,847 ± 0,02	0,925 ± 0,01	0,896 ± 0,06
		15	0,843 ± 0,04	0,829 ± 0,04	0,927 ± 0,01	0,905 ± 0,07
		20	0,826 ± 0,05	0,841 ± 0,06	0,929 ± 0,00	0,904 ± 0,04
	1E-05	5	0,830 ± 0,04	0,841 ± 0,04	0,923 ± 0,01	0,917 ± 0,03
		10	0,817 ± 0,05	0,826 ± 0,03	0,928 ± 0,01	0,872 ± 0,04
		15	0,826 ± 0,06	0,857 ± 0,02	0,928 ± 0,01	0,904 ± 0,04
		20	0,808 ± 0,05	0,847 ± 0,04	0,924 ± 0,01	0,910 ± 0,05
	1E-04	5	0,805 ± 0,04	0,836 ± 0,06	0,921 ± 0,03	0,900 ± 0,04
		10	0,832 ± 0,04	0,868 ± 0,05	0,925 ± 0,01	0,908 ± 0,05
		15	0,841 ± 0,02	0,831 ± 0,08	0,925 ± 0,01	0,915 ± 0,03
		20	0,826 ± 0,07	0,834 ± 0,05	0,924 ± 0,01	0,906 ± 0,03
128	1E-06	5	0,835 ± 0,05	0,829 ± 0,05	0,924 ± 0,01	0,913 ± 0,04
		10	0,842 ± 0,02	0,841 ± 0,05	0,928 ± 0,01	0,906 ± 0,03
		15	0,841 ± 0,04	**0,878 ± 0,04**	0,932 ± 0,01	0,906 ± 0,03
		20	0,831 ± 0,05	0,822 ± 0,07	0,919 ± 0,02	0,906 ± 0,03
	1E-05	5	0,841 ± 0,07	0,854 ± 0,05	0,925 ± 0,01	0,899 ± 0,04
		10	0,849 ± 0,04	0,856 ± 0,05	0,925 ± 0,01	0,906 ± 0,05
		15	0,827 ± 0,04	0,856 ± 0,05	0,927 ± 0,03	0,893 ± 0,03
		20	0,825 ± 0,05	0,812 ± 0,06	0,927 ± 0,02	0,903 ± 0,06
	1E-04	5	0,823 ± 0,04	0,857 ± 0,03	0,929 ± 0,01	0,912 ± 0,02
		10	0,831 ± 0,04	0,854 ± 0,04	0,923 ± 0,01	0,917 ± 0,01
		15	0,841 ± 0,05	0,854 ± 0,02	**0,934 ± 0,01**	0,908 ± 0,02
		20	0,816 ± 0,03	0,843 ± 0,04	0,925 ± 0,02	0,895 ± 0,03

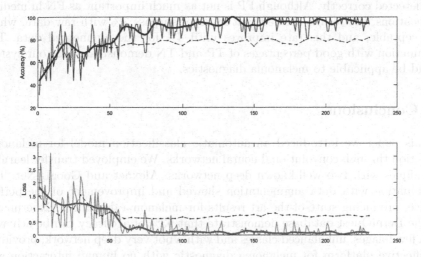

Fig. 4. Training progress of the best configuration of the augmented Alexnet.

Table 3. Best configuration of the proposed models.

Configuration	Non-augmented		Augmented	
	Alexnet	GoogleNet	Alexnet	GoogleNet
Batch size	32	128	128	16
Learning rate	1E-05	1E-06	1E-04	1E-04
Validation frequency	20	15	15	15

Table 4. Average matrix confusion of the proposed models for the best configuration.

		Predicted	
		Melanoma	Nevus
Real	Melanoma	0.52	0.09
	Nevus	0.07	0.32

(a) Non-Augmented Alexnet

		Predicted	
		Melanoma	Nevus
Real	Melanoma	0.58	0.03
	Nevus	0.04	0.35

(b) Augmented Alexnet

		Predicted	
		Melanoma	Nevus
Real	Melanoma	0.54	0.08
	Nevus	0.07	0.31

(c) Non-Augmented GoogleNet

		Predicted	
		Melanoma	Nevus
Real	Melanoma	0.56	0.06
	Nevus	0.04	0.34

(d) Augmented GoogleNet

are detected correctly. Although FP is not as much important as FN in medical applications, the results showed that in never surpass 7% with raw data, which is acceptable, and this rate improves until 4% for the augmented data. The conjunction with good percentages of TP and TN demonstrate that our system would be applicable to melanoma diagnostics.

4   Conclusions

In this paper we introduced an automatic classification model for melanoma detection through convolutional neural networks. We employed transfer learning techniques with two well-known deep networks: Alexnet and GoogLeNet. The combination with data augmentation showed and improvement of the performance, producing state-of-the-art results for melanoma detection. Experiments on the DermQuest database demonstrate that 93% of accuracy can be achieved with few images, unbalanced classes and with a not very deep network, providing an effective platform for melanoma diagnostic with no human interaction and avoiding the segmentation step.

Further works include the use of the other datasets, which would improve the performance and reduce the false negative detections even more. An extensive statistical analysis with more precision measures applied to a detailed optimization of the network model would allow us to achieve better results for the following cancer diagnosis and healing processes.

Acknowledgements. This work is partially supported by the Ministry of Economy and Competitiveness of Spain under grant TIN2014-53465-R, project name Video surveillance by active search of anomalous events. It is also partially supported by the Autonomous Government of Andalusia (Spain) under grant TIC-657, project name Self-organizing systems and robust estimators for video surveillance. All of them include funds from the European Regional Development Fund (ERDF). The authors thankfully the grant of the University of Málaga and acknowledge the computer resources, technical expertise and assistance provided by the SCBI (Supercomputing and Bioinformatics) center of the University of Málaga. Karl Thurnhofer-Hemsi (FPU15/06512) is funded by a PhD scholarship from the Spanish Ministry of Education, Culture and Sport under the FPU program.

References

1. American Cancer Society, I. (ed.): Cancer Facts & Figures 2016. American Cancer Society, Atlanta (2016)
2. Asha Gnana Priya, H., Anitha, J., Poonima Jacinth, J.: Identification of melanoma in dermoscopy images using image processing algorithms. In: 2018 International Conference on Control, Power, Communication and Computing Technologies, ICCPCCT 2018, pp. 553–557 (2018)
3. Bakheet, S.: An SVM framework for malignant melanoma detection based on optimized hog features. Computation 5(1), 4 (2017)

4. Devassy, B.M., Yildirim-Yayilgan, S., Hardeberg, J.Y.: The impact of replacing complex hand-crafted features with standard features for Melanoma classification using both hand-crafted and deep features. In: Arai, K., Kapoor, S., Bhatia, R. (eds.) IntelliSys 2018. AISC, vol. 868, pp. 150–159. Springer, Cham (2019). https://doi.org/10.1007/978-3-030-01054-6_10
5. Hussain, Z., Gimenez, F., Yi, D., Rubin, D.: Differential data augmentation techniques for medical imaging classification tasks. In: AMIA Annual Symposium Proceedings, vol. 2017, p. 979. American Medical Informatics Association (2017)
6. Jafari, M.H., et al.: Skin lesion segmentation in clinical images using deep learning. In: 2016 23rd International Conference on Pattern Recognition (ICPR), pp. 337–342, December 2016
7. Jafari, M.H., Nasr-Esfahani, E., Karimi, N., Soroushmehr, S.M.R., Samavi, S., Najarian, K.: Extraction of skin lesions from non-dermoscopic images for surgical excision of melanoma. Int. J. Comput. Assist. Radiol. Surg. **12**(6), 1021–1030 (2017). https://doi.org/10.1007/s11548-017-1567-8
8. Jerant, A.F., Johnson, J.T., Sheridan, C.D., Caffrey, T.J.: Early detection and treatment of skin cancer. Am. Fam. Phys. **62**(2), 357–368 (2000)
9. Krizhevsky, A., Sutskever, I., Hinton, G.E.: Imagenet classification with deep convolutional neural networks. In: Proceedings of the 25th International Conference on Neural Information Processing Systems, NIPS 2012, vol. 1, pp. 1097–1105. Curran Associates Inc., USA (2012). http://dl.acm.org/citation.cfm?id=2999134.2999257
10. Litjens, G., et al.: A survey on deep learning in medical image analysis. Med. Image Anal. **42**, 60–88 (2017). http://www.sciencedirect.com/science/article/pii/S1361841517301135
11. Nachbar, F., et al.: The ABCD rule of dermatoscopy: high prospective value in the diagnosis of doubtful melanocytic skin lesions. J. Am. Acad. Dermatol. **30**(4), 551–559 (1994)
12. Nida, N., Irtaza, A., Javed, A., Yousaf, M., Mahmood, M.: Melanoma lesion detection and segmentation using deep region based convolutional neural network and fuzzy c-means clustering. Int. J. Med. Inf. **124**, 37–48 (2019)
13. Ruela, M., Barata, C., Marques, J., Rozeira, J.: A system for the detection of melanomas in dermoscopy images using shape and symmetry features. Comput. Methods Biomech. Biomed. Eng.: Imag. Vis. **5**(2), 127–137 (2017)
14. Szegedy, C., et al.: Going deeper with convolutions. In: 2015 IEEE Conference on Computer Vision and Pattern Recognition (CVPR), vol. 0, pp. 1–9, June 2015. https://doi.org/10.1109/CVPR.2015.7298594
15. Victor, A., Ghalib, M.: Automatic detection and classification of skin cancer. Int. J. Intell. Eng. Syst. **10**(3), 444–451 (2017)
16. Yadav, V., Kaushik, V.: Detection of melanoma skin disease by extracting high level features for skin lesions. Int. J. Adv. Intell. Paradigms **11**(3–4), 397–408 (2018)
17. Yu, L., Chen, H., Dou, Q., Qin, J., Heng, P.A.: Automated melanoma recognition in dermoscopy images via very deep residual networks. IEEE Trans. Med. Imag. **36**(4), 994–1004 (2017)

BatchNorm Decomposition for Deep Neural Network Interpretation

Lucas Y. W. Hui[(✉)] and Alexander Binder

ISTD, Singapore University of Technology and Design, Changi, Singapore
lucas.hui@hotmail.com

Abstract. Layer-wise relevance propagation (LRP) has shown potential for explaining neural network classifier decisions. In this paper, we investigate how LRP is to be applied to deep neural network which makes use of batch normalization (BatchNorm), and show that despite the functional simplicity of BatchNorm, several intuitive choices of published LRP rules perform poorly for a number of frequently used state of the art networks. Also, we show that by using the ε-rule for BatchNorm layers we are able to detect training artifacts for MobileNet and layer design artifacts for ResNet. The causes for such failures are analyzed deeply and thoroughly. We observe that some assumptions on the LRP decomposition rules are broken given specific networks, and propose a novel LRP rule tailored for BatchNorm layers. Our quantitatively evaluated results show advantage of our novel LRP rule for BatchNorm layers and its wide applicability to common deep neural network architectures. As an aside, we demonstrate that one observation made by LRP analysis serves to modify a ResNet for faster initial training convergence.

Keywords: Layer-wise relevance propagation · Batch normalization · Convolutional neural networks

1 Introduction

This paper examines the topic of deep neural network (DNN) decision explanations for input samples. Important approaches to this problem, which has attracted considerable interest, include among others: sensitivity [13] and SmoothGrad [15] which computes gradient of image classification score with respect to input pixels, deconvolutional neural nets [18] and guided backpropagation [16] which can be derived by applying certain zeroing-out-rules to gradients and ReLU layers, deepLIFT [12], and layer-wise relevance propagation [1] which is based on the deep-Taylor methodology [9]. While layer-wise relevance propagation (LRP) has been demonstrated effective in interpreting various convolutional neural networks [8,11,14], most of its applications so far were shown on such networks without batch normalization. There remains a question how it performs in state-of-the-art deep neural networks [4–6,17] which employ batch normalization [3,7] for improved gradient flow.

© Springer Nature Switzerland AG 2019
I. Rojas et al. (Eds.): IWANN 2019, LNCS 11507, pp. 280–291, 2019.
https://doi.org/10.1007/978-3-030-20518-8_24

Batch normalization at test time amounts to a simple affine transformation. Yet we can show that several natural choices of treating these layers in an backward pass for LRP, such as the ε-rule which is common for fully connected layers or the identity rule, results for some popular deep neural networks in very poor explanations. Given the strategic function and positions of BatchNorm layers, we are motivated to explore the roles of different BatchNorm decomposition rules as well as various BatchNorm statistics to the interpretability of various network architectures, and to ultimately achieve better network explanations and network learning performance.

2   Layer-Wise Relevance Propagation

Features of an input sample are computed and propagated through layers of neurons within a convolutional network during classification process. When a final decision is made by the classifier, the decision score can be back propagated using LRP reversely through the layers and neurons. Each neuron in the process accumulates the relevant contributions from the decision score, and hence gives insight to how the network derives the decision. At input layer, the back propagated relevance reaches pixel resolution and forms a heat-map representing importance of each pixel in contribution to the classifier decision. Detail explanation of LRP using deep Taylor decomposition may be found in [1].

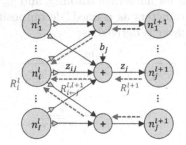

Fig. 1. LRP relevance back-propagation

As illustrated in Fig. 1 with two layers of neurons within a network, an example forward propagation of input features following the solid arrows is given as:

$$z_j = \sum_i z_{ij} + b_j = \sum_i x_i w_{ij} + b_j \tag{1}$$

where x_i are outputs of layer l neuron, w_{ij} and b_j are network weight and bias parameters, and z_j is input to activation function of layer $l+1$ neuron. The corresponding back-propagation of relevance R following the dotted red arrows with conversation property is given as:

$$R_i^l = \sum_j R_{i \leftarrow j}^{l,l+1} \tag{2}$$

In addition, a number of back propagation or decomposition rules have been developed [2,10] for different interpretation or functional purposes. Some of the basic decomposition rules are shown in Table 1.

Table 1. A few published LRP decomposition rules.

Identity	$R_i^l = R_i^{l+1}$
ε-rule	$R_i^l = \sum_j R_j^{l+1} \left(\dfrac{x_i w_{ij}}{\sum_{i'} x_{i'} w_{i'j} + b_j + \varepsilon \cdot \text{sign}(\sum_{i'} x_{i'} w_{i'j} + b_j)} \right)$
w^+-rule	$R_i^l = \sum_j R_j^{l+1} \left(\dfrac{x_i w_{ij}^+}{\sum_{i'} x_{i'} w_{i'j}^+} \right)$
$\alpha\beta$-rule	$R_i^l = \sum_j R_j^{l+1} \left(\alpha \dfrac{(x_i w_{ij})^+}{\sum_{i'} (x_{i'} w_{i'j})^+ + b_j^+} - \beta \dfrac{(x_i w_{ij})^-}{\sum_{i'} (x_{i'} w_{i'j})^- + b_j^-} \right)$

Using these basic rules, we can derive decomposition accordingly for batch normalization layers. Given the definition of batch normalization process [7] as:

$$y_i = \frac{x_i - \mu_c}{\sqrt{\sigma_c^2 + \epsilon}} \gamma_c + \beta_c \tag{3}$$

where μ_c and σ_c are channel-wise running-batch average and standard deviation of input x_i, ϵ is a constant for numerical stability, and γ_c and β_c are channel-wise affine transformation parameters as trained, decomposition according to ε-rule for batch normalization layer is derived as:

$$R_i^l = \frac{x_i w_i}{x_i w_i + b_i + \varepsilon \cdot \text{sign}(x_i w_i + b_i)} R_i^{l+1} \tag{4}$$

$$\text{with } w_i = \frac{\gamma_c}{\sqrt{\sigma_c^2 + \epsilon}} \tag{5}$$

$$b_i = \beta_c - \frac{\mu_c \gamma_c}{\sqrt{\sigma_c^2 + \epsilon}} \tag{6}$$

and similarly according to $\alpha\beta$-rule is derived as:

$$R_i^l = \left[\alpha \frac{(x_i w_i)^+}{(x_i w_i)^+ + b_i^+} - \beta \frac{(x_i w_i)^-}{(x_i w_i)^- + b_i^-} \right] R_i^{l+1} \tag{7}$$

For network interpretation according to LRP, Preset is used to define a set of decomposition rules for specific function layers within the network. In this paper, we define LRP Presets according to Table 2 with specific intention to interpret in depth the various state-of-art deep neural networks using their batch normalization layer statistics.

Resulting heat-maps of LRP, according to Presets with the three basic and commonly used rules (ε, Identity, and $\alpha 1\beta 0$) on batch normalization layers for a

Table 2. LRP rules as applied to neural network layers in this paper.

DNN functional layers	LRP decomposition rules
Convolution	$\alpha\beta$-rule with $\alpha = 1$ and $\beta = 0$, hence $\alpha 1 \beta 0$
Batch normalization	Identity, ε-rule, $\alpha 1 \beta 0$, or newly proposed
Fully connected/Linear	ε-rule

selected set of existing images, are shown in Fig. 2 for pre-trained MobileNet-V2[1], Fig. 3 for ResNet-50[2], Fig. 4 for DenseNet-121[3], and Fig. 5 for InceptionResNet-V2[4]. From figures shown, we can clearly notice that (1) the ε-rule, although widely known as the default rule especially for linear functions, fails quite badly in case of MobileNet, (2) different rules perform or fail differently in different networks, and (3) none of the three commonly known basic rules produces consistent results across all tested networks.

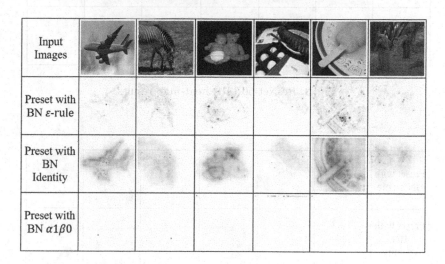

Fig. 2. MobileNet-V2 LRP heat-map results

3 Observations and Analysis

3.1 BatchNorm Decomposition with ε-Rule

The ε-rule has been widely used as the default decomposition rule for many linear functions; nevertheless, while applying to the BatchNorm layers in MobileNet,

[1] https://github.com/tonylins/pytorch-mobilenet-v2.
[2] https://pytorch.org/docs/stable/torchvision/models.html#id3.
[3] https://pytorch.org/docs/stable/torchvision/models.html#id5.
[4] https://github.com/Cadene/pretrained-models.pytorch/blob/master/pretrained models/models/inceptionresnetv2.py.

resulting heat-maps (Fig. 2) give little or no useful information. To explain, we examine in more details the statistics from this pre-trained MobileNet-V2 Batch-Norm layers. Figure 6 shows an example set of this BatchNorm layer input and bias parameters starting from top-layer (output) to input-layer, where Bp-Mean and Bn-Mean are the means of positive and negative bias respectively (Avg. of b^+ and b^-), and Xp-Mean and Xn-Mean are the means of positive and negative input respectively (Avg. of x^+ and x^-).

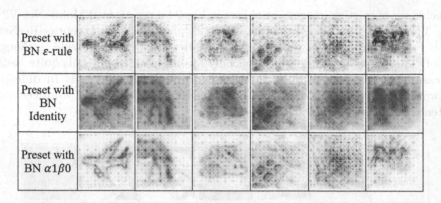

Fig. 3. ResNet-50 LRP heat-map results

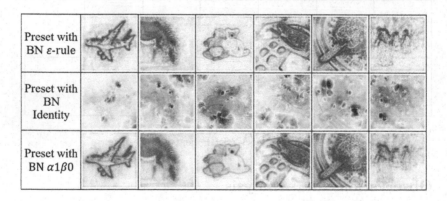

Fig. 4. DenseNet-121 LRP heat-map results

From Fig. 6, there is clearly a huge peak on the 49th BatchNorm layer (hence near to the input layer). This exceptionally huge peak is highly unexpected given the many machine learning techniques such learning rate, regularization, and dropout. It may be due to a peculiarity in the design of the MobileNet. Furthermore, according to specific statistics of this BatchNorm layer as shown in Fig. 7, we can observe a large negative peak input x^- combined with a large

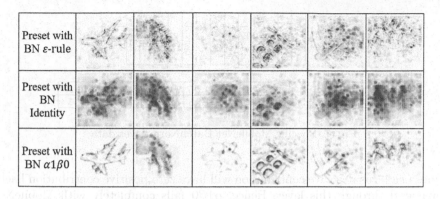

Fig. 5. InceptionResNet-V2 LRP heat-map results

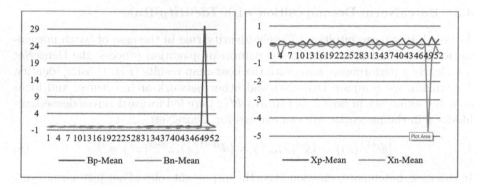

Fig. 6. MobileNet BatchNorm layer statistics

positive peak bias b^+ to generate a much smaller or normalized output $z^{\pm}$. This observation highlights a fact that bias values may not only be used for changing input bias but also for changing output magnitude. Another way to interpret, the bias has a huge cancelling effect to the input which causes ε-rule to fail in this situation. This observation serves to explain the poorer heat-map performance in the case of ResNet (Fig. 3) and InceptionResNet (Fig. 5).

3.2 BatchNorm Decomposition with $\alpha\beta$-Rule

Also commonly used, the $\alpha\beta$-rule has the ability to separate positive and negative contributions for example to indicate a pixel's positive or negative impact to a classifier decision. However, the $\alpha\beta$-rule fails badly in the case of MobileNet (Fig. 2) too. According to data from Fig. 7, the negative only input (hence Avg. $x^+ = 0$) to the BatchNorm layer along with the positive only weight (or Avg. w^- = 0) is another unexpected observation. This fact changes the assumptions about how relevance can be back-propagated in LRP; specifically, the $\alpha\beta$-rule which

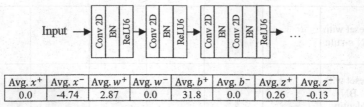

Avg. x^+	Avg. x^-	Avg. w^+	Avg. w^-	Avg. b^+	Avg. b^-	Avg. z^+	Avg. z^-
0.0	-4.74	2.87	0.0	31.8	0.0	0.26	-0.13

Fig. 7. Specific MobileNet BatchNorm layer data

defines $(xw)^-$ as negative contribution will have zero positive contribution back-propagated through this layer. Hence, $\alpha 1 \beta 0$ fails completely with MobileNet since this specific BatchNorm is within the only branch in the network.

3.3 BatchNorm Decomposition with *Identity*-Rule

Although it is more intuitive to apply identity rule in the case of batch normal-ization given that it is a one-to-one neuron propagation process, the DenseNet has clearly a performance issue with the heat-map results (Fig. 4) using identity. To explain, we compare DenseNet and other network architectures, and focus on a key difference in how data features $\Phi^d(x)$ are fed forward across dense-layer blocks d in the networks. We can express for DenseNet:

$$\{\Phi^{d+1}(x)\} = \{\Phi^d(x), z^d\} = \{\Phi^{d-1}(x), z^{d-1}, z^d\} = \ ... \tag{8}$$

In this case, batch normalization plays the important roles of not just normalizing typically results from a convolution layer, but also results from multiple dense-layer blocks. To illustrate, Fig. 8 shows the bias (b_j and β_c) parameters of last batch normalization layer in DenseNet. Stronger bias is accordingly applied to features from latest few dense-layer blocks (values on right side of the plot).

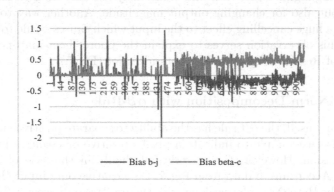

Fig. 8. Specific DenseNet bias data

Similarly, it can be observed from the weight parameters (w_i and γ_c) of the same batch normalization layer in Fig. 9 that much higher weights are applied to later dense-layer features. Given the importance of bias as well as weight, we explain the poorer heat-map performance of DenseNet with identity rule where bias is ignored.

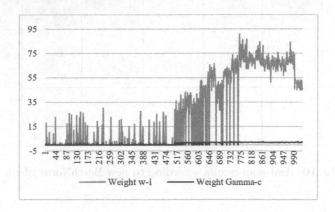

Fig. 9. Specific DenseNet weight data

3.4   BatchNorm Decomposition with New $|z|$-Rule

Based on observations and analysis so far, a new batch normalization decomposition rule can be defined to cater to different network characteristics, and to give consistent performance across network architectures hence improving the LRP analysis process. This new batch normalization decomposition rule possess three properties: (a) include the bias term, (b) support negative input term as contribution, and (c) avoid cancellation by the bias. Hence, a first implementation of the rule, called the $|z|$-rule, is defined as:

$$R_i^l = \frac{|x_i w_i|}{|x_i w_i| + |b_i| + \varepsilon} \cdot R_i^{l+1} \tag{9}$$

According to (9), the negative input and bias cancellation effects are avoided by considering only the absolute magnitude of weighted input term and bias term. The $|z|$-rule can also be rewritten with the help of sign function as:

$$R_i^l = \frac{x_i w_i}{x_i w_i + b_i \cdot sign(x_i w_i) \cdot sign(b_i)} \cdot R_i^{l+1} \tag{10}$$

In Fig. 10, the LRP heat-map results with new $|z|$-rule for BatchNorm layers are shown consistent across all four tested networks. In terms of perturbation performance [11], the $|z|$-rule is very much aligned with $\alpha 1 \beta 0$ rule for DenseNet-121 (Fig. 11), and in between $\alpha 1 \beta 0$ and Identity rule for ResNet-50 (Fig. 12).

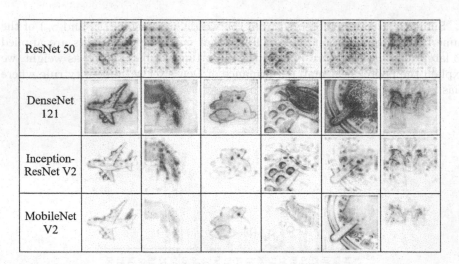

Fig. 10. Heat-map results according to new BatchNorm $|z|$-rule

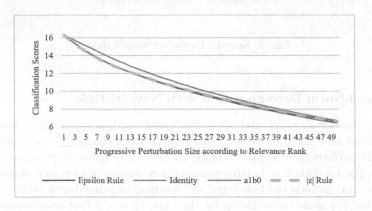

Fig. 11. DenseNet-121 perturbation test

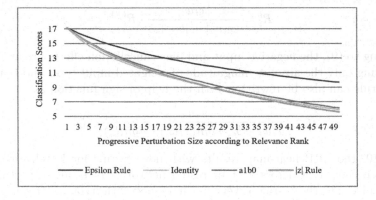

Fig. 12. ResNet-50 perturbation test

3.5   Peculiarity About ResNet Heat-Maps

A key observation from the LRP heat-maps of ResNet in Fig. 3 is that all heat-maps have a special dot-pattern artifact regardless of LRP rule used. Also noted is that the dots have fixed positions and fixed distance in between; more importantly, the dots don't seem to suggest efficient use of pixels from all image location. We investigate this according to ResNet Architecture illustrated in Fig. 13 where there are three 2:1 down-sampling layers involving the residual signal (dotted arrows) using convolution of 1×1 kernel with 2×2 stride. This decimation process explains the heat-map dot-pattern as relevance cannot be propagated to decimated locations. The three red numbers in Fig. 13 represent the total amount of relevance back propagated through the decimated residual path of the network according to one example image. The red numbers are significantly lower compared to their neighbor hence indicating a loss of relevance in such residual decimation path.

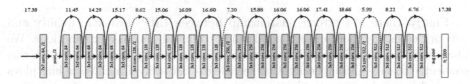

Fig. 13. ResNet architecture and Relevance propagation [4] (Color figure online)

To validate, an experiment was done comparing two models: (1) original reference ResNet-50 model and (2) modified ResNet-50 with the three convolution layers upgraded to 3×3 kernel with 2×2 stride. The two models were trained using ImageNet dataset for a limited number of epochs for illustration purpose, and results are shown in Fig. 14. Observations are: (1) with modified down-sampler the dot-pattern is removed from LRP heat-maps of given example, (2) the top-5 prediction scores are spread wider, and (3) the top-1 validation accuracy (Ex Acc@1) after each epoch is around 3% higher than the reference model (Ref Acc@1) hence showing a potential of better model accuracy or faster convergence rate. We can derive accordingly the minimum distance between the dots as $2^3 = 8$ pixels, which corresponds to the heat-map observation.

4   Discussion

We presented LRP with focus on BatchNorm layers and analyzed several state-of-art deep neural network classifiers using multiple decomposition rules. When deriving our BatchNorm decomposition rules, we took the approach of modeling BatchNorm as a simplified affine transformation. According to LRP heap-map results, we observed different performance issues with the multiple decomposition rules across the tested classifiers. Based on further analysis, we uncovered several network classifier peculiarities.

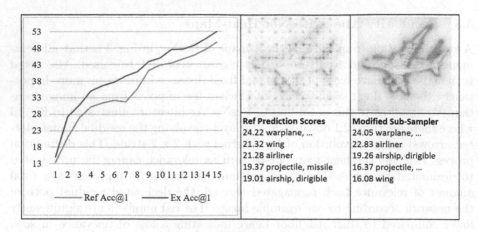

Fig. 14. Modified ResNet down-sampler

Firstly, there is a strong relationship between LRP heat-map quality and network efficiency; hence it is a good guide to initial network explanations. We illustrated this especially with the ResNet LRP heat-map peculiarity which was connected to sub-optimal decimation kernel size. Secondly, BatchNorm bias plays an much more important role than expected. In our case, there are two key factors (6) that determine this BatchNorm bias: the running batch mean μ_c and the network bias β_c. We highlighted with DenseNet the additional function of normalizing at different levels the different features from previous Dense-Blocks. On the other hand, this bias became exceptionally huge at one BatchNorm layer within MobileNet. This unexpected peak may be explained by the fact that μ_c, although a key factor to BatchNorm bias, is not regularized directly. Lastly, the entire set of features to a single network (BatchNorm) layer can be negative. This again can be explained by the fact that there are two contributors to bias in case of BatchNorm. Nevertheless, this changes some assumptions on LRP decomposition; specifically, the positive and negative contributions related to the $\alpha\beta$-rule are challenged. As a side note, this peculiarity can impact other explanation approaches (e.g. gradient times input).

In concluding the presented observations and analysis results, a new batch normalization decomposition rule ($|z|$-rule) is proposed to provide more robust performance across various deep network architectures. It resolves inefficiencies of conventional decomposition rules on specific network architectures. While LRP has been proven effective compared to other explanation methods via perturbation tests [11], its applications such as to computer vision or efficient deep network design may be enhanced with additional rules and presets.

References

1. Bach, S., Binder, A., Montavon, G., Klauschen, F., Müller, K.R., Samek, W.: On pixel-wise explanations for non-linear classifier decisions by layer-wise relevance propagation. PLoS ONE **10**(7), 1–46 (2015). https://doi.org/10.1371/journal.pone.0130140
2. Binder, A., Montavon, G., Lapuschkin, S., Müller, K.-R., Samek, W.: Layer-wise relevance propagation for neural networks with local renormalization layers. In: Villa, A.E.P., Masulli, P., Pons Rivero, A.J. (eds.) ICANN 2016. LNCS, vol. 9887, pp. 63–71. Springer, Cham (2016). https://doi.org/10.1007/978-3-319-44781-0_8
3. Bjorck, J., Gomes, C., Selman, B., Weinberger, K.Q.: Understanding batch normalization (2018)
4. He, K., Zhang, X., Ren, S., Sun, J.: Deep residual learning for image recognition. In: CVPR, pp. 770–778 (2016)
5. Howard, A.G., et al.: Mobilenets: efficient convolutional neural networks for mobile vision applications (2017)
6. Huang, G., Liu, Z., Weinberger, K.Q., van der Maaten, L.: Densely connected convolutional networks. arXiv preprint arXiv:1608.06993 (2016)
7. Ioffe, S., Szegedy, C.: Batch normalization: accelerating deep network training by reducing internal covariate shift (2015)
8. Krizhevsky, A., Sutskever, I., Hinton, G.E.: Imagenet classification with deep convolutional neural networks. In: Advances in Neural Information Processing Systems, pp. 1106–1114 (2012)
9. Montavon, G., Bach, S., Binder, A., Samek, W., Müller, K.R.: Explaining nonlinear classification decisions with deep Taylor decomposition. Pattern Recogn. **65**, 211–222 (2017). https://doi.org/10.1016/j.patcog.2016.11.008
10. Montavon, G., Lapuschkin, S., Binder, A., Samek, W., Mller, K.R.: Explaining nonlinear classification decisions with deep Taylor decomposition. Pattern Recogn. **65**, 211–222 (2017). https://doi.org/10.1016/j.patcog.2016.11.008
11. Samek, W., Binder, A., Montavon, G., Lapuschkin, S., Müller, K.R.: Evaluating the visualization of what a deep neural network has learned. IEEE Trans. Neural Netw. Learn. Syst. **28**(11), 2660–2673 (2017)
12. Shrikumar, A., Greenside, P., Kundaje, A.: Learning important features through propagating activation differences. CoRR abs/1704.02685 (2017). http://arxiv.org/abs/1704.02685
13. Simonyan, K., Vedaldi, A., Zisserman, A.: Deep inside convolutional networks: visualising image classification models and saliency maps. arXiv preprint arXiv:1312.6034 (2013)
14. Simonyan, K., Zisserman, A.: Very deep convolutional networks for large-scale image recognition (2014)
15. Smilkov, D., Thorat, N., Kim, B., Vigas, F., Wattenberg, M.: Smoothgrad: removing noise by adding noise (2017)
16. Springenberg, J.T., Dosovitskiy, A., Brox, T., Riedmiller, M.A.: Striving for simplicity: the all convolutional net. CoRR abs/1412.6806 (2014), http://arxiv.org/abs/1412.6806
17. Szegedy, C., Ioffe, S., Vanhoucke, V., Alemi, A.: Inception-v4, inception-resnet and the impact of residual connections on learning (2016)
18. Zeiler, M.D., Fergus, R.: Visualizing and understanding convolutional networks. In: Fleet, D., Pajdla, T., Schiele, B., Tuytelaars, T. (eds.) ECCV 2014. LNCS, vol. 8689, pp. 818–833. Springer, Cham (2014). https://doi.org/10.1007/978-3-319-10590-1_53

Video Categorisation Mimicking Text Mining

Cristian Ortega-León[1], Pedro A. Marín-Reyes[1], Javier Lorenzo-Navarro[1(✉)],
Modesto Castrillón-Santana[1], and Elena Sánchez-Nielsen[2]

[1] Instituto Universitario SIANI, Universidad de las Palmas de Gran Canaria,
Las Palmas de Gran Canaria, Spain
pedro.marin102@alu.ulpgc.es, javier.lorenzo@ulpgc.es
[2] Departamento de Ingeniería Informática y de Sistemas, Universidad de la Laguna,
Santa Cruz de Tenerife, Spain

Abstract. With the rapid growth of online videos on the Web, there is
an increasing research interest in automatic categorisation of videos. It is
essential for multimedia tasks in order to facilitate indexing, search and
retrieval of available video files on the Web. In this paper, we propose a
different technique for the video categorisation problem using only visual
information. Entity labels extracted from each frame using a deep learning network, mimic words giving rise to manage the video classification
task as a text mining problem. Experimental evaluation on two widely
used datasets confirms that the proposing approach fits perfectly to video
classification problems. Our approach achieves 64.30% in terms of Mean
Average Precision (mAP) in CCV dataset, above other approaches that
make use of both visual and audio information.

Keywords: Video classification · Text classification · Text mining ·
Semantic video tagging

1 Introduction

Online video is responsible for more than 58% of Internet traffic, with an upward
trend, as suggested in a recent report [5]. The main reasons for this trend
are the growth and consolidation of social networks and streaming platforms,
such as YouTube, Vimeo, Viddler or Netflix. For these platforms, due to the
huge amounts of uploaded videos, the problem of classifying them into gen-
res/categories is an essential issue. At the high structure level, film or video sets
are categorised into different subcategories, such as fiction, horror, thriller, etc.

Currently, most of the digital content uploaded to web pages comprises meta-
information which is assigned manually by the user when s/he shares the video.
However, this approach is time-consuming for users; being also affected by the
different users' understanding on video categories. Therefore, a reviewing process
is needed to verify video labels. In this context, automatic video categorisation
is a key approach to solve these limitations. Intelligent systems can in fact help

© Springer Nature Switzerland AG 2019
I. Rojas et al. (Eds.): IWANN 2019, LNCS 11507, pp. 292–301, 2019.
https://doi.org/10.1007/978-3-030-20518-8_25

users to tag videos, suggesting the most accurate labels. Thus, resulting categories will be more coherently assigned than allowing users to choose manually the categories for the uploaded videos.

Commonly, video categorisation is developed mainly by using models based on meta-textual, visual Region of Interest (ROI) features, audio or their combination, as presented by [22–24,27,30]. Nowadays, authors keep on using the same cues, but making use of Deep Neural Networks (DNNs) as [9,31]. However, any focus based on low level features is not robust enough for online video categorisation because videos from the same category are frequently diverse in term of features. Based on this limitation, this work aims to propose a methodology that allows automatic video categorisation making use of text mining based strategy to extract features. The adopted approach is evaluated using different models over various public datasets.

The paper is organised as follows. Section 2 presents the related work. Section 3 describes the proposed approach. The experimental design is illustrated in Sect. 4. Section 5 presents the obtained results of the experiments. Finally, Sect. 6 concludes with a summary.

2 Related Work

2.1 Non Deep Learning Based Approaches

Video categorisation has received the attention of the computer vision community since years. In [3] an extended survey about the literature of video classification is presented where the authors highlighted three types of modalities: text, audio and visual features. The authors explored an extensive number of approaches using single modality and combinations of them. Another work [4], proposed a systematic study of automatic categorisation of consumer videos, dividing them into a set of classes with diverse semantic concepts, which have been carefully selected based on user studies. In this sense, they manually annotated over approximately 1,300 recordings from real users. Their goals are summarised in: (i) evaluation of the state of the art in multimedia analytics, (ii) evaluation of different approaches in consumer video classification; and (iii) to discover new research opportunities. The work summarised in [12] described a generic video classification algorithm that detects object of interest. They used online user-submitted recordings and aimed to categorise videos into six broad categories. Recently, the approach described in [32] proposed an improved K-means algorithm to categorise video fragments.

2.2 Deep Learning Based Approaches

An extensive empirical evaluation of Convolutional Neural Networks (CNNs) on large-scale video classification is described in [11], using a dataset of one million YouTube videos belonging to 487 categories. Another work [31] proposed and evaluated different Deep Neural Network (DNN) models to combine image information across a video over longer time periods. The authors proposed two methods capable of handling full length videos. One of the proposed methods explored various convolutional temporal feature pooling architectures. The second proposed method explicitly modeled the video as an ordered sequence of frames, applying Long Short Term Memory (LSTM) units which are connected to the output of the underlying CNN. Moreover, [21] proposed to leverage high-level semantic features to improve the state of the art of temporal model in video categorisation. A LSTM network was used with the aim to understand what is learned by the network. Firstly, object features were extracted from a CNN model that was trained to recognise 20K objects. These features feeded the LSTM to capture the video temporal dynamics. The work described in [28] presented a novel approach to combine multiple layers and modalities of DNNs for video classification. In [9], the authors studied the challenging problem of categorising videos according to high-level semantics such as the existence of a particular human action or a complex event. The authors proposed a novel unified approach that jointly exploits the feature and class relationships to improved categorisation performance. Particularly, these two types of relationships are estimated and applied by rigorously imposing regularisation in the learning process of a DNN.

All the above mentioned works share a similar point of view to extract features, which are later used to train a classifier. In this paper, we propose a different focus, i.e. to apply a text-mining approach to the problem of video categorisation, deriving high-quality information from semantic information extracted over visual objects of the video.

3   Methodology

The proposal of this paper is based on mimicking the text mining pipeline for the video categorisation problem. Firstly, it is necessary to establish the analogies between the elements in which a video can be broken down; and the elements that make up a set of documents. In fact, a focus to find a equivalence between a video, which is composed of frames, and a document, which is made up of paragraphs or pages, may be text classification. Thus, a frame is formed by a set of visual entities, that are equivalent to the set of words that appear in the paragraph or page. The above analogies are clarified in Fig. 1. As a result, a relationship between the concepts used in text classification and those used in video classification can be established.

The first process in text mining is called tokenization. Similarly, in video categorisation firstly it is needed to extract the visual entities from the frames. Visual entities are extracted using an object detector based on a deep neural network. In this way, labels such as *person, dog, car* are obtained for the video frames. The semantic information of each visual entity is then considered as a word in our approach This information is extracted from the corresponding class of the detected entity. These are the basics to manage video categorisation as text classification.

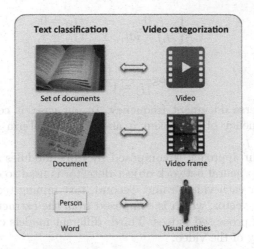

Fig. 1. Equivalence among the elements used in text classification and those used in video categorisation.

Fig. 2. Approach overview. Semantic visual entities are extracted from the video frames to apply text mining. Then, a classification process is carry out to categorise the video.

At this point, it is necessary to transform a video, $V = \{f_1, f_2, ..., f_j, ..., f_{|V|}\}$, into a feature vector of fixed dimension for each video. This problem is similar to transform a document into a feature vector. For this purpose, we employ the concept Bag of Words (BoW), which allows to encode documents, in our case a video, in features vectors regardless of the number of frames of the video. Applying a BoW technique consists in converting the token sets of each video frame into sparse vectors of the size of the vocabulary ($Voc = \{t_1, t_2, ..., t_i, ..., t_{|Voc|}\}$) built from the tokens that compose each frame. Formally, $\mathbb{R}^l \to \mathbb{R}^{|Voc|}$ where l is the number of tokens in the video. Thus, we obtain as many vectors as videos in which the values that are not 0 depend on the chosen method to weight the presence of vocabulary tokens in each frame (w_{ij}). The simplest technique is the Boolean model,

$$w_{ij} = \begin{cases} 0 \to & t_i \quad \neg\text{appears}(f_j) \\ 1 \to & t_i \quad \text{appears}(f_j) \end{cases} \tag{1}$$

where 0 is assigned if a token t_i does not appears in a frame f_j, otherwise, 1. However, the boolean model does not take into account the relevance of each token. In order to consider the relevance of each token, the model, proposed by [20], Term Frequency - Inverse Document Frequency (TF-IDF) is used; and it is defined as follows:

$$w_{ij} = \begin{cases} 0 & \rightarrow & tf_{ij} = 0 \\ tf_{ij} \times idf_i & \rightarrow & tf_{ij} \geq 1 \end{cases} \tag{2}$$

where

$$idf_i = \log \frac{|V|}{|f_j \in V : t_i \in f_j|} \tag{3}$$

being idf_i the inverse document frequency. In our case, it corresponds to the inverse video frequency of the token i; and t_f is the Term Frequency of the token i in a frame j.

In summary, our approach is composed by three modules as summarised in Fig. 2. First, a deep neural network object detector is used to extract the visual entities (words) for each video frame. Second, text mining techniques are used to obtain a feature vector, which is composed from the extracted visual entities of the whole set of processed videos. Third, different models can be adopted to obtain the category of the video.

4 Experiments

The experimental evaluation is performed on two widely used benchmarks. First, a subset of videos from YouTube-8M, see [1], is selected. This dataset consists of 200 videos corresponding to ten different categories. These are related to sports (Basketball, Bowling, Cycling, Football, Jumping, Parachuting, Rallying, Surfing, Tennis and WinterSport). Second, Columbia Consumer Video (CCV) dataset is selected, see [10]. This collection consists of 9,317 videos. However, just 7,578 of them could be downloaded due to broken links. Either the user who uploaded the video has deleted or blocked it, or YouTube has deleted the video due to copyright infringement. Once both datasets were obtained, we followed a protocol regarding training and testing. A repeated holdout validation was adopted, including 10 repetitions with re-sampling of the samples, making use of two third of the samples to train and the rest to test.

Table 1. Results in terms of mAP for the subsets built from YouTube-8M and CCV with different classifiers.

	NB	SVM	KNN	C4.5	RF
YouTube-8M	94.00	**98.50**	90.60	82.40	94.00
CCV	61.20	**64.30**	53.10	46.30	58.60

As general purpose object detector to get the visual entities, YOLO9000 [19] is used. Unlike other approaches that are based on a sliding window approach, this network divides the image into regions that corresponds to bounding boxes and for each one the network assigns a probability, obtaining a confidence score (α) for each object class and bounding box. The version used in this work is the YOLOv3 trained with 9,000 objects. Thus, for each video, a vector $\mathbf{X}_v$ of 9,000

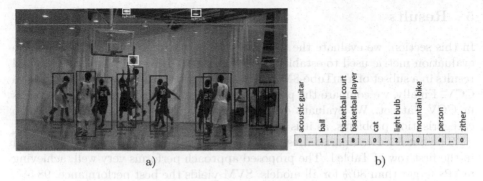

Fig. 3. (a) Example of yolo9000 detections. (b) Resulting translation from a frame into a vector where each column represents the number of occurrences of the entities.

elements is obtained as follows. Each frame of the video is processed with YOLO, making a vector $\mathbf{X}_f$ where each position represents how many entities of each type appear in the frame (Fig. 3), considering those entities with an $\alpha \geq 0.1$. Then, $\mathbf{X}_v$ is computed as the sum of all the $\mathbf{X}_f$ of the video. So, $\mathbf{X}_v$ gives how many object of each class are detected in the video. After that, a matrix M where each row corresponds to the $\mathbf{X}_v$ of each video is obtained. To note that M is the TF matrix of the video but as it was explained before in this work TF-IDF is going to be used. So, a matrix M_{tf-idf} is computed for M using (3) with $|V| = 9,000$. From now on, the problem of video classification can be considered as a traditional classification problem because each video is represented with a 9,000 element feature vector independently of the number of frames the video comprises, being the M_{tf-idf} the dataset for training and testing the different classifiers under consideration using a holdout approach.

To classify the samples, classifiers with different kinds of heuristics were used: Naïve Bayes (NB) proposed by [2]; Decision Tree (C4.5) with pruning is used to avoid overfitting, it is introduced by [18]; the ensemble method proposed by [14], denominated as Random Forest (RF) is used with 10 trees; Support Vector Machine (SVM) presented by [6] with polynomial kernel ($C = 1$); and K-Nearest Neighbors (KNN), proposed by [7], is applied with K set to 10.

A first experiment is carried out to assess the validity of the proposal in the subset of YouTube-8M with only 10 video categories as it was explained before. After this first experiment, the CCV dataset is used to evaluate the performance of the proposal that is a real challenge dataset with 20 video categories. The approaches taken for comparison purposes are: [8–10, 13, 15–17, 25, 26, 29]. These are the state of the art in the last years in CCV dataset.

The software/hardware requirements to carry out the experiments comprise an i7 core with a Nvidia GeForce GTX 960 graphic card, where YOLO9000 was executed with Tensorflow and Keras library. The classification methods used Weka library running over Java programming language.

5 Results

In this section, we evaluate the validity of our approach. First, we describe the evaluation metric used to establish a comparison. Then, we detail the obtained results in a subset of YouTube-8M. Furthermore, we present the results related to CCV. Finally, we compare the performance of our approach and other methods in CCV dataset. We evaluate our methodology using a well known in video categorisation problems, as it is Mean Average Precision (mAP).

The obtained comparison results with for YouTube-8M subset are presented in the first row of Table 1. The proposed approach performs very well, achieving mAPs larger than 80% for all models. SVM yields the best performance, 98.5%, which is 16 % points better than the worst classifier (C4.5) and four over the second classifiers (NB and RF). With this toy example, we verify the validity of our methodology for video categorisation.

In relation to CCV, the achieved results are summarizes in the second row of Table 1. Again the best mAP is obtained for SVM, but just 64.3%, three percentage points larger than NB, the second best classifier, and above 18 points larger than C4.5.

Table 2. Results obtaining by different methods in CCV.

Method	mAP
∝ SVM [13]	43.60
SIFT + STIP + MFCC [10]	59.50
Feature Weighting via Optimal Thresholding (FWOT) [26]	60.30
Reduced Analytic Dependency Model (RADM) [16]	63.04
Robust Late Fusion (RLF) [29]	64.00
Regularized Multi-modality Auto-Encoders (RMAE) [8]	64.00
Our (SVM)	**64.30**
Sample Specific Late Fusion (SSLF) [15]	68.20
Student's-t Mixture Model + Temporal Pyramids (StMM+TP) [17]	71.70
Regularized Deep Neural Network (rDNN) [9]	73.50
Multi-Stream Multi-class Fusion (MSMF) [25]	84.90
Labelling by humans	77.40

Finally, Table 2 presents the results obtained from different authors over CCV, including the results reported by our approach using a SVM classifier. Rank 5 is reached with respect to the state of the art methods in CCV. Most of the approaches that have a lower score make use of Scale-Invariant Feature Transform (SIFT), Spatial-Temporal Interest Points (STIP) as features. These features are not significant to extract independent features from different categories. Our result is achieved using just BoW features, contrary to other approaches which combines multiple features and modalities. Furthermore, it is interesting to compare with the reported human performance, 77.40%. The latter suggests the complexity of categorising the CCV dataset. It is awe-inspiring that some methods over-match the understanding of the humans about video categorisation. For instance, in the case of [25] is 7.5% over labelling by humans. Labelling by humans is presented by [10].

6 Conclusions

This paper has focused on the use of a novel technique for the problem of video categorisation. The adopted approach, based on text mining, extract visual entities from videos, which are represented textually. This gives rise to manage such textual information following the standard text mining protocol, applying tokenization and BoW.

Results are evaluated over two datasets, a subset of YouTube-8M to evaluate at first step whether our hypothesis is feasible. Reaching a 98.5% in terms of mAP with SVM classifier. Moreover, CCV dataset is used to test in a challenging categorisation set where humans are not capable to classify correctly the different categories, reaching a 77.40% of mAP. Our approach achieves 64.30% using SVM as classifier.

Those results suggest the possibility to make use of *forgotten* features in video categorisation, which may in a close future be combined with state of the art approaches to boost the overall performance.

Acknowledgements. This research work has been partially supported by the Spanish Ministry of Economy and Competitiveness (TIN2015-64395-R MINECO/FEDER), by the Office of Economy, Industry, Commerce and Knowledge of the Canary Islands Government (CEI2018-4), and the Computer Science Department at the Universidad de Las Palmas de Gran Canaria.

References

1. Abu-El-Haija, S., et al.: Youtube-8M: a large-scale video classification benchmark. arXiv Preprint arXiv:1609.08675 (2016)
2. Bayes, T., Price, R., Canton, J.: An Essay Towards Solving a Problem in the Doctrine of Chances (1763)
3. Brezeale, D., Cook, D.J.: Automatic video classification: a survey of the literature. IEEE Trans. Syst. Man Cybern. Part C Appl. Rev. **38**(3), 416–430 (2008)
4. Chang, S.F., et al.: Large-scale multimodal semantic concept detection for consumer video. In: Proceedings of the International Workshop on Multimedia Information Retrieval, pp. 255–264. ACM (2007)
5. Convivia (2018). https://www.conviva.com/. Accessed 13 Dec 2018
6. Cortes, C., Vapnik, V.: Support-vector networks. Mach. Learn. **20**(3), 273–297 (1995)
7. Fix, E., Hodges Jr., J.L.: Discriminatory analysis-nonparametric discrimination: consistency properties. California Univ Berkeley, Technical report (1951)
8. Jhuo, I.H., Lee, D.: Video event detection via multi-modality deep learning. In: International Conference on Pattern Recognition, pp. 666–671. IEEE (2014)

9. Jiang, Y.G., Wu, Z., Wang, J., Xue, X., Chang, S.F.: Exploiting feature and class relationships in video categorization with regularized deep neural networks. IEEE Trans. Pattern Anal. Mach. Intell. **40**(2), 352–364 (2018)
10. Jiang, Y.G., Ye, G., Chang, S.F., Ellis, D., Loui, A.C.: Consumer video understanding: A benchmark database and an evaluation of human and machine performance. In: Proceedings of the ACM International Conference on Multimedia Retrieval, p. 29. ACM (2011)
11. Karpathy, A., Toderici, G., Shetty, S., Leung, T., Sukthankar, R., Fei-Fei, L.: Large-scale video classification with convolutional neural networks. In: Proceedings of the IEEE Conference on Computer Vision and Pattern Recognition, pp. 1725–1732 (2014)
12. Kowdle, A., Chang, K.W., Chen, T.: Video categorization using object of interest detection. In: IEEE International Conference on Image Processing, pp. 4569–4572. IEEE (2010)
13. Lai, K.T., Felix, X.Y., Chen, M.S., Chang, S.F.: Video event detection by inferring temporal instance labels. In: IEEE Conference on Computer Vision and Pattern Recognition, pp. 2251–2258. IEEE (2014)
14. Liaw, A., Wiener, M., et al.: Classification and regression by randomForest. R News **2**(3), 18–22 (2002)
15. Liu, D., Lai, K.T., Ye, G., Chen, M.S., Chang, S.F.: Sample-specific late fusion for visual category recognition. In: IEEE Conference on Computer Vision and Pattern Recognition, pp. 803–810. IEEE (2013)
16. Ma, A.J., Yuen, P.C.: Reduced analytic dependency modeling: robust fusion for visual recognition. Int. J. Comput. Vis. **109**(3), 233–251 (2014)
17. Nagel, M., Mensink, T., Snoek, C.G., et al.: Event fisher vectors: robust encoding visual diversity of visual streams. In: British Machine Vision Conference, vol. 2, p. 6 (2015)
18. Quinlan, J.R.: Induction of decision trees. Mach. Learn. **1**(1), 81–106 (1986)
19. Redmon, J., Farhadi, A.: Yolo9000: Better, Faster, Stronger (2016)
20. Salton, G., Buckley, C.: Term-weighting approaches in automatic text retrieval. Inf. Process. Manage. **24**(5), 513–523 (1988)
21. Sun, Y., Wu, Z., Wang, X., Arai, H., Kinebuchi, T., Jiang, Y.G.: Exploiting objects with LSTMs for video categorization. In: Proceedings of the ACM on Multimedia Conference, pp. 142–146. ACM (2016)
22. Truong, B.T., Dorai, C.: Automatic genre identification for content-based video categorization. In: Proceedings of the International Conference on Pattern Recognition, vol. 4, pp. 230–233. IEEE (2000)
23. Wang, J., Duan, L., Xu, L., Lu, H., Jin, J.S.: TV ad video categorization with probabilistic latent concept learning. In: Proceedings of the International Workshop on Multimedia Information Retrieval, pp. 217–226. ACM (2007)
24. Wu, X., Zhao, W.L., Ngo, C.W.: Towards Google challenge: combining contextual and social information for web video categorization. In: Proceedings of the ACM International Conference on Multimedia, pp. 1109–1110. ACM (2009)
25. Wu, Z., Jiang, Y.G., Wang, X., Ye, H., Xue, X.: Multi-stream multi-class fusion of deep networks for video classification. In: Proceedings of the ACM on Multimedia Conference, pp. 791–800. ACM (2016)
26. Xu, Z., Yang, Y., Tsang, I., Sebe, N., Hauptmann, A.G.: Feature weighting via optimal thresholding for video analysis. In: IEEE International Conference on Computer Vision, pp. 3440–3447. IEEE (2013)

27. Yang, L., Liu, J., Yang, X., Hua, X.S.: Multi-modality web video categorization. In: Proceedings of the International Workshop on Multimedia Information Retrieval, pp. 265–274. ACM (2007)

28. Yang, X., Molchanov, P., Kautz, J.: Multilayer and multimodal fusion of deep neural networks for video classification. In: Proceedings of the ACM on Multimedia Conference, pp. 978–987. ACM (2016)

29. Ye, G., Liu, D., Jhuo, I.H., Chang, S.F.: Robust late fusion with rank minimization. In: 2012 IEEE Conference on Computer Vision and Pattern Recognition (CVPR), pp. 3021–3028. IEEE (2012)

30. Yuan, X., Lai, W., Mei, T., Hua, X.S., Wu, X.Q., Li, S.: Automatic video genre categorization using hierarchical SVM. In: IEEE International Conference on Image Processing, pp. 2905–2908. IEEE (2006)

31. Yue-Hei Ng, J., Hausknecht, M., Vijayanarasimhan, S., Vinyals, O., Monga, R., Toderici, G.: Beyond short snippets: deep networks for video classification. In: Proceedings of the IEEE Conference on Computer Vision and Pattern Recognition, pp. 4694–4702 (2015)

32. Zhou, Y., Song, W.: Video classification algorithm based on improved k-means. Tech. Bull. 55(1), 138–144 (2017). www.scopus.com

Trainable Thresholds for Neural Network Quantization

Alexander Goncharenko[1,2(✉)], Andrey Denisov[1,2], Sergey Alyamkin[1(✉)], and Evgeny Terentev[3]

[1] Expasoft LLC, Novosibirsk 630090, Russia
{a.goncharenko,s.alyamkin}@expasoft.ru
[2] Novosibirsk State University, Novosibirsk 630090, Russia
[3] Microtech, Moscow, Russia
https://expasoft.com/, https://english.nsu.ru/, https://microtech.ai

Abstract. Embedded computer vision applications for robotics, security cameras, and mobile phone apps require the usage of mobile neural network architectures like MobileNet-v2 or MNAS-Net in order to reduce RAM consumption and accelerate processing. An additional option for further resource consumption reduction is 8-bit neural network quantization. Unfortunately, the known methods for neural network quantization lead to significant accuracy reduction (more than 1.2%) for mobile architectures and require long training with quantization procedure.

To overcome this limitation, we propose a method that allows to quantize mobile neural network without significant accuracy loss. Our approach is based on trainable quantization thresholds for each neural network filter, that allows to accelerate training with quantization procedure up to 10 times in comparison with the standard techniques.

Using the proposed technique, we quantize the modern mobile architectures of neural networks with the accuracy loss not exceeding 0.1%. Ready-for-use models and code are available at:
https://github.com/agoncharenko1992/FAT-fast-adjustable-threshold.

Keywords: Distillation · Machine learning · Neural networks · Quantization

1 Introduction

Mobile neural network architectures [1–3] allow running AI solutions on mobile devices due to the small size of models, low memory consumption, and high processing speed while providing a relatively high level of accuracy in image recognition tasks. In spite of their high computational efficiency, these networks continuously undergo further optimization to meet the requirements of edge devices. One of the promising optimization directions is to use quantization to int8, which is natively supported by mobile processors, either with or without training. Both methods have certain advantages and disadvantages.

© Springer Nature Switzerland AG 2019
I. Rojas et al. (Eds.): IWANN 2019, LNCS 11507, pp. 302–312, 2019.
https://doi.org/10.1007/978-3-030-20518-8_26

Quantization of the neural network without training is a fast process as in this case a pre-trained model is used. However, the accuracy of the resultant network is particularly low compared to the one typically obtained in commonly used mobile architectures of neural networks [4]. On the other hand, quantization with training is a resource-intensive task which results in low applicability of this approach.

Current article suggests a method which allows speeding up the procedure of training with quantization and at the same time preserves a high accuracy of results for 8-bit discretization.

2   Related Work

In general case the procedure of neural network quantization implies discretization of weights and input values of each layer. Mapping from the space of float32 values to the space of signed integer values with n significant digits is defined by the following formulae:

$$S_w = \frac{2^n - 1}{T_w} \tag{1}$$

$$T_w = max|W| \tag{2}$$

$$W_{int} = \lfloor S_w \cdot W \rceil \tag{3}$$

$$W_q = clip(W_{int}, -(2^{n-1} - 1), 2^{n-1} - 1)$$
$$= min(max(W_{int}, -(2^{n-1} - 1)), 2^{n-1} - 1) \tag{4}$$

Here $\lfloor \rceil$ is rounding to the nearest integer number, W – weights of some layer of neural network, T – quantization threshold, max calculates the maximum value across all axes of the tensor. Input values can be quantized both to signed and unsigned integer numbers depending on the activation function on the previous layer.

$$S_i = \frac{2^n - 1}{T_i} \tag{5}$$

$$T_i = max|I| \tag{6}$$

$$I_{int} = \lfloor S_i \cdot I \rceil \tag{7}$$

$$I_q^{signed} = clip(I_{int}, -(2^{n-1} - 1), 2^{n-1} - 1) \tag{8}$$

$$I_q^{unsigned} = clip(I_{int}, 0, 2^n - 1) \tag{9}$$

After all inputs and weights of the neural network are quantized, the procedure of convolution is performed in a usual way. It is necessary to mention that the result of operation must be in higher bit capacity than operands. For example, in Ref. [5] authors use a scheme where weights and activations are quantized to 8-bits while accumulators are 32-bit values.

Potentially quantization threshold can be calculated on the fly, which, however, can significantly slow down the processing speed on a device with low system resources. It is one of the reasons why quantization thresholds are usually calculated beforehand in calibration procedure. A set of data is provided to the network input to find desired thresholds (in the example above - the maximum absolute value) of each layer. Calibration dataset contains the most typical data for the certain network and this data does not have to be labeled according to procedure described above.

2.1 Quantization with Knowledge Distillation

Knowledge distillation method was proposed by Hinton [6] as an approach to neural network quality improvement. Its main idea is training of neural networks with the help of pre-trained network. In Refs. [7,8] this method was successfully used in the following form: a full-precision model was used as a model-teacher, and quantized neural network - as a model-student. Such paradigm of learning gives not only a higher quality of the quantized network inference, but also allows reducing the bit capacity of quantized data while keeping an acceptable level of accuracy.

2.2 Quantization Without Fine-Tuning

Some frameworks allow using the quantization of neural networks without fine-tuning. The most known examples are TensorRT[1], Tensorflow [9] and Distiller framework from Nervana Systems[2]. However, in the last two models calculation of quantization coefficients is done on the fly, which can potentially slow down the operation speed of neural networks on mobile devices. In addition, to the best of our knowledge, TensorRT framework does not support quantization of neural networks with the architectures like MobileNet.

2.3 Quantization with Training/Fine-Tuning

One of the main focus points of research publications over the last years is the development of methods that allow to minimize the accuracy drop after neural network quantization. The first results in this field were obtained in Refs. [10–13]. The authors used the Straight Through Estimator (STE) [14] for training the weights of neural networks into 2 or 3 bit integer representation. Nevertheless, such networks had substantially lower accuracy than their full-precision analogs.

---

[1] https://developer.nvidia.com/tensorrt - NVIDIA TensorRT[TM] platform, 2018.
[2] https://github.com/NervanaSystems/distiller.

The most recent achievements in this field are presented in Refs. [15,16] where the quality of trained models is almost the same as for original architectures. Moreover, in Ref. [16] the authors emphasize the importance of the quantized networks ensembling which can potentially be used for binary quantized networks. In Ref. [5] authors report the whole framework for modification of network architecture allowing further launch of learned quantized models on mobile devices.

In Ref. [17] the authors use the procedure of threshold training which is similar to the method suggested in our work. However, the reported approach has substantial shortcomings that prevent its usage for fast conversion of pretrained neural network on mobile devices. First of all there is a requirement to train the threshold on the full ImageNet dataset [18], and second of all there are no examples demonstrating the accuracy of networks which are considered to be the standards for mobile platforms.

In current paper we propose a novel approach to set the quantization threshold with fast fine-tuning procedure on a small set of unlabeled data that allows to overcome the main drawbacks of known methods. We demonstrate performance of our approach on modern mobile neural network architectures (MobileNet-v2, MNAS).

3   Method Description

Under certain conditions (see Figs. 1 and 2) the processed model can significantly degrade during the quantization process. The presence of outliers for weights distribution shown in Fig. 1 forces to choose a high value for thresholds that leads to accuracy degradation of quantized model.

Outliers can appear due to several reasons, namely specific features of calibration dataset such as class imbalance or non-typical input data. They also can be a natural feature of the neural network, that are, for example, weight outliers formed during training or reaction of some neurons on features with the maximum value.

Overall it is impossible to avoid outliers completely because they are closely associated with the fundamental features of neural networks. However, there is a chance to find a trade-off between the value of threshold and distortion of other values during quantization, and thus get a better quality of the quantized neural network.

3.1   Quantization with Threshold Fine-Tuning

Differentiable Quantization Threshold. In Refs. [11,13,14] it is shown that the Straight Through Estimator (STE) can be used to define a derivative of a function which is non-differentiable in the usual sense (*round*, *sign*, *clip*, etc.). Therefore, the value which is an argument of this function becomes differentiable and can be trained with the method of steepest descent, also called the gradient descent method. Such variable is a quantization threshold and its training can

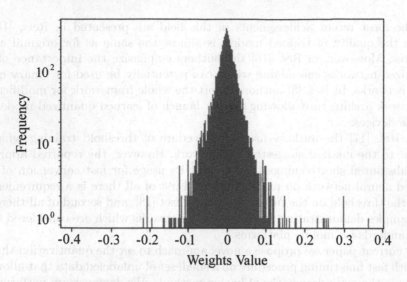

Fig. 1. Distribution of weights of ResNet-50 neural network before the quantization procedure.

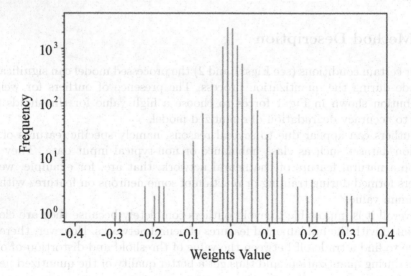

Fig. 2. Distribution of weights of ResNet-50 neural network after the quantization procedure (the number of values appeared in bins near zero increased significantly).

directly lead to the optimal quality of the quantized network. This approach can be further optimized through some modifications as described below.

Batch Normalization Folding. Batch normalization (BN) layers play an important role in training of neural networks because they speed up train procedure convergence [19]. Before making quantization of neural network weights, we suggest to perform batch normalization folding with the network weights similar to method described in Ref. [5]. As a result we obtain the new weights calculated by the following formulae:

$$W_{fold} = \frac{\gamma W}{\sqrt{\sigma^2 + \varepsilon}} \tag{10}$$

$$b_{fold} = \beta - \frac{\gamma \mu}{\sqrt{\sigma^2 + \varepsilon}} \tag{11}$$

We apply quantization to weights which were fused with the BN layers because it simplifies discretization and speeds up the neural network inference. Further in this article the folded weights will be implied (unless specified otherwise).

Threshold Scale. All network parameters except quantization thresholds are fixed. The initial value of thresholds for activations is the value calculated during calibration. For weights it is the maximum absolute value. Quantization threshold T is calculated as

$$T = clip(\alpha, min_\alpha, max_\alpha) \cdot T_{max} \tag{12}$$

where α is a trained parameter which takes values from min_α to max_α with saturation. The typical values of these parameters are found empirically, which are equal to 0.5 and 1.0 correspondingly. Introducing the scale factor simplifies the network training since the update of thresholds is done with different learning rates for different layers of neural network as they can have various orders of values. For example, values on the intermediate layers of VGG network may increase up to 7 times in comparison with the values on the first layers.

Therefore the quantization procedure can be formalized as follows:

$$T_{adj} = clip(\alpha, 0.5, 1) \cdot T_i \tag{13}$$

$$S_I = \frac{2^n - 1}{T_{adj}} \tag{14}$$

$$I_q = \lfloor I \cdot S_I \rceil \tag{15}$$

The similar procedure is performed for weights. The current quantization scheme has two non-differentiable functions, namely *round* and *clip*. Derivatives of these functions can be defined as:

$$I_q = \lfloor I \rceil \tag{16}$$

$$\frac{dI_q}{dI} = 1 \tag{17}$$

$$X_c = clip(X, a, b) \tag{18}$$

$$\frac{dX_c}{dX} = \begin{cases} 1, if X \in [a, b] \\ 0, otherwise \end{cases} \tag{19}$$

Bias quantization is performed similar to Ref. [5]:

$$b_q = clip(\lfloor S_i \cdot S_w \cdot b \rfloor, -(2^{31} - 1), 2^{31} - 1) \tag{20}$$

Training of Asymmetric Thresholds. Quantization with symmetric thresholds described in the previous sections is easy to implement on certain devices, however it uses an available spectrum of integer values inefficiently which significantly decreases the accuracy of quantized models. Authors in Ref. [5] effectively implemented quantization with asymmetric thresholds for mobile devices, so it was decided to adapt the described above training procedure for asymmetric thresholds.

For asymmetric thresholds there are left (T_l) and right (T_r) range limits. However, for quantization procedure it is more convenient to use other two values: left limit and width, and train these parameters. If the left limit is equal to 0, then scaling of this value has no effect. That is why a shift for the left limit is introduced. It is calculated as:

$$R = T_r - T_l \tag{21}$$

$$T_{adj} = T_l + clip(\alpha_T, min_{\alpha_T}, max_{\alpha_T}) \cdot R \tag{22}$$

The coefficients min_{α_T}, max_{α_T} are set empirically. They are equal to -0.2 and 0.4 in the case of signed variables, and to 0 and 0.4 in the case of unsigned. Range width is selected in a similar way. The values of min_{α_R}, max_{α_R} are also empiric and equal to 0.5 and 1.

$$R_{adj} = clip(\alpha_R, min_{\alpha_R}, max_{\alpha_R}) \cdot R \tag{23}$$

Vector Quantization. Sometimes due to high range of weight values it is possible to perform the discretization procedure more softly, using different thresholds for different filters of the convolutional layer. Therefore, instead of a single quantization factor for the whole convolutional layer (scalar quantization) there is a group of factors (vector quantization). This procedure does not complicate the realization on devices, however it allows increasing the accuracy of the quantized model significantly. Considerable improvement of accuracy is observed for models with the architecture using the Depth-wise separable convolutions. The most known networks of this type are MobileNet-v1 [1] and MobileNet-v2 [2].

3.2   Training on the Unlabeled Data

Most articles related to neural network quantization use the labeled dataset for training discretization thresholds or directly the network weights. In the proposed approach it is recommended to discard initial labels of train data which significantly speeds up transition from a trained non-quantized network to a quantized one as it reduces the requirements to the train dataset. We also suggest to optimize root-mean-square error (RMSE) between outputs of quantized and original networks before applying the softmax function, while leaving the parameters of the original network unchanged.

Suggested above technique can be considered as a special type of quantization with distillation [7] where all components related to the labeled data are absent.

The total loss function L is calculated by the following formula:

$$L(x; W_T, W_A) = \alpha H(y, z^T) + \beta H(y, z^A) + \gamma H(z^T, z^A) \tag{24}$$

In our case α and β are equal to 0, and

$$H(z^T, z^A) = \sqrt{\sum_{i=1}^{N} \frac{(z_i^T - z_i^A)^2}{N}} \tag{25}$$

where:

- z^T is the output of non-quantized neural network,
- z^A is the output of quantized neural network,
- N is batch size,
- y is the label of x example.

4   Experiments and Results

4.1   Experiments Description

Researched Architectures. The procedure of quantization for architectures with high redundancy is practically irrelevant because such neural networks are hardly applicable for mobile devices. Current work is focused on experiments on the architectures which are actually considered to be a standard for mobile devices (MobileNet-v2 [2]), as well as on more recent ones (MNasNet [3]). All architectures are tested using 224×224 spatial resolution.

Training Procedure. As it is mentioned above in the Sect. 3.2 ("Training on the unlabeled data"), we use RMSE between the original and quantized networks as a loss function. Adam optimizer [20] is used for training, and cosine annealing with the reset of optimizer parameters - for learning rate. Training is carried out on approximately 10% part of ImageNet dataset [18]. Testing is done on the validation set. 100 images from the training set are used as calibration data. Training takes 6–8 epochs depending on the network.

4.2   Results

The quality of network quantization is represented in the Tables 1 and 2.

Table 1. Quantization in the 8-bit scalar mode.

Architecture	Symmetric thresholds, %	Asymmetric thresholds, %	Original accuracy, %
MobileNet v2	8.1	19.86	71.55
MNas-1.0	72.42	73.46	74.34
MNas-1.3	74.92	75.30	75.79

Experimental results show that the scalar quantization of MobileNet-v2 has very poor accuracy. A possible reason of such quality degradation is the usage of ReLU6 activation function in the full-precision network. Negative influence of this function on the process of network quantization is mentioned in Ref. [21]. In case of using vector procedure of thresholds calculation, the accuracy of quantized MobileNet-v2 network and other researched neural networks is almost the same as the original one.

For implementation the Tensorflow framework [9] is chosen because it is rather flexible and convenient for further porting to mobile devices. Pre-trained networks are taken from Tensorflow[3] repository. To verify the results, the program code and quantized scalar models in the .lite format, ready to run on mobile phones are presented in the repository specified in the abstract.

Table 2. Quantization in the 8-bit vector mode.

Architecture	Symmetric thresholds, %	Asymmetric thresholds, %	Original accuracy, %
MobileNet v2	71.11	71.39	71.55
MNas-1.0	73.96	74.25	74.34
MNas-1.3	75.56	75.72	75.79

5   Conclusion

This paper demonstrates the methodology of neural network quantization with fine-tuning. Quantized networks obtained with the help of our method demonstrate a high accuracy that is proved experimentally. Our work shows that setting a quantization threshold as multiplication of the maximum threshold value and trained scaling factor, and also training on a small set of unlabeled data allow using the described method of quantization for fast conversion of pre-trained models to mobile devices.

---

[3] https://github.com/tensorflow/tensorflow/blob/master/tensor-flow/lite/g3doc/models.md - Image classification (Quantized Models).

References

1. Howard, A.G., et al.: Mobilenets: Efficient convolutional neural networks for mobile vision applications. arXiv preprint arXiv:1704.04861 (2017)
2. Sandler, M., Howard, A., Zhu, M., Zhmoginov, A., Chen, L.: Inverted residuals and linear bottlenecks: mobile networks for classification, detection and segmentation. In: IEEE Conference on Computer Vision and Pattern Recognition CVPR (2018)
3. Tan, M., Chen, B., Pang, R., Vasudevan, V., Le, Q.V.: MnasNet: platform-aware neural architecture search for mobile. arXiv preprint arXiv:1807.11626 (2018)
4. Lee, J.H., Ha, S., Choi, S., Lee, W., Lee, S.: Quantization for rapid deployment of deep neural networks. arXiv preprint arXiv:1810.05488 (2018)
5. Jacob, B., et al.: Quantization and training of neural networks for efficient integer-arithmetic only inference. In: Conference on Computer Vision and Pattern Recognition CVPR (2018)
6. Hinton, G., Vinyals, O., Dean, J.: Distilling the knowledge in a neural network. arXiv preprint arXiv:1503.02531 (2015)
7. Mishra, A., Marr, D.: Apprentice: Using knowledge distillation techniques to improve low-precision network accuracy. arXiv preprint arXiv:1711.05852 (2017)
8. Mishra, A., Nurvitadhi, E., Cook, J.J., Marr, D.: WRPN: wide reduced-precision networks. arXiv preprint arXiv:1709.01134 (2017)
9. Abadi, M., et al.: Tensorflow: Largescale machine learning on heterogeneous distributed systems. arXiv preprint arXiv:1603.04467 (2016)
10. Courbariaux, M., Bengio, Y., David, J.: Training deep neural networks with low precision multiplications. In: International Conference on Learning Representations ICLR (2015)
11. Hubara, I., Courbariaux, M., Soudry, D., El-Yaniv, R., Bengio, Y.: Binarized neural networks. In: Advances in Neural Information Processing Systems (NIPS), pp. 4107–4115 (2016)
12. Rastegari, M., Ordonez, V., Redmon, J., Farhadi, A.: XNOR-Net: imagenet classification using binary convolutional neural networks. In: Leibe, B., Matas, J., Sebe, N., Welling, M. (eds.) ECCV 2016. LNCS, vol. 9908, pp. 525–542. Springer, Cham (2016). https://doi.org/10.1007/978-3-319-46493-0_32
13. Zhou, S., Wu, Y., Ni, Z., Zhou, X., Wen, H., Zou, Y.: DoReFa-Net: Training low bitwidth convolutional neural networks with low bitwidth gradients. arXiv preprint arXiv:1606.06160 (2016)
14. Bengio, Y., Leonard, N., Courville, A.C.: Estimating or propagating gradients through stochastic neurons for conditional computation. arXiv preprint arXiv:1308.3432 (2013)
15. McDonnell, M.D.: Training wide residual networks for deployment using a single bit for each weight. In: International Conference on Learning Representations ICLR (2018)
16. Zhu, S., Dong, X., Su, H.: Binary ensemble neural network: More bits per network or more networks per bit? arXiv preprint arXiv:1806.07550 (2018)
17. Baskin, C., et al.: Nice: Noise injection and clamping estimation for neural network quantization. arXiv preprint arXiv:1810.00162 (2018)
18. Russakovsky, O., et al.: Imagenet large scale visual recognition challenge. arXiv preprint arXiv:1409.0575 (2014)
19. Ioffe, S., Szegedy, C.: Batch normalization: accelerating deep network training by reducing internal covariate shift. In: International Conference on Machine Learning ICML (2015)

20. Kingma, D.P., Ba, J.L.: Adam: a method for stochastic optimization. In: International Conference on Learning Representations ICLR (2015)
21. Sheng, T., Feng, C., Zhuo, S., Zhang, X., Shen, L., Aleksic, M.: A quantization-friendly separable convolution for mobilenets. arXiv preprint arXiv:1803.08607 (2018)

Integration of CNN into a Robotic Architecture to Build Semantic Maps of Indoor Environments

D. Chaves[1,2(✉)], J. R. Ruiz-Sarmiento[1], N. Petkov[2], and J. Gonzalez-Jimenez[1]

[1] Machine Perception and Intelligent Robotics group (MAPIR),
Department of System Engineering and Automation,
Biomedical Research Institute of Malaga (IBIMA),
University of Malaga, Málaga, Spain
{davfercha,jotaraul,javiergonzalez}@uma.es
[2] Johann Bernoulli Institute of Mathematics and Computing Science,
University of Groningen, Groningen, The Netherlands
n.petkov@rug.nl

Abstract. In robotics, semantic mapping refers to the construction of a rich representation of the environment that includes high level information needed by the robot to accomplish its tasks. Building a semantic map requires algorithms to process sensor data at different levels: geometric, topological and object detections/categories, which must be integrated into an unified model. This paper describes a robotic architecture that successfully builds such semantic maps for indoor environments. For this purpose, within a ROS-based ecosystem, we apply a state-of-the-art Convolutional Neural Network (CNN), concretely YOLOv3, for detecting objects in images. The detection results are placed within a geometric map of the environment making use of a number of modules of the architecture: robot localization, camera extrinsic calibration, data form a depth camera, etc. We demonstrate the suitability of the proposed framework by building semantic maps of several home environments from the Robot@Home dataset, using Unity 3D as a tool to visualize the maps as well as to provide future robotic developments.

Keywords: Semantic map · CNN · Object detection · YOLO · Unity 3D · Robotic architecture · Robot@Home · ROS

1 Introduction

A cornerstone task for the operation of mobile robots in human environments (like homes or offices) is the recognition of the objects. This permits the robot to acquire knowledge about its surroundings, enabling it to interact with humans and to carry out high-level tasks [1–3]. Object recognition has been a hot topic and a challenge within the computer vision field for decades, mainly due to its complexity: the number of objects classes to categorize can be high (including

© Springer Nature Switzerland AG 2019
I. Rojas et al. (Eds.): IWANN 2019, LNCS 11507, pp. 313–324, 2019.
https://doi.org/10.1007/978-3-030-20518-8_27

high intra-class variability and similarities between different classes), difficulties in the observation of objects like lighting conditions, occlusions, different poses/points of view, etc [4–7]. In recent years this issue has experimented a great advance provided by the rise of *Neural Network* based approaches, mainly *Convolutional Neural Networks* (CNNs) [1,8] (*e.g.* Faster R-CNN [9], YOLOv3 [10], SSD [11] or R-FCN [12]). These *Deep Learning* techniques take advantage of the computational power of modern *Graphics Processing Units* (GPUs) and the availability of huge repositories of labeled data, like Pascal VOC [13], COCO [14], or ImageNet [15], to produce models showing a remarkable success.

In order to be exploited by a mobile robot, the objects recognized by CNNs, usually coming in the form of bounding boxes over an intensity image, must be adequately incorporated in the robot's world representation. This poses a number of challenges, including linking the detected objects in an image to the geometrical map, which consists of low-level entities (*e.g.* points, segments, planes, etc.) referred to a common coordinates frame. This type of representation including geometric information as well as instances of detected objects is known as *objects' map* or *semantic map* [16], and are necessary for the execution of high-level tasks like efficient object search (*e.g.* bring me the cup from the kitchen table), reasoning (*e.g.* if a bed is detected within a room, it must be a bedroom and the robot must avoid it during the night), gas source localization (*e.g.* this gas is emitted by ovens, where are there?), etc [3,17].

A simple way of building such semantic maps is by annotating the RGB images containing the detected objects with the robot poses from where they were first observed [18,19]. A more elaborated alternative, as the one we present here, is to place the detected objects in the geometric map [20,21]. This requires the relative position of the camera w.r.t. the robot, the robot pose, and some way to infer depth from the image.

In this paper we describe a specific robotic architecture aimed to build semantic maps. This architecture, based on the Robot Operating System (ROS) framework [22], orchestrates modules that handle the object detection in the images using the YOLOv3 CNN [10], robot localization, object 3D pose estimation and a map representation that combines a geometric map and the detected objects. The later module, which is in charge of the integration and maintenance of the semantic map, feeds a Unity 3D [23] based one that offers a number of possibilities, like a powerful visualization of the built semantic map to share with an user in immersive telepresence applications [24] or algorithms for task planning [25], or simulation in augmented environments [23], among others.

To assess the suitability of the proposed architecture we carried out experiments with the Robot@Home [26] dataset. This dataset contains information collected from indoor environments, concretely homes, captured by a laser scanner and four RGB-D cameras mounted on a mobile robot. It also includes geometric maps of such houses. We illustrate how our proposal is able to successfully build and show semantic maps using such data.

2   Achitecture Description

This section describes the proposed architecture for the building of semantic maps. Figure 1 illustrates its main components and their connections that, as commented, are implemented within the ROS framework [22]. ROS is an open-source collection of tools, libraries and conventions that aims to simplify the task of creating a complex and robust robotic behavior, being the default choice in the robotics community for developing software.

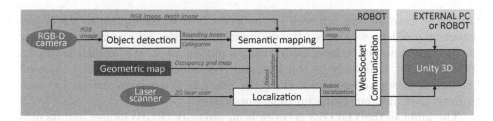

Fig. 1. Overview of the proposed architecture showing its main components and the connections between them. White boxes are modules, the green one represents previously generated data, while ovals stand for sensors capturing data. (Color figure online)

Briefly, the localization module relies on an (previously built) geometric map of the environment and the observations coming from a laser scanner to localize the robot within it (see Sect. 2.1), while the object detection one watches for objects in images coming from an RGB-D camera (Sect. 2.2). Then, the semantic mapping module is in charge of computing the 3D pose of the detected objects w.r.t. the camera frame and, using the robot localization and the pose of the camera w.r.t. the robot frame, positioning these objects in the geometric map (see Sect. 2.3). Finally, this information is propagated by the ROSBridge websocket to the Unity 3D visualizer, which can be running on the same robot or in any other external computer for the user convenience (Sect. 2.4). In the next sections we describe these components and their interactions with more detail.

2.1   Robot Localization: AMCL

For localizing the robot within the given geometric map we resort to a well-known third party development for ROS, namely the *AMCL* package[1]. This package implements the popular Adaptive Monte Carlo Localization method proposed by Fox [27], a probabilistic localization approach that employs a particle filter to track the position of a robot. The ROS implementation is fed with a previously computed geometric map, as well as with information coming from

---

[1] http://wiki.ros.org/amcl.

2D laser scanners, and yields the location of the robot within such a map. For the construction of the geometric map we rely on the also popular *Gmapping* package[2], which contains a ROS wrapper for *OpenSlam's Gmapping*. It provides a laser-based Simultaneous Localization and Mapping (SLAM) algorithm for creating 2D geometric maps in the form of occupancy grids. In this work we launch it in advance to produce the map, which is used for localizing the robot as well as a component of the semantic map (recall Fig. 1).

2.2 Object Detecion: YOLO CNN

Given the great success achieved by modern CNNs when applied to the object detection problem, we studied the different state of the art alternatives looking for the best suited to be integrated within a mobile robot. We decided to rely on *YOLO* [28,29], concretely its third version [10], which exhibits a good trade-off between accuracy and execution time. For example, it reaches a similar accuracy to RetinaNet [30], but being up to 4 times faster.

The input to the YOLOv3 network is an intensity (RGB) image, and its output is a set of bounding boxes containing the detected objects, annotated with their object categories and confidence values. These values tell us how confident the network is about that results, information that can be lately propagated to the semantic map [3]. To perform these computations in short times, considering that its architecture it of medium size (53 convolution layers distributed in successive layers of 3×3, 1×1 and some shortcut connections (residuals)), the network incorporates a number of ingenious ideas. To name a few, it has a convolutional implementation of sliding windows to efficiently detect objects at any region of the image, a novel way to carry out bounding box predictions and non-max suppression, aiming to avoid multiple detections of the same object, or the utilization of anchor boxes for the identification of overlapping objects. The interested reader can find further information in the YOLO papers [10,28,29].

Since we are interested in the detection of objects within indoor environments, the chosen YOLO CNN has been pre-trained with the COCO dataset [14], which includes everyday object categories like chair, sofa, potted plant, bed, dining-table, toilet, tv monitor, etc. Figure 2 shows some intensity images annotated with the output of this network, including bounding boxes and objects categories. We can check its good performance even with partially occluded objects (*e.g.* the chairs or the spoon), or with those observed from a challenging point of view (*e.g.* the teddy bear). Finally, to integrate the CNN into the ROS ecosystem we use the `Darcknet_ros` [31] package. This package implements a wrapper for YOLOv3 that is feed with RGB images from the camera/s mounted on the robot and, as commented, yields detected objects in the form of bounding boxes, category annotations and confidence values.

---

[2] http://wiki.ros.org/gmapping.

2.3 Semantic Mapping

The semantic map implementation consists of a geometric map and a number of detected objects positioned with respect to it (recall Fig. 1). Since we are using a geometric map provided by *Gmapping*, we focused here in the description of how the objects are placed within that map by the semantic mapping module.

Fig. 2. Detected objects by YOLOv3-tiny in the form of bounding boxes with category annotations. On the left, a microwave, a refrigerator and a spoon; in the center, a bed with a teddy bear overlap; on the right, a potter plant and some chairs.

As commented in the previous section, the CNN outcome is, among others, a set of bounding boxes of detected objects. The first step for placing them in the geometric map is the retrieval of their 3D pose w.r.t. the camera frame. For that we rely on RGB-D cameras, which provide both intensity and depth information of the scene. These cameras come with an internal extrinsic calibration that permits us, given a bounding box in the intensity image, obtain the depth measurements associated with it by propagating such bounding box to the depth one. We then build a point cloud representation from the information within that bounding box using the camera intrinsic parameters, and perform a simple but effective segmentation to extract the points belonging to the object. Concretely, we compute the center of mass of the point cloud and, using it as the seed, perform an euclidean segmentation where points closer than a certain threshold are added until there are no more points to include. This procedure removes both points belonging to other objects in the foreground/background as well as spurious points.

Let's define P_{C_j} as the set of points in such a segmented point cloud expressed in the reference frame of the C_j camera, and P_{k,C_j} the k-th point. Therefore, such a point can be placed in the global, geometric map frame through the following composition of poses and points:

$$P_{k,G} = (R_G \oplus C_{j,R}) \oplus P_{k,C_j} \tag{1}$$

being $R_G = [x, y, \theta]$ the robot localization provided by the AMCL package and $C_{j,R}$ the extrinsic calibration of the C_j camera, that is, its rigid transformation w.r.t. the robot frame. This allows us to propagate the information provided by the CNN, in the form of bounding boxes in the intensity image, to a (colored) point cloud representation placed in the geometric map. Such a point cloud, along with the object category and confidence value, are then incorporated to the semantic map in order to be further exploited during the robot operation. Depending on the application, this map could also incorporate the intensity image where the object was detected.

2.4   Semantic Map Visualization: Unity

During the robot operation, the semantic map being built is shown to the user by means of Unity 3D[3]. This is a 3D video game development platform with a number of interesting features. In a nutshell, it permits us to show the information within the semantic map in order to share the robot knowledge, providing a number of primitives for displaying information of different nature. These primitives greatly simplify the development of software by providing a way to easily analyze its outcome in a three-dimensional representation, as well as permit the user to visualize it through a clean, uncluttered interface.

Besides its strengths as a visualization tool, Unity 3D provides a range of possibilities for robotics research. In fact, there is a long-term trend to use video game development frameworks to advance in research fields tightly related to robotics, as is the case of Artificial Intelligence (AI) [23]. To name some of these possibilities, Unity 3D enables the execution of path planning packages for robot navigation [25], the utilization of reinforcement learning techniques [23], or the immersive teleoperation of the robot by the user [24]. An additional feature especially relevant for this work is that the framework allows to annotate the detected objects with meta-information, like their category, recognition confidence, utilities, etc. These annotations are codified in such a way that enable the user to efficiently search for objects with specific characteristics. For example, if the robot or the user need to warm up the food, the framework can search through the detected objects and retrieve those with that utility [3].

In our implementation, the generated semantic map can be visualized in the robot screen (if available) or in an external computer. The second option might be profitable by constrained robotic platforms, avoiding an overload of the processes being executed. In both cases, to communicate Unity 3D with ROS we resort to the *rosbridge_suite* package[4], which provides communication mechanisms between systems via *WebSocket* [32].

3   Results

To demonstrate the suitability of our proposal, we present in this section the obtained results when executing it within real indoor environments. For that, we first describe the used dataset (see Sect. 3.1), and then discuss the produced semantic maps by the proposed robotic architecture (Sect. 3.2).

---

[3] https://unity3d.com.
[4] http://wiki.ros.org/rosbridge_suite.

3.1   Dataset: Robot@Home

Robot@Home [26] is a repository of data gathered through the inspection of a number of houses by a Giraff robot [33], a mobile robotic platform equipped with a rig of 4 RGB-D cameras placed vertically and a 2D laser scanner mounted on its base (see Fig. 3-left). It was collected aimed at serving as a benchmark for semantic mapping algorithms, and contains diverse raw (2D laser scans and RGB-D observations) as well as processed data (3D reconstructions and 2D geometric maps, both annotated with the ground truth categories of the surveyed rooms and objects, as well as topological maps). It is publicly available at: http:// mapir.isa.uma.es/work/robot-at-home-dataset.

Of particular interest to this work are the ~70k provided RGB-D observations capturing a large variety of rooms and objects, summing up ~1,900 object instances belonging to categories like bottle, sink, toilet, caibinet, book, bed, pillow, cushion, microwave, bowl, etc. Figure 3-right shows an example of observations coming from the 4 cameras, including both intensity (RGB) and depth images. Aiming to show the generality and applicability of the proposed architecture, for the detection of objects we have only considered the information coming from the RGB-D camera looking ahead, since it supposes a more common sensory configuration.

Fig. 3. Left, Giraff robot used to collect the Robot@Home dataset, including a rig with 4-RGB-D cameras among its sensors. Right, example of intensity (top) and depth (bottom) images from the dataset captured simultaneously by these cameras.

3.2   Produced Semantic Maps

The proposed architecture for semantic mapping has been executed with the information collected from four houses within Robot@Home, namely *alma*, *anto*, *pare* and *rx2*. Figure 4 reports the obtained semantic maps, visualized using Unity 3D, while Fig. 5 shows some details of those maps. These maps are progressively built by processing the information coming from the robot sensors as described in Sect. 2, being the reported maps the outcome of such process

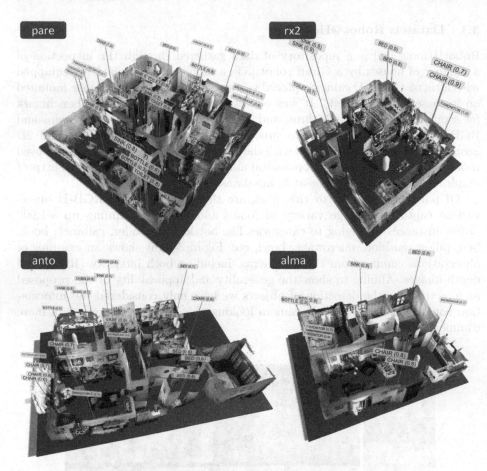

Fig. 4. Visual 3D representations of the obtained semantic maps. The green boxes represent the detected centroids of the detected objects, which are annotated with their categories and confidence values. (Color figure online)

after visiting all the rooms in the different houses. For improving visualization, we have replaced the occupancy grid maps used for localization by point cloud based ones. This permits us to recreate virtual spaces similar to the real ones, more appealing to the eye and informative for the user. It has been also introduced an avatar (a white oval sphere) synchronized with the position of the robot on the map. When the CNN detects an object, if its associated confidence value is greater than a threshold (higher than 70% in these experiments), we create a box representing such object placed at its centroid, and annotate it with its label and confidence value. Recall that all this knowledge comes from the semantic map being built, which also contains the information relative to detected objects with lower confidence values. The interested reader can check the following video, where we illustrate the building of one of these maps: https://youtu.be/uF0VXeY2_Oo.

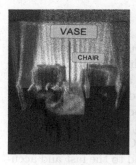

Fig. 5. Some details of the built semantic maps illustrating object detections.

From the available versions of the YOLO CNN, in these experiments we opted for YOLOv3-tiny, a reduced version designed for constrained platforms like mobile robots. It exhibits very short execution times (220 FPS running in GPU), but at the expense of a lower recognition accuracy. As commented, this network is trained to detect the 80 different object categories annotated in the COCO dataset. Some of the most frequently appearing categories within homes are table, handbag, suitcase, bottle, cup, knife, bed, chair, etc, most of them also appearing in the houses explored in Robot@Home.

The robot spent different amounts of time building the semantic maps, which is mainly due to the variable houses' size. Concretely, it needed $172s$. to map *alma*, $201s$. for *anto*, it spent $247s$. in *pare*, and $94s$. in *rx2*. Concerning the accuracy of the object detection component, it achieved a success of $\sim 78\%$. We analyzed the reported wrong detections and notice that more of them are due to false positives of the *bottle* class. Thereby, a way to improve such accuracy could be to further process the bottle detections, or to use a different, higher threshold to consider than the detection of a bottle is reliable.

It is worth mentioning some limitations that we found in our current implementation and possible ways to overcome them. On the one hand, some objects are prone to be partially detected due to their size, spatial position, or occlusions. We found it especially challenging with chairs and beds. This implies that we are only locating in the map a portion of them, which can be an issue for some tasks. A way to tackle this issue could be the utilization of active perception techniques, launched when an object has been detected at the border of the image. On the other hand, the same object can be detected multiple times if it is large or appears in different (not necessarily consecutive) images. In our implementation we use a simple procedure to fuse the point clouds from object detections sharing category and location in the geometric map, however, it is needed a more sophisticated mechanism to associate to the same object detections from different points of views. The utilization of active perception could help to disambiguate if two detections belong to the same object, which can be also useful in case of occlusions. Deep Learning techniques might be also applied here to decide if different detections refer to the same object, in a similar way to its application to person identity verification [34].

4 Conclusions

In this work we have presented a robotic architecture for building semantic maps of indoor environments. Such a semantic map contains information about the robot workspace at different levels of abstraction, that is, a geometric map, and a number of objects detected. Its implementation is based on the Robot Operating System (ROS), which orchestrates modules for robot localization, object detection, semantic map building and map visualization. Such an implementation resorts to Adaptive Monte Carlo Localization (AMCL) for retrieving the robot position in a geometric map using 2D laser scans, and to the fast and accurate YOLOv3 CNN to detect objects in intensity images from a camera. The information coming from these modules is fused by a third one that retrieves and stores the 3D poses of the detected objects w.r.t. the geometric map. Unity 3D has been presented as the chosen tool to visualize and interact with the semantic maps, and its possible applications has been also discussed, including path planning, reinforcement learning, or immersive teleoperation. The suitability of the proposal has been demonstrated by constructing semantic maps from the information about houses contained in the Robot@Home dataset. We also provided a video illustrating those maps: https://youtu.be/uF0VXeY2_Oo.

Despite the obtained promising results, the proposed architecture has significant room to improve. First, we plan to incorporate active perception techniques for facing issues like partial observability of objects, or to disambiguate detection results with low confidence values. At this point it can be also useful applying techniques to exploit the contextual relations within the environment, *e.g.* that it is more probable to find a toilet closer to a sink than to a table, as is the case of Conditional Random Fields [35, 36]. Second, since there are solutions for reaching a noticeable computational power that can be integrated into mobile robot platforms (*e.g.* NVIDIA Jetson TX1 or TX2), our aim is to change YOLOv3-tiny by YOLOv3, which should exhibit also low execution times but a higher accuracy. We will also explore the possibility of processing the intensity images in a external computer to alleviate the computational load. Finally, we also plan to take advantage of the built semantic maps to efficiently perform robotic tasks [3, 17].

Acknowledgments. This work has been supported by the research projects *WISER* (DPI2017-84827-R), funded by the Spanish Government and financed by the European Regional Development's funds (FEDER), *MoveCare* (ICT-26-2016b-GA-732158), funded by the European H2020 program, and by a postdoc contract from the I-PPIT program of the University of Málaga, and the UG PHD scholarship program from the University of Groningen.

References

1. Garcia-Garcia, A., Orts-Escolano, S., Oprea, S., Villena-Martinez, V., Garcia-Rodriguez, J.: A review on deep learning techniques applied to semantic segmentation. arXiv preprint arXiv:1704.06857 (2017)
2. Chen, L.C., Papandreou, G., Kokkinos, I., Murphy, K., Yuille, A.L.: Semantic image segmentation with deep convolutional nets and fully connected CRFs. In: International Conference on Learning Representations (ICLR) (2015)

3. Ruiz-Sarmiento, J.R., Galindo, C., González-Jiménez, J.: Building multiversal semantic maps for mobile robot operation. Knowl.-Based Syst. **119**, 257–272 (2017)
4. Pinto, N., Cox, D.D., DiCarlo, J.J.: Why is real-world visual object recognition hard? PLOS Comput. Biol. **4**(1), 1–6 (2008)
5. Ruiz-Sarmiento, J.R., Galindo, C., Gonzalez-Jimenez, J.: A survey on learning approaches for undirected graphical models. application to scene object recognition. Int. J. Approximate Reasoning **83**(C), 434–451 (2017)
6. Kasaei, S.H., Oliveira, M., Lim, G.H., Seabra Lopes, L., Tomé, A.M.: Interactive open-ended learning for 3D object recognition: an approach and experiments. J. Intell. Robot. Syst. **80**, 537–553 (2015)
7. Ruiz-Sarmiento, J.R., Galindo, C., Gonzalez-Jimenez, J.: UPGMpp: a software library for contextual object recognition. In: 3rd Workshop on Recognition and Action for Scene Understanding (REACTS) (2015)
8. Han, J., Zhang, D., Cheng, G., Liu, N., Xu, D.: Advanced deep-learning techniques for salient and category-specific object detection: a survey. IEEE Signal Process. Mag. **35**(1), 84–100 (2018)
9. Ren, S., He, K., Girshick, R., Sun, J.: Faster R-CNN: towards real-time object detection with region proposal networks. IEEE Trans. Pattern Anal. Mach. Intell. **39**(6), 1137–1149 (2017)
10. Redmon, J., Farhadi, A.: Yolov3: An incremental improvement. arXiv preprint arXiv:1804.02767 (2018)
11. Liu, W., et al.: SSD: single shot multibox detector. In: Leibe, B., Matas, J., Sebe, N., Welling, M. (eds.) ECCV 2016. LNCS, vol. 9905, pp. 21–37. Springer, Cham (2016). https://doi.org/10.1007/978-3-319-46448-0_2
12. Dai, J., Li, Y., He, K., Sun, J.: R-FCN: object detection via region-based fully convolutional networks. In: Proceedings of the 30th International Conference on Neural Information Processing Systems NIPS2016, pp. 379–387. Curran Associates Inc., USA (2016)
13. Everingham, M., Eslami, S.M., Gool, L., Williams, C.K., Winn, J., Zisserman, A.: The pascal visual object classes challenge: a retrospective. Int. J. Comput. Vis. **111**(1), 98–136 (2015)
14. Lin, T.Y., et al.: Microsoft COCO: common objects in context. In: Fleet, D., Pajdla, T., Schiele, B., Tuytelaars, T. (eds.) ECCV 2014. LNCS, vol. 8693, pp. 740–755. Springer, Cham (2014). https://doi.org/10.1007/978-3-319-10602-1_48
15. Russakovsky, O., et al.: Imagenet large scale visual recognition challenge. Int. J. Comput. Vis. **115**(3), 211–252 (2015)
16. Kostavelis, I., Gasteratos, A.: Semantic mapping for mobile robotics tasks: a survey. Robot. Auton. Syst. **66**, 86–103 (2015)
17. Monroy, J., Ruiz-Sarmiento, J.R., Moreno, F.A., Melendez-Fernandez, F., Galindo, C., Gonzalez-Jimenez, J.: A semantic-based gas source localization with a mobile robot combining vision and chemical sensing. Sensors **18**(12), 4174 (2018)
18. Zender, H., Mozos, O.M., Jensfelt, P., Kruijff, G.J., Burgard, W.: Conceptual spatial representations for indoor mobile robots. Robot. Auton. Syst. **56**(6), 493–502 (2008)
19. Pronobis, A., Jensfelt, P.: Large-scale semantic mapping and reasoning with heterogeneous modalities. In: 2012 IEEE International Conference on Robotics and Automation (ICRA), pp. 3515–3522, May 2012
20. Pangercic, D., Pitzer, B., Tenorth, M., Beetz, M.: Semantic object maps for robotic housework - representation, acquisition and use. In: 2012 IEEE/RSJ International Conference on Intelligent Robots and Systems, pp. 4644–4651, October 2012

21. Günther, M., Ruiz-Sarmiento, J.R., Galindo, C., Gonzalez-Jimenez, J., Hertzberg, J.: Context-aware 3D object anchoring for mobile robots. Robot. Auton. Syst. **110**, 12–32 (2018)
22. Quigley, M., et al.: ROS: an open-source robot operating system. In: ICRA Workshop on Open Source Software, vol. 3, p. 5. Kobe, Japan (2009)
23. Juliani, A., et al.: Unity: A general platform for intelligent agents. arXiv preprint arXiv:1809.02627 (2018)
24. Codd-Downey, R., Forooshani, P.M., Speers, A., Wang, H., Jenkin, M.: From ROS to unity: leveraging robot and virtual environment middleware for immersive teleoperation. In: 2014 IEEE International Conference on Information and Automation (ICIA), pp. 932–936, July 2014
25. Hu, Y., Meng, W.: ROSUnitySim: development and experimentation of a real-time simulator for multi-unmanned aerial vehicle local planning. Simulation **92**(10), 931–944 (2016)
26. Ruiz-Sarmiento, J.R., Galindo, C., González-Jiménez, J.: Robot@home, a robotic dataset for semantic mapping of home environments. Int. J. Robot. Res. **36**(2), 131–141 (2017)
27. Fox, D.: KLD-sampling: adaptive particle filters. In: Advances in Neural Information Processing Systems, pp. 713–720 (2002)
28. Redmon, J., Divvala, S., Girshick, R., Farhadi, A.: You only look once: unified, real-time object detection, pp. 779–788 (2016)
29. Redmon, J., Farhadi, A.: Yolo9000: Better, faster, stronger. arXiv preprint arXiv:1612.08242 (2016)
30. Lin, T.Y., Goyal, P., Girshick, R., He, K., Dollár, P.: Focal loss for dense object detection, pp. 2980–2988 (2017)
31. Bjelonic, M.: YOLO ROS: real-time object detection for ROS (2016–2018). https://github.com/leggedrobotics/darknet_ros
32. Wang, V., Salim, F., Moskovits, P.: The Definitive Guide to HTML5 WebSocket, vol. 1. Springer, Heidelberg (2013). https://doi.org/10.1007/978-1-4302-4741-8
33. González-Jiménez, J., Galindo, C., Ruiz-Sarmiento, J.: Technical improvements of the giraff telepresence robot based on users' evaluation. In: 2012 IEEE ROMAN: The 21st IEEE International Symposium on Robot and Human Interactive Communication, pp. 827–832. IEEE (2012)
34. Schroff, F., Kalenichenko, D., Philbin, J.: FaceNet: a unified embedding for face recognition and clustering. In: Proceedings of the IEEE Computer Society Conference on Computer Vision and Pattern Recognition, pp. 815–823, 07–12 June 2015
35. Ruiz-Sarmiento, J.R., Galindo, C., Monroy, J., Moreno, F.A., Gonzalez-Jimenez, J.: Ontology-based conditional random fields for object recognition. Knowl.-Based Syst. **168**, 100–108 (2019)
36. Ruiz-Sarmiento, J.R., Galindo, C., González-Jiménez, J.: Scene object recognition for mobile robots through semantic knowledge and probabilistic graphical models. Expert Syst. Appl. **42**(22), 8805–8816 (2015)

Tandem Modelling Based Emotion Recognition in Videos

Salma Kasraoui[1]($\boxtimes$), Zied Lachiri[1]($\boxtimes$), and Kurosh Madani[2]($\boxtimes$)

[1] LR-SITI Laboratory National Engineering School of Tunis,
University of Tunis el Manar, BP. 37, Le Blvedre, 1002 Tunis, Tunisia
{salma.kasraoui,zied.lachiri}@enit.utm.tn
[2] Université Paris-Est Créteil, LISSI EA-3956 Laboratory,
Sénart-FB Institute of Technology, 36-37 rue Charpak, 77127 Lieusaint, France
madani@u-pec.fr

Abstract. The work presented in this paper introduces a new model for emotion recognition from videos, Tandem Modelling (TM). The core of the proposed system consists of a hybrid neural network model that joins two feed-forward neural net models with a bottle-neck connection layer (BNL). Specifically, appearance and motion of each video sequence are encoded using a hand-crafted spatio-temporal descriptor. The obtained features are propagated through a not fully-connected neural net (NFCN) and a new tandem features are generated from the BNL. In a second level, a fully connected network (FCN) is trained with the so-extracted features to encode one of the six basic emotional states (anger, disgust, fear, happiness, sadness and surprise) with the neutral state. The classification results reached by the proposed TM show superiority over state-of-the-art approaches.

Keywords: Neural networks · Tandem Modelling ·
Emotion recognition · Hand-crafted features

1 Introduction and Background

Automatic sensing and analysis of human behavior and in particular human facial behavior has been one of the most attractive research areas in artificial intelligence over the last two decades. Facial expressions are the human preminent mean to recognize emotions, attitudes and intentions. They convey rich and multimodal informations and if we could establish a framework with the ability to detect faces in realistic scenes and analyse facial expressions, we could therefore deploy this technology in a wide range of real-life applications. The state-of-the-art in the field is relatively advanced. Impressive results have been witnessed in the last few years with the development of cost-effective approaches to tracking, detecting and recognizing facial expressions from static and dynamic images [1,2]. However, the maturity of these approaches is not fully reached. There still are great issues in handling facial occlusions, head movements and

© Springer Nature Switzerland AG 2019
I. Rojas et al. (Eds.): IWANN 2019, LNCS 11507, pp. 325–336, 2019.
https://doi.org/10.1007/978-3-030-20518-8_28

the large illumination changes. To this end, and with the emergence of deep learning, recent attempts rely on learning various facial representations from raw data by fusing multiple deep neural networks typically convolutional neural networks (CNN) and recurrent neural networks (RNN). For instance, Fan et al. [3] proposed two features' sets extracted from CNN-RNN and 3D CNN models. The improvement was registered after adding an audio module. The combination of CNN with different depths and RNN models to capture high-level textural and temporal features has been scrutinized in an important amount of works [4–6]. More recently, a number of hybrid architectures was proposed. Hereof, Yao et al. [7] put forward Holo-Net, a stack of multiple CNN blocks allowing multiscale features generation. In [8], the authors introduced the Supervised Scoring Ensemble (SSE), a learning approach that constructs a connection layer from all the probabilities of the fully-connected layers of a CNN to extract high-level representations. In a follow-up work [9], deep semantic CNN features with highresolution were proposed for a more comprehensive depiction of features and for a better performance. In [10], the authors learned facial landmarks trajectory and added them to CNN-RNN based features. Another hybrid approach with facial landmarks was proposed in [11] in which landmarks euclidean distances (LMED) was computed and combined with CNN features. Although the combination of deep features seem to offer a slight improvement in the classification accuracy with realistic data, these approaches are vulnerable to over-fitting due to the small amount of training data and the learning process is complex since different deep models are trained.

In this paper, we propose a novel hybrid framework, Tandem Modelling (TM), based on two feed-forward neural networks of the same type hierarchically combined for visual emotion classification. The proposed framework consists of a two-stage neural network with a bottleneck connection layer. The first stage performs a mapping of the input video features into seven different sub-spaces. This step is achieved by a neural network with not fully-connected units (NFCN) that trains seven emotional classes independently in order to generate more discriminative features from a narrow-size hidden units in each of the connected units of the NFCN. At this level, a bottleneck connection layer (BNL) is created by chaining all the outputs of the BNL. The second stage trains a fully connected network (FCN) with the new data. The input to the first stage are low-level features extracted from video data. The idea of Tandem Modelling stems, in part, from the tandem architecture proposed by Hermansky et al. [12] for automatic speech recognition (ASR). The approach trains a neural network based on hand-crafted audio features and uses the posterior probabilities after further transformations as features in a Gaussian mixture based Hidden Markov Model (HMM) classifier. Later, in [13], a bottle-neck layer was created within a neural net to extract bottleneck features from the outputs of that hidden layer. Taking advantages from the effectiveness of probabilistic and bottleneck features in representing the acoustic space, we studied in [14] the behaviour of these tandem features when applied in facial expressions recognition task and showed their capability of improving the classification accuracy for the examined classes. We extend this work here by adding a not fully-connected network (NFCN) instead of the standard fully connected net (FCN) in the first stage and FCN instead of support vector machines in the second stage. The rationale of this framework is threefolds:

1. We rely on engineered features as they are based on strong mathematical conceptions and remain straightforward for tremendous video classification tasks.
2. The first NFCN allows a separate classes' learning which regularizes their confusions. Besides, the fully-connected architecture in the second stage allows the exploration of interclass relationships.
3. TM uses the same classifier in each stage which alleviates the computation complexity. Furthermore, the nets are easy to train.

To the best of our knowledge, the proposed TM is a novelty in emotion recognition from videos. We show through our experiments the relevance of our TM over state-of-the-art and deep learning based techniques.

In what follows, we describe the overall architecture of the proposed system in Sect. 2. Experiments are detailed in Sect. 3. We report and discuss our system performance in Sect. 4. In Sect. 5, we conclude the paper and we give insights about our coming work.

2   Tandem Modelling

2.1   Overview

Our automatic video emotion recognition framework involves a pre-processing step in which the face region is localized in each video frame, a spatio-temporal representation of the so-detected faces and a classification step. This latter is carried-out with the proposed Tandem Modelling where an overview is shown in Fig. 1. Tandem Modelling takes as input the extracted spatio-temporal feature vectors of the faces. These inputs are propagated through a not-fully connected network (NFCN) with a bottleneck hidden layer. As can be seen, the NFCN hidden units are divided into groups whithin each of them the elements of one class are propagated and the dimensionality of their corresponding vectors is reduced through a bottleneck units group. A subsequent classification is then performed with a fully-connected network (FCN) trained with the outputs of the bottleneck units concatenated together to construct a BNL.

2.2   Network Learning

Notation. Let $\{D_n \in D\}_{n=0}^{N-1}$ be our database of N video clips linked with K classes. After pre-processing and low-level features extraction, a training sample from $\{D_n \in D\}_{n=0}^{N-1}$ can be represented as a m-tuple (X_n^k, y_n), $n \in [0, ..., N-1]$, $k \in [1, ..., K]$; where $X_n^k = (x_n^{(1k)}, x_n^{(2k)}, ..., x_n^{(mk)})$ is the $n-th$ video sample feature vector of dimension m associated with the class k and y_n is its ground-truth label.

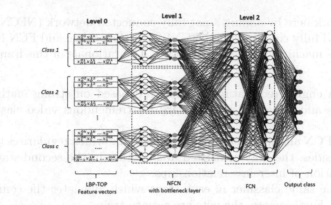

Fig. 1. Tandem Modelling framework design.

NFCN Learning. A standard feed-forward multi-layered network provides a bio-inspired architecture in which the information spread is feed-forward from the input layer to the output layer. Between the two, one or multiple hidden layers with undefined number of hidden neurons are interconnected by adaptive weights. These connectionnist weights construct a complex computational model of L layers and $s - j$ units in each layer l. In general, Given N training samples X_n, the transition from a layer $l - 1$ to a layer l ($l > 1$) can be written as:

$$a_l = g(\theta_{l-1}^T a_{l-1} + b_{l-1}) \tag{1}$$

where θ_{l-1}^T denotes the weight corresponding to the layer, running from the input i to the hidden unit j, b_{l-1} is the bias of that layer and $g(.)$ is an activation function which is, in our case, the hyperbolic tangent given by the Eq. 2:

$$g(a_l) = \frac{1 - \exp(-2a_l)}{1 + \exp(-2a_l)} \tag{2}$$

To adapt the weights within each layer, one can formulate the following optimization problem:

$$\min_\theta \sum_{n=1}^N \mathcal{L}(\hat{y_n}, y_n) + \frac{\lambda}{2} \sum_{l=1}^{l-1} ||\theta_l||^2 \tag{3}$$

where the first part of the equation $\mathcal{L}(.)$ corresponds to the empirical loss over all the samples of the training data by computing the discrepancy between the outputs of the net $\hat{y_n} = a_L$ and the ground-truth y_n. In our case, the empirical loss is the softmax activation function summerized in the Eq. 4:

$$\mathcal{L}(x^{(i)}) = \frac{\exp(x^{(i)})}{\sum_{k=1}^K \exp(x_k^{(i)})} \tag{4}$$

Where $x^{(i)}$ denotes the ith element of the input to $\mathcal{L}$ corresponding to the class k. This results in a vector of all the probabilities that a sample $x^{(i)}$ belongs to a class k. The output is the class with the highest probability. The second part is a L_2 regularization term added to prevent overfitting. In practice, the bias is included to the feature vector for simplicity with a fixed value equal to one. NFCN can be seen as a multi-label hierarchical model where the connectivity of the hidden layers is divided into K unit groups $\{G_k\}_{k=1}^{K}$ as depicted in Fig. 2. Each group G_k is considered as a sub-network where the training process follows the Eq. 3 with the empirical parameters λ_k and θ_l^k and $\mathcal{L}_k(\hat{y}_{n,k}, y_{n,k})$ where $\hat{y}_{n,k}$ and $y_{n,k}$ are the ground-truth and the output of the sub-net G_k respectively. With such hierarchy, a small number of weights is involved compared with the fully-connected architecture. This prevents from overfitting issues which might be caused by the small number of training data. The independent class training ignores the knowledge sharing between the different categories which reduces the co-linearity between the samples of different classes and consequently, the confusion between the classes would be regularized.

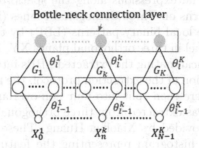

Fig. 2. Topology of the NFCN

FCN Learning. The FCN performs a simultaneous multi-class training with a learning procedure equivalent to a standard multi-layered network. This step allows the exploration of the eventual commonalities shared between the emotional classes. The input to the FCN in our case are the outputs of the bottleneck connection layer (BNL) of the NFCN updated during the NFCN training. The FCN training matrix is defined by the Eq. 5:

$$\Psi = \bigcup_{k=1}^{K} \{a_{BNL}\}_k \tag{5}$$

where $\{a_{BNL}\}_k$ are the activations of the bottleneck layer in each sub-network G_k. The weights are optimized according to the Eq. 3 for a standard network. The output layer of K units derives the final emotional class $\hat{y}_n$ of a sample X_n according to the softmax function given in the Eq. 4. The trick behind choosing a bottleneck layer to train the FCN is to reduce the dimensionality of the initial input and to employ the outputs of hidden layers as a connection between the two networks as using the posterior probabilities to train the FCN seem to be suboptimal [14].

3 Experiments

3.1 Data and Features

For our experiments, we adopt two standard databases in discrete video emotion recognition: The Surrey Audio-Visual Expressed Emotion (SAVEE) [15] and the Acted Facial Emotion in the Wild (AFEW) [16] where illustrations of the samples are given in Figs. 3 and 4. The pre-processing step starts with the face detection and alignment. For this, we employed Viola and Jones algorithm to localize faces in each video sequence from the two databases. However, this algorithm doesn't handle the face rotation. While in SAVEE database all the faces were correctly detected, the AFEW frames detection could not be performed with Viola and Jones algorithm. In this case, we refered to the IntraFace Library Tracker following the work in [17] to track the faces from each video segment. Then, all detected faces were resized into 128×128. One straightforward descriptor for facial expressions along the spatio-temporal dimensions is the Local Binary Patterns on Three Orthogonal Planes (LBP-TOP) descriptor. LBP-TOP extracts the local binary patterns (LBP) by thresholding the neighborhood of a central pixel in three orthogonal planes XY, XT and YT. The LBP-TOP code is got by concatenating the extracted LBPs into one single histogram. For a detailed explanation of the extraction protocol, the reader can refer to [18]. The resulted frames were then split into four non-overlapping blocks. For each bloc, LBP histogram was extracted from three orthogonal planes accordingly to the implementation provided by Xiaohua Huang. These histograms were then concatenated into one histogram representing the feature vector of the given video.

Fig. 3. Illustration of SAVEE frames.

Fig. 4. Illustration of AFEW frames.

3.2 Implementation and Evaluation Metric

Before the training phase, all inputs were normalized in the range of $[0, 1]$. The training was performed in two stages. In the first stage, the LBP-TOP matrices of each class were independently propagated through a 5-layers MLP, the NFCN. The first and the third hidden layers were of size 1000 units and the bottleneck layers size varied from 25 to 150 units with a step of 5 units. The network was trained with the log-loss function and the tanh activation function. The weights were optimized with the Broyden-Fletcher-Goldfarb-Shanno (BFGS) solver which is potent for small training sets. The learning rate and the weight decay were fine-tuned until no improvement is reached in the training accuracy. Then, the outputs of the bottleneck connection layer were propagated through the FCN in the second stage. The number of hidden layers varied from one to four hidden layers and the number of hidden units was in the range of $[150, 1000]$. For the training, the tanh activation function, the log-loss function and the BFGS solver were used. In order to select the best hyperparameters for each sub-network in the NFCN and the FCN, a 5-fold cross validation was performed on the training sets of SAVEE and AFEW databases. For SAVEE database, the training and the test sets were selected according to previous works by randomly choosing 70% of the data as a training set and the remaining 30% were defined as a test set. As to AFEW database, the data is provided with a training set, a validation set and a test set. As the labels of the test set are not available, we evaluated the framework performance on the validation set. The framework was implemented with python language programming. For the network training, we used sklearn for machine learning [19]. To evaluate the system performance, we used the precision metric which is calculated by the Eq. 6:

$$precision = \frac{TP}{TP + FP} \tag{6}$$

Where TP (True Positive) is the number of testing examples correctly classified as class c and FP (False Positive) is the number of testing examples not belonging to the class c and were classified as c.

4 Results and Discussions

In our experiments, we first started by studying the performance of our proposed TM on profile view images from SAVEE database. The results displayed in Table 1 indicate the relevance of TM with a mean accuracy 100% which

Table 1. Classification results on SAVEE database.

Model	Accuracy (%)
Standard MLP	95.08
Optical Flow + 3D-CNN [1]	97.92
CNN + LBP + LPQ-TOP [2]	85.83
Tandem Modelling	100

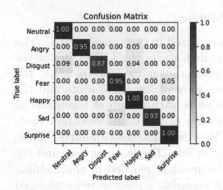

Fig. 5. Confusion matrix with a standad MLP on SAVEE database

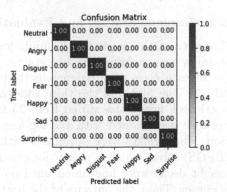

Fig. 6. Confusion matrix with Tandem Modelling on SAVEE database

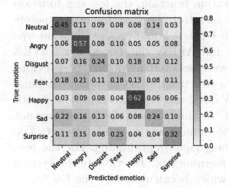

Fig. 7. Confusion matrix with a standad MLP

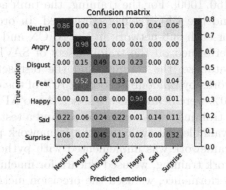

Fig. 8. Confusion matrix with Tandem Modelling

outperforms the baseline results (standard MLP) with a margin of 5%. A comparison with state-of-the-art works on this database is given in the same Table 1 where TM consistently shows its superiority over these works. From the confusion matrices given in Figs. 5 and 6, one can observe that TM fully recognize the seven facial expressions from the SAVEE database. The per-class comparison with state-of-the-art works on SAVEE database is not given in this work as the confusion matrices in the referenced works are not provided. In the second part of the experiments, we evaluated the framework on challenging facial images from the AFEW database where some samples are shown in Fig. 4. The AFEW database has unequal distribution of classes. From the outcomes of works achieved on this database and given the confusion between classes, we hypothesized in the beginning of our experiments that by equilibrating the samples distribution, the training would be improved. Thus, we followed the Synthetic Minority Over-sampling Technique (SMOTE) for class equilibrating. Table 2 reports the overall accuracies obtained with the two data distributions. One can first note that TM dramatically outperforms the baseline by more than 20%.

This implies the relevance of TM over a standard neural net. To highlight the robustness of TM, we detail in Figs. 7 and 8 the performances of a standard neural net and TM on each emotional class as well as the classes confusion. TM proves a discriminant effect in classifying neutral, anger and happy expressions with mean accuracies of 83.61%, 96.67% and 88.52% respectively. The classification of the disgust expression was augmented with an overall accuracy of 48.72%. The classification of fear, sad and surprise expressions was improved; however, these expressions are still poorly recognized. Fear and surprise expressions are mainly confused with anger and disgust expressions respectively. This is simply justified by the small amount of training data for these classes. Regarding the classes' balancing with the SMOTE class equilibrating technique, the classification of neutral, anger and happy expressions is not increased. This could be explained by the fact that these classes are majorities and the number of training samples for these expressions is not augmented. The improvement of fear, sad and surprise classifications is not noteworthy and the disgust expression accuracy is augmented by 72.13%. One can conclude that the unequal distribution of the dataset classes is not the main issue of the emotion recognition system failure in recognizing the minority classes. This observation holds true only when the difference between the amounts of data in each class is considerable. This is confirmed by the classification result of the disgust expression after equilibrating the classes. Disgust expression is the most minority class in the dataset and by applying SMOTE, the classification accuracy was significantly increased. In our experiments, we also employed support vector machines with the NFCN, however, no similar accuracy improvement was observed. This indicates that TM is better performing with the same structure (NFCN-FCN). A comparison with the top-performing frameworks on AFEW is given in Table 3. Obviously, TM outperforms these works which were based, as stated in the introduction, on deep visual features fusion. Figure 9 provides a per-class comparison with the available top-performing achievements on AFEW database [3,7,8]. TM yields better performance for almost all the classes. Even though the sad expression is worst classified by TM which is explained by the small amount of training samples of the sad expression that decreased after face detection. It is important to mention here that the works with which our framework was compared employed the speech channel to help boosting the recognition results which fills the gap caused by the missing visual information. From the experimental results, the proposed TM offered numerous advantages. First, TM helped denoising the data of the minority class and improved its accuracy. The confusions between the classes were optimized . Second, TM constructs a robust classification model with a performance that is suitable to small databases and capable of learning high-level features. This is offered by the NFCN which allowed features learning and non-linear dimensionality reduction instead of using the conventional linar methods like principal component analysis (PCA) and linear discriminant analysis (LDA). Tandem Modelling is an efficient model for facial emotion classification from profile view images. However, this finding can not be totally generalized to the realistic context as the classification accuracy of some classes is still needing more investigations. The explanation of this can be depicted from Figs. 3 and 4. Under uncontrolled environment where images are taken randomly with bad illumination conditions, blur, absence of some facial emotion components, the generalization capability of a model can not reach 100% as in the acted context where images of the same class are highly correlated.

Table 2. Classification results on AFEW database with different networks topologies.

Model	Balanced data	Unbalanced data
Standard MLP	35.2	36.92
Tandem Modelling	56.67	**59.03**

Table 3. Comparison of TM with state-of-the-art approaches on AFEW database.

State-of-the-art	Accuracy (%)
Baseline [17]	38.81
Holo-Net [7]	51.96
2 CNN-RNNs + 1 C3D [3]	51.96
SSE [8]	46.47
Textures + CNN-BRNN [10]	44.46
Deeply supervised CNN [9]	57.43
4 CNNs + LMED [11]	55.09
Proposed framework	**59.03**

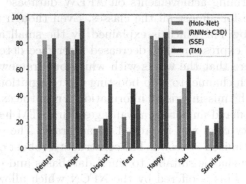

Fig. 9. Per-class comparison of TM with state-of-the-art works on AFEW database. Holo-Net, RNN-C3D and SSE are referenced in [3,7,8] respectively.

5 Conclusion

In this paper, we proposed Tandem Modelling, a novel framework for facial emotion recognition. The investigated model was built upon two levels feed-forward neural networks (NCFN and FCN). NFCN was added to train hand-crafted features and reduce their dimensionality through its bottlenck layer which was then used to train the FCN to classify the seven basic emotions from the SAVEE and the AFEW databases. TM achieved significant results in the classification accuracy compared with state-of-the-art approaches which investigated multi cues for video sequence representations and deep neural networks. Given the current results, one can think about combining hybrid features sets with TM.

References

1. Chen, J., Chen, Z., Chi, Z., Fu, H.: Facial expression recognition based on multiple feature fusion in video. In: ICCPR 2018 (2018)
2. Zhao, J., Mao, X.: Learning deep facial expression features from image and optical ow sequences using 3D CNN. Visual Comput. **34**(10), 1461–1475 (2018)
3. Fan, Y., et al.: Video-based emotion recognition using CNN-RNN and C3D hybrid networks. In: Proceedings of the 18th ACM ICMI 2016, Tokyo, Japan, pp. 445–450 (2016)
4. Kim, D.H., Lee, M.K., Choi, D.Y., Song, B.C.: Multi-modal emotion recognition using semi-supervised learning and multiple temporal models. In: Proceedings of the 19th ACM ICMI 2017, Glasgow, UK, pp. 569–576 (2017)
5. Bargal, S.A., Barsoum, E., Ferrer, C.C., Zhang, C.: Emotion recognition in the wild from videos using images. In: Proceedings of the 18th ACM ICMI 2016, Tokyo, Japan, pp. 433–436 (2016)
6. Liu, C., et al.: Multiple spatio-temporal feature learning for video-based emotion recognition in the wild. In: Proceedings of the 20th ACM ICMI 2018, Boulder, CO, USA, pp. 646–652 (2018)
7. Yao, A., Cai, D., Hu, P., Wang, S., Sha, L., Chen, Y.: HoloNet: towards robust emotion recognition in the wild. In: Proceedings of the 18th ACM ICMI 2016, Tokyo, Japan, pp. 472–478 (2016)
8. Hu, P., Cai, D., Wang, S., Yao, A., Chen, Y.: Learning supervised scoring ensemble for emotion recognition in the wild. In: Proceedings of the 19th ACM ICMI17, Glasgow, UK, pp. 553–560 (2017)
9. Fan, Y., Lam, J.C.K., Li, V.O.K.: Networks, video-based emotion recognition using deeply-supervised neural. In: Proceedings of the 20th ACM ICMI18, Boulder, CO, USA, pp. 584–588 (2018)
10. Yan, J., et al.: Multi-clue fusion for emotion recognition in the wild. In: Proceedings of the 18th ACM ICMI16, Tokyo, Japan, pp. 458–463 (2016)
11. Liu, C., Tang, T., Lv, K., Wang, M.: Multi-features based emotion recognition for video clips. In: Proceedings of the 20th ACM ICMI18, Boulder, CO, USA, pp. 630–634 (2018)
12. Hermansky, H., Ellis, D.P.W., Sharma, S.: Tandem connectionist feature extraction for conventional HMM systems. In: Proceedings of ICASSP, Istanbul, Turkey, pp. 1635–1638 (2000)

13. Grezl, F., Karafiat, M., Kontav, S., Cernocky, J.: Probabilistic and bottle-neck features for LVCSR of meetings. In: Proceedings of ICASSP, Honolulu, HI, USA, pp. 757–760 (2007)
14. Kasraoui, S., Lachiri, Z., Madani, K.: Probabilistic and bottleneck tandem features for emotion recognition from videos. In: Proceedings of the International Conferences Interfaces and Human Computer Interaction, Game and Entertainment Technologies and Computer Graphics, pp. 387–391. IADIS Press, Madrid (2018)
15. Haq, S., Jackson, P.J.B., Edge, J.: Speaker-dependent audio-visual emotion recognition. In: AVSP (2009)
16. Dhall, A., Gedeon, L.T., Frampton, T.D., et al.: Acted Facial Expressions in the Wild Database, Technical report (2011)
17. Dhall, A., et al.: Emotiw 2016: video and group-level emotion recognition challenges. In: Proceedings of the 18th ACM ICMI16, Tokyo, Japan, pp. 427–432 (2016)
18. Zhao, G., Pietikainen, M.: Dynamic texture recognition using local binary patterns with an application to facial expressions. IEEE Trans. PAMI 29(6), 915–928 (2007)
19. Pedregosa, F., et al.: Scikit-learn: machine learning in Python. J. Mach. Learn. Res. 12, 2825–2830 (2011)

System Identification, Process Control, and Manufacturing

Computational Intelligence Approach for Liquid-Gas Flow Regime Classification Based on Frequency Domain Analysis of Signals from Scintillation Detectors

Robert Hanus[1($\boxtimes$)] ⓘ, Marcin Zych[2] ⓘ, and Marek Jaszczur[2] ⓘ

[1] Rzeszów University of Technology,
12 Powstańców Warszawy Ave., 35-959 Rzeszów, Poland
rohan@prz.edu.pl
[2] AGH – University of Science and Technology,
30 Mickiewicz Ave., 30-059 Kraków, Poland
zych@geol.agh.edu.pl, jaszczur@agh.edu.pl

Abstract. Liquid-gas flows frequently occur in the mining, energy, chemical, and oil industry. One of the well-known non-contact method applied for measurement of parameters for such flows is the gamma-ray absorption technique. An analysis of the signals from scintillation detectors allows us to determine the flow parameters and to identify the flow structure. In this work, four types of liquid-gas flow regimes known as a slug, plug, bubble, and transitional plug – bubble were evaluated using selected computational intelligence methods. The experiments were carried out for two-phase water-air flow in horizontal pipe with internal diameter equal to 30 mm using a sealed Am-241 gamma-ray sources and a NaI(Tl) scintillation detectors. Based on the signal analysis in the frequency domain, eight features for the fluid flow were extracted and then were used at the input of the classifier. Three computational intelligence methods: single decision tree, multilayer perceptron, and radial basis function neural network were used for the flow structure identification. It was found that all the methods give good classification results for the types of analysed liquid-gas flow.

Keywords: Computational intelligence · Two-phase flow ·
Gamma absorption · Pattern recognition · Artificial neural networks

1 Introduction

Two-phase liquid-gas flow commonly occurs in nature and in industry, e.g. in mining, nuclear, chemical, food, thermal and petrochemical engineering. This type of flow may be analysed using several measurement methods, such as computer tomography, optical equipment, Coriolis flowmeters, Particle Image Velocimetry (PIV), Laser Doppler Anemometry (LDA), Magnetic Resonance Imaging (MRI) and nuclear techniques [1–6].

Knowledge of a two-phase flow regime is significant in order to conduct a number of industrial processes properly. Therefore flow regime identification motivates many studies. Recent works show applications of computational intelligence methods as artificial neural networks (ANN) for this purpose [7–12]. The flow structures

© Springer Nature Switzerland AG 2019
I. Rojas et al. (Eds.): IWANN 2019, LNCS 11507, pp. 339–349, 2019.
https://doi.org/10.1007/978-3-030-20518-8_29

classification with these methods is comprised of three steps: data acquisition and signal processing, signal feature extraction, and flow regime classification. Generally, computational intelligence methods exploit various features of signals in the time, frequency and state-space domain. Many works discuss the application of gamma-ray densitometry and ANN to the determination of flow regime and individual phase transportation in multiphase flow in pipelines [13–17]. The vast majority of these works concentrate on static conditions only.

This article shows how the signals from scintillation detectors, analysed in the frequency domain, can be used to recognise the structure of the water-air flow in a horizontal pipeline using selected computational intelligence methods: single decision tree, multilayer perceptron, and radial basis function neural network. The experiments described in this paper were conducted in dynamic conditions. This work continues the research presented in [18], where time domain analysis were used for signals from scintillation detectors.

2 Gamma-Absorption Method and Experimental Set-Up

The gamma absorption technique is based on exponential decreasing of a gamma beam in function of composition and geometry of the absorbent. The variation in the intensity of radiation are registered by the scintillation probes and converted into output electrical impulses [19].

The typical gamma-absorption set for two-phase water-air flow measurement is presented in Fig. 1.

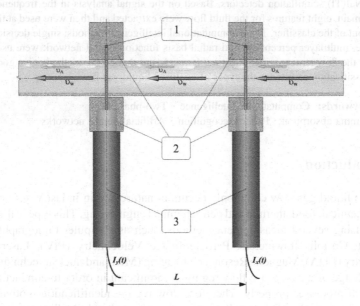

Fig. 1. The gamma-absorption measuring set: 1 – gamma radiation source, 2 – collimators, 3 – scintillation detector, v_A, v_W – velocities of air and water respectively.

Two sealed radioactive sources (1) emit gamma radiation beams with an energy of 59.5 keV shaped by collimators (2). Photons pass through the pipeline with flowing compound and detector's collimators (2) before achieving probes (3). The single gamma-absorption measuring set consist of linear Am-241 source with an activity of 100 mCi and scintillation probe with 2" NaI(Tl) crystal. For flow rate measurement, the sets are placed at a distance of $L = 97$ mm between them [19]. Count signals $I_x(t)$ and $I_y(t)$ are recorded at the outputs of scintillation probes.

Measuring equipment described above was applied in the experimental set-up, built in Sedimentological Laboratory of the AGH University of Science and Technology in Krakow, Poland. Figure 2 shows the diagram of the hydraulic testing installation.

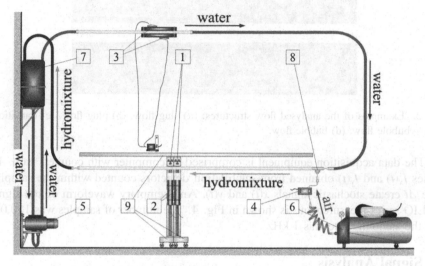

Fig. 2. Diagram of experimental set-up: 1 – gamma-ray sealed source, 2 – scintillation probe, 3 – ultrasonic flow meter, 4 – mass flow meter of air, 5 – pump, 6 – compressor, 7 – air-removing container, 8 – air nozzle, 9 – travers system of the gamma-absorption set.

The test section consists of a transparent acrylic glass pipe with internal diameter of 30 mm and a length of 4.5 m. Water is pumped by a pump (5), and the air is provided by the compressor (6) by a mass flow meter (4) and injector nozzle (8). Due to this, the controlled mixture of water and air fills the measuring pipe and flows up to the air removing tank (7). The measuring system consisting of two gamma-radioactive sources (1), two scintillation probes (2) and the system of data acquisition and analysis. The measuring set is mounted on a unique shifting system (9), which allows moving the set along the pipeline. The water flow rate is measured by use Uniflow 990 ultrasonic flowmeter (3). The pump (5) controlled by an inverter enables selection of water rate in the measuring pipe with a velocity between 0.5–2.5 m/s [18].

The transparent measuring section of the pipe allows photographic documentation of the observed structures of two-phase flow. Figure 3 shows examples of analysed water-air flow regimes.

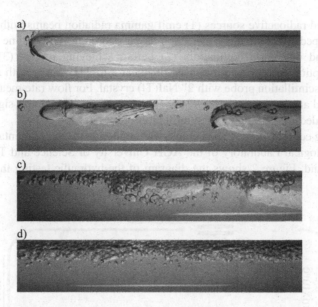

Fig. 3. Examples of the analysed flow structures: (a) slug flow, (b) plug flow, (c) transitional plug – bubble flow, (d) bubble flow.

The data acquisition equipment is comprised of computer with counters card. The pulses $I_x(t)$ and $I_y(t)$ obtained from scintillation detectors counted within the sampling time Δt create stochastic signals $x(t)$ and $y(t)$. An exemplary waveform of $x(t)$ signals for LIQ series measurements is shown in Fig. 4. The number of samples was 180,000, and the sampling rate was 1 kHz.

3 Signal Analysis

3.1 Determination of Signal Parameters in the Frequency Domain

The signals obtained from the scintillation probes are generally stochastic. For such signals in the frequency domain, the power spectral density (ASDF) and the cross-spectral density (CSDF) are usually determined. The calculated (for the positive range of frequencies) single-sided ASDF $G_{xx}(f)$, and CSDF $G_{xy}(f)$ are given by the formulas [20]:

$$ASDF = G_{xx}(f) = 2 \int_{-\infty}^{\infty} R_{xx}(\tau)e^{-j2\pi f\tau}d\tau \tag{1}$$

$$CSDF = \underline{G}_{xy}(f) = 2 \int_{-\infty}^{\infty} R_{xy}(\tau)e^{-j2\pi f\tau}d\tau \tag{2}$$

where $R_{xx}(\tau)$, and $R_{xy}(\tau)$ are autocorrelation and cross-correlation function respectively, τ – time delay, f – frequency.

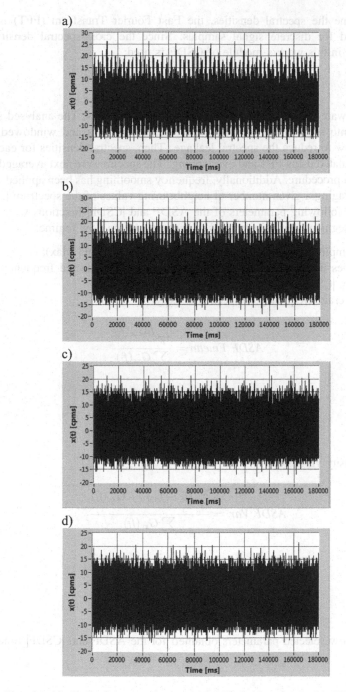

Fig. 4. Signals $x(t)$ obtained in experiments: (a) LIQ 1 (slug flow), (b) LIQ 4 (plug flow), (c) LIQ 7 (transitional plug – bubble flow), and (d) LIQ 10 (bubble flow).

To determine the spectral densities, the Fast Fourier Transform (FFT) is most commonly used for discrete signal samples. Since the cross-spectral density is a complex value, in this work a module of CSDF is used:

$$|CSDF| = \{\mathrm{Re}[\underline{G}_{xy}(f)]^2 + \mathrm{Im}[\underline{G}_{xy}(f)]^2\}^{0.5} \tag{3}$$

Matlab software was used to calculate the spectral densities. The analysed signals were divided into segments with a length of 10,000 samples and windowed using Hanning window to reduce the spectral leakage. Then, spectral densities for each segment were calculated using FFT and overlapping. The spectra were next averaged in the set using Welch procedure. Additionally, frequency smoothing has been applied, which consists of averaging a given number of neighbouring values of the spectrum [20].

Finally, the following parameters of the ASDF and |CSDF| functions were calculated for each segment to obtain features to recognition the flow regime:

- maximum amplitude of spectral density (ASDF Max, |CSDF| Max);
- sum of values under the spectral density graph for the selected frequency range (ASDF Sum, |CSDF| Sum);
- weighted average frequency of spectral density:

$$ASDF\ Fmean = \frac{\sum\limits_i f_i G_{xx}(f_i)}{\sum\limits_i G_{xx}(f_i)} \tag{4}$$

$$|CSDF|\ Fmean = \frac{\sum\limits_i f_i G_{xy}(f_i)}{\sum\limits_i G_{xy}(f_i)} \tag{5}$$

- spectral density variance:

$$ASDF\ Var = \frac{\sum\limits_i (f_i - \bar{f}_{ASDF})^2 G_{xy}(f_i)}{\sum\limits_i G_{xy}(f_i)} \tag{6}$$

$$|CSDF|\ Var = \frac{\sum\limits_i (f_i - \bar{f}_{|CSDF|})^2 G_{xy}(f_i)}{\sum\limits_i G_{xy}(f_i)} \tag{7}$$

Figure 5 show selected parameters obtained for the ASDF and |CSDF| functions.

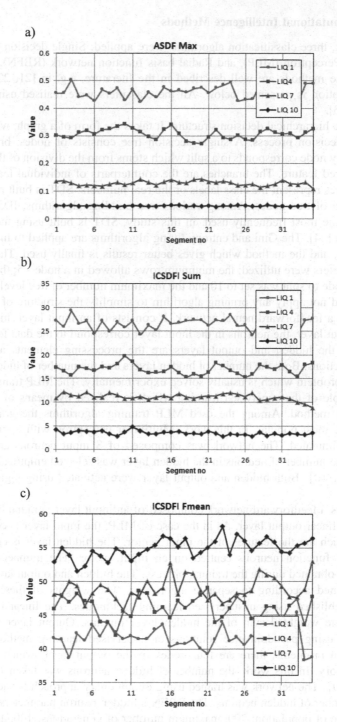

Fig. 5. The values of selected ASDF and |CSDF| parameters for LIQ experiments: (a) *ASDF Max*, (b) *|CSDF| Sum*, (c) *|CSDF| Fmean*.

3.2   Computational Intelligence Methods

In this work, three classification algorithms were applied: Single decision tree (SDT), Multilayer Perceptron (MLP), and Radial basis function network (RBFN).

Since the methods are well described in the literature, e.g. in [21, 22], only the brief description is presented below. All the classifiers were realised using DTREG software [23].

SDT is a hierarchical decision structure. It takes the form of a graph, which is next used in a decision process. A single decision tree consists of nodes, branches and leaves. Every node corresponds to a split which stems from the division of the values of the considered feature. The branches are the counterparts of individual feature values and the leaves represent the class labels or the real numbers. SDT is built by means of the selection of tree nodes. Among various tree building algorithms, ID3, C4.5 and CART are the most frequently used. In this study, SDT is built using the algorithm presented in [24]. The Gini and entropy fitting algorithms are applied to find variables for the split, and the method which gives better results is finally used. The following SDT parameters were utilized: the minimum rows allowed in a node - 5; the minimum size for a node to split was set to 10 and the maximum number of tree levels was equal to 6. We did not apply any pruning algorithm to simplify the structure of the tree.

MLP is a feedforward neural network. It consists of an input layer, hidden layers and an output layer, the neurons in the input layer correspond to the data features. The neurons in the hidden and output layers are the processing elements activated by transfer functions. Both the number of hidden layers and the number of hidden neurons is an open problem which is usually solved experimentally. The MLP training process is an example of the supervised learning which is realized by means of error back-propagation method. Among the used MLP training algorithms the gradient-based methods are most popular. In this work, Multilayer perceptron with a single hidden layer was simulated. The network was composed of 8 input neurons and 4 output neurons. The number of neurons in the hidden layer was selected empirically from the set $\{2, 3, \ldots, 10\}$. Both hidden and output layer were activated using logistic transfer function.

RBFN is a feedforward network composed of an input layer, a radial basis hidden layer and a linear output layer. As in the case of MLP, the input layer is composed of neurons which are the features of the input vector. The hidden layer is composed of radial basis function neurons centred on an input vector. The numbers of hidden neurons are obtained during the training process. The hidden and output layers weights are determined according to separate algorithms. The weights of these layers are usually established using unsupervised learning techniques. The linear output layer calculates the weighted sum of the hidden layer outputs. Output layer weights are determined using pseudoinverse solutions or supervised learning methods. For the classification problems, there are four nodes in the output layer, which represent a target category. In this work, the number of hidden neurons was taken from the set $\{1, 2, \ldots, 20\}$. The network was trained using evolutionary approach for achieving the optimal number of hidden neurons. The RBFN's hidden neuron parameters were set as follows: size of population: 200, maximum number of generations: 100, and boosting iterations: 50. The optimal RBF structure was obtained for 16 neurons in hidden layer.

4   Classification Results

For the analysis using computational intelligence methods, a total of 1120 data were used (140 data for each of the eight features) [25]. As predictors parameters (4)–(7), the maximum amplitude of spectral density (ASDF Max, |CSDF| Max), and the sum of values under the spectral density graph for the selected frequency range (ASDF Sum, |CSDF| Sum) were applied. The following classes are defined: slug flow (Class 1), plug flow (Class 2), transitional plug -bubble flow (Class 3), bubble flow (Class 4). The prediction ability is determined using a cross-validation procedure by computing the accuracy (Acc), sensitivity (Sen) and specificity (Spe) defined as follows:

$$Acc = \frac{TP + TN}{TP + FP + TN + FN} \tag{8}$$

$$Sen = \frac{TP}{TP + FN} \tag{9}$$

$$Spe = \frac{TN}{TN + FP} \tag{10}$$

where TP, TN, FP and FN denote true positive, true negative, false positive and false negative counts, respectively. The calculations were repeated ten times with random subset selection. The averaged results are presented in Table 1.

Table 1. The accuracy, sensitivity and specificity of each class determined by considered computational intelligence methods

	Class 1			Class 2			Class 3			Class 4		
	Acc	Sen	Spe	Acc	Sen	Spe	Acc	Sen	Spe	Acc	Sen	Spe
SDT	0,992	1	0,99	0,985	0,971	0,99	0,985	0,971	0,99	0,992	0,971	1
MLP	1	1	1	1	1	1	1	1	1	1	1	1
RBFN	1	1	1	1	1	1	1	1	1	1	1	1

As shown from Table 1, all classes are ideally predicted by MLP and RBFN ANN since the measured indices Acc, Sen and Spe are equal to 1 in each case. The least effective results (but still very good) were obtained by SDT.

5   Conclusion

This article presents the problem of identifying the liquid-gas flow regimes in a horizontal pipeline using gamma-ray absorption, frequency domain analysis of obtained signals and computational intelligence methods. The results of the experiments were conducted under dynamic conditions for four types of liquid-gas flow regimes as a slug, plug, bubble, and transitional plug – bubble.

In this work three well known computational intelligence methods as STD, MLP, RBFN were applied. The accuracy, sensitivity and specificity were used to evaluate the performance of the algorithms. The values of these indicators were very high for all the methods. They were equal to 1 for MLP and RBFM ANN except for the single decision tree, where the values were slightly smaller but still very good.

It was found that it is possible to identify four water-air flow regimes in the horizontal pipeline based on the frequency-domain analysis of signals from scintillation detectors. Obtained results confirm the possibility of application of gamma-ray absorption in combination with computational intelligence methods for liquid-gas flow regime classification.

The results of this work correspond well with the results obtained in [18], where time domain analysis and computational intelligence methods were used for signals from scintillation detectors.

Acknowledgements. The authors would like to thank dr Leszek Petryka for his cooperation during the measurements.

This project is financed by the Minister of Science and Higher Education of the Republic of Poland within the "Regional Initiative of Excellence" program for years 2019–2022. Project number 027/RID/2018/19, amount granted 11 999 900 PLN.

Ministry of Science and Higher Education
Republic of Poland

References

1. Falcone, G., Hewitt, G.F., Alimonti, C.: Multiphase Flow Metering: Principles and Applications. Elsevier, Amsterdam (2009)
2. Powell, R.L.: Experimental techniques for multiphase flows. Phys. Fluids **20**, 040605 (2008)
3. Pusppanathan, J., et al.: Single-plane dual-modality tomography for multiphase flow imaging by integrating electrical capacitance and ultrasonic sensors. IEEE Sens. J. **17**(19), 6368–6377 (2017)
4. Rahim, A.R., et al.: Optical tomography: velocity profile measurement using orthogonal and rectilinear arrangements. Flow Meas. Instrum. **23**, 49–55 (2012)
5. Tamburini, A., Cipollina, A., Micale, G., Brucato, A.: Particle distribution in dilute solid liquid unbaffled tanks via a novel laser sheet and image analysis based technique. Chem. Eng. Sci. **87**, 341–358 (2013)
6. Xue, T., Qu, L., Cao, Z., Zhang, T.: Three-dimensional feature parameters measurement of bubbles in gas–liquid two-phase flow based on virtual stereo vision. Flow Meas. Instrum. **27**, 29–36 (2012)
7. Xie, T., Ghiaasiaan, S.M., Karrila, S.: Artificial neural network approach for flow regime classification in gas–liquid–fiber flows based on frequency domain analysis of pressure signals. Chem. Eng. Sci. **59**, 2241–2251 (2004)
8. Sun, T., Zhang, H.: Neural networks approach for prediction of gas–liquid two-phase flow pattern based on frequency domain analysis of vortex flowmeter signals. Meas. Sci. Technol. **19**, 015401 (2008)

9. Santoso, B., Indarto, Deendarlianto, Thomas, S.W.: The identification of gas-liquid co-current two phase flow pattern in a horizontal pipe using the power spectral density and the Artificial Neural Network (ANN). Mod. Appl. Sci. **6**(9), 56–67 (2012)

10. Rosa, E.S., Salgado, R.M., Ohishi, T., Mastelari, N.: Performance comparison of artificial neural networks and expert systems applied to flow pattern identification in vertical ascendant gas–liquid flows. Int. J. Multiph. Flow **36**, 738–754 (2010)

11. Zhou, Y., Chen, F., Sun, B.: Identification method of gas-liquid two-phase flow regime based on image multi-feature fusion and support vector machine. Chin. J. Chem. Eng. **16**(6), 832–840 (2008)

12. Abbagoni, B.M., Yeung, H.: Non-invasive classification of gas–liquid two-phase horizontal flow regimes using an ultrasonic Doppler sensor and a neural network. Meas. Sci. Technol. **27**, 084002 (2016)

13. Roshani G.H., Nazemi, E., Roshani, M.M.: Flow regime independent volume fraction estimation in threephase flows using dual-energy broad beam technique and artificial neural network. Neural Comput. Appl. (2016). https://doi.org/10.1007/s00521-016-2784-8

14. Salgado, C.M., Pereira, C., Schirru, R., Brandão, L.E.B.: Flow regime identification and volume fraction prediction in multiphase flows by means of gamma-ray attenuation and artificial neural networks. Prog. Nucl. Energy **52**, 555–562 (2010)

15. Roshani, G.H., Nazemi, E., Feghhi, S.A.H., Setayeshi, S.: Flow regime identification and void fraction prediction in two-phase flows based on gamma ray attenuation. Measurement **62**, 25–32 (2015)

16. Roshani, G.H., Nazemi, E., Roshani, M.M.: A novel method for flow pattern identification in unstable operational conditions using gamma ray and radial basis function. Appl. Radiat. Isot. **123**, 60–68 (2017)

17. Roshani, G.H., Nazemi, E., Feghhi, S.A.H.: Investigation of using 60Co source and one detector for determining the flow regime and void fraction in gas-liquid two-phase flows. Flow Meas. Instrum. **50**, 73–79 (2016)

18. Hanus, R., Zych, M., Kusy, M., Jaszczur, M., Petryka, L.: Identification of liquid-gas flow regime in a pipeline using gamma-ray absorption technique and computational intelligence methods. Flow Meas. Instrum. **60**, 17–23 (2018)

19. Johansen, G.A., Jackson, P.: Radioisotope Gauges for Industrial Process Measurements. Wiley, New York (2004)

20. Bendat, J.S., Piersol, A.G.: random data. Analysis and Measurement Procedures, 4th edn. Wiley, New York (2010)

21. Larose, D.T.: Discovering Knowledge in Data: An Introduction to Data Mining. Wiley, New York (2005)

22. Wu, X., Kumar, V., Ross Quinlan, J., et al.: Top 10 algorithms in data mining. Knowl. Inf. Syst. **14**, 1–37 (2008)

23. http://www.dtreg.com. Accessed 11 Jan 2019

24. Sherrod, P.H: DTREG predictive modelling software. http://www.dtreg.com

25. Hanus, P.: Research of fluid-gas two-phase flow by means of artificial intelligence methods. MS thesis, Rzeszów University of Technology, Faculty of Electrical and Computer Engineering (2018)

Waste Classification System Using Image Processing and Convolutional Neural Networks

Janusz Bobulski⬛ and Mariusz Kubanek(✉)⬛

Institute of Computer and Information Sciences,
Czestochowa University of Technology,
Dabrowskiego Street 69, 42-201 Czestochowa, Poland
{janusz.bobulski,mariusz.kubanek}@icis.pcz.pl

Abstract. Image segmentation and classification is more and more being of interest for computer vision and machine learning researchers. Many systems on the rise need accurate and efficient segmentation and recognition mechanisms. This demand coincides with the increase of computational capabilities of modern computer architectures and more effective algorithms for image recognition. The use of convolutional neural networks for the image classification and recognition allows building systems that enable automation in many industries. This article presents a system for classifying plastic waste, using convolutional neural networks. The problem of segregation of renewable waste is a big challenge for many countries around the world. Apart from segregating waste using human hands, there are several methods for automatic segregation. The article proposes a system for classifying waste with the following classes: polyethylene terephthalate, high-density polyethylene, polypropylene and polystyrene. The obtained results show that automatic waste classification, using image processing and artificial intelligence methods, allows building effective systems that operate in the real world.

Keywords: Deep learning · Waste management · Image processing

1 Introduction

In many European countries waste segregation has already been introduced at the beginning of the recycling path, i.e. at home. Just people divide waste into groups such as plastic, metal, glass and organic/bio. The use of selective automatic techniques for these groups is easier than for municipal solid waste (MSW). Unfortunately, a large part of the waste is still collected in the form of the MSW, which is why the countries strive for the most effective reprocessing of waste materials. In order to do this, the rubbish should be effectively sort into individual factions and materials. Therefore, an important task is to isolate individual

© Springer Nature Switzerland AG 2019
I. Rojas et al. (Eds.): IWANN 2019, LNCS 11507, pp. 350–361, 2019.
https://doi.org/10.1007/978-3-030-20518-8_30

types of materials from the MSW. Therefore, techniques and procedures for segregating waste are used for the main groups of materials such as paper, glass, metal, wood, plastic and biomass by property system [1,2].

The biggest challenge, however, is the separation of various types of materials within a given group, i.e. sorting different color of glass or different types of plastic. The problem of plastic garbage is interesting and at the same time important due to the possibility of recycling only some types of plastic (e.g. PET). To simplify the recycling process, international labeling of various types of plastics was introduced Fig. 1. These are: 01 - PE (polyethylene terephthalate), 02 - HDPE (high-density polyethylene), 03 - PVC (polyvinyl chloride), 04 - LDPE (low-density polyethylene), 05 - PP (polypropylene), 06 - PS (polystyrene) and 07 (other).

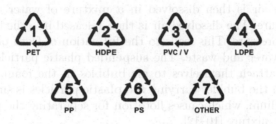

Fig. 1. Labeling of plastic waste

In domestic waste, a significant part is occupied by plastic elements, and within it may be distinguished four dominant types: PET, HDPE, PP, PS. Unfortunately, at this stage of recycling, there is no division into individual types of plastic and they often end up in a group of plastic waste or MSW. There is a problem with the separation of different types of plastics, some of which can be re-used. One of the possibilities is the use of computer vision techniques, in particular image recognition.

2   Review of Methods for Separating Plastic Waste

The whole process of automatic sorting of materials suitable for reprocessing from MSW is complicated. First, the dry waste is separated from the wet, and then the dry waste fraction is subjected to a grinding process. In order to sort materials containing iron, magnetic drum techniques are used. Subsequently, non-ferrous metals are sorted using various indirect sorting techniques, such as eddy current or X-ray radiation. However, for the separation of plastic waste, one of the following methods can be used - direct or indirect [3].

2.1   Direct Methods

The hydrocyclone uses centrifugal force to separate materials of different densities. This method can be used to separate materials such as ABS (acrylonitrile butadiene styrene), PE (polyethylene), HIPS (high impact polystyrene) and PVC (polyvinyl chloride). Different factors affect the buoyancy of a given material, e.g. different density, shape and level of separation from other materials, which is used to separate various fractions from MSW [4–6]. Jigging is using gravity to separate materials that works based on the interaction of buoyancy, resistance, gravity and acceleration. In this process, a mixture of solids and water is placed in a perforated vessel called a pulse bed. The sedimentary bed is shaken to induce vertical currents in the water column. This causes the particles to rise. The currents can be ascending or descending. Materials with higher density settle at the bottom. Segregation takes place according to the density, size and shape of the material [7–9].

Froth flotation technique uses the hydrophobicity of the plastic to separate it from the rest of the waste. The waste is ground into small particles and mixed with water. The air is then dissolved in a mixture of water and waste pulp under high pressure. The dissolved air is then released into the flotation section at atmospheric pressure. This leads to the formation of froth on the surface of the mixture of water and waste. The suspended plastic particles, due to their hydrophobicity, attach themselves to air bubbles in the foam. The combined specific weight of the bubbles carrying the plastic particles is smaller compared to the liquid medium, which causes flotation for separating the plastic from the waste and water mixture [10–12].

2.2 Indirect Sorting

X-ray transmission (XRT) is a fast indirect sorting technique, because X-ray image capture takes a few milliseconds. The imaging module uses a high intensity beam of radiation. Part of this beam penetrates into the material and is absorbed, and a part is passed to the detector below the test material. The radiation captured by the detector is analysed to obtain information about the atomic density of the material. X-ray sorting techniques can be divided into two types: double x-ray energy (DE-XRT) and X-ray fluorescence (XRF) [13,14].

The XRF technician may be used to recover plastic waste fractions. Unfortunately, this technique is only applicable to the recovery of PVC from other types of plastic [15]. The basis of the XRF technique is the induction of individual atoms by an external laser source, which leads to the emission of X-ray photons. The emitted photons form a unique spectral signature corresponding to the atomic weight, which allows to determine the type of material. In the case of plastic, its spectral signature is a superposition of spectral signatures of components that can be identified using machine learning techniques.

Another technique for sorting waste is EDXRF (Energy Dispersive X-ray Fluorescence) that uses markers added to the polymer matrix to sort plastic particles. These labels are formed by many substances dispersed in the material thereby increasing the selectivity of the polypropylene sorting [16]. The X-ray is focused on a small area of material and goes to the detector. The signal from the detector is then sent to the processing unit. This unit controls the radiation source whose spectral signal is analysed and then used to separate materials containing specific quantities of markers. XRF is a non-destructive technique capable of identifying black polymers as well as dirty waste [17].

2.3   Optical Based Sorting

Commonly techniques often used physical features but ignored visual properties like colour, shapes, texture and size for the sorting of waste. In optical sorting, camera based sensors are used for the identification of waste fractions. In this section we present optical sorting techniques.

Sorting technique based on features like shape and colour was proposed by Huang et al. [18]. This method combines a 3D colour camera and laser beam over the conveyor belt. This technique formed triangles over the image from the camera on the base laser beam, so is called triangulation scanning. The technique achieves an accuracy of 99 % for plastic fractions.

Spectral imaging is a combination of spectral reflectance measurement and image processing technologies. We may found several spectral imaging methods using NIR (near infrared), VIS (visual image spectroscopy) and HSI (hyperspectral imaging) [19,20]. A hyperspectral sensor produces images over a continuous range of narrow spectral bands and next system analysis the spectroscopic data. The conveyor system moves the waste fractions beneath the spectral camera acquires images. At the second stage data is pre-processing and reduction. Next to perform material classification a special algorithm is applied. A set of compressed air nozzles is mounted at the end of the conveyor belt and depending upon the classifier decision, one of nozzles are triggered the waste into particular bins [21,22].

Waste classification is also possible using artificial intelligence. In work [23] RecycleNet is carefully optimized deep convolutional neural network architecture for classification of selected recyclable object classes. This novel model reduced the number of parameters in a 121 layered network from 7 million to about 3 million. In paper [24] they provided the concept of automatic processing of plastic and metal waste combined with another mechanism for economic motivation of end users. In work [25] authors propose a multilayer hybrid deep-learning system (MHS) to automatically sort waste disposed of by individuals in the urban public area. This system deploys a high-resolution camera to capture waste image and sensors to detect other useful feature information.

All the methods cited have their advantages, but they also generate problems with the use of appropriate specific and often expensive technologies (direct sorting, indirect sorting), or demanding high computational efficiency (artificial intelligence). The use of artificial intelligence to classify waste became the main motivator for the authors. The possibilities obtained by neural networks in the field of image classification and recognition show that it is possible to build effective systems also in the field of waste selection, but the complexity of the whole system must be simplified.

3 Proposed System

After extracting plastic garbage from the MSW, a computer system based on image processing can be used to divide it into different types Fig. 2. The method we propose uses an RGB digital camera and a computer with software for classifying plastic waste. In contrast, an air stream is used to direct the waste to a specific container, assuming that the waste will be transported separately on the conveyor belt. The software used in this system uses image processing techniques in the process of image pre-processing. Convolution neural network and deep learning [26, 28, 29] are used to recognize objects.

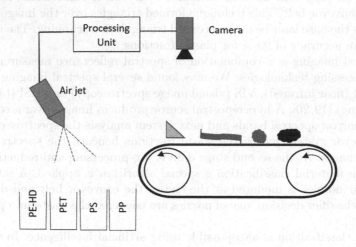

Fig. 2. Proposed system for plastic waste sorting

4 Convolutional Neural Network

Convolutional neural network (CNN) is a feed-forward artificial neural network in which the organization of neurons is similar to the animal visual cortex [27]. In order to recognize the shape of an object, the local arrangement of pixels is important. CNN starts with recognition of smaller local patterns on the image and concatenate them into more complex shapes. CNN was proved to be efficient especially in object recognition on an image [28–30]. CNNs might be an effective solution to the waste sorting problem. CNN explicitly assumes the input is an image and reflects it onto its architecture. CNN usually contains Convolutional layer, Pooling layer and Fully-connected layer. Convolutional layers and Pooling layers are stacked on each other, fully-connected layers at the top of the network outputs the class probabilities.

5 Structure of Research Networks

A number of important factors had to be taken into account when working on the appropriate selection of the network structure. First of all, the size of the input image was an important element. Too high resolution resulted in increasing the number of calculations, which resulted in fairly frequent overload of memory available computing unit. On the other hand, too low resolution of the input data could have prevented the achievement of the expected performance. It was decided to conduct research for images with a resolution of 120×120 pixels and 227×227 pixels. The assumption that the resolution of input images is 227×227 pixels results from the established AlexNet structure [30]. The assumption that the resolution of input images is 120×120 is an experimental assumption. This choice allowed testing for images almost twice smaller than in the case of AlexNet.

Another important element was the selection of the number and types of layers of the CNN network. Two CNN networks were tested, differing in the number of layers and the number and size of convolution filters. The first type of network tested (based on the AlexNet network) contained 23 layers. For this structure, 9×9 convolution filters were used in the first convolution layer. A total of six weave layers were responsible for the appropriate encoding of the input image, which was then fed into a three-layer fully connected layer.

The second type of network, based on the authors' proposals, preceded by preliminary tests (other structures tested achieved worse results, or it was hard to adapt them to the adopted resolution.), contained 15 layers. The first convolution layer consisted of 64 convolution filters with dimensions 9×9. Three layers of convolution encode information, transferred to a two-layer fully connected layer. The network diagram for 120×120 pixel images is shown in Fig. 3. For CNN prepared in this way, tests were carried out with images of various input resolutions.

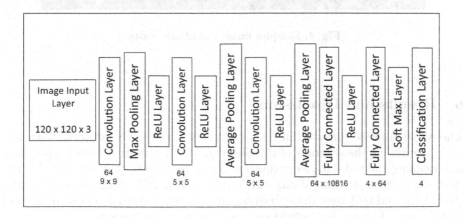

Fig. 3. The structure of our networks - 15 layers

Preparation of input data for the learning and testing phase was important in the context of correct classification of objects in natural working conditions. In the case of deep neural networks, as many data as possible should be collected for each identified class. In our case, it was necessary to collect photos of classified waste. We adopted a simplified model where there could be only one waste within the camera lens. This approach does not reflect the natural working conditions, but for research needs it gives sufficient opportunities to generate a properly functioning network. All collected images represented objects classified into four considered classes: PS, PP, PE-HD and PET. These images came from the WaDaBa database [31], and their samples may be seen on Fig. 4. Due to the availability of research facilities, it was easiest to create a PET class due to the largest amount of this type of waste that can be recycled. This resulted in the fact that a different number of images assigned to individual classes was collected. To increase the number of images in individual classes and to obtain a similar number of images for each class in the final stage, it was necessary to make copies of images rotated by different angles of rotation within each class. Such set of data (about 33 000 simulated images per class) was sufficient to properly teach the prepared network structures.

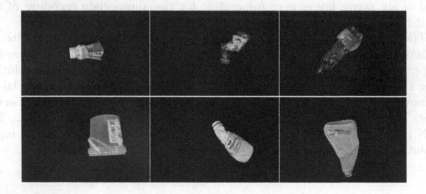

Fig. 4. Samples images of plastic waste

6   Results and Discussion

The research phase consisted in training two prepared CNN network structures and determining the accuracy of classification at different divisions of input data into training and test data. Four divisions were used: - 90% (training data) - 10% (test data), 80% (training data) - 20% (test data), 70% (training data) - 30% (test data) and 60% (test data) training data) - 40% (test data). The numerical distribution of photos for individual classes is shown in Table 1.

For the learning data prepared in this way, the CNN network learning process was carried out (parameters of the test computer: Intel (R) Core (TM) i7-5820K

Table 1. Numerical division of data into training data and test data

Number of division	Type of division	PE-HD Training/Test	PET Training/Test	PP Training/Test	PS Training/Test
1	90% - 10%	32400/3600	29700/3300	29952/3328	33696/3744
2	80% - 20%	28800/7200	26400/6600	26624/6656	29952/7488
3	70% - 30%	25200/10800	23100/9900	23296/9984	26208/11232
4	60% - 40%	21600/14400	19800/23200	19968/13312	22464/14976

CPU 3.30 GHz, RAM 64 GB, GPU NVIDIA GeForce GTX 960). Training was carried out for two pre-prepared network structures, with input image resolutions of 120×120 pixels and 227×227 pixels. Learning was carried out for a variable value of learning coefficient, starting from 0.001 and decreasing every subsequent 4 epoch. 1064 iterations for one epoch were considered.

Table 2 summarizes all tests carried out for a stratified network using images with a resolution of 120×120 pixels.

Table 2. Learning results of a 15-layer network (image resolution 120×120 pixels)

Number of division	Type of division	2 epochs accuracy [%]	Time [min]	4 epochs accuracy [%]	Time [min]	10 epochs accuracy [%]	Time [min]
1	90% - 10%	93,27	29	97,43	61	**99,92**	217
2	80% - 20%	92,71	27	96,97	57	98,69	203
3	70% - 30%	90,74	24	93,68	52	97,78	184
4	60% - 40%	86,57	20	90,25	49	92,77	167

Table 3 presents all tests carried out for a stratified network using images with a resolution of 227×227 pixels.

Table 3. Learning results of a 15-layer network (image resolution 227×227 pixels)

Number of division	Type of division	2 epochs accuracy [%]	Time [min]	4 epochs accuracy [%]	Time [min]	10 epochs accuracy [%]	Time [min]
1	90% - 10%	69,43	79	80,25	183	**91,72**	540
2	80% - 20%	66,76	76	77,80	174	88,34	527
3	70% - 30%	63,89	70	73,69	171	84,64	504
4	60% - 40%	60,70	69	70,45	159	80,23	498

Table 4 summarizes all tests conducted for a 23-layer network using images with a resolution of 120×120 pixels.

Table 4. Learning results of a 23-layer network (image resolution 120 × 120 pixels)

Number of division	Type of division	2 epochs accuracy [%]	Time [min]	4 epochs accuracy [%]	Time [min]	10 epochs accuracy [%]	Time [min]
1	90% - 10%	73,29	63	92,31	125	**96,41**	364
2	80% - 20%	70,08	61	90,83	121	93,39	347
3	70% - 30%	67,13	57	86,24	114	90,29	311
4	60% - 40%	62,44	55	83,04	106	88,46	301

Table 5 summarizes all tests conducted for a 23-layer network using images with a resolution of 227 × 227 pixels.

Table 5. Learning results of a 23-layer network (image resolution 227 × 227 pixels)

Number of division	Type of division	2 epochs accuracy [%]	Time [min]	4 epochs accuracy [%]	Time [min]	10 epochs accuracy [%]	Time [min]
1	90% - 10%	62,38	73	86,83	214	**99,23**	725
2	80% - 20%	60,44	71	84,21	199	97,51	707
3	70% - 30%	59,21	69	80,49	192	97,92	642
4	60% - 40%	58,94	64	72,84	165	93,45	549

Analyzing the obtained results, it can be seen that in the case of a stratified network, for our training data composed of images with a resolution of 120 × 120 pixels, 4 epochs are sufficient to obtain an acceptable level. Further learning, even with a reduced learning rate, no longer significantly affects accuracy. Accuracy at 4 epochs of 97.43% is a very good result as a fairly small number of iterations. After 10 epochs, this accuracy is still increasing and amounts to almost 100%. Of course, this is for the division, where up to 90% of images are used to teach the network. For images with a resolution of 227 × 227 pixels, the computation time is almost doubled. In addition, accuracy at 91.72% at the first assumed division into training and test data is not acceptable for the proper functioning of the system in real conditions.

The training process is slightly different in the case of a 23-layer network, for which the same division into training and test data was used, as in the case of a 15-layer network. It is able to achieve an accuracy index of 99.23 for the first training data sharing and 227 × 227 pixel image resolution, but a learning time of 725 min compared to 217 min makes the learning process inconvenient. This is quite important information in the context of the learning process of the system, which is to function in real conditions. In the case of training images with a resolution of 120 × 120 pixels, the 23-layer network after 10 epochs did not reach the level that was obtained for the 15-layer network.

7 Conclusion and Future Works

The conducted research has shown that the 15-layer network proposed by us allows achieving high efficiency for images with a resolution more than twice lower than the available 23-layer network with a dedicated resolution of 227×227 pixels. Classification of segregated waste into four main classes takes place in most cases without error. Of course, this is to a certain extent caused by the artificially increased number of individual class representatives. Further work will mainly consist of extending the database of segregated waste images with photos of waste in more realistic conditions. Hence, efforts to obtain recordings of waste on a conveyor belt from enterprises dealing with waste segregation. Another noticeable thing is the definitely shorter learning time for a 15-layer network compared to a 23-layer network, especially for 120×120 pixel images. Our research in the future will assume the possibility of training the network while working in real conditions, which is possible to implement with our proposal. After introducing modifications to the training database, we also want to determine the accuracy for real images of waste taken from the conveyor belt during the segregation process.

Acknowledgements. The project financed under the program of the Minister of Science and Higher Education under the name "Regional Initiative of Excellence" in the years 2019–2022 project number 020/RID/2018/19, the amount of financing 12,000,000 PLN.

References

1. Kumar, P., Sikder, P.S., Pal, N.: Biomass fuel cell based distributed generation system for Sagar Island. Bull. Pol. Acad. Sci.: Tech. Sci. **66**(5), 665–674 (2018)
2. Kaczorek, T.: Responses of positive standard and fractional linear systems and electrical circuits with derivatives of their inputs. Bull. Pol. Acad. Sci.: Tech. Sci. **66**(4), 419–426 (2018)
3. Gundupalli, S.P., Hait, S., Thakur, A.: A review on automated sorting of source-separated municipal solid waste for recycling. Waste Manag. **60**, 56–74 (2017)
4. Al-Salem, S.M., Lettieri, P., Baeyens, J.: Recycling and recovery routes of plastic solid waste (PSW): a review. Waste Manag. **29**(10), 2625–2643 (2009)
5. Richard, G.M., Mario, M., Javier, T., Susana, T.: Optimization of the recovery of plastics for recycling by density media separation cyclones. Resour. Conserv. Recycl. **55**(4), 472–482 (2011)
6. Yuan, H., Fu, S., Tan, W., He, J., Wu, K.: Study on the hydrocyclonic separation of waste plastics with different density. Waste Manag. **45**, 108–111 (2015)
7. De Jong, T.P.R., Dalmijn, W.L.: Improving jigging results of non-ferrous car scrap by application of an intermediate layer. Int. J. Miner. Process. **49**(1), 59–72 (1997)
8. Pita, F., Castilho, A.: Influence of shape and size of the particles on jigging separation of plastics mixture. Waste Manag. **48**, 89–94 (2016)
9. Li, J., Xu, Z., Zhou, Y.: Application of corona discharge and electrostatic force to separate metals and nonmetals from crushed particles of waste printed circuit boards. J. Electrostat. **65**(4), 233–238 (2007)

10. Vajna, B., et al.: Complex analysis of car shredder light fraction. Open Waste Manag. J. **2**(53), 2–50 (2010)
11. Patachia, S., Moldovan, A., Tierean, M., Baltes, L.: Composition determination of the Romanian municipal plastics wastes. In: Proceeding of the 26th International Conference on Solid Waste Technology and Management (2011)
12. Wang, C.Q., Wang, H., Fu, J.G., Liu, Y.N.: Flotation separation of waste plastics for recycling a review. Waste Manag. **41**, 28–38 (2015)
13. De Jong, T.P.R., Dalmijn, W.L.: X-ray transmission imaging for process optimisation of solid resources. In: Proceedings R: 02 Congress (2002)
14. De Jong, T.P.R., Dalmijn, W.L., Kattentidt, H.U.R.: Dual energy X-ray transmission imaging for concentration and control of solids. In: Proceedings of IMPC-2003 XXII International Minerals Processing Conference, Cape Town (2003)
15. Brunner, S., Fomin, P., Kargel, C.: Automated sorting of polymer flakes: fluorescence labeling and development of a measurement system prototype. Waste Manag. **38**, 49–60 (2015)
16. Bezati, F., Massardier, V., Balcaen, J., Froelich, D.: A study on the dispersion, preparation, characterization and photo-degradation of polypropylene traced with rare earth oxides. Polym. Degrad. Stab. **96**(1), 51–59 (2015)
17. Bezati, F., Froelich, D., Massardier, V., Maris, E.: Addition of tracers into the polypropylene in view of automatic sorting of plastic wastes using X-ray fluorescence spectrometry. Waste Manag. **30**(4), 591–596 (2010)
18. Huang, J., Pretz, T., Bian, Z.: Intelligent solid waste processing using optical sensor based sorting technology. In: 3rd International Congress on Image and Signal Processing (CISP), vol. 4, pp. 1657–1661. IEEE (2010)
19. Kreindl, G.: Sorting of mixed commercial waste for material recycling. In: Proceeding of TAKAG 2011 Deutsch-Trkische Abfalltage, Suttgart (2011)
20. Pieber, S., Meirhofer, M., Ragossnig, A., Brooks, L., Pomberger, R., Curtis, A.: Advanced waste-splitting by sensor based sorting on the example of the MTPlant Oberlaa. In: Tagungsband zur 10, DepoTech Conference, pp. 695–698 (2010)
21. Picn, A., Ghita, O., Whelan, P.F., Iriondo, P.M.: Fuzzy spectral and spatial feature integration for classification of nonferrous materials in hyperspectral data. IEEE Trans. Ind. Inform. **5**(4), 483–494 (2009)
22. Picn, A., Ghita, O., Bereciartua, A., Echazarra, J., Whelan, P.F., Iriondo, P.M.: Real-time hyperspectral processing for automatic nonferrous material sorting. J. Electron. Imaging. **21**(1), 013018 (2012)
23. Bircanoglu, C., Atay, M., Beser, F., Genc, O., Kizrak, M.A.: RecycleNet: intelligent waste sorting using deep neural networks (2018)
24. Kokoulin, A.N., Tur, A.I., Yuzhakov, A.A.: Convolutional neural networks application in plastic waste recognition and sorting. In: IEEE Conference of Russian Young Researchers in Electrical and Electronic Engineering (EIConRus), pp. 1094–1098 (2018)
25. Chu, Y., Huang, C., Xie, X., Tan, B., Kamal, S., Xiong, X.: Multilayer hybrid deep-learning method for waste classification and recycling. Comput. Intell. Neurosci. (2018)
26. Wang, M., Wang, Z., Li, J.: Convolutional neural network applies to face recognition in small and medium databases. In: 4th International Conference on Systems and Informatics, ICSAI 2017, pp. 1368–1372, January 2018
27. Gua, J., et al.: Recent advances in convolutional neural networks. Pattern Recognit. **77**, 354–377 (2018)

28. Wang, L., Ouyang, W., Wang, X., Lu, H.: Visual tracking with fully convolutional networks. In: Proceedings of International Conference on Computer Vision (ICCV), pp. 3119–3127 (2015)
29. Zhao, Z.Q., Zheng, P., Xu, S.T., Wu, X.: Object detection with deep learning: a review. IEEE Trans. Neural Netw. Learn. Syst. **PP**, 1–21 (2019)
30. Krizhevsky, A., Sutskever, I., Hinton, G.E.: ImageNet classification with deep convolutional neural networks. Commun. ACM **60**(6), 84–90 (2012)
31. Bobulski, J., Piatkowski, J.: PET waste classification method and plastic waste database - WaDaBa. In: Choraś, M., Choraś, R. (eds.) IP&C 2017. Advances in Intelligent Systems and Computing, vol. 681, pp. 57–64. Springer, Cham (2018). https://doi.org/10.1007/978-3-319-68720-9_8

Artificial Neural Networks for Bottled Water Demand Forecasting: A Small Business Case Study

Israel D. Herrera-Granda[1]([✉]), Joselyn A. Chicaiza-Ipiales[1],
Erick P. Herrera-Granda[1], Leandro L. Lorente-Leyva[1],
Jorge A. Caraguay-Procel[1], Iván D. García-Santillán[1],
and Diego H. Peluffo-Ordóñez[2]

[1] Facultad de Ingeniería en Ciencias Aplicadas, Universidad Técnica del Norte,
Av. 17 de Julio, 5-21, y Gral. José María Córdova, Ibarra, Ecuador
idherrera@utn.edu.ec
[2] Escuela de Ciencias Matemáticas y Tecnología Informática, Yachay Tech,
Hacienda San José s/n, San Miguel de Urcuquí, Ecuador
dpeluffo@yachaytech.edu.ec

Abstract. This paper shows a neural networks-based demand forecasting model designed for a small manufacturer of bottled water in Ecuador, which currently doesn't have adequate demand forecast methodologies, causing problems of customer orders non-compliance, inventory excess and economic losses. However, by working with accurate predictions, the manufacturer will have an anticipated vision of future needs in order to satisfy the demand for manufactured products, in other words, to guarantee on time and reasonable use of the resources. To solve the problems that this small manufacturer has to face a historic demand data acquisition process was done through the last 36 months costumer order records. In the construction of the historical time series, that was analyzed, demand dates and volumes were established as input variables. Then the design of forecast models was done, based on classical methods and multi-layer neural networks, which were evaluated by means of quantitative error indicators. The application of these methods was done through the R programming language. After this, a stage of training and improvement of the network is included, it was evaluated against the results of the classic forecasting methods, and the next 12 months were predicted by means of the best obtained model. Finally, the feasibility of the use of neural networks in the forecast of demand for purified water bottles, is demonstrated.

Keywords: Long-term demand forecasting · Small business ·
Artificial neural networks

1 Introduction

Within the business approach, small and large manufacturing businesses have to face uncertain environments daily, one of the most representative is the demand study and forecast of products that customers require from the manufacturer, ergo, if the latter

© Springer Nature Switzerland AG 2019
I. Rojas et al. (Eds.): IWANN 2019, LNCS 11507, pp. 362–373, 2019.
https://doi.org/10.1007/978-3-030-20518-8_31

were able to know in advance what their customers require, he would be able to stock up with raw materials in advance without falling into unnecessary acquisitions and economic resources waste. This implies a competitive advantage and is the basis for an adequate management of supply chains.

The proper use of resources in a small business is an aspect of great importance, because this kind of business are still in development stages and an inadequate decision could cause great damage being in some cases the cause of the closure of its operations.

Within the field of demand forecasting and more specifically the demand forecasting for finished products, several studies have been carried out, among which Aburto's work developed in 2017 can be mentioned, which develops a hybrid approach for demand forecasting of finished products based on the analysis of an ARIMA model and introducing it into a neural network which tries to reproduce the patterns of a historical time series obtained on the daily demand of products in a supermarket [1].

In 2014 Saha proposes the implementation of a generic demand forecasting model of servers based on neural networks (NN) which uses a historical demand of 52 weeks and its parameters were modified in order to guaranty an accuracy range from 84 to 89% [2].

In 2015 Slimani's work focuses on finding the optimal multilayer perceptron structure (MPL) for demand forecasting of the products of a supermarket in Morocco and establishes the idea of implementing this forecast with NN in a short supply chain with focused on game theory [3].

In 2017 Abraham et al. propose the design, training and simulation of an NN for demand forecasting of soybean produced and exported in Mato Grosso - Brazil, this model was implemented in Matlab software. The results of this forecast indicated in advance that the demand would increase by about 26% compared to the previous year, which meant that producers increased their production, mobilizing around 6 million tons of soybeans in the first 5 months of 2017 [4].

Currently the applications of the NN in the forecasting of product demand have diversified their field of action. For example, in 2018 Fu et al. proposed an integrating approach between NNs and a parametric method for the forecasting of the intermittent demand of semiconductors in Taiwan [5]. On the other hand, Ezekwesili et al. tested forecast models based on NNs and Bootstrap methods to improve the demand forecast accuracy for commercial aircraft parts [6]. While Yang et al. in the year of 2018 in China, collected sales information in more than 10 million sale points of a bakery products franchise to build historical time series and used this data to forecast the demand of products by short term or by hour, using neural networks [7].

Approach of the Case Study

The studied company in this work is a small manufacturer and distributor of bottled water that operates in the north of Ecuador. In relation to its customer service policies, an indicator of delivery punctuality has been implemented, and its value was found at 80% initially. However, it is intended to maximize this value with the use of a forecast model that allows planning the company's resources to consequently increase the punctuality in the delivery of orders.

The objective of this study is maximizing the indicator of punctuality in costumer deliveries to consequently raise customer's satisfaction, through the neural network-forecasting model.

The study of the demand of water bottles through time, seeks to analyze and predict their behavior in the future using artificial neural networks methodology, in addition to the diagnosis of the forecasting process that the company used to do initially. Finally, a framework of demand forecasting for the studied company was designed.

2   Materials and Methods

Multi-Layer Perceptrons Neuronal Networks

Artificial neural networks can be used as forecasting methods that are based on simple mathematical models inspired by the human brain. These allow complex non-linear relationships between the response variable and its predictors [6].

The neural network prediction model will be established according to the influence factors, considering within the inputs the factors that affect the demand, and the predicted demand as output [7].

Multi-Layer Perceptrons (MLP) Neural Networks are networks formed by interconnected nodes or neurons in various types of configurations and with a hierarchical organization, which try to interact with the real world objects in the same way that a biological nervous system does. An scheme of the architecture and learning process of the MLP is shown in Fig. 1. The MLP are constituted by interconnected neurons and arranged in three or more layers. The data enters through the input layer, pass through the hidden layers and exit through the output layer. It should be mentioned that the hidden layers can be constituted by several layers [6, 8].

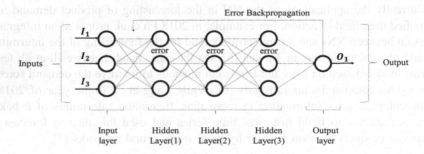

Fig. 1. MLP neural network with back-propagation learning methodology [6, 8–10].

Multi-Layer Perceptron (MLP) for Time Series Forecasting

Let data vectors of independent values, X_t or $Y \forall Y = \{Y_t, = Y_{t-1}, \ldots, Y_{t-i+1}\}$, be the historic time series t, in addition take its predicted values as $\widehat{y}_{t+1}$. Within the present work, the implementation of a typical MLP network specially designed for time series forecast of non-linear autoregressive model of order p NAR(p), is proposed, being p the delayed positions of the traditional forecast models, hence the future forecast can be defined by the variable $\widehat{y}_{t+h}$. Consider also $n = p = I$ for the total number of entries for the MLP, $g(f)$ is a non-linear transfer function and let H be the total number of hidden units in the network [1].

Within the network, the weights of each neuron are defined as w, being $w = (\beta, \gamma)$, the output layers $\beta = \{\beta_1, \beta_2, \ldots, \beta_H\}$, and the hidden layers $\gamma = \{\gamma_{11}, \gamma_{12}, \ldots, \gamma_{21}, \ldots \gamma_{hI}\}$. The variables β_0, γ_{0i} represent the biases of each neuron. Then, the function that represents a single layer MLP with a single output is shown in Eq. 1.

$$f(Y, w) = \beta_0 + \sum_{h=1}^{H} \beta_h g\left(\gamma_{0i} + \sum_{i=1}^{I} \gamma_h Y_i\right) \tag{1}$$

Within the neural dynamics the sigmoidal logistic function or the hyperbolic tangent function are used, so that as mentioned above each hidden node is calculated by means of the NAR (p) model in all the input nodes, which are combined by means of a weighted sum of one or more output nodes, as shown in the Fig. 2.

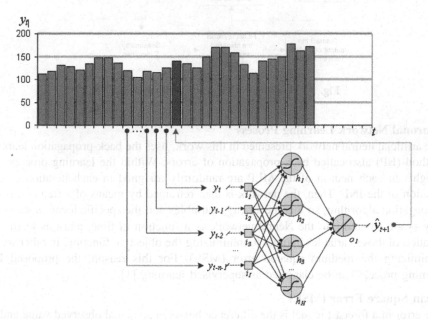

Fig. 2. Architecture of a autoregressive MLP for time series forecasting [11].

Seasonality Detection in the Historical Data Series

One of the main benefits of the NNFOR package lies in the detection of seasonality in a historical data series, where S is the seasonal distance underlying the total seasonal distance, and their individual seasonal distances d_p are calculated based on the Euclidean distance formula. In this way, if a seasonality takes the value $s \neq 0$ then $d_p > 0$, and analogously if $s = 12$ then $d_p = 0$. Consider also an ordered pair of vectors $P, Q(p_i \in P, q_i \in Q; \forall i = 1, 2, \ldots, s)$. Seasonality detection is achieved by a penalty τ applied to relatively large vectors, as shown in the Eq. 2.

$$d_p(P, Q)_s = \log(d(p, Q)_s + 1) - \tau \log(S) \tag{2}$$

Additionally, by means of an Iterative Neural Filter (INF) inspired by the traditional ARIMA iterative model, overlapping seasons which could interact with other seasonal frequencies, are identified. In this way, less dominant patterns can be considered for the future forecast. By means of the variables $x_{s,1}$ and $x_{s,2}$, seasonality is coded in the data series, as shown in the Eq. 3. In addition, a scheme of the operation of the INF is shown in Fig. 3.

$$x_{s,1}(t) = \sin\frac{2\pi t}{S} \ y \ x_{s,2}(t) = \cos\frac{2\pi t}{S} \tag{3}$$

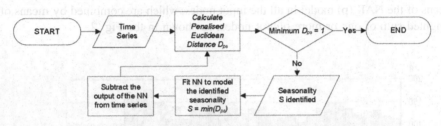

Fig. 3. Flow chart of the iterative filter (INF) [11].

Neuronal Network Learning Process

The artificial neural network presented in this work, uses the back-propagation learning method (BP) also called back-propagation of errors. Within the learning process, the weights in each neuron of the MLP are randomly assigned in each iteration or calculation of the INF. Then, the network is then retrained by means of a standard back-propagation algorithm, so that the $x_{s,1}, x_{s,2}$ variables and the specific location variables z_1 y z_2 are entered into the Neural network as a function of time, and this learns the location of those variables in order to minimizing the objective function, in other words minimizing the medium squared error (MSE). For this reason, the proposed NN learning process can be classified as supervised learning [1].

Mean Square Error (MSE)

The error in a forecast model is the difference between an actual observed value and its predicted value, so it is possible to estimate the accuracy of a model. In the MSE indicator, each error is raised to square, then these errors are summed and divided among the number of observations, thus major errors are penalized, this indicator is used to measure the accuracy of a forecast model. Since through this, it is possible to differentiate between a technique that produces moderate errors, to another in which the errors are minor but that occasionally throw some extremely big errors [12].

$$MSE = \frac{\sum_{t=1}^{n}(Y_t - \widehat{Y}_t)^2}{n} \tag{4}$$

Where:

Y_t = Real value in t period.
$\widehat{Y}_t$ = Predicted value in t period.
n = Number of observations.

Square Root of Mean Error (RMSE)

Both the RMSE and the MSE penalize larger errors but the RMSE has the same unit of measurement as the original series and is interpreted more easily [12].

$$RMSE = \sqrt{MSE} \tag{5}$$

Applied Methodology to the Case Study

For the solution of the studied problem, it began with the characterization of the studied business. So that, it is possible to collect the historical data of the demand for purified water bottles. It was initially detected that the company had a policy of costumer orders recording, whose validity in forecasting matters is greater than the data of the invoices of finished products [13].

It could also be detected that the studied company does not analyze the behavior of the product in the market, in the long term, within the present work is defined daily time intervals as short-term planning period, medium-term includes time intervals quarterly or semi-annual and long-term is understood as annual time periods [14].

Through a checklist, it was possible to characterize the company's current forecasting system, which actually is cataloged as qualitative and visionary [15], ergo, a forecast is made based on the experience of the manager of the company, without the use of quantitative methods or classical statistics, which makes it difficult to measure the accuracy of the initial forecast methodology. Additionally, it can be mentioned that the company does not have an adequate definition of the annual production goals, and control problems were noticed at different levels of its supply chain.

For the present article, we work with a historical time series of purified water bottle orders of the last three years. The literature concerning to business forecasts recommends the use of orders instead of invoices of finished products, since in this way it avoids falling into forecasting errors and at inventory level, avoids errors produced by backorders that were not historically recorded, which potentially could affect modifying the inventory levels of the studied company [14, 16].

The adequate construction of the historical demand series allows to carry out quantitative analyzes, which allows to have a clue to determine the most suitable model for the forecast. In the present case study, a historical series of high variability is treated. The study of the literature made in the first section indicates the benefits of demand forecasting through neural networks, especially when dealing with high variability demands. However, the approach of neural networks is also contrasted against classical methods of forecasting through the use of specific software such as Forecast Pro v5.0 [17].

Subsequently, a training phase of the network allows to improve the accuracy of the forecast, as well as the main indicator which is the Medium Square Error (MSE).

In this way, by means of several modifications in the architecture and parameters of the neural multilayer network (MLP) MSE is gradually improved. Having the correctly trained network the next stage can be started, as shown in Fig. 4.

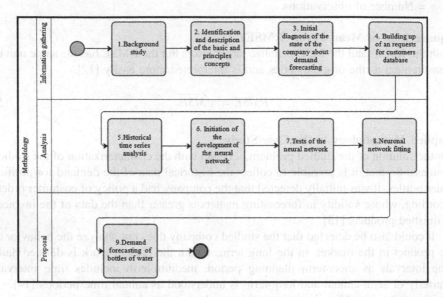

Fig. 4. Methodological framework for forecasting demand in the case study.

3   Results and Discussion

Analysis of the Historical Time Series

The water bottling company manages a manual register and a diary of order, sales and return of empty bottles. The register dates from June 01, 2015 to January 15, 2019. The sample we work with corresponds to 3 years of productive activity and data recording of the company. This amount is considered well-timed for forecasts development. Figure 4 shows historical order data of the last 3 years spread over 12 months, on which a multiplicative decomposition is made analyzing its tendency, seasonality and randomness. As shown in Fig. 5 according to the seasonality section of multiplicative decomposition, an annual seasonality is obtained that starts from each half year and is repeated annually, the trend is decreasing and begins from the second half of 2016 until the second half of 2018.

Design and Validation of the Neuronal Network

Initially, through the R programming language, and the NNFOR packet, time series were introduced to test an automated multi-layer neural network analysis by the NNFOR package. The output indicates that the resulting network has five hidden nodes, was trained 20 times and the different forecasts were combined using the median operator. The MLP function automatically generates sets of networks, which training begins with different random initial weights. In addition, it provides the entries that

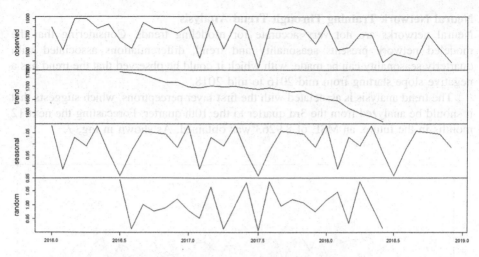

Fig. 5. Multiplicative decomposition of the time series.

were included in the network, resulting in a MSE of 44.0968, which, based on the related works literature review can be catalogued as very high [3], and therefore it was decided to move on new phases of training, in order to improve this network indicator.

First Layer NN Training

The analysis of the time series decomposition, according to seasonality section, indicates that there is an annual seasonality starting from every half year. There is also evidenced, the existence of 3 patterns, but in order to take advantage of the analysis of neural network patterns, each annual pattern was divided into 4 parts, so that 12 initial perceptrons were used, which means that a quarterly pattern analysis is carried out. Then the number of perceptrons in the first layer is equal to 36/3 = 12.

To save time in network training in each test the default value of 20 iterations is used. This value implies a compromise between training speed and performance, as shown below. Forecasting the next 12 months an MSE of 7.3677 was obtained, as shown in Fig. 6.

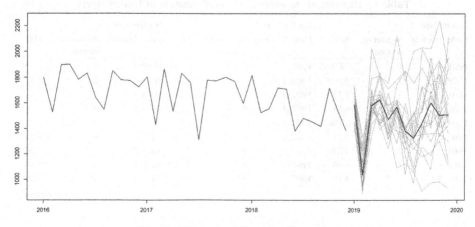

Fig. 6. Forecast with seasonality setting.

Neural Network Training Through Trend Analysis

Neural networks are not very accurate for modeling trends. Considering that the modeled network presents seasonality and trend, differentiations associated with quarterly seasonality can be made, with which it could be observed that the trend has a negative slope starting from mid 2016 to mid 2018.

The trend analysis is associated with the first layer perceptrons, which suggests that it should be analyzed from the 3rd quarter to the 10th quarter. Forecasting the next 12 months in the future, an MSE of 8.9263 was obtained. As shown in Fig. 7.

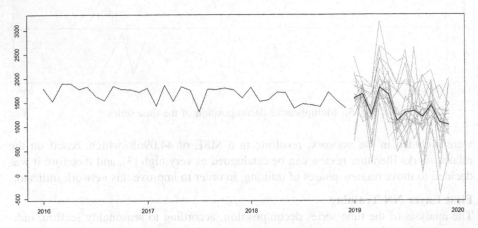

Fig. 7. Forecast with trend setting.

Computational Experiments to Improve NN Hidden Layers

Based on the proper configuration of the first layer or input layer, it was decided to make modifications to the configurations of the next three hidden layers of the network, with the aim of improving the MSE forecast indicator for the next 12 months, as shown in Table 1.

Table 1. Experiments to improve the configuration of hidden layers.

Hidden layer 1			Hidden layer 2			Hidden layer 3		
Code_Training	Number of neurons	MSE	Code_Training	Number of neurons	MSE	Code_Training	Number of neurons	MSE
Fit01_1	1	7473,15	Fit02_1	1	27,70	Fit03_1	1	26,40
Fit01_2	2	358,99	Fit02_2	2	1,92	Fit03_2	2	1,97
Fit01_3	3	7,12	Fit02_3	3	0,44	Fit03_3	3	0,24
Fit01_4	4	6,63	Fit02_4	4	1,23	Fit03_4	4	1,50
Fit01_5	5	10,98	Fit02_5	5	0,84	Fit03_5	5	1,12
Fit01_6	6	6,21	Fit02_6	6	1,00	Fit03_6	6	0,85
Fit01_7	7	11,40	Fit02_7	7	1,07	Fit03_7	7	1,26
Fit01_8	8	6,06	Fit02_8	8	2,25	Fit03_8	8	0,88
Fit01_9	9	6,18	Fit02_9	9	1,22	Fit03_9	9	0,59

(continued)

Table 1. (*continued*)

Hidden layer 1			Hidden layer 2			Hidden layer 3		
Code_Training	Number of neurons	MSE	Code_Training	Number of neurons	MSE	Code_Training	Number of neurons	MSE
Fit01_10	10	6,91	Fit02_10	10	1,60	Fit03_10	10	0,84
Fit01_11	11	6,48	Fit02_11	11	1,20	Fit03_11	11	0,72
Fit01_12	12	8,14	Fit02_12	12	1,97	Fit03_12	12	0,13
Fit01_13	13	3,71	Fit02_13	13	1,58	Fit03_13	13	0,67
Fit01_14	14	6,95	Fit02_14	14	1,63	Fit03_14	14	1,05
Fit01_15	15	6,53	Fit02_15	15	1,48	Fit03_15	15	0,90
Fit01_16	16	4,74	Fit02_16	16	1,71	Fit03_16	16	0,86
Fit01_17	17	9,41	Fit02_17	17	1,25	Fit03_17	17	0,91
Fit01_18	18	6,32	Fit02_18	18	1,18	Fit03_18	18	0,97
Fit01_19	19	5,42	Fit02_19	19	0,86	Fit03_19	19	1,97
Fit01_20	20	6,24	Fit02_20	20	1,16	Fit03_20	20	0,53
Fit01_21	21	6,16	Fit02_21	21	1,10	Fit03_21	21	0,72
Fit01_22	22	2,71	Fit02_22	22	1,17	Fit03_22	22	0,56
Fit01_23	23	4,99	Fit02_23	23	1,36	Fit03_23	23	0,68
Fit01_24	24	5,84	Fit02_24	24	1,13	Fit03_24	24	0,47
Fit01_25	25	5,47	Fit02_25	25	0,86	Fit03_25	25	0,94
Fit01_26	26	6,37	Fit02_26	26	0,62	Fit03_26	26	1,46
Fit01_27	27	5,38	Fit02_27	27	0,99	Fit03_27	27	0,78
Fit01_28	28	6,87	Fit02_28	28	0,93	Fit03_28	28	0,67
Fit01_29	29	5,54	Fit02_29	29	1,04	Fit03_29	29	1,01
Fit01_30	30	4,55	Fit02_30	30	0,60	Fit03_30	30	0,68
Fit01_31	31	3,84	Fit02_31	31	1,19	Fit03_31	31	0,92
Fit01_32	32	2,52	Fit02_32	32	1,33	Fit03_32	32	0,81

Optimized demand forecasting network

After experimentation process on the configurations of the neural network layers, a network of 12 input perceptrons was obtained. It was analyzed every 12 patterns and from this first layer, 3 hidden layers are deployed with 22 nodes in the first layer, 26 nodes in the second and 12 nodes in the last layer. The models were executed 20 times selecting the best of them as optimum and trained 10 times to the point where the error was transformed into a constant. The obtained MSE was 0.95. The optimal network configuration and forecast result for the next 12 months are shown in Fig. 8. Table 2 shows a comparison of the optimal network with traditional forecasting method suggested by Forecast ProV5.0.

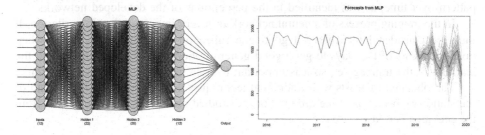

Fig. 8. Optimal network for future demand forecasting.

Table 2. Results comparison of optimal neural network vs method traditional

Parameters	Optimal neural network		Forecast Pro V5.0
Characteristics of forecast model	Neural networks with multilayer perceptrons (MLP)		Exponential smoothing
	Learning method with backpropagation (BP) and use of an algorithm with Iteration Neural Filter (INF)		
	Package: NNFOR		
	Input layer: 12 neurons Hidden layer 1:22 neurons Hidden layer 2:26 neurons Hidden layer 3:12 neurons		
Decomposition method for time series	Multiplicative method		Winters multiplicative
Forecast for 2019	January	1663	1601
	February	1310	1359
	March	1461	1552
	April	1352	1519
	May	1626	1563
	June	1217	1453
	July	1012	1332
	August	1471	1417
	September	1059	1466
	October	1326	1550
	November	1600	1484
	December	834	1383
MSE		0,96	14400
RMSE		0,98	120

4 Conclusions

The company's information management of the orders data allows doing accurate predictions according to the time of the year, as a preliminary step to supply chain planning.

Due to the learning capacity of neural networks, in this case, seasonality, in addition to its tendency, it is possible to make use of them in prediction problems, where the variable to be estimated has a varying behavior, but with a similar behavior pattern over time, as was identified in the perceptrons of the developed networks.

In the training process of a neural network it is very important the stop criterion. It must be considered to achieve a good generalization of the neural network when operating. One of the ways to get a good generalization is not to be so rigid with the learning of the training set, so it is important to train the network through iterations.

The obtained forecasts with traditional techniques do not show a realistic trend of the stationary behavior of the orders. For the studied time series only some data from the previous periods is used.

The neural networks learn and memorize the dynamics of the behavior of the historical series and take time as a variable that affects the cycle of orders. In this way they offer more robust results that are closer to reality.

Acknowledgments. The authors acknowledge to the research project "Optimización de la Distribución Física y en Planta en la Cadena de Suministro aplicando Técnicas Heurísticas" supported by Agreement HCD Nro. UTN-FICA-2019-0149 by Facultad de Ingeniería en Ciencias Aplicadas from Universidad Técnica del Norte. As well, authors thank the valuable support given by the SDAS Research Group (https://www.sdas-group.com).

References

1. Aburto, L., Weber, R.: A sequential hybrid forecasting system for demand prediction. In: Perner, P. (ed.) MLDM 2007. LNCS (LNAI), vol. 4571, pp. 518–532. Springer, Heidelberg (2007). https://doi.org/10.1007/978-3-540-73499-4_39
2. Saha, C., Lam, S.S., Boldrin, W.: Demand forecasting for server manufacturing using neural networks. In: Proceedings of the 2014 Industrial and Systems Engineering Research Conference. State University of New York at Binghamton (2015)
3. Slimani, I., El Farissi, I., Achchab, S.: Artificial neural networks for demand forecasting: application using Moroccan supermarket data. In: International Conference on Intelligent Systems Design and Applications, ISDA, June 2016, pp. 266–271 (2016). https://doi.org/10.1109/isda.2015.7489236
4. Abraham, E.R., dos Reis, J.G.M., Colossetti, A.P., de Souza, A.E., Toloi, R.C.: Neural network system to forecast the soybean exportation on Brazilian port of Santos. In: Lödding, H., Riedel, R., Thoben, K.-D., von Cieminski, G., Kiritsis, D. (eds.) APMS 2017. IAICT, vol. 514, pp. 83–90. Springer, Cham (2017). https://doi.org/10.1007/978-3-319-66926-7_10
5. Fu, W., Chien, C.-F., Lin, Z.-H.: A hybrid forecasting framework with neural network and time-series method for intermittent demand in semiconductor supply chain. In: Moon, I., Lee, G.M., Park, J., Kiritsis, D., von Cieminski, G. (eds.) APMS 2018. IAICT, vol. 536, pp. 65–72. Springer, Cham (2018). https://doi.org/10.1007/978-3-319-99707-0_9
6. Hyndman, R., Athnasopoulos, G.: Forecasting: Principles and Practice. OTexts, Australia (2018)
7. Hu, Y., Sun, S., Wen, J.: Agricultural machinery spare parts demand forecast based on BP neural network. In: Applied Mechanics and Materials, pp. 1822–1825 (2014)
8. Matich, D.: Redes Neuronales: Conceptos Básicos y Aplicaciones, Rosario (2001)
9. Ramasubramanian, K., Singh, A.: Machine Learning Using R. Apress, Berkeley (2019)
10. Scenio: ¿Qué es una Red Neuronal? Parte 2 : La Red | DotCSV, (2018)
11. Crone, S.F., Kourentzes, N.: Feature selection for time series prediction – a combined filter and wrapper approach for neural networks. Neurocomputing **73**, 1923–1936 (2010). https://doi.org/10.1016/j.neucom.2010.01.017
12. Hanke, J.: Pronósticos en los negocios, México (2010)
13. Chopra, S., Meindl, P.: Supply Chain Management: Strategy, Planning, and Operation (2013)
14. Heizer, J., Render, B., Munson, C.: Operations management (2016)
15. Montemayor, E.: Métodos de pronósticos para negocios, Mexico (2013)
16. Hanke, J., Wichern, D.: Business forecast. Pearson Educación (2010)
17. Business Forecast Systems, Inc.: Better decisions demand forecast accuracy - forecast pro. https://www.forecastpro.com/

Image and Signal Processing

Detection of Cancerous Lesions
with Neural Networks

Hassan El-khatib[1], Dan Popescu[1(✉)], and Loretta Ichim[1,2]

[1] Faculty of Control and Computers,
University POLITEHNICA Bucharest, Bucharest, Romania
hassan.elk@yahoo.com,
{dan.popescu, loretta.ichim}@upb.ro
[2] "Stefan S. Nicolau" Institute of Virology, Bucharest, Romania

Abstract. The paper presents two methods for automatic identification of skin cancer in forms of melanoma. In the first method we design a Neural Network that help us to classify the skin lesions. The design of the Neural Network is discussed by analyzing the performance of the training process of the network and the number of target classes. The sensitivity, specificity and accuracy are then determined. The second method uses GoogleNet Convolutional Neural Network (CNN), which is pretrained with the large image database ImageNet. The CNN model is then fine-tuned to classify skin lesions using transfer learning. The classification accuracy is also calculated. The results obtained using the two methods were then compared. The experimental results on a free database demonstrates that the second method can provide high accuracy if some conditions are respected when designing the neural networks.

Keywords: Image processing · Neural Network ·
Convolutional Neural Network · Medical diagnosis · Dermatoscopic images ·
Melanoma detection

1 Introduction

Melanoma is a form of skin cancer that begins in melanocytes, the cells that produce skin pigment, and are located deep within the epidermis or in the surface moss. Even though it is not the most common form of skin cancer, melanoma is responsible for most of the deaths from this type of cancer. The main problem with this disease is that if it is not detected early and treated, melanoma can fast spread (metastasize) to surrounding tissues and other internal organs such as lungs and liver. Therefore, early detection of melanoma is more than necessary taking into account that in most of cases this disease is curable if the melanoma is removed in a short time since its appearance. In other words, it is necessary to create a precise and easy-to-use melanoma detection system to be used by everyone considering that anyone today has a smart phone that can be equipped with such a system.

Melanomas often looks like nevi and is caused mainly by UV exposure. Generally, in order to differentiate a non-cancerous lesion from a cancerous lesion the dermatologist analyzes by visual examination more attributes such as color (if the lesion has

© Springer Nature Switzerland AG 2019
I. Rojas et al. (Eds.): IWANN 2019, LNCS 11507, pp. 377–389, 2019.
https://doi.org/10.1007/978-3-030-20518-8_32

uniform or non-uniform color), dimensions, analysis of lesion edges and so on. In some cases, however, where a more detailed examination is needed, the visual analysis is followed by a dermatoscopic examination that gives detailed images of lesions which are taken with a high-resolution and high-performance camera [1]. As mentioned in Argenziano et al. [2], without the help of dermatoscop device, dermatologists can have just a 65%–80% accuracy rate in case of melanoma detection. According to Ara et al. [3] and to Fabbrocini et al. [4], adding dermatoscopic images to visual inspection results in a melanoma detection accuracy of 75%–84% by the dermatologists.

Over the years, many studies have been conducted to build a system that can detect skin lesions, using machine learning, so that, until 2016, most studies proposed the classical method: preprocessing of the image for the first step, followed by segmentation, feature extraction, and, in the end, the classification [1, 4, 5]. The problem of this type of method is that, the feature extraction needs a high-level expertise being also time consuming in the attempt to choose the suitable and useful features. In 2018, Jianu et al. [6] added to the classic features, such as geometric features (area, perimeter and so on) and texture descriptors (HOG, the fractal dimension and lacunarity). The fractal analysis features gave the best results in the way to classify the benign and malignant nevi.

Starting with 2016, the attention of the most researchers regarding the automatic detection of melanoma, turned to deep learning technique such as Convolutional Neural Network (CNN) [1, 7, 8]. This is because CNN have been successfully used in image recognition and classification area buy giving better results in identifying, for example, faces or other small things such as pencils, food, or traffic signs]. In most of the studies regarding the automatic detection of skin cancer, the authors use CNN pretrained to a large dataset such as ImageNet, applied as a feature extractor [1].

In [9] the authors used only a 399 images dataset for melanoma or benign nevi classification. The images were acquired with a standard camera. As usually, the first step, was a preprocessing step to which they added data augmentation. Then, the most specific features where extracted by using a pretrained AlexNet. The classification step was then done by using a k-nearest-neighbor classifier. As the authors mentioned, the algorithm achieved an accuracy of 93.64%, a specificity of 95.18%, and a sensitivity of 92.1% [1, 9].

Codella et al. [10], also extracted the features with the help of an AlexNet model. The authors used the International Skin Imaging Collaboration (ISIC) database which is a larger database with 2624 dermatoscopic images for the classification of melanoma versus nonmelanoma nevi or melanoma versus atypical lesions. The authors modified the outputs of AlexNet and used features from sparse coding, low-level handcrafted features and, a convolutional U-network and a deep residual network. The final classification was then done by using a Support Vector Machine. The obtained results for classification of melanoma versus nonmelanoma showed an accuracy, sensitivity and specificity of, 93.1%, 94.9%, respectively 92.8%.

There are also some studies in which the authors used the benefits of pretrained Convolutional Neural Network by using end-to-end learning. For example, in Esteva et al. [11], the classification was done by using GoogLeNet Inception v3 model, which was pretrained with ImageNet database. The transfer learning was used by fine tuning the Convolutional Neural Network model to classify the skin lesions.

In this paper we proposed two methods for automatic identification of skin cancer in forms of melanoma: a Neural Network designed to classify the skin lesions and a GoogleNet Convolutional Neural Network, fine-tuned to classify skin lesions using transfer learning.

2 Methodology for Skin Lesion Detection Using Neural Network

The first step in developing an information system, able to correctly diagnose melanoma to preprocess the input image (Fig. 1).

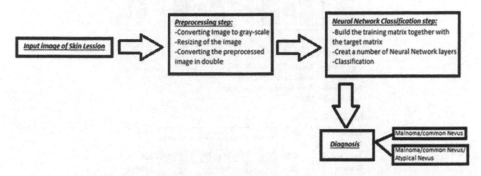

Fig. 1. Flow chart of the proposed system using a NN as classifier.

The preprocessing step consists in converting the image in gray-scale, resizing the image in order to have a correct matrix. The second step is the neural network classification that consists in building the training matrix together with the target matrix and after that, creating a few desired hidden layers. At least, the classification is done. We choose to test the network through two types of classification. First the classification of skin lesion in melanoma or common nevus and, after that, the classification of skin lesion in three classes, by adding also the atypical nevus class.

The algorithm was implemented in Matlab using the network function and "nntraintool". This last mentioned function makes the training GUI appears before the training ends. Through the training window, we can see details about training algorithms, status including network accuracy. The training window also offers us useful plots. Subsequent paragraphs, however, are indented.

3 Experimental Results for Skin Lesion Detection Using Neural Network

In the experiments, there were used 69 images for the Neural Network with two targets in the output layer: 43 images of common nevus and 26 images of melanoma which were used for Neural Network training. In order to test the neural network, 20 images were used: 10 images of common nevus and 10 images of melanoma. In case of Neural Network with three targets in the output layer we added to the training process 49 images of Atypical nevus. All the images were extracted from the International Skin Imaging Collaboration and PH₂ databases [12].

We made several attempts until we achieved a good result of classification. We designed a neural network with 70 hidden layers and 2 targets in output layer (Fig. 2). This combination had the best result in classifying the skin lesion in melanoma or common nevus.

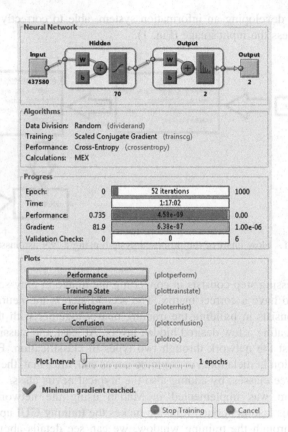

Fig. 2. Neural Network training tool.

Due to useful Matlab neural network toolbox we were able to analyze the training process. As it can be seen the Neural Network training tool allow us to see details regarding training algorithms and status including training accuracy. In our case the number of epochs achieved by the training process is 59 and the elapsed time was 1 h 17 min and 02 s. We also were able to analyze the confusion matrix, as the Neural Network training tool allows us to see useful plots. Matlab computes Confusion Matrix for training, validation and test.

The rows are associated to the Output Class or predicted class while the columns are associated with the Target Class. The diagonal of the matrix coincides to the correctly classified images. The incorrectly classified images are represented through the secondary diagonal. In all cells, the percentage of total number of images and number of images are shown. The right column of each matrix shows the percentages of all images belonging to a class that are correctly and incorrectly classified. The red color is for incorrectly classified images while the green color is for the correct classified ones. The row at the bottom of the matrix shows the percentages of all images belonging to a class that are correctly and incorrectly classified. The overall accuracy is represented in the bottom tight.

As we can see from Fig. 3, the training confusion matrix shows us a good training process, having 20 images classified as melanoma and 27 as common nevus. This means that the system hadn't any errors in the training process. The same thing we can tell about the validation confusion matrix, but not the same for testing the training. In this case, one image that supposed to be common nevus, was classified as melanoma. All these results take us think that the neural network with 70 hidden layers could be a good one for skin lesion detection if we would have more images as training data.

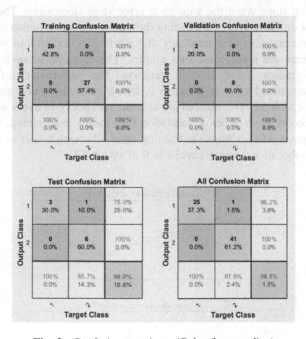

Fig. 3. Confusion matrices. (Color figure online)

Another useful plot is the performance plot, that shows how the network mean square error drops rapidly as it learns. In Fig. 4 we can see that the best validation performance was 0.0019928 at epoch 52.

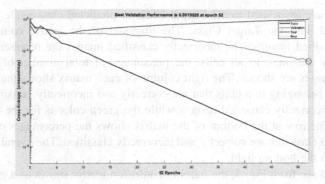

Fig. 4. Training performance. (Color figure online)

The blue line indicates lower error on the training data if the number of epochs increases. The green line shows the validation error. The training curve decreases more than the validation curve and this way, represents that, the performance of the trained network with learning data is better than with the data not involved in the learning process. Training stops when the validation error stops decreasing. The red line represents the error on the data used for testing the training process. In this case we can see that the error increases, so we cannot say that we have a perfect training.

In Fig. 5 there are represented two graphs. The first one is associated with the results of the gradient value in each iteration. The performance of our network increases as the gradient value is closer to 0. In our case, the lowest gradient value is 6.3788e–07 at epoch 52. The number of validations checks in every epoch is represented in the second graph. A lot of validation fails means overtraining, but in our case the results are good. Matlab stops the training after achieving 6 fails in a row. We can see that the number of validation checks is 0 at epoch 52.

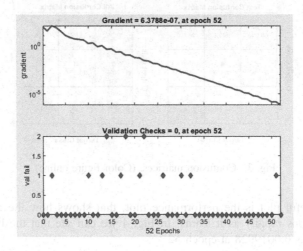

Fig. 5. Neural Network train state plot.

The receiver operating characteristic (ROC) curve, shows us the results of our network simulation after training. The area that is placed under the ROC curve is associated with the measure of validity of a diagnostic test. In Fig. 6, the ROC graphs for, training, validation and testing the classifier are represented. In the last graph, the global ROC of the system is represented.

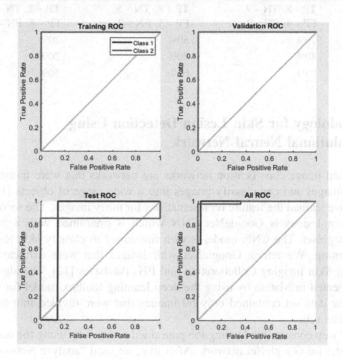

Fig. 6. Neural Network training, testing and validation states ROC plot.

In case of a test without any errors the sensitivity and specificity should be 100%, that means that the graph must show points in the upper-left side, which is the case of training and validation ROC, but not the case of test ROC where we can see that the value of 100% for sensitivity and specificity for class 1 is reached very fast while for the second class it takes a little bit more, but in most part the results are just as good. The NN performances are presented in Table 1 for different cases (TP - True positive: melanoma image correctly identified as melanoma, TN - True negative: common nevus correctly identified as common nevus, FP - False positive: common nevus incorrectly identified as melanoma, and FN - False negative: melanoma incorrectly identified as common nevus).

Table 1. Experimental results.

Performances	Cases		
	Neural Network with 70 layers and 2 targets in the output layer	Neural Network with 30 layers and 2 targets in the output layer	Neural Network with 15 layers and 3 targets in the output layer
Results	TP - 8, TN - 7, FP - 3, FN - 2	TP - 8, TN - 5, FP - 5, FN - 2	TP - 2, TN - 2, FP - 8, FN - 8
Specificity	70%	50%	20%
Sensitivity	80%	80%	20%
Accuracy	75%	65%	50%

4   Methodology for Skin Lesion Detection Using Convolutional Neural Network

The pretrained image classification networks are networks that were trained on a big number of images and can classify images into a wide range of objects [13]. So, the networks have learned the feature representations for many images. The second method to detect skin lesions is GoogleNet CNN which is pretrained with a large image database ImageNet. The CNN model is then fine-tuned to classify skin lesions using transfer learning. We retrain GoogleNet with images that were extracted from the International Skin Imaging Collaboration and PH$_2$ databases [12]. The algorithm was also implemented in Matlab by using the deep learning toolbox model for GoogleNet network. Our data set contained only 69 images that were divided into training and validation data sets.

The first step consisted in loading the pretrained network by using the deep learning toolbox model for GoogleNet network. After that, we used "analyze Network" Matlab function to display a window that showed us the network architecture and some information about the network layers [13]. We saw that the first network layer is the image input layer which requires images of $224 \times 224 \times 3$, size. So, in our algorithm, one requirement consists in resizing the images.

The features extracted by the convolutional layers are used by the last learnable layer and the final classification layer to classify the input image. In GoogleNet, these two last layers, encapsulates the information on how to associate the features into class probabilities, predicted labels and loss value. We replaced the two layers with new layers fitted to our data set, in order to retrain the pretrained network to classify our images. We first, find the names of the two layers to replace by using Matlab function "findLayersToReplace". We replace then the last layer with learnable weights, with a new fully connected layer with two outputs, because two is the desired number of classes [13].

Next step we set the learning rates of earlier layers to zero, to "freeze" the weights of these layers. This is done because we don't want the parameters of the frozen layers to be updated. If we let them to be updated, the network training will be time consuming. In our case, the data set is small and freezing earlier layers can prevent the overfitting. The final step consists in validation images classification using the fine-tuned network, and calculation of the classification accuracy.

The flow chart of the proposed system is presented in Fig. 7.

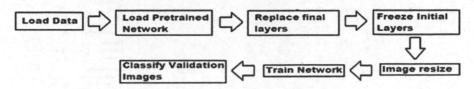

Fig. 7. Flow chart of the proposed system.

5 Experimental Results for Skin Lesion Detection

In the experiments, there were used 69 images for the Convolutional Neural Network with two targets in the output layer: 43 images of common nevus and 26 images of melanoma. Approximately 70% of the images were used for training and 30% for validation. In case of Neural Network with three targets in the output layer we added 49 images of Atypical nevus. All the images were extracted from the International Skin Imaging Collaboration and PH₂ databases [12].

The results obtained in case of CNN with two targets can be seen in Fig. 8(a, b, c, and d), Figs. 9 and 11a. In case of CNN with three targets the results can be seen in Fig. 8(e, f, g, and h), Figs. 10 and 11b.

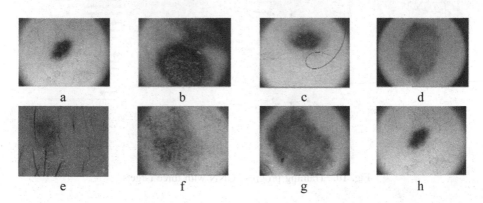

Fig. 8. Validation images with predicted labels and predicted probabilities (a) training "Common Nevus" 99.9%, (b) training "Melanom" 99.8%, (c) training "Common Nevus" 99.8%, (d) training "Common Nevus" 100%. Validation images with predicted labels and predicted probabilities with three targets: probabilities (e) training "Melanom" 48%, (f) training "Atypical Nevus" 57.7%, (g) training "Common Nevus" 35.5%, (h) training "Atypical Nevus" 57%.

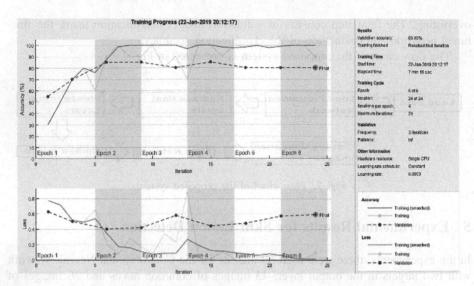

Fig. 9. Training progress CNN with two targets.

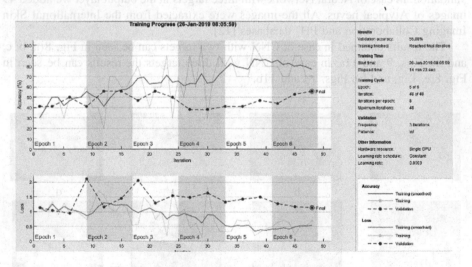

Fig. 10. Training progress CNN with three targets.

So, in Fig. 8 we displayed four sample validation images with predicted labels and the predicted probabilities of the images having those labels.

In the two targets case, the obtained validation accuracy was 80% and the elapsed time was 7 min and 15 s. The number of epochs was 6. In the three targets case we obtained a validation accuracy of 55.88% and an elapsed time of 14 min and 23 s.

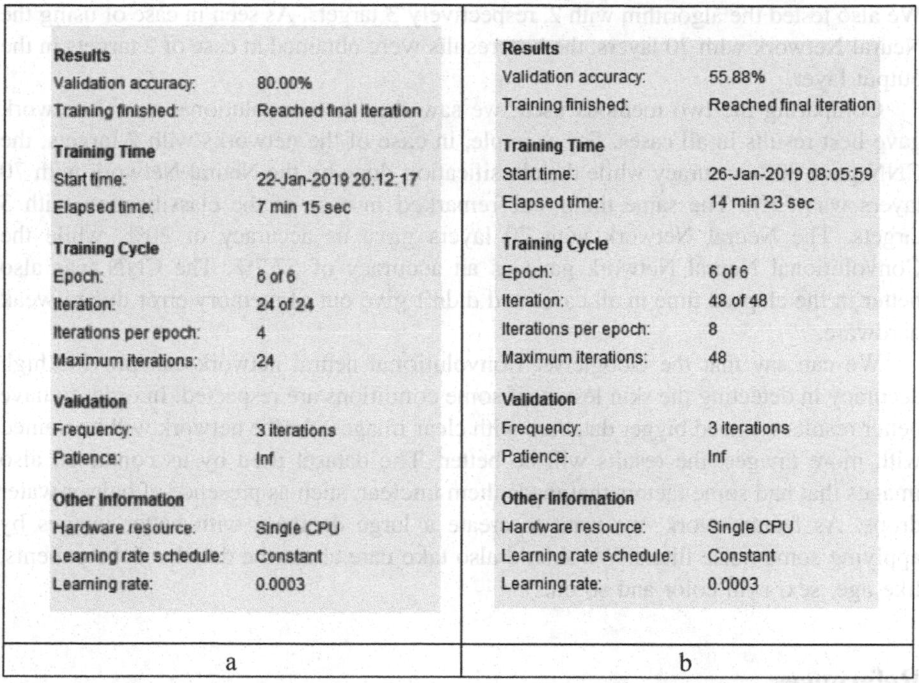

Results		Results	
Validation accuracy:	80.00%	Validation accuracy:	55.88%
Training finished:	Reached final iteration	Training finished:	Reached final iteration
Training Time		**Training Time**	
Start time:	22-Jan-2019 20:12:17	Start time:	26-Jan-2019 08:05:59
Elapsed time:	7 min 15 sec	Elapsed time:	14 min 23 sec
Training Cycle		**Training Cycle**	
Epoch:	6 of 6	Epoch:	6 of 6
Iteration:	24 of 24	Iteration:	48 of 48
Iterations per epoch:	4	Iterations per epoch:	8
Maximum iterations:	24	Maximum iterations:	48
Validation		**Validation**	
Frequency:	3 iterations	Frequency:	3 iterations
Patience:	Inf	Patience:	Inf
Other Information		**Other Information**	
Hardware resource:	Single CPU	Hardware resource:	Single CPU
Learning rate schedule:	Constant	Learning rate schedule:	Constant
Learning rate:	0.0003	Learning rate:	0.0003
a		b	

Fig. 11. Characteristics of the training process: (a) CNN with two targets, (b) CNN with three targets.

6   Conclusions and Discussions

In this paper, we presented the reliability of use of Neural Network and convolutional Neural Network for detecting skin lesions.

First, we created a neural network using Matlab functions and toolboxes in order to observe the performance, accuracy, specificity and sensitivity of the Neural network. In order to train and test our system, we used images from PH_2 dataset, which is a free database with 200 dermatoscopic images.

The best result was obtained by using a neural network having 70 hidden layers and two targets in the output layer. We also tested the effectiveness of the neural network by inserting 3 outputs to the system. We observed that the result is much weaker than the result given by the system when using 2 targets in the output layer. Due to our weak hardware components of our pc, we couldn't test a big number of images in neural network with a big number of layers. For example, for the neural network with 2 outputs and 75 layers, the out of memory error appeared.

Second, we detected skin lesions using GoogleNet convolutional neural network which was pretrained with the large image database ImageNet. The Convolutional Neural Network was fine-tuned in order to classify skin lesions by using transfer learning. We retrained GoogleNet with the same images that we used in the first case.

We also tested the algorithm with 2, respectively 3 targets. As seen in case of using the Neural Network with 70 layers, the best results were obtained in case of 2 targets in the output layer.

Comparing the two methods used, we saw that the convolutional neural network gave best results in all cases. For example, in case of the networks with 2 targets, the CNN gave 80% accuracy while the classification done by the Neural Network with 70 layers was 75%. The same thing was remarked in case of the classification with 3 targets. The Neural Network with 70 layers gave us accuracy of 20%, while the Convolutional Neural Network gave us an accuracy of 57,7%. The CNN was also better in the elapsed time in all cases and didn't give out of memory error due to weak hardware.

We can say that the GoogleNet Convolutional neural network can provide high accuracy in detecting the skin lesions if some conditions are respected. In order to have better results we need bigger database with clear images. As the network will be trained with more images, the results will be better. The dataset used by us contained also images that had some factors that made them unclear, such as presence of hair or water drops. As feature work, we want to create a large database with better images by applying some noise filters. We should also take care about the details of the patients, like age, sex, skin color and so on.

References

1. Brinker, T.J., et al.: Skin cancer classification using convolutional neural networks: systematic. Rev. J. Med. Internet Res. **20**(10), e11936 (2018)
2. Argenziano, G., Soyer, H,P.: Dermoscopy of pigmented skin lesions: a valuable tool for early diagnosis of melanoma. Lancet Oncol. **2**(7), 443–449 (2001)
3. Ara, A., Deserno, T.M.: A systematic review of automated melanoma detection in dermatoscopic images and its ground truth data. Proc. SPIE Int. Soc. Opt. Eng. **8318**, 1–6 (2012)
4. Fabbrocini, G., et al.: Teledermatology: from prevention to diagnosis of nonmelanoma and melanoma skin cancer. Int. J. Telemed. Appl. **2011**(17), 125762 (2011)
5. Hart, P.E., Stork, D.G., Duda, R.O.: Pattern Classification, 2nd edn. Wiley, Hoboken (2000)
6. Jianu, S.R.S., Ichim, L., Popescu, D., Chenaru, O.: Advanced processing techniques for detection and classification of skin lesions. In: 22nd International Conference on System Theory, Control and Computing (ICSTCC), pp. 498–503. Sinaia, Romania, 10–12 October 2018
7. Oliveira, R.B., Papa, J.P., Pereira, A.S., Tavares, J.M.R.S.: Computational methods for pigmented skin lesion classification in images: review and future trends. Neural Comput. Appl. **29**(3), 613–636 (2016)
8. LeCun, Y., Bengio, Y., Hinton, G.: Deep learning. Nature **521**(7553), 436–444 (2015)
9. Pomponiu, V., Nejati, H., Cheung, N.M.: Deepmole: deep neural networks for skin mole lesion classification. In: Proceedings of the IEEE International Conference on Image Processing (ICIP), Phoenix, AZ, 25–28 September 2016
10. Codella, N., Cai, J., Abedini, M., Garnavi, R., Halpern, A., Smith, J.R.: Deep learning, sparse coding, and SVM for melanoma recognition in dermoscopy images. In: 6th International Workshop on Machine Learning in Medical Imaging, pp. 118–126, Munich, Germany, 5–9 October 2015

11. Esteva, A., et al.: Dermatologist-level classification of skin cancer with deep neural networks. Nature **542**(7639), 115–118 (2017)
12. Gutman, D., et al.: Skin lesion analysis toward melanoma detection: a challenge. In: The International Symposium on Biomedical Imaging (ISBI) 2016, Hosted by the International Skin Imaging Collaboration (ISIC) (2016). arXiv, 04 May 2016
13. https://www.mathworks.com/help/deeplearning/examples/train-deep-learning-network-to-classify-new-images.html. Accessed 20 Jan 2019

A Deep Learning Approach to Anomaly Detection in the Gaia Space Mission Data

Alessandro Druetto[1], Marco Roberti[1], Rossella Cancelliere[1(✉)],
Davide Cavagnino[1], and Mario Gai[2]

[1] Computer Science Department, University of Turin,
Via Pessinetto 12, 10149 Torino, Italy
rossella.cancelliere@unito.it
[2] National Institute for Astrophysics, Astrophysical Observatory of Turin,
V. Osservatorio 20, 10025 Pino Torinese, Italy

Abstract. The data reduction system of the Gaia space mission generates a large amount of intermediate data and plots for diagnostics, beyond practical possibility of full human evaluation. We investigate the feasibility of adoption of deep learning tools for automatic detection of data anomalies, focusing on convolutional neural networks and comparing with a multilayer perceptron. The results evidence very good accuracy ($\sim$99.7%) in the classification of the selected anomalies.

Keywords: Deep learning · Astronomical data · Diagnostics · Big data

1 Introduction

In recent years supervised learning's popularity increased dramatically, focusing on deep learning algorithms that are able to exploit large datasets [18]: in particular, deep learning has shown outstanding performance in spatio-temporal sequences processing (such as text or speech) and image processing. This led to two different – but equally successful – deep architectures: recurrent neural networks (RNNs, [29]) for tasks characterized by the presence of time sequences, and convolutional neural networks (CNNs) for imaging.

CNNs were firstly introduced in 1989 to recognize handwritten ZIP codes [23] and then used over the next ten years even if the relatively small datasets at the time were not suitable to proper training of CNNs with a huge number of parameters.

Only the advent of modern, much larger datasets makes the training of very deep CNNs effective: the breakthrough came in 2012, when Krizhevsky et al. [22] achieved the highest classification accuracy in the ILSVRC 2012 competition, using a CNN trained on the images of ImageNet dataset.

This radical shift would not have been possible without two factors that have been a real booster for its success, albeit not directly related to the field: on the

© Springer Nature Switzerland AG 2019
I. Rojas et al. (Eds.): IWANN 2019, LNCS 11507, pp. 390–401, 2019.
https://doi.org/10.1007/978-3-030-20518-8_33

one hand, the already mentioned impressive growth of data availability, mostly due to the development and the widespread diffusion of the internet technology; on the other hand, a computing capability hardly conceivable before, that takes advantage of recent years' parallelisation trend thanks to GPU technologies. These ideas and practices are now recognised as the foundation for modern CNNs. Since then, CNNs have been successfully applied, inter alia, in the recognition and classification of various items, such as hand-written digits (as in the MNIST [24] dataset), traffic signs [30], and more recently the 1000-category ImageNet dataset [22]; other important applications include object detection, image segmentation and motion detection.

CNNs have also recently found increasing usage in astrophysical applications, for better exploitation and inter-calibration of the large datasets produced from modern sky surveys' information. Some relevant examples include the development of CNNs for:

- derivation of fundamental stellar parameters (i.e. effective temperature, surface gravity and metallicity) [21];
- studies of galaxy morphology [32];
- high-resolution spectroscopic analysis using APO Galactic Evolution Experiment data [25];
- determination of positions and sizes of craters from Lunar digital elevation maps [31].

Also, in [34] ExoGAN (Exoplanet Generative Adversarial Network) is presented, a new deep-learning algorithm able to recognize molecular features, atmospheric trace-gas abundances, and planetary parameters using unsupervised learning.

In this paper we investigate the use of CNNs in the framework of the Gaia mission [28] of the European Space Agency (ESA), which will provide an all-sky catalogue of position, proper motion and parallax of about 1.7 billion objects among Milky Way stars and bright galaxies.

Our research concerns initial exploration of deep learning tools for data mining on the huge set of as yet unexploited Gaia plots, with the goal of improving on the identification of transients and peculiar operating conditions.

The current preliminary study is focused on two specific areas: (i) identification of runaway conditions on the Gaia plots (with parameters drifting beyond appropriate limiting values), and (ii) identification of one or more missing data in the plots.

In Sect. 2 we recall the main features of the Gaia mission and of the data used in this work; Sect. 3 includes a description of the deep architecture we use and of the experimental framework; we also analyse and evaluate the achieved results. Finally, in Sect. 4 we draw our conclusions, also outlining options for future work.

2   An Overview of Gaia

The Gaia mission will observe every object in the sky brighter than its limiting magnitude $V = 20$ mag, with unprecedented astrometric accuracy [16,26] and astrophysical potential [12,14,17]. The current version of the output catalogue, based on the first half of the mission, is the Data Release 2 [15] (described also in https://cosmos.esa.int/web/gaia/dr2), available e.g. through the user interface https://gea.esac.esa.int/archive/, and it is already widely used by the astronomical community. The catalogue is materialised in the astrometric parameters (position, proper motion and parallax) of the sample of observed objects in DR2, mainly stars in our Galaxy. Gaia was launched on 2013, Dec. 19 from the French Guyana space center, and it reached its operating site (the L2 Lagrange point) about two weeks later. The five year mission lifetime has recently been extended by one plus one years.

The Gaia concept [28] relies on simultaneous observation of two fields of view, by two nominally equal telescopes, separated by a large basic angle ($\sim$106°), repeatedly covering the full sky by the combination of orbital revolution, satellite spin and precession. The Gaia focal plane has 7×9 large format astrometric Charge Coupled Device (CCD) sensors, complemented by specialised CCDs for detection of object as they enter the field, and sensors for photometric and spectroscopic measurements, for a grand total of about 1 Gpixel.

The data reduction scheme [26] derives, by self-consistency of the set of measurements, the kynematic information on celestial objects throughout the mission lifetime, factoring out the instrument parameters and their evolution by calibration.

The Gaia data reduction is managed by the Data Processing and Analysis Consortium (DPAC), including more than 450 European scientists. The DPAC is composed of Coordination Units (CUs), each in charge of specific parts of the overall reduction chain. Initial processing (Initial Data Treatment, IDT) performs preliminary estimate of several parameters (star position and magnitude, and initial attitude), which are then fed to different computing chains, all accessing the main database which includes raw and intermediate data, as well as the final results. Processing are split in many layers, operating on different data amounts and complexity, from daily operations (which must keep up with the continuous data inflow), up to the six-month cycle related to full sphere solution. Reduction software updates, required to account for unforeseen aspects of operating conditions, and for a progressively improving understanding of the instrument response, are also synchronised to the six month cycle.

The CU3, in particular, takes care of the so-called Core Processing, operating on the unpacked, filtered and pre-processed data of a large subset of well-behaved stars, and reconstructing their astrometric parameters, which are progressively improved for each object as the measurements pile up, by means of increasingly accurate (and more computer intensive) algorithms provided by the DPAC scientists.

The iterative astrometric core solution provides the calibration data and attitude reconstruction needed for all the other treatments, in addition to the astrometric solution of several million primary sources and the overall reference system of coordinates.

The whole Gaia database, including raw, intermediate and final data for the nominal five year mission, is estimated to exceed 1 Petabyte.

The processing chain includes a number of diagnostics functions, implemented in each unit at different levels, which have been used to monitor the mission behaviour in normal periods, the progressive evolution of many instrument parameters, and the insurgence of critical conditions related either to excessive variation of the payload (e.g. optical transmission degradation by contamination), or to external disturbances (e.g. solar flares), requiring modification of on-board setup. Moderate instrument variations must be taken into account by the data reduction system, by appropriate update of the parameters used, or by further algorithm development.

Diagnostics [9,10] is based on a large number of intermediate quantities, whose monitoring is often automated, but for cases where human assessment is considered as potentially necessary the data reduction system includes automatic generation of several thousand plots on a daily timeframe, which are stored in the database as intermediate data in graphical file format.

During critical periods, many such plots are studied by the payload experts for better understanding of the disturbances in play, and to define corrective measures if needed. Over most of the Gaia lifetime, fortunately, good operating conditions have been experienced so that most of the plots were not further considered. However, it is becoming clear that the payload is in a state of continual evolution, at the targeted precision level, and that the final mission accuracy will benefit of further improvements in the data processing taking into account the instrument response at a more detailed level.

In this paper, we evaluate the feasibility of using CNNs for processing of the huge set of as yet unexploited Gaia plots, with the goal of improving on the identification of transients and peculiar operating conditions.

We focus on two specific areas: identification of **runaway conditions** on the Gaia plots (with parameters drifting beyond appropriate limiting values), and **missing data** in the plots (ignored in the normal processing if it has small duration).

2.1   Selected Input Data

The first feasibility tests have been performed on the family of plots showing the daily statistics of along scan photo-center estimate. Such plots are generated on every day of operation for each of the 7×9 CCDs in the astrometric focal plane, and for each telescope. Since the electro-optical instrument response changes over the field of view, they are similar but appreciably different at the level of precision of Gaia. The data of each 30 min observation segment provide an average value and an error bar, due not only to photon statistics fluctuations, but also to "cosmic scatter", i.e. different characteristics of the many thousand detected celestial sources. Each plot includes therefore 48 points with errors. The abscissa is the mission running time, in satellite revolutions, whereas the vertical axis is in micro-meters with respect to the center of the readout window of each object. One detector pixel is 10 μm; a slip by more than one pixel in either direction of the average photo-center requires re-adjustment of the on-board parameters used for computation of the placement of the read-out windows.

To issue an alert and trigger the corrective action it is therefore necessary to implement an automatic detection of such conditions, so far managed mostly by human supervision. Similar checks, with different thresholds, can be applied to other relevant quantities summarising the quality of the measured data, e.g. the statistics of image moments like root mean square width, skewness and kurtosis.

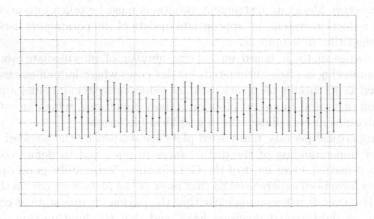

Fig. 1. A sample Gaia plot.

The original plots, generated for human evaluation, have format 1500×927 pixels, therefore imposing a large computational load due to their size. However, human readability requires a large amount of white space between points, which are placed in fixed positions (every half hour, over one day). Besides, the vertical labels are the same within each family of plots, and the horizontal labels (corresponding to the running time, a known quantity) are different but irrelevant to the test goals. Therefore, we decided to alleviate the data size, and correspondingly the computational load, by implementing an automated procedure which "squeezes" the initial plots by cutting the image strips associated to the labels, and removing most of the white space between useful data points. An example is shown in Fig. 1. We applied the same pre-processing also to a large number of plots expressly generated for simulating additional anomalous data; this is necessary for proper supervised CNN training, because "unfortunately" Gaia works in good operating conditions for most of the time.

Furthermore, in order to ease the detection of runaway conditions and missing data, we generated a new set of **difference plots** by subtracting from each of them the reference zero-offset case. This operation also removes the grid and axes of the original plots. The resulting images have much smaller format (128×128), but retain the initial information, providing a compression which is not strictly

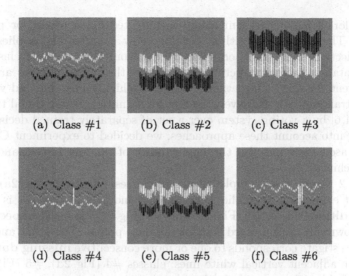

(a) Class #1   (b) Class #2   (c) Class #3

(d) Class #4   (e) Class #5   (f) Class #6

Fig. 2. Examples for the six classes of input images.

required by the deep learning tools used for subsequent computation, but follows good general practices of economy.

These final images, shown in Fig. 2, are the inputs to the CNN models used for our diagnostic task.

3   Experimentation and Results

In this section we present in detail the neural architectures used during the experimentation and the obtained results.

3.1   Our Proposed Model

Our principal aim is to classify GAIA images with respect to the presence or absence of certain kind of anomalies. In the task of supervised classification, there exist a lot of successful approaches: two widespread examples are support vector machines and random forests.

A support vector machine [13], a binary classifier in its standard formulation, builds a special kind of separation rule, called a linear classifier, with theoretical guarantees of good predictive performance. Statistical learning theory [33] gives theoretical basis for this family of methods. To work even with non-linear data, the so-called kernel trick can be used to construct special kinds of non-linear rules. Also, many approaches exist to build a non-binary classifier system from a set of binary classifiers (one-vs-all, one-vs-one, error correcting output codes (ECOC) [11], Directed Acyclic Graph (DAG) [27]). In all of these approaches we combine prediction results from multiple previously trained binary classifiers.

A random forest [5] is a machine learning technique useful for prediction problems. The seminal algorithm, developed by Breiman [8], applies random feature selection [2, 19] and bootstrap aggregation [7] to a set of classification trees to obtain a better prediction. It is known that decision trees are not the most efficient classification method, as they are highly unstable and very likely to overfit training data. However, random forest mitigates individual trees overfitting [3, 4, 6] by a voting system over a set of separately trained decision trees.

Taking into account these approaches, we decided to experiment CNNs over GAIA dataset, verifying also the applicability of random forests and support vector machines.

In Fig. 2 we can see examples of all six classes. Class #1 (Fig. 2a) contains images not evidencing anomalies. One typical anomaly in the plot is the **runaway condition**: classes #2 (Fig. 2b) and #3 (Fig. 2c) represent respectively the cases of downward and upward shift of the data points. The other anomaly we want to investigate corresponds to one or more consecutive **missing data** points, resulting in adjacent vertical white lines: classes #4 (Fig. 2d), #5 (Fig. 2e) and #6 (Fig. 2f) identify one, two or three consecutive missing data points respectively. Wider gaps, corresponding to more relevant on-board failures, are detected by other subsystems triggering suitable corrective actions.

The difference plots are composed by vertical strokes in greyscale; those are comparable to handwritten strokes produced when someone draws or writes something. This fact gives us a hint to analyse our data with an architecture similar to the one used as state-of-art for the MNIST dataset of handwritten digits, i.e. the CNN with 2 convolutional blocks (*double-convolution CNN*) described in the following.

Recalling briefly how a CNN is usually built, its principal constituents are a set of convolutional blocks, followed by a set of fully connected layers. A convolutional block is composed by some layers of convolutional filters, followed by an activation function and a pooling.

A convolution acts as a filter that multiplies each pixel in the $N \times N$ subregion for the corresponding value, summing up all values to get a single scalar.

The previously computed weighted sum is then fed to an activation function, usually the Rectified Linear Unit, or ReLU.

After application of both the filter and the activation, a pooling is performed: every square block of size $M \times M$ (typically $M < N$) is represented by its maximum or average value.

The set of fully connected layers behaves as a classical multilayer perceptron; all neurons of the last convolutional layer are connected to one (or more) layers of hidden neurons, and from here to the output layer.

Our proposed double-convolution CNN is shown in Fig. 3.

In the perspective of possible reduction of the computational effort in operation, we also explored a *single-convolution CNN* containing only one convolutional block.

We decided to compare our CNNs also to a simpler network structure, in particular our third model is a *multilayer perceptron* (MLP).

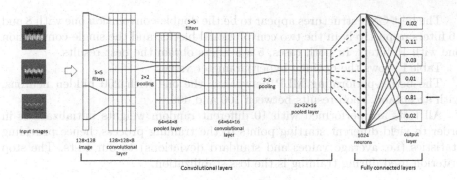

Fig. 3. CNN structure.

For analogy with the previously built CNN, the MLP structure is extracted directly from such network. In particular, we used the final fully connected part of the CNNs. All models share the choice of the ReLU activation function.

For the entire validation and testing processes, we generated a total of 9000 images: 6000 for the training set, 1500 for the validation set and 1500 for the test set. All such images are equally spread among the 6 different classes, thus providing a balanced distribution.

Since the possible classes are 6, the typical representation of labels for each image is a one-hot mono-dimensional array. This array can represent, in fact, a probability distribution and works well with the cross-entropy loss function. We minimize such loss using the ADAM optimizer [20].

The previously introduced validation set guided the tuning of hyperparameters in our models. In particular, we grid-searched for best values over the tunable parameters: convolutional filters size; number of convolutional filters; max-pooling filter size; dropout rate; batch size; and, both in the MLP and in the fully connected layer on the CNNs, number of hidden neurons.

Table 1. Final hyperparameter choice.

Hyperparameter	CNN model	
	Single-convolution	Double-convolution
1st conv. filters size	5×5	5×5
2nd conv. filters size	-	5×5
1st conv. number of filters	16	8
2nd conv. number of filters	-	16
Max-pooling filter size	2×2	2×2
Dropout rate	0.5	0.5
Batch size	50	50
Number of hidden neurons	1024	1024

The best CNN structures appear to be the double-convolution one with 8 and 16 filters respectively in the two convolutional blocks, and the single-convolution one with 16 filters. In both cases, 5 × 5 filters obtain the best results.

Table 1 shows the optimal hyperparameter values.

The better-performing MLP network is the one with 200 hidden neurons, with acceptably good results between 100 and 400.

All tests are performed with 10 different random weights initializations in order to yield different starting points to the training process, hence providing statistics (i.e. average values and standard deviations) to our tests. The stop criterion used during training is the loss stabilization.

3.2   Experimental Results

All the experiments are run using a software framework based on TensorFlow [1] and written in Python 3.6.

The three models, selected during validation procedure, are tested on a GeForce GTX 1070 with 1920 CUDA cores.

Table 2. Test results: statistics over 10 different random weights initializations.

	Accuracy		Time (s)	
	Avg	Std	Avg	Std
Single-convolution CNN	0.9967	0.0045	158.64	0.12
Double-convolution CNN	0.9965	0.0013	127.56	0.07
MLP	0.9655	0.0110	18.54	0.05

The results achieved on the test set are summarised in Table 2. We remark that both CNNs achieve almost the same accuracy, over 99%; however, the double-convolution CNN seems more stable, since the standard deviation is smaller. MLP reaches an accuracy ~3% lower than CNNs: this might suggest that a non-convolutional network is less able to evince structure and discrepancies between classes than a convolutional one. Besides, MLP standard deviation is more than two times higher than CNN's worst case. We also remark that its training time is smaller, as expected.

Notwithstanding the excellent performance achieved by the CNN approach, for the sake of completeness we also explored the aforementioned methods, i.e. random forest and support vector machine. The former reached a test accuracy of 0.9387 with 4096 trees; the latter, instead, provided a test accuracy of 0.5093 with RBF kernel and $C = 1$. We suppose that the poor performance of support vector machine is due to the extremely high dimensionality of input data, since each image pixel is an independent dimension.

In order to assess the statistical significance of our results, we decided to perform a Student's t-test. It provides evidence, at a significance level of 99.9%, that the MLP performance is worse than either CNN architectures. Similarly, the performance of single-convolution CNN and double-convolution CNN is equal, at the same significance level.

An interesting fact is that the CNN training time is smaller in the double-convolution case. This behavior is due to the presence, in this case, of a second pooling layer further reducing the input size of the fully connected layer.

4 Conclusions

We deal with the issue of detection and classification of anomalous data on automatically generated images from the intermediate processing in the data reduction system of the Gaia space mission.

We investigate the application of convolutional neural networks, evidencing very good classification accuracy and quite acceptable training time.

Single- and double-convolution CNNs have comparable performance, with better stability and shorter training time in the latter case. However, the former, lighter architecture would result in faster runtime on operation. MLP still provides good classification performance, but significantly lower than either CNN (in spite of faster training and running time). Random forests and support vector machines achieve, respectively, acceptable and poor results.

The results are promising with respect to possible adoption of CNNs and deep learning tools in the Gaia data reduction system. Further investigation may be devoted to increasing the range of target anomalies, thus refining the diagnostic class definition.

Acknowledgements. We acknowledge the contribution of sample plots and discussion on the requirements from D. Busonero and E. Licata (INAF-OATo). The activity has been partially funded by the Italian Space Agency (ASI) under contracts Gaia Mission, The Italian Participation to DPAC, 2014-025-R.1.2015 and 2018-24-HH.0.

References

1. Abadi, M., et al.: TensorFlow: a system for large-scale machine learning. In: Keeton, K., Roscoe, T. (eds.) 12th USENIX Symposium on Operating Systems Design and Implementation, OSDI 2016, Savannah, GA, USA, 2–4 November 2016, pp. 265–283. USENIX Association (2016)
2. Amit, Y., Geman, D.: Shape quantization and recognition with randomized trees. Neural Comput. 9(7), 1545–1588 (1997)
3. Banfield, R.E., Hall, L.O., Bowyer, K.W., Bhadoria, D., Kegelmeyer, W.P., Eschrich, S.: A comparison of ensemble creation techniques. In: Roli, F., Kittler, J., Windeatt, T. (eds.) MCS 2004. LNCS, vol. 3077, pp. 223–232. Springer, Heidelberg (2004). https://doi.org/10.1007/978-3-540-25966-4_22
4. Bernard, S., Heutte, L., Adam, S.: On the selection of decision trees in Random Forests. In: 2009 International Joint Conference on Neural Networks, pp. 302–307 (2009)
5. Bharathidason, S.: Improving classification accuracy based on random forest model with uncorrelated high performing trees (2014)

6. Boinee, P., Angelis, R.D., Foresti, G.L.: Ensembling classifiers – an application to image data classification from cherenkov telescope experiment (2005)
7. Breiman, L.: Heuristics of instability and stabilization in model selection. Ann. Stat. **24**(6), 2350–2383 (1996)
8. Breiman, L.: Random forests. Mach. Learn. **45**(1), 5–32 (2001)
9. Busonero, D., Lattanzi, M., Gai, M., Licata, E., Messineo, R.: Running AIM: initial data treatment and μ-arcsec level calibration procedures for Gaia within the astrometric verification unit. In: Modeling, Systems Engineering, and Project Management for Astronomy VI, p. 91500K (2014)
10. Busonero, D., Licata, E., Gai, M.: Astrometric instrument model software tool for Gaia real-time instrument health monitoring and diagnostic. Revista Mexicana de Astronomía y Astrofísica **45**, 39–42 (2014)
11. Dietterich, T.G., Bakiri, G.: Solving multiclass learning problems via error- correcting output codes. CoRR cs.AI/9501101 (1995)
12. Evans, D., et al.: Gaia data release 2-photometric content and validation. Astron. Astrophys. **616**, A4 (2018)
13. Fradkin, D., Muchnik, I.: Support Vector Machines for Classification. DIMACS Series in Discrete Mathematics and Theoretical Computer Science (2006)
14. Gaia Collaboration, Babusiaux, C., et al.: Gaia data release 2. Observational Hertzsprung-Russell diagrams. Astron. Astrophys. **616**, A10 (2018)
15. Gaia Collaboration, Brown, A.G.A., et al.: Gaia data release 2. Summary of the contents and survey properties. Astron. Astrophys. **616**, A1 (2018)
16. Gaia Collaboration, Mignard, F., et al.: Gaia data release 2. The celestial reference frame (Gaia-CRF2). Astron. Astrophys. **616**, A14 (2018)
17. Gaia Collaboration, Spoto, F., et al.: Gaia data release 2. Observations of solar system objects. Astron. Astrophys. **616**, A13 (2018)
18. Goodfellow, I.J., Bengio, Y., Courville, A.C.: Deep Learning. MIT Press, Cambridge (2016)
19. Ho, T.K.: The random subspace method for constructing decision forests. IEEE Trans. Pattern Anal. Mach. Intell. **20**(8), 832–844 (1998)
20. Kingma, D.P., Ba, J.: Adam: a method for stochastic optimization. CoRR abs/1412.6980 (2014)
21. Kou, R., Petit, P., Paletou, F., Kulenthirarajah, L., Glorian, J.-M.: Deep learning determination of stellar atmospheric fundamental parameters. In: Proceedings of the Annual meeting of the French Society of Astronomy and Astrophysics, SF2A-2018, pp. 167–169 (2018)
22. Krizhevsky, A., Sutskever, I., Hinton, G.E.: ImageNet classification with deep convolutional neural networks. In: Proceedings of the 25th International Conference on Neural Information Processing Systems - Volume 1, NIPS 2012, Lake Tahoe, Nevada, pp. 1097–1105. Curran Associates Inc. (2012)
23. LeCun, Y., et al.: Backpropagation applied to handwritten zip code recognition. Neural Comput. **1**(4), 541–551 (1989)
24. Lecun, Y., Bottou, L., Bengio, Y., Haffner, P.: Gradient-based learning applied to document recognition. Proc. IEEE **86**(11), 2278–2324 (1998)
25. Leung, H.W., Bovy, J.: Deep learning of multi-element abundances from high resolution spectroscopic data. Mon. Not. R. Astron. Soc. **483**, 3255–3277 (2019)
26. Lindegren, L., et al.: Gaia data release 2. The astrometric solution. Astron. Astrophys. **616**, A2 (2018)
27. Platt, J.C., Cristianini, N., Shawe-Taylor, J.: Large margin DAGs for multiclass classification (2000)

28. Prusti, T., et al.: The Gaia mission. Astron. Astrophys. **595**, A1 (2016)
29. Rumelhart, D.E., Hinton, G.E., Williams, R.J.: Learning representations by back-propagating errors. Nature **323**(6088), 533–536 (1986)
30. Sermanet, P., LeCun, Y.: Traffic sign recognition with multi-scale Convolutional Networks. In: The 2011 International Joint Conference on Neural Networks, pp. 2809–2813 (2011)
31. Silburt, A., et al.: Lunar crater identification via deep learning. Icarus **317**, 27–38 (2019)
32. Tuccillo, D., Huertas-Company, M., Decencière, E., Velasco-Forero, S., Domínguez Sánchez, H., Dimauro, P.: Deep learning for galaxy surface brightness profile fitting. Mon. Not. R. Astron. Soc. **475**, 894–909 (2018)
33. Vapnik, V.N.: The Nature of Statistical Learning Theory. Springer, Heidelberg (1995). https://doi.org/10.1007/978-1-4757-2440-0
34. Zingales, T., Waldmann, I.P.: ExoGAN: retrieving exoplanetary atmospheres using deep convolutional generative adversarial networks. Astron. J. **156**, 268 (2018)

On Possibilities of Human Head Detection for Person Flow Monitoring System

Petr Dolezel[1]([⊠])(iD), Dominik Stursa[1](iD), and Pavel Skrabanek[2](iD)

[1] Faculty of Electrical Engineering and Informatics, University of Pardubice,
Pardubice, Czech Republic
petr.dolezel@upce.cz
[2] Institute of Automation and Computer Science, Brno University of Technology,
Brno, Czech Republic
skrabanek@fme.vutbr.cz

Abstract. Along with the development of human society, economy, industry and engineering, as well as with growing population in the world's biggest cities, various approaches to person detection have become the subject of great interest. One approach to developing a person detection system is proposed in this paper. A high-angle video sequence is considered as the input to the system. Then, three classification algorithms are considered: support vector machines, pattern recognition neural networks and convolutional neural networks. The results showed very little difference between the classifiers, with the overall accuracy more than 95% over a testing set.

Keywords: Person flow monitoring · Support vector machines ·
Pattern recognition neural network · Convolutional neural network ·
Histograms of oriented gradients

1 Introduction

Along with development of human society, economy, industry and engineering, as well as with growing population in the world's biggest cities, various approaches to person detection have become the subject of great interest. Monitoring of person flow, as a branch of the person detection problem, has an indispensable importance in the public transport system safety surveillance, passenger flow prediction, transport planning, or transport vehicle load monitoring. Apparently, the possibility of a precise person flow detection has a great positive effect on public transport systems, station control and management, and cost optimization.

The work was supported from ERDF/ESF "Cooperation in Applied Research between the University of Pardubice and companies, in the Field of Positioning, Detection and Simulation Technology for Transport Systems (PosiTrans)" (No. CZ.02.1.01/0.0/0.0/17_049/0008394).

I. Rojas et al. (Eds.): IWANN 2019, LNCS 11507, pp. 402–413, 2019.
https://doi.org/10.1007/978-3-030-20518-8_34

The monitoring of the person flow problem is constantly getting more and more focus from both academic and corporate experts. Various approaches to person flow detection are based on infra-red sensors [1], radar sensors [7], lasers [6] or 3D laser scanners [2]. However, these approaches often suffer problems of counting every object that passes through, and also are not able to track the objects precisely. So, in present days, person flow monitoring systems are often implemented using video processing algorithms and computer vision techniques. Generally, several ways of image and video processing can be considered as tools for person flow detection. In particular, statistical methods based on model learning, shape feature, skin color feature or area estimation are widely used [27].

Focusing on shape feature evaluation, human appearance, pose, orientation, and movement are typically considered as inputs for further processing [19]. However, if the monitoring is going to be used in public areas, it is appropriate to avoid collecting data that will enable a specific identification of the person (especially faces). Thus, a high-angle video acquisition tends to be a natural solution of the mentioned difficulty - see Fig. 1.

Fig. 1. High-angle shot, persons cannot be identified.

When dealing with a person flow problem using a high-angle video acquisition, only few approaches have been proposed. Gao et al. [10] provide a technique combining convolutional neural networks and cascade Adaboost methods. In [9], the authors use a depth camera along with a classical RGB camera. Both articles provide a method for head and shoulder detection, they do not consider a strict high-angle video acquisition, though. Still, a head itself can provide a strong feature due to its almost circular shape. Then, the Hough transform can be applied to human head detection for getting the flow monitoring result [21]. However, authors of this contribution propose another approach for feature extraction - histograms of oriented gradients. In the previous research, we dealt with a very specific problem of white wine grape detection and counting using visual data. Although a totally different problem, the grape shape is similar to a head. And in the research summarized in [23, 24], histograms of oriented gradients have proven to be an optimal tool for feature extraction.

Therefore, an approach for a person's head detection is derived and comprehensively tested in this paper. This approach is intended to be used as a key part of a person flow monitoring system, which is going to be implemented in various means of transport for passenger counting. The paper is structured as follows. The problem is properly formulated in next section. Then, the used methods are described and the dataset acquisition is illustrated. The experiments are presented as the subsequent section and the paper is finished with the conclusions.

2   Problem Formulation

The aim of this work is to develop a person detector in real-life RGB images. The images are supposed to be derived from a video sequence acquired orthogonally - from above. In the computer vision, the detection process is usually compounded of four steps. During the first step, an object image is acquired from a large real-life image; the second step performs image preprocessing; the third one provides extraction of features; and the final step represents the classification of the object image using the feature vector. In this particular approach, the inputs of the detector are size normalized RGB object images cropped from a real-life video. The outputs are classes of the object images - see Fig. 2 for a basic illustration of the functionality.

Fig. 2. Person detector functionality.

The structure of the detector is based on our previous work [23–25]. Nevertheless, each part of the detector is redesigned in order to fit to the new purpose. The necessary details about all the parts are summarized in the following subsections.

2.1   Image Preprocessing

The image preprocessing consists of two steps. The first step deals with the conversion of an input RGB object image from RGB to the grayscale format according to the ITU-R recommendation BT.601 [12]. The resulting grayscale image is obtained by eliminating the hue and saturation information, while retaining the luminance.

The second step of the preprocessing normalizes the contrast of the grayscale image. Each pixel of the resulting image can acquire values from $[0, 1]$. The output of the image preprocessing is the contrast normalized grayscale image.

2.2 Feature Extraction

As mentioned above, histograms of oriented gradients (HOGs) [8] are considered as a suitable tool for feature extraction. In simple words, HOG feature descriptor provides distribution (histograms) of directions of gradients (i.e. oriented gradients) of the image. Thus, HOGs encode local shape information from regions within an image.

In order to get beneficial information, HOG cell size and a number of bins need to be set properly. HOG cell size represents the sub-frames of the image under examination. The number of bins affects the sensitivity of gradient directions - all the gradient directions in the sub-frame are divided into the particular number of bins. The volume of each bin consequently provides the information about the dominant gradient directions. It is widely recommended to use 9 bins, but there is no explicit recommendation for the HOG cell size - see Fig. 3, where HOG features are extracted using various cell sizes.

Fig. 3. HOG features for 9 bins and cell size $[16, 16]$ px, $[8, 8]$ px and $[6, 6]$ px. Original size of the object image is $[51, 51]$ px. The length of white abscissae is related to the gradients in the image.

2.3 Classification Techniques

The aim of a classification technique is to decide a category of an object captured in an object image. In this contribution, two categories of objects, 'head' and 'not head', are assumed. The class 'head' is called 'positive' and the class 'not head' is called 'negative'.

Based on previous authors' experience, several approaches are considered as possible classification algorithms. Support vector machines (SVMs), which traditionally provide good results with HOGs, are suggested as a first possibility [18].

Except SVMs, feedforward multilayer neural networks seem to be a decent choice for classification in combination with HOGs [26]. For pattern recognition in input data, hyperbolic tangent activation functions are recommended to use in hidden layers and softmax activation functions in output layer. See [20] for detailed information. Such a topology of feedforward network is then called the pattern recognition network (PRN).

Convolutional neural networks (CNNs) are selected as the third approach to be tested. With a rapidly growing possibilities of parallel computing, CNNs became a leading methodology for image processing and analyzing [14]. Compared to other approaches, CNNs use relatively little preprocessing due to the usage of convolutional and pooling layers, traditionally implemented as anterior layers. Therefore, convolutional neural networks leverage spatial information of the object images and a separate feature extraction technique is not necessary to be employed. During the last decade or two, many successful architectures of CNNs were introduced. A brief list of the most popular ones is available in [3].

Therefore, a schematic representation of the detector is shown in Fig. 4.

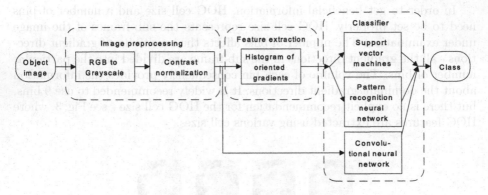

Fig. 4. Flow chart of the person detector.

3   Dataset Creation

The important step for a person detector design is a preparation of appropriate training and evaluation sets. The source data should be apparently acquired within the conditions as close to the real situation as possible. Therefore, in this case, the initial video sequences were acquired in the public places both inside and outside under various light conditions. Then, a number of object images was cropped from those sequences. Eventually, 1562 original object images were acquired with the size normalized to 51 px × 51 px. The data were divided into four subsets according to Table 1. Some examples are illustrated in Fig. 5.

In addition, in order to support the generalization of the detector, the training set was artificially enhanced - each object image was transformed to provide three more descendants using 90, 180 and 270° rotation.

Table 1. Dataset

Training set		Testing set	
Positive	Negative	Positive	Negative
375	406	379	402

Fig. 5. Three positive (left) and three negative examples from training set.

4   Experiments with Classifiers

As mentioned in Sect. 2.3, three approaches are supposed to be tested in this contribution. In addition, each approach provides a number of variants. Thus, the particular conditions of the testing experiments are defined in the following subsections. Note that the conditions are set after a huge set of blind experiments, which are not described here in detail.

4.1   Extraction of Histograms of Oriented Gradients

The following setting of the descriptor has demonstrated to be sufficient. Specifically, a linear gradient direction dividing into 9 bins in 0°–180°; cells of size 8×8 px; blocks of 2×2 cells; and 1 overlapping cell between adjacent blocks in both directions. Therefore, each object image, which consists of 2601 px, provides 900 elements in the feature vector.

4.2   Support Vector Machine Classifier

SVM classifiers provide various results depending especially on the selected kernel. Linear, polynomial or radial basis function (RBF) kernels are implemented in the most of the cases. Beside the applied kernel function, the performance of a SVM classifier is also influenced by a regularization constant C. Performance of a classifier with the RBF kernel is further influenced by a kernel width σ. In order to tune these parameters, a grid search algorithm [4] combined with the 10-fold cross-validation is used.

Therefore, a set of experiments is performed in order to design an optimal SVM classifier. SVM classifiers with a linear kernel, RBF kernel and polynomial kernel with order equal to 2, 3 and 4 are considered. The training set described in Sect. 3 is used. Since the grid search algorithm belongs to a family of stochastic algorithms, the SVM classifier optimization is performed a hundred times for each classifier and the resulting values of a loss function obtained by cross-validated SVM classifier are observed. The loss function is defined as follows.

$$E_{SVM} = \sum_{j=1}^{n} w_j I \{\hat{y}_j \neq y_j\}. \tag{1}$$

Loss function represents the weighted fraction of misclassified observations, where y_j is the class label, $\hat{y}_j$ is the class label corresponding to the class with the maximal posterior probability, w_j is the weight for the observation j, $I\{.\}$ is the indicator function and n is the sample size. In our case, all the weights are equal and $\sum_j w_j = 1$.

The observed values are shown in Fig. 6 for all the selected kernel functions. The central lines in the box graphs, shown in the figure, are medians of loss function resulting values; the edges of the boxes are 25th and 75th percentiles; and the whiskers extend to the most extreme data points (except outliers).

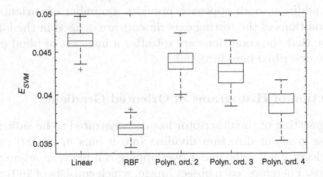

Fig. 6. Resulting values of loss function for SVM classifiers with various kernel functions.

The values pictured in Fig. 6 indicate, that the SVM classifier with the RBF kernel function provides the most suitable behavior. However, the more important quantity would be obtained by the evaluation of the classifiers using the testing set (see Sect. 3). This evaluation is provided in the next section.

4.3   Pattern Recognition Neural Network Classifier

Beside training set acquisition, the procedure of PRNs design also involves training, pruning and testing. The essential information about this procedure is described here. More information about the process can be found e.g. in [11].

While a training of PRNs means to find suitable weights and biases of the network, the pruning converts the net into a simpler one while the performance is kept close to the original one. In our approach, a topology search is performed in the following way: PRNs of various topologies are trained using a scaled conjugate gradient algorithm [17] (random 85% of the training data set - see Sect. 3 - is used for training, 15% for validation) and the performance is observed. Similarly to the previous experiment, PRN training is a stochastic process. Therefore, the experiments are performed a hundred times and the results are statistically evaluated. A loss function for the evaluation is defined using a binary cross entropy function.

$$E_{PRN} = -\frac{1}{n} \sum_{j=1}^{n} [o_j \ln(y_j) + (1 - o_j) \ln(1 - y_j)], \tag{2}$$

where o_j is the desired output, y_j is the actual output of the neural network and n is the number of samples.

The observed resulting values of E_{PRN} are shown in Fig. 7 for various topologies of PRN beginning with two neurons in one hidden layer and ending with two hidden layers, each witch five neurons.

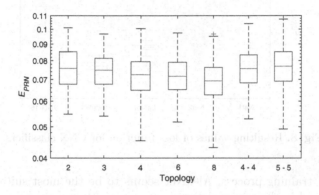

Fig. 7. Resulting values of loss function for PRN classifiers. Labels along the X-axis represent number of neurons in hidden layers.

Looking at Fig. 7, all the topologies provide similar results. Again, the more important tests using testing set are provided in the next section.

4.4   Convolutional Neural Network Classifier

As mentioned above, a number of architectures of CNNs is available for testing, these days. According to some literature research as well as previous authors' experience, five specific architectures are selected for implementation. The first two architectures are relatively simple. Net1 consists of one convolutional layer, one max-pooling and one fully connected dense layer with 512 neurons. The layer with 2 neurons and softmax activation function is implemented as the output layer. Net2 is similar, but it contains a more complex sequence of anterior layers. In particular, it is convolutional layer + convolutional layer + max-pooling layer + convolutional layer + max-pooling layer. Both networks are adapted from [16]. The third one is one of pioneering architectures - LeNet [5], while the fourth one is probably the most cited topology - AlexNet [15]. This architecture was originally designed to win the ImageNet Large-Scale Visual Recognition Challenge, but it spread to a huge number of industrial and engineering applications. The last selected architecture is called VGG-16 net, based on the a large number of simple and repetitive layers, which is, in some cases, effective to implement [22].

Similarly to the previous cases, the mentioned architectures are trained in order to classify correctly the dataset described in Sect. 3. ADAM algorithm is implemented as optimizer [13]. Again, the experiments are performed a hundred times due to a stochastic character of training and a binary cross entropy function is used as loss function - see (2). The resulting values are shown in Fig. 8.

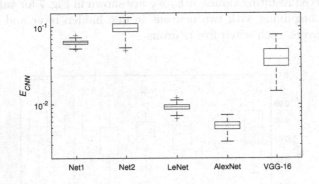

Fig. 8. Resulting values of loss function for CNN classifiers.

After the training process, AlexNet seems to be the most suitable CNN to be implemented in the person detector.

In the next section, all the variants designed here are tested using the testing set. The testing procedure should denote the best possibility among of all.

5   Results and Discussion

The aim of this section is to evaluate all the proposed classifiers. A good practice for the evaluation is to determine the accuracy of the classifiers over the testing set. However, two additional metrics, precision and recall, are proposed to evaluate the classifiers comprehensively. The metrics are described by following equations.

$$\text{Accuracy} = \frac{TP + TN}{TP + FP + TN + FN}, \tag{3}$$

$$\text{Precision} = \frac{TP}{TP + FP}, \tag{4}$$

$$\text{Recall} = \frac{TP}{TP + FN}, \tag{5}$$

where TP (true positive) is the number of correctly classified positive images, FN (false negative) is the number of misclassified positive images, FP (false positive) is the number of misclassified negative images, and TN (true negative) is the number of correctly classified negative images.

The resulting values of the metrics for all the classifiers are summarized in Table 2.

Table 2. Testing results

Classifier	Accuracy	Precision	Recall
SVM with linear kernel function	0.9539	0.9561	0.9485
SVM with RBF kernel function	0.9645	0.9718	0.9545
SVM, polynomial order 2	0.9539	0.9525	0.9525
SVM, polynomial order 3	0.9545	0.9544	0.9518
SVM, polynomial order 4	0.9581	0.9589	0.9545
PRN, [2]	0.9539	0.9567	0.9479
PRN, [3]	0.9529	0.9536	0.9492
PRN, [4]	0.9549	0.9532	0.9538
PRN, [6]	0.9529	0.9653	0.9367
PRN, [8]	0.9542	0.9654	0.9393
PRN, [4 4]	0.9504	0.9474	0.9505
PRN, [5 5]	0.9542	0.9531	0.9525
Net1	0.9129	0.9356	0.8813
Net2	0.9206	0.9249	0.9103
LeNet	0.9501	0.9359	0.9631
AlexNet	0.9040	0.8671	0.9472
VGG_16	0.8886	0.8544	0.9288

The testing results, shown in Table 2, indicate several interesting outcomes. First of all, best accuracy and precision along all the classifiers is provided by the SVM model with a RBF kernel function, which may be a surprising fact, considering the list of classifiers. Then, the tested CNNs provide generally the worst performance. And thirdly, all the performances are very similar, accuracy between 95% and 96%. This feature could indicate, that although the classifiers are generally trained sufficiently, some samples in the testing set can be outside the regular position. Comprehensive check of the results shows, that if a testing sample is misclassified by one classifier, it is misclassified by at least 8 other classifiers in more than 60% of cases.

6 Conclusion

In this contribution, a set of classifiers for person detection from a high-angle image is introduced, designed and tested. According to the results presented above, the image feature extraction using a histogram of oriented gradients in combination with pattern recognition network or support vector machine as a classifier looks like an effective solution for such issues. Apparently, not only the accuracy, but also computation time is necessary to be tuned in order to provide a suitable tool for the monitoring of person flow in real-life conditions. Hence, computational efficiency and classifier implementation using special hardware for parallel processing will be the next subject of interest.

References

1. Ahmed, A., Siddiqui, N.A.: Design and implementation of infra-red based computer controlled monitoring system. In: 2005 Student Conference on Engineering Sciences and Technology, pp. 1–5, August 2005. https://doi.org/10.1109/SCONEST.2005. 4382890
2. Akamatsu, S., Shimaji, N., Tomizawa, T.: Development of a person counting system using a 3D laser scanner. In: 2014 IEEE International Conference on Robotics and Biomimetics (ROBIO 2014), pp. 1983–1988, December 2014. https://doi.org/10. 1109/ROBIO.2014.7090627
3. Aloysius, N., Geetha, M.: A review on deep convolutional neural networks. In: 2017 International Conference on Communication and Signal Processing (ICCSP), pp. 0588–0592, April 2017. https://doi.org/10.1109/ICCSP.2017.8286426
4. Bergstra, J., Bengio, Y.: Random search for hyper-parameter optimization. J. Mach. Learn. Res. 13, 281–305 (2012)
5. Bottou, L., et al.: Comparison of classifier methods: a case study in handwritten digit recognition. In: Proceedings of the 12th IAPR International Conference on Pattern Recognition, Vol. 3 - Conference C: Signal Processing (Cat. No. 94CH3440-5), vol. 2, pp. 77–82, October 1994. https://doi.org/10.1109/ICPR.1994.576879
6. Chen, Z., Yuan, W., Yang, M., Wang, C., Wang, B.: SVM based people counting method in the corridor scene using a single-layer laser scanner. In: 2016 IEEE 19th International Conference on Intelligent Transportation Systems (ITSC), pp. 2632–2637, November 2016
7. Choi, J.W., Quan, X., Cho, S.H.: Bi-directional passing people counting system based on IR-UWB radar sensors. IEEE Internet Things J. 5(2), 512–522 (2018). https://doi.org/10.1109/JIOT.2017.2714181
8. Dalal, N., Triggs, B.: Histograms of oriented gradients for human detection. In: 2005 IEEE Computer Society Conference on Computer Vision and Pattern Recognition (CVPR 2005), vol. 1, pp. 886–893, June 2005. https://doi.org/10.1109/CVPR. 2005.177
9. Fu, H., Ma, H., Xiao, H.: Real-time accurate crowd counting based on RGB-D information. In: 2012 19th IEEE International Conference on Image Processing, pp. 2685–2688, September 2012. https://doi.org/10.1109/ICIP.2012.6467452
10. Gao, C., Li, P., Zhang, Y., Liu, J., Wang, L.: People counting based on head detection combining Adaboost and CNN in crowded surveillance environment. Neurocomputing 208, 108–116 (2016). https://doi.org/10.1016/j.neucom.2016.01.097, http://www.sciencedirect.com/science/article/pii/S0925231216304660. sI: BridgingSemantic
11. Haykin, S.: Neural Networks: A Comprehensive Foundation. Prentice Hall, Upper Saddle River (1999)

12. ITU-R Recommendation BT.601: Studio encoding parameters of digital television for standard 4:3 and wide screen 16:9 aspect ratios, March 2011
13. Kingma, D.P., Ba, J.: Adam: a method for stochastic optimization. CoRR abs/1412.6980 (2014). http://arxiv.org/abs/1412.6980
14. Kizuna, H., Sato, H.: The entering and exiting management system by person specification using deep-CNN. In: 2017 Fifth International Symposium on Computing and Networking (CANDAR), pp. 542–545, November 2017. https://doi.org/10.1109/CANDAR.2017.40
15. Krizhevsky, A., Sutskever, I., Hinton, G.: Imagenet classification with deep convolutional neural networks. In: Advances in Neural Information Processing Systems, vol. 2, pp. 1097–1105 (2012)
16. Millstein, F.: Deep Learning with Keras. CreateSpace Independent Publishing Platform (2018)
17. Moller, M.: A scaled conjugate gradient algorithm for fast supervised learning. Neural Netw. **6**(4), 525–533 (1993)
18. Paisitkriangkrai, S., Shen, C., Zhang, J.: Performance evaluation of local features in human classification and detection. IET Comput. Vis. **2**(4), 236–246 (2008). https://doi.org/10.1049/iet-cvi:20080026
19. Pore, S.D., Momin, B.F.: Bidirectional people counting system in video surveillance. In: 2016 IEEE International Conference on Recent Trends in Electronics, Information Communication Technology (RTEICT), pp. 724–727, May 2016. https://doi.org/10.1109/RTEICT.2016.7807919
20. Resch, C., Pineda, F., Wang, J.J.: Automatic recognition and assignment of missile pieces in clutter. In: 1999 International Joint Conference on Neural Networks, IJCNN 1999, vol. 5, pp. 3177–3181 (1999). https://doi.org/10.1109/IJCNN.1999.836162
21. Shang, H., Wang, T.: Bus passenger counting based on frame difference and improved Hough transform. In: 2012 2nd International Conference on Consumer Electronics, Communications and Networks (CECNet), pp. 3132–3135, April 2012. https://doi.org/10.1109/CECNet.2012.6201616
22. Simonyan, K., Zisserman, A.: Very deep convolutional networks for large-scale image recognition. arXiv:1409.1556, October 2014
23. Skrabanek, P., Dolezel, P.: Robust grape detector based on SVMs and HOG features. Comput. Intell. Neurosci. **2017** (2017). https://doi.org/10.1155/2017/3478602
24. Skrabanek, P., Majerik, F.: Evaluation of performance of grape berry detectors on real-life images, pp. 217–224 (2016)
25. Skrabanek, P., Runarsson, T.P.: Detection of grapes in natural environment using support vector machine classifier. In: Proceedings of the 21st International Conference on Soft Computing MENDEL 2015, 23–25 June 2015, pp. 143–150. Brno University of Technology, Brno (2015)
26. Taskiran, M., Cam, Z.G.: Offline signature identification via HOG features and artificial neural networks. In: 2017 IEEE 15th International Symposium on Applied Machine Intelligence and Informatics (SAMI), pp. 83–86, January 2017. https://doi.org/10.1109/SAMI.2017.7880280
27. Wu, X.: Design of person flow counting and monitoring system based on feature point extraction of optical flow. In: 2014 Fifth International Conference on Intelligent Systems Design and Engineering Applications, pp. 376–380, June 2014. https://doi.org/10.1109/ISDEA.2014.92

Performance of Classifiers on Noisy-Labeled Training Data: An Empirical Study on Handwritten Digit Classification Task

Irfan Ahmad[✉]

Information and Computer Science Department,
King Fahd University of Petroleum and Minerals, Dhahran, Saudi Arabia
irfanics@kfupm.edu.sa

Abstract. Machine learning is an important area of Artificial Intelligence. It has applications in almost all the fields of science. Supervised machine learning, for classification problems, involves training the classifiers with labeled data. There are many classifiers, each having its own strengths and weaknesses in terms of classification accuracy and the ability of dealing with noisy class labels in the training data. There is limited work reported in the literature on investigating the performance of classifiers under different levels of class noise in the training data. The current work aims to presents a thorough investigation on the effects of class mislabeling on the performance of different classifiers. Five commonly used classifiers; SVM, random forest, ANN, naïve Bayes, and KNN were investigated on a benchmark database of handwritten digit images. Classifiers were trained with different levels of labeling noise, ranging from low, to medium, to very high, and their recognition performances were evaluated and compared. The study led to some interesting observations which are presented in this paper.

Keywords: Classifier · Training · Noisy labels · Random forest · SVM · KNN · ANN · Naïve Bayes · Supervised learning

1 Introduction and Related Work

Machine learning is an important part of pattern recognition applications. In the case of supervised machine learning, classifiers are trained with large amounts of labeled training data. In general, the more the training data, the better is the performances of the recognition systems. Both, the quantity and the quality of labeled data is as important as developing powerful classifiers and discriminating features (cf. e.g., [4,19]). To compare different classifiers and techniques, benchmark databases are created and, many a times, made publicly available (e.g., [10,11]).

Gathering and manually annotating data is a tedious and time consuming activity. Researchers have presented techniques to alleviate some of these

© Springer Nature Switzerland AG 2019
I. Rojas et al. (Eds.): IWANN 2019, LNCS 11507, pp. 414–425, 2019.
https://doi.org/10.1007/978-3-030-20518-8_35

problems by performing semi-supervised or unsupervised labeling of data (e.g., [9,13]). Some other researchers have looked into alternate ways of training the classifiers which does not require time consuming preparation and annotation of data (e.g., [1,8]). Most of these approaches rely on iteratively training the system such that a classifier starts with quite a noisy training data and iteratively tries to improve the training annotations using some heuristics and other approaches. For these techniques to work, there is an underlying assumption that the systems are initialized using data having some acceptable level of noise in the annotations. If the degree of noise exceeds these assumptions, the system may not initialize robustly in order to benefit from the iterative improvements.

According to the best knowledge of the author, the existing literature lacks thorough investigations on performance of different classifiers on varying levels of noise in training annotations. The current study is an attempt to explore this issue using handwritten digit recognition as a case study in order to investigate how, many of the popular, machine learning based classifiers perform under different degrees of noise in data annotation. The aim of the study was not to present yet another system for handwritten digit classification. A lot of work has already been done to address this problem with authors reporting very high recognition rates of close to 99% and above on benchmark databases (cf. e.g., [2,18]). The goal of the present work was to investigate how different levels of class-labeling noise can effect the performance of classifiers.

The rest of the paper is organized as follows: In Sect. 2, we present the related work. In Sect. 3, we will present our methodology on how we investigate the classifiers' performances using noisy annotations. The experiments and the results are presented in Sect. 4. Some discussions on the results are presented in Sect. 5. Finally, we present the conclusions and possible future works in Sect. 6.

2   Related Work

Zhu and Wu [22] studied the impact of class noise and attribute noise on the performance of different classifiers. For the class noise, the authors presented the impact of noisy labels on a classification problem involving 4 classes. Moreover, there were only 6 predictor attributes and all of them were categorical with no numerical attributes. The performance of 5 different classifiers were studied; namely C4.5, C4.5 rules, HCV, IR, and Prism. Noise level of up to 50% was investigated. The experiments demonstrated that the classifiers' accuracy decrease linearly with increase of noise level. The authors suggest that the best approach in dealing with noisy labels is to remove the noisy training instances. A number of approaches are proposed in the literature to deal with removal of noisy training data (e.g., [14,21]).

Nazari et al. [12] investigated the impact of class noise on performance of 3 different classifiers; Decision Tree, SVM, and K-Nearest Neighbor. Most of the datasets experimented had only 2 or 3 classes with only one dataset having 6 classes. In addition, the number of samples in the datasets are only few hundreds with the dataset having 6 classes having only 214 samples. Moreover, the authors did not mention how they fine tune the classifiers when they increase the noise level. The study dealt with two level of noises–10% and 20%–and did not investigate higher levels of noise. The authors concluded that SVM is more robust to noise based on the results of the experiments although it seems the results are statistically insignificant considering the small size of the dataset.

Sáez et al. [16] presented an interesting study on the effects of class noise on the performance of C4.5 decision tree and SVM classifier. The paper investigated the effect on classifiers' performances when 10% and 20% class-labeling noise is added to the training set. The authors also presented a new measure to compare the performances of different classifiers keeping into perspective a classifier's performance on the noise-free training data.

Based on the literature survey, we see that dealing with noisy training data is a common problem in many machine learning applications and, thus, a thorough investigation on how different levels of noise impact the performance of various, commonly used, classifiers is very important. We investigated this topic with 5 commonly used classifiers in machine learning. Moreover, we tried with high variations in noise levels starting with 0% noise to 'nearly' random labels. In situations where high levels of noise are present, even the robust classifiers face issues and the techniques to deal with removing noisy data reach their limitations, too (cf. [16]). In the following sections, we present our investigation on this topic.

3   Methodology

In this section, we present the methodology of our investigation. We first present the database used for the experimentations. Next, we explain how different levels of label noise was introduced to the training sets. We follow this with brief details on the different classifiers used for experimentation. Finally, we present the measures we used to report the results.

3.1   Dataset

To study the impact of noisy-labeled training data, we selected a dataset of handwritten Arabic digits which was collected and published by CENPARMI [3]. These handwritten digits were extracted from real-bank checks written by many writers. The CENPARMI Arabic bank checks is a popular database in pattern recognition research (e.g., [7,8]). There are a total of 10423 digits in the database divided into the training set and the test set. The training set consists of 7388 digits while the test set contains the remaining 3035 digits.

In our experiments, we further split the training set by randomly selecting 1113 digit images from the training set for system development. This was done in order to calibrate the hyper parameters of the system so that we do not optimally train the system using the test set, directly. Thus, the training set consists of 6275 digits in our experiments.

We performed a number of experiments with different levels of label noise in the training set. As a first experiment, we trained and tested our system on the original data without adding any noise. The results we get from this system serves as a baseline to allow us to compare how the system performs when noise is added by modifying the class labels randomly. We term this training set as *train–0* signifying that 0% noise is added to this set. We created a number of additional training sets whereby we kept increasing the noise by 10% in every successive set. This led to training sets having 0%, 10%, 20%, 30%, 40%, 50%, 60%, 70%, and 80% class-labeling noise. In addition, we also created two sets with 75% and 85% noise, respectively. Training sets with 10% to 30% noise are termed as 'low' noise. Sets with 40% to 60% noise are termed as 'medium' noise, and sets with 70% or higher noise are termed as 'high' noise sets. It should be noted that the class labels are modified by randomly (uniform) selecting new labels for the classes of the selected digit images. Thus, *train–10* means that 10% of the digit images in the training set have incorrect class labels.

3.2   Classifiers

To study the effects of noise on different classifiers, we selected five of the commonly used classifiers: Support Vector Machine (SVM), Random Forest (RF), K-Nearest Neighbor (KNN), Multi-Layer Perceptron (MLP), and Naïve Bayes (NB) classifier. In this section, we will briefly describe the five classifiers used in the experimentation.

Support Vector Machine (SVM): SVMs are one of the most powerful and widely used classifiers. It works by selecting support vectors from the input training data and defining a hyper-plane that separates the classes as far as possible based on the selected support vectors. The selected support vectors are mapped to a higher dimension features space in order to define the decision boundaries in complex real-word scenarios where the classes seem difficult to be separated in the original feature space. In our experiments, we use the *radial basis function* (RBF) as the kernel with the parameters, *gamma* (the kernel coefficient) and *cost* (penalty for misclassification), selected using the grid search.

Random Forest (RF): RF is another powerful algorithm which is used for classification problems. It is an ensemble classifier that works by creating a number of decision trees by selecting a random subset of features and training data for each decision tree. The final output of the classifier is based on majority voting on the output of individual decision trees. RF is, in general, quite robust to over-fitting. Parameters like *forest size* and the *criteria for splitting* were calibrated based on the performance on the development set and independent of the test set.

K-Nearest Neighbor (KNN): KNN is another simple and powerful non-parametric algorithm used widely in classification problems. It is also known as a lazy classifier that delays all the computation to the classification stage. The idea of KNN is quite simple. It basically looks at the 'K' closest examples in the feature space of the training data to the sample that needs to be classified. Euclidean distance is one the most common distance metric used to decide the nearness in the feature space. The classifier output the test sample class based on the most frequent class among the K nearest neighbors of it. An as alternative to majority voting, weighted majority voting is also employed where the weight of an example is inversely proportional to its distance from the test sample. The value of the hyper-parameter K was empirically decided in our experiments based on the results on the development set.

Multi-layer Perceptron (MLP): An MLP is an example of feed-forward artificial neural networks. It consists of an input layer, one or more hidden layers, and an output layer. Each node in a layer connects to all the nodes in the next layer and its contribution to the connecting node is weighted. The last hidden layer is connected to the nodes of the output layer. Each node combines the weighted input of the values from the incoming nodes and passes them through the activation function. This value, in-turn, becomes the output of the node. The discriminative learning is achieved, mainly, by adjusting the weights to minimize the classification error. Hyper-parameters like number of hidden layers, the number of nodes in the hidden layer, the number of training iterations were decided using the cross-validation results.

Naïve Bayes (NB): Naïve Bayes is a probabilistic and a generative classifier. It uses the Bayes' theorem as a basis for making the probabilistic decisions. The classifier makes a strong assumption that the input features are independent of each other and hence the term 'naïve'. Although, the assumption of independence between the features are strong, the naïve Bayes classifier performs reasonably good in practical situations. Moreover, this classifier is scalable as the number of parameters grow linearly to the number of predictor variables. The probabilities for each class given the input features can be learned using the maximum-likelihood training. Smoothing of probabilities need to be performed so as to deal with situations where a specific value of a feature, given a class, is zero based on the training data. This avoids over-fitting.

We used KNIME [5] and DTREG [17] for experimentation. DTREG was used for the SVM classifier while KNIME was used for the other four classifiers.

3.3 Measures

We used the following measures to report and compare the results:

Error Rate in Classification: This measure calculates the error in classification. Error rate can be state as the following:

$$Error(\%) = \frac{S}{N} \times 100 \qquad (1)$$

where; S is the number of miss-classification and N is the total number of instances to be classified.

Error rate is a commonly used measure to report the performance of the classifiers. The smaller the error, the better a classifier's performance. It is also reported in percentage. Many researchers prefer to use the complimentary measure which is termed as classification rate, recognition rate, or accuracy.

Equalized Loss of Accuracy (ELA): Equalized Loss of Accuracy (ELA) was presented in [16]. It was designed with the goal of studying and comparing classifiers under scenarios where class labeling is noisy. This measure is an enhancement of an earlier measure proposed in [15] and takes into consideration the robustness of a classifier under noisy-labeled training data as well as under noise-free training data. ELA is defined as follows:

$$ELA = \frac{100 - A_x}{A_0} \qquad (2)$$

where; A_x = Accuracy of the classifier with $x\%$ noise and A_0 = Accuracy of the classifier with 0% noise.

Significance Interval at 95% Confidence Level: In addition to the above two presented measure, we also report the significance interval of the error rates as 95% confidence level. Reporting the significance interval is important when comparing the performances of two or more classifiers as small differences may not be statistically significant and can be attributed to 'chance'. Moreover, sometimes a classifier performs slightly better with some noisy labels as compared to when no noise was present in training. Checking the significance interval helps assert if the difference was insignificant, which generally may be the case, and should not be taken seriously. The statistical test for the difference of two proportions as presented in [6] was used to report the significance interval of the errors.

4 Experiments and Results

In this section, we will present the experiments we conducted and the results we obtained. Before training the classifiers, we first normalized the size of the digit images and extracted features from them. These features were then used, along with the labels, to train the classifiers. For size normalization, we resized the handwritten digit images such that each digit is of width 64 pixels without changing the original aspect ratio. Each image was segmented into 15 columns vertically and 15 rows horizontally and 9 different geometrical features were computed from each column and row segment of the images, respectively. These features were adapted from [20] by setting the bottom of the image segment as the baseline. The computed features were concatenated to form a feature vector having a dimension of 270 ($15 \times 9 + 15 \times 9$). These features were used to train the classifiers.

Training with no Noise: As a first set of experiments, each of the five classifiers was trained using the original class labels, i.e., with no label noise. A separate development set containing 1113 images were used to optimally calibrate the hyper-parameters of each of the classifiers. Finally, the trained classifiers were used to predict the digit classes of the test-set images. In Table 1, we summarize the error rates and ELA for the five classifiers. As we can see from the table, all the five classifiers have pretty good performance and very low error rates (under 4%). The best performing classifier is the SVM classifier with error rate of 1.35% followed by random forest classifier with error rate of 1.45%. The significance interval of the results is ±0.40 at 95% confidence level. Thus, the difference in results for both the classifiers are insignificant.

Training with Low Noise: Our next set of experiments are related to training the classifiers with 'low' noises in class labeling, i.e., 10%, 20%, and 30% noise, respectively. The steps of image size normalization and feature extraction are exactly the same. The only different is in class labels of the training data. Again, classifiers' hyper-parameters were optimally calibrated on a development set which is independent of the test set. In Table 2, we present the results in terms of error rates and ELA for different classifiers under different 'low' noise levels. The increase in error rates for the SVM classifier is very low. This shows that the SVM classifier is quite robust to low noises in class labels. Also, the ELA for the SVM classifier is quite low. The error rate for random forest is also very low. In fact, the degradation of performance due to low-noise is even less than SVM which shows that random forest is more robust to noise than SVM although the difference is performance is very small. ELA values for random forest are better than SVM. Error rates for KNN classifier is also quite low for low-noise training but are slightly higher that SVM and random forest classifiers. Error rates and ELA values for the naïve Bayes and the ANN classifiers are significantly higher than SVM, random forest, and KNN. Naïve Bayes performs the worst in low-noise conditions.

Training with Medium Noise: Our next set of experiments are related to training the classifiers with 'medium' noises in class labelling, i.e., 40%, 50%, and 60% noise, respectively. As was the case before, the steps for image size normalization and feature extraction are exactly the same. The only difference is in class labels for the training data. In Table 3, we present the results in terms of error rates and ELA for the classifiers under different medium-noise levels. From the table we can observe that the performance starts deteriorating

Table 1. Results in error rate and ELA for different classifiers when no noise in class-labeling. The significance interval of the results is ±0.40 at 95% confidence level.

Classifier	Error rate	ELA
SVM	**1.35**	**0.0137**
Random forest	**1.45**	**0.0147**
KNN	2.04	0.0209
Naïve Bayes	3.46	0.0358
ANN	3.53	0.0365

Table 2. Results in error rate and ELA for different classifiers when 'low' noise in class-labeling. The significance interval of the results is ±0.40 at 95% confidence level.

Classifier	train-10		train-20		train-30	
	Error rate	ELA	Error rate	ELA	Error rate	ELA
SVM	**1.78**	**0.0180**	1.91	0.0194	**1.98**	**0.0200**
Random forest	**1.45**	**0.0471**	1.38	0.0140	1.61	0.0164
KNN	2.34	0.0239	2.70	0.0276	2.64	0.0269
Naïve Bayes	6.79	0.0703	6.56	0.0679	7.25	0.0751
ANN	3.95	0.0410	4.05	0.0420	4.91	0.0509

for all the classifiers but the performances of the classifiers are still, relatively speaking, robust. The worst results in terms of error rates and SLA is for the ANN classifier suggesting that it is less robust to medium noise among the five classifiers. An interesting observation is related to random forest classifier as it shows the best performance in medium-noise situations. SVM and KNN classifiers are performing almost similar and naïve Bayes comes just above the ANN classifier. So, the most robust classifier to medium amount of noise in class labels is the random forest classifier and the least robust is the ANN.

Training with High Noise: Our next set of experiments are related to training the classifiers with 'high' noises in class labelling, i.e., 70%, 75%, 80%, and 85% noise, respectively. As the classifiers' performance starts deteriorating sharply in these experiments, we experiment with smaller noise steps of 5%. In Table 4, we present the results in terms of error rates and ELA for the classifiers under, different, high-noise levels. Performances of all the classifiers deteriorate significantly under high noise conditions but there are some interesting observations. The SVM classifier performs better than the random forest under high noise conditions. ANN has the worst deterioration in the performance. Naïve Bayes classifier, interestingly, seems to be more robust than SVM and random forest classifiers. The most interesting results are for the KNN classifier. It shows the

Table 3. Results in error rate and ELA for different classifiers when 'medium' noise in class-labeling. The significance interval of the results is ±0.44 at 95% confidence level.

Classifier	train-40		train-50		train-60	
	Error rate	ELA	Error rate	ELA	Error rate	ELA
SVM	2.83	0.0287	3.39	0.0344	4.58	0.0464
Random forest	**1.78**	**0.0181**	2.18	0.0221	**3.10**	**0.0314**
KNN	3.03	0.0330	3.23	0.0454	4.45	0.0454
Naïve Bayes	6.95	0.0720	6.43	0.0666	7.58	0.0785
ANN	5.83	0.0605	7.48	0.0775	13.51	0.1400

best performance under high noise conditions. It can correctly recognize about 70% of the test set digits with only, about, 15% correct labels in the training data.

Table 4. Results in error rate and ELA for different classifiers when 'high' noise in class-labeling. The significance interval of the results is ±0.74 at 95% confidence level.

Classifier	train-70		train-75		train-80		train-85	
	Error rate	ELA	Error rate	ELA	Error rate	ELA	Error rate	ELA
SVM	**5.80**	**0.059**	10.69	0.108	22.67	0.230	44.94	0.456
Random forest	7.68	0.078	14.40	0.146	30.28	0.307	61.81	0.627
KNN	**5.93**	**0.061**	**8.07**	**0.082**	**13.25**	**0.135**	**29.59**	**0.302**
Naïve Bayes	11.43	0.118	15.88	0.165	15.65	0.162	35.09	0.364
ANN	22.47	0.233	39.84	0.413	42.08	0.436	65.96	0.684

5   Discussions

In this section, we would like to summarize the results of the experiments and would like to also discuss some other aspects of training and recognition. When it comes to noise-free labeling of classes in the training data, all the five classifiers performed quite good but the best performance was recorded for SVM and random forest. This confirms that these two classifiers are among the most powerful classifiers when no noise is present in the training data. Under low-noise conditions, SVM and random forest classifiers are still the two best classifiers but random forest seems to be slightly more robust than SVM. KNN came third followed by the naïve Bayes classifier. ANN was the most sensitive classifier to low noise in training labels. For the medium-noise experiments, random forest was the most robust classifier followed by SVM and KNN. Again, ANN performs the worst. Random forest, although performing robustly under low and medium noise, was outperformed by SVM under high-noise conditions. Interestingly, naïve Bayes performed significantly better than SVM and random forest classifiers under high-noise conditions. The most robust classifier under high-noise conditions was the KNN classifier and the results are significantly better than any other classifier. Figure 1 (best viewed in color) plots the error rates for the five classifiers under different levels of noise in the training data.

Hyper Parameters and Training Time: In this section, we would like to discuss some of the observations regarding the hyper parameters of different classifiers under different amount of noise in training labels. We will also discuss the difference in training time for different classifiers. For the SVM classifier, the hyper parameters were selected using the grid search. It was observed that the grid search takes significantly more time to select the optimal hyper parameters when noise is present in the training labels and the time taken was proportional to the amount of noise added. For the random forest classifier, as we have more

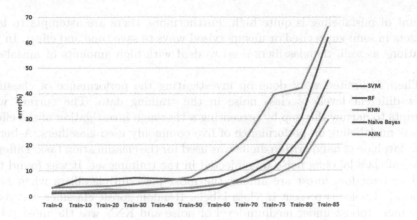

Fig. 1. Plot of error rates for different classifiers under different amounts of class miss-labels in training data. (Color figure online)

noise in the training labels, increasing the size of the forest (number of trees) does have some impact on the classifier's performance but this also meant it took more time to train and classify. The differences in the training times were not as significant as in the case of grid search for SVM. For the naïve Bayes classifier, we optimally calibrated the smoothing probability based on the results on the development set. It was observed that increasing the smoothing probability was better as we increase the noise but beyond a certain noise level (around 50% noise), the smoothing probability need to be decreased significantly for optimal results. For the KNN classifier, we need to keep increasing the value of K as we deal with noisier training data. This is clearly understandable because having noisy neighbors will force KNN to look for more neighbors before deciding the class based on the majority. As long as the number of true labels are higher than the false labels, the classifier can predict the correct label. Also, when medium to high noise is present in the training labels, it is better not to use weighted distance as it can be misleading. A very 'close' neighbor can have a wrong label and may influence the overall score. So, it is better to just perform a simple majority voting among the neighbors.

6   Conclusions and Future Work

Supervised machine learning, for classification problems, involves training the classifiers with labeled data. There are many classifiers, each having its own strength and weakness in terms of classification accuracy and dealing with noisy class labels in the training sets. The training data, inevitably, includes some class mislabeling due to a number of reasons like human error or corrupted data. Additionally, in some classification problems, attempts are made to use the training data in semi-supervised or unsupervised manner. In these situations, the training data is not only noisy, in terms of class mislabeling, but also the amount of mislabeling is quite high. Furthermore, there are attempts to label datasets in semi-supervised or unsupervised ways to save time and effort. In this situations as well, the classifiers need to deal with high amounts of mislabeled data.

There is limited work done on investigating the performance of classifiers under different levels of class noise in the training data. The current work attempts to narrow this gap by presenting a thorough investigation of the effects of class mislabeling on performance of five commonly used classifiers. A benchmark database of handwritten digits was used for the classification task. Different degree of class-labeling noise was injected in the training set. It was found that SVM and random forest are among the most powerful classifiers when no or low level of noise is present in class labels. Random forest classifier was found to be more robust under medium-level of noise and KNN was the most robust classifier when high amount class mislabeling was present.

Some of the possible future works include investigating this topic using other datasets having different characteristics. This work was performed on only one dataset and, thus, further experimentation on other datasets need to be performed to support the results of the current work and to confirm if the behavior is consistent across different datasets. Also, studying the impact of class mislabeling in the training set from a theoretical perspective seems interesting. This can also help the researchers to investigate ways of dealing with noisy training data especially in situations where high levels of noise are present.

Acknowledgment. The author would like to thank King Fahd University of Petroleum and Minerals (KFUPM) for supporting this work.

References

1. Ahmad, I., Fink, G.A.: Training an Arabic handwriting recognizer without a handwritten training data set. In: 2015 13th International Conference on Document Analysis and Recognition (ICDAR), pp. 476–480. IEEE (2015)
2. Ahmad, I., Mahmoud, S.A.: Arabic bank check processing: state of the art. J. Comput. Sci. Technol. **28**(2), 285–299 (2013)
3. Al-Ohali, Y., Cheriet, M., Suen, C.: Databases for recognition of handwritten arabic cheques. Pattern Recognit. **36**(1), 111–121 (2003)
4. Baird, H.S.: The state of the art of document image degradation modelling. In: Chaudhuri, B.B. (ed.) Digital Document Processing. ACVPR, pp. 261–279. Springer, London (2007). https://doi.org/10.1007/978-1-84628-726-8_12
5. Berthold, M.R., et al.: KNIME-the Konstanz information miner: version 2.0 and beyond. ACM SIGKDD Explor. Newsl. **11**(1), 26–31 (2009)
6. Dietterich, T.G.: Approximate statistical tests for comparing supervised classification learning algorithms. Neural Comput. **10**(7), 1895–1923 (1998)
7. Gimenez, A., Andrés-Ferrer, J., Juan, A., et al.: Discriminative Bernoulli mixture models for handwritten digit recognition. In: 2011 International Conference on Document Analysis and Recognition, pp. 558–562. IEEE (2011)
8. Helali, M., Alneghaimish, A., Ahmad, I.: Handwritten digit recognition under constrained training conditions. IET Conference Proceedings pp. 35–36 (2017)

9. Kozielski, M., Nuhn, M., Doetsch, P., Ney, H.: Towards unsupervised learning for handwriting recognition. In: 2014 14th International Conference on Frontiers in Handwriting Recognition, pp. 549–554. IEEE (2014)

10. Mahmoud, S.A., et al.: KHATT: an open Arabic offline handwritten text database. Pattern Recognit. **47**(3), 1096–1112 (2014)

11. Marti, U.V., Bunke, H.: The IAM-database: an English sentence database for offline handwriting recognition. Int. J. Doc. Anal. Recognit. **5**(1), 39–46 (2002)

12. Nazari, Z., Nazari, M., Danish, M.S.S., Kang, D.: Evaluation of class noise impact on performance of machine learning algorithms. IJCSNS **18**(8), 149 (2018)

13. Richarz, J., Vajda, S., Grzeszick, R., Fink, G.A.: Semi-supervised learning for character recognition in historical archive documents. Pattern Recognit. **47**(3), 1011–1020 (2014)

14. Sabzevari, M., Martínez-Muñoz, G., Suárez, A.: A two-stage ensemble method for the detection of class-label noise. Neurocomputing **275**, 2374–2383 (2018)

15. Sáez, J.A., Luengo, J., Herrera, F.: Fuzzy rule based classification systems versus crisp robust learners trained in presence of class noise's effects: a case of study. In: 2011 11th International Conference on Intelligent Systems Design and Applications, pp. 1229–1234. IEEE (2011)

16. Sáez, J.A., Luengo, J., Herrera, F.: Evaluating the classifier behavior with noisy data considering performance and robustness: the equalized loss of accuracy measure. Neurocomputing **176**, 26–35 (2016)

17. Sherrod, P.H.: DTREG predictive modeling software (2003). http://www.dtreg.com

18. Tabik, S., Peralta, D., Herrera-Poyatos, A., Herrera, F.: A snapshot of image preprocessing for convolutional neural networks: case study of mnist. Int. J. Comput. Intell. Syst. **10**(1), 555–568 (2017)

19. Varga, T., Bunke, H.: Perturbation models for generating synthetic training data in handwriting recognition. In: Marinai, S., Fujisawa, H. (eds.) Machine Learning in Document Analysis and Recognition. SCI, vol. 90, pp. 333–360. Springer, Heidelberg (2008). https://doi.org/10.1007/978-3-540-76280-5_13

20. Wienecke, M., Fink, G.A., Sagerer, G.: Toward automatic video-based whiteboard reading. Int. J. Doc. Anal. Recognit. (IJDAR) **7**(2–3), 188–200 (2005)

21. Yuan, W., Guan, D., Zhu, Q., Ma, T.: Novel mislabeled training data detection algorithm. Neural Comput. Appl. **29**(10), 673–683 (2018)

22. Zhu, X., Wu, X.: Class noise vs. attribute noise: a quantitative study. Artif. Intell. Rev. **22**(3), 177–210 (2004)

Combination of Multiple Classification Results Based on K-Class Alpha Integration

Gonzalo Safont[(⊠)], Addisson Salazar, and Luis Vergara

Institute of Telecommunications and Multimedia Applications,
Universitat Politècnica de València, Valencia, Spain
gonsaar@upvnet.upv.es,
{asalazar,lvergara}@dcom.upv.es

Abstract. This work introduces vector score integration (VSI), a novel alpha integration method to perform soft fusion of scores in K-class classification problems. The parameters of the method are optimized to achieve the least mean squared error between the fused scores and the ideal scores over a set of training data. VSI was applied to perform soft fusion of multiple classifiers working on two sets of real polysomnographic data from subjects with sleep disorders. In both sets, the signal is automatically staged in three classes: wake, rapid eye movement (REM) sleep, and non-REM sleep. Four single classifiers were considered: linear discriminant analysis, naive Bayes, classification trees, and random forests. VSI was able to successfully combine the scores from the considered classifiers, outperforming all of them and a classical fusion technique (majority voting).

Keywords: Soft fusion · Alpha integration · Classification · EEG · Score integration

1 Introduction

Decision fusion is one form of data fusion that combines the decisions of multiple classifiers into a common decision about the activity that has occurred [1, 2]. Instead of combining features [3], also known as "early fusion", many decision fusion algorithms combine multiple classifiers to make use of their different properties [4, 5] or because of the difficulties in combining features from multiple modalities [6]. This fusion can be performed by combining the scores yielded by each classifier, also known as "soft" fusion (see [4, 7–11]), or by combining the final decisions of the classifiers, also known as "hard" fusion (see [12–14]).

The optimal combination of multiple classifiers requires consideration of the statistical dependencies among their outputs, resulting in complex methods (e.g., [15–17]). A practical approach to this is defining combination functions whose parameters are then optimized over a training set, thus implicitly considering any dependencies between the outputs of the classifiers. One such approach is alpha integration, which was originally conceived as a method to integrate stochastic models [18, 19] and was recently proposed as a soft fusion method for binary problems [20]. In that work, the fusion improved the automatic detection of microarousals in patients with obstructive

© Springer Nature Switzerland AG 2019
I. Rojas et al. (Eds.): IWANN 2019, LNCS 11507, pp. 426–437, 2019.
https://doi.org/10.1007/978-3-030-20518-8_36

sleep apnea. Several optimality criteria were developed for the estimation of the parameters of alpha integration, such as least mean squared error (LMSE) and minimum probability of error (MPE). Other works have centered on estimating the parameters of alpha integration using the LMSE criterion [21, 22].

This paper presents a novel alpha integration method to perform soft fusion of the scores in multi-class classification problems, which we have called vector score integration (VSI). The scores produced by the classifiers are combined multi-dimensionally, accounting for possible interactions among the scores of different classes and classifiers. A training algorithm based on the optimization of the LMSE criterion is also presented. The performance of VSI was tested on the task of performing automatic staging of polysomnographic data from subjects with sleep disorders. Three classes were considered: wake, rapid eye movement (REM) sleep, and non-REM sleep. The data were classified with the following four commonly-used single classifiers: linear discriminant analysis, naive Bayes, classification trees, and random forests. The scores yielded by these four single classifiers were combined using a classical fusion technique (majority voting) and using VSI. In all cases, performance was considered using the accuracy and kappa coefficient of the results.

The rest of this paper is organized as follows. Section 2 includes a review of two-class alpha integration, which is extended to soft fusion of multiple classes in Sect. 3. Section 4 presents the results of the proposed methods on several sets of real data. Conclusions end the paper.

2   Two-Class Alpha Integration

Let us review the concepts of two-class alpha integration. We will assume that there is a group of D detectors with a set of scores $s_i, i = 1 \ldots D$, normalized between 0 and 1, with higher values of s_i indicating that hypothesis H_1 (detection is true) is more likely than hypothesis H_0 (detection is not true). The goal of alpha integration is optimally integrating these scores into a unique score s_α, given by

$$s_\alpha\big(\mathbf{s} = [s_1 \ldots s_D]^T\big) = \begin{cases} \left\{\sum_{i=1}^{D} w_i \cdot s_i^{\frac{1-\alpha}{2}}\right\}^{\frac{2}{1-\alpha}} & , \alpha \neq 1 \\ \exp\left\{\sum_{i=1}^{D} w_i \cdot \log(s_i)\right\} & , \alpha = 1 \end{cases} \tag{1}$$

where α and the coefficients $\mathbf{w} = [w_1 \ldots w_D]^T$ are the parameters to be optimized, subject to $w_i \geq 0$, $\sum_{i=1}^{D} w_i = 1$. Due to these constraints, s_α is bound between 0 and 1.

2.1   Least Mean Squared Error Criterion

Let us assume we have a set of couples $\{\mathbf{s}^j, y^j\}, j = 1 \ldots N$, where $\mathbf{s}^j = \left[s_1^j \ldots s_i^j \ldots s_D^j\right]^T$ is the vector of scores provided by the D detectors when y^j is the corresponding known binary decision ($y^j = 1$ if H_1 is true and $y^j = 0$ if H_0 is true). The mean squared error of alpha integration is defined as

$$\varepsilon = \frac{1}{N}\sum_{j=1}^{N}\left(y^j - s_\alpha(\mathbf{s}^j)\right)^2 \tag{2}$$

To learn the parameters of alpha integration, the derivatives of (2) with respect to α and $\mathbf{w}$ are to be calculated:

$$\frac{\partial \varepsilon}{\partial \alpha} = -\frac{2}{N}\sum_{j=1}^{N}\left(y^j - s_\alpha(\mathbf{s}^j)\right)\frac{\partial s_\alpha(\mathbf{s}^j)}{\partial \alpha} \tag{3.a}$$

$$\frac{\partial s_\alpha(\mathbf{s}^j)}{\partial \alpha} = \frac{2s_\alpha(\mathbf{s}^j)}{1-\alpha}\left\{\frac{\log\left(\sum_{i=1}^{D}w_i f_\alpha(s_i^j)\right)}{1-\alpha} + \frac{\sum_{i=1}^{D}w_i\frac{\partial f_\alpha(s_i^j)}{\partial \alpha}}{\sum_{i=1}^{D}w_i f_\alpha(s_i^j)}\right\} \tag{3.b}$$

$$\frac{\partial f_\alpha(s_i^j)}{\partial \alpha} = -\frac{1}{2}\log(s_i^j)(s_i^j)^{\frac{1-\alpha}{2}} \tag{3.c}$$

where, according to (1), $f_\alpha(\cdot)$ is a differentiable monotone function given by:

$$f_\alpha(z) = \begin{cases} z^{\frac{1-\alpha}{2}} & , \alpha \neq 1 \\ \log z & , \alpha = 1 \end{cases} \tag{4}$$

Moreover,

$$\frac{\partial \varepsilon}{\partial w_i} = -\frac{2}{N}\sum_{j=1}^{N}\left(y^j - s_\alpha(\mathbf{s}^j)\right)\frac{\partial s_\alpha(\mathbf{s}^j)}{\partial w_i} \tag{5.a}$$

$$\frac{\partial s_\alpha(\mathbf{s}^j)}{\partial w_i} = \begin{cases} \frac{2}{1-\alpha}\left(\frac{s_\alpha(\mathbf{s}^j)f_\alpha(s_i^j)}{\sum_{l=1}^{D}w_l f_\alpha(s_i^j)}\right) & , \alpha \neq 1 \\ s_\alpha(\mathbf{s}^j)\log(s_i^j) & , \alpha = 1 \end{cases} \tag{5.b}$$

Equations (3.a, 3.b and 3.c) to (5.a and 5.b) can be used to optimize the parameters of the model, for instance, using gradient descent to update α and $\mathbf{w}$ as:

$$\alpha(l+1) = \alpha(l) - \eta_\alpha\frac{\partial \varepsilon}{\partial \alpha}(l) \tag{6.a}$$

$$\mathbf{w}(l+1) = \mathbf{w}(l) - \eta_w\frac{\partial \varepsilon}{\partial w_i}(l) \tag{6.b}$$

where values η_α and η_w are the learning rate constants that control the speed of convergence.

3   K-Class Alpha Integration

In this section, we propose an extension of the two-class alpha integration method to the problem of K-class classification. Given K classes, indexed by $k = 1\ldots K$, and D classifiers, the ith classifier will produce a vector of scores $\mathbf{s}_i = [s_{1i}\ldots s_{Ki}]^T$, $i = 1\ldots D$. We will assume the scores are normalized to unit sum, $\sum_{k=1}^K s_{ki} = 1$. The true class identifier vector is defined as $\mathbf{y} = [y_1\ldots y_K]^T$, where

$$y_k = \begin{cases} 1, & \text{if the true class is } k \\ 0, & \text{otherwise} \end{cases} \tag{7}$$

We propose a vector score integration (VSI) method that accounts for cross dependencies among scores from different classes.

Given a set of scores from D classifiers for K classes, $\mathbf{S} = [\mathbf{s}_1\ldots \mathbf{s}_D]$, we can obtain a vector of integrated scores $\mathbf{s}_{\alpha_k} = [s_{\alpha_k 1}\ldots s_{\alpha_k K}]^T$ using the alpha integration function (1):

$$s_{\alpha_k m}(\mathbf{r}_m) = \begin{cases} \left(\sum_{i=1}^D w_{ki} \cdot r_{mi}^{\frac{1-\alpha_k}{2}}\right)^{\frac{2}{1-\alpha_k}} & , \; \alpha_k \neq 1 \\ \exp\left(\sum_{i=1}^D w_{ki} \cdot \log(r_{mi})\right) & , \; \alpha_k = 1 \end{cases} \tag{8}$$

where $m = 1\ldots K$, and $\mathbf{r}_m$ is the mth row of $\mathbf{S}$. The alpha parameters of each class are applied on the scores for all classes, resulting in K fused scores per class, for a total number of K^2 fused scores. Thus, the parameters of VSI are the following: α_k and $w_{ki}, k = 1\ldots K, i = 1\ldots D$.

Once we have the vectors of integrated scores $\mathbf{s}_{\alpha_k}, k = 1\ldots K$, classification is performed by choosing the vector that is closest to an ideal output $\mathbf{y}_{(k)}$ (1 in class k and 0 otherwise). In this work, we considered the Euclidean distance, thus arriving to

$$\hat{k} = \min_k \left\| \mathbf{y}_{(k)} - \mathbf{s}_{\alpha_k} \right\| \tag{9}$$

The fused scores provided by the method are those corresponding to the chosen class, $\mathbf{s}_{\alpha_{\hat{k}}}$.

3.1   K-Class LMSE Criterion

The optimization of the alpha integration parameters of class k will be performed using the subset of the whole training set where the true class is k. We denote this subset by $\left\{ \mathbf{S}_{(k)}^j, \mathbf{y}_{(k)}^j \right\}, j = 1\ldots N_k$, where N_k is the number of training couples in the subset. As per the definition of $\mathbf{y}$ in (7), since all values in this subset belong to class $k, y_{(k)k}^j = 1$ and $y_{(k)m}^j = 0, m \neq k$.

Given this subset, the LMSE cost function for class k is given by

$$\varepsilon_k = \frac{1}{N_k} \sum_{n=1}^{N_k} \sum_{m=1}^{K} \left(y_{(k)m} - s_{\alpha_k m} \left(\mathbf{r}_{m(k)}^n \right) \right)^2 \tag{10}$$

The derivatives of ε_k with respect to α_k are:

$$\frac{\partial \varepsilon_k}{\partial \alpha_k} = -\frac{2}{N_k} \sum_{n=1}^{N_k} \sum_{m=1}^{K} \left(y_{(k)m} - s_{\alpha_k m} \left(\mathbf{r}_{m(k)}^n \right) \right) \frac{\partial s_{\alpha_k m} \left(\mathbf{r}_{m(k)}^n \right)}{\partial \alpha_k} \tag{11}$$

where

$$\frac{\partial s_{\alpha_k m} \left(\mathbf{r}_{m(k)}^n \right)}{\partial \alpha_k} = \frac{2 s_{\alpha_k m} \left(\mathbf{r}_{m(k)}^n \right)}{1 - \alpha_k} \left\{ \frac{\sum_{i=1}^{D} w_{ki} \cdot \frac{\partial f_{\alpha_k m} \left(r_{mi(k)}^n \right)}{\partial \alpha_k}}{\sum_{i=1}^{D} w_{ki} \cdot f_{\alpha_k m} \left(r_{mi(k)}^n \right)} + \frac{\log \left(\sum_{i=1}^{D} w_{ki} \cdot f_{\alpha_k m} \left(r_{mi(k)}^n \right) \right)}{1 - \alpha_k} \right\} \tag{12.a}$$

$$\frac{\partial f_{\alpha_k m} \left(r_{mi(k)}^n \right)}{\partial \alpha_k} = -\frac{1}{2} \log \left(r_{mi(k)}^n \right) \left(r_{mi(k)}^n \right)^{\frac{1-\alpha_k}{2}} \tag{12.b}$$

And the derivatives with respect to the weights w_{ki} are the following:

$$\frac{\partial \varepsilon_k}{\partial w_{ki}} = -\frac{2}{N_k} \sum_{n=1}^{N_k} \sum_{m=1}^{K} \left(y_{(k)m} - s_{\alpha_k m} \left(\mathbf{r}_{m(k)}^n \right) \right) \frac{\partial s_{\alpha_k m} \left(\mathbf{r}_{m(k)}^n \right)}{\partial w_{ki}} \tag{13.a}$$

$$\frac{\partial s_{\alpha_k m} \left(\mathbf{r}_{m(k)}^n \right)}{\partial w_{ki}} = \begin{cases} \frac{2}{1-\alpha_k} \left(\frac{s_{\alpha_k m} \left(\mathbf{r}_{m(k)}^n \right) f_{\alpha_k} \left(r_{mi(k)}^n \right)}{\sum_{l=1}^{D} w_{kl} \cdot f_{\alpha_k} \left(r_{ml(k)}^n \right)} \right) & , \alpha_k \neq 1 \\ s_{\alpha_k m} \left(\mathbf{r}_{m(k)}^n \right) \log \left(r_{mi(k)}^n \right) & , \alpha_k = 1 \end{cases} \tag{13.b}$$

Using these derivatives, we can estimate the parameters that optimize the LMSE criterion, for instance, using a gradient descent algorithm similar to (6.a and 6.b).

4   Experiments and Results

In order to test the proposed method, we considered the classification of physiological signals from patients with sleep disorders.

4.1   Analysis of EEG Data from Subjects with Sleep Disorders

The proposed alpha integration method was tested on a set of real polysomnograms publicly available from the St Vincent's University Hospital/University College Dublin Sleep Apnea Database in Physionet [23]. The database contains polysomonograms (PSG) from 25 adult subjects (21 male, 4 female) with suspected sleep disorders. The PSG is a multimodal biomedical record that includes multiple physiological signals. A full PSG was available for each subject, but in this work we only considered two bipolar electroencephalographic (EEG) channels, C3-A2 and C4-A1. The EEG signals were sampled at 128 Hz and band-pass filtered between 0.5 and 30 Hz, and scoring was available for every 30-s epoch. An example of such signals is shown in Fig. 1.

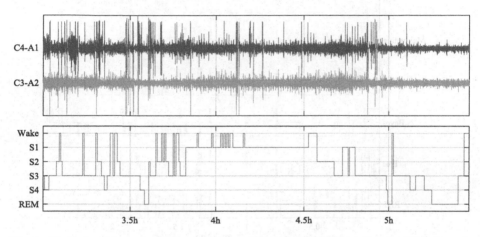

Fig. 1. Example of the extracted data from one of the subjects: (a) the two bipolar EEG channels considered, C3-A2 and C4-A1; (b) hypnogram for that time frame.

For this experiment, three classes were considered: wake, REM sleep, and non-REM sleep (sleep stages 1 through 4). For classification, the following features were extracted from each EEG channel in 30 s epochs: power in frequency bands delta (0–4 Hz), theta (5–7 Hz), alpha (8–12 Hz), sigma (13–15 Hz) and beta (16–30 Hz); and the activity, mobility and complexity of the signal [24]. These features are typically used in the literature on sleep staging [25].

The classification methods considered were the following: four single classifiers (LDA, naive Bayes, classification trees and RDF), a classical fusion technique (majority voting), and the proposed alpha based integration method (VSI). Each subject was classified independently from the rest. The data were divided into three datasets: training, validation and testing. In order to preserve the prior probabilities, the observations of each class were distributed as evenly as possible across the three datasets. The considered single classifiers were trained on the training dataset, the proposed alpha integration methods were trained on the scores of the single classifiers of the validation dataset, and the performance of all methods was estimated on the testing dataset. The results for each subject were obtained as the average of 100 iterations.

An example of the classification obtained by each of the compared methods for one of the patients is shown in Fig. 2. It can be seen that alpha integration method yielded classes that were more in line with the actual labels provided by the expert. This was particularly true for classes after 1.5 h, where the variability of the labels of the single classifiers and the classical fusion technique was larger than that of the proposed alpha integration based method (VSI).

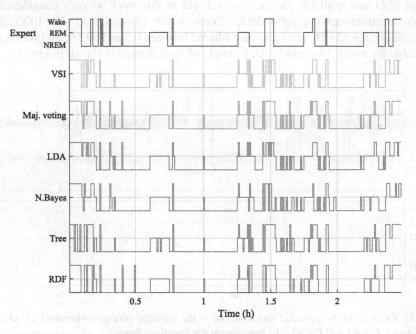

Fig. 2. Resulting classifications on the testing dataset for one of the subjects.

In this work we considered two indicators of classification performance: accuracy and kappa coefficient. Both are obtained from the confusion matrix, $\mathbf{C}$, where c_{ij} is the number of samples that belong to class i and where classified as class j. The accuracy was defined as the amount of correctly-classified samples, $acc = \frac{1}{N}\sum_{i=1}^{K} c_{ii}$, where N is the number of samples. For the kappa coefficient, we considered the following extension for multiclass classification [26]:

$$\kappa = \frac{N \cdot \sum_{i=1}^{K} c_{ii} - \sum_{i=1}^{K} c_{i+} \cdot c_{+i}}{N^2 - \sum_{i=1}^{K} c_{i+} \cdot c_{+i}} \tag{14}$$

where c_{i+} is the number of samples that belong to class i, and c_{+i} is the number of samples classified as class i.

The average accuracy and kappa values for all 25 patients are shown in Fig. 3. These values are similar to those in the literature for this dataset, e.g., [27, 28]. The proposed alpha integration method yielded better results than considered single classifiers and classical fusion technique. In particular, alpha integration using VSI method yielded the best results, returning similar values of accuracy and kappa. The VSI method achieved an average 2.38% more accuracy and 3.88% more kappa than the best performing single classifier (RDF), and 4.09% more accuracy and 4.65% more kappa than the classical fusion technique (majority voting).

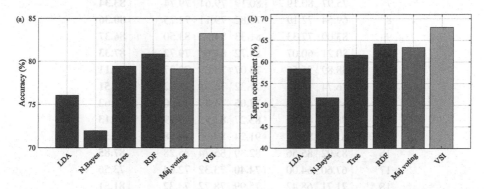

Fig. 3. Average classification results on subjects with sleep disorders: (a) accuracy; (b) kappa values.

The accuracy yielded by each method for each patient is shown in Table 1, with bold fonts indicating the best result for each patient. In accordance with the values in Fig. 3, the proposed alpha integration method yielded the best results in 20 of the 25 subjects, with the classical fusion technique yielding the best result for one subject. In the subjects where the best performance was achieved by one of the competitive methods (subjects 2, 13, 17, 19, and 24), the difference with the results yielded by alpha integration was small (less than 1% in all cases). These results indicate that the improvement in performance yielded by the proposed alpha integration method was consistent across subjects.

In order to further verify the performance of VSI, a second experiment was performed using different data. This second experiment considered six subjects diagnosed with obstructive sleep apnea, and the data were captured with the help of the Neurology and Neurophysiology Units of Hospital Universitari i Politècnic La Fe, Valencia (Spain). A full PSG is present for each subject, but in this work we only considered channels C3-A2 and C4-A1. The EEG signals were sampled at 256 Hz, and scoring was provided by a physician at irregular intervals for every night. These signals were then converted to the same 30 s epochs that were considered for the first experiment on real data, and the same classes were considered (wake, REM sleep, and non-REM sleep). Any classes containing insufficient observations were removed from the data prior to classification. Performance was evaluated using the same Monte Carlo experiments and performance indicators considered for the previous experiment.

Table 1. Accuracy obtained for each method for each subject of the first database.

Subject	LDA	N. Bayes	Tree	RDF	Maj. voting	VSI
1	79.52	59.44	79.92	80.84	79.20	**84.26**
2	85.37	**89.12**	85.03	87.72	88.50	88.57
3	71.27	77.82	79.64	80.07	79.09	**83.02**
4	75.84	72.86	75.84	80.56	76.88	**82.16**
5	80.07	71.22	84.50	82.44	80.70	**85.02**
6	68.25	70.63	82.94	83.41	78.45	**86.59**
7	75.97	80.19	80.19	79.61	79.74	**83.34**
8	69.54	72.19	76.82	79.11	72.25	**80.36**
9	83.00	77.33	81.33	84.83	83.50	**86.37**
10	80.21	60.07	84.72	84.65	79.72	**87.33**
11	58.89	55.56	74.07	75.89	72.37	**79.11**
12	76.74	70.54	71.32	76.94	75.62	**81.51**
13	65.25	60.33	**77.05**	74.49	70.00	76.62
14	88.21	85.55	83.27	87.72	87.79	**89.13**
15	81.39	81.39	91.24	89.96	90.58	**92.55**
16	85.92	85.56	82.39	85.74	85.28	**87.85**
17	67.60	64.00	**74.40**	73.32	72.64	73.56
18	71.71	68.42	75.99	78.72	73.32	**81.51**
19	85.50	89.31	91.60	91.64	**92.90**	92.63
20	74.56	43.90	74.56	77.21	78.85	**81.11**
21	76.41	75.42	66.45	74.49	75.95	**78.90**
22	70.04	66.24	77.22	76.16	73.04	**79.32**
23	81.72	80.29	84.23	84.84	84.41	**87.20**
24	80.47	81.14	82.49	**83.60**	82.90	83.40
25	67.92	60.42	69.17	67.79	65.33	**69.88**

The average classification results on the second dataset are shown in Fig. 4 and Table 2. Results are consistent with those of the first experiment (see Fig. 3). The considered classical fusion technique was generally unable to improve on the single classifiers, whereas alpha integration methods were capable of optimally combining the available results from the different classifiers. Numerically, the proposed alpha integration method (VSI) obtained an average improvement of 1.85% more accuracy and 3.91% more kappa than the best performing single classifier (RDF), and 4.71% more accuracy and 5.24% more kappa than the best performing of the classical fusion technique. In accordance with the values in Fig. 4, the values of Table 2 show the proposed alpha integration method (VSI) yielded the best result for all the 6 subjects of this database.

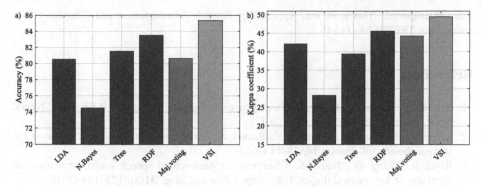

Fig. 4. Average classification results on the second dataset of subjects with sleep disorders: (a) accuracy; (b) kappa values.

Table 2. Accuracy obtained for each method for each subject of the second database.

Subject	LDA	N. Bayes	Tree	RDF	Maj. voting	VSI
1	84.43	76.72	88.44	88.52	85.65	**89.72**
2	87.15	79.64	88.15	91.14	89.55	**92.28**
3	64.24	54.10	65.78	67.25	61.22	**71.13**
4	76.01	76.46	78.47	81.79	74.57	**82.97**
5	78.07	71.36	77.12	78.44	78.80	**81.09**
6	93.34	87.23	91.15	93.87	94.06	**94.93**

5 Conclusion

This work has proposed a new method based on alpha integration to perform soft fusion of scores in K-class classification problems: vector score integration (VSI). A training algorithm has been developed to optimize the parameters of VSI with respect to the least mean squared error criterion. The proposed method was tested on several sets of real data (EEG signals). In these experiments, alpha integration was used to fuse the results from four single classifiers: LDA, naive Bayes, classification trees, and RDF. A classical fusion technique (majority voting) was also included for comparison.

The experiments on real data performed automatic staging of polysomnographic data into three classes: wake, REM sleep, and non-REM sleep. Two different datasets were considered: a public database with records from 25 patients with sleep disorders, and a private study with six patients with obstructive sleep apnea. The proposed alpha integration method yielded better results than the single classifiers and a classical fusion technique across all experiments. In the cases where the classical fusion technique was unable to improve significantly on the considered single classifiers, the proposed alpha integration method was still able to exploit information from all single classifiers in order to yield optimally fused scores.

Acknowledgment. This work was supported by Spanish Administration and European Union under grant TEC2017-84743-P.

References

1. Yuksel, S., Wilson, J., Gader, P.: Twenty years of mixture of experts. IEEE Trans. Neural Netw. Learn. Syst. **23**, 1177–1193 (2012)
2. Khaleghi, B., Khamis, A., Karray, F., Razavi, S.: Multisensor data fusion: a review of the state-of-the-art. Inf. Fusion **14**, 28–44 (2013)
3. Rivet, B., Wang, W., Naqvi, S., Chambers, J.: Audiovisual speech source separation: an overview of key methodologies. IEEE Signal Process. Mag. **31**(3), 125–134 (2014)
4. Wang, S., et al.: Fusion of machine intelligence and human intelligence for colonic polyp detection in CT colonography. In: International Symposium on Biomedical Imaging: From Nano to Macro, Chicago, IL, USA, pp. 160–164 (2011)
5. Mohandes, M., Deriche, M., Aliyu, S.: Classifiers combination techniques: a comprehensive review. IEEE Access **6**, 19626–19639 (2018)
6. Lahat, D., Adali, T., Jutten, C.: Multimodal data fusion: an overview of methods, challenges and prospects. Proc. IEEE **103**, 1449–1477 (2015)
7. Fattah, M.: New term weighting schemes with combination of multiple classifiers for sentiment analysis. Neurocomputing **167**, 434–442 (2015)
8. Abellán, J., Mantas, C.: Improving experimental studies about ensembles of classifiers for bankruptcy prediction and credit scoring. Expert Syst. Appl. **41**, 3825–3830 (2014)
9. Salazar, A., Safont, G., Soriano, A., Vergara, L.: Automatic credit card fraud detection based on non-linear signal processing. In: International Carnahan Conference on Security Technology (ICCST), Boston, MA, USA, pp. 207–212 (2012)
10. Salazar, A., Safont, G., Vergara, L.: Surrogate techniques for testing fraud detection algorithms in credit card operations. In: International Carnahan Conference on Security Technology (ICCST), Rome, Italy (2014). Article no. 6986987
11. Salazar, A., Safont, G., Vergara, L.: Semi-supervised learning for imbalanced classification of credit card transaction. In: International Joint Conference on Neural Networks (IJCNN), Rio de Janeiro, Brazil (2018). Article no. 8489755
12. Zhang, J., Wu, Y., Bai, J., Chen, F.: Automatic sleep stage classification based on sparse deep belief net and combination of multiple classifiers. Trans. Inst. Meas. Control **38**(4), 435–451 (2015)
13. Kevric, J., Jukic, S., Subasi, A.: An effective combining classifier approach using tree algorithms for network intrusion detection. Neural Comput. Appl. **28**(suppl1), 1051–1058 (2017)
14. Safont, G., Salazar, A., Soriano, A., Vergara, L.: Combination of multiple detectors for EEG based biometric identification/authentication. In: International Carnahan Conference on Security Technology (ICCST), Boston, MA, USA, pp. 230–236 (2012)
15. Poh, N., Bengio, S.: How do correlation and variance of base experts affect fusion in biometric authentication tasks. IEEE Trans. Signal Process. **53**, 4384–4396 (2005)
16. Vergara, L., Soriano, A., Safont, G., Salazar, A.: On the fusion of non-independent detectors. Digit. Signal Process. **50**, 24–33 (2016)
17. Safont, G., Salazar, A., Bouziane, A., Vergara, L.: Synchronized multi-chain mixture of independent component analyzers. In: Rojas, I., Joya, G., Catala, A. (eds.) IWANN 2017. LNCS, vol. 10305, pp. 190–198. Springer, Cham (2017). https://doi.org/10.1007/978-3-319-59153-7_17

18. Amari, S.: Integration of stochastic models by minimizing α-divergence. Neural Comput. **19**, 2780–2796 (2007)
19. Wu, D.: Parameter estimation for α-GMM based on maximum likelihood criterion. Neural Comput. **21**, 1776–1795 (2009)
20. Soriano, A., Vergara, L., Bouziane, A., Salazar, A.: Fusion of scores in a detection context based on alpha-integration. Neural Comput. **27**, 1983–2010 (2015)
21. Choi, H., Choi, S., Katake, A., Choe, Y.: Learning α-integration with partially-labeled data. In: IEEE International Conference on Acoustics, Speech, and Signal Processing, Dallas, TX, USA, pp. 2058–2061 (2010)
22. Choi, H., Choi, S., Choe, Y.: Parameter learning for alpha integration. Neural Comput. **25**, 1585–1604 (2013)
23. Heneghan, C.: St. Vincent's University Hospital/University College Dublin Sleep Apnea Database. https://www.physionet.org/pn3/ucddb/. Accessed 08 Mar 2019
24. Hjorth, J.: The physical significance of time domain descriptors in EEG analysis. Electroencephalogr. Clin. Neurophysiol. **34**(3), 321–325 (1973)
25. Motamedi-Fakhr, S., Moshrefi-Torbati, M., Hill, M., Hill, C., White, P.: Signal processing techniques applied to human sleep EEG signals – a review. Biomed. Signal Process. Control **10**, 21–33 (2014)
26. Carletta, J.: Assessing agreement on classification tasks: the kappa statistic. Comput. Linguist. **22**(2), 249–254 (1996)
27. Xie, B., Minn, H.: Real-time sleep apnea detection by classifier combination. IEEE Trans. Inf. Technol. Biomed. **16**(3), 469–477 (2012)
28. Wang, S., Hua, G., Hao, G., Xie, C.: A cycle deep belief network model for multivariate time series classification. Math. Probl. Eng. **2017**, 1–7 (2017)

Acceleration of Online Recognition of 2D Sequences Using Deep Bidirectional LSTM and Dynamic Programming

Dmytro Zhelezniakov[✉], Viktor Zaytsev, and Olga Radyvonenko

Samsung R&D Institute Ukraine (SRK), 57, L'va Tolstogo Str, Kyiv 01032, Ukraine
{d.zheleznyak,v.zaytsev,o.radyvonenk}@samsung.com

Abstract. In this work, the approach for online recognition of 2D sequences using deep bidirectional LSTM was proposed. One of the complex cases of online sequence recognition is handwritten mathematical expressions (HME). In spite of many achievements in this area, it is a still challenging task as, in addition to character segmentation and recognition, the tasks of structure, relations, and grammar analysis should be resolved. Such a combination of recognizers could lead to an increase in computational complexity for large expressions, which is unacceptable for on-device recognition in mobile applications. As end-to-end neural systems do not achieve plausible accuracy and recognition speed for on-device calculations so far, to overcome this problem we proposed a deep-learning solution that employs recurrent neural networks (RNNs) for structure and character recognition in combination with re-ordering and modified CYK algorithm for expression construction. Also, we explored a variety of structural and optimization enhancements to CYK algorithm that significantly improved the performance in terms of the recognition speed while the recognition accuracy remained at the same level. The ablation study for the introduced optimization techniques demonstrated significant improvement of recognition speed keeping the recognition accuracy comparable with the existing state-of-the-art approaches.

Keywords: Deep learning · Recurrent neural networks ·
Recognition acceleration · Dynamic programming ·
Human computer interfaces · Handwritten mathematical expression ·
Online recognition

1 Introduction

The recent development of sequence-to-sequence model-based deep learning approaches has led to drastic accuracy improvements in the sequence recognition and modeling tasks. Such technology established as the state-of-the-art on complex tasks like automatic speech recognition [6], machine translation [3], question answering, caption generation [24]. Deep sequence-to-sequence models could integrate many separate aspects and features of the training sequence

© Springer Nature Switzerland AG 2019
I. Rojas et al. (Eds.): IWANN 2019, LNCS 11507, pp. 438–449, 2019.
https://doi.org/10.1007/978-3-030-20518-8_37

(acoustic, touch, language) into a single neural system and do not assume the conditional independence as in standard hidden Markov models.

One of the examples of successful application of RNN-based sequence-to-sequence modeling is online handwriting recognition [8], where the sequence of touch or pen strokes is considered as an input of the recognition system. A stroke contains point coordinates from pen-down to pen-up. Applying advanced RNNs to the online handwritten text recognition has shown outperforming results [4, 18], while recognition of more complex online structures requires more research. For example, handwritten mathematical expressions (HME) is typically more difficult and challenging for employing RNNs directly because of the presence of complex two-dimensional sequential structures.

With the advent of computationally powerful touch-screen mobile devices various application came up with online handwriting recognition for note taking and educational purposes [17]. Rolling out such technologies on-device puts forward new requirements for recognition performance and speed [7,13].

In this paper, we present an effective RNN-based approach for online recognition of 2D sequences by the example of HME. Also, we addressing the task of the recognition acceleration and explore a variety of optimization improvements to the algorithm for expression construction [1] based on dynamical programming, which significantly speed up the recognition. Finally, we characterize the performance of the proposed RNN architecture and decoding techniques via evaluation using open source CROHME dataset [15].

The structure of the paper is organized as follows: after this introduction and related works analysis, Sect. 3 describes the proposed approach in detail. The experimental validation is presented in Sect. 4. The conclusion and future work are presented at the end of the paper.

2 Related Work

The task of *mathematical expression recognition* system could be considered as the translation of handwritten strokes into sequences of LaTeX expressions.

Generally, recognition of HME includes three independent tasks [1]:

1. Segmentation of characters – a grouping of the strokes that belong to the same character.
2. Recognition of characters – recognizing characters and labeling them into a form that the computer can use, for example in LaTeX or MathML representation.
3. Structural analysis – identifying the spatial relations between characters to determine the most likely interpretation as a mathematical expression in terms of grammar.

In the classical approaches based on structural recognition these tasks can be addressed either sequentially or in parallel [1,2,9,19]. Recognition errors made at an earlier level has a significant contribution to the next levels.

In contrast to the traditional approaches, deep learning allows to build end-to-end recognition system. In [22] the end-to-end system based a Bidirectional Long Short-Term Memory neural network (BLSTM) was proposed. The model achieved Correct Expression Recognition Rate (ExpRate) 30.57% on CROHME2014, which is low comparing to the structural approaches. It describes the inability of BLSTMs to memorize the mapping of the complex 2D structures on the expression graph directly, due to stroke order. In [21] the attention-based encoder-decoder has been applied to implement end-to-end system for offline HME recognition and demonstrated ExpRate 44.55% on CROHME2014. The initial results based on the convolutional encoder and LSTM an attention-based parser for composing of LaTeX sequences was improved in [20] by applying multi-scale attention with DenseNet encoder, which allowed to enhance ExpRate up to 52.8%. The main drawback of this system is high computational complexity, which makes it harder to be applied for on-device computing in mobile applications.

Thus, despite the large amount of related works [12,16,22,23], the complete problem of RNN-based mathematical expression recognition containing segmentation, character recognition and expression construction tasks, with respect to calculation speed, which is very important for mobile applications, is rarely analyzed in the literature.

In this paper, we present the system employing the advantages of BLSTM and fast optimized parsing suitable for using in on-device calculations.

3   Proposed Solution

We propose a combined BLSTM-based approach for online recognition of HME. As shown in Fig. 1, it consists of four general steps:

1. Preprocessing of input handwritten strokes.
2. Reordering of the strokes.
3. Character recognition and segmentation.
4. Expression building.

As a first step, the preprocessing methods based on geometric heuristics and strokes' statistics is used [10]. As a result, they perform baseline detection, size normalization, slant/slope correction, and resampling. Then for preliminary structure analysis and further preparation of strokes for character recognition, the strokes' reordering and detection of fractions are performed. The method for reordering the sequence of strokes helps to unify handwriting expressions structure in two special cases of interest: the horizontal bar of the character – a fraction and a radical sign. Our method exploits shape information for selecting the strokes that most likely correspond to the features of interest, and the information about layout and topology for locating the strokes representing the expression underneath the detected fraction and the radical sign. The performance of our method, evaluated on in-house test dataset, has shown a correct reordering of the sequence 95.3% of the cases. Thus, the proposed method allows

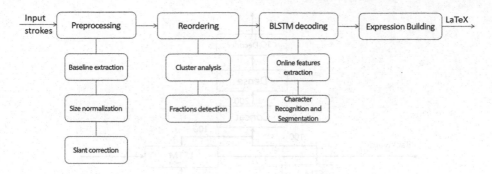

Fig. 1. General overview of the proposed approach

obtaining a more stable and invariant description of the HME in terms of elementary stroke sequences, and therefore helpful as a preprocessing step for the following RNN-based character recognition step.

A BLSTM model used in this paper is given by the next equations [14]:

$$
\begin{aligned}
i_t &= \sigma_i(x_t W_{xi} + h_{(t-1)} W_{hi} + b_i) \\
f_t &= \sigma_f(x_t W_{xf} + h_{(t-1)} W_{hf} + b_f) \\
c_t &= f_t c_{t-1} + i_t \tau_c(x_t W_{xc} + h_{(t-1)} W_{hc} + b_c) \\
o_t &= \sigma_o(x_t W_{xo} + h_{(t-1)} W_{ho} + b_o) \\
h_t &= o_t \tau_c,
\end{aligned}
\tag{1}
$$

where x_t is the input at the time stamp t; i_t and f_t are input and forget gate vectors; c_t is cell activation vector; o_t is cell output vector; h_t is hidden states vector; W denote weight matrices and b – bias vectors. The activation functions σ is logistic sigmoid, and τ is hyperbolic tangent.

The proposed approach relies on using BLSTM simultaneously for segmentation and character classification. It allows to simplify the recognition workflow and decrease recognition time. The *input* layer receives normalized feature vectors, containing 3 features: delta x coordinates, delta y coordinates, penup/pen-down. *Dropout* layers ($p = 0.5$) are active only during training and are included to improve the networks' generalization performance. The class conditional probabilities are obtained at the output of the *softmax* layer. The objective train function is the categorical cross-entropy. The prediction result is decoded by the *argmax* function.

Training is performed using mini-batch gradient descent, the model parameters update rule is ADADELTA ensuring robust parameter convergence. The architecture of RNN for character recognition and segmentation is presented at Fig. 2. The structure of the trained neural network is next: an input layer, two forward and backward layers with the layer which combines their outputs, dense layer, and CTC decoding layer. Total segmentation network consists of 8 layers and about 300 000 weights.

Utilizing of BLSTM for segmentation of 2D sequences with respect to classification of sequences' elements introduces the requirements to the training dataset

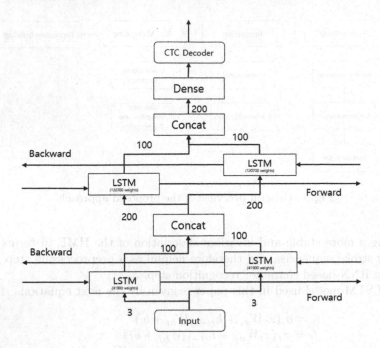

Fig. 2. Architecture of neural network for the character recognition and segmentation

composition. From our experience, it is desirable that the single sequences' elements be present in the training dataset, as, without stand-alone elements of sequences, BLSTM could not perform segmentation task with high performance [11].

3.1 Recognition Based on Decoding of BLSTM and Expressions Construction Using Dynamic Programming

The introduction of the RNN at the first step has reduced the need for segmentation variation search. After BLSM max-decoding we obtain the characters score matrix and the segmentation information. To achieve better performance and provide the system with balanced recognition accuracy and processing speed, we proposed a modular expression builder, that employs ideas of spatial relations, dominance, character mutual visibility, and probabilistic context-free grammar (PCFG) [5].

To construct the model of spatial relations we used ten well-known relation types Fig. 3b [1].

Also, a concept of body box [13] instead of bounding boxes was employed. The eleven types of characters were introduced: (1) tiny symbols (dot, comma, prime, cdot, ldots, etc.); (2) horizontal lines (minus, frac line, tilde, etc.); (3) large operators (sum, prod, int, etc.); (4) ascent; (5) descent; (6) ascent + descent; (7) common; (9) radical; (10) small operators (+, *, ring); (11) brackets.

Body box is calculated depending on the type of the character, then the features are extracted using updated body box information. The description of spatial relation features is given in the Table 1. Body box information is updating with respect to adjacent characters (Fig. 4b). After calculation of the spatial relation features for each character candidate (Table 1), they are distributed to the 3 types – subsequent, top/bottom, and sub/superscript. Each type represented as a special bigram in the language model. Different classification approaches for spatial relation classification were examined. The most accurate results were demonstrated by Decision Trees, Random Forest, and Gaussian Mixture Models. Currently, Decision Trees are used because of acceptable accuracy (95.96%) with relatively low computational complexity.

Table 1. Features for spatial relation model

Feature	Description
F_{CX}	A distance between centers of characters' bounding boxes along X coordinate
F_{CY}	A distance between centers of characters' bounding boxes along Y coordinate
F_{TT}	A distance between upper coordinates of characters' bounding boxes
F_{BB}	A distance between lower coordinates of characters' bounding boxes
F_{BT}	A distance between the lower coordinate of the previous and the upper coordinate of the next characters bounding box
F_{LL}	A distance between left coordinates of characters' bounding boxes
F_{RL}	A distance between the right coordinate of the previous and the left coordinate of the next characters bounding box
F_{LR}	A distance between the left coordinate of the previous and the right coordinate of the next characters bounding box

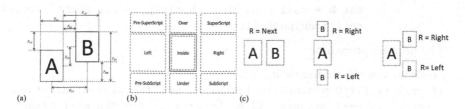

Fig. 3. (a) Features for spatial relation model; (b) Spatial relations; (c) Relation types for the language model

In order to cover the complex 2D structure of HME, we have introduced two different bigram language models for expression construction:

1. Language sequence model: We added a scoring heuristic score S_{LS} – the probability of the character B after character A for specific spatial relation type R.
2. Language relation model: Calculation of S_{LR} – the probability of spatial relation type between character A and B.

$$S_{LS} = P(B|AR) \tag{2}$$
$$S_{LR} = P(R|AB) \tag{3}$$

PCFG contains rules for each relation type and produces two score values: binary grammar score S_{GB} and terminal grammar score S_{GT}. For parsing HME we introduced optimizing heuristics into bottom-up parsing and dynamic programming CYK-aglorithm [1]. The first one is the concepts of dominance and mutual visibility for characters' candidates. The second is the dynamic pruning the search area depending on found characters candidates. As a result, the parsing time per expression was reduced on average four times, and for large expressions – tens of times.

We use the following notation: H[level] is a set of hypothesis; P[p, l, r] is a set of productions rules; PL is a list of all production types.

The pseudo-code of the our modification of CYK-aglorithm for HME recognition acceleration and pruning is presented below:

```
procedure constructExpression()
for each level = 1 to length - 1
  for each leftLen = 0 to level - 1
    set rightLen = length - leftLen - 1
    for each left to H[leftLen]
      for pt left to PL
        // search of the canidates according to the search area
        set rightCandidate = call findCandidates(rightLen, pt)
        for each right in rightCandidate
          if call !isVisible(left, right, pt) continue
            for each p in P[pt, left.S, right.S]
            set score = call calculateScore(p, left, right)
            set h = call createHypothesis(p, left, right, score)
            call addHypothesis(h)
```

The pruning procedure:

```
procedure addHypothesis(h)
if call !verifyDominantRelation(h) // verification of dominance
  return // exit without adding hypothesis for the next steps
// verification of new hypothesis
set minScore = HUGE, set counter = 0, set worst = nil
set maxCounter = max(3, min(10, h.level)) // pruning criteria
```

```
for each candidate to H[h.level]
  if candidate.symbols = h.symbols // in the one node
    set counter = counter + 1
    if candidate.score < minScore then
      set worst = candidate
      set minScore = candidate.score
    if counter > maxCounter and h.score > minScore then
      *worst = h // replace the worst hypothesis by new one
      return
  call H[h.level].add(h)
```

Symbol dominance helps us to skip creation of redundant hypothesis during expression construction by defining the minimal requirements for complex expressions with radicals, fractions, arrows, etc. Specified geometric areas are used for dominance verification (Fig. 4a).

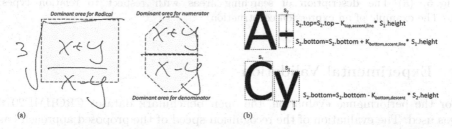

(a)   (b)

Fig. 4. (a) Illustration of the dominance areas for the radical and the fraction; (b) The example of the characters' body boxes calculation with respect to adjacent characters

For the expression construction two score metrics were introduced: the terminal hypothesis score (H_T) and the binary hypothesis score (H_B):

$$H_T = K_C \cdot \log(S_C) + K_{TG} \cdot \log(S_{GT}). \tag{4}$$

H_T is used on the first level of expression construction when there is no information for H_B calculation. At the next levels, H_T is transforming to the left hypothesis score H_L or right hypothesis score H_R depends on position in the parsing tree.

H_B is calculated as follows:

$$H_B = K_R \cdot \log(S_R) + K_{LS} \cdot \log(S_{LS})$$
$$+K_{LR} \cdot \log(S_{LR}) + K_{GB} \cdot \log(S_{GB}) + H_L + H_R + P, \tag{5}$$

where S_R – relation score, S_{GB} – binary grammar score, S_{GT} – terminal grammar score, S_{LS} – language sequence score, S_{LR} – language relation score, S_C – character score, H_R – left hypothesis score, H_L – right hypothesis score, P – hypothesis penalty.

The problem of setting weights K_C, K_{TG}, K_R, K_{LS}, K_{LR}, $K_G B$ were resolved with use of genetic algorithm.

P – obtained by decision tree heuristically, based on expert opinion.

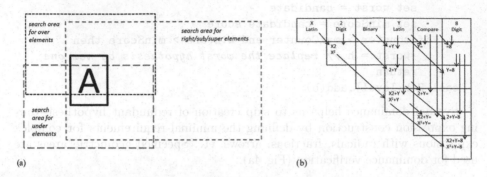

(a)   (b)

Fig. 5. (a) The description of searching areas with respect to relation types; (b) The example of an expression construction

4   Experimental Validation

For the performance evaluation, the open benchmark dataset CROHME2016 was used. The evaluation of the recognition speed of the proposed approach has been conducted on Intel(R) Core(TM) i7-2600 CPU @ 3.40 GHz. The BLSTM calculation step average time is about 60 ms per expression. The increase of calculation time for mobile device was estimated as 5 times – about 300 ms. The evaluation results on CROHME2016 dataset for the overall approach in comparison with the state-of-the-art accuracy [15] are given in Table 3. The ablation study for the introduced optimization techniques is given in the Table 2. The performance of the proposed approach is 10 times magnitude better keeping the recognition quality comparable with the state-of-the-art approaches.

Table 2. Evaluation of recognition speed of the proposed accelerated algorithm for expression construction

Approach	Average ERT (ms)	Expression Rate (%)
CYK	188	65.76
CYK + visibility	140	65.93
CYK + prunung	113	65.59
CYK + visibility + pruning	60	65.76
CYK + dominance	49	65.93
CYK + dominance + pruning	33	65.76
Our approach	**19**	**65.76**

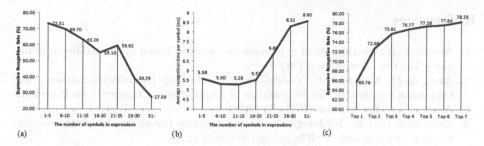

Fig. 6. Performance study of the proposed approach analyzed in terms of the length and complexity of input sequences: (a) Expression rate with respect to the number of symbols in expressions; (b) Dependency of average recognition time per symbol on the number of symbols in the expression; (c) Evaluation of Top–N accuracy

Table 3. Performance evaluation results

Approach	Expression rate, %	Character recognition rate, %	Recognition time, ms
Our approach	65.76	94.33	79
MyScript	67.62	92.82	–

The study of dependency of recognition performance on the HME length and complexity is given in Fig. 6.

5   Conclusion and Outlook

In this work, we present a fast and highly accurate system for online recognition of 2D structures such as handwritten mathematical expressions.

The novelty in our approach is using BLSTM model for segmentation, character recognition and relation analysis. The input strokes are reordered that improve the classification result. Also, we presented the variety of structural and optimization enhancements to CYK algorithm which significantly improved the performance in terms of the recognition speed. The proposed approach was evaluated using CROHME dataset and demonstrated the recognition quality comparable (ExpRate 65.76%) with the existing state-of-the-art approaches [15].

To further improvement, we plan to incorporate a new sequence re-ordering techniques, study another RNN architectures and make experiments with different scoring functions for better segmentation. In addition, we are going to examine other deep learning approaches for spatial relation classification.

References

1. Álvaro, F.M.: Mathematical expression recognition based on probabilistic grammars, Doctoral dissertation (2015)
2. Awal, A.-M., Mouchére, H., Viard-Gaudin, C.: Towards handwritten mathematical expression recognition. In: Proceedings of International Conference on Document Analysis and Recognition (ICDAR) (2009)
3. Bahdanau, D., Cho, K., Bengio, Y.: Neural machine translation by jointly learning to align and translate. CoRR, arXiv:1409.0473 (2014)
4. Carbune, V. et al.: Fast multi-language LSTM-based online handwriting recognition. arXiv:1902.10525 (2019)
5. Celik, M., Yanikoglu, B.: Probabilistic mathematical formula recognition using a 2D context-free graph grammar. In: Proceedings of International Conference on Document Analysis and Recognition (ICDAR), pp. 161–166 (2011)
6. Chiu, C., et al.: State-of-the-art speech recognition with sequence-to-sequence models. In: Proceedings of IEEE International Conference on Acoustics, Speech and Signal Processing (ICASSP), Calgary, Canada, pp. 4774–4778 (2018)
7. Degtyarenko, I., Radyvonenko, O., Bokhan, K., Khomenko, V.: Text/shape classifier for mobile applications with handwriting input. Int. J. Doc. Anal. Recogn. (IJDAR) **19**(4), 369–379 (2016)
8. Graves, A., Liwicki, M., Fernandez, S., Bertolami, R., Bunke, H., Schmidhuber, J.: A novel connectionist system for unconstrained handwriting recognition. IEEE Trans. Pattern Anal. Mach. Intell. **31**(5), 855–868 (2009)
9. Hu, L., Zanibbi, R.: MST-based visual parsing of online handwritten mathematical expressions. In: Proceedings of 15th International Conference on Frontiers in Handwriting Recognition (ICFHR) (2016)
10. Huang, B., Zhang, Y., Kechadi, T.: Preprocessing techniques for online handwriting recognition. Stud. Comput. Intell. **164**, 25–45 (2008)
11. Khomenko, V., Shyshkov, O., Radyvonenko O., Bokhan K.: Accelerating recurrent neural network training using sequence bucketing and multi-GPU data parallelization. In: Proceedings of IEEE International Conference on Data Stream Mining and Processing (DSMP) (2016). arXiv:1708.05604
12. Le, A.D., Indurkhya, B., Nakagawa, M.: Pattern generation strategies for improving recognition of handwritten mathematical expressions. arXiv:1901.06763 (2019)
13. Le, A. D., Nakagawa, M.: Speedup of parsing for recognition of online handwritten mathematical expressions. In: 14th International Conference on Document Analysis and Recognition (ICDAR) (2017)
14. Le, A. D., Nakagawa, M.: Training an end-to-end system for handwritten mathematical expression recognition by generated patterns. In: 14th International Conference on Document Analysis and Recognition (ICDAR) (2017)
15. Mouchére, H., Zanibbi, R., Garain, U., Viard-Gaudin, C.: Advancing the state of the art for handwritten math recognition: the CROHME competitions (2011–2016)
16. Nguyen, H.D., Le, A.D., Nakagawa M.: Deep neural networks for recognizing online handwritten mathematical symbols. In: 3rd IAPR Asian Conference on Pattern Recognition (ACPR) (2015)
17. Ormanci, U., Cepni, S., Deveci, I., Ozhan, A.: A thematic review of interactive whiteboard use in science education: rationales, purposes, methods and general knowledge. J. Sci. Educ. Technol. **24**(5), 532–548 (2015)

18. Volkova, V., Deriuga, I., Osadchiy, V., Radyvonenko, O.: Improvement of character segmentation using recurrent neural networks and dynamic programming. In: IEEE Second International Conference Data Stream Mining and Processing 2018 (DSMP), Lviv, Ukraine, pp. 218–222 (2018)

19. Tapia, E., Rojas, R.: Recognition of on-line handwritten mathematical expressions using a minimum spanning tree construction and symbol dominance. In: Lladós, J., Kwon, Y.-B. (eds.) GREC 2003. LNCS, vol. 3088, pp. 329–340. Springer, Heidelberg (2004). https://doi.org/10.1007/978-3-540-25977-0_30

20. Zhang, J., Du, J., Dai, L.: Multi-scale attention with dense encoder for handwritten mathematical expression recognition. In: ICPR, Beijing, China (2018)

21. Zhang, J., et al.: Watch, attend and parse: an end-to-end neural network based approach to handwritten mathematical expression recognition. Pattern Recogn. **71**, 196–206 (2017)

22. Zhang, T., Mouchére, H., Viard-Gaudin, C.: Using BLSTM for interpretation of 2D languages - case of handwritten mathematical expressions. Doc. Numérique **19**, 135–157 (2016)

23. Zhang, T., Mouchére, H., Viard-Gaudin, C.: Tree-based BLSTM for mathematical expression recognition. In: International Conference on Document Analysis and Recognition (ICDAR) (2017)

24. Xu, K. et al.: Show, attend and tell: neural image caption generation with visual attention. In: ICML, pp. 2048–2057 (2015)

A New Graph Based Brain Connectivity Measure

Addisson Salazar$^{(\boxtimes)}$, Gonzalo Safont, and Luis Vergara

Institute of Telecommunications and Multimedia Applications,
Universitat Politècnica de València, Valencia, Spain
{asalazar, lvergara}@dcom.upv.es,
gonsaar@upvnet.upv.es

Abstract. This paper presents a new measure of brain connectivity based on graphs. The method to estimate connectivity is derived from the set of transition matrices obtained by multichannel hidden Markov modeling and graph connectivity theory. Analysis of electroencephalographic signals from epileptic patients performing neuropsychological tests with visual stimuli was approached. Those tests were performed as clinical procedures to evaluate the learning and short-term memory capabilities of the patients. The proposed method was applied to classify the stages (stimulus display and subject response) of the Barcelona and the Wechsler Memory Scale - Figural Memory tests. To evaluate the capabilities of the proposed method, commonly used brain connectivity measures: correlation, partial correlation, and coherence were implemented for comparison. Results show the proposed method clearly outperforms the other ones in terms of classification accuracy and brain connectivity structures.

Keywords: Brain connectivity · Signal processing on graphs · HMM · EEG · Neuropsychological tests

1 Introduction

The subject considered in brain connectivity is the organization of anatomical and functional connections or interactions between brain regions to process information. The complexity and extension of this area joint with prevalent neurological diseases, and the number of neurodegenerative diseases coming from aging of population make this area has been increasingly studied. Brain connections determine spatially distributed networks formed by local and distant brain regions returning synchronized activity. Brain connectivity can be roughly organized into three classes (anatomical, functional, and effective connectivity) depending on the assumptions considered [1]. There are several methods proposed to measure brain connectivity such as coherence, partial coherence, correlation, independent component analysis-based methods, and graphical models, see for instance [1, 2] and the references within.

Brain connectivity has been measured mainly on image datasets, e.g., magnetic resonance imaging (MRI), diffusion MRI, and functional MRI (fMRI) [3]. However, brain source imaging to reconstruct electrical activity everywhere in the brain based on surface electroencephalography (EEG) and intracranial EEG signals has been recently

I. Rojas et al. (Eds.): IWANN 2019, LNCS 11507, pp. 450–459, 2019.
https://doi.org/10.1007/978-3-030-20518-8_38

approached [4, 5]. Thus, EEG signals has been employed for localization of source cortical areas as well as connectivity estimates of their interactions [6].

In this paper, we propose a new method to measure brain connectivity based on graph signal processing (GSP), see [7] and the references within. One key aspect of GSP is defining the graph connectivity. This can be made considering typical inter-actions from the application where the graph signal is defined, e.g., time or space proximity between two nodes. GSP has been successfully applied in quite different problems such as detection of arousals in apnea patients [8, 9] and credit card fraud detection [10]. In this work, a state transition probability matrix constructed from the parameters of a Multichannel Hidden Markov Model (MHMM) [11, 12] is considered to define the proposed graph connectivity measure. The application approached here was the classification of the stages (stimulus display and subject response) of two visual neuropsychological tests from EEG signals. Those tests were made as part of the clinical analysis to evaluate the learning and short-term memory capabilities of epileptic patients.

The rest of the paper is organized as follows. Section 2 includes the theoretical derivation of the proposed method. Section 3 describes the methods used in compar-isons. Section 4 contains experimental results and discussion. An overall conclusion is given in Sect. 5.

2 Proposed Connectivity Measure

The proposed method to measure connectivity is presented in this section.

2.1 State Transition Probability Matrix

Let us consider the state transition probability from the parameters of a MHMM,

$$\pi_{\mathbf{k}_t \mathbf{k}_{t-1}} = P[\mathbf{k}(t)/\mathbf{k}(t-1)], \tag{1}$$

which the probability of having a state vector $\mathbf{k}(t) = [k_1(t) \ldots k_L(t)]^T$ ($l = 1 \ldots L$ is the number of channels) in time instant t conditioned to having a vector of states $\mathbf{k}(t-1) = [k_1(t-1) \ldots k_L(t-1)]^T$ at instant $t-1$.

We could try to estimate the transition probability between the state k_m of one of the channels at instant $t-1$ and the state k_l of the other channel at instant t and vice versa $P[k_l(t)/k_m(t-1)] = P[k_m(t)/k_l(t-1)]$. Let us name i to the state of channel l at instant t or at instant $t-1$, and j to the state of channel m at instant t or at instant $t-1$, we can write,

$$P[k_l(t)/k_m(t-1)] = P[i/j] = P[k_m(t)/k_l(t-1)] = P[j/i]. \tag{2}$$

We may define a symmetric (KxK) state transition probability matrix between any pair of channels, whose generic element is given by $\mathbf{P}(i,j) = P[i/j] = P[j/i]$. The elements of that matrix can be estimated by integration of $\pi_{\mathbf{k}_t \mathbf{k}_{t-1}}$

$$P(i,j) = \frac{1}{2} \left(\sum_{k_t=a_t} \sum_{k_{t-1}=a_{t-1}} \pi_{k_t k_{t-1}} + \sum_{k_t=b_t} \sum_{k_{t-1}=b_{t-1}} \pi_{k_t k_{t-1}} \right), \qquad (3)$$

where $\mathbf{a}_t$ represent the subset of vectors $\mathbf{k}_t$ such that $k_l(t) = i$, $\mathbf{a}_{t-1}$ represents the subset of vectors $\mathbf{k}_{t-1}$ such that $k_m(t-1) = j$, $\mathbf{b}_t$ represents the subset of vectors $\mathbf{k}_t$ such that $k_m(t) = j$, y $\mathbf{b}_{t-1}$ represents the subset of vectors $\mathbf{k}_{t-1}$ such that $k_l(t-1) = i$.

2.2   Graph Model Based Method

Let us consider an undirected graph G $(K, E, \mathbf{W})$. K represents the set of nodes of the graph, E represents the set of edges connecting every two nodes, $\mathbf{W}$ is the adjacency matrix of weights associated to every edge, thus matrix element $\mathbf{W}(i,j)$ is a measure of the interaction or influence of node i in node j and vice versa. In an undirected graph $\mathbf{W}$ is symmetric. Normally, a transition matrix is defined by normalizing $\mathbf{W}$ in the form $= \mathbf{D}^{-1}\mathbf{W}$, where $\mathbf{D}$ is a diagonal matrix having element $\mathbf{D}(i,i) = \sum_{j=1}^{K} \mathbf{W}(i,j)$, the degree of node i. Then $\mathbf{P}(i,j)$ is a value between 0 and 1, which in the context of graph theory measures the probability of transit from node i to node j (or vice versa) in a one step of a random walk across the graph nodes.

We may consider a graph of K nodes, having a transition matrix equal to the state transition probability matrix previously defined. Every node corresponds to a possible state $k = 1 \ldots K$ of any of the two entities, and every edge connects the state of the first entity at time $t - 1$ with the state of the second entity at state t and vice versa.

Measures of the graph connectivity are mostly related with the graph spectral information: the eigenvalues of the selected graph matrix. In particular, the eigenvalues of a symmetric transition matrix satisfy

$$1 = \lambda_0 \geq \lambda_1 \ldots \lambda_{K-1} \geq -1. \qquad (4)$$

We will consider that a graph is maximally connected when every node is connected to all the other nodes and, in a step of a random walk, it is possible to go from one node to another (or to stay at the same node) with the same probability. This implies that

$$P(i,j) = \frac{1}{K} \ \forall i,j \Rightarrow \ \lambda_1 = 1 \ \lambda_2 = 0 \ldots \lambda_K = 0. \qquad (5)$$

Conversely, a graph is minimally connected when every node is isolated, i.e., $\mathbf{P}(i,i) = 1 \ \forall i$, or connected to only one another node with transition probability 1 in a step of a random walk. This implies that only one element of every row (or column) of $\mathbf{P}$ is equal to 1 and the rest elements are equal to 0.

$$\text{given } i, \ \mathbf{P}(i,j) = \begin{cases} 1 & \text{for some } j \neq i \\ 0 & \text{rest} \end{cases} \Rightarrow, \ \lambda_1 = 1 \ \lambda_k = \pm 1 \ k = 2 \ldots K. \qquad (6)$$

Notice that maximal graph connectivity coincides with lack of connectivity between the two entities and minimal graph connectivity is equivalent to total connectivity of the two entities. We can take advantages of the effect of connectivity in the eigenvalues and propose a general connectivity measure as a function of the eigenvalues

$$c_\lambda = f(\lambda), \tag{7}$$

where $\lambda = [\lambda_2 \ldots \lambda_K]^T$ is the eigenvalue vector, excluding λ_1 which is non-informative (always equals 1). We can define a normalized coefficient of connectivity as

$$c_C = \frac{1}{(K-1)^{1/2}} \|\lambda\|_2 = c_{\|\lambda\|_2}, \tag{8}$$

where $\|.\|_2$ is the Euclidean norm. But other options are possible for $f(\lambda)$. This can be considered a particular case of general coefficient of connectivity using the p-norm of vector λ

$$c_{\|\lambda\|_p} = \frac{1}{(K-1)^{1/p}} \cdot \|\lambda\|_p. \tag{9}$$

Notice that $0 \le c_{\|\lambda\|_p} \le 1$.

The Simplest Case
Let us consider the case of $L = 2$ and $K = 2$. Thus, we have two channels in MHMM evolving with time in two possible states, i.e., $k_1(t) = 1 \, or \, 2$ and $k_2(t) = 1 \, or \, 2$. Then $\mathbf{k}(t) = [k_1(t) \, k_2(t)]^T$ and the transition parameters estimated from the model will be, in this case

$$\pi_{\mathbf{k}_t \mathbf{k}_{t-1}} = P[k_1(t) \, k_2(t)/k_1(t-1) \, k_2(t-1)]. \tag{10}$$

From these parameters we can estimate the (2×2) state transition matrix $\mathbf{P}$ using (3).

$$\mathbf{P}(i,j) = \frac{1}{2} \left(\sum_{m=1}^{2} \sum_{m'=1}^{2} P[i,m/m',j] + P[m,i/j,m'] \right). \tag{11}$$

$\mathbf{P}$ is a symmetric doubly stochastic matrix, hence for the (2×2) case it must be $\mathbf{P}(1,1) = \mathbf{P}(2,2)$ and $\mathbf{P}(1,2) = \mathbf{P}(2,1) = 1 - \mathbf{P}(1,1)$. The (two) eigenvalues of $\mathbf{P}$ are the solutions of the equation

$$det(\mathbf{P} - \lambda\mathbf{I}) = 0 \iff (\mathbf{P}(1,1) - \lambda)^2 - (1 - \mathbf{P}(1,1))^2 = 0, \tag{12}$$

solving for λ we find the two roots, namely

$$\lambda_1 = 1 \quad \lambda_2 = 2\mathbf{P}(1,1) - 1. \tag{13}$$

There are two possibilities of total connectivity: $\mathbf{P}(1,1) = 1$, (if the first entity is in a given state at instant $t - 1$, the other will necessarily be at the same state at instant t), or $\mathbf{P}(1,1) = 0$ (if the first entity is in a given state at instant $t - 1$, the other will necessarily be at the opposite state at instant t. When $(1,1) = 1 \Rightarrow \lambda_2 = 1$. When $\mathbf{P}(1,1) = 0 \Rightarrow \lambda_2 = -1$. In both cases $c_{\|\lambda\|_p} = \frac{1}{(2-1)^{1/p}} \cdot (|\lambda_2|^p)^{1/p} = |\lambda_2| = 1$. Lack of connectivity implies $\mathbf{P}(1,1) = \frac{1}{2} \Rightarrow 1 \quad \lambda_2 = 0$ and $c_{\|\lambda\|_p} = |\lambda_2| = 0$. Figure 1 shows the variation of the connectivity parameter with $\mathbf{P}(1,1)$.

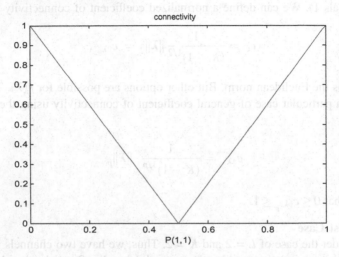

Fig. 1. Variation of the connectivity parameter with $\mathbf{P}(1,1)$.

3 Methods for Comparison

This section includes definition of methods that are commonly used to estimate brain connectivity and were employed for comparison with the proposed one.

In the following definitions, we will assume that the observations at time t are denoted by random vector $\mathbf{x}(t) = [x_1(t) \ldots x_M(t)]^T$, with M being the number of channels. The coherence of two channels $x_l(t)$ and $x_m(t)$ at frequency f is defined as

$$c_{lm}(f) = \frac{|P_{lm}(f)|^2}{P_{ll}(f)P_{mm}(f)}, \tag{14}$$

where $P_{lm}(f)$ is the cross spectral density of channels $x_l(t)$ and $x_m(t)$ at frequency f; and $P_{ll}(f)$ and $P_{mm}(f)$ are the power spectral densities of $x_l(t)$ and $x_m(t)$, respectively, at frequency f. The coherence is symmetrical and bound within 0 and 1, with higher values indicating more similar frequency contents. In this work, we average the coherence within each of the following frequency bands: delta (0.5–4 Hz), theta (4–7 Hz), lower alpha (8–10 Hz), higher alpha (10–12 Hz), beta (13–30 Hz) and gamma (30–40 Hz).

The correlation of two channels $x_l(t)$ and $x_m(t)$ is defined as

$$r_{lm} = \frac{E[(x_l - \mu_l)(x_m - \mu_m)]}{\sigma_l \sigma_m},\tag{15}$$

where $E[]$ is the expectation operator; μ_l, μ_m are the average values of channels $x_l(t)$, $x_m(t)$; and σ_l, σ_m are the standard deviation of channels $x_l(t)$, $x_m(t)$. The correlation is symmetrical and bound within the interval $[-1, 1]$. Values close to zero indicate low correlation between the channels, whereas values close to 1 (-1) indicate a linear direct (inverse) relationship between the channels.

Both the correlation and the coherence consider pairs of channels in isolation, without considering possible relationships with other channels. Conversely, the partial correlation measures the association of two channels after controlling for the effect of the rest of channels:

$$\rho_{lm} = \frac{E\left[(x_l' - E[x_l'])(x_m' - E[x_m'])\right]}{\sqrt{E\left[(x_l' - E[x_l'])^2\right]E\left[(x_m' - E[x_m'])^2\right]}},\tag{16}$$

where x_l', x_m' are the channels x_l, x_m after removing the contribution of the other channels. In practice, the partial correlation is estimated from the inverse of the covariance matrix,

$$\rho_{lm} = -\frac{\upsilon_{lm}}{\sqrt{\upsilon_{ll}\upsilon_{mm}}},\tag{17}$$

where υ_{lm} is the lm element of matrix Υ, being $\Upsilon^{-1} = E\left[(\mathbf{x} - E[\mathbf{x}])(\mathbf{x} - E[\mathbf{x}])^T\right]$. The partial correlation has the same limits and interpretation as the correlation.

4   Results

The proposed connectivity measure was tested on two datasets of EEG signals from six epileptic patients while performing two neuropsychological tests. The tests were carried out in a clinical environment to evaluate the learning and short-term memory capabilities of the patients. During the tests, multichannel EEG signals were captured with the help of the Neurology and Neurophysiology Units at Hospital Universitari i Politècnic La Fe, Valencia (Spain). The analysis of the tests is an essential area of clinical neurophysiology, since the evaluation of the learning and memory functions of the patient is a critical part of their neuropsychological condition assessment. Recently, we have demonstrated the automatic classification, based on change detection using a probabilistic distance, of the stages of neuropsychological tests from EEG signals [13].

The following two visual neuropsychological tests were implemented. The first test was the Barcelona test [14]. During each trial of the Barcelona test, the subject is shown a probe item for 10 s, and after a 10-s retention interval, they must be able to pick the probe item from a set of four similar items. The second test was the Wechsler Memory

Scale (WMS) - Figural Memory [15]. This is an immediate recognition test of abstract designs. The participant is shown a set of three abstract figures, rectangles with a pattern of shapes inside them, for 10 s. Afterward, the examinee is shown a set of nine similar figures from which they have to select the three figures they were shown before. There are three trials of increasing difficulty. Scoring is calculated from the number of correctly-selected figures.

Each trial was divided in two stages: stimulus display and subject response. The considered connectivity methods were then used to estimate the connectivity of the EEG channels on each stage of each trial. Each set of EEG signals was captured on $L = 18$ bipolar EEG channels set according to the 10–20 system, sampled at 500 Hz. A band pass filter (0.5 Hz and 50 Hz) and a notch filter (50 Hz) were applied on the acquired signals.

Besides, a preprocessing step consisting of quantizes the signal of each channel into Q signals levels was implemented; the quantized values of the signal were used to calculate a transition matrix like in HMM [16]. All the L transition matrices obtained for the EEG channels were used to estimate the joint connectivity for the MHMM using the proposed procedure explained in Sect. 2.

The connectivity returned by the proposed method, Eq. (8), was compared with those returned by the following commonly used connectivity measures: correlation, partial correlation, and coherence. As was commented in Sect. 3, the coherence was averaged for the following frequency bands: delta (0.5–4 Hz), theta (4–7 Hz), lower alpha (8–10 Hz), higher alpha (10–12 Hz), beta (13–30 Hz) and gamma (30–40 Hz).

Several classification experiments were performed in order to test the connectivity matrix stability and its capability to capture meaningful brain connectivity structures. The EEG samples were split into two labeled datasets corresponding to the stages of the tests (display and response). A different connectivity matrix was obtained for each of those datasets. This process was repeated for all the trials in each test, obtaining between 10 and 20 connectivity matrices per subject and test. Then, a leave-one-out classification experiment was performed with a 1-nearest neighbor classifier, using the matrices as features and attempting to classify the two stages of each trial. This process was repeated using each of the considered methods; in the case of the proposed method, we considered a range of values for Q.

The average results of the experiment are shown in Figs. 2 and 3 shows the results for Barcelona test and WMS - Figural Memory, respectively. The results shown correspond to the better-performing band in terms of classification accuracy for each test (delta band for the Barcelona test and gamma band for the WMS - Figural Memory test).

It can be seen that performance was higher for the proposed method once the number of considered levels is high enough. Besides, notice that the result of the proposed method depends on the value of Q, meanwhile the results of the other methods are estimated on the signal without quantization, and thus, they do not depend on the value of Q. The obtained performance demonstrates that the connectivity matrices returned by the proposed method were more stable (and thus, more similar across trials) than those returned by the other competitive methods.

The connectivity matrices returned by each method for the Barcelona test are compared in Fig. 4. In order to show the graphs better, only the strongest 10% (highest

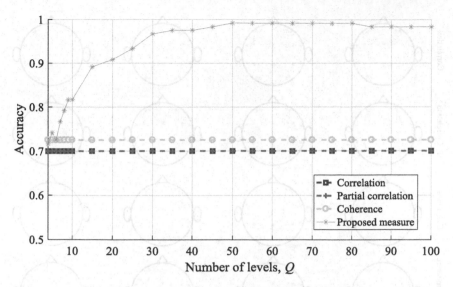

Fig. 2. Classification accuracy results for Barcelona test.

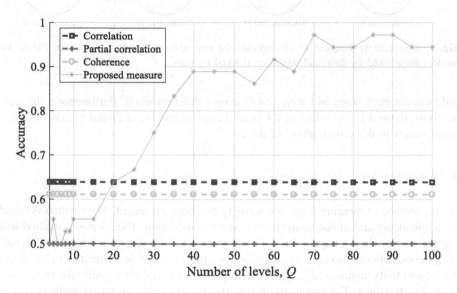

Fig. 3. Classification accuracy results for WMS test - Figural Memory.

connectivity values) of the connections have been shown. It can be seen that the connectivity matrices returned by the correlation and partial correlation were very similar for the two stages of the trial, thus, it was difficult to find differences between display and response. Conversely, the connectivity yielded by the coherence was different for each stage, but it was also different for the same stage across different trials. In contrast, the proposed connectivity measure yielded matrices that were

458   A. Salazar et al.

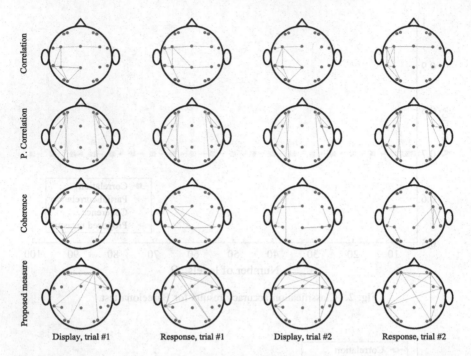

Fig. 4. Connectivity matrices of all methods for two trials of the Barcelona test. Electrodes (nodes) are marked by dots, and edges are marked by lines.

different for each stage and were stable across different trials. Furthermore, the connectivity showed involvement of the frontal regions and the occipital region, which is consistent with the visual nature of the test.

5 Conclusion

A new method to measure brain connectivity has been presented. The method is based on graph theory and multichannel hidden Markov modeling. The proposed method was applied to classify the stages of two visual neuropsychological tests from EEG signals of epileptic patients. Several classification experiments were performed in order to test the connectivity matrix stability and its capability to capture meaningful brain connectivity structures. The connectivity patterns found by the proposed method clearly allows to discriminate better the two stages of the tests (stimulus display and subject response) compared with the ones obtained by the coherence, correlation, and partial correlation. Besides, brain region activation highlighted in the connectivity matrices estimated by the proposed method are consistent with clinical basis of the tests.

Acknowledgements. This work was supported by Spanish Administration and European Union under grants TEC2014-58438-R and TEC2017-84743-P.

References

1. Lang, E.W., Tomé, A.M., Keck, I.R., Górriz-Sáez, J.M., Puntonet, C.G.: Brain connectivity analysis: a short survey. Comput. Intell. Neurosci. **2012**, 1–21 (2012). Article ID 412512
2. Straathof, M., Sinke, M.R., Dijkhuizen, R.M., Otte, W.M.: A systematic review on the quantitative relationship between structural and functional network connectivity strength in mammalian brains. J. Cereb. Blood Flow Metab. **39**(2), 189–209 (2019)
3. Frau-Pascual, A., et al.: Quantification of structural brain connectivity via a conductance model. Neuroimage **189**, 485–496 (2019)
4. Becker, H., et al.: SISSY: an efficient and automatic algorithm for the analysis of EGG sources based on structured sparsity. Neuroimage **157**, 157–172 (2017)
5. Wendling, F., Chauvel, P., Biraben, A., Bartolomei, F.: From intracerebral EEG signals to brain connectivity: identification of epileptogenic networks in partial epilepsy. Front. Syst. Neurosci. **4**, 1–13 (2010). Article 154
6. Mahjoory, K., Nikulin, V.V., Botrel, L., Linkenkaer-Hansen, K., Fato, M.M., Haufe, S.: Consistency of EEG source localization and connectivity estimates. Neuroimage **152**, 590–601 (2017)
7. Ortega, A., Frossard, P., Kovacevic, J., Moura, J.M.F., Vandergheynst, P.: Graph signal processing: overview, challenges and applications. Proc. IEEE **106**, 808–828 (2018)
8. Belda, J., Vergara, L., Salazar, A., Safont, G.: Estimating the Laplacian matrix of Gaussian mixtures for signal processing on graphs. Signal Process. **148**, 241–249 (2018)
9. Belda, J., Vergara, L., Safont, G., Salazar, A.: Computing the partial correlation of ICA models for Non-Gaussian graph signal processing. Entropy **21**(1), 22 (2019). https://doi.org/10.3390/e21010022. 1–16
10. Vergara, L., Salazar, A., Belda, J., Safont, G., Moral, S., Iglesias, S.: Signal processing on graphs for improving automatic credit card fraud detection. In: 51st International Carnahan Conference on Security Technology, pp. 1–6 (2017)
11. Safont, G., Salazar, A., Vergara, L., Gomez, E., Villanueva, V.: Multichannel dynamic modeling of non-Gaussian mixtures. To appear in Pattern Recognition (2019)
12. Safont, G., Salazar, A., Bouziane, A., Vergara, L.: Synchronized multi-chain mixture of independent component analyzers. In: Rojas, I., Joya, G., Catala, A. (eds.) IWANN 2017, Part I. LNCS, vol. 10305, pp. 190–198. Springer, Cham (2017). https://doi.org/10.1007/978-3-319-59153-7_17
13. Safont, G., Salazar, A., Vergara, L., Gomez, E., Villanueva, V.: Probabilistic distance for mixtures of independent component analyzers. IEEE Trans. Neural Netw. Learn. Syst. **29**(4), 1161–1173 (2018)
14. Quintana, M., et al.: Spanish multicenter normative studies (Neuronorma project): norms for the abbreviated Barcelona test. Arch. Clin. Neuropsychol. **26**(2), 144–157 (2010)
15. Strauss, E.: A Compendium of Neuropsychological Tests. Oxford University Press, Oxford (2006)
16. Campanharo, A.S.L.O., Doescher, E., Ramos, F.M.: Automated EEG signals analysis using quantile graphs. In: Rojas, I., Joya, G., Catala, A. (eds.) IWANN 2017, Part II. LNCS, vol. 10306, pp. 95–103. Springer, Cham (2017). https://doi.org/10.1007/978-3-319-59147-6_9

Soft Computing

Many-Objective Cooperative
Co-evolutionary Feature Selection:
A Lexicographic Approach

Jesús González[✉][iD], Julio Ortega[iD], Miguel Damas[iD],
and Pedro Martín-Smith[iD]

Department of Computer Architecture and Technology, CITIC,
University of Granada, Granada, Spain
{jesusgonzalez,jortega,mdamas,pmartin}@ugr.es
http://atc.ugr.es/

Abstract. This paper presents a new wrapper method able to optimize
simultaneously the parameters of the classifier while the size of the subset
of features that better describe the input dataset is also being minimized.
The search algorithm used for this purpose is based on a co-evolutionary
algorithm optimizing several objectives related with different desirable
properties for the final solutions, such as its accuracy, its final number
of features, and the generalization ability of the classifier. Since these
objectives can be sorted according to their priorities, a lexicographic
approach has been applied to handle this many-objective problem, which
allows the use of a simple evolutionary algorithm to evolve each one of
the different sub-populations.

Keywords: Many-objective evolutionary algorithm ·
Cooperative co-evolutionary algorithm · Lexicographic optimization ·
Feature selection · Wrapper approach

1 Motivation

Wrapper methods are intrinsically simple, what has made them quite popular.
They basically consist of a classifier, a search algorithm, and way to assess the
prediction accuracy of the learning machine in order to guide the search towards
good feature subsets [19]. Besides, since wrapper methods select the features
subset according to the classifier that will be applied later on to the test set,
they usually achieve better accuracy than filter methods, although they are also
more computationally expensive [15].

This work was supported by projects TIN2015-67020-P (Spanish "Ministerio de
Economía y Competitividad") and PGC2018-098813-B-C31 (Spanish "Ministerio de
Ciencia, Innovación y Universidades"), and by European Regional Development Funds
(ERDF).

© Springer Nature Switzerland AG 2019
I. Rojas et al. (Eds.): IWANN 2019, LNCS 11507, pp. 463–474, 2019.
https://doi.org/10.1007/978-3-030-20518-8_39

Regarding the classifier, Support Vector Machines (SVM) have been widely used within wrappers [22]. However, the functioning of this classifier rely on some parameters that must be fine tuned, according to the dataset, in order to achieve a good accuracy. SVM relies on the regularization parameter C and also on the set of parameters that define the type of kernel used. The correct initialization of these parameters is of capital importance, mainly because the final result of the wrapper method will depend on them. The problem here is that the value of these parameters depend on the final dataset defined by the selected features, which is a priori unknown.

Some approaches fix the parameters of the classifier heuristically before the wrapper method is applied. For example, in [22] these parameters were optimized with the whole training set (containing all the features) before the application of the wrapper procedure. However, these parameters might not be optimal for the final feature subset found by the wrapper algorithm. Besides, different values for the parameters could have provided a different feature subset.

Since the parameters of the classifier depend on the final feature subset, and this subset is the result of the search algorithm applied within the wrapper method, this paper proposes that the classifier parameters should be simultaneously optimized while the search algorithm is finding the best subset of features. That is, two interdependent problems should be simultaneously optimized: the parameters of the classifier, which depend on the subset of features used, and the best subset of features, which depend on the classifier used within the wrapper method. Cooperative Co-Evolutionary algorithms (CCEAs) are particularly well suited to this scenario, since they have been designed to evolve different species of solutions simultaneously [27].

On the other hand, some objectives should be taken into account in order to guide the search towards a good couple of classifier and subset of features. First of all, the classification error and generalization capability should be optimized, since the aim of this work is to find the subset of features that best describe the original dataset. Another objective that many approaches take into account is the number of features, that should also be minimized. Lastly, and since the classifier parameters are also going to be optimized, some objectives could also be defined. For example, in the case of the SVM classifier, the minimization of C is preferred, since lower values of C avoid over-fitting and also speed up the training and test processes. Thus, this is a Many-Objective Optimization Problem (MaOP) too, that is, a Multi-Objective Problem (MOP) with more than tree objectives [10].

Evolutionary algorithms (EAs) have also been widely applied to solve MaOPs [20]. However, and although there are several approaches to design Many-Objective EAs (MaOEAs), such as Pareto-based, indicator-based or aggregation-based approaches, all of them have been designed to treat all the objectives equally. That is, all the objectives in the problem have the same level of importance. Although this is the case in most MaOPs, not all the objectives have the same priority in the case of feature selection problems. For example, given two possible solutions for the problem, the one with a lower classification error is preferred, and only when the classification error for both is similar, the number of features should be considered. Thus, a priority-based scheme should be incorporated into the algorithm to guide the search towards adequate solutions.

Although some attempts have been made to support objective priorities in EAs [9, 29], a much simpler approach is possible if the problem allows to set a different priority level for each objective. This approach, which was initially proposed in 1975 [25], is lexicographic optimization, and problems meeting this restriction are also known as Lexicographic MOPs (LMOPs) [18]. Lexicographic optimizers try to meet all the objectives sequentially. First, the most important objective is optimized. Then, among the solutions meeting this objective, a smaller set of solutions is chosen to satisfy the second objective, and so on until all the objectives have been considered [1]. Although it may seem a quite simple approach, there are relevant LMOPs that have been successfully solved with it, even at the present time, such as the design and optimization of integrated vehicle control systems [17] or the design of autonomous vehicles [28].

There exist many analytical algorithms for LMOPs, but all of them impose restrictions, such as the differentiability and convexity of the objective functions [33]. Thus, for any general problem not meeting these constraints a more robust optimization technique is necessary. On the other hand, Multi-Objective EAs (MOEAs) have been applied successfully to MOPs where analytical approaches have failed [5]. Therefore they can also be used to solve any kind of LMOP, even with not convex and not continuous objectives. This is the approach presented in this paper, a Lexicographic Many-Objective Cooperative Co-Evolutionary Algorithm (LeMaOCCEA) to simultaneously optimize the parameters of the classifier within a wrapper method while the number of features is also being minimized.

The rest of the paper is organized as follows. Section 2 describes the lexicographic relation for MaOEAs, a relation that allows the full ranking of possible solutions for a MaOP where a different level of priority can be defined for each objective. Then, Sect. 3 describes the wrapper method proposed in this paper, a CCEA based approach using this lexicographic relation. After that, Sect. 4 presents some experimental results obtained with the proposed approach and compares them with those obtained with other wrapper methods, and finally, Sect. 5 concludes this work.

2 A Lexicographic Relation for MaOEAs

Assuming that n_o objectives have been defined for the problem, and that these objectives have been sorted according to their priority, the fitness for any solution for the problem can be expressed as:

$$f = \left[f^0, f^1, \ldots, f^{n_o-1}\right]^T \in \mathbb{R}^{n_o} \tag{1}$$

Given two fitness evaluations, f_1 and f_2, and a precision threshold t, the lexicographic relations between them, noted as $\prec_l$ and $\preceq_l$, can be defined as:

$$f_1 \prec_l f_2 \;\Leftrightarrow\; \exists k \in [0, n_o) \cap \mathbb{N} : f_1^k < f_2^k \;\wedge\; |f_1^k - f_2^k| \geq t$$
$$\wedge \; |f_1^i - f_2^i| < t \;\forall i < k \tag{2}$$
$$f_1 =_l f_2 \;\Leftrightarrow\; |f_1^i - f_2^i| < t \;\forall i \in [0, n_o) \cap \mathbb{N} \tag{3}$$
$$f_1 \preceq_l f_2 \;\Leftrightarrow\; f_1 \prec_l f_2 \;\vee\; f_1 =_l f_2 \tag{4}$$

As can be seen, this formulation differs from the pure mathematical lexicographic relation because a threshold t has been introduced to let the Decision Maker (DM) decide the precision used to make the comparison. Regarding the behavior of the algorithm using this relation, it is quite similar to classical lexicographic optimization techniques. It optimizes the objectives in order, but with an important difference. Since an EA is used, local optima can now be avoided [24].

The use of this relation allows the full ranking of the solutions of a MaOP, an thus, a simple EA can be used (not a MOEA or a MaOEA) with smaller populations, since the algorithm will now provide only one optimal solution in each population instead of a Pareto set of not comparable solutions.

3   Proposal

This section describes the main components of the LeMaOCCEA wrapper method presented in this paper, a wrapper method able to optimize the parameters of the classifier while the number of features taken into account is also being minimized. Figure 1 shows its flowchart. The contributions of this paper have been highlighted, indicating also the section that describes each one of them. The remaining steps of the method are taken from the original EA.

3.1   Cooperative Co-evolutionary Approach

Since the wrapper method is based on an CCEA, potential solutions of the problem will be evolved within populations of different species. In this case, a hybrid single-level and two-level approach is proposed [16]. One species will be used to evolve the parameters of the classifier while the features of the input dataset will also be split into several species. The number of species used to minimize the input features subset is not fixed a priori, and should be chosen for each experiment according to the number of features in the dataset, in order to balance the search spaces of every population.

With respect to the many-objective optimization part of the feature selection problem, the use of the lexicographic relation described above allows the use of a simple EA scheme to evolve each one of the species defined within the CCEA.

Finally, regarding the selection of representatives, the shuffle-and-pair method [26] will be used. This method shuffles the indexes to access individuals in each population and then combines all the individuals having the same index to form and evaluate a complete solution. Since only one evaluation per individual may seem a poor estimation of its fitness, the process is repeated r times, and then the best evaluation for each individual is chosen as its fitness. The only restriction for this method is that all the populations must be of the same size m.

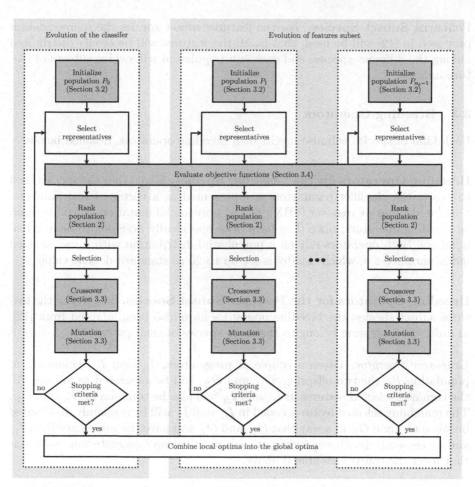

Fig. 1. Flowchart of the proposed wrapper method. The steps that are not highlighted are taken from the original EA

3.2   Species Representation

Since the proposed LeMaOCCEA is based on a hybrid single-level and two-level approach, different representations for the species are needed. Specifically, one representation for the parameters of the classifier, that will evolve in one population, and another one for the subsets of features that will be co-evolved in the remaining populations.

Classifier Species. The SVM parameters will be encoded as a vector of floating point numbers in the first population (P_0).

Features Subset Species. For the features subset species the representation proposed in [12] will be used, that is, all the features will be evenly distributed among the different species, and each sub-population will evolve subsets of feature indexes.

3.3 Breeding Operators

Each kind of species will also use its own breeding operators, detailed below.

Breeding Operators for the Classifier Species. Given that the species used to evolve the classifier parameters is represented as a vector of real numbers, Simulated Binary Crossover (SBX) [6] and polynomial mutation [7] will be used as breeding operators, since they have been specifically designed to handle real numbers. Both operators rely on a polynomial distribution with a user-defined index parameter ν, which usually is set up to 20 as standard default value.

Breeding Operators for the Features Subset Species. Regarding the features subset species, the breeding operators have also been adapted from [12], in order that offspring belong to the same species of their parents.

Crossover Operator. Given a couple of progenitors, I_{j_k} and I_{j_l}, belonging to population P_j, the two offspring, O_{j_k} and O_{j_l}, will be generated as follows. All the common selected features in I_{j_k} and I_{j_l} will also be common in O_{j_k} and O_{j_l}. The remaining selected features coded in I_{j_k} and I_{j_l} will be randomly distributed between O_{j_k} and O_{j_l} in a way that O_{j_k} and O_{j_l} will have the same sizes than I_{j_k} and I_{j_l} respectively. This crossover procedure will always generate valid solutions that meet the constraints stated above.

Mutation Operator. This operator affects each individual gene or selected feature within an individual separately. If a determined gene is chosen to be mutated, then it may be deleted (reducing the number of selected features within the individual) or randomly altered, being assigned a new value that should be a valid feature index for the problem and should not be repeated within the individual.

Once all the genes have been processed, the mutation operator may also add a new feature index to the individual, increasing the number of selected features, but only if the new feature index is not already selected.

3.4 Lexicographic Objectives for the Feature Selection Problem

It seems sensible that the most prioritary objective should be related to the accuracy of the classifier with the reduced features subset. Many works propose the use of the cross-validation error to avoid over-fitting when training classifiers [19]. However, and although this approach has proven successful, it has an important drawback. It is quite computationally expensive, since all the potential solutions tried by the search algorithm must be evaluated several times. This inconvenient affects even more to CCEA approaches, since all the solutions in each sub-population must be re-evaluated every generation because their fitness also depend on the solutions belonging the remaining sub-populations.

To overcome this problem, a new MOP-based distributed cross-validation approach has been proposed in [12]. For each potential solution (subset of input features) to be evaluated, the original training dataset D is randomly split into two subsets, D_{tr} and D_{val} according to parameter p_{val}, which indicates the percentage of solutions used to validate the solution. Although D is split into two different random subsets for each individual evaluation, the division procedure always assures that a percentage p_{val} of samples of each class in D are included in D_{val}. Then, the classifier is trained only with the samples belonging to D_{tr}, and later, two objectives are evaluated: the accuracy obtained by the classifier using D_{tr} and also the accuracy using D_{val}. For the accuracy estimation, the Kappa index [4] is proposed in [12], since it not only takes into account the accuracy of the classifier, but also the per class error distribution.

The following objective in importance should be the size of the final features subset, which should also be minimized, since the problem being solved is a feature selection problem. Finally, the last objective in importance is related to the regularization parameter C used by SVM classifiers. High values for this parameter allows the SVM to use more training examples for the definition of the hyperplane, which may tend to over-fitting, while lower values of C relax this condition. Since the accuracy of the classifier has been included as one of the higher priority objectives, setting the minimization of C as the lower priority objective will guide the algorithm towards solutions with a higher generalization ability without sacrificing the classifier accuracy.

Taking these reflections into account, the following objectives, sorted according to their priority, will be taken into account:

1. Minimize the error rate estimated using D_{val} (SVM is trained with D_{tr}).
2. Minimize the error rate estimated using D_{tr}
3. Minimize the number of features
4. Minimize the regularization parameter C of the SVM.

This approach considers the validation accuracy as the most prioritary objective, in order to avoid overfitting. Then the training accuracy is used to untie solutions having similar validation accuracies. For those solutions with similar validation and training accuracies, the number of features will be considered, and finally, if there are sill tied solutions, the regularization parameter of the SVM will be taken into account.

4   Experimentation

This section describes all the details related to the experimentation process carried out to evaluate the proposed LeMaOCCEA wrapper method. Experiments have been performed using the Zoo dataset [11], a well known dataset belonging to the UCI machine learning repository [8]. For all the experiments, each dataset has been randomly split into two sets as proposed in [32]: a training set containing 70% of all samples in each class, and a test set formed by the remaining samples. Adopting the same methodology will provide a fair comparison between the proposed wrapper method and those described in [32].

4.1 Other Wrapper Alternatives

The results of the proposed LeMaOCCEA wrapper method will be compared to those obtained by the following wrapper methods. All of them use KNN with $k = 5$ to simplify the evaluation process [32].

Linear Forward Selection (LFS): This wrapper method [13] is derived from the classical Sequential Forward Selection (SFS) [30], with the main difference that LFS restricts the number of features that are considered in each step of the forward selection, which can reduce the number of evaluations, optimizing the overall computation time.

Greedy Stepwise Backward Selection (GSBS): It is based on the traditional Sequential Backward Selection (SBS) method [23]. It starts with all the available features and removes sequentially one feature per iteration until the elimination of any remaining feature worsens the accuracy of the classifier [2].

Commonly Used PSO Algorithm (ErFS): This method applies the Particle Swarm Optimization (PSO) search algorithm [14] to minimize the error rate of the classifier. The implementation presented in [32] fixes the inertia weight $w = 0.7298$ and the acceleration constants $c_1 = c_2 = 1.49618$.

PSO with a Two-Stage Fitness Function (2SFS): This wrapper method, also based on the PSO algorithm, divides the evolutionary process into two stages. The first one only minimizes the error rate, whereas the second one also takes into account the number of features in the fitness function [31]. Since the results of this algorithm have also been taken from [32], the parameters of the PSO algorithm are the same than in ErFS.

4.2 Implementation Details and Parameterization of the Proposed LeMaOCCEA Method

The implementation of the co-evolutionary algorithm, along with the breeding operators for the population evolving the parameters of the SVM classifier have been taken from ECJ [21], a research Evolutionary Computation (EC) system written in *Java* and developed within the *Evolutionary Computation Laboratory* at the George Mason University, VA, USA. Moreover, for the SVM classifiers

Table 1. Parameters for the wrapper method

Parameter	Value
Number of populations (n_p)	5
Population size (m)	150
Number of generations (n)	300
Mutation probability for the feature selection species ($p_{m_{fs}}$)	0.01
Mutation probability for the SVM parameters species ($p_{m_{svm}}$)	0.05
Rate of training samples used for validation (p_{val})	0.33
Number of executions of the wrapper method (l)	40
Lexicographic precision threshold (t)	0.001
Co-evolutionary evaluation number of shuffles (r)	2

LibSVM has been used [3]. The rest of the code has been written by the authors of this work.

Since the LeMaOCCEA wrapper method presented in this paper relies on a CCEA, there are some parameters that must be chosen to make it work. Table 1 shows the values used for the experiments presented in this section. The number of species (populations) has been fixed in a way that the number of features assigned to each population should be 4 or 5.

4.3   Results

Table 2 shows the results obtained by the proposed LeMaOCCEA wrapper method, along with those achieved by the other wrapper methods introduced in Sect. 4.1. Although at first sight it seems that the proposed wrapper method outperforms the others, a Kruskal-Wallis statistical test has been applied to both, the test error rates and also the number of features of the results provided by each alternative. Table 3 shows the pairwise comparison of the proposed LeMaOCCEA wrapper method with all the other wrapper methods. As can be seen, the proposed wrapper method is able to achieve statistically significant better error rates and also smaller subsets of features than the other alternatives.

Table 2. Results obtained by the different wrapper alternatives

Method	Test error rate	# Features
LFS	20.950	8
GSBS	20	7
ErFS	4.500 ± 0.009	9.180
2SFS	4.500 ± 0.009	9.180
LeMaOCCEA	2.832 ± 1.609	4.800 ± 0.791

Table 3. p-values obtained from multiple pairwise comparison of the test error rate and the number of features achieved by the different wrapper alternatives using the Kruskal-Wallis statistical test

Wrapper 1	Wrapper 2	Test error rate	# Features
LFS	LeMaOCCEA	0.000	0.000
GSBS	LeMaOCCEA	0.000	0.000
ErFS	LeMaOCCEA	0.000	0.000
2SFS	LeMaOCCEA	0.000	0.000

5 Conclusions

This paper has presented a new wrapper approach which hybridized ideas of CCEAs and lexicographic optimization in order to be able to optimize simultaneously the parameters of a SVM classifier and the subset of features that better represent a dataset. The lexicographic approach allows to introduce any number of objectives easily, allowing to solve even MaOPs with a simple EA scheme as the search algorithm for each species. Another advantage of the proposed LeMaOCCEA wrapper method is that it produces only one solution per execution, instead of a set of Pareto optimal solutions, which simplifies the work of the DM.

The proposed LeMaOCCEA wrapper method is also able to obtain statistically significant better results than the rest of wrapper methods it has been compared to, taking into account both the test error rate and also the number of features finally selected.

References

1. Ben-Tal, A.: Characterization of Pareto and lexicographic optimal solutions. In: Fandel, G., Gal, T. (eds.) Multiple Criteria Decision Making Theory and Application. Lecture Notes in Economics and Mathematical System, vol. 177, pp. 1–11. Springer, Berlin (1979). https://doi.org/10.1007/978-3-642-48782-8_1
2. Caruana, R., Freitag, D.: Greedy attribute selection. In: Cohen, W.W., Hirsh, H. (eds.) Proceedings of the Eleventh International Conference on International Conference on Machine Learning, ICML 1994, pp. 28–36. Morgan Kaufmann, New Brunswick, July 1994
3. Chang, C.C., Lin, C.J.: LIBSVM: a library for support vector machines. ACM Trans. Intell. Syst. Technol. **2**(3), 27 (2011). http://www.csie.ntu.edu.tw/~cjlin/libsvm
4. Cohen, J.: A coefficient of agreement for nominal scales. Educ. Psychol. Measure. **20**(1), 37–46 (1960). https://doi.org/10.1037/h0026256
5. Deb, K.: Multi-Objective Optimization Using Evolutionary Algorithms. Wiley Interscience Series in Systems and Optimization. Wiley, Chichester (2001)
6. Deb, K., Agrawal, R.B.: Simulated binary crossover for continuous search space. Complex Syst. **9**(2), 115–148 (1995). https://www.complex-systems.com/abstracts/v09_i02_a02/

7. Deb, K., Agrawal, S.: A niched-penalty approach for constraint handling in genetic algorithms. In: Dobnikar, A., Steele, N.C., Pearson, D.W., Albrecht, R.F. (eds.) Artificial Neural Networks and Genetic Algorithms, pp. 235–243. Springer, Portorož (1999). https://doi.org/10.1007/978-3-7091-6384-9_40

8. Dheeru, D., Karra Taniskidou, E.: UCI machine learning repository. School of Information and Computer Sciences, University of California, Irvine, CA, USA (2017). http://archive.ics.uci.edu/ml

9. Drechsler, N., Sülflow, A., Drechsler, R.: Incorporating user preferences in many-objective optimization using relation ε-preferred. Natural Comput. 14(3), 469–483 (2015). https://doi.org/10.1007/s11047-014-9422-0

10. Farina, M., Amato, P.: On the optimal solution definition for many-criteria optimization problems. In: Keller, J., Nasraoui, O. (eds.) Proceedings of the 2002 Annual Meeting of the North American Fuzzy Information Processing Society, pp. 233–238. IEEE, New Orleans, June 2002. https://doi.org/10.1109/NAFIPS.2002.1018061

11. Forsyth, R.S.: Zoo dataset. Mapperley Park, Nottingham, UK (1990). https://archive.ics.uci.edu/ml/datasets/Zoo

12. González, J., Ortega, J., Damas, M., Martín-Smith, P., Gan, J.Q.: A new multi-objective wrapper method for feature selection - accuracy and stability analysis for BCI. Neurocomputing 333(14), 407–418 (2019). https://doi.org/10.1016/j.neucom.2019.01.017

13. Gutlein, M., Frank, E., Hall, M., Karwath, A.: Large-scale attribute selection using wrappers. In: Smith-Miles, K., Keogh, E., Lee, V.C. (eds.) Proceedings of the 2009 IEEE Symposium on Computational Intelligence and Data Mining, CIDM 2009. IEEE, Nashville, March 2009. https://doi.org/10.1109/CIDM.2009.4938668

14. Kennedy, J., Eberhart, R.: Particle swarm optimization. In: Proceedings of the IEEE International Conference on Neural Networks, ICNN 1995, vol. 6, pp. 1942–1948. IEEE, Perth, November 1995. https://doi.org/10.1109/ICNN.1995.488968

15. Khalid, S., Khalil, T., Nasreen, S.: A survey of feature selection and feature extraction techniques in machine learning. In: Arai, K., Mellouk, A. (eds.) Proceedings of the 2014 Science and Information Conference, pp. 372–378. The Science and Information (SAI) Organization, London, August 2014. https://doi.org/10.1109/SAI.2014.6918213

16. Khare, V.R., Yao, X., Sendhoff, B.: Credit assignment among neurons in co-evolving populations. In: Yao, X., et al. (eds.) PPSN 2004. LNCS, vol. 3242, pp. 882–891. Springer, Heidelberg (2004). https://doi.org/10.1007/978-3-540-30217-9_89

17. Khosravani, S., Jalali, M., Khajepour, A., Kasaiezadeh, A., Chen, S.K., Litkouhi, B.: Application of lexicographic optimization method to integrated vehicle control systems. IEEE Trans. Ind. Electron. 65(12), 9677–9686 (2018). https://doi.org/10.1109/TIE.2018.2821625

18. Klepikova, M.G.: The stability of lexicographic optimization problem. USSR Comput. Math. Math. Phys. 25(1), 21–29 (1985). https://doi.org/10.1016/0041-5553(85)90037-0

19. Kohavi, R., John, G.H.: Wrappers for feature subset selection. Artif. Intell. 97(1–2), 273–324 (1997). https://doi.org/10.1016/S0004-3702(97)00043-X

20. Li, B., Li, J., Tang, K., Yao, X.: Many-objective evolutionary algorithms: a survey. ACM Comput. Surv. 48(1), 13 (2015). https://doi.org/10.1145/2792984

21. Luke, S., et al.: ECJ 26. A Java-based evolutionary computation research system. https://cs.gmu.edu/~eclab/projects/ecj/

22. Madonado, S., Weber, R.: A wrapper method for feature selection using support vector machines. Inf. Sci. **179**(13), 2208–2217 (2009). https://doi.org/10.1016/j.ins.2009.02.014
23. Marill, T., Green, D.: On the effectiveness of receptors in recognition systems. IEEE Trans. Inf. Theory **9**(1), 11–17 (1963). https://doi.org/10.1109/TIT.1963.1057810
24. Michalewicz, Z.: Genetic Algorithms + Data Structures = Evolution Programs, 3rd edn. Springer, Berlin (1998)
25. Podinovskii, V.V., Gavrilov, V.M.: Optimization with respect to successive criteria (Optimizatsiya po posledovatel'no primenyaemym kriteriyam). Sovetskoe Radio, Moscow, Russia (1975)
26. Popovici, E., Bucci, A., Wiegand, R.P., De Jong, E.D.: Coevolutionary principles. In: Rozenberg, G., Bäck, T., Kok, J.N. (eds.) Handbook of Natural Computing, pp. 987–1033. Springer, Berlin (2012). https://doi.org/10.1007/978-3-540-92910-9_31
27. Potter, M.A., De Jong, K.A.: Cooperative coevolution: an architecture for evolving coadapted subcomponents. Evol. Comput. **8**(1), 1–29 (2000). https://doi.org/10.1162/106365600568086
28. Rasekhipour, Y., Fadakar, I., Khajepour, A.: Autonomous driving motion planning with obstacles prioritization using lexicographic optimization. Control Eng. Pract. **77**, 235–246 (2018). https://doi.org/10.1016/j.conengprac.2018.04.014
29. Schmiedle, F., Drechsler, N., Große, D., Drechsler, R.: Priorities in multi-objective optimization for genetic programming. In: Spector, L., Goodman, E.D., Wu, A., Langdon, W.B., Voigt, H.M. (eds.) Proceedings of the 3rd Annual Conference on Genetic and Evolutionary Computation, GECCO 2001, pp. 129–136. Morgan Kaufmann Publishers Inc., San Francisco, July 2001. https://dl.acm.org/citation.cfm?id=2955256
30. Whitney, A.W.: A direct method of nonparametric measurement selection. EEE Trans. Comput. **C–20**(9), 1100–1103 (1971). https://doi.org/10.1109/T-C.1971.223410
31. Xue, B., Zhang, M., Browne, W.N.: New fitness functions in binary particle swarm optimisation for feature selection. In: Abbass, H., Essam, D., Sarker, R. (eds.) Proceedings of the 2012 IEEE Congress on Evolutionary Computation, CEC 2012. IEEE, Brisbane, June 2012. https://doi.org/10.1109/CEC.2012.6256617
32. Xue, B., Zhang, M., Browne, W.N.: Particle swarm optimization for feature selection in classification: a multi-objective approach. IEEE Trans. Cybern. **43**(6), 1656–1671 (2013). https://doi.org/10.1109/TSMCB.2012.2227469
33. Zykina, A.V.: A lexicographic optimization algorithm. Autom. Remote Control **65**(3), 363–368 (2004). https://doi.org/10.1023/B:AURC.0000019366.84601.8e

An Online Tool for Unfolding Symbolic Fuzzy Logic Programs

Ginés Moreno$^{(\boxtimes)}$ and José Antonio Riaza

Department of Computing Systems, UCLM, 02071 Albacete, Spain
{Gines.Moreno,JoseAntonio.Riaza}@uclm.es

Abstract. In many declarative frameworks, unfolding is a very well-known semantics-preserving transformation technique based on the application of computational steps on the bodies of program rules for improving efficiency. In this paper we describe an online tool which allows us to unfold a symbolic extension of a modern fuzzy logic language where program rules can embed concrete and/or symbolic fuzzy connectives and truth degrees on their bodies. The system offers a comfortable interaction with users for unfolding symbolic programs and it also provides useful options to navigate along the sequence of unfolded programs. Finally, the symbolic unfolding transformation is connected with some fuzzy tuning techniques that we previously implemented on the same tool.

Keywords: Fuzzy logic programming · Symbolic execution · Unfolding

1 Introduction

During the last decades, several fuzzy logic programming systems have been developed, where the classical SLD resolution principle of logic programming has been replaced by a fuzzy variant with the aim of dealing with partial truth and reasoning with uncertainty in a natural way. Most of these systems implement (extended versions of) the resolution principle introduced by Lee [8], such as Elf-Prolog [3], F-Prolog [9], Fril [1], MALP [10], R-fuzzy [2], and FASILL [6].

In this paper we focus on the so-called *multi-adjoint logic programming* approach MALP [10], a powerful and promising approach in the area of fuzzy logic programming. Intuitively speaking, logic programming is extended with a *multi-adjoint lattice* L of truth values (typically, a real number between 0 and 1), equipped with a collection of *adjoint pairs* $\langle \&_i, \leftarrow_i \rangle$ and connectives: implications, conjunctions, disjunctions, and other operators called aggregators, which are interpreted on this lattice. Consider, for instance, the following MALP rule: "$good(X) \leftarrow_\mathsf{P} @_\mathsf{aver}(nice(X), cheap(X))$ *with* 0.8", where the adjoint pair $\langle \&_\mathsf{P}, \leftarrow_\mathsf{P} \rangle$ is defined as shown in the first line of Fig. 1, and the aggregator $@_\mathsf{aver}$

This work has been partially supported by the EU (FEDER), the State Research Agency (AEI) and the Spanish *Ministerio de Economía y Competitividad* under grant TIN2016-76843-C4-2-R (AEI/FEDER, UE).

© Springer Nature Switzerland AG 2019
I. Rojas et al. (Eds.): IWANN 2019, LNCS 11507, pp. 475–487, 2019.
https://doi.org/10.1007/978-3-030-20518-8_40

$$\&_P(x,y) \triangleq x * y \qquad \leftarrow_P (x,y) \triangleq \begin{cases} 1 & \text{if } y \leq x \\ x/y & \text{if } 0 < x < y \end{cases} \qquad \textit{Product logic}$$

$$\&_G(x,y) \triangleq \min(x,y) \qquad \leftarrow_G (x,y) \triangleq \begin{cases} 1 & \text{if } y \leq x \\ x & \text{otherwise} \end{cases} \qquad \textit{Gödel logic}$$

$$\&_L(x,y) \triangleq \max(0, x+y-1) \qquad \leftarrow_L (x,y) \triangleq \min(x-y+1, 1) \qquad \textit{Łukasiewicz logic}$$

Fig. 1. Adjoint pairs of three different fuzzy logics over $\langle [0,1], \leq \rangle$.

is typically defined as $@_{\text{aver}}(x_1, x_2) \triangleq (x_1 + x_2)/2$. Therefore, the rule specifies that X is good—with a truth degree of 0.8—whenever X be nice and cheap enough. Assuming that X is nice and cheap with, e.g., truth degrees n and c, respectively, then X is good with a truth degree of $0.8 * ((n + c)/2)$.

When specifying a MALP program, it might sometimes be difficult to assign weights—truth degrees—to program rules, as well as to determine the right connectives.[1] This is a common problem with fuzzy control system design, where some trial-and-error is often necessary. In our context, a programmer can develop a prototype and repeatedly execute it until the set of answers is the intended one. Unfortunately, this is a tedious and time consuming operation.

In order to overcome this drawback, in [15] we have recently introduced a symbolic extension of MALP programs called *symbolic multi-adjoint logic programming* (sMALP). Here, we can write rules containing *symbolic* weights and *symbolic* connectives, i.e., truth degrees and operators which are not defined on its associated multi-adjoint lattice. In order to evaluate these programs, we introduce a symbolic operational semantics that delays the evaluation of symbolic expressions. Therefore, a *symbolic answer* could now include symbolic (unknown) truth values and connectives. The approach is correct in the sense that using the symbolic semantics and then replacing the unknown values and connectives by concrete ones gives the same result as replacing these values and connectives in the original sMALP program and, then, applying the concrete semantics on the resulting MALP program. Furthermore, in [15,16][2] it is shown how sMALP programs can be used to tune a program w.r.t. a given set of test cases, thus easing what is considered the most difficult part of the process: the specification of the right weights and connectives for each rule.

On the other hand, in [4,5,11,12] we successfully adapted the unfolding transformation to MALP and FASILL programs, and in [13,14] we introduced its symbolic extension for sMALP programs, which motivates the present work. The main goal of this paper is to describe the online implementation of the symbolic unfolding technique in FLOPER (*Fuzzy LOgic Programming Environment for Research*), which is freely available from http://dectau.uclm.es/malp/sandbox.

---

[1] For instance, we have typically several adjoint pairs as shown in Fig. 1: *Łukasiewicz logic* $\langle \&_L, \leftarrow_L \rangle$, *Gödel logic* $\langle \&_G, \leftarrow_G \rangle$ and *product logic* $\langle \&_P, \leftarrow_P \rangle$, which might be used for modeling *pessimist, optimist* and *realistic scenarios*, respectively.

[2] The online tuning tool is available through http://dectau.uclm.es/malp/sandbox.

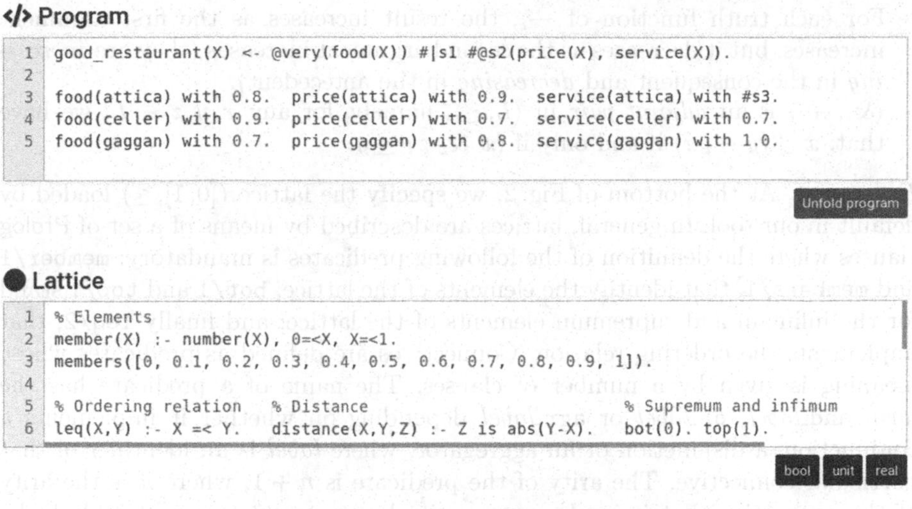

</> Program

```
1  good_restaurant(X) <- @very(food(X)) #|s1 #@s2(price(X), service(X)).
2
3  food(attica) with 0.8.   price(attica) with 0.9.   service(attica) with #s3.
4  food(celler) with 0.9.   price(celler) with 0.7.   service(celler) with 0.7.
5  food(gaggan) with 0.7.   price(gaggan) with 0.8.   service(gaggan) with 1.0.
```

Unfold program

● Lattice

```
1  % Elements
2  member(X) :- number(X), 0=<X, X=<1.
3  members([0, 0.1, 0.2, 0.3, 0.4, 0.5, 0.6, 0.7, 0.8, 0.9, 1]).
4
5  % Ordering relation   % Distance                          % Supremum and infimum
6  leq(X,Y) :- X =< Y.   distance(X,Y,Z) :- Z is abs(Y-X).   bot(0). top(1).
```

bool unit real

Fig. 2. Screenshot of the online tool showing a loaded program and lattice.

The structure of this paper is as follows. Sections 2 and 3 focus on the syntax and operational semantics of the framework of symbolic multi-adjoint logic programming by showing how such kind of programs can be loaded and executed into the online tool. Then, in Sect. 4, we describe the capability of the tool for unfolding symbolic programs. Finally, Sect. 5 concludes and points out some directions for further research.

2 Symbolic Multi-adjoint Logic Programs

We assume the existence of a multi-adjoint lattice $\langle L, \preceq, \&_1, \leftarrow_1, \ldots, \&_n, \leftarrow_n \rangle$, equipped with a collection of *adjoint pairs* $\langle \&_i, \leftarrow_i \rangle$—where each $\&_i$ is a conjunctor which is intended to be used for the evaluation of *modus ponens* [10]—. In addition, on each program rule, we can have a different adjoint implication ($\leftarrow_i$), conjunctions (denoted by $\land_1, \land_2, \ldots$), adjoint conjunctions ($\&_1, \&_2, \ldots$), disjunctions ($|_1, |_2, \ldots$), and other operators called aggregators (usually denoted by $@_1, @_2, \ldots$); see [17] for more details. More exactly, a multi-adjoint lattice fulfills the following properties:

- $\langle L, \preceq \rangle$ is a (bounded) complete lattice.[3]
- For each truth function of $\&_i$, an increase in any of the arguments results in an increase of the result (they are *increasing*).

---

[3] A complete lattice is a (partially) ordered set $\langle L, \preceq \rangle$ such that every subset S of L has infimum and supremum elements. It is bounded if it has bottom and top elements, denoted by $\bot$ and $\top$, respectively. L is said to be the *carrier set* of the lattice, and $\preceq$ its ordering relation.

- For each truth function of $\leftarrow_i$, the result increases as the first argument increases, but it decreases as the second argument increases (they are *increasing* in the consequent and *decreasing* in the antecedent).
- $\langle \&_i, \leftarrow_i \rangle$ is an *adjoint pair* in $\langle L, \preceq \rangle$, namely, for any $x, y, z \in L$, we have that: $x \preceq (y \leftarrow_i z)$ if and only if $(x \&_i z) \preceq y$.

Example 1. At the bottom of Fig. 2, we specify the lattice $([0, 1], \leq)$ loaded by default in our tool. In general, lattices are described by means of a set of Prolog clauses where the definition of the following predicates is mandatory: member/1 and members/1, that identify the elements of the lattice; bot/1 and top/1 stand for the infimum and supremum elements of the lattice; and finally leq/2, that implements the ordering relation. Connectives are defined as predicates whose meaning is given by a number of clauses. The name of a predicate has the form and_*label*, or_*label* or agr_*label* depending on whether it implements a conjunction, a disjunction or an aggregator, where *label* is an identifier of that particular connective. The arity of the predicate is $n + 1$, where n is the arity of the connective that it implements, so its last parameter is a variable to be unified with the truth value resulting of its evaluation.

In this work, given a multi-adjoint lattice L, we consider a first order language $\mathcal{L}_L$ built upon a signature Σ_L, that contains the elements of a countably infinite set of variables $\mathcal{V}$, function and predicate symbols (denoted by $\mathcal{F}$ and Π, respectively) with an associated arity—usually expressed as pairs f/n or p/n, respectively, where n represents its arity—, and the truth degree literals Σ_L^T and connectives Σ_L^C from L. Therefore, a well-formed formula in $\mathcal{L}_L$ can be either:

- A *value* $v \in \Sigma_L^T$, which will be interpreted as itself, i.e., as the truth degree $v \in L$.
- $p(t_1, \ldots, t_n)$, if $t_1, \ldots, t_n$ are terms over $\mathcal{V} \cup \mathcal{F}$ and p/n is an n-ary predicate. This formula is called *atomic* (or just an atom).
- $\varsigma(e_1, \ldots, e_n)$, if $e_1, \ldots, e_n$ are well-formed formulas and ς is an n-ary connective with truth function $[\![\varsigma]\!] : L^n \mapsto L$.

As usual, a *substitution* σ is a mapping from variables in $\mathcal{V}$ to terms over $\mathcal{V} \cup \mathcal{F}$ such that $Dom(\sigma) = \{x \in \mathcal{V} \mid x \neq \sigma(x)\}$ is its domain. Substitutions are usually denoted by sets of mappings like, e.g., $\{x_1/t_1, \ldots, x_n/t_n\}$. Substitutions are extended to morphisms from terms to terms in the natural way. The identity substitution is denoted by *id*. The composition of substitutions is denoted by juxtaposition, i.e., $\sigma\theta$ denotes a substitution δ such that $\delta(x) = \theta(\sigma(x))$ for all $x \in \mathcal{V}$.

A MALP *rule* over a multi-adjoint lattice L is a formula $H \leftarrow_i \mathcal{B}$, where H is an *atomic formula* (usually called the *head* of the rule), $\leftarrow_i$ is an implication symbol belonging to some adjoint pair of L, and $\mathcal{B}$ (which is called the *body* of the rule) is a well-formed formula over L without implications. A *goal* is a body submitted as a query to the system. A MALP program is a set of expressions R *with* v, where R is a rule and v is a *truth degree* (a value of L) expressing the confidence of a programmer in the truth of rule R. By abuse of the language, we often refer to R *with* v as a rule (see, e.g., [10] for a complete formulation of the MALP framework).

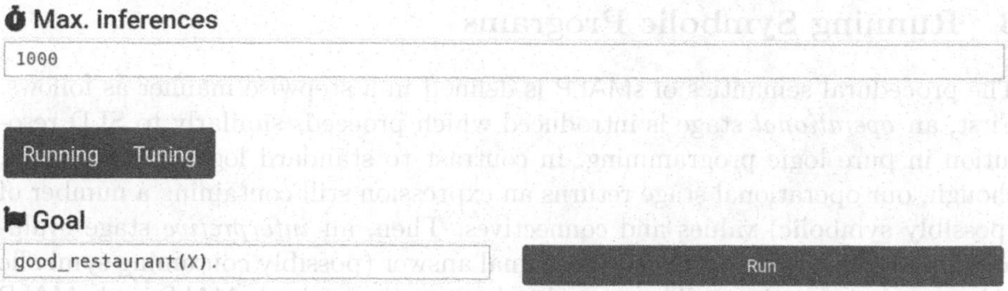

⏱ **Max. inferences**

```
1000
```

Running   Tuning

⚐ **Goal**

```
good_restaurant(X).
```
Run

Fig. 3. Screenshot of the online tool showing the run input area.

We are now ready for summarizing the *symbolic* extension of multi-adjoint logic programming initially presented in [15] where, in essence, we allow some undefined values (truth degrees) and connectives in the program rules, so that these elements can be systematically computed afterwards according to a given set of test cases thanks to the use of our tuning techniques implemented in [16]. In the following, we will use the abbreviation sMALP to refer to programs belonging to this setting. Here, given a multi-adjoint lattice L, we consider an augmented language $\mathcal{L}_L^s \supseteq \mathcal{L}_L$ which may also include a number of symbolic values, symbolic adjoint pairs and symbolic connectives which do not belong to L. Symbolic objects are usually denoted as o^s with a superscript s and, in our online tool, their identifiers always start with #.

Definition 1 (sMALP program). *Let L be a multi-adjoint lattice. An sMALP program over L is a set of symbolic rules, where each symbolic rule is a formula $(H \leftarrow_i \mathcal{B} \text{ with } v)$, where the following conditions hold:*

- *H is an atomic formula of $\mathcal{L}_L$ (the head of the rule);*
- *$\leftarrow_i$ is a (possibly symbolic) implication from either a symbolic adjoint pair $\langle \&^s, \leftarrow^s \rangle$ or from an adjoint pair of L;*
- *$\mathcal{B}$ (the body of the rule) is a symbolic goal, i.e., a well-formed formula of $\mathcal{L}_L^s$;*
- *v is either a truth degree (a value of L) or a symbolic value.*

Example 2. At the top of Fig. 2, we can see a sMALP program loaded into our online tool. Here, we consider a travel guide that offers information about three restaurants, named *attica*, *celler* and *gaggan*, where each one of them is featured by three factors: the restaurant services (cleanliness, parking, etc), the quality of its food, and the price, denoted by predicates *service*, *food* and *price*, respectively. Here, we assume that all weights can be easily obtained except for the weight of the fact *service*(*attica*), which is unknown, so we introduce a symbolic weight #$s3$. Also, the programmer has some doubts on the connectives used in the first rule, so she introduces two symbolic connectives, i.e., the disjunction and aggregator symbols #|$s1$ and #@$s2$.

3   Running Symbolic Programs

The procedural semantics of sMALP is defined in a stepwise manner as follows. First, an *operational* stage is introduced which proceeds similarly to SLD resolution in pure logic programming. In contrast to standard logic programming, though, our operational stage returns an expression still containing a number of (possibly symbolic) values and connectives. Then, an *interpretive* stage evaluates these connectives and produces a final answer (possibly containing symbolic values and connectives). The procedural semantics of both MALP and sMALP programs is based on a similar scheme. The main difference is that, for MALP programs, the interpretive stage always returns a value, while for sMALP programs we might get an expression containing symbolic values and connectives that should be first instantiated in order to compute a value.

In the following, $C[A]$ denotes a formula where A is a sub-expression which occurs in the—possibly empty—context $C[]$. Moreover, $C[A/A']$ means the replacement of A by A' in context $C[]$, whereas $Var(s)$ refers to the set of distinct variables occurring in the syntactic object s, and $\theta[Var(s)]$ denotes the substitution obtained from θ by restricting its domain to $Var(s)$. An sMALP *state* has the form $\langle Q; \sigma \rangle$ where Q is a symbolic goal and σ is a substitution. We let $\mathcal{E}^s$ denote the set of all possible sMALP states.

Definition 2 (Admissible step). *Let L be a multi-adjoint lattice and $\mathcal{P}$ an sMALP program over L. An* admissible step *is formalized as a state transition system, whose transition relation $\rightarrow_{AS} \subseteq (\mathcal{E}^s \times \mathcal{E}^s)$ is the smallest relation satisfying the following transition rules:*[4]

1. $\langle Q[A]; \sigma \rangle \rightarrow_{AS} \langle (Q[A/v \&_i B])\theta; \sigma\theta \rangle$,
   if $\theta = mgu(\{H = A\}) \neq fail$, $(H \leftarrow_i B \text{ with } v) \ll \mathcal{P}$ *and* B *is not empty.*[5]
2. $\langle Q[A]; \sigma \rangle \rightarrow_{AS} \langle (Q[A/\bot]); \sigma \rangle$,
   if there is no rule $(H \leftarrow_i B \text{ with } v) \ll \mathcal{P}$ *such that* $mgu(\{H = A\}) \neq fail$.

Here, $(H \leftarrow_i B \text{ with } v) \ll \mathcal{P}$ *denotes that* $(H \leftarrow_i B \text{ with } v)$ *is a renamed apart variant of a rule in* $\mathcal{P}$ *(i.e., all its variables are fresh). Note that symbolic values and connectives are not renamed.*

Observe that the second rule is needed to cope with expressions like $@_{aver}(p(a), 0.8)$, which can be evaluated successfully even when there is no rule matching $p(a)$ since $@_{aver}(0, 0.8) = 0.4$. We sometimes call *failure steps* to this kind of admissible steps.

In the following, given a relation $\rightarrow$, we let $\rightarrow^*$ denote its reflexive and transitive closure. Also, an L^s-*expression* is now a well-formed formula of $\mathcal{L}^s_L$ which is composed by values and connectives from L as well as by symbolic values and connectives.

---

[4] Here, we assume that A in $Q[A]$ is the selected atom. Furthermore, as it is common practice, $mgu(E)$ denotes the *most general unifier* of the set of equations E [7].

[5] For simplicity, we consider that facts $(H \text{ with } v)$ are seen as rules of the form $(H \leftarrow_i \top \text{ with } v)$ for some implication $\leftarrow_i$. Furthermore, in this case, we directly derive the state $\langle (Q[A/v])\theta; \sigma\theta \rangle$ since $v \&_i \top = v$ for all $\&_i$.

Definition 3 (Admissible derivation). *Let L be a multi-adjoint lattice and $\mathcal{P}$ be an* sMALP *program over L. Given a goal $\mathcal{Q}$, an admissible derivation is a sequence $\langle \mathcal{Q}; id \rangle \rightarrow_{AS}^{*} \langle \mathcal{Q}'; \theta \rangle$. When $\mathcal{Q}'$ is an L^s-expression, the derivation is called* final *and the pair $\langle \mathcal{Q}'; \sigma \rangle$, where $\sigma = \theta[\mathcal{V}ar(\mathcal{Q})]$, is called a symbolic admissible computed answer (*saca, *for short) for goal $\mathcal{Q}$ in $\mathcal{P}$.*

Given a goal $\mathcal{Q}$ and a final admissible derivation $\langle \mathcal{Q}; id \rangle \rightarrow_{AS}^{*} \langle \mathcal{Q}'; \sigma \rangle$, we have that $\mathcal{Q}'$ does not contain atomic formulas. Now, $\mathcal{Q}'$ can be *solved* by using the following interpretive stage:

Definition 4 (Interpretive step). *Let L be a multi-adjoint lattice and $\mathcal{P}$ be an* sMALP *program over L. Given a* saca *$\langle \mathcal{Q}; \sigma \rangle$, the interpretive stage is formalized by means of the following transition relation $\rightarrow_{IS} \subseteq (\mathcal{E}^s \times \mathcal{E}^s)$, which is defined as the least transition relation satisfying:*

$$\langle \mathcal{Q}[\varsigma(r_1, \ldots, r_n)]; \sigma \rangle \rightarrow_{IS} \langle \mathcal{Q}[\varsigma(r_1, \ldots, r_n)/r_{n+1}]; \sigma \rangle$$

where ς denotes a connective defined on L and $[\![\varsigma]\!](r_1, \ldots, r_n) = r_{n+1}$.

An interpretive derivation of the form $\langle \mathcal{Q}; \sigma \rangle \rightarrow_{IS}^{*} \langle \mathcal{Q}'; \theta \rangle$ such that $\langle \mathcal{Q}'; \theta \rangle$ cannot be further reduced, is called a final *interpretive derivation. In this case, $\langle \mathcal{Q}'; \theta \rangle$ is called a symbolic fuzzy computed answer (*sfca, *for short). Also, if $\mathcal{Q}'$ is a value of L, we say that $\langle \mathcal{Q}'; \theta \rangle$ is a fuzzy computed answer (*fca, *for short).*

In Fig. 3, we can see the run area of our online tool. After introducing a goal and clicking on the Run button, the system executes the goal and generates both the whole set of its sfca's as well as its associated derivation tree (in plain text and also graphically), as seen in Figs. 4 and 7. Each sfca appears on a different leaf of the tree, where each state contains its corresponding goal and substitution components and they are drawn inside yellow rectangles. Computational steps, colored in blue, are labeled with the program rule they exploit in the case of *admissible* steps or with the word "IS", corresponding to *interpretive* steps.

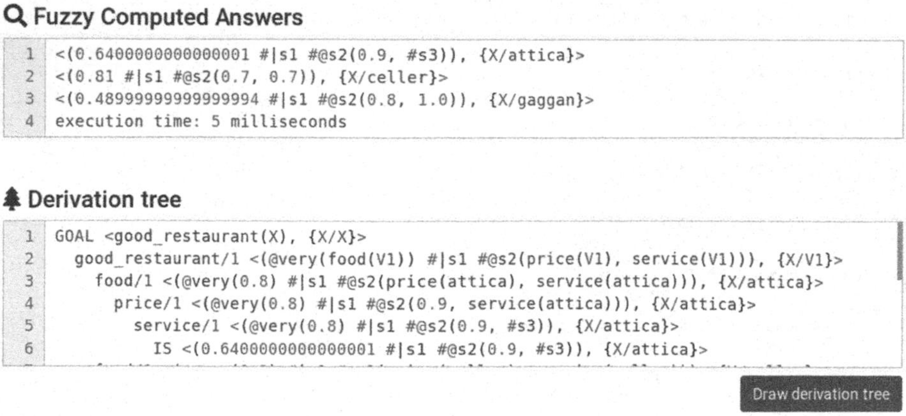

Q Fuzzy Computed Answers

```
1   <(0.6400000000000001 #|s1 #@s2(0.9, #s3)), {X/attica}>
2   <(0.81 #|s1 #@s2(0.7, 0.7)), {X/celler}>
3   <(0.48999999999999994 #|s1 #@s2(0.8, 1.0)), {X/gaggan}>
4   execution time: 5 milliseconds
```

♣ Derivation tree

```
1   GOAL <good_restaurant(X), {X/X}>
2     good_restaurant/1 <(@very(food(V1)) #|s1 #@s2(price(V1), service(V1))), {X/V1}>
3       food/1 <(@very(0.8) #|s1 #@s2(price(attica), service(attica))), {X/attica}>
4         price/1 <(@very(0.8) #|s1 #@s2(0.9, service(attica))), {X/attica}>
5           service/1 <(@very(0.8) #|s1 #@s2(0.9, #s3)), {X/attica}>
6             IS <(0.6400000000000001 #|s1 #@s2(0.9, #s3)), {X/attica}>
```

Draw derivation tree

Fig. 4. Screenshot of the symbolic derivation tree generated by the online tool.

Example 3. Consider again the multi-adjoint lattice L and the sMALP program $\mathcal{P}$ of Example 2. Figure 4 shows the admissible and interpretive derivations for the goal $good\_restaurant(X)$ in $\mathcal{P}$, where we can see that the first associated saca is $\langle\#|s1(@_{very}(0.8), \#@s2(0.9, \#s3)); \{X/attica\}\rangle$, and its sfca is $\langle\#|s1(0.64, \#@s2(0.9, \#s3)); \{X/attica\}\rangle$, since it cannot be further reduced.

4   Unfolding Symbolic Multi-adjoint Logic Programs

Unfolding is a well-known, widely used, semantics-preserving program transformation rule. In essence, it is usually based on the application of operational steps on the body of program rules [18]. The unfolding transformation is able to improve programs, generating more efficient code, and is the basis for developing sophisticated and powerful programming tools, such as fold/unfold transformation systems or partial evaluators, etc.

Definition 5. (Symbolic Unfolding). *Let $\mathcal{P}$ be an* sMALP *program and let $R : A \leftarrow \mathcal{B} \in \mathcal{P}$ be a program rule with no empty body. Then, the symbolic unfolding of rule R in program $\mathcal{P}$ is the new* sMALP *program $\mathcal{P}' = (\mathcal{P} - \{R\}) \cup \{A\sigma \leftarrow \mathcal{B}' \mid \langle\mathcal{B}; id\rangle \rightarrow \langle\mathcal{B}'; \sigma\rangle\}$.*

Example 4. We consider again the sMALP program of Example 2. Below the input program of our online tool (see Fig. 2), we have the option to unfold the loaded program. By clicking on the Unfold program button, all rules of the program will be listed as shown in Fig. 5. If we unfold its first rule (with selected atom $food(X)$) by applying a $\rightarrow_{AS}$ step with the three facts defining predicate

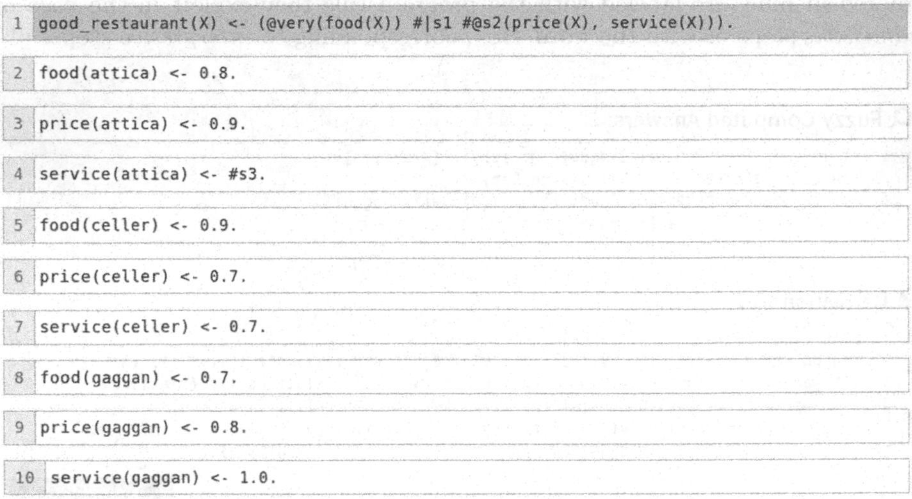

```
1  good_restaurant(X) <- (@very(food(X)) #|s1 #@s2(price(X), service(X))).

2  food(attica) <- 0.8.

3  price(attica) <- 0.9.

4  service(attica) <- #s3.

5  food(celler) <- 0.9.

6  price(celler) <- 0.7.

7  service(celler) <- 0.7.

8  food(gaggan) <- 0.7.

9  price(gaggan) <- 0.8.

10 service(gaggan) <- 1.0.
```

Fig. 5. Screenshot of the online tool showing the unfolding area.

```
1   good_restaurant(attica) <- (@very(0.8) #|s1 #@s2(price(attica), service(attica))).
2   good_restaurant(celler) <- (@very(0.9) #|s1 #@s2(price(celler), service(celler))).
3   good_restaurant(gaggan) <- (@very(0.7) #|s1 #@s2(price(gaggan), service(gaggan))).
4
5   food(attica) with 0.8.   price(attica) with 0.9.   service(attica) with #s3.
6   food(celler) with 0.9.   price(celler) with 0.7.   service(celler) with 0.7.
7   food(gaggan) with 0.7.   price(gaggan) with 0.8.   service(gaggan) with 1.0.
```

(a) Program $\mathcal{P}_1$ after first unfolding based on a $\rightarrow_{AS}$ step.

```
1   good_restaurant(attica) <- (@very(0.8) #|s1 #@s2(0.9, #s3)).
2   good_restaurant(celler) <- (@very(0.9) #|s1 #@s2(0.7, 0.7)).
3   good_restaurant(gaggan) <- (@very(0.7) #|s1 #@s2(0.8, 1.0)).
4
5   food(attica) with 0.8.   price(attica) with 0.9.   service(attica) with #s3.
6   food(celler) with 0.9.   price(celler) with 0.7.   service(celler) with 0.7.
7   food(gaggan) with 0.7.   price(gaggan) with 0.8.   service(gaggan) with 1.0.
```

(b) Program $\mathcal{P}_7$ after unfoldings based on several $\rightarrow_{AS}$ steps.

```
1   good_restaurant(attica) <- (0.6400000000000001 #|s1 #@s2(0.9, #s3)).
2   good_restaurant(celler) <- (@very(0.9) #|s1 #@s2(0.7, 0.7)).
3   good_restaurant(gaggan) <- (@very(0.7) #|s1 #@s2(0.8, 1.0)).
4
5   food(attica) with 0.8.   price(attica) with 0.9.   service(attica) with #s3.
6   food(celler) with 0.9.   price(celler) with 0.7.   service(celler) with 0.7.
7   food(gaggan) with 0.7.   price(gaggan) with 0.8.   service(gaggan) with 1.0.
```

(c) Program $\mathcal{P}_8$ after first unfolding based on a $\rightarrow_{IS}$ step.

```
1   good_restaurant(attica) <- (0.6400000000000001 #|s1 #@s2(0.9, #s3)).
2   good_restaurant(celler) <- (0.81 #|s1 #@s2(0.7, 0.7)).
3   good_restaurant(gaggan) <- (0.48999999999999994 #|s1 #@s2(0.8, 1.0)).
4
5   food(attica) with 0.8.   price(attica) with 0.9.   service(attica) with #s3.
6   food(celler) with 0.9.   price(celler) with 0.7.   service(celler) with 0.7.
7   food(gaggan) with 0.7.   price(gaggan) with 0.8.   service(gaggan) with 1.0.
```

(d) Final unfolded program $\mathcal{P}_{10}$.

Fig. 6. Screenshots of the online tool showing the unfolded program.

food, we obtain the program shown in Fig. 6a, which removes the unfolded rule and contains three new rules.[6]

After 6 more unfolding operations based on $\rightarrow_{AS}$ steps, we have exploited all the atoms, as we can see in Fig. 6(b). Then, if we unfold its first rule (with selected expression @$_{very}(0.8)$) by applying a $\rightarrow_{IS}$ step, we obtain the program shown in Fig. 6(c). Finally, after 2 unfolding operations based on $\rightarrow_{IS}$ steps, we reach the final program shown in Fig. 6(d), that cannot be further unfolded. It is easy to see that the new program produces the same set of sfca's for a

---

[6] It is important to note that, at execution time, each implication symbol belonging to a concrete adjoint pair is replaced by its adjoint conjunction. So, given a rule $A \leftarrow_i \mathcal{B}$ *with* v, it is converted to $A \leftarrow v \;\&_i \mathcal{B}$ before unfolding.

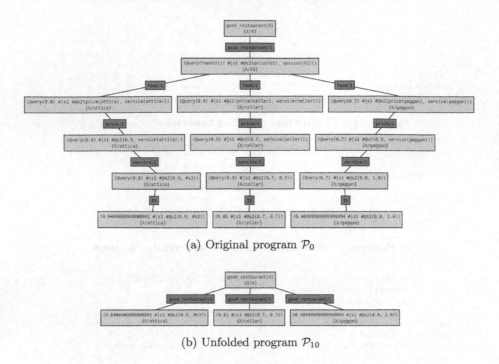

(a) Original program $\mathcal{P}_0$

(b) Unfolded program $\mathcal{P}_{10}$

Fig. 7. Screenshots of the graphical symbolic trees generated by the online tool.

given goal but reducing the length of derivations. For instance, the derivation performed w.r.t. the original program illustrated in Example 3 can be emulated in the final program with just one $\rightarrow_{AS}$ step, as we can see in Fig. 7, which shows the graphical trees generated by the online tool for the goal good_restaurant(X) with the initial program (a) and its final unfolded version (b).

Since the sMALP language was initially conceived for tuning fuzzy logic programs, we finish this section by exploring the synergies between this technique and the unfolding transformation described and implemented in this paper.

Given a multi-adjoint lattice L and a symbolic language $\mathcal{L}_L^s$, in the following we consider *symbolic substitutions* that are mappings from symbolic values and connectives to expressions over $\Sigma_L^T \cup \Sigma_L^C$. Symbolic substitutions are denoted by $\Theta, \Gamma, \ldots$ Furthermore, for all symbolic substitution Θ, we require the following condition: $\leftarrow^s/\leftarrow_i \in \Theta$ iff $\&^s/\&_i \in \Theta$, where $\langle \&^s, \leftarrow^s \rangle$ is a symbolic adjoint pair and $\langle \&_i, \leftarrow_i \rangle$ is an adjoint pair in L. Intuitively, this is required for the substitution to have the same effect both on the program and on an L^s-expression.

Given an sMALP program $\mathcal{P}$ over L, we let $\mathsf{sym}(\mathcal{P})$ denote the symbolic values and connectives in $\mathcal{P}$. Given a symbolic substitution Θ for $\mathsf{sym}(\mathcal{P})$, we denote by $\mathcal{P}\Theta$ the program that results from $\mathcal{P}$ by replacing every symbolic symbol e^s by $e^s\Theta$. Trivially, $\mathcal{P}\Theta$ is now a MALP program.

The following theorem formally proved in [15] is a key result in order to use sMALP programs for tuning the components of a MALP program:

Theorem 1. *Let L be a multi-adjoint lattice and $\mathcal{P}$ be an sMALP program over L. Let $\mathcal{Q}$ be a goal. Then, for any symbolic substitution Θ for* sym$(\mathcal{P})$*, we have that $\langle v; \theta \rangle$ is a fca for Q in $\mathcal{P}\Theta$ iff there exists a sfca $\langle Q'; \theta' \rangle$ for $\mathcal{Q}$ in $\mathcal{P}$ and $\langle \mathcal{Q}'\Theta; \theta' \rangle \rightarrow^{*}_{IS} \langle v; \theta' \rangle$, where θ' is a renaming of θ.*

Example 5. Consider again the multi-adjoint lattice L and the sMALP program $\mathcal{P}$ in Fig. 2. Let $\Theta = \{\#|s_1/|_{\texttt{godel}}, \#@s_2/@_{\texttt{aver}}, \#s_3/0.6\}$ be a symbolic substitution. Given the sfca from Example 3, we have:

$$\langle \#|s1(0.64, \#@s2(0.9, \#s3))\rangle\Theta; \{X/attica\}\rangle \equiv \langle |_{\texttt{godel}}(0.64, @_{\underline{\texttt{aver}}}(0.9, 0.6)); \{X/attica\}\rangle$$

So, we have the following interpretive final derivation for the instantiated sfca:

$$\langle |_{\texttt{godel}}(0.64, @_{\underline{\texttt{aver}}}(0.9, 0.6)); \{X/attica\}\rangle \rightarrow_{IS}$$
$$\langle |_{\texttt{godel}}(0.64, \overline{0.75}); \{X/attica\}\rangle \qquad \rightarrow_{IS}$$
$$\langle 0.75; \{X/attica\}\rangle$$

By Theorem 1 we have that, for *good_restaurant*(X) in $\mathcal{P}\Theta$, $\langle 0.75; \{X/attica\}\rangle$ is also a fca.

Note that the previous theorem establishes two equivalent ways for obtaining the same sets of fca's for a given goal, which in both cases require the same computational effort as illustrated in the previous example, where the final fca $\langle 0.8; \{X/attica\}\rangle$ is reached after applying 4 admissible steps plus 3 interpretive steps. However, in [15,16] we proved that for tuning purposes, the computational complexity of our techniques can be largely improved only by the method based on the application of symbolic computational steps to find saca's and then instantiating them to reach final fca's (instead of instantiating the original sMALP program an then proceed with the classical operational semantics of MALP). It is easy to see that efficiency of tuning a sMALP program $\mathcal{P}$ could be increased once again if the same method would receive as input an unfolded version of $\mathcal{P}$.

5   Conclusions and Future Work

In this paper, we have collected from [13,14] our initial formulation of an unfolding transformation for optimizing sMALP programs, which represents a symbolic extension of fuzzy logic programs belonging to the so-called *multi-adjoint logic programming* approach. Here, we have implemented such technique and described the online tool freely available via URL http://dectau.uclm.es/malp/sandbox.

Since in [6], we have designed a new fuzzy language extending MALP with *similarity relations*, as future work we plan to enrich the present implementation of the symbolic unfolding technique to cope with such similarity relations according to some preliminary guidelines provided in [12]. Moreover, a pending task related to the tuning techniques that we have combined here with symbolic unfolding, consists in exploring their synergies with machine learning strategies and fuzzy SMT/SAT solvers.

References

1. Baldwin, J.F., Martin, T.P., Pilsworth, B.W.: Fril- Fuzzy and Evidential Reasoning in Artificial Intelligence. Wiley, Hoboken (1995)
2. Guadarrama, S., Muñoz, S., Vaucheret, C.: Fuzzy prolog: a new approach using soft constraints propagation. Fuzzy Sets Syst. **144**(1), 127–150 (2004)
3. Ishizuka, M., Kanai, N.: Prolog-ELF incorporating fuzzy logic. In: Joshi, A.K. (ed.) Proceedings of the 9th International Joint Conference on Artificial Intelligence, IJCAI 1985, pp. 701–703. Morgan Kaufmann (1985)
4. Julián-Iranzo, P., Medina-Moreno, J., Morcillo, P.J., Moreno, G., Ojeda-Aciego, M.: An unfolding-based preprocess for reinforcing thresholds in fuzzy tabulation. In: Rojas, I., Joya, G., Gabestany, J. (eds.) IWANN 2013. LNCS, vol. 7902, pp. 647–655. Springer, Heidelberg (2013). https://doi.org/10.1007/978-3-642-38679-4_65
5. Julián-Iranzo, P., Moreno, G., Penabad, J.: On fuzzy unfolding. A multi-adjoint approach. Fuzzy Sets Syst. **154**, 16–33 (2005)
6. Julián-Iranzo, P., Moreno, G., Penabad, J.: Thresholded semantic framework for a fully integrated fuzzy logic language. J. Log. Algebr. Methods Program. **93**, 42–67 (2017)
7. Lassez, J.L., Maher, M.J., Marriott, K.: Unification revisited. In: Minker, J. (ed.) Foundations of Deductive Databases and Logic Programming, pp. 587–625. Morgan Kaufmann, Los Altos (1988)
8. Lee, R.C.T.: Fuzzy logic and the resolution principle. J. ACM **19**(1), 119–129 (1972)
9. Li, D., Liu, D.: A Fuzzy Prolog Database System. Wiley, Hoboken (1990)
10. Medina, J., Ojeda-Aciego, M., Vojtáš, P.: Similarity-based unification: a multi-adjoint approach. Fuzzy Sets Syst. **146**, 43–62 (2004)
11. Morcillo, P.J., Moreno, G.: Improving multi-adjoint logic programs by unfolding fuzzy connective definitions. In: Rojas, I., Joya, G., Catala, A. (eds.) IWANN 2015. LNCS, vol. 9094, pp. 511–524. Springer, Cham (2015). https://doi.org/10.1007/978-3-319-19258-1_42
12. Moreno, G., Penabad, J., Riaza, J.A.: On similarity-based unfolding. In: Moral, S., Pivert, O., Sánchez, D., Marín, N. (eds.) SUM 2017. LNCS (LNAI), vol. 10564, pp. 420–426. Springer, Cham (2017). https://doi.org/10.1007/978-3-319-67582-4_32
13. Moreno G., Penabad J., Riaza J.A.: Symbolic unfolding of multi-adjoint logic programs. In: 9th European Symposium on Computational Intelligence and Mathematics, ESCIM 2017, pp. 1–8 (2017). http://escim2017.uca.es/proceedings/ (extended version published by Springer)
14. Moreno, G., Penabad, J., Riaza, J.A.: Symbolic unfolding of multi-adjoint logic programs. In: Cornejo, M.E., Kóczy, L.T., Medina, J., De Barros Ruano, A.E. (eds.) Trends in Mathematics and Computational Intelligence. SCI, vol. 796, pp. 43–51. Springer, Cham (2019). https://doi.org/10.1007/978-3-030-00485-9_5
15. Moreno, G., Penabad, J., Riaza, J.A., Vidal, G.: Symbolic execution and thresholding for efficiently tuning fuzzy logic programs. In: Hermenegildo, M.V., Lopez-Garcia, P. (eds.) LOPSTR 2016. LNCS, vol. 10184, pp. 131–147. Springer, Cham (2017). https://doi.org/10.1007/978-3-319-63139-4_8

16. Moreno, G., Riaza, J.A.: An online tool for tuning fuzzy logic programs. In: Costantini, S., Franconi, E., Van Woensel, W., Kontchakov, R., Sadri, F., Roman, D. (eds.) RuleML+RR 2017. LNCS, vol. 10364, pp. 184–198. Springer, Cham (2017). https://doi.org/10.1007/978-3-319-61252-2_13
17. Nguyen, H.T., Walker, E.A.: A First Course in Fuzzy Logic. Chapman & Hall, Boca Ratón (2006)
18. Pettorossi, A., Proietti, M.: Rules and strategies for transforming functional and logic programs. ACM Comput. Surv. **28**(2), 360–414 (1996)

Ensemble of Attractor Networks for 2D Gesture Retrieval

Carlos Dávila[1], Mario González[1]([⊠]), Jorge-Luis Pérez-Medina[1],
David Dominguez[2], Ángel Sánchez[3], and Francisco B. Rodriguez[2]

[1] Intelligent and Interactive Systems Lab (SI2 Lab), Universidad de Las Américas,
Quito, Ecuador
{carlos.davila.armijos,mario.gonzalez.rodriguez,
jorge.perez.medina}@udla.edu.ec
[2] Grupo de Neurocomputación Biológica, Dpto. de Ingeniería Informática,
Escuela Politécnica Superior, Universidad Autónoma de Madrid, 28049 Madrid, Spain
{david.dominguez,f.rodriguez}@uam.es
[3] ETSII, Universidad Rey Juan Carlos, 28933 Móstoles, Madrid, Spain
angel.sanchez@urjc.es

Abstract. This work presents an Ensemble of Attractor Neural
Networks (EANN) model for gesture retrieval. 2D single-stroke gestures
were captured and tested offline by the ensemble. The ensemble was
compared to a single attractor with the same complexity, i.e. with equal
connectivity. We show that the ensemble of neural networks improves
the gesture retrieval in terms of capacity and quality of the gestures
retrieval, regarding the single network. The ensemble was able to improve
the retrieval of correlated patterns with a random assignment of pattern
subsets to the ensemble modules. Thus, optimizing the ensemble input is
a possibility for maximizing the patterns retrieval. The proposed EANN
proved to be robust for gesture recognition with large initial noise promising to be robust for gesture invariants.

Keywords: Single-stroke gestures · Offline recognition ·
Gesture encoding · Hopfield network · Synaptic dilution ·
Storage capacity

1 Introduction

Recent works have dealt with an EANN increasing the network storage capacity
[1,2]. Using a divide-and-conquer strategy, each diluted module of the ensemble
stores and retrieves a subset of patterns, increasing the capacity respect to a
single densely connected module of similar connectivity. The aforementioned
works have processed random patterns to measure the model limits in terms of
storage capacity. This set of attractor networks whose operation is based on the
increase in capacity retrieval through the dilution of connections has never been
tested with real structured patterns. Thus, in the present work, the ensemble is
used to deal with real patterns, i.e. a 2D on-screen gestures dataset.

© Springer Nature Switzerland AG 2019
I. Rojas et al. (Eds.): IWANN 2019, LNCS 11507, pp. 488–499, 2019.
https://doi.org/10.1007/978-3-030-20518-8_41

Attractor networks as memory machines for pattern storing and retrieval have been recently used in many applications such as pattern recognition [3–5] and socioeconomics modeling [6], to mention a few. The attractor dynamical properties such as pattern denoising and completion, are desirable for the aforementioned applications, as well as, for gesture retrieval where the attractor robustness to initial conditions is desirable.

Gesture recognition has become largely used in many domains of interactivity [7]. The introduction of new technologies and the new exigences of users lead to incorporate more and more natural interfaces in interactive systems [8]. A gesture is generally defined as a sequence of points between a starting and an ending point. A gesture is characterized by having a scale, a direction, a position among other aspects. The process of the recognition of a gesture is summarized in a comparison of the similarity of a sequence of points of a candidate gesture with a reference gesture that is part of a datasets. Normally, the comparison is based on a principle of obtaining the sum of the distances of each point of the gestures to be compared. In this sense, several distances have been proposed, such as: the Euclidean distance used by \$1 algorithm [9], Dynamic Time Warping (DTW) used by [10], dynamic programming in combination with a Hierarchical Temporal Memory [11], and the Levenshtein distance used in [12,13].

The present work proposes an EANN model for 2D single-stroke gesture retrieval. The attractor modules perform a holistic processing of the gesture not needing to extract individual features (i.e., distances between points). The gestures were encoded as binary patterns from the (x, y) points captured, codifying the gesture strokes directions as increasing, decreasing, or not changing. The EANN improved the retrieval capacity of the single module and demonstrated to be robust to initial conditions with a gesture dissimilarity/noise as high of 40% of the pattern sites. A random assignment of the pattern subsets to the ensemble modules was necessary to improve the retrieval quality of correlated patterns. This opens the possibility of a combinatorial optimization of the input of patterns to the ensemble modules.

The rest of the paper is organized as follows. Section 2 describes the ensemble model and the learning and retrieval dynamics of the modules, as well as the performance measures of the model. Section 3 describes the gesture encoding process, and the gestures dataset. Section 4 presents the results in terms of the network retrieval performance, storage capacity, comparing the EANN with a single attractor. Also, examples of the gestures retrieval are presented. Section 5 concludes the paper and discusses future applications of the ensemble model.

2   The Model

In this section the base module of the ensemble, that is, the Attractor Neural Network (ANN) model is described. Starting with the neural coding, the network topology, learning and retrieval dynamics. Then, a schematic representation of the modularized EANN system is illustrated for the gesture retrieval dataset.

2.1   Modules Coding, Topology and Dynamics

The state of a neuron in a network of N units is defined, at any discrete time t, by a set of N binary variables $\tau^t = \{\tau_i^t \in 0, 1; i = 1, \ldots, N\}$, where 1 and 0 represent, respectively, active and inactive states. The target of network will be to recover a set of patterns, that in this work will be 2D gestures, $\{\eta^\mu, \ \mu = 1, \ldots, P\}$ that have been stored by a learning process. Each pattern corresponds to a stable fixed point attractor and the network retrieval state satisfies $\tau^t = \eta^\mu$, for large enough time t. The patterns are encoded as a set of binary variables $\eta^\mu = \{\eta_i^\mu \in 0, 1; i = 1, \ldots, N\}$, with resulting probability

$$p(\eta_i^\mu = 1) = a, \quad p(\eta_i^\mu = 0) = 1 - a, \tag{1}$$

where $a \in (0, 1)$ stands for the corresponding average activity ratio of the patterns dataset, as treated in biased coded networks [5, 14].

The synaptic couplings between the neurons i and j are given by the adjacency matrix

$$J_{ij} \equiv C_{ij} W_{ij}, \tag{2}$$

where the topology matrix $C = \{C_{ij}\}$ describes the connectivity structure of the neural network and $W = \{W_{ij}\}$ is the matrix with the learning weights. The topology matrix corresponds to a symmetric random network [15], with a network degree of K, that is each node is connected to K other nodes. The network is then characterized by the dilution parameter $\gamma = K/N$, which represents the connectivity ratio of the ANN.

The retrieval of a gesture pattern is achieved through the noiseless neuron dynamics

$$\tau_i^{t+1} = \Theta(h_i^t - \theta(a)), \quad i = 1, \ldots, N, \tag{3}$$

where

$$h_i^t \equiv \frac{1}{K} \sum_j J_{ij} \frac{\tau_j^t - q_j^t}{\sqrt{Q_j^t}} \tag{4}$$

denotes the local field at neuron i and time t and $\theta(a)$ is the threshold of firing. In Eq. (3) is used the step function: $\Theta(x) = 1$, $x \geq 0$, $\Theta(x) = 0$, $x < 0$.

In Eq. (4) is introduced the average activity of the neighborhood of neuron i, $q_i^t = \langle \tau^t \rangle_i$, and its corresponding variance, $Q_i^t = Var(\tau^t)_i = \langle (\tau^t)^2 \rangle_i - \langle \tau^t \rangle_i^2$. The neighborhood average is defined as $\langle f^t \rangle_i \equiv \sum_j C_{ij} f_j^t / K$.

The normalized variables are used, the site and time dependence being implicit:

$$\sigma \equiv \frac{\tau - q}{\sqrt{Q}}, \ q \equiv \langle \tau \rangle, \ Q \equiv Var(\tau) = q(1 - q) \tag{5}$$

$$\xi \equiv \frac{\eta - a}{\sqrt{A}}, \ a \equiv \langle \eta \rangle, \ A \equiv Var(\eta) = a(1 - a), \tag{6}$$

where a and q are the pattern and neuronal activities, respectively. The averages done in this work run over different groups, and are indicated in each case. The uniform binary neuronal model is recovered when $a = 1/2$ [16].

In terms of these normalized variables, the neuron dynamics [14,16] can be written as

$$\sigma_i^{t+1} = g(h_i^t - \theta(a), q_i^t), \tag{7}$$

$$h_i^t \equiv \frac{1}{K} \sum_j J_{ij}\sigma_j^t, \ i = 1, \ldots, N, \tag{8}$$

where the gain function is given by

$$g(x, y) \equiv [\Theta(x) - y]/\sqrt{y(1 - y)}. \tag{9}$$

The weight matrix W is updated according to the Hebb's learning rule,

$$W_{ij}^{\mu+1} = W_{ij}^{\mu} + \xi_i^{\mu}\xi_j^{\mu} \ (\text{online}), \quad W_{ij} = \sum_{\mu=1}^{P} \xi_i^{\mu}\xi_j^{\mu} \ (\text{offline}). \tag{10}$$

It is worth to note that the offline learning rule is non-iterative, that is, all patterns must be learned before the beginning of the retrieval process (Eq. 8). In contrast, the online rule is iterative and the retrieve process is repeated for each new pattern learned by the network [6]. Weights start at $W_{ij}^0 = 0$ and after P learning steps, they reach the value $W_{ij} = \sum_{\mu}^{P} \xi_i^{\mu}\xi_j^{\mu}$. The network learns $P = \alpha K$ patterns, where α is the load ratio, which measures the capacity retrieval of the network. A threshold is necessary to keep the neural activity close to that of the learned patterns, $\theta(a) = \frac{1-2a}{2\sqrt{A}}$ is used, where a corresponds to the mean activity of the learning pattern set, and $A = a(1 - a)$ to the activity variance [5,14].

In order to characterize the retrieval ability of the network modules, the overlap is used,

$$m \equiv \frac{1}{N} \sum_i^N \xi_i\sigma_i, \tag{11}$$

which is the statistical correlation between the learned pattern ξ_i and the neural state σ_i. For $m = 1$, one has a perfect retrieval of the pattern by the network, for $m = 0$ no retrieval is achieved, and for intermediate values the pattern is retrieved with noise. Hence, the value of m is a measure of the retrieval quality of the pattern performed by the network [5,16].

The initial state of the network $\tau^{t=0}$ and the corresponding internal transformation $\sigma_i^{t=0}$, is a noisy state loaded from a test pattern set where $m^{t=0} < 1$, i.e., initial noise is allowed at the start of the retrieval phase. That is, the test set represents the same gestures than the learned set, but a different realization by the user [4].

2.2   Ensemble of Attractor Neural Networks (EANN)

A schematic representation of the single ANN is presented in Fig. 1-left. The connectivity ratio γ is diluted with $K < N$. A set of P gestures $\boldsymbol{\xi}$ is presented to the network in a learning phase, represented with the red dashed arrow.

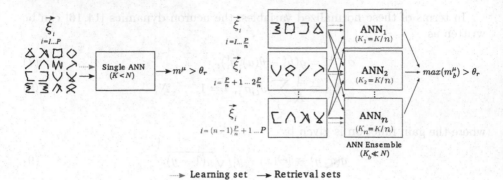

Fig. 1. Schematic representation of the model. Left: single attractor network with whole dataset. Right: ensemble of attractor networks with pattern subsets.

Then, this set of gestures is presented in a retrieval phase in order to test the recall abilities of the network in terms of the retrieved patterns load α, and the quality of the retrieval m. This is represented with the solid black arrow.

In Fig. 1-right, a schematic representation of an ensemble of ANN modules with a number of n components is presented. The connectivity in each ANN_b module b is highly diluted with $K_b \ll N$, $b \in \{1, ..., n\}$, and connectivity ratio $\gamma_b = K_b/N$. In this work the single network $n = 1, K = 200, \gamma = 0.2$ is compared with an EANN of $n = 4, K_b = 50, \gamma_b = 0.05$ modules. The computational cost of the single ANN and the ANN ensemble are the same, one uses $K = K_b \times n$. The set of patterns is divided into disjoint subsets of uniform size $P_b = P/n$, and each pattern subset is learned by its corresponding ANN_b module as represented with the red dashed arrows. E.g. $\{\xi^\mu, \ \mu = 1, \ldots, P/n\}$ for the first module $ANN_b, b = 1$, as shown in Fig. 1-right. The solid black arrows in Fig. 1-right, represent the retrieval stage, in which all the pattern subsets are presented to all ANN modules in order to test the discrimination among them. The target patterns are considered as retrieved by the ANN module with the higher overlap value over the retrieval threshold λ, i.e. $max(m_b^\mu) > \lambda$. For comparison purpose, the retrieval threshold is assumed to take the same value $\lambda = 0.5$ for each component in the ensemble, as well as, for the single ANN system.

3   Pattern Encoding

This section describes the process of the raw gesture encoding to a binary pattern, and the resulting binarized dataset is presented and described.

3.1   Gesture Binarization

The raw gestures can be expressed as a tuple of continuous points $\pi^\mu = \{(x_i, y_i) \in \mathbb{R}; i = 1, \ldots, S\}$, where $S = 1500$ is the number of points captured for each gesture. To be processed by the network the gesture signal π, is resampled to be length $S = N/2 + 1$, where $N = 1000$ is the network size, and the

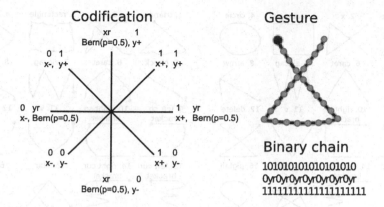

Fig. 2. Left: binary codification. Right: binarized gesture example.

binarization occurs as described schematically in Fig. 2. The gesture encoding returns a binary chain of size $N = 1000$, where half of the sites corresponds to the resampled x and the other half to the y interleaved in a binarized tuple. The value of $N = 1000$ was chosen for the network/pattern size (nodes/sites) as a minimum value to allow diluted ensemble modules.

Figure 2 left panel shows the change in stroke directions and their respective binary encoding. If x_i or y_i increase, the resulting codification is $\eta_i = 1$, otherwise $\eta_i = 0$. A threshold ϕ is used to encode strokes that are changing in x or y but not both, that is, vertical or horizontal strokes. If the change (δx or δy) is smaller than ϕ, $\eta_{yi} = yr$, $yr \sim Bern(p = 0.5)$ will encode vertical strokes, and $\eta_{xi} = xr$, $xr \sim Bern(p = 0.5)$, will encode horizontal ones. In this work $\phi = 0.01$ was found empirically to work well, for the \$1 dataset [9]. Here, $xr, yr \sim Bern(p)$ stands for the variable follows a Bernoulli distribution with probability parameter p. For changes δx or δy that are larger than ϕ, the stroke (x_i, y_i) is considered either increasing or decreasing respect to the previous point (x_{i-1}, y_{i-1}) and are encoding accordingly, $x+, y+ \Rightarrow 1, x-, y- \Rightarrow 0$. Figure 2 right panel depicts an example of gesture and the resulting codification. The black dot, indicates the starting point of the gesture. Note that the first point (black dot) is used as reference of the stroke changes in the binarization process. One can observe that when the stroke is increasing in x and decreasing in y, the resulting chain is a set of 10. When the stroke is decreasing in x but not increasing nor decreasing in y the resulting chain is $0yr$, where $yr \sim Bern(p = 0.5)$. Finally, when the gesture is increasing in x, y the resulting chain is 11 for each point.

3.2 Gesture Dataset

The gestures dataset is composed of $P = 16$ gestures as depicted in Fig. 3. The left panel shows the continuous points (π^μ) gestures acquired using the methodology described in [7]. The right panel corresponds to the binarized patters that will be stored by the network ensemble. Note that the binarized gestures in

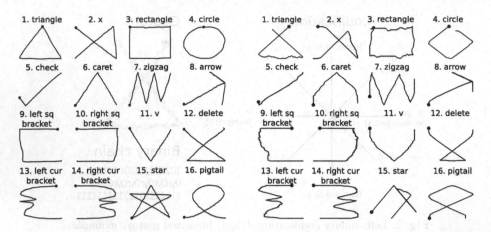

Fig. 3. Left: continuous gestures dataset [9]. Right: Binarized patterns stored by the EANN. Binarized set with $P = 16$ patterns with size $N = 1000$.

Fig. 3-right are depicted as an xy-plot of the cumulative sum for each binary pattern $\eta^\mu \times 2 - 1$. Again, the black initial dot, indicates the starting point of the gesture. The $1 dataset [9] was used because its gestures have similar characteristics. For example, gestures 2 and 12 differ, basically, in the dynamics as the gesture was created. That is, they differ in their orientation and rotation. Thus, this is a standard dataset that can test the ensemble attractor retrieval abilities. The binarized dataset has a mean activity of $a = 0.5803125$, which results in $\theta(a) = -0.17093$ for the retrieval dynamics in Eq. 3. The dataset has a mean overlap between gestures of $\langle m \rangle_{\mu\nu} = 0.1895$. This is a high correlation between patterns, thus the cross-talk noise is large for the present dataset. The ensemble might be helpful in such scenarios, given that one can optimize the pattern subsets assignments, to minimize the subset overlap and improve the retrieval.

4   Results

This section presents the retrieval performance of the network, comparing a single module attractor with an EANN. The performance is shown in terms of the retrieval parameter m for each pattern in the dataset. The retrieval quality of each of the gestures from a noisy initial state is presented and the final retrieved state of the network depicted as gesture strokes.

4.1   Network Retrieval Performance

Figure 4 top panels depict the performance of the single module $n = 1, K = 200, \gamma = 0.2$ vs. the ensemble of attractors of similar connectivity $n = 4$, $K_b = 50, \gamma_b = 0.05$ in Fig. 4 bottom panels. The performance is measured using

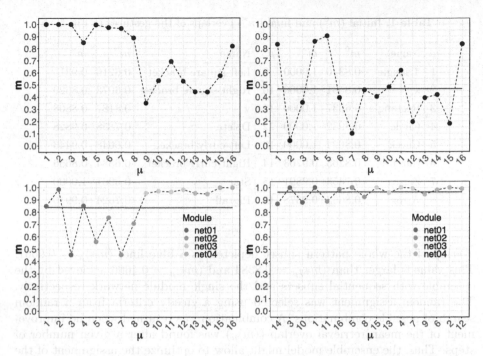

Fig. 4. Top left: Online retrieval performance, $n = 1, K = 200, \gamma = 0.2$ single module. Top right: Offline retrieval performance, $n = 1, K = 200, \gamma = 0.2$ single module. Bottom left: Offline retrieval performance, $n = 4, K_b = 50, \gamma_b = 0.05$ ensemble. Bottom right: Offline retrieval performance, $n = 4, K_b = 50, \gamma_b = 0.05$ ensemble with random assignment of patterns to modules. Blue line: mean retrieval overlap $\langle m* \rangle_\mu$ of the systems. Dashed lines depicted as a visual guide. (Color figure online)

the overlap parameter, where a value of $m = 1$ means perfect retrieval, $m = 0$ no retrieval, and intermediate values, retrieval with a level of noise of $1 - m$. Figure 4 top left panel depicts the online retrieval performance where patterns are learned (Eq. 10-left) and tested one by one. One can appreciate that the critical point between high and low retrieval (m) is around half the dataset (8 patterns). Figure 4 top right panel shows the offline retrieval for the whole dataset where all $P = 16$ patterns are learned at once (see Eq. 10-right) and tested. Only 4 patterns ($\mu = \{1, 4, 5, 16\}$) are retrieved with a high overlap $m^\mu > 0.8$. Figure 4 bottom left panel shows the offline retrieval of a $n = 4, K_b = 50, \gamma_b = 0.05$ ensemble. The $P = 16$ patterns are divided in $n = 4$ sequential subsets and $P_b = 4$ are learned by each of the modules. Compared with the single network, the performance improved from 4 to 11 patterns with an overlap $m_b^\mu > 0.8$.

Figure 4 bottom right panel shows the offline retrieval of a $n = 4, K_b = 50, \gamma_b = 0.05$ ensemble with a random assignment of the patterns, instead of sequential subsets as in Fig. 4 bottom left. Such subset arrangement allows to retrieve the whole dataset ($P = 16$) with a high overlap $m_b^\mu > 0.8$. The improvement of this random assignment can also be observed by the mean retrieval

Table 1. Initial (m^0) and final (m^*) overlaps of the gestures in Fig. 5.

μ	Name	m^0	m^*	μ	Name	m^0	m^*
1	Triangle	0.8346	1.0000	9	Left-square bracket	0.6216	0.9315
2	x	0.8561	1.0000	10	Right-square bracket	0.6073	0.8629
3	Rectangle	0.6337	1.0000	11	v	0.8187	0.8808
4	Circle	0.8932	1.0000	12	Delete	0.7858	0.9828
5	Check	0.8624	1.0000	13	Left-curly bracket	0.6048	0.9443
6	Caret	0.8263	0.9703	14	Right-curly bracket	0.6584	0.8656
7	Zigzag	0.7454	0.9891	15	Star	0.5961	1.0000
8	Arrow	0.8663	0.9378	16	Pigtail	0.7624	0.9916

overlap of the whole pattern subset depicted as a blue line, $\langle m* \rangle_\mu = 0.9631$. This value is larger than $\langle m* \rangle_\mu = 0.8384$ and $\langle m* \rangle_\mu = 0.4666$ achieved by the ensemble with sequential subsets and the single module network respectively. The random assignment was selected using a greedy criteria from a random search of possible combinations. A combination was returned when no improvement of the mean retrieval overlap ($\langle m \rangle_\mu$) was found after a given number of steps. Thus, the ensemble model might allow to optimize the assignment of the patterns to the modules to improve the retrieval performance of the network, both in terms of the number of retrieved patterns (capacity), and the quality of the retrieval (overlap).

4.2 Gesture Retrieval Quality

Figure 5 depicts the gestures learned and retrieved by the $n = 4, K_b = 50, \gamma_b = 0.05$ ensemble with random assignment of patterns to modules. The columns with label "Stored" correspond to the patterns presented to the network during the learning phase. The columns with label "Tested" correspond to the initial state of the network at the beginning of the retrieval phase m^0. Such initial states corresponds to a different capture of the gestures by the user, that is two different instances of the gestures. The overlaps between the tested and stored patterns are depicted in Table 1 in the m^0 columns. The mean overlap value between stored and tested gestures is $\langle m^0 \rangle_\mu = 0.7483$ with a minimum and maximum initial overlaps of 0.5961 and 0.8932 respectively. That is the initial (tested) states have a considerable level of noise compared with the learned (stored) patterns. After $t = 100$ steps of the ensemble modules updating the final states of the network overlaps m^* are depicted in the columns labeled as "Retrieved" in Fig. 5. A high quality retrieval is performed by the ensemble modules, as can be visually inspected. The retrieval overlaps are shown in Table 1, in the m^* columns, this values corresponds to the depiction in Fig. 4 bottom right panel. The gestures are effectively retrieved with a mean overlap of $\langle m^* \rangle_\mu = 0.9631$, with a minimum and maximum retrieval overlaps of 0.8629 and 1 respectively.

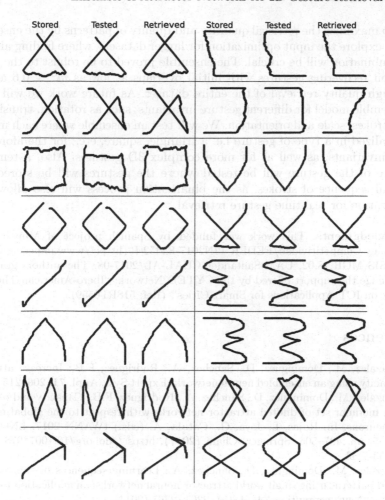

Fig. 5. Two retrieval examples: delete (top) and right curly bracket (bottom) gestures.

As observed, the ensemble modules are robust to the initial noisy conditions with a high attractor basin, starting with an initial condition as low as 0.5961 for the star pattern ($\mu = 15$), achieving a perfect retrieval of the gesture ($m^* = 1.0$).

5 Conclusions

We evaluated the EANN model [1,2] performance for learning and retrieval of a 2D single-stroke gesture dataset. The EANN outperformed the single network offline retrieval of the gestures. A random assignment of the gestures to the ensemble modules improved the retrieval of the highly correlated gestures, allowing the EANN to recover all gestures with a high quality retrieval ($m^* > 0.9$), contrary to the single network. Thus, an input optimization can be carried out in order to maximize the retrieval quantity and quality of patterns by the ensemble. We will explore the input optimization for larger datasets, where finding an optimal combination will be crucial. The ensemble proved to be robust to the initial condition recovering gestures with initial overlaps as low as $m^0 \sim 0.6$ achieving a high quality retrieval of the entire dataset. As future work we will check the ensemble model for different gesture invariants, such as rotation, translation, multi-strokes, scale and orientation. We will test an ensemble where each module is specialized in a type of gesture i.e. a triangle, square, etc., for the aforementioned invariants, as well as for more complex 3D gestures. Also, a temporal sequence of the gesture will be tested where the gestures will be stored as a temporal sequence of strokes, as the binarization process will also allow such configuration for real-time gesture retrieval.

Acknowledgments. This work was funded by Spanish project of Ministerio de Economía y Competitividad/FEDER TIN2017-84452-R (http://www.mineco.gob.es/), UDLA SIS MGR.18.02, UAM-Santander CEAL-AL/2017-08. The authors gratefully acknowledge the support offered by the CYTED Network: "Ibero-American Thematic Network on ICT Applications for Smart Cities" (Ref: 518RT0559).

References

1. Gonzalez, M., Dominguez, D., Sanchez, A., Rodriguez, F.B.: Increase attractor capacity using an ensembled neural network. Expert Syst. Appl. **71**, 206–215 (2017)
2. González, M., Dominguez, D., Sánchez, Á., Rodríguez, F.B.: Capacity and retrieval of a modular set of diluted attractor networks with respect to the global number of neurons. In: Rojas, I., Joya, G., Catala, A. (eds.) IWANN 2017. LNCS, vol. 10305, pp. 497–506. Springer, Cham (2017). https://doi.org/10.1007/978-3-319-59153-7_43
3. González, M., Dominguez, D., Sánchez, Á.: Learning sequences of sparse correlated patterns using small-world attractor neural networks: an application to traffic videos. Neurocomputing **74**(14–15), 2361–2367 (2011)
4. González, M., Dominguez, D., Rodríguez, F.B., Sanchez, A.: Retrieval of noisy fingerprint patterns using metric attractor networks. Int. J. Neural Syst. **24**(07), 1450025 (2014)
5. Doria, F., Erichsen Jr., R., González, M., Rodríguez, F.B., Sánchez, Á., Dominguez, D.: Structured patterns retrieval using a metric attractor network: application to fingerprint recognition. Phys. A: Stat. Mech. Appl. **457**, 424–436 (2016)
6. González, M., del Mar Alonso-Almeida, M., Avila, C., Dominguez, D.: Modeling sustainability report scoring sequences using an attractor network. Neurocomputing **168**, 1181–1187 (2015)
7. Vanderdonckt, J., Roselli, P., Pérez-Medina, J.L.: !FTL, an articulation-invariant stroke gesture recognizer with controllable position, scale, and rotation invariances. In: Proceedings of the 20th ACM International Conference on Multimodal Interaction, ICMI 2018, pp. 125–134. ACM, New York (2018)
8. Rozado, D., Moreno, T., San Agustin, J., Rodriguez, F., Varona, P.: Controlling a smartphone using gaze gestures as the input mechanism. Hum.-Comput. Interact. **30**(1), 34–63 (2015)

9. Wobbrock, J.O., Wilson, A.D., Li, Y.: Gestures without libraries, toolkits or training: a \$1 recognizer for user interface prototypes. In: Proceedings of the 20th Annual ACM Symposium on User Interface Software and Technology, UIST 2007, pp. 159–168. ACM, New York (2007)
10. Taranta, II, E.M., LaViola Jr., J.J.: Penny pincher: a blazing fast, highly accurate \$-family recognizer. In: Proceedings of the 41st Graphics Interface Conference, GI 2015, pp. 195–202. Canadian Information Processing Society, Toronto (2015)
11. Rozado, D., Rodriguez, F.B., Varona, P.: Low cost remote gaze gesture recognition in real time. Appl. Soft Comput. **12**(8), 2072–2084 (2012)
12. Beuvens, F., Vanderdonckt, J.: Designing graphical user interfaces integrating gestures. In: Proceedings of the 30th ACM International Conference on Design of Communication, SIGDOC 2012, pp. 313–322. ACM, New York (2012)
13. Coyette, A., Schimke, S., Vanderdonckt, J., Vielhauer, C.: Trainable sketch recognizer for graphical user interface design. In: Baranauskas, C., Palanque, P., Abascal, J., Barbosa, S.D.J. (eds.) INTERACT 2007. LNCS, vol. 4662, pp. 124–135. Springer, Heidelberg (2007). https://doi.org/10.1007/978-3-540-74796-3_14
14. Dominguez, D., González, M., Rodríguez, F.B., Serrano, E., Erichsen Jr., R., Theumann, W.: Structured information in sparse-code metric neural networks. Phys. A: Stat. Mech. Appl. **391**(3), 799–808 (2012)
15. Albert, R., Barabási, A.L.: Statistical mechanics of complex networks. Rev. Mod. Phys. **74**(1), 47 (2002)
16. Dominguez, D., González, M., Serrano, E., Rodríguez, F.B.: Structured information in small-world neural networks. Phys. Rev. E **79**(2), 021909 (2009)

Sparse Least Squares Support Vector Machines Based on Genetic Algorithms: A Feature Selection Approach

Pedro Hericson Machado Araújo and Ajalmar R. Rocha Neto[✉]

Computer Science Department, Federal Institute of Ceará, IFCE, Fortaleza, Brazil
hericson.araujo@ppgcc.ifce.edu.br, ajalmar@ifce.edu.br

Abstract. This paper presents a new approach for pruning dataset features (i.e., feature selection) based on genetic algorithms (GAs) and sparse least squares support vector machines (LSSVM) for classification tasks. LSSVM is a modified version of standard Support Vector Machine (SVM), which is in general faster to train than SVM since the training process of SVM requires the solution of a quadratic programming problem while the LSSVM demands only the solution of a linear equation system. GAs are applied to solve optimization problems without the assumption of linearity, differentiability, continuity or convexity of the objective function. There are some works where GAs and LSSVM work together, however, mostly to find the LSSVM kernel and/or classifier parameters. Nevertheless, our new proposal combines LSSVM and GAs for achieving sparse models, in which each support vector has just a few features in a feature selection sense. The idea behind our proposal is to remove non-relevant features from the patterns by using GAs. Removing a pattern has less impact than removing a feature since the training dataset has in general more patterns than features. On the basis of the results, our proposal leaves non-relevant features out of the set of features and still maintains or even improves the classification accuracy.

Keywords: Least Squares Support Vector Machines ·
Pruning methods · Genetic Algorithms

1 Introduction

There are many works that relies on Least Square Support Vector Machines (LSSVM) [17] with Evolutionary Computation (EC), especially Genetic Algorithms (GAs). As known, GAs are mostly applied to optimize the LSSVM kernel parameters, as well as the classifier parameters [10,20]. Genetic Algorithms (GAs) are optimization methods and therefore they can be used to generate useful solutions to search problems. By reasons of the underlying features of GAs, some optimization problems can be solved without the assumption of linearity, differentiability, continuity or convexity of the objective function. GAs as a meta-heuristic method are also used to handle classification tasks even with LSSVM.

© Springer Nature Switzerland AG 2019
I. Rojas et al. (Eds.): IWANN 2019, LNCS 11507, pp. 500–511, 2019.
https://doi.org/10.1007/978-3-030-20518-8_42

A theoretical advantage of kernel methods such as Support Vector Machines [19] concerns the empirical and structural risk minimization which balances the complexity of the model against its success at fitting the training data, along with the production of sparse solutions [16]. Support Vector Machines (SVM) are the most popular kernel methods. The LSSVM is an alternative to the standard SVM formulation [19]. A solution for the LSSVM is achieved by solving linear KKT systems[1] in a least square sense. In fact, the solution follows directly from solving a linear equation system, instead of a QP optimization problem. On the one hand, it is in general easier and less computationally intensive to solve a linear system than a QP problem. On the other hand, the resulting solution is far from sparse, in the sense that it is common to have all training samples being used as SVs.

To handle the lack of sparseness in SVM and LSSVM solutions, several *reduced set* (RS) and pruning methods have been proposed, respectively. These methods comprise a bunch of techniques aiming at simplifying the internal structure of those models, while keeping the decision boundaries as similar as possible to the original ones. They are very useful in reducing the computational complexity of the original models, since they speed up the decision process by reducing the number of SVs. They are particularly important for handling large datasets, when a great number of data samples may be selected as support vectors, either by pruning less important SVs [8] or by constructing a smaller set of training examples [11].

It is worth noticing that removing patterns is a way of achieving sparse models, however, one can also achieve sparse LSSVM models by eliminating features from the dataset so that each removed feature avoids L values being stored, where L is the number of patterns in the training dataset. Therefore, removing a pattern has, in general, less impact than removing a feature since the training dataset has more patterns than features.

In order to combine the aforementioned advantages of LSSVM classifiers and genetic algorithms, this work puts both of them to work together for selecting features while achieving equivalent (in some cases, higher) performances than standard full-set LSSVM classifiers. Our proposal finds out sparse classifiers based on a single objective genetic algorithm which takes into account the classification accuracy. Differently from previous works, genetic algorithms are used here to obtain less dataset features, not to find out the optimal values of the kernel and classifier parameters. The remaining part of this paper is organized as follows. In Sect. 2, the related work. In Sect. 3 we review the fundamentals of the LSSVM classifiers. After that, in Sect. 4 we discuss the use of Principal Component Analysis (PCA). In Sect. 5, we introduce genetic algorithms that will help to understand the proposal and then in Sect. 6 we demonstrate the proposal of this work called FPGAS-LSSVM. In Sect. 7 we present our simulations and discussions. After all, the paper is concluded in Sect. 8.

---

[1] Karush-Kuhn-Tucker systems.

2   Related Work

Many works have been applying to achieve models that works only with a few attributes (feature selection methods) or are built based only on a few patterns (pruning methods for sparse models). With respect to LSSVMs, feature selection or pruning methods yield sparse models. In [9] the authors use Principal Component Analysis (PCA) to analyze posts of twitter users that have been retweeted in a short period, by first analyzing the specific features of the tweet. A qualitative study was used to select the attributes which allow the best classification performance, such a method was called the PCA-LSSVM.

In [14], a pruning method named Genetic Algorithms for Sparse LSSVM (GAS-LSSVM) based on GAs with single objective function for sparse LSSVMs was proposed. Similarly, in [13], sparse LSSVMs based on multiobjective GAs were proposed by pruning patterns. In such a proposal named MOGAS-LSSVM, the multiobjective function takes into account sparsity and accuracy. In [12], a new method based on GAs with a single objective function to achieve sparse LSSVMs with fixed size of support vectors was proposed. Differently from the previous works, our proposal named FPGAS-LSSVM handles the number of features not the number of patterns in LSSVM. In Table 1, we present some information about the above-mentioned methods and our proposal.

Table 1. Comparison of pruning methods.

Reference	Name	Classifier	Pruning	Method
[14]	GAS-LSSVM	LSSVM	Patterns	Genetic Algorithm
[13]	MOGAS-LSSVM	LSSVM	Patterns	Genetic Algorithm
[12]	FSGAS-LSSVM	LSSVM	Patterns	Genetic Algorithm
[7]	PCA-LSSVM	LSSVM	Features	Principal Component Analysis
Proposed method	FPGAS-LSSVM	LSSVM	Features	Genetic Algorithm

3   LSSVM Classifiers

The primal problem formulation for LSSVM classifiers [17] is given by

$$\min_{\mathbf{w},\xi_i,b} \left\{ \tfrac{1}{2}\mathbf{w}^T\mathbf{w} + \gamma\tfrac{1}{2} \sum_{i=1}^{L} \xi_i^2 \right\} \tag{1}$$
$$\text{subject to } y_i[(\mathbf{w}^T x_i) + b] = 1 - \xi_i, i = 1, ..., L$$

where γ is a positive cost parameter, $\{\xi_i\}_{i=1}^{L}$ are the slack variables and b is the bias. We reach the Lagrangian function for LSSVM classifiers by rearranging Eq. (1).

$$L(\mathbf{w}, b, \boldsymbol{\xi}, \alpha) = \frac{1}{2}\mathbf{w}^T\mathbf{w} + \gamma\frac{1}{2} \sum_{i=1}^{L} \xi_i^2 - \sum_{i=1}^{L} \alpha_i(y_i(\mathbf{x}_i^T\mathbf{w} + b) - 1 + \xi_i), \tag{2}$$

where $\{\alpha_i\}_{i=1}^{L}$ are the Lagrange multipliers. The conditions for optimality are given by the partial derivatives

$$\frac{\partial L(\mathbf{w}, b, \boldsymbol{\xi}, \alpha)}{\partial \mathbf{w}} = 0 \Rightarrow \mathbf{w} = \sum_{i=1}^{L} \alpha_i y_i \mathbf{x}_i,$$

$$\frac{\partial L(\mathbf{w}, b, \boldsymbol{\xi}, \alpha)}{\partial b} = 0 \Rightarrow \sum_{i=1}^{L} \alpha_i y_i = 0,$$

$$\frac{\partial L(\mathbf{w}, b, \boldsymbol{\xi}, \alpha)}{\partial \alpha_i} = 0 \Rightarrow y_i(\mathbf{x}_i^T \mathbf{w} + b) - 1 + \xi_i = 0, \tag{3}$$

$$\frac{\partial L(\mathbf{w}, b, \boldsymbol{\xi}, \alpha)}{\partial \xi_i} = 0 \Rightarrow \alpha_i = \gamma \xi_i.$$

It is possible to formulate a linear system to represent the problem, based on Eq. (3).

$$\mathbf{Dz} = 1, \tag{4}$$

where

$$\mathbf{D} = \begin{bmatrix} 0 & y^T \\ y & \Omega + \gamma^{-1}\mathbf{I} \end{bmatrix}, \mathbf{z} = \begin{bmatrix} b \\ \alpha \end{bmatrix}, 1 = \begin{bmatrix} 0 \\ 1 \end{bmatrix} \tag{5}$$

$\Omega \in \mathbb{R}^{L \times L}$ is a matrix whose entries are given by $\Omega_{i,j} = y_i y_j x_i^T x_j, i, j = 1, ..., L$. In addition, $y = [y_1 \ldots y_L]^T$ and the symbol 1 denotes a vector of ones with dimension L. The solution of this linear system can be computed by direct inversion of matrix $\mathbf{D}$ as follows

$$\mathbf{z} = \mathbf{D}^{-1}1 = \begin{bmatrix} 0 & y^T \\ y & \Omega + \gamma^{-1}\mathbf{I} \end{bmatrix}^{-1} \begin{bmatrix} 0 \\ 1 \end{bmatrix}. \tag{6}$$

Once we have the values of the Lagrange multipliers and the bias, the output can be calculated based on the classification function described as

$$f(\mathbf{x}) = sign\left(\sum_{i=1}^{L} \alpha_i y_i \mathbf{x}^T \mathbf{x}_i + b\right) \tag{7}$$

It is simple to visualize in Eq. (7) which straightforward the usage of the kernel trick, which is applied to generate non-linear versions of the standard linear SVM classifier. This procedure works by replacing the dot product $\mathbf{x}^T \mathbf{x}_i$ with the kernel function $k(\mathbf{x}, \mathbf{x}_i)$.

4   Principal Component Analysis (PCA)

Principal Component Analysis are being widely used in biology, image recognition and signal processing [4,18,21]. It is used to make a classifier more efficient and, in addition, it is a useful statistical technique [5]. The feature extraction is one of most important part of the pattern recognition [2,3].

Çalisir and Dogantekin [6] justify the use of PCA in their work by addressing that features must be reduced (from the original data set) to a lower dimension by a feature extractor. The reduced feature vector obtained includes most useful information from the original feature vector. They use the LSSVM classifier and call the PCA-LSSVM method. The main advantage of PCA is that once you have found patterns in the data, you can compress the data, i.e., by reducing the number of dimensions, without much loss of information [15].

5   Genetic Algorithms

The Genetic Algorithm is a meta-heuristic search method inspired by natural evolution of the species, such as inheritance, crossing, natural selection and mutation, in which each possible solution is represented by an individual, composed of one or more chromosomes, being coded as a sequence of bits, that is, strings of 0s and 1s. In this work the gene vector is represented in binary form as a string of 0s and 1s, other models of representation are possible. Genetic algorithms are a form of local search that uses evolution-based methods to make small changes in a population of chromosomes in an attempt to identify an optimal solution. Due to the characteristics of GA, it is easier to solve various problems by GA than other mathematical methods which do have to rely on the assumption of linearity, continuity, differentiability, or convexity of the objective function [1].

In GAs, a set of individuals (population) is evolved similar as the evolution of the species, by natural selection. The population contains candidate individuals to better solve the optimization problem involving the best solution, i.e., fitness function. Each Individual of the population has genes (gene vector) called the chromosome, so that it can be modified by mutation operations and combined with other individuals to generate new individuals using the crossover operator in the reproduction. The initial population contains randomly generated individuals, which we call generation. The value of the fitness function indicates how good the solution is. This value is used to guide the reproduction process, in which individuals with high aptitude value are more likely to distribute their genes in the new population. The evaluation value is calculated after the decoding of the chromosome (individual), and the best solution for the problem to be solved is therefore the one with the highest value for the fitness function.

The best individuals in the population will be the one with the greatest fitness value, these individuals remain in the generations, with an operation called elitism. Standard implementation of genetic algorithm for natural selection is the roulette wheel but there is also the selection by tournament.

6 Proposal: Feature Pruning Genetic Algorithm for Sparse LSSVM (FPGAS-LSSVM)

Our proposal called FPGAS-LSSVM is based on a single genetic algorithm objective that aims to maximize the accuracy of the classifier by pruning features from the original dataset. The individual is represented by a binary vector, where one ('1') stands for the presence of a certain feature in the training (validation and test) dataset while zero ('0') stands for the absence of such a feature. Our fitness function is the value of the resulting accuracy of a classifier when we take into account the genes that were set to "one". In this approach, each individual has as many features as the number of genes set up to "one", if the gene has a value of "zero" then the feature will be deleted. The proposal is based on the accuracy obtained by the classifier after pruning the features, in order to remove features that are not relevant to the problem and still maintains or even improves the classification accuracy.

6.1 Individuals or Chromosomes

In our proposal, the individual is described by a vector $s = [g_1 \ g_2 \cdots g_i \cdots g_{q-1} \ g_q]$, where $g_i \in \{0,1\}$ and q is the number of dataset features. Whenever $g_i = 0$, it means the i-th feature is out of the training dataset. This way, we have as many features as the number of "ones" in the individual. We can also say that we have the number of features pruned like the number of "zeros" in the vector s. As stated, each pruned feature avoids saving L values. Note that LSSVM requires to save all the patterns as support vectors, so each removed feature helps save L dataset values from such a feature. To demonstrate the proposed approach, we present in an simplified way an example of how to model an individual for the FPGAS-LSSVM. Let us consider the matrix $\mathbf{B}$ like a sample dataset with 7 features as presented in Fig. 1, this means an individual stood for a gene vector equals to [1111111]. In our model, the following gene vector [1010111] stands for the removal of the second and fourth features (columns) as shown in Fig. 2.

$\mathbf{B} =$

0.4378	0.1326	0.2037	0.8998	0.4694	0.8705	0.4578
0.2802	0.5450	0.5444	0.2179	0.4138	0.6030	0.7222
0.9852	0.8278	0.8749	0.0770	0.5027	0.2653	0.3390
0.6088	0.8370	0.1210	0.4742	0.1254	0.8648	0.4012
0.2537	0.8333	0.8564	0.8350	0.1323	0.0581	0.5270

Fig. 1. Matrix $\mathbf{B}$ with the 7 features from the original dataset.

The matrix $\mathbf{B}_1$ represents the result of the pruning of the features on the original dataset, as follows in the Fig. 2. In general, in order to eliminate a certain feature from the dataset which is described by the i-th gene (g_i) in the vector $s = [g_1 \ g_2 \ g_3 \ g_4 \ g_5 \ g_6 \ g_7]$, it is necessary to remove the i column.

B1 =

0.4378	0.2037	0.4694	0.8705	0.4578
0.2802	0.5444	0.4138	0.6030	0.7222
0.9852	0.8749	0.5027	0.2653	0.3390
0.6088	0.1210	0.1254	0.8648	0.4012
0.2537	0.8564	0.1323	0.0581	0.5270

Fig. 2. Matrix $\mathbf{B}_1$ obtained after pruning from the original dataset

6.2 Fitness Function for FPGAS-LSSVM

The main idea of the fitness function is to maximize the accuracy of an individual who has a candidate solution s, while pruning the features of the original dataset, then we define fitness as an objective function given by:

$$fitness(s) = accuracy(s) \tag{8}$$

6.3 FPGAS-LSSVM Algorithm

FPGAS-LSSVM algorithm for training a classifier can be described as follows.

1   Initiate $t = 0$, where t stands for the generation;
2   Generate initial population $P(t)$, i.e., a set of $\{s_j\}_{j=1}^{p}$, where p is the number of individuals at the generation t, randomly;
3   For each individual $s_j \in P(t)$
4   Train LSSVM Linear taking into account the removed features by Eq. (6);
5   Evaluate the fitness, i.e., the validation set accuracy using Eq. (8);
6   While $t \le t_{max}$, where t_{max} is the maximum number of generations
7   Select some individuals from $P(t)$;
8   Apply crossover operator to selected individuals;
9   Apply mutation operator to selected individuals;
10   Apply elitism operator to selected individuals;
11   Compute $t = t + 1$;
12   Evaluate validation set accuracy for new population $P(t)$ using Eq. (8);
13   Select the best individual or solution, i.e., s^o;

7   Simulation and Discussion

We carried out some simulations taking into account a dataset Z, which was randomly divided into Z_1, Z_2 and Z_3 with 60%, 20%, 20% of the samples for training, validation and testing, respectively. This way, the dataset $Z = Z_1 \cup Z_2 \cup Z_3$. Real world datasets were used in our simulations from UCI Machine Learning Repository[2] (Bupa Liver Disorders, Australian Credit Approval, Pima Indians

---

[2] http://archive.ics.uci.edu/ml/.

Diabets, Sonar, Hepatitis) and the binary version of the vertebral column pathology (VCP) dataset. Information about datasets such as name, abbreviation, and quantity of patterns and features can be viewed in the Table 2. The LSSVM and PCA-LSSVM classifiers were used for comparison purposes. It can be observed in Table 3 the average accuracy along with the standard deviation, mean pruning rate (# PR) and the mean time of classification in seconds (# CT) referring to each classifier and respective dataset, executed in the test set after 20 independent runs.

Table 2. List of datasets classified

Dataset	Abbreviation	Patterns	Features
Pima Indians Diabets	PID	768	8
Bupa Liver Disorders	BLD	345	6
Sonar	SON	208	60
Hepatitis	HEP	142	19
Australian Credit Approval	AUS	690	14
Vertebral Column Pathologies	VCP	310	6

Each GA execution starts by generating a population of possible solutions in a random basis. For the next generation we keep 10% of the best individuals selected due to elitist selection, 80% of new individuals from reproduction using crossover operator and finally, 10% of new individuals are obtained by mutation. The algorithm stops when the average relative change in the best fitness function value is less than or equal to 10^{-5} or by reaching the number of generations. After reaching the stopping criterion, we choose the best individual i.e., the one with the best accuracy. We present some results in Table 3. By analysing the Table 3, it is possible to observe that the performance of the classifiers with the set of reduced features were similar or even higher than the conventional LSSVM and PCA-LSSVM. Since we carried out simulations for our proposal (FPGAS-LSSVM) and use the resulting average of selected features as the goal for the PCA-LSSVM, both of them have the same pruning rate. This way, the number of ones ('1') in a solution s^o for each realization will be the number of principal components present in our simulations. This was done in order to have a fair comparison among the models.

Besides that, we present the results for PCA-LSSVM when using different degree of information (i.e., number of principal components) in Figs. 3 and 4. These figures have the accuracy versus number of principal components in the first column and percentage of the information versus number of principal components in the second.

Table 3. Results for the LSSVM, PCA-LSSVM and FPGAS-LSSVM classifiers with 60%, 20% and 20% of the full dataset for training, validation and testing, respectively.

Dataset	Model	γ	Accuracy(%)	# PR(%)	# CT
PID	LSSVM	0.05	75.48 ± 3.3	-	0.00157
	PCA-LSSVM		73.93 ± 3.2	32.5	0.00120
	FPGAS-LSSVM		75.80 ± 3.2	32.5	0.00134
BLD	LSSVM	0.05	66.01 ± 5.9	-	0.00063
	PCA-LSSVM		59.05 ± 5.9	30.0	0.00023
	FPGAS-LSSVM		66.81 ± 7.0	30.0	0.00054
SON	LSSVM	0.05	74.65 ± 7.6	-	0.00009
	PCA-LSSVM		74.76 ± 6.4	51.7	0.00008
	FPGAS-LSSVM		76.16 ± 5.7	51.7	0.00007
HEP	LSSVM	0.05	64.65 ± 8.7	-	0.00008
	PCA-LSSVM		66.80 ± 7.1	52.8	0.00006
	FPGAS-LSSVM		66.20 ± 6.4	52.8	0.00005
AUS	LSSVM	0.05	86.59 ± 2.3	-	0.00117
	PCA-LSSVM		84.52 ± 2.3	37.1	0.00091
	FPGAS-LSSVM		86.30 ± 2.4	37.1	0.00094
VCP	LSSVM	0.05	83.71 ± 5.7	-	0.00046
	PCA-LSSVM		81.12 ± 6.5	25.0	0.00039
	FPGAS-LSSVM		83.06 ± 5.9	25.0	0.00038

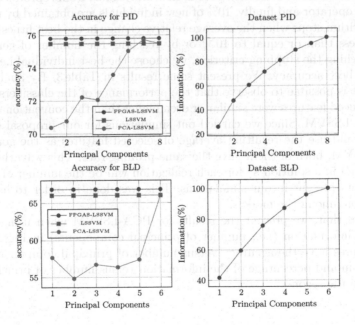

Fig. 3. Accuracy, number of principal components and percentage of information present in datasets PID and BLD.

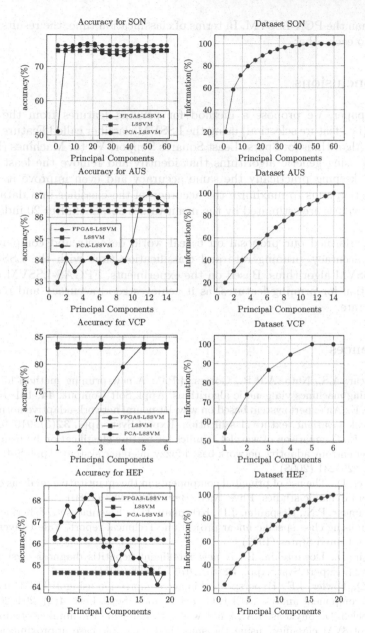

Fig. 4. Accuracy, number of principal components and percentage of information present in datasets SON, AUS, VCP and HEP.

By observing the Table 3 one can realize that our proposal obtains better accuracies for PID, BLD and SON and similar for the remaining data sets, being the accuracy very close to the conventional LSSVM and, in most cases, better than the PCA-LSSVM. In terms of classification time, the results are very similar to each other.

8   Conclusions

In this paper, we propose a method for pruning features from the original dataset (i.e., feature selection) using the LSSVM classifier called Feature Pruning Genetic Algorithms for Sparse Least Squares Support Vector Machines (FPGAS-LSSVM) using genetic algorithms that identify and remove the least relevant features, keeping practically the same accuracy and even improve it. The fitness function aims to maximize the accuracy of the classifier in a dataset. The accuracy values are estimated by the mean values obtained after 20 independent runs.

We performed our proposal on 6 real world datasets and compared three criteria: accuracy, pruning rate and classification time with the LSSVM and PCA-LSSVM algorithms. Based on the experiments, FPGAS-LSSVM can be a good option for pruning features, as it achieves good accuracy and reasonable pruning rate.

References

1. Alencar, A.S., Neto, A.R.R., Gomes, J.P.P.: A new pruning method for extreme learning machines via genetic algorithms. Appl. Soft Comput. **44**, 101–107 (2016)
2. Avci, E.: An expert system based on wavelet neural network-adaptive norm entropy for scale invariant texture classification. Expert Syst. Appl. **32**(3), 919–926 (2007)
3. Avci, E.: Comparison of wavelet families for texture classification by using wavelet packet entropy adaptive network based fuzzy inference system. Appl. Soft Comput. **8**(1), 225–231 (2008)
4. Barber, D.: The use of principal components in the quantitative analysis of gamma camera dynamic studies. Phys. Med. Biol. **25**(2), 283 (1980)
5. Belhumeur, P.N., Hespanha, J.P., Kriegman, D.J.: Eigenfaces vs. fisherfaces: recognition using class specific linear projection. Technical report, Yale University New Haven United States (1997)
6. Çalişir, D., Dogantekin, E.: A new intelligent hepatitis diagnosis system: PCA-LSSVM. Expert Syst. Appl. **38**(8), 10705–10708 (2011)
7. Du, Q., Fowler, J.E.: Hyperspectral image compression using JPEG2000 and principal component analysis. IEEE Geosci. Remote Sens. Lett. **4**(2), 201–205 (2007)
8. Geebelen, D., Suykens, J.A., Vandewalle, J.: Reducing the number of support vectors of SVM classifiers using the smoothed separable case approximation. IEEE Trans. Neural Netw. Learn. Syst. **23**(4), 682–688 (2012)
9. Morchid, M., Dufour, R., Bousquet, P.M., Linares, G., Torres-Moreno, J.M.: Feature selection using principal component analysis for massive retweet detection. Pattern Recogn. Lett. **49**, 33–39 (2014)
10. Mustafa, M., Sulaiman, M., Shareef, H., Khalid, S.A.: Reactive power tracing in pool-based power system utilising the hybrid genetic algorithm and least squares support vector machine. IET Gener. Transm. Distrib. **6**(2), 133–141 (2012)

11. Peres, R.T., Pedreira, C.E.: Generalized risk zone: selecting observations for classification. IEEE Trans. Pattern Anal. Mach. Intell. **31**(7), 1331–1337 (2009)
12. Silva, D.A., Rocha Neto, A.R.: A genetic algorithms-based LSSVM classifier for fixed-size set of support vectors. In: Rojas, I., Joya, G., Catala, A. (eds.) IWANN 2015. LNCS, vol. 9095, pp. 127–141. Springer, Cham (2015). https://doi.org/10.1007/978-3-319-19222-2_11
13. Silva, D.A., Silva, J.P., Neto, A.R.R.: Novel approaches using evolutionary computation for sparse least square support vector machines. Neurocomputing **168**, 908–916 (2015)
14. Silva, J.P., Neto, A.R.D.R.: Sparse least squares support vector machines via genetic algorithms. In: 2013 BRICS Congress on Computational Intelligence and 11th Brazilian Congress on Computational Intelligence (BRICS-CCI & CBIC), pp. 248–253. IEEE (2013)
15. Smith, L.I.: A tutorial on principal components analysis. University of Otago, Technical report (2002)
16. Steinwart, I.: Sparseness of support vector machines. J. Mach. Learn. Res. **4**(Nov), 1071–1105 (2003)
17. Suykens, J.A., Vandewalle, J.: Least squares support vector machine classifiers. Neural Process. Lett. **9**(3), 293–300 (1999)
18. Thomaz, C.E., Giraldi, G.A.: A new ranking method for principal components analysis and its application to face image analysis. Image Vis. Comput. **28**(6), 902–913 (2010)
19. Vapnik, V.: Statistical Learning Theory, vol. 3. Wiley, New York (1998)
20. Yu, L., Chen, H., Wang, S., Lai, K.K.: Evolving least squares support vector machines for stock market trend mining. IEEE Trans. Evol. Comput. **13**(1), 87–102 (2009)
21. Zhao, W., Krishnaswamy, A., Chellappa, R., Swets, D.L., Weng, J.: Discriminant analysis of principal components for face recognition. In: Wechsler, H., Phillips, P.J., Bruce, V., Soulié, F.F., Huang, T.S. (eds.) Face Recognition, pp. 73–85. Springer, Heidelberg (1998). https://doi.org/10.1007/978-3-642-72201-1_4

Mathematics for Neural Networks

A Neural Network-Based Approach to Sensor and Actuator Fault-Tolerant Control

Marcin Pazera[ID], Marcin Mrugalski[✉][ID], Marcin Witczak[ID], and Mariusz Buciakowski[ID]

Institute of Control and Computation Engineering, University of Zielona Góra, 65-246 Zielona Góra, Poland
{M.Pazera,M.Mrugalski,M.Witczak,M.Buciakowski}@issi.uz.zgora.pl

Abstract. The paper is devoted to the problem of design of robust estimator and controller on the basis of the neural-network model represented in a linear parameter-varying form. In particular the fault-tolerant controller for multiple sensor and actuator faults is developed. The proposed approach is able to minimise the influence of the multiple faults of sensor as well as actuator on the controlled system. The robust estimator and robust controller procedure boil down to solving a set of linear matrix inequalities. The illustrative part of the paper is devoted to the application of the proposed approach to fault tolerant control of the laboratory multi-tank system.

Keywords: Recurrent Neural Network ·
Actuator and sensors fault estimation · Fault Tolerant Control

1 Introduction

Over the decades of the development of Artificial Neural Networks (ANNs) theory, many architectures of neural models and training algorithms have been proposed [8,11,13]. The research focused on ANNs, in addition to a purely theoretical and cognitive character, also had a practical and application dimension. It is possible to differentiate several interesting applications of ANNs in the areas of modelling and identification of static and dynamic systems, signal prediction, artificial intelligence, image and speech recognition, feature classification, fault diagnosis and control [8,11,14,18]. Due to the scale and scope of the practical applications, ANNs stand out especially in the area of industrial system fault detection, estimation of remaining useful life and control of industrial processes and systems. It results from the specific ANNs features such as the ability to model complex non-linear dynamic systems, generalisation abilities and measurements-based learning properties [8]. Unfortunately, ANNs also have some important weaknesses that limit the area of their application in the model-based Fault Tolerant Control (FTC) [16], which is particularly important at the moment. It is due to the rarely available simple description of the

© Springer Nature Switzerland AG 2019
I. Rojas et al. (Eds.): IWANN 2019, LNCS 11507, pp. 515–526, 2019.
https://doi.org/10.1007/978-3-030-20518-8_43

neural model in the state space representation. The system description in this form allows the use of a wide range of well-known control system design tools. In addition, ANNs perfectly find themselves in the tasks of fault detection, but they are difficult to apply in fault estimation, what is extremely important in the case of their usage in FTC systems. Another important limitation is that existing ANNs enable system modelling while they do not provide a description of the model's uncertainty along with the description of the disturbances affecting the system. All the above-mentioned disadvantages causes that the analytic methods such as EKF [6], a sliding mode observers and an high-gain observers [10], a minimum-variance estimators [7] and $\mathcal{H}_\infty$ approaches [3,15,17] opposite to ANNs are applied. Unfortunately, these methods have the main disadvantage that they can be only used for a limited class of non-linear systems. It should be also underlined that the most of the analytic fault estimation and FTC approaches are dedicated to either actuator or sensor faults only. To overcome all the above-mentioned problems, in the paper a new fault estimation method with the application of the Recurrent Neural Network (RNN) represented in a Linear Parameter-Varying (LPV) form is proposed. Such method enables simultaneous estimation of several sensors and actuators faults. In order to make the designed estimator robust to the model and measurement uncertainties the Quadratic Boundedness (QB) approach is applied [1,2]. In the proposed technique it is assumed that all unknown uncertainties are bounded within an ellipsoid and they are taken into account during the design of the robust estimator. In the subsequent part of the paper a novel FTC scheme is developed. On the basis of the proposed RNN based fault estimator a new control strategy allowing for the compensation of the effect of faults is developed.

The paper is organised as follows: Sect. 2 delivers description of the RNN in the LPV form. Section 3 presents the design procedure of the robust fault estimator for the multiple actuators and sensors. Section 4 is devoted to the robust controller design, which minimises the influence of actuator and sensor faults. Section 5 shows an illustrative example of the proposed approach. Finally, Sect. 6 is devoted to conclusions.

2   Preliminaries

Let us consider the following representation of the a non-linear discrete-time system:

$$x_{k+1} = g\left(x_k, u_k\right), \tag{1}$$
$$y_k = Cx_k. \tag{2}$$

where $u_k \in \mathbb{R}^r$ represents the system control inputs and $y_k \in \mathbb{R}^m$ denotes its output. Vector $x_k \in \mathbb{X} \subset \mathbb{R}^n$ represents the state vector and unknown function $g\left(\cdot\right)$ describes a non-linear relation between the system inputs u_k and its state x_k. As it was preciously mentioned the ANNs and especially the RNN can be effectively applied in the non-linear and dynamic systems identification tasks.

Taking into consideration possible actuator and sensor faults and external exogenous disturbances, the system (1)–(2) can be redefined in the following LPV-like form [12]:

$$x_{k+1} = A\left(\gamma_k\right) x_k + B\left(\gamma_k\right) u_k + B\left(\gamma_k\right) f_{a,k} + W_1 w_{1,k}, \tag{3}$$

$$y_k = C x_k + f_{s,k} + W_2 w_{2,k}, \tag{4}$$

where γ_k represents an appropriate scheduling parameter. Furthermore, vectors $f_{a,k} \in \mathbb{F}_a \subset \mathbb{R}^r$ and $f_{s,k} \in \mathbb{F}_s \subset \mathbb{R}^m$ represent the actuator and sensor fault. The vectors $w_{1,k}$ and $w_{2,k}$ represent exogenous disturbance vectors representing process and measurement uncertainties. Furthermore, W_1 and W_2 stand for their distribution matrices. It has to be noticed that there is no sensor fault distribution matrix associated with the sensor fault vector $f_{s,k}$ which means that all measurements can be biased by the faults.

The primary goal of the paper relies on the development of the estimator which allow for simultaneous estimation of the sensor and actuator faults. The subsequent goal is to develop the control scheme allowing for minimisation of the influence of the faults on the controlled system. For this purpose an appropriate Lyapunov candidate function V_k has to be assumed which allow to perform the design of the fault estimator and controller but also proves its convergence. On the beginning let us remind the following definition and lemma [4] for $u_k = 0$ and $f_{a,k} = 0$:

Definition 1. *The system* (3) *is strictly quadratically bounded for all allowable* $w_k \in \mathcal{E}$, $k \geq 0$, *if* $V_k > 1 \implies V_{k+1} - V_k < 0$ *for any* $w_k \in \mathcal{E}$.

Lemma 1. *[4] The following statements are equivalent:*

1. *There exist* $X \prec 0$ *and* $W \prec 0$ *such that* $V^T X V - W \prec 0$,
2. *There exist* $X \prec 0$ *and* $W \prec 0$ *such that*

$$\begin{bmatrix} -W & V^T U^T \\ UV & X - U - U^T \end{bmatrix} \prec 0. \tag{5}$$

3   Estimation of the Actuator and Sensor Faults

The primary goal of this section is to propose a novel actuator and sensor fault estimation scheme. In order to achieve such goal the following form of the estimator is developed:

$$\hat{x}_{k+1} = A\left(\gamma_k\right) \hat{x}_k + B\left(\gamma_k\right) u_k + B\left(\gamma_k\right) \hat{f}_{a,k} + K_x \left(y_k - C\hat{x}_k - \hat{f}_{s,k}\right), \tag{6}$$

$$\hat{f}_{a,k+1} = \hat{f}_{a,k} + K_a \left(y_k - C\hat{x}_k - \hat{f}_{s,k}\right), \tag{7}$$

$$\hat{f}_{s,k+1} = \hat{f}_{s,k} + K_s \left(y_k - C\hat{x}_k - \hat{f}_{s,k}\right), \tag{8}$$

where matrices K_x, K_a and K_s represent the estimator gain matrices. The proposed scheme enables to estimate the state $\hat{x}_k$ of the system along with the sensor $\hat{f}_{s,k}$ and actuator $\hat{f}_{a,k}$ faults.

It should be underlined that by subtracting $\hat{f}_{s,k}$ from the Eq. 6 the state estimate $\hat{x}_k$ free from sensor faults can be obtained. Thus, the state estimate $\hat{x}_k$ converges to the real state x_k instead of the output measurements y_k encumber by the fault. In order to calculate the gain matrices of the estimator (6)–(8) it is necessary to define the state estimation error:

$$e_{k+1} = x_{k+1} - \hat{x}_{k+1} = [A(\gamma_k) - K_x C] e_k + W_1 w_{1,k}$$
$$+ B(\gamma_k) e_{a,k} - K_x e_{s,k} - K_x W_2 w_{2,k}, \tag{9}$$

where $e_{a,k} = f_{a,k} - \hat{f}_{a,k}$ and $e_{s,k} = f_{s,k} - \hat{f}_{s,k}$. Moreover, the dynamics of the actuator fault estimation error can be defined as:

$$e_{a,k+1} = e_{a,k} + \varepsilon_{a,k} - K_a C e_k - K_a e_{s,k} - K_a W_2 w_{2,k}, \tag{10}$$

and adequate the sensor fault estimation error is defined as:

$$e_{s,k+1} = \varepsilon_{s,k} + [I - K_s] e_{s,k} - K_s C e_k - K_s W_2 w_{2,k}. \tag{11}$$

By assuming the following form of vectors $\bar{e}_{k+1} = [e_{k+1}{}^T, e_{a,k+1}{}^T, e_{s,k+1}{}^T]^T$, and $\bar{w}_k = [w_{1,k}^T, w_{2,k}^T \varepsilon_{a,k}^T, \varepsilon_{s,k}]^T$ the following form of the actuator and sensor faults estimation error can be defined:

$$\bar{e}_{k+1} = X(\gamma_k) \bar{e}_k + Z \bar{w}_k, \tag{12}$$

with: $X(\gamma_k) = \bar{A}(\gamma_k) - \bar{K}\bar{C}$ and $Z = \bar{W} - \bar{K}\bar{V}$, where:

$$\bar{W} = \begin{bmatrix} W_1 & 0 & 0 & 0 \\ 0 & 0 & I & 0 \\ 0 & 0 & 0 & I \end{bmatrix}, \bar{A}(\gamma_k) = \begin{bmatrix} A(\gamma_k) & B(\gamma_k) & 0 \\ 0 & I & 0 \\ 0 & 0 & I \end{bmatrix},$$
$$\bar{C} = [C \; 0 \; I], \quad \bar{V} = [0 \; W_2 \; 0 \; 0], \quad \bar{K} = [K_x^T, K_a^T, K_s^T]^T. \tag{13}$$

In order to design the estimator and perform its convergence analysis the QB approach [1] is applied and the following assumptions are formulated:

Assumption 1: $\varepsilon_{a,k} = f_{a,k+1} - f_{a,k}$, $\varepsilon_{a,k} \in \mathcal{E}_{\varepsilon_a}$, $\mathcal{E}_{\varepsilon_a} = \{\varepsilon_a : \varepsilon_a^T Q_{\varepsilon_a} \varepsilon_a \le 1\}$ and $Q_{\varepsilon_a} \succ 0$.
Assumption 2: $\varepsilon_{s,k} = f_{s,k+1} - f_{s,k}$, $\varepsilon_{s,k} \in \mathcal{E}_{\varepsilon_s}$, $\mathcal{E}_{\varepsilon_s} = \{\varepsilon_s : \varepsilon_s^T Q_{\varepsilon_s} \varepsilon_s \le 1\}$, and $Q_{\varepsilon_s} \succ 0$.

Such assumptions are necessary to obtain fault estimation conditions. $\varepsilon_{a,k}$ and $\varepsilon_{s,k}$ are unknown but bounded within the ellipsoidal set [19]:

Assumption 3: $w_{1,k} \in \mathcal{E}_{w_1}$, $\mathcal{E}_{w_1} = \{w : w_1^T Q_{w_1} w_1 \le 1\}$ and $Q_{w_1} \succ 0$,
Assumption 4: $w_{2,k} \in \mathcal{E}_{w_2}$, $\mathcal{E}_{w_2} = \{w : w_2^T Q_{w_2} w_2 \le 1\}$, and $Q_{w_2} \succ 0$,

Assumption 5: $\mathcal{E}_{\bar{w}} = \{\bar{w} : \bar{w}^T Q_v \bar{w} \leq 1\}$, $Q_v = \frac{1}{4}\text{diag}\left(Q_{w_1}, Q_{w_2} Q_{\varepsilon_a}, Q_{\varepsilon_s}\right)$ and $Q_v \succ 0$.

The mentioned assumptions and definitions provided by [1,5] enable to formulate the following Lemma for (12):

Lemma 2. *The following statements are equivalent:*

1. *The system given by (12) is strictly quadratically bounded for all* $\bar{w}_k \in \mathcal{E}_{\bar{w}}$,
2. *There is a scalar* $\alpha \in (0,1)$ *such that*

$$\begin{bmatrix} X\left(\gamma_k\right)^T PX\left(\gamma_k\right) - P + \alpha P & X\left(\gamma_k\right)^T PZ \\ Z^T PX\left(\gamma_k\right) & Z^T PZ - \alpha Q_v \end{bmatrix} \preceq 0. \tag{14}$$

and consequently the subsequent theorem:

Theorem 1. *The system given by (12) is strictly quadratically bounded for* $\bar{w}_k \in \mathcal{E}_{\bar{w}}$ *if there exist* U, $P \succ 0$ *as well as* $0 < \alpha < 1$, *such that:*

$$\begin{bmatrix} -P + \alpha P & * & * \\ 0 & -\alpha Qv & * \\ P\bar{A}\left(\gamma_k\right) - U\bar{C} & P\bar{W} - UV & -P \end{bmatrix} \prec 0. \tag{15}$$

Proof. On the basis of (12) and (14) it can be seen that QB is equivalent to:

$$V_{k+1} - (1-\alpha)V_k - \alpha \bar{w}_k^T Q_v \bar{w}_k < 0, \tag{16}$$

what results:

$$\bar{e}_{k+1} P\bar{e}_{k+1} - \bar{e}_k^T P\bar{e}_k + \alpha \bar{e}_k^T P\bar{e}_k - \alpha \bar{w}_k^T Q_v \bar{w}_k < 0. \tag{17}$$

By assuming that:

$$v_k = \begin{bmatrix} \bar{e}_k \\ \bar{w}_k \end{bmatrix}, \tag{18}$$

the relation (17) can be rewritten as:

$$v_k^T \begin{bmatrix} X\left(\gamma_k\right)^T PX\left(\gamma_k\right) - P + \alpha P & X\left(\gamma_k\right)^T PZ \\ Z^T PX\left(\gamma_k\right) & Z^T PZ - \alpha Q_v \end{bmatrix} v_k \prec 0. \tag{19}$$

By applying the Schur complement and then multiplying by $\text{diag}\,(I, I, P)$ the following inequality is obtained:

$$\begin{bmatrix} -P + \alpha P & 0 & X\left(\gamma_k\right)^T P \\ 0 & -\alpha Q_v & Z^T P \\ PX\left(\gamma_k\right) & PZ & -P \end{bmatrix} \prec 0. \tag{20}$$

By substituting $PX\left(\gamma_k\right) = P\bar{A}\left(\gamma_k\right) - UC$ and $PZ = P\bar{W} - U\bar{V}$. completes the proof.

It is also worth highlighting that the multiplication of α and P in (15) causes that it is a set of Bilinear Matrix Inequalities (BMIs). To avoid such problem a set of LMIs with iterative changing $\alpha \in (0,1)$ is solved and in consequence the following gain matrices can be calculated:

$$\bar{K} = [K_x^T, K_a^T, K_s^T]^T = P^{-1} U. \tag{21}$$

4   Robust Neural Network-Based Controller Design

The main goal of this section is to develop the controller which allows to minimise the influence of the actuator and sensor faults into the system. Thus, let us define the control strategy combining the sensor and actuator fault compensation:

$$u_k = -K_c \hat{x}_k - \hat{f}_{a,k}. \tag{22}$$

During design procedure the sensor fault-free state estimate $\hat{x}_k$ is used. It follows from the fact that it converges to the real state regardless of occurrence of the sensor faults. The design of the controller boils down to the determination of the gain matrix K_c. On the basis of the defined control strategy (22) and assuming estimation error as:

$$e_k = x_k - \hat{x}_k, \tag{23}$$

the equation defining the system state (3) can be expressed as:

$$x_{k+1} = [A(\gamma_k) - B(\gamma_k) K_c] x_k + B(\gamma_k) K_c e_k - B(\gamma_k) e_{a,k} + W_1 w_{1,k}. \tag{24}$$

Similarly as it was in Sect. 3 devoted to the fault estimation, by assuming that

$$\bar{v}_k = \begin{bmatrix} \bar{e}_k \\ \bar{w}_k \end{bmatrix}, \tag{25}$$

the closed-loop system (24) can be redefined to the following form

$$x_{k+1} = \tilde{X}(\gamma_k) x_k + \tilde{Z} \bar{v}_k, \tag{26}$$

where $\tilde{X}(\gamma_k) = \tilde{A}(\gamma_k) - B(\gamma_k) K_c$, $\tilde{Z}(\gamma_k) = \begin{bmatrix} \tilde{B}(\gamma_k) \tilde{W}_1 \end{bmatrix}$, $\tilde{W}_1 = \begin{bmatrix} W_1 \ 0 \ 0 \ 0 \end{bmatrix}$, $\tilde{B}(\gamma_k) = \begin{bmatrix} B(\gamma_k) K_c - B(\gamma_k) K_c 0 \end{bmatrix}$.

On the basis of the above system description, a candidate of a Lyapunov function can be defined:

$$V_{c,k} = x_k^T U^{-T} P U^{-1} x_k. \tag{27}$$

and the controller design procedure can be summarised in the form of the following theorem:

Theorem 2. *The system given by* (26) *is strictly quadratically bounded for* $\bar{w}_k \in \mathcal{E}_{\bar{w}}$ *if there exist* U, N, P *as well as a scalar* $0 < \alpha < 1$ *such that the following inequality is fulfil*

$$\begin{bmatrix} -P + \alpha P & * & * \\ 0 & -\alpha Q_c & * \\ \tilde{A}(\gamma_k) U - B(\gamma_k) N & \tilde{Z}(\gamma_k) & P - U - U^T \end{bmatrix} \prec 0, \tag{28}$$

for $\tilde{Z}(\gamma_k) = \begin{bmatrix} \tilde{B}(\gamma_k) \tilde{W}_1 \end{bmatrix}$.

Proof. The problem of the QB controller design boils down to obtain matrices N, U and P. Thus, it is necessary to find a Lyapunov function such that:

$$V_{c,k+1} - (1 - \alpha) V_{c,k} - \alpha \bar{v}_k^T Q_c \bar{v}_k < 0, \qquad (29)$$

which leads to:

$$x_k^T (S_1(\gamma_k)) x_k + x_k^T (S_2(\gamma_k)) \bar{v}_k + \bar{v}_k^T (S_3(\gamma_k)) x_k + \bar{v}_k^T (S_4(\gamma_k)) \bar{v}_k < 0, \qquad (30)$$

with

$$S_1(\gamma_k) = \tilde{X}(\gamma_k)^T U^{-T} P U^{-1} \tilde{X}(\gamma_k) - U^{-T} P U^{-1} + \alpha U^{-T} P U^{-1}, \qquad (31)$$

$$S_2(\gamma_k) = \tilde{X}(\gamma_k)^T U^{-T} P U^{-1} \tilde{Z}(\gamma_k), \qquad (32)$$

$$S_3(\gamma_k) = \tilde{Z}(\gamma_k)^T U^{-T} P U^{-1} \tilde{X}(\gamma_k), \qquad (33)$$

$$S_4(\gamma_k) = \tilde{Z}(\gamma_k)^T U^{-T} P U^{-1} \tilde{Z}(\gamma_k) - \alpha Q_c. \qquad (34)$$

By substituting:

$$\tilde{v}_k = \begin{bmatrix} x_k \\ \bar{v}_k \end{bmatrix}, \qquad (35)$$

the relation (30) can be redefined as follows:

$$\tilde{v}_k^T \begin{bmatrix} S_1(\gamma_k) & S_2(\gamma_k) \\ S_3(\gamma_k) & S_4(\gamma_k) \end{bmatrix} \tilde{v}_k \prec 0. \qquad (36)$$

By multiplying the relation (36) by the expression $\left(U^T, I \right)$ and (U, I), respectively, and using Lemma 1 the following inequality is obtained:

$$\begin{bmatrix} -P + \alpha P & 0 & U^T \tilde{X}(\gamma_k)^T \\ 0 & -\alpha Q_c & \tilde{Z}(\gamma_k) \\ \tilde{X}(\gamma_k) U & \tilde{Z}(\gamma_k) & P - U - U^T \end{bmatrix} \prec 0. \qquad (37)$$

Finally, by substituting:

$$\tilde{X}(\gamma_k) U = \tilde{A}(\gamma_k) U - B K_c U = \tilde{A}(\gamma_k) U - B N, \qquad (38)$$

into (37), completes the proof.

It should be noticed that (28) is a set of BMIs for the same reason as it was in Sect. 3 devoted to the fault estimation. Let us remind that such problem boils down to iterative solving of a set of LMIs (28) for a fixed values of α. Finally, the controller design procedure relies on calculation of the gain matrix:

$$K_c = U^{-1} N. \qquad (39)$$

5 Illustrative Example

To illustrate the effectiveness of the proposed approach it was validated with the application of the non-linear Multi-Tank (M-T) system provided by Inteco [9]. Such a system (c.f. Fig. 1) is designed to simulate the developed FD and FTC strategies for the real M-T systems in the safe laboratory environment. It is built with three separated tanks which are connected as a cascade. Each tank has a different shape what gives a whole system non-linear character. Moreover, each tank is equipped with a liquid level sensor. The water is pumped into the upper tank by 12[V] DC pump and then, with the gravity, outflows to the subsequent tanks. The digital I/O board is used to control the system via Simulink in Matlab.

Fig. 1. The laboratory M-T system by Inteco

The implementation of the proposed adaptive estimator as well as the fault-tolerant controller requires neural model of the M-T system. Thus, at the beginning the RNN consisted of 13 neurons in hidden layer was trained with the application of the Levenberg-Marquardt algorithm on the basis of fault-free data.

Moreover, during the experiments the following fault scenario has been explored comprising the sensor $f_{s,k} = [f_{s,1,k}, f_{s,2,k}, f_{s,3,k}]$ and actuator $f_{a,k}$

faults, respectively:

$$\boldsymbol{f}_{s,1,k} = 0,$$

$$\boldsymbol{f}_{s,2,k} = \begin{cases} y_{2,k} + 0.04, & 600[s] \leq t \leq 800[s], \\ 0, & \text{otherwise}, \end{cases}$$

$$\boldsymbol{f}_{s,3,k} = \begin{cases} y_{3,k} - 0.1, & 700[s] \leq t \leq 900[s], \\ 0, & \text{otherwise}, \end{cases} \tag{40}$$

$$\boldsymbol{f}_{a,k} = \begin{cases} -0.45, & 55[s] \leq t \leq 85[s], \\ 0, & \text{otherwise}. \end{cases} \tag{41}$$

As it can be seen the first fault relies on the temporarily lost of the 45% of the effectiveness of the pump. The second fault boils down to the damage of the two sensors which appears partially in the same time. During the experiment the initial state for the system and estimator are assumed as $x_0 = [0, 0, 0]$ and $\hat{x}_0 = [0.01, 0.01, 0.01]$. The actuator as well as the sensor faults estimates are initiated by $\hat{\boldsymbol{f}}_{a,0} = 0$ and $\hat{\boldsymbol{f}}_{s,0} = [0, 0, 0, 0]$, respectively.

As a result of solving a set of LMIs (15), presented in Sect. 3, the following gain matrices for the state and for the actuator and sensor faults estimation has been calculated:

$$\boldsymbol{K}_x = \begin{bmatrix} 0.3757 & -0.0244 & -0.0001 \\ -0.0481 & 0.0061 & -0.0006 \\ 0.0000 & -0.0014 & 0.0017 \end{bmatrix}, \tag{42}$$

$$\boldsymbol{K}_a = \begin{bmatrix} 0.3109 & -0.0197 & -0.0002 \end{bmatrix}, \tag{43}$$

$$\boldsymbol{K}_s = \begin{bmatrix} 0.3738 & 0.0470 & -0.0008 \\ 0.0025 & 0.9885 & 0.0039 \\ -0.0002 & -0.0002 & 1.0021 \end{bmatrix}. \tag{44}$$

Similarly, the controller feedback gain matrix has been obtained by solving a set of LMIs defined by (28), and it has value:

$$\boldsymbol{K}_c = \begin{bmatrix} 0.8747 & 0.0199 & 0.0004 \end{bmatrix}. \tag{45}$$

It should be mentioned that the control goal was to achieve and keep the appropriate water level in the middle tank of the M-T system.

Figures 2(a), (b) and (c) present the water levels in the upper, middle and lower tank respectively, obtained with the fault-tolerant control provided by the proposed control strategy. The state estimates are pursuing quickly to the real water level, despite the fact of occurrence of the actuator fault. Moreover, the water levels in the other two tanks were measured incorrectly by the faulty sensors. It should be underlined that opposite to the sensor faults the state estimates have not been proceeding the faulty output but they estimate the real state. Furthermore, the set-point has been changed during the experiment, though the system tried to reach and keep the water level at the desired level as close as possible. AS it can be seen the sensors fault estimate have been obtained with a very high accuracy which is shown on the left side of Fig. 3.

a)   b)

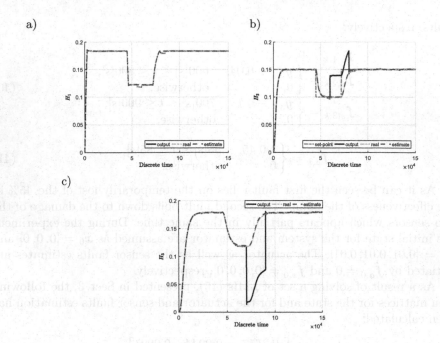

c)

Fig. 2. The response of the system

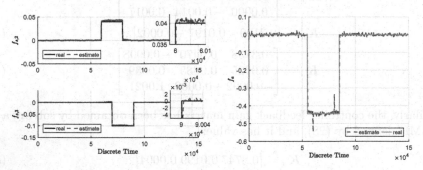

Fig. 3. The sensor faults $\boldsymbol{f}_{s,k}$ with their estimates $\hat{\boldsymbol{f}}_{s,k}$ (left) and the actuator fault $\boldsymbol{f}_{a,k}$ with its estimate $\hat{\boldsymbol{f}}_{a,k}$ (right)

Finally, on the right Fig. 3 the real actuator fault along with its estimate are presented. Despite the sensor faults, the actuator fault has been identified and estimated properly. Only two small spikes can be observed at the time instances related to the beginning and ending of the abrupt sensor fault in the middle tank. As it was already mentioned, the control goal was to reach and keep the appropriate water level in the middle tank. The system keeps the reference as close as possible. As a result of estimating the faults quite good and then by employing them into the suitable control strategy in order to compensate the effect of faults, the system is controlled in a very proper way. It can be concluded that the obtained results prove the effectiveness of the developed approach.

6 Conclusions

In the paper a novel methodology for the FTC under the presence of simultaneous actuator and sensor faults is developed. To tackle the simultaneous estimation problem for the state and sensor as well as actuator faults a novel observer has been proposed. Such approach guarantees a predefined disturbance attenuation level and convergence of the obtained estimator. Subsequently, on the basis of the sensors faults estimate it was possible to estimate the real state of the system output without faults. In the next part of the paper the FTC control approach was proposed. The robust controller has been developed using similar criteria. The design procedure for both, estimator as well as controller, boils down to solving a set of LMIs. The last part of the paper contains an illustrative example devoted to the application of the proposed approach in FTC task of the non-linear M-T system. The obtained results show the high effectiveness of the proposed approach.

Acknowledgements. The work was supported by the National Science Centre, Poland under grant: UMO-2017/27/B/ST7/00620.

References

1. Alessandri, A., Baglietto, M., Battistelli, G.: Design of state estimators for uncertain linear systems using quadratic boundedness. Automatica **42**(3), 497–502 (2006)
2. Cayero, J., Rotondo, D., Morcego, B., Puig, V.: Optimal state observation using quadratic boundedness: application to UAV disturbance estimation. Int. J. Appl. Math. Comput. Sci. **29**(1), 99–109 (2019)
3. Chen, L., Patton, R., Goupil, P.: Robust fault estimation using an LPV reference model: addsafe benchmark case study. Control Eng. Pract. **49**, 194–203 (2016)
4. de Oliveira, M.C., Bernussou, J., Geromel, J.C.: A new discrete-time robust stability condition. Syst. Control Lett. **37**(4), 261–265 (1999)
5. Ding, B.: Dynamic output feedback predictive control for nonlinear systems represented by a Takagi-Sugeno model. IEEE Trans. Fuzzy Syst. **19**(5), 831–843 (2011)
6. Foo, G.H.B., Zhang, X., Vilathgamuwa, D.M.: A sensor fault detection and isolation method in interior permanent-magnet synchronous motor drives based on an extended Kalman filter. IEEE Trans. Ind. Electron. **60**(8), 3485–3495 (2013)
7. Gillijns, S., De Moor, B.: Unbiased minimum-variance input and state estimation for linear discrete-time systems. Automatica **43**(1), 111–116 (2007)
8. Haykin, S.: Neural Networks and Learning Machines, vol. 3. Pearson, Upper Saddle River (2009)
9. INTECO. Multitank System - User's manual (2013). www.inteco.com.pl
10. Khalil, H.K., Praly, L.: High-gain observers in nonlinear feedback control. Int. J. Robust Nonlinear Control **24**(6), 993–1015 (2014)

11. Mrugalski, M.: Advanced Neural Network-Based Computational Schemes for Robust Fault Diagnosis. Springer, Cham (2014). https://doi.org/10.1007/978-3-319-01547-7
12. Mrugalski, M., Luzar, M., Pazera, M., Witczak, M., Aubrun, C.: Neural network-based robust actuator fault diagnosis for a non-linear multi-tank system. ISA Trans. **61**, 318–328 (2016)
13. Nelles, O.: Non-linear Systems Identification. From Classical Approaches to Neural Networks and Fuzzy Models. Springer, Berlin (2001). https://doi.org/10.1007/978-3-662-04323-3
14. Nguyen, A., Yosinski, J., Clune, J.: Deep neural networks are easily fooled: high confidence predictions for unrecognizable images. In: Proceedings of the IEEE Conference on Computer Vision and Pattern Recognition, pp. 427–436 (2015)
15. Nobrega, E.G., Abdalla, M.O., Grigoriadis, K.M.: Robust fault estimation of uncertain systems using an LMI-based approach. Int. J. Robust Nonlinear Control **18**(18), 1657–1680 (2008)
16. Noura, H., Theilliol, D., Ponsart, J.C., Chamseddine, A.: Fault-tolerant Control Systems: Design and Practical Applications. Springer, London (2009). https://doi.org/10.1007/978-1-84882-653-3
17. Pazera, M., Buciakowski, M., Witczak, M.: Robust multiple sensor fault-tolerant control for dynamic non-linear systems: application to the aerodynamical twin-rotor system. Int. J. Appl. Math. Comput. Sci. **28**(2), 297–308 (2018)
18. Russell, S.J., Norvig, P.: Artificial Intelligence: A Modern Approach. Pearson Education Limited, Malaysia (2016)
19. Witczak, M., Buciakowski, M., Puig, V., Rotondo, D., Nejjari, F.: An LMI approach to robust fault estimation for a class of nonlinear systems. Int. J. Robust Nonlinear Control **26**(7), 1530–1548 (2015)

Estimating Supervisor Set Using Machine Learning and Optimal Control

Konrad Kosmatka and Andrzej Nowakowski[✉]

Faculty of Mathematics and Computer Science, University of Łódź,
Banacha 22, 90-238 Łódź, Poland
{konrad.kosmatka,andrzej.nowakowski}@wmii.uni.lodz.pl

Abstract. The paper deals with the problem of finding an estimation of supervisor set and content of machine learning block. We propose a construction of a probability distribution of the learning set using the empirical risk functional defined by Vapnik and applying a new dual dynamic programming ideas to formulate a new optimization problem. As a consequence we state and prove a verification theorem for an approximate probability distribution defining the approximation of the supervisor set.

1 Introduction

One of the most popular application of neural networks is pattern recognition (discriminant analysis). Usually, a kind of supervised learning method to feedforward networks is applied. There exist several types of neural networks which can realize the pattern recognition. However, the neural networks have some disadvantages. We do not know whether the network learned (in fact some function) on the given supervised set really is that we are looking for, i.e. if we change the data, is the answer of neural network (for this new data) correct? That problem was studied wildly by Vapnik and published in several books. Vapnik and his coauthor propose statistical tools to solve this problem and add one block to the neural network set: machine learning block (see Fig. 1). Moreover, for the set of the given data (the supervised set) he introduced a probability distribution function (unknown), which should measure a quality of the supervised set for demanded function from learning block. Thus, an estimation of the probability distribution of learning set is a challenging problem. The Vapnik model [1] (see Fig. 1) consists of three components: (i) a generator (G) of random vectors $x \in \mathbb{R}^n$, drawn independently from a fixed but unknown probability distribution function $F(x)$, (ii) supervisor (S) that returns an output value y to every input vector x, according to a conditional distribution unknown function $F(y \mid x)$, (iii) a learning machine LM (LM-block) capable of implementing a set of functions $f(x, \alpha)$, $\alpha \in \Lambda$, where Λ is a set of parameters. The problem of learning is that of choosing from the given set of functions $f(x, \alpha)$, $\alpha \in \Lambda$, the one that best approximates the supervisor's response.

The selection of the desired function is based on a training set of l independent and identically distributed (i.i.d.) observations: $(x_1, y_1), \ldots, (x_l, y_l)$, drawn

© Springer Nature Switzerland AG 2019
I. Rojas et al. (Eds.): IWANN 2019, LNCS 11507, pp. 527–538, 2019.
https://doi.org/10.1007/978-3-030-20518-8_44

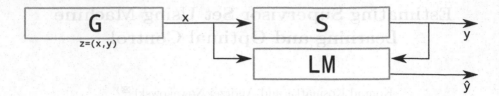

Fig. 1. Vapnik's learning model

according to $F(x,y) = F(x)F(y \mid x)$ (compare [2]). In order to choose the best available approximation to the supervisor's response, one measures the loss, or discrepancy, $L(y, f(x, a))$ between the response y of the supervisor to a given input x and the response $f(x, \alpha)$ provided by the learning machine. In [1] (see also [3]) the expected value of the loss is approximated by the empirical risk functional

$$R(\alpha) = \frac{1}{l} \sum_{i=1}^{l} L(z_i, f(z_i, \alpha)).$$

As we interest mostly in approximate estimation of the probability distribution of learning set, in order to take it into account more explicitly, we modify that functional to the following

$$R(\alpha) = \sum_{i=1}^{l} L(z_i, f(z_i, \alpha)) p(z_i), \tag{1}$$

where $z_i = (x_i, y_i)$, $i = 1, \ldots, l$ and $p(z_i)$ is a probability mass function corresponding to the discrete probability distribution function $F(z)$. The probability distribution function $F(z)$ defines how good is the supervisor's set z_i to characterize unknown (but existing!) supervisor's function (operator). The goal is to find the function $f(x, \alpha_0)$ that minimizes the risk functional $R(\alpha)$ over the class of functions $f(x, \alpha)$, $\alpha \in \Lambda$, in the situation where the joint probability distribution function $F(z)$ is unknown and the only available information is contained in the training set. The problems considered up to now using statistical tools, were consistency of learning processes, bounds on the rate of convergence of learning processes as well constructing learning algorithms. However, we do not find papers with algorithm allowing to find the probability measure $F(z)$ as well to find an approximate function $f(z, \alpha_\varepsilon)$ to the empirical risk functional (1) at $f(x, \alpha_0)$. From the practical point of view just finding suitable set of points $z_1, \ldots, z_l$, i.e. the function $F_\varepsilon(z)$ and a function $f(z, \alpha_\varepsilon)$ such that

$$|R(\alpha_0) - R(\alpha_\varepsilon)| < \varepsilon \tag{2}$$

are the most important. But, we do not know a priori $f(z, \alpha_0)$ and $F(z)$ hence to measure (2) is not an easy job. The main tools applied in the machine learning are from statistics and probability theory. However the problem of the minimization of the empirical risk functional (1) can be reformulated in terms of the optimal control theory treating parameter $\alpha \in \Lambda$ as a control of the function $f(z, \alpha)$. Then choosing the set of controls Λ in a suitable way and the corresponding family of functions $f(z, \alpha)$ are particularly essential. In the optimal control theory usually Λ is a set of functions defined on some domain over which integral in (1) is taken and the set of functions $f(z, \alpha)$, $\alpha \in \Lambda$ is described by differential equations. As $z = (x, y)$ is an element of some space $\mathbb{R}^{n+1}$, thus a domain Ω of the observations for $z_1, \ldots, z_l$ should be from $\mathbb{R}^{n+1}$. This fact determines that differential equations defining the set of functions $f(x, \alpha)$, $\alpha \in \Lambda$ must be of a partial differential equation (PDE) type. We use a system of elliptic partial differential equations. For such a general optimal control problem we have tools in optimal control theory which provide conditions in a form of verification type theorem allowing to find approximate solutions. Therefore, the aim of this paper is to formulate in the rigorous way the optimal control problem and to prove a verification theorem for an approximate minimum of the empirical risk functional (1) related to pattern recognition in a family of functions $f(z, \alpha)$, $\alpha \in \Lambda$ defined by a system of elliptic PDE.

2   The Optimal Control Problem for Functional (1)

Let Ω be an open domain in $\mathbb{R}^{n+1}$ of the variables $z = (x, y)$, $x \in \mathbb{R}^n$, $y \in \mathbb{R}$. Let us fix any finite interval $I \subset \mathbb{R}^m$, $m \geq 1$, in general, m is different than $n + 1$. In Ω we consider a family

$$\Lambda = \left\{ \alpha(z), \ z \in \Omega : \alpha \in L^2(\Omega), \ \alpha(z) \in I \right\}$$

of controls. We consider the functions $f : \Omega \times I \to [-1, 1]$, forming the machine learning block (LM-block). We define them as being solutions of semilinear elliptic equation. In the sequel we omit dependence of $f(z, \alpha)$ on α and we simply write $f(z)$, $z \in \Omega$ and then the function $H(z, f(z), \alpha(z), p(z))$ below depends on $\alpha(z)$. This relates to notations in optimal control theory. Thus denote by $\mathbf{Q}$ (LM-block) a family of functions $f(z)$ being defined by controls $\alpha \in \Lambda$ and (being solutions of) the following elliptic equation

$$-\Delta f(z) = H(z, f(z), \alpha(z), p(z)), \ z \in \Omega, \tag{3}$$

where $H : \mathbb{R}^{n+1} \times \mathbb{R} \times \mathbb{R}^m \times \mathbb{R} \to \mathbb{R}$, is continuous and sufficiently regular functions ensuring existence of solutions to (3) for each control $\alpha \in \Lambda$ belonging to the Sobolev space $W^{2,2}(\Omega)$. We do not impose any boundary conditions for f. However we assume that the interval I and the function H are chosen in such a way that $-1 \leq f(z) \leq 1$, $z \in \Omega$, $\alpha(\cdot) \in \Lambda$. We use (3) to define only a certain family of functions generated by $\alpha \in \Lambda$. Denote by $\mathcal{P}$ a set of $p(z)$, probability mass functions corresponding to the discrete probability distribution functions $F(z)$ whose domain of definition is contained in Ω, i.e. the set of measurable functions $p : \Omega \to [0, 1]$ with $\int_\Omega p(z) dz = 1$, which are different from zero at fixed mass points $z_i \in \Omega$, $i = 1, \ldots, l$ and $(1/l) \sum_{i=1}^{l} p(z_i) \leq 1$. We do not assume that $(1/l) \sum_{i=1}^{l} p(z_i)$ equals 1 as we admit that the observation $z_i \in \Omega$,

$i = 1, \ldots, l$ may be not enough good to characterize a function we are looking for approximation of (2). According to the above new notations we rewrite the empirical risk functional (1) to the form

$$R(\alpha, p) = \sum_{i=1}^{l} L(z_i, f(z_i), \alpha(z_i)) p(z_i), \tag{4}$$

where $z_i \in \Omega$, $i = 1, \ldots, l$ are given fixed, as they form supervisor's known set. We assume that $L : \Omega \times \mathbb{R} \times \mathbb{R}^m \to \mathbb{R}$ is continuous. The optimal control problem **R** for (3) is as follows:

$$\text{minimize } R(\alpha, p)$$

subject to

$$-\Delta f(z) = H(z, f(z), \alpha(z), p(z)), \ z \in \Omega, \tag{5}$$

$$\int_{\Omega} p(z) dz = 1, \ 0 < \delta < \sum_{i=1}^{l} p(z_i) \le 1 \tag{6}$$

with $\alpha \in \Lambda$, $p \in \mathcal{P}$ and some fixed $0 < \delta < 1$. The set of all trio (f, α, p) satisfying (5), (6) we denote by Ad.

Note that the problem **R** is a classical optimal control problem with distributed parameters α and p. Notice also that we included in H the probability mass functions p as they should have influence on the definition of the machine learning block. In order to derive verification conditions, in fact, sufficient optimality conditions for **R** we shall apply ideas from [4,5].

3 Dual Approach to R

In this paper we want to construct a new dual dynamic programming theory for the control problem **R**. The dual approach to the dynamic programming was first introduced in [4] and then developed in several papers to different optimal control problems governed by: elliptic, parabolic and wave equations (see e.g. [5–7]). Essential point in this dual approach is that we do not deal directly with a value function, but with some auxiliary function, defined in a dual set, satisfying a dual dynamic equation and then we derive sufficient optimality conditions for the primal value function. Such an approach has some advantages: we do not need any properties of the value function such as smoothness or convexity. In the control problem **R** we deal with sufficiently regular state equation and in consequence, the dual dynamic equations we consider also in strong form, i.e. we look for solutions which are strong solutions. However, in this paper we want to construct a new dual dynamic programming equations. Thus, let us start first with the definition of a dual set. Let $\mathbf{P} \subset \mathbb{R}$ be an open set of the variables $\mathbf{p}$. We should stress that $\mathbf{P}$ is a set which is chosen by us. Let $P \subset \mathbb{R}^{n+2}$ be an open set of the variables $(z, \mathbf{p})$, $z \in \Omega$, $\mathbf{p} \in \mathbf{P}$, i.e.

$$P = \{(z, \mathbf{p}) \in \mathbb{R}^{n+2} : z \in \Omega, \ \mathbf{p} \in \mathbf{P}\}. \tag{7}$$

Why we extend our primal space of z variables? In classical approach to necessary optimality conditions of Pontryagin maximum principle, as well in one variable as with distributed parameters, we deal with the space of variables z and with so called conjugate variable $(y^0, \mathbf{y})$-multiplier. In our case $\mathbf{p}$ is nothing more as just multiplier $\mathbf{y}$ staying by our constraints. However, novelty in our approach is that we move all our study to that extended space $(z, \mathbf{p})$. This is in a spirit of machine learning introduced by Vapnik [1] in seventies of the former century, i.e. we jump to a larger space, do some work and come back to our primal space with new results.

Denote by $W^{1;2}(P)$ the specific Sobolev space of functions with real values of the variables $(z, \mathbf{p})$, having the second order weak or generalized derivative (in the sense of distributions) with respect to z. Our notation for the function space is used for the function depending on the primal variable z, and the dual variable $\mathbf{p}$. The primal and dual variables are independent and the functions in the space $W^{1;2}(P)$ enjoy different properties with respect to z and $\mathbf{p}$. We use it to derive optimality properties for a value function known in classical calculus of variation and optimal control theory as dynamic programming equality. In order to formulate verification theorem for concrete optimal control problem usually as a first step, in the classical approach, a Hamiltonian is constructed, the value function is defined and then a kind of the Hamilton-Jacobi equation is posed which the value function should satisfy (also in Crandall-Lions viscosity approach). In the dual dynamic programming the strategy is similar in spite that all notions are built in the dual space. However, when we deal with optimal control problems with constraints being partial differential equations two questions appear: what is the Hamiltonian and what is a differential operator imposed on the value function or in dual approach on an auxiliary function. Usually the Hamiltonian is built taking the integrand from our functional adding right hand side of constraints multiplied by multipliers and as a differential operator the linear differential operator defining the type of PDE is taken. In the case of our problem (elliptic equation) the answer for both questions is not easy and not unique: on the left hand side of (5) we have a linear differential operator, it concerns a state f. Certainly the auxiliary function V has to be real valued, as it must relate somehow to a value function. This implies that dynamic equations have to consist of one equation only. The main problem is to choose in a proper way differential operator for the auxiliary function V and a correct Hamiltonian, as these choices depend on themselves. We decided that for the dual approach of dynamic programming to our problem it is better to apply for V the elliptic operator $-\Delta$ only. We pose that equation is in a strong form (see (8)). We should stress that this equation is considered in the set P, i.e. in the set of the variables $(z, \mathbf{p})$.

In order to prove the verification theorem, we require that the function $V(z, \mathbf{p})$ satisfies, in P, for some $y^0 \in L^2(\Omega)$, the second order partial differential equation of dual dynamic programming of the form:

$$-\Delta_z V(z, \mathbf{p}) - \sup\{H(z, V(z, \mathbf{p}), \alpha, p), \ \alpha \in I, \ p \in [0,1]\}$$
$$= -\Delta_z V(z, \mathbf{p}) - H(z, V(z, \mathbf{p}), \alpha(z, \mathbf{p}), p(z, \mathbf{p})) \ = y^0(z), \ (z, \mathbf{p}) \in P, \quad (8)$$

$$y^0(z_i) = L(z_i, V(z_i, \mathbf{p}), \alpha(z_i, \mathbf{p}))p(z_i, \mathbf{p}), \ i = 1, \ldots, l, \ \mathbf{p} \in \mathbf{P}, \tag{9}$$

where $\alpha(z, \mathbf{p}), p(z, \mathbf{p})$ are functions in P for which a supremum is attained. Since the function H is continuous and I, $[0, 1]$ are compact sets thus $\alpha(z, \mathbf{p})$, $p(z, \mathbf{p})$ exist and are continuous. One may wonder how to find that equation? The answer is: we apply to the auxiliary function V the differential operator $-\Delta$ and take the Hamiltonian augmented by the right hand side of elliptic equation plus the cost functional. That is different to a Hamiltonian in classical mechanic. We would like to stress that in classical mechanics appears by H the dual variable $\mathbf{p}$ known as momentum. In our new approach we do not use it. Denote by $\mathbf{p}(z)$, $z \in \Omega$ the new trajectory being a solution to the following equation, for some $\alpha(z)$, $p(z)$, $z \in \Omega$ and fixed $y(\cdot) \in L^2(\Omega)$

$$-\Delta_z V(z, \mathbf{p}(z)) - H(z, V(z, \mathbf{p}(z)), \alpha(z), p(z)) = y(z) \tag{10}$$

with $z \in \Omega$, while $V(z, \mathbf{p})$, $(z, \mathbf{p})$, $\in P$ is a solution to (8). Now, we will call $\mathbf{p}(z)$, $z \in \Omega$, the dual trajectory, while $f(z)$, $z \in \Omega$ stands for the primal trajectory. Having the function V we can define a relation between the set P of $(z, \mathbf{p})$ and primal set of (z, f), which in our case is unknown. Thus put

$$\mathbf{V} = \{(z, f) : z \in \Omega, \ f = V(z, \mathbf{p}), (z, \mathbf{p}) \in P\}. \tag{11}$$

We tell that $\mathbf{p}(z)$ is dual to $f(z)$ if both are generated by the same controls $\alpha(z)$, $p(z)$. Further, we confine ourselves only to those admissible trajectories $f(\cdot)$ for which there exist functions $\mathbf{p}(z), (z, \mathbf{p}(z)) \in P$, satisfying (10) such that $f(z) = V(z, \mathbf{p}(z))$ for $z \in \Omega$. Thus denote

$$Ad_\mathbf{V} = \{(f, \alpha, p) \in Ad : \quad \text{there exist} \quad \mathbf{p}(z), \ \mathbf{p} \in L^2(\Omega)$$
$$\text{satisfying (10) such that } f(z) = V(z, \mathbf{p}(z)), \ z \in \Omega\}.$$

Actually, it means that we are going to study problem $\mathbf{R}$ possibly in some smaller set $Ad_\mathbf{V}$, which is determined by the V. All the above was simply precise description of the family $Ad_\mathbf{V}$. This means we must reformulate the problem $\mathbf{R}$ to the following

$$R^V = \inf_{(f, \alpha, p) \in Ad_\mathbf{V}} \sum_{i=1}^l L(z_i, f(z_i), \alpha(z_i))p(z_i). \tag{12}$$

We named R^V dual optimal value in contrast to optimal value

$$R = \inf_{(f, \alpha, p) \in Ad} R(\alpha, p)$$

as R^V depends strongly upon dual trajectories $\mathbf{p}(z)$, which in fact determine the set $Ad_\mathbf{V}$. Moreover, essential point is that the set $Ad_\mathbf{V}$ is, in general, smaller than Ad i.e. $Ad_\mathbf{V} \subset Ad$ and thus the dual optimal value R^V is greater than the optimal value R, i.e. $R^V \geq R$. In order to find the set $Ad_\mathbf{V}$, first we must find the function V i.e. to solve Eq. (8) and then to define the set of admissible dual trajectories (see below). It is not easy work, but then we will have a possibility to assert that suspected trajectory is really optimal with respect to all trajectories lying in Ad_V. This fact for our problem is presented in literature for the first time. Of course, one can wonder are we able to find V or is the set Ad_V nonempty? The answer is not simple. In some cases we can solve that problem, in many cases we cannot do it, similarly as in the classical calculus of variation.

4 Sufficient Optimality Conditions for Problem (12)

The optimal value for the problem formulated in (12), in general, we can not calculate directly. An existence result, as well as necessary optimality conditions for that problem are also not available in literature. In this section we formulate and prove sufficient optimality conditions for the existence of the optimal value R^V as well for the optimal pair but optimal relative to the set Ad_V.

Theorem 1. *Assume that there exists a $W^{1,2}(P)$ solution V of (8) on P, i.e. there exists $y^0 \in L^2(\Omega)$ such that V satisfies (8). Let $\bar{\mathbf{p}}(\cdot) \in L^2(\Omega)$ with the corresponding $\bar{\alpha}(z)$, $\bar{p}(z)$ satisfy (accordingly to (10))*

$$-\Delta_z V(z, \bar{\mathbf{p}}(z)) - H(z, V(z, \bar{\mathbf{p}}(z)), \bar{\alpha}(z), \bar{p}(z)) = y(z).$$

Let $\bar{f}(z)$, $z \in \Omega$, together with the corresponding controls $\bar{\alpha}(z)$, $\bar{p}(z)$, $z \in \Omega$ belong to Ad_V, i.e. $(\bar{f}, \bar{\alpha}, \bar{p}) \in Ad_V$ and suppose that for $(\bar{f}, \bar{\alpha}, \bar{p})$ the following equality holds

$$- \Delta_z V(z, \bar{\mathbf{p}}(z)) - H(z, V(z, \bar{\mathbf{p}}(z)), \bar{\alpha}(x), \bar{p}(z)) = y^0(z), \tag{13}$$

$$y^0(z_i) = L(z_i, \bar{f}(z_i), \bar{\alpha}(z_i))\bar{p}(z_i), \ i = 1, \ldots, l.$$

Then $(\bar{f}(\cdot), \bar{\alpha}(\cdot), \bar{p}(\cdot))$ is an optimal trio relative to all $(f(\cdot), \alpha(\cdot), p(\cdot)) \in Ad_V$.

Proof. Let us take any $(f(\cdot), \alpha(\cdot), p(\cdot)) \in Ad_V$ and corresponding to it $\mathbf{p}(z)$, $z \in \Omega$, i.e. $(\alpha(z), \mathbf{p}(z), p(z))$, $z \in \Omega$ satisfy (10) with fixed $y \in L^2(\Omega)$, defining Ad_V. Thus we have

$$- \Delta_z V(z, \mathbf{p}(z)) - H(z, V(z, \mathbf{p}(z)), \alpha(z), p(z)) = y(z). \tag{14}$$

Now, we apply along $(\mathbf{p}(\cdot), \alpha(\cdot), p(\cdot))$ the dynamic inequality (8) and (9). We get

$$- \Delta_z V(z, \mathbf{p}(z)) - H(z, V(z, \mathbf{p}(z)), \alpha(z), p(z)) \leq y^0(z), \tag{15}$$

$$y^0(z_i) = L(z_i, V(z_i, \mathbf{p}(z_i)), \alpha(z_i))p(z_i), \ i = 1, \ldots, l, \tag{16}$$

where $f(z) = V(z, \mathbf{p}(z))$, $z \in \Omega$ is generating by the controls $\alpha(\cdot)$, $p(z)$. Joining (14)–(16) we come to

$$L(z_i, f(z_i), \alpha(z_i))p(z_i) \geq (y(z_i) - y^0(z_i)), \ i = 1, \ldots, l. \tag{17}$$

Similarly, we follow with the pair $(\bar{\alpha}(\cdot), \bar{p}(\cdot))$ and corresponding to it dual trajectory $\bar{\mathbf{p}}(\cdot)$ and primal trajectory $\bar{f}(\cdot)$. Then we come to the following equality

$$L(z_i, \bar{f}(z_i), \bar{\alpha}(z_i))\bar{p}(z_i) = (y(z_i) - y^0(z_i)), \ i = 1, \ldots, l$$

and then,

$$\sum_{i=1}^{l} L(z_i, \bar{f}(z_i), \bar{\alpha}(z_i))\bar{p}(z_i) \leq \sum_{i=1}^{l} L(z_i, f(z_i), \alpha(z_i))p(z_i),$$

i.e. the assertion of the theorem.

5 Verification Theorem for Approximate Optimality

As we are mostly interested in approximative problem, therefore we will also consider instead of dual dynamic equality (8) a certain type of an inequality

$$-\varepsilon + y_\varepsilon^0(z) \leq -\Delta_z V(z, \mathbf{p}) - \sup\{H(z, V(z, \mathbf{p}), \alpha, p),\ \alpha \in I,\ p \in [0, 1]\}$$

$$= -\Delta_z V(z, \mathbf{p}) - H(z, V(z, \mathbf{p}), \alpha_\varepsilon(z, \mathbf{p}), p_\varepsilon(z, \mathbf{p})) \leq y_\varepsilon^0(z),\ (z, \mathbf{p}) \in P, \quad (18)$$

$$y_\varepsilon^0(z_i) = L(z_i, V(z_i, \mathbf{p}), \alpha_\varepsilon(z_i, \mathbf{p}))p_\varepsilon(z_i, \mathbf{p}),\ i = 1, \ldots, l,\ \mathbf{p} \in \mathbf{P}, \quad (19)$$

where $\alpha_\varepsilon(z, \mathbf{p})$, $p_\varepsilon(z, \mathbf{p})$ are functions in P for which a supremum is attained. Next, denoting its solution by V_ε, we write

$$\mathbf{V}_\varepsilon = \{(z, f):\ z \in \Omega,\ f = V_\varepsilon(z, \mathbf{p}),\ (z, \mathbf{p}) \in P\}, \quad (20)$$

i.e. we similarly as in (10) require that dual trajectories satisfy

$$-\Delta_z V_\varepsilon(z, \mathbf{p}(z)) - H(z, V_\varepsilon(z, \mathbf{p}(z)), \alpha(z), p(z)) = y_\varepsilon(z), \quad (21)$$

with some fixed $y_\varepsilon(\cdot) \in L^2(\Omega)$. Next define, for this fixed y_ε, the set

$$Ad_{V_\varepsilon} = \{(f, \alpha, p) \in Ad:\ \text{exist } \mathbf{p}(z) \in L^2(\Omega)$$
$$\text{satisfying (21) such that } f(z) = V_\varepsilon(z, \mathbf{p}(z)),\ z \in \Omega\}.$$

Put

$$R^{V_\varepsilon} = \inf_{(f, \alpha, p) \in Ad_{V_\varepsilon}} \sum_{i=1}^{l} L(z_i, f(z_i), \alpha(z_i))p(z_i).$$

An ε-optimal value for problem $\mathbf{R}$ we call each value $R_\varepsilon^{V_\varepsilon}$ satisfying the inequality, for the above fixed $\varepsilon > 0$

$$R^{V_\varepsilon} \leq R_\varepsilon^{V_\varepsilon} \leq R^{V_\varepsilon} + \varepsilon.$$

The above means that we are looking for such controls α_ε, p_ε which will lead corresponding to them state f_ε to such a value that

$$R(\alpha_\varepsilon, p_\varepsilon) \leq R(\alpha, p) + \varepsilon.$$

for all $(f, \alpha, p) \in Ad_{V_\varepsilon}$.

Having the above notions and the inequality, below, we formulate and prove our approximate theorem so called a verification theorem being, in fact, sufficient ε-optimality conditions for our problem.

In order to formulate verification theorem about ε-optimality conditions we need to consider a solution V_ε to dynamic programming inequality (18) as well the set $Ad_{\mathbf{V}_\varepsilon}$ defined by V_ε. Thus assume that we have any solution V_ε of (18) and the set $Ad_{\mathbf{V}_\varepsilon}$.

Theorem 2. *Assume that there exists a $W^{1:2}(P)$ solution V_ε of (18) on P with $y_\varepsilon^0 \in L^2(\Omega)$. Let $f_\varepsilon(z)$, $z \in \Omega$, together with corresponding controls $\alpha_\varepsilon(z)$, $p_\varepsilon(z)$, $z \in \Omega$ belong to $Ad_{\mathbf{V}_\varepsilon}$ i.e. $(f_\varepsilon, \alpha_\varepsilon, p_\varepsilon) \in Ad_{\mathbf{V}_\varepsilon}$ and there exists corresponding to it $\mathbf{p}_\varepsilon(z)$. Suppose that for $(f_\varepsilon, \alpha_\varepsilon, p_\varepsilon)$ and $\mathbf{p}_\varepsilon(z)$ the following inequality holds*

$$\varepsilon + y_\varepsilon^0(z) \le -\Delta_z V_\varepsilon(z, \mathbf{p}_\varepsilon(z)) - H(z, V_\varepsilon(z, \mathbf{p}_\varepsilon(z)), \alpha_\varepsilon(z), p_\varepsilon(z)) \le y_\varepsilon^0(z), \quad (22)$$

with

$$y_\varepsilon^0(z_i) = L(z_i, V_\varepsilon(z_i, \mathbf{p}_\varepsilon(z_i)), \alpha_\varepsilon(z_i))p_\varepsilon(z_i), \quad i = 1, \ldots, l.$$

Then $(f_\varepsilon(\cdot), \alpha_\varepsilon(\cdot), p_\varepsilon(\cdot))$ is an ε-optimal trio relative to all $(f(\cdot), \alpha(\cdot), p(\cdot)) \in Ad_{\mathbf{V}_\varepsilon}$.

Proof. Let us take any $(f(\cdot), \alpha(\cdot), p(\cdot)) \in Ad_{\mathbf{V}_\varepsilon}$ and corresponding to it $\mathbf{p}(z)$, $z \in \Omega$, i.e $(f(z), \alpha(z), p(z))$, $\mathbf{p}(z)$, $z \in \Omega$ satisfy (21). Thus, as $f(z) = V_\varepsilon(z, \mathbf{p}(z))$ we have

$$-\Delta V_\varepsilon(z, \mathbf{p}(z)) - H(z, V_\varepsilon(z, \mathbf{p}(z)), \alpha(z), p(z)) = y_\varepsilon(z). \quad (23)$$

Applying along $(f(\cdot), \alpha(\cdot), p(\cdot))$ and $\mathbf{p}(z)$ the inequality (18) we infer that

$$\Delta_z V_\varepsilon(z, \mathbf{p}(z)) - H(z, V_\varepsilon(z, \mathbf{p}(z)), \alpha(z), p(z)) \ge -\varepsilon + y_\varepsilon^0(z) \quad (24)$$

with

$$y_\varepsilon^0(z_i) = L(z_i, -V_\varepsilon(z_i, \mathbf{p}(z_i)), \alpha(z_i))p(z_i), \quad i = 1, \ldots, l. \quad (25)$$

Joining (24)–(25) we come to

$$-\varepsilon + y_\varepsilon^0(z_i) - y_\varepsilon(z_i) - \le L(z_i, -V_\varepsilon(z_i, \mathbf{p}(z_i)), \alpha(z_i))p(z_i), \quad i = 1, \ldots, l. \quad (26)$$

Similarly, we follow with the trio $(f_\varepsilon(\cdot), \alpha_\varepsilon(\cdot), p_\varepsilon(\cdot))$ and corresponding to it trajectory $\mathbf{p}_\varepsilon(\cdot)$ but now, instead of (18) using (22). Then we come to the following inequality

$$L(z_i, -V_\varepsilon(z_i, \mathbf{p}_\varepsilon(z_i)), \alpha_\varepsilon(z_i))p_\varepsilon(z_i) \le y_\varepsilon^0(z_i) - y_\varepsilon(z_i) \quad (27)$$

and then we get

$$\sum_{i=1}^l L(z_i, f_\varepsilon(z_i), \alpha_\varepsilon(z_i))p_\varepsilon(z_i) \le \sum_{i=1}^l L(z_i, f(z_i), \alpha(z_i))p(z_i) + \varepsilon,$$

i.e. the assertion of the theorem.

6 Computational Algorithm

The main task is to find the LM-block and probability mass function that gives adequate approximation of the problem. Our observations are described as a set of l elements $z_i = (x, y)$, $i = 1, \ldots, l$. The sufficient conditions formulated for ε-value function allow us to build numerical approach to calculate suboptimal trio $(f_\varepsilon, \alpha_\varepsilon, p_\varepsilon)$.

In general, we do not know whether the learning set contains enough information that will allow to build a correct approximation. As an example, we can consider a periodic function like $\sin x$. Having all the observations only at axis intercepts, i.e. for $x = k\pi$, $y = 0$, $k \in \mathbb{Z}$, our first intuition improperly concludes a constant function $y = 0$ as a result of the approximation. However, quite often we have some additional information about the nature of resulting function. Therefore, we may want to define or amplify the information between successive elements of the set of observations z_i. This can be accomplished by defining N probability mass functions $p(z)$ and adequately rich set of N functions $f(z)$ within the machine learning LM-block.

The algorithm, we present below, ensures that we find in finite number of steps the suboptimal trio.

Algorithm:

1. Let the observation will be given as a finite set z_i, $i = 1, \ldots, l$ (i.e. learning points).
2. Consider a finite set of N probability mass functions p_i, $i = 1, \ldots, N$ from $\mathcal{P}$.
3. Consider a finite family α_i, $i = 1, \ldots, N$ of controls belonging to Λ

$$\Lambda = \left\{ \alpha(z),\ z \in \Omega : \alpha \in L^2(\Omega),\ \alpha(z) \in I \right\}.$$

4. Choose an appropriate $H(z, f, \alpha, p)$.
5. Solve the semilinear elliptic Eq. (3) for each pair (α_i, p_i) from 3. and 4. in order to obtain a finite family $\mathbf{Q}$ of functions $f_i(z)$, $i = 1, \ldots, N$ from the set Λ.
6. Minimize the functional (4), i.e. $R(\alpha, p)$, with respect to $p_i, \alpha_i, i = 1, \ldots, N$, in order to obtain the $(f_\varepsilon, \alpha_\varepsilon, p_\varepsilon)$ trio.
7. Define an open set $\mathbf{P} \subset \mathbb{R}$ of variables $\mathbf{p}$.
8. Solve (18)–(19) to obtain $V_\varepsilon(z, \mathbf{p})$.
9. Solve (21) to obtain $\mathbf{p}_\varepsilon(z)$ corresponding to $(f_\varepsilon, \alpha_\varepsilon, p_\varepsilon)$.
10. With the trio $(f_\varepsilon, \alpha_\varepsilon, p_\varepsilon)$, function $V_\varepsilon(z, p)$ and $\mathbf{p}_\varepsilon(z)$, we can refer to Theorem 2 to verify if (22) for given $\varepsilon \geq 0$ holds.
    If this verification is positive, we obtain that trio $(f_\varepsilon, \alpha_\varepsilon, p_\varepsilon)$ is an ε-optimal trio relative to all $(f, \alpha, p) \in Adv$ and $R(\alpha_\varepsilon, p_\varepsilon)$ is an ε-optimal value with given ε. Otherwise, go to step 2.

Ad_V is relatively compact in $\mathbf{Q} \times \Lambda \times \mathcal{P}$. Therefore, there exists finite number of trio (f_j, α_j, p_j), $j = 1, \ldots, M$, such that balls with the center at (f_j, α_j, p_j) and radius ε cover the Ad_V. That denotes the above algorithm is convergence for properly chosen trio (f, α, p) in 2., 3., 4. and 5. We would like to stress that the family of functions $\mathbf{Q}$ is uniquely determined by Eq. (3), i.e. by the function H which we choose in 4. The quality and type of $\mathbf{Q}$ depends on the additional information of the given problem. Similarly, the set $\mathcal{P}$, as a control set occurring in our control problem $\mathbf{R}$, may be restricted to some smaller set depending on the kind of the information we have at our dispose. The obtained p_ε from the algorithm above estimates the quality of learning points z_i, $i = 1, \ldots, l$ for the function f_ε.

7 Conclusion

We present absolutely new approach to the problem of an approximation of a function defined by the neural network learned on a given supervised set (set of observations). To this effect we apply ideas of Vapnik model [1] i.e. we add, following him, the machine learning block. However Vapnik [1] continues his studying developing statistical tools and thus the result he received are of statistical type. Our approach is different. The problem is how to construct the LM-block and how to calculate the minimum of the functional $R(\alpha, p)$ on the set of functions defined in LM-block. The only available tools are in the optimal control theory. But then we need to have the LM-block defined by any differential equations. Because the learning set belongs at least to two dimensional space thus we need a kind of PDE as the differential equation. We decided to use semi-linear elliptic equation of which non-linearity (the right hand side) contain the control function α and the control (mass function) p. Then we formulate the optimal control problem $\mathbf{R}$. Next, we build a new dual approach for that problem and construct sufficient optimality criteria for it. However, the most important results from the practical point of view are in Sect. 5 where the verification conditions are given of the approximate minimum for the problem $\mathbf{R}$ in order that the condition (2) to be satisfied i.e. suitable conditions are formulated which allow to check whether the pair $(\alpha_\varepsilon, p_\varepsilon)$ satisfies criteria to be ε-optimal for $\mathbf{R}$. Just these verification conditions are base to construct the computational algorithm in Sect. 6. As the consequence of this algorithm, we are able to calculate a pair $(\alpha_\varepsilon, p_\varepsilon)$ which satisfies the verification criteria to be ε-optimal. This means, using that algorithm, we find the mass function p_ε and thus the probability distribution function F_ε which estimates the quality of the learning set. But we also find the approximate function $f_\varepsilon(z)$ from the LM-block which approximates our neural network generated by supervised set. The numerical computations and experimental results will be presented in the next paper.

References

1. Vapnik, V.: The Nature of Statistical Learning Theory. Springer, New York (2000). https://doi.org/10.1007/978-1-4757-3264-1
2. Vapnik, V., Izmailov, R.: Knowledge transfer in SVM and neural networks. Ann. Math Artif. Intell. **81**(1), 3–19 (2017)
3. Vapnik, V., Braga, I., Izmailov, R.: Constructive setting for problems of density ratio estimation. Stat. Anal. Data Min. **8**(3), 137–146 (2015)
4. Nowakowski, A.: The dual dynamic programming. Proc. Am. Math. Soc. **116**, 1089–1096 (1992)
5. Nowakowski, A.: ε-value function and dynamic programming. J. Optim. Theory Appl. **138**, 85–93 (2008)
6. Galewska, E., Nowakowski, A.: A dual dynamic programming for multidimensional elliptic optimal control problems. Numer. Funct. Anal. Optim. **27**, 279–289 (2006)
7. Nowakowski, A., Sokolowski, J.: On dual dynamic programming in shape control. Commun. Pure Appl. Anal. **11**, 2473–2485 (2012)

Application of Artificial Neural Network Model for Cost Optimization in a Single-Source, Multi-destination System with Non-deterministic Inputs

Modestus O. Okwu[1][(✉)], Vitalian U. Chukwu[2],
and Onyewuchi Oguoma[3]

[1] Department of Mechanical Engineering,
Federal University of Petroleum Resources Effurun, Warri, Delta, Nigeria
okwu.okechukwu@fupre.edu.ng,
mechanicalmodestus@yahoo.com
[2] DGS, Federal University of Technology, Owerri, Imo, Nigeria
[3] Niger Delta Development Commission, Port Harcourt, Nigeria

Abstract. Decision making, especially at the strategic level of corporate management has continued to grow complex by the day arising from its connection with the dynamism being experienced in technology, innovation and information carriage in recent time. Consequently, decision makers at this level can no longer depend on the power of intuition but require creative algorithms. The major thrust of this paper, therefore, is to propose an imaginative approach for solving a linear programming problem (LPP) for multi-product distribution by creating a transshipment system and analysis using artificial neural network (ANN) technique for effective product allocation and cost prediction. Data obtained from a bottling company in Nigeria were used for the analysis. The results of the analyses indicate that the company spent ₦6,332,304.00 (17,590.00USD) on product distribution within the period of survey. However, from the established creative model, the total cost of distribution of product to available depots equaled ₦4,170,500.00 (11585.00.USD). The current operational cost was saved by roughly 34% using the ANN model. The study, therefore, concludes that ANN model provides a good prediction response in solving transportation/transshipment problems, thus offers an appropriate distribution strategies for cost optimization.

Keywords: Artificial neural network · Multi-echelon · LPP

1 Background

On a daily basis, multi-product manufacturing companies are challenged by the need for an efficient mode of managing inventory of products and resources in the form of raw materials, component parts and finished goods [1]. This, especially in multi-echelon systems, is quite a complex task compared to a simple system of single-echelon coordination [2]. Prior to the boost offered by technology and information management strategies, suppliers relied on intuition during allocation or distribution of

© Springer Nature Switzerland AG 2019
I. Rojas et al. (Eds.): IWANN 2019, LNCS 11507, pp. 539–554, 2019.
https://doi.org/10.1007/978-3-030-20518-8_45

products, either in the form of raw materials to be distributed to factories, or finished products ready for distribution to retail outlets. There is no doubt that in complex and bleary systems, allocation of products require efficient and robust computational algorithms which can unravel these challenges arising in different fields [3]. Such complex challenges can be addressed using heuristic techniques [4]. Transportation model is concerned with the movement of goods from several supply locations to several customer locations. The structure of a transportation problem involves a large number of shipping routes from several supply origins to several demand destinations [5]. Thus, determining optimal routes to minimise costs associated with physical distribution management has been a serious challenge to decision makers. However, linear programming can be used to generate practical applications to the model, which often serves more as a theoretical framework than offering empirical solution. [6] investigated an integrated supplier selection and inventory control problems in supply chain and design by developing a mathematical model for a multi-echelon systems. [7] studied a two-stage, multi-item inventory system with stochastic demand using heuristic search techniques based on the trade-off between the inventory cost and setup cost. Other research on Supply chain and distribution have been considered like: Mixed-integer programming model for supplier selection [8], multi-objective mixed integer LP model [9], synchronized supplier assortment and customers demand for single-source multi-source strategies [10]. The objective of this research is to propose a model with the potential for optimization of cost of product distribution in a single source, multi-product distribution system using LP model and training the network using ANN model.

1.1 Introduction to Artificial Neural Network

Using creative algorithm in solving problems offers a potent alternative to classical solution methods in solving problems. A good example is the use of artificial neural network. As a computation and learning model, they are presented as unique approach for solving day- to-day complex problems. They have been used to solve stochastic problems in various areas, such as engineering, business, manufacturing, medicine, military etc. [11]. Neural networks are modeled in the like of human brain to process real time data or information. The network is composed of a large number of highly interconnected processing elements called neurons which work in parallel to solve imprecise problems. Neural networks are different from the conventional or traditional methods in that they take a diverse approach in solving a problem [12]. Soft computing techniques are quite easy to use as they require refining of weights and continuous training of dataset [13].

1.2 Training Artificial Neural Network

Neural network training is a complex task. It is highly significant for supervised learning. Multi-layer Perceptron Networks (MLP) are feed-forward ANN models for machine learning. MLPs are type of NN used for training and used as classifier system [14].

Weight adjustment is an important process of training an ANN. Such training is usually carried out with the help of Back-Propagation (BP) algorithm [15]. One of the common disadvantage of the BP algorithm is its slow convergence [16]. Due to this well known demerit of BP, meta-heuristic techniques have been proposed as a better replacement [17]. Weight refinement of input and output dataset is necessary during ANN training [18]. [19] observed the classification of ANN into self organizing maps, feed-forward NN (FFNN), feedback NN (FBNN). [20] studied application of ANN during thermal analysis of heat exchangers, [21] further applied ANN in refrigeration, air conditioning and heat pump systems. Other areas where ANN have been used include: Forecasting of accident victims and highway casualties [22–24]; Student achievement prediction in E-learning courses [25]; Road traffic prediction [26, 27]; Gas turbine diagnosis [28]; Urban taxi demand forecasting [29].

1.3   Learning Algorithm

The neural network (NN) is a predictive technique which require training of network by adjusting the appropriate weight values or connections. The learning process in ANN are categorized into three namely: Reinforced learning, supervised and unsupervised learning. According to [30], the most popular of the three categories of ANN are the supervised and unsupervised learning. The architecture and classification of the learning algorithm is shown in Fig. 1.

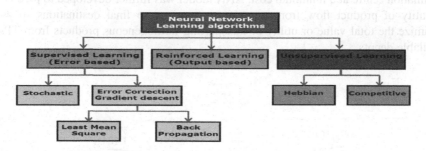

Fig. 1. Classification of learning algorithms (Chakraborty 2010)

2   Model Development

The classical mathematical model is developed and presented considering physical distribution of products in a bottling plant from a supply centre to several demand centres in a single-source, multi-product to multi-destination transshipment problem. The structure or architecture of the network is shown in Fig. 2 which represents a system of source showing diverse shipping routes and demand centres or destination points.

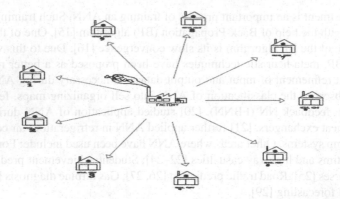

Fig. 2. Structure of the distribution network from source to destinations

In modelling the system, three transshipment points (TPs) were created as intermediate points to allow easy flow of products from source to TPs and from TPs to other available depots, see Fig. 3. The linear programming (LP) model is presented, with the demand and cost input variables. Figure 4 shows the quantity of products needed at the ten destination points and the unit cost of product distribution from source to destinations. The objective is to determine shipping routes between supply centres and demand centres so as to satisfy the required quantity of bottled products at each destination centre at a minimum cost. ANN model was further developed to predict the quantity of product flow from transshipment points to final destinations so as to optimize the total value or utility of transporting homogeneous products from TPs to available depots.

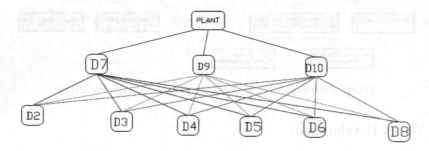

Fig. 3. Architecture of the system with transshipment at D_7, D_9 and D_{10}

(a)

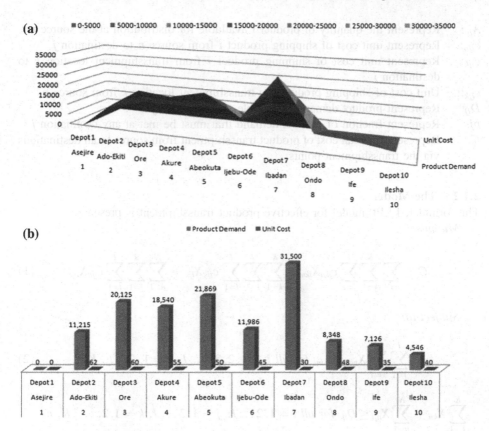

(b)

Fig. 4. (a) Location point and demand requirements (b) Product demand and unit cost of distribution from destination point

2.1 Mathematical Model Development

2.1.1 Notations

The notations used for modelling purposes are presented:

I: Total products available at the source, ready for distribution.

N: Total number of products sources to be distributed

J: Number of destinations/depots to be served

K: Available transshipment points.

i: Index for proper identification of products with $i = 1, 2 \ldots I$

n: Index for proper identification of products sources with $n = 1, 2 \ldots N$

j: Index denoting destinations/depots with $j = 1, 2, 3, \ldots J$

k: Index representing transshipment points with $k = 1, 2, \ldots, K$

X_{ijn}:   Product quantity i shipped from source n to available destination j from source n

X_{ikn}:   Product quantity i shipped from source n to transshipment point k

X_{ijk}:   Product quantity i shipped from transshipment location k to destination j

Z_k: Represent the capacity of the transshipment center, k

A_{in}:   Represent the quantity of product i available for distribution at the source n
c_{ijn}:   Represent unit cost of shipping product i from source n to destination j
c_{ijk}:   Represent unit cost of shipping product i from transshipment location k to destination j
c_{ijk}:   Unit cost of shipping product i to transshipment location k from source n
D_{ij}:   Represent product demand i available at destination j
bj: Represent fraction of product demand that must be met at any destination j
C: Represent the total cost of product transshipment from source to all destinations via the transshipment points

2.1.2 The Model

The formulated LPP model for effective product transshipment is presented:

Minimise

$$C = \sum_{n=1}^{N}\sum_{k=1}^{K}\sum_{j=1}^{J} c_{jkn}X_{jkn} + \sum_{k=1}^{K}\sum_{j=1}^{J}\sum_{i=1}^{I} c_{ijk}X_{ijk} + \sum_{n=1}^{N}\sum_{j=1}^{J}\sum_{i=1}^{I} c_{ijn}X_{ijn} \tag{1}$$

Subject to:

$$\sum_{i=1}^{I}\sum_{k=1}^{K} X_{nik} \le A_{in} \ \text{for all } i = 1,2,3\ldots..I, \, k = 1, 2, \ldots, K \tag{2}$$

$$\sum_{k=1}^{K} X_{kji} + \sum_{n=1}^{N} X_{nji} \le D_{ij} \ \text{for all } i = 1,2\ldots I; \, j = 1,2\ldots J, \, k = 1,2, \ldots, K, \, n$$
$$= 1, 2\ldots, N \tag{3}$$

$$\sum_{n=1}^{N} X_{nji} + \sum_{k=1}^{K} X_{kji} \ge b_j D_{ij} \ \text{for all } i = 1, 2\ldots I; \, j = 1, 2\ldots J, \, k = 1, 2, \ldots, K$$
$$\tag{4}$$

$$\sum_{j=1}^{N} X_{ik} \le Z_k \ \ \forall k = 1, 2, \ldots, K \tag{5}$$

$$X_{ij} \ge 0 \ \ \forall i = 1, 2\ldots I; \, j = 1, 2\ldots J \tag{6}$$

3 Solution Technique/Procedure

The objective is to create the optimal transshipment route for easy flow from source to transshipment points (TP) for onward distribution to the final destination given the fuzziness of the input variables (demand and cost). It becomes difficult to solve the problem using the classic linear programming solutions and popular solvers. A novel approach to solving this type of complex distribution problem is by using combined classical technique and artificial neural network (ANN) model which works based on the principle of neural weight update algorithms to determine the optimal weight vectors that minimize the objective function of the LP problem presented by considering solution with low mean square error (MSE) value and high correlation coefficient (R^2) value.

3.1 Company of Case Study

The case presented here is that of a bottling company with a single production plant, $n = 1$, a range of products represented by the single product family, $i = 1$, servicing the south western part of Nigeria and some neighboring cities via 10 depots, $j = 10$, and 3 transshipment points, $k = 3$, which also serve as depots. Three-years data collected from the bottling plant are used for analysis. Matrices were developed for the distances and the associated cost for effective manipulation of ANN toolbox. The current practice in the company, however, is to ship directly to all depots from the plants. The system architecture is shown in Fig. 2. For a better and robust solution, three transshipment points (TPs) were created (D7, D9 and D10) as shown in Fig. 3. The transshipment points serve as intermediate points for onward and easy movement of products to the final destinations. The product quantity available within the period of investigation was 135,255. The demand of product at different destinations and unit costs of product delivery are clearly shown in Figs. 4a and b.

3.2 Model Implementation

The ANN architecture with input, hidden and output characteristics are shown in Fig. 5. They demonstrate a close similitude of the inbound and outbound bottling system. The input corresponds to the cost and demand set variables and the weighted strength of the ANN network represent the route to different destinations from the sources. Output layer is set as the availability at different TPs. D represents product demand at different destinations. The dataset for training the network is clearly shown in Fig. 6. The Y_k values represent the products shipped from source to TPs for onward movement to other destinations. The accuracy of the training process is a function of the R and MSE values.

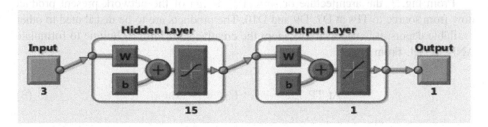

Fig. 5. ANN architecture with input, hidden and output layers

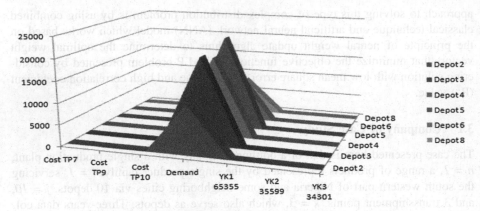

Fig. 6. Product at transshipment point to be distributed to depots

Total sum of the neuron U_k is shown in Eq. 9

$$U_k = \sum_{j=1}^{n} x_j w_{kj} \qquad (7)$$

ANN Model Notations:

W_{kn}:	Connection weight of neurons
X_j:	Signal flow from source/input through synaptic weights to summing junction i.e. $X_1, X_2, X_3, \ldots\ldots X_j$
j:	Index identifying the number of input
n:	Index identifying destinations
k:	Index identifying process element
V_k:	Linear combiner with bias to support input signal
Y_k:	Total output
U_k:	Linear combiner output due to input signal
b_k:	Weight of the biases between layers
USD:	United State dollar
NGN:	Nigerian Naira

From Fig. 3, the architecture or structural design of the network present product flow from source to TPs at D7, D9 and D10. The products are to be distributed to other available depots, it is necessary to adopt the creative algorithm technique to formulate ANN model. From Eq. 7.

$$\text{At TP 7,} \qquad U_{7k} = \sum_{j=1}^{6} x_j w_{7kj} \qquad (8)$$

$$\text{At TP 9,} \qquad U_{9k} = \sum_{j=1}^{6} x_j w_{9kj} \qquad (9)$$

$$\text{At TP 10,} \qquad U_{10k} = \sum_{j=1}^{6} x_j w_{10kj} \qquad (10)$$

A parameter of external function or bias (b_k), of ANN is introduced and the combined input of the network (V_k) is further expressed as:

$$V_k = U_{7k} + U_{9k} + U_{10k} + bk \qquad (11)$$

Substituting the value, U_k from Eqs. (8, 9 & 10) into (11), the combined input is further computed with Eq. (12).

$$V_k = \sum_{j=1}^{6} x_j w_{7kj} + \sum_{j=1}^{6} x_j w_{9kj} + \sum_{j=1}^{6} x_j w_{10kj} \qquad (12)$$

Y_k which is the total input of ANN is further expressed by defining the threshold function as a sigmoid function, which is further represented in Eqs. (13, 14 & 15)

$$Y_{7k} = \varphi(.) \sum_{j=1}^{6} x_j w_{7kj} + b_{7k} \qquad (13)$$

$$Y_{9k} = \varphi(.) \sum_{j=1}^{6} x_j w_{9kj} + b_{9k} \qquad (14)$$

$$Y_{10k} = \varphi(.) \sum_{j=1}^{6} x_j w_{10kj} + b_{10k} \qquad (15)$$

Figure 6, shows the network training data for products at TPs to be allocated to destinations. This was achieved by mimicking the network using the sourced data, by considering the input elements of cost and demand denoted by W and X respectively and the output data which represent product availability at the TPs, that is Y_k. D is demand of products at various locations. The iteration was performed several times with a goal of minimizing the error values.

3.3 Results and Discussion

In the study, the goal is to predict the optimum cost of product distribution considering a single source with three transshipment points and multi-destinations. The ANN was trained using different layers. Figure 5 represent the training process using three input, fifteen hidden and one output layer (3-15-1). The retraining process was done several times by varying different network layers (input, hidden and output). successful assessment of the accuracy of the model involve uninterrupted training and weight

modification with focus on MSE and R value for valuable determination of the degree
of association between the predicted and expected value.

$$MSE = \frac{1}{n} \sum_{m=1}^{n} \left(Y_{exp.m} - Y_{pred.m}\right)^2 \tag{16}$$

$$R = \left(\frac{\sum_{m=1}^{n} \left(Y_{pred.m} - y_{pred}\right)\left(Y_{exp.m} - y_{exp}\right)}{\sqrt{\sum_{m=1}^{n} \left(Y_{pred.m} - y_{pred}\right)^2 \sum_{m=1}^{n} \left(Y_{exp.m} - y_{exp}\right)^2}} \right) \tag{17}$$

Equation representing the MSE and R value is clearly stated in system of Eqs. 16
and 17. Where Y_{pred} is the predicted value obtained from the neural network model and
Y_{exp} is the experimental value. y_{exp} and y_{pred} are also the average of the experimental
and predicted values, respectively. The performance results and best possible iteration
result of the ANNs for testing data with different number of neuron in the hidden layer
are shown in Table 1. According to the testing performance of the networks, the
network model having MSE value = 1.57×10^{-5} was found to be more accurate with
performance shown (see Table 1 in bold item number 11).

Table 1. Performance results of MLP network for different numbers of neurons in the hidden
layer for testing data set

No.	MSE	R value			
		Training	Testing	Validation	Performance
1	6514260	0.9688	−0.9990	0.5950	0.37153
2	2940735	0.4768	−1.0000	1.0000	0.41513
3	3505	0.9994	−0.5558	−0.19681	0.63866
4	2800	1.0000	0.1035	−0.5057	0.69111
5	2851	1.0000	0.11657	0.80388	0.68865
6	3586	1.0000	0.10346	−0.5057	0.69111
7	3357	1.0000	−0.55358	0.71808	0.56428
8	2676	0.6810	0.4638	0.81260	0.65247
9	2280	0.9821	0.7218	0.56580	0.73260
10	1950	0.9784	0.7136	0.6842	0.71216
11	**1.57×10^{-5}**	**1.0000**	**0.77089**	**0.95504**	**0.80369**

MSE = Mean   Squared   Error,   R = Regression   Coefficient,
R^2 = Average determination

In Fig. 7, MSE against epochs during the training process for network prediction is
plotted in which the best result for validation data set was achieved at epoch 5 with
minimum MSE of 1.57×10^{-5} (Table 1 in bold item 11). Figure 8 illustrate the plots
of the ANN predicted output after series of training. The performance value obtained is
established as shown in Table 2. The recital value for the ANN sensitivity analysis

denote a satisfactory result. The MSE value at the 11th iteration process for validation and testing are quite low with R value is close to unity (Table 2).

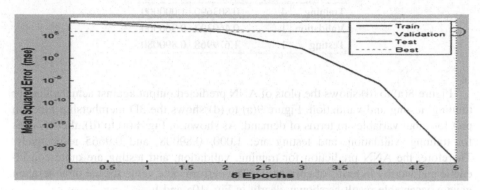

Fig. 7. MSE Variations for training, testing and validation data sets against epoch during the training of the data set

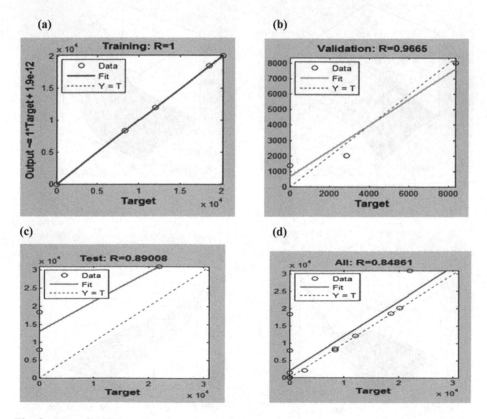

Fig. 8. Plot of ANN predicted output against actual value for (A) Training (B) Validation (C) Testing (D) Target.

Table 2. Performance value for ANN modelling (Final iteration)

Results	Samples	MSE	R
Training	4	0.803690	1.000000
Validation	1	0.460307	0.966500
Testing	1	1.679968	0.890080

Figure 8(a) to (d) shows the plots of ANN predicted output against actual value for training, testing and validation. Figure 9(a) to (d) shows the 3D membership function plot for input variables in terms of demand. As shown in Fig. 8(a) to (d), the R values for training, validation, and testing are: 1.000, 0.89008, and 0.9665 respectively. Therefore, the ANN prediction for training, validation, and testing are quite considerable and worthy in terms of correlation. The performance value for the final iteration gave a acceptable result as shown clearly in Fig. 10a and b.

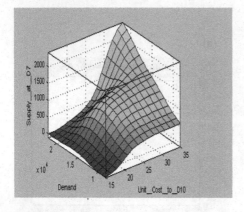
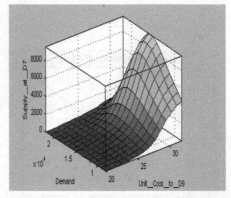
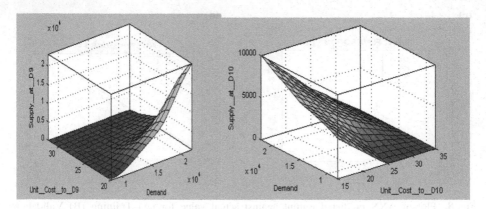

Fig. 9. 3D input-output surface plots

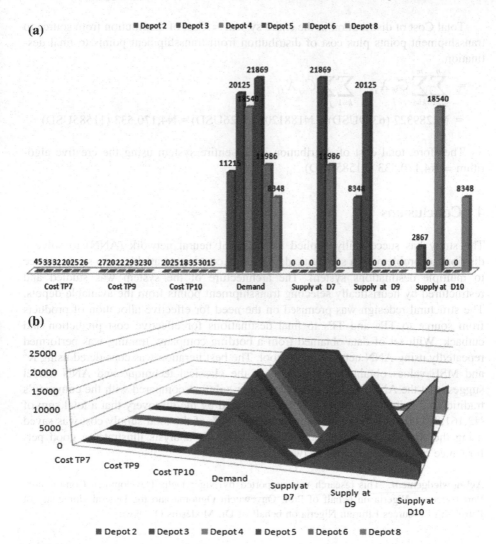

Fig. 10. (a) Final products allocated to destinations from TPs (b) ANN solution graph for product supply from TPs

The final allocation is presented in Figs. 10a and b. The figures show quantity of products supplied from TPs to each of the destinations using classical technique and effective prediction using ANN. The total distribution cost per week obtained in Nigerian Naira, NGN (1USD = 360NGN) is shown in Figs. 10a and b. The ANN solution is marginally cheaper than the company's traditional demands. Total cost of transshipment is equal to the total cost of distribution at D7 plus total cost of distribution at D9 plus total cost of distribution at D10. Total transshipment cost is ₦1881206 (5,226USD).

Total Cost of distribution in the entire system = Cost of distribution from source to transshipment points plus cost of distribution from transshipment points to final destination.

$$= \sum_{i=1}^{n}\sum_{k=1}^{m}C_{ik}X_{ik} + \sum_{k=1}^{n}\sum_{j=1}^{m}C_{kj}X_{kj}$$

$$= ₦2289327 \ (6359USD) + ₦1881206 \ (5226USD) = ₦4,170,533 \ (11585USD)$$

Therefore, total cost of distribution in the entire system using the creative algorithm = ₦4,170,533 (11585USD).

4 Conclusions

The study has successfully applied the artificial neural network (ANN) to solve a distribution problem with stochastic demand and cost input variables in a single source to multiple destinations system. The architecture of the system was studied and restructured by heuristically selecting transshipment points from the available depots. The structural redesign was premised on the need for effective allocation of products from source to TPs and TPs to final destinations for effective cost prediction and cutback. With set of data obtained from a bottling company, training was performed repeatedly using ANN network fitting tool. The best iteration was considered using R^2 and MSE values obtained. Application of the classical technique and ANN model suggest that the ANN model produced a better solution compared with the company's traditional distribution blueprint. This is indicated by the discovery that a total sum of ₦2,161,804.00NGN (6,005USD) representing 34% of the total supply cost was saved using the ANN model developed. Conclusively, the analysis illustrates a good performance of proposed Neural Network.

Acknowledgement. This research was supported by Niger Delta Development Commission, Port Harcourt, Nigeria on behalf of Prof. Onyewuchi Oguoma and the Federal University of Petroleum Resources Effurun, Nigeria on behalf of Dr. Modestus O. Okwu.

References

1. Bijvank, M.: Service inventory management: solution techniques for inventory systems without backorders (2009)
2. Alev, T.G., Ali, F.G., Fusun, U.: A new methodology for multi-echelon inventory management in stochastic and neuro-fuzzy environments. Int. J. Prod. Econ. **128**(2), 248–260 (2010)
3. Okwu, M.O., Adetunji, O.: A comparative study of artificial neural network (ANN) and adaptive neuro-fuzzy inference system (ANFIS) models in distribution system with nondeterministic inputs. Int. J. Eng. Bus. Manag. **10**, 1–17 (2018). https://doi.org/10.1177/1847979018768421
4. Azmi, A., Lewis, H.W.: A new optimization algorithm for combinatorial problems. (IJARAI) Int. J. Adv. Res. Artif. Intell. **2**(5), 63–77 (2013)

5. Sharma, J.K.: Operations Research: Theory and Applications, 5th edn. Macmillian Publishers, India Ltd., Delhi (2013). ISBN 978-9350-59336-3
6. Cong, G., Xueping, L.: A multi-echelon inventory system with supplier selection and order allocation under stochastic demand. Int. J. Prod. **151**, 37–47 (2014)
7. Yuli, Z., Shiji, S., Zhang, H., Cheng, W., Yin, W.: A hybrid genetic algorithm for two-stage multi-item inventory system with stochastic demand. Neural Comput. Appl. **21**, 1087–1098 (2010)
8. Hammami, R., Frein, Y., Hadj-Alouane, A.B.: An international supplier selection model with inventory and transportation management decisions. Flex. Serv. Manuf. J. **24**, 4–27 (2012)
9. Demirtas, E.A., Ustun, O.: Analytic network process and multi-period goal programming integration in purchasing decisions. Comput. Ind. Eng. **56**(2), 677–690 (2009)
10. Sawik, T.: Optimization of cost and service level in the presence of supply chain disruption risks: single vs. multiple sourcing. Comput. Oper. Res. **51**, 11–20 (2014)
11. Rasit, A.: Artificial neural networks applications in wind energy systems: a review. Renew. Sustain. Energy Rev. J. **49**, 534–562 (2015)
12. Okwu, M.O., Oreko, B.U., Okiy, S., Uzorh, A.C., Oguoma, O.: Artificial neural network model for cost optimization in a dual-source multi-destination outbound system. J. Prod. Manuf. Cogent Eng. **5**, 1447774 (2018). https://doi.org/10.1080/23311916.2018.1447774
13. Kartalopoulous, S.V.: Understanding Neural Networks and Fuzzy Logic. Prentice Hall, Upper Saddle River (2003)
14. Heaton, J.: Introduction to Neural Networks for Java, 2nd edn. Heaton Research, Inc. (2008)
15. Valian, E., Mohanna, S., Tavakoli, S.: Improved cuckoo search algorithm for feed-forward neural network training. Int. J. Artif. Intell. Appl. (IJAIA) **2**(3), 36–43 (2011)
16. Mingguang, L., Gaoyang, L.: Artificial neural network co-optimization algorithm based on differential evolution. In: Second International Symposium on Computational Intelligence and Design (2009)
17. Espinal, A., et al.: Comparison of PSO and DE for training neural networks. In: 10th Mexican International Conference on Artificial Intelligence (2011)
18. Popoola, L.T., Babagana, G., Susu, A.A.: A review of an expert system design for crude oil distillation column using the neural networks model and process optimization and control using genetic algorithm framework. Adv. Chem. Eng. Sci. **3**, 164–170 (2013)
19. Beale, M.H., Hagan, M.T., Demuth, H.B.: Neural Network Toolbox^{TM}7 User's Guide (2010)
20. Mohanraj, M., Jayaraj, S., Muraleedharan, C.: Applications of artificial neural networks for thermal analysis of heat exchangers – a review. Int. J. Ther. Sci. **90**, 150–172 (2015)
21. Mohanraj, M., Jayaraj, S., Muraleedharan, C.: Applications of artificial neural networks for refrigeration, air conditioning and heat pump systems. Renew. Sustain. Energy Rev. J. **16**, 1340–1358 (2012)
22. Vasavi, S.: Extracting hidden patterns within road accident data using machine learning techniques. In: Mishra, D., Azar, A., Joshi, A. (eds.) Information and Communication Technology. AISC, vol. 625, pp. 13–22. Springer, Singapore (2018). https://doi.org/10.1007/978-981-10-5508-9_2
23. Chen, Q., Song, X., Yamada, H., Shibasaki, R.: Learning deep representation from big and heterogeneous data for traffic accident inference. In: Proceedings of the Thirtieth AAAI Conference on Artificial Intelligence, Phoenix, AZ, USA, 12–17 February 2016, pp. 338–344 (2016)
24. Ren, H., Song, Y., Wang, J., Hu, Y., Lei, J.: A deep learning approach to the citywide traffic accident risk prediction. In: Proceedings of the 2018 21st International Conference on Intelligent Transportation Systems (ITSC), Maui, HI, USA, 4–7 November 2018

25. Lykourentzou, I., Giannoukos, I., Mpardis, G., Nikolopoulos, V., Loumos, V.: Early and dynamic student achievement prediction in E-learning courses using Neural Networks. J. Am. Soc. Inform. Sci. Technol. **60**(2), 372–380 (2009)
26. More, R., Mugal, A., Rajgure, S., Adhao, R.B., Pachghare, V.K.: Road traffic prediction and congestion control using artificial neural networks. In: Proceedings of the International Conference on Computing, Analytics and Security Trends (CAST), Pune, India, 19–21 December 2016, pp. 52–57 (2016)
27. Wu, Y., Tan, H., Qin, L., Ran, B., Jiang, Z.: A hybrid deep learning based traffic flow prediction method and its understanding. Transp. Res. Part C Emerg. Technol. **90**, 166–180 (2018)
28. Loboda, I.: Neural Networks for Gas Turbine Diagnosis; Machine Learning. Springer, New York (2016). Chapter 8
29. Liao, J., Zhou, S., Di, L., Yuan, X., Xiong, B.: Large-scale short-term urban taxi demand forecasting using deep learning. In: Proceedings of the 23rd Asia and South Pacific Design Automation Conference, Jeju, Korea, 22–25 January 2018, pp. 428–433 (2018)
30. Chakraborty, R.C.: Fundamental of Neural Networks: Soft Computing Course Lecture Notes (2010). http://www.myreaders.info/html/soft_computing.html

A New Online Class-Weighting Approach with Deep Neural Networks for Image Segmentation of Highly Unbalanced Glioblastoma Tumors

Mostefa Ben Naceur[1,2]([✉]), Rostom Kachouri[1], Mohamed Akil[1],
and Rachida Saouli[2]

[1] Gaspard Monge Computer Science Laboratory, A3SI, ESIEE Paris, CNRS,
University Paris-Est, Champs-sur-Marne, France
{mostefa.bennaceur,rostom.kachouri,mohamed.akil}@esiee.fr
[2] Smart Computer Sciences Laboratory,
Computer Sciences Department, Exact.Sc, and SNL, University of Biskra,
Biskra, Algeria
rachida.saouli@esiee.fr

Abstract. The most common problem among image segmentation methods is unbalanced data, where we find a class or a label of interest has the minority of data compared to other classes. This kind of problems makes Artificial Neural Networks, including Convolutional Neural Networks (CNNs), bias toward the more frequent label. Thus, training a CNNs model with such kind of data, will make predictions with low sensitivity, where the most important part in medical applications is to make the model more sensitive toward the lesion-class, i.e. tumoral regions. In this work, we propose a new Online Class-Weighting loss layer based on the Weighted Cross-Entropy function to address the problem of class imbalance. Then, to evaluate the impact of the proposed loss function, a special case study is done, where we applied our method for the segmentation of Glioblastoma brain tumors with both high- and low-grade. In this context, an efficient CNNs model called OcmNet is used. Our results are reported on BRATS-2018 dataset where we achieved the average Dice scores 0.87, 0.75, 0.73 for whole tumor, tumor core, and enhancing tumor respectively compared to the Dice score of radiologist that is in the range 74%–85%. Finally, the proposed Online Class-Weighting loss function with a CNNs model provides an accurate and reliable segmentation result for the whole brain in 22 s as inference time, and that make it suitable for adopting in research and as a part of different clinical settings.

Keywords: Online Class-Weighting · Weighted Cross-Entropy ·
Convolutional Neural Networks · Deep learning ·
Glioblastoma tumors · Image segmentation

© Springer Nature Switzerland AG 2019
I. Rojas et al. (Eds.): IWANN 2019, LNCS 11507, pp. 555–567, 2019.
https://doi.org/10.1007/978-3-030-20518-8_46

1   Introduction and Related Work

Current state-of-the-art image segmentation in the field of Deep Learning are based on Convolutional Neural Networks, where in general we find a feature extractor with a bank of convolution layers (i.e. trainable parameters), then pooling layers to make the images less sensitive and invariant to small translations (i.e. resisting to local translation), the last stage in CNNs models is a classifier which classifies each pixel (or voxel) into one of many classes. Since 2012, the tool of CNNs model led to a big breakthrough in all computer vision applications such as image classification, segmentation, e.g. LeCun et al. [1], Krizhevsky et al. [2] also in object detection Redmon et al. [3]. In the field of medical image classification and segmentation, we can classify methods-based CNNs into two categories: 2D-CNNs and 3D-CNNs models. Many works are proposed in the first category such as Davy et al. [4], Pereira et al. [5], Chang [6], Ben Naceur et al. [7], Zhao et al. [8] and Havaei et al. [9] who extended the previous work of Davy et al. [4], where these methods use 2 dimensional patches as an input to train and classify medical image data. On the other hand, in the second category, we can cite works such as Urban et al. [10], Kamnitsas et al. [11,12], where these methods use 3 dimensional image patches as an input to CNNs models. In the context of our ongoing work [7], we have faced and solved many Deep learning issues: (1) we proposed new three CNNs architectures for brain tumor segmentation problem, (2) we proposed a non-parametric fusion function that merges different brain tumor predictions, acting as a voting function which elects only the tumor region with a high probability, (3) we proposed a new optimizer to overcome the problem of vanishing gradient; where the gradient's signal becomes almost zero (vanishingly small) at the front layers of deep CNNs model. The proposed optimizer adds a new bloc after each training phase and so on until we obtain the satisfied segmentation results in terms of Dice coefficient.

The issue of data imbalance is common across multi-label image segmentation tasks, where there are many proposed methods in state-of-the-art. In overall there are two categories of solutions to this issue: some methods try to mitigate this problem by proposing equal sampling of training images patches [9,13], on the other hand, some methods propose a new loss functions: cross entropy-based median frequency balancing [14], cross entropy-based weight map [15], combination of sensitivity and specificity [16], asymmetric similarity loss function [17], and many others [18–22]. To our knowledge, the existing methods are usually a specific function dedicated for well-defined applications however, these methods could be efficient in its corresponding applications but do not work well in others.

The aim of this paper is to overcome the issue of classical loss functions and define a generic one. To this end, we propose a new method called Online Class-Weighting. Then, to evaluate the impact and the effectiveness of our new proposed Weighted Cross-Entropy loss function, we train a CNNs model with our Online Class-Weighting for the problem of fully automatic brain tumor segmentation. For achieving this goal, our main contributions are divided into two folds:

1. To address the unbalanced data issue, we present a new Online Class-Weighting method that is based on the Weighted Cross-Entropy loss function,
2. To demonstrate the Online Class-Weighting performance, we evaluate our proposition within a case study applied to a fully automatic brain tumor segmentation of Highly Unbalanced Glioblastoma tumors.

2   Proposed Online Class-Weighting Method

Our proposed Online Class-Weighting method is based on the Weighted Cross-Entropy loss function (see Eq. 3) which is used in most image-related applications [14,15].

$$p_j = \frac{\exp^{z_j}}{\sum\limits_{k=1}^{n} \exp^{z_k}} \quad for \; j \in \{1,..,n\} \tag{1}$$

$$Loss(p,q)_j = -\frac{1}{n}(\times \sum_j q_j \times \log(p_j)) \tag{2}$$

$$Loss(p,q)_j = -\frac{1}{n}(\sum_j W_j \times q_j \times \log(p_j)) \quad w_j \in [0,1] \, \& \sum_{j=1}^{n} W_j = 1 \tag{3}$$

Where Loss $(q,p)_j$ is the loss function (i.e. Eq. 3) that represents the error between the estimated probability $p_j \in [0,1]$ (i.e. Eq. 1) and the ground truth class $q_j \in \{0,1\}$, n is the number of classes, $w_j \in [0,1]$ is the weighting factor assigned to the class j.

In this work, to make the prediction of the estimated probability accurate and faster, we have used one-hot encoding, where this encoding gives all probabilities of the ground truth to one class (i.e. the correct class), and the other classes become zero, e.g. $q = [1,0,0,0]$, this vector q indicates that the first class is the correct class. In this case, to calculate the overall error we need only one operation instead of many operations for all classes.

The estimated probability is computed using Softmax function (see Eq. 1) which is used in the last layer of Neural Network after a forward propagation. Thus, this function squash the output to become between 0 and 1. Then, these probabilities are fed into Weighted Cross-Entropy function (see Eq. 3), where this function defines a weighting factor $w_j \in [0,1]$ assigned to the class j compared to classical Cross-Entropy loss function (see Eq. 2). The main issue of the Weighted Cross-Entropy loss function is how to find the weighting factors w_j. To solve this issue, we propose here a new method called Online class-weighting. It allows to find w_j for each class during the training phase by searching for the best parameters using the training rate progress of each class with respect to different training iterations. Indeed, Online Class-Weighting initializes the weighting factors based on some computed measurements such as an evaluation metric, or an error's change. At the beginning, it measures the performance by evaluating the training results with and without the initialization of each class. This evaluation helps to measure the training accuracy of each class. After that, a new training is launched while rewarding each class with an adapted weighting factor w_j until the algorithm reaches a defined epoch iteration. w_j is computed regarding to the previous accuracy of the corresponding class. We note that our proposed Online Class-Weighting algorithm is generic and able to be applied with any number of classes and suitable for different possible weighting factor initialization, and evaluation metrics. In the following, the Online Class-Weighting algorithm describes the proposed method.

Algorithm 1. Online Class-Weighting

Data: training dataset of m instances $(X_{1:m}, Y_{1:n})$, where an input image $X_t \in \mathbb{R}^{d \times d}$, and a target image $Y_j \in \mathbb{N}^{d \times d}$, n is the number of labels and $j \in \{1, ..., n\}$. Z, S are respectively the total number of epoches and a set of epoches. $Z, S, \in \mathbb{N}^+$, where $0 < S < Z$.

Result: set of weighting factors $\{w_j, / \sum_{j=1}^{n} w_j = 1, w_j \in [0, 1]\}$, and the final evaluation results M_j for each class j.

1 Initialization of the n weighting factors $w_{1,j}$, where $w_{1,j} = 1$;
2 Launch training for S epoches with the initialized weighting factors $w_{1,j}$;
3 Evaluation of the first training results for each class $(M_{1,j})$;
4 Weighting factors re-initialization $w_{2,j}$ where $w_{2,j} = \mathrm{f}(X)$, $\sum_{j=1}^{n} w_{2,j} = 1$;
5 Launch second training for S epoches with the initialized $w_{2,j}$;
6 Re-evaluation of the second training results for each class $(M_{2,j})$;
7 Computing stopping criteria: $C = \mathrm{g}(\mathrm{Error}_2(\mathrm{training}), \mathrm{Error}_2(\mathrm{testing}))$;
8 Initialization of the training iteration index: $i = 3$;
9 **while** *stopping criteria is not satisfied* **do**
10 | Updating the weighting factors: $w_{i,j} = \mathrm{h}(M_{i-1,j}, M_{i-2,j})$, where $\sum_{j=1}^{n} w_{i,j} = 1$ and i is the iteration index of the current S epoches;
11 | Launch training with the updated weighting factors $w_{i,j}$;
12 | Re-evaluation of the training results for each class $(M_{i,j})$;
13 | Computing stopping criteria: $C = \mathrm{g}(\mathrm{Error}_i(\mathrm{training}), \mathrm{Error}_i(\mathrm{testing}))$;
14 | Training iteration index i equals to: $i = i + 1$;

The algorithm of Online Class-Weighting works as a policy for the assessment of a Deep Learning model performance. Lines from 1 to 6 attempt to evaluate the behavior of a Deep learning model without and with applying the weighting factors. This assessment consists of computing the evaluation metrics (e.g. Dice coefficient, area under precision-recall curve [17]) of the training results for each class, for example Healthy, Necrosis and Non-enhancing tumor, Edema, and Enhancing tumor class. After that, we use these measurements to calculate the change in these weighting factors over the training phase. Line 7 computes the stopping criteria based on the training and testing errors. Lines 10 and 11 calculate the new weighting factors based on the evaluation metrics between the current training iteration and the two previous ones. To demonstrate the Online

Class-Weighting performance, a case study for Glioblastoma tumor segmentation is described in the next section.

3   A Case Study: Online Class-Weighting Approach for Glioblastoma Tumor Segmentation

In this section, we evaluate the proposed Online Class-Weighting algorithm with our CNNs model on BRATS-2018 dataset. For that, we present the used CNNs architecture that is designed for the segmentation of Glioblastoma brain tumors. Moreover, we introduce simple but effective settings (e.g. the weighting factors initialization) used in the Online Class-Weighting algorithm. Finally, we illustrate different experiments applied with different evaluation metrics.

3.1   Used Neural Networks Architecture

To demonstrate the performance and the efficiency of our loss function, we trained a fully convolutional neural networks architecture (see Fig. 1), where the design of this network is based on:

1. The extraction of multiple hierarchical representations (i.e. multiple feature maps) over the entire input image.
2. It is known that the use of pooling layer allows images to lose some information, which leads to reducing the features' size, thus we designed this network architecture to use the minimum number of pooling layers.
3. Instead of using fully connected layers as the default classifier in many CNNs architectures [1,2,23], we used a $1 \times 1 \times 4$ convolution Softmax layer to save the memory and to speed up the inference time.
4. Our used CNNs architecture called OcmNet which is built based on the rule of using many interconnected modules, where this technique is implemented in known CNNs architectures such as GoogLeNet [24].

Convolutional Neural Networks are known for its ability to extract many complicated and hierarchical features from input images, but one of the issues known in CNNs algorithm is: it labels each pixel (or voxel) separately from the others (Havaei et al. [9], Pereira et al. [25]). Thus, to make CNNs take into account the influence of pixel's neighborhood, we concatenate the output of different levels from the first module (see Fig. 1) with the second module to encourage the OcmNet to make a second prediction of the pixel's label. For the pre-processing step, we apply three normalization steps as follows:

1. Removing the 1% highest and lowest intensities: this technique helps to remove some noise at the tail of the histogram.
2. Subtracting the mean and dividing by the standard deviation of non zero values in all channels.
3. Isolating the background from the tumoral regions by assigning the minimum values to -9. This normalization step helps CNNs to differentiate easily between the background and the tumoral regions.

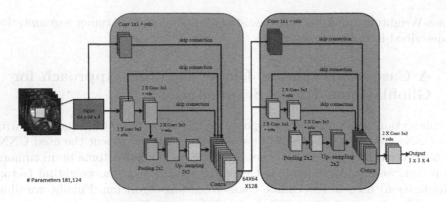

Fig. 1. The OcmNet architecture with $64 \times 64 \times 4$ four channel input patches, consisting of two dense modules. The output $1 \times 1 \times 4$ is a 1×1 convolution layer with 4 classes of BRATS dataset where we have 4 sub-regions: Healthy tissue, Necrosis and Non-enhancing tumor, Edema, Enhancing core. Also, this architecture has 181,124 parameters.

For the post-processing step, we remove some mis-classified non-tumor regions by applying two post-processing techniques:

1. Using a global threshold equals to 110 for each 2D MRI image to remove small non-tumoral regions based on connected-components. We refer to this Post-processing as Post-processing 1.
2. Using a Morphological opening operation, where we noticed that this operator improves the results of the first post-processing. We refer to this Post-processing as Post-processing 2.

3.2   Training and Testing Datasets

BRATS's dataset has a training set of 210 patients with high-grade Glioblastoma and 75 low-grade Glioblastoma. Glioblastoma tumors have 4 classes: Healthy tissue in addition to 3 sub-regions (Necrosis and Non-Enhancing tumor, Edema and Enhancing tumor). Each patient's brain image comes with 4 MRI sequences (i.e., T1, T1c, T2, flair) and the Ground truth labels that are made by radiologist. We split BRATS dataset into 70% for training (i.e. first phase), 30% for testing (i.e. second phase), then in the validation phase (i.e. third phase), we used BRATS 2018 validation set which contains 66 MRI images of patients with unknown grade. For the evaluation of our CNNs architectures on the validation dataset, we used the online evaluation system[1]. Moreover, OcmNet model is trained from scratch using a large number of MRI Overlapping Patches where we applied 25% overlap among patches.

---

[1] Center for Biomedical Image Computing and Analytics University of Pennsylvania, Url:https://ipp.cbica.upenn.edu/.

3.3   Online Class-Weighting Settings

After the definition of the Online Class-Weighting algorithm (see Sect. 2), here we introduce different settings used in this algorithm:

1. Starting from the weighting factors: we initialize the weighting factors using the mean distribution, i.e. each weighting factor equals to $1/n$ (n is the number of classes), where n in our case study equals to 4, i.e. Healthy regions, Necrosis and Non-enhancing tumor, Edema and Enhancing tumor. The algorithm of Online Class-Weighting, line 4 (see Algorithm 1), initializes the weighting factors $w_{2,j}$ as $w_{2,j} = f(X)$, thus $w_{2,j} = 1/n = 1/4 = 0.25$.
2. The second point consists of selecting the best candidates for the current iteration (i) during the training phase[2] as a function of distance between the training results of the iteration ($i-1$) and the iteration ($i-2$). After each S epoches (we found $S = 5$ provides the best results in our experiments) the algorithm changes the weighting factor according to $w_{i,j} = h(M_{i-1,j}, M_{i-2,j})$ (see Algorithm 1, line 10), where $M_{i-1,j}$ and $M_{i-2,j}$ are the Dice coefficient (see Sect. 3.4) of the training iteration ($i-1$) and ($i-2$) respectively. The function h selects two candidates in the following formulas then in the next step (see step 3), calculates the new weighting factors:

$$Candidate(class)_u = \underset{j \in [1:n]}{Argmax}(M_{i-1,1} - M_{i-2,1}), ...(M_{i-1,4} - M_{i-2,4}) \quad (4)$$

$$Candidate(class)_v = \underset{j \in [1:n]}{Argmin}(M_{i-1,1} - M_{i-2,1}), ...(M_{i-1,4} - M_{i-2,4}) \quad (5)$$

The algorithm chooses two candidates using $Argmax$ and $Argmin$ functions, where Eq. (4) returns the class that showed improvement for the last S epoches, thus the algorithm rewards this class. Equation (5) returns the class that is stucked in a valley of the sub-optimal point at the local mimina of loss function. In general, this class is the more frequent label, thus the algorithm penalizes this class.
3. After selecting the best candidates using $Argmax$ and $Argmin$, the algorithm computes the new weighting factors as the following:

$$W_{i,u} = W_{i-1,u} + \beta \qquad W_{i,v} = W_{i-1,v} - \beta \quad (6)$$

where $W_{i,u}$, and $W_{i,v}$ in Eq. (6) are the best candidates, β is the rate of weight's change, where β in our experiments is equal to 0.02 for obtaining a soft weight's change between two successive weighting factors.
4. The last point is stopping criteria (see Algorithm 1, lines 7 and 13):
   1. If the error's change (EC) between training and testing sets is greater than a threshold equals to 0.1: $EC = |$ Error(testing) - Error(training) $|$.
   2. Or testing set accuracy does not improve for a number of epoches K. We found $K = 20$ provides the best results in our experiments, (see Fig. 2).
   3. The last parameters is: if the above two criteria (i.e. a and b) are always false, thus the algorithm will stop after a defined number of epoches Z, where in our case study $Z = 240$.

---

[2] From this technique we inspired the name of Online Class-Weighting loss function.

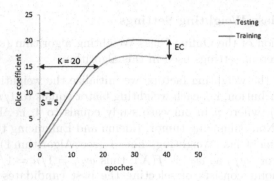

Fig. 2. The curve of stopping criteria that is based on three parameters: Error's change (EC), K epoches, S epoches.

3.4   Evaluation Metrics

For the evaluation of our tumor segmentation method, we use the most pertinent evaluation metrics in state-of-the-art that are used in BRATS[3]: complete (i.e., necrosis and non-enhancing tumor, edema, enhancing tumor), core (i.e. necrosis and non-enhancing tumor, enhancing tumor), enhancing (i.e. enhancing tumor), the evaluation metrics are calculated as follows:

Dice= $\frac{|P_1 \wedge T_1|}{(|P_1| + |T_1|)/2}$, Sensitivity = $\frac{|P_1 \wedge T_1|}{|T_1|}$, Specificity= $\frac{|P_0 \wedge T_0|}{|T_0|}$,

$$Hausdorff = max \left\{ \sup_{p \in \partial P_1} \inf_{t \in \partial T_1} d(p, t), \sup_{t \in \partial T_1} \inf_{p \in \partial P_1} d(t, p) \right\}$$

where $\wedge$ is the logical AND operator, P is the model predictions and T is the ground truth labels. T_1 and T_0 represent the true lesion region and the remaining normal region respectively. P_1 and P_0 represent the predicted lesion region and the predicted to be normal respectively. ($|.|$) is the number of pixels. In addition, $'p'$ and $'t'$ are two points of the surface ∂P_1 and ∂T_1 respectively, and $d(p,t)$, $d(t,p)$ are the shortest least-squares distance (i.e. euclidean distance) between point $'p'$ and $'t'$ and vice versa for $d(t,p)$.

3.5   Experimental Results and Discussion

Online Class-Weighting Versus Cross-Entropy, Focal Loss Function:
To demonstrate the performance of the Online Class-Weighting algorithm, we compare our loss function with state-of-the art Cross-entropy and Focal loss [21]. Cross-entropy and Focal loss are known and showed a high performance in many computer vision applications. Table 1 shows the results of our proposed CNNs architecture with 3 loss functions: Focal loss, Cross-Entropy and Online Class-Weighting. As we can see from Table 1, according to the Dice score of Healthy, NCR/NET and Edema classes, we have competitive results with Cross-Entropy

---

[3] https://www.med.upenn.edu/sbia/brats2018/data.html.

Table 1. Healthy, Necrosis and Non-enhancing tumor (NCR/NET), Edema, Enhancing tumor represent the different Glioblastoma tumor sub-regions segmentation results of OcmNet architecture with 3 different loss functions: Focal loss, Cross-Entropy, and our proposed Online Class-Weighting. Moreover, for each experiment we show the Dice score of 10 MRI images from BRATS-2018 dataset. Values in bold are the best results.

Methods	Dice score					
	Healthy	NCR/NET	Edema	Enhancing	Complete	Core
OcmNet + Focal loss	0.99	0.38	0.70	0.60	0.79	0.62
OcmNet + Cross-Entropy loss	0.99	**0.63**	0.80	0.72	0.87	0.81
OcmNet + Online Class-weighting	**0.99**	0.61	**0.80**	**0.75**	**0.88**	**0.83**

function. However, according to the Dice Complete, Core and Enhancing score, our OcmNet provides the best results with Online Class-Weighting loss. Thus, OcmNet architecture with Online Class-Weighting provides the best results on 5 over 6 metrics. With these results, Online Class-Weighting method outperformed the results of Cross-Entropy and Focal loss function.

Comparison to State-of-the-Art: Table 2 illustrates the results of the segmentation performance of different methods. As we can see, our proposed Ocm-Net with Online Class-Weighting model improved the segmentation results in terms of Dice, sensitivity, specificity, Hausdorff distance. Moreover, OcmNet with Online Class-Weighting performs better than the latest methods applied to Glioblastoma brain tumor segmentation such as [4,8,12] in terms of Dice score and Specificity. Also, OcmNet with Online Class-Weighting architecture obtained competitive results with [5,6,9,10] in terms of Dice coefficient and Hausdorff distance. Also, as we can see from Table 2 and according to the standard deviation, OcmNet with Online Class-Weighting could achieve 100% prediction on some patients MRI images in terms of Dice score and specificity, i.e. the prediction of OcmNet with Online Class-Weighting model corresponds to the manual segmentation of radiologist experts. Havaei et al. [9], Zhao et al. [8] proposed a method based on equal sampling of training images patches to solve the problem of class-imbalance. However, we tried many experiments on this method and we found that it mitigates the impact of class-imbalance problem but it does not solve it (class-imbalance). Thus, loss function-based methods are the best choice and more robust. Cross entropy-based median frequency balancing [14], cross entropy-based weight map [15] propose a modified cross-entropy loss function, but each of which adapts this function to a specific application, which is not the case with our proposed Online Class-Weighting loss function. It does not assume any distribution or *prior knowledge* about the application fields before the training phase. In addition, the most important metric for a clinical decision support system [17] is sensitivity (recall), where OcmNet with Online

Table 2. Segmentation results of our proposed OcmNet and Online Class-Weighting method with the state-of-the-art CNNs methods. WT, TC, ET denote Whole Tumor (complete), Tumor Core, Enhancing Tumor core respectively. (±) is the standard deviation. Post 1 and Post 2 denote post-processing 1 and post-processing 2 respectively. Fields with (-) are not mentioned in the published work.

Methods	Dice score			Sensitivity			Specificity			Hausdorff		
	WT	TC	ET	WT	TC	ET	WT	TC	ET	WT	TC	ET
Urban et al. [10]	0.88	0.83	0.72	-	-	-	-	-	-	-	-	-
Axel et al. [4]	0.79	0.68	0.57	-	-	-	0.79	0.67	0.63	-	-	-
Pereira et al. [5]	0.87	0.73	0.68	-	-	-	0.86	0.77	0.70	-	-	-
Chang [6]	0.87	0.81	0.72	-	-	-	-	-	-	9.1	10.1	6.0
Havaei et al. [9]	0.88	0.79	0.73	0.89	0.79	0.68	0.87	0.79	0.80	-	-	-
Kamnitsas et al. [12]	0.847	0.67	0.629	-	-	-	0.876	0.607	0.662	-	-	-
Zhao et al. [8]	0.84	0.73	0.62	-	-	-	0.82	0.76	0.67	-	-	-
OcmNet + Online Class-weighting	0.863 (±0.1)	0.752 (±0.2)	0.71 (±0.3)	0.863 (±0.2)	0.763 (±0.3)	0.774 (±0.3)	0.992 (±0.01)	0.996 (±0.01)	0.998 (±0.003)	23.37 (±28.2)	18.55 (±23.7)	11.97 (±24.6)
OcmNet + Online Class-weighting + Post 1	0.87 (±0.1)	0.753 (±0.2)	0.73 (±0.3)	0.86 (±0.2)	0.76 (±0.3)	0.776 (±0.3)	0.994 (±0.01)	0.996 (±0.01)	0.998 (±0.004)	14.71 (±22.5)	16.33 (±21.8)	9.83 (±20.7)
OcmNet + Online Class-weighting + Post 1 + Post 2	0.864 (±0.1)	0.75 (±0.2)	0.72 (±0.3)	0.853 (±0.2)	0.76 (±0.3)	0.76 (±0.3)	0.994 (±0.01)	0.996 (±0.01)	0.998 (±0.003)	14.63 (±22.4)	16.37 (±21.9)	8.14 (±20.04)

Class-Weighting achieved a high accuracy on this metric (i.e. sensitivity), see Table 2. In summary, OcmNet with Online Class-Weighting model obtained high segmentation results, and so did with the Post-processing techniques, where the Post-processing 2 improved a lot the Hausdorff distance metric. The first conclusion from these segmentation results is that Gioblastoma brain tumors contain in most cases one connected region, i.e. we have demonstrated this hypothesis after applying the Post-processing 1, where we have seen an improvement in the segmentation performance. The second conclusion is that the class of Enhancing tumor does not have much border with the healthy tissue, i.e. this is demonstrated after applying the Post-processing 2, where we have seen a decrease in the surface of mis-classified Enhancing tumor region and an improvement in the Hausdorff distance metric. Finally, in this study, we have demonstrated that our proposed method has achieved the state-of-the-art segmentation performance in terms of Dice score, sensitivity, specificity, Hausdorff distance.

Inference Time: To deploy a deep learning model on a large scale or on a real-time system, it is necessary to improve the deep learning inference time. Figure 3 shows the speed (seconds) versus CNNs-based methods. Our proposed OcmNet with Online Class-Weighting method is the fastest method among all other models.

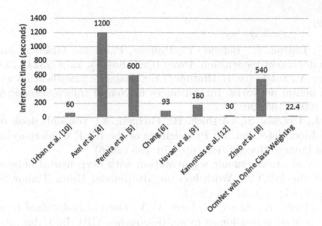

Fig. 3. Evaluation results of inference time (seconds) with the state-of-the-art CNNs methods versus our proposed OcmNet with Online Class-Weighting model.

4 Conclusion

In this work, we presented a new generic Online Class-Weighting algorithm, where the performance and the efficiency of this algorithm is demonstrated through a brain tumor segmentation case study with OcmNet model. In addition, the training of OcmNet with the Online Class-Weighting showed a high performance against Cross-Entropy and Focal loss functions. Moreover, the advantage of Online Class-Weighting is: it allows to find the weighting factors for each class during the training by searching for the best parameters using the rate of progress of each class with respect to the training phase. Moreover, our experimental results show that the OcmNet with Online Class-Weighting architecture improved the segmentation results. Also, these results are demonstrated through a separated validation dataset with real MRI images where the average Dice scores are 0.87, 0.75, 0.73, for whole tumor, tumor core, and enhancing tumor respectively. OcmNet with Online Class-Weighting and Post-processing 1 and Post-processing 2 is the best architecture in terms of Dice, sensitivity, specificity and Hausdorff distance. This architecture has only 181,124 parameters and that make it suitable for adopting in research and as a part of different clinical settings. In this work, we proposed a generic loss function that can be applied to other applications. Based on the preliminary results, our OcmNet with Online Class-Weighting algorithm provides a very promising result. As a perspective of this research, we intend to investigate different scenarios and parameters for an optimal use of the proposed Online Class-Weighting method.

References

1. LeCun, Y., Bottou, L., Bengio, Y., Haffner, P., et al.: Gradient-based learning applied to document recognition. Proc. IEEE **86**(11), 2278–2324 (1998)
2. Krizhevsky, A., Sutskever, I., Hinton, G.E.: ImageNet classification with deep convolutional neural networks. In: Advances in Neural Information Processing Systems, pp. 1097–1105 (2012)
3. Redmon, J., Divvala, S., Girshick, R., Farhadi, A.: You only look once: unified, real-time object detection. In: Proceedings of the IEEE Conference on Computer Vision and Pattern Recognition, pp. 779–788 (2016)
4. Davy, A., et al.: Brain tumor segmentation with deep neural networks. In: Proceedings of the MICCAI Workshop on Multimodal Brain Tumor Segmentation Challenge BRATS, pp. 01–05 (2014)
5. Pereira, S., Pinto, A., Alves, V., Silva, C.A.: Deep convolutional neural networks for the segmentation of gliomas in multi-sequence MRI. In: Crimi, A., Menze, B., Maier, O., Reyes, M., Handels, H. (eds.) BrainLes 2015. LNCS, vol. 9556, pp. 131–143. Springer, Cham (2016). https://doi.org/10.1007/978-3-319-30858-6_12
6. Chang, P.D., et al.: Fully convolutional neural networks with hyperlocal features for brain tumor segmentation. In: Proceedings MICCAI-BRATS Workshop, pp. 4–9 (2016)
7. Ben Naceur, M., Saouli, R., Akil, M., Kachouri, R.: Fully automatic brain tumor segmentation using end-to-end incremental deep neural networks in MRI images. Comput. Methods Programs Biomed. **166**, 39–49 (2018)
8. Zhao, X., Yihong, W., Song, G., Li, Z., Zhang, Y., Fan, Y.: A deep learning model integrating FCNNs and CRFs for brain tumor segmentation. Med. Image Anal. **43**, 98–111 (2018)
9. Havaei, M., et al.: Brain tumor segmentation with deep neural networks. Med. Image Anal. **35**, 18–31 (2017)
10. Urban, G., Bendszus, M., Hamprecht, F., Kleesiek, J.: Multi-modal brain tumor segmentation using deep convolutional neural networks. In: Proceedings of hte MICCAI BraTS (Brain Tumor Segmentation) Challenge, Winning Contribution, pp. 31–35 (2014)
11. Kamnitsas, K., et al.: DeepMedic for brain tumor segmentation. In: Crimi, A., Menze, B., Maier, O., Reyes, M., Winzeck, S., Handels, H., et al. (eds.) BrainLes 2016. LNCS, vol. 10154, pp. 139–149. Springer, Cham (2016). https://doi.org/10.1007/978-3-319-55524-9_14
12. Kamnitsas, K., et al.: Efficient multi-scale 3D CNN with fully connected CRF for accurate brain lesion segmentation. Med. Image Anal. **36**, 61–78 (2017)
13. Lai, M.: Deep learning for medical image segmentation. arXiv preprint arXiv:1505.02000 (2015)
14. Badrinarayanan, V., Kendall, A., Cipolla, R.: SegNet: a deep convolutional encoder-decoder architecture for image segmentation. IEEE Trans. Pattern Anal. Mach. Intell. **39**(12), 2481–2495 (2017)
15. Ronneberger, O., Fischer, P., Brox, T.: U-Net: convolutional networks for biomedical image segmentation. In: Navab, N., Hornegger, J., Wells, W.M., Frangi, A.F. (eds.) MICCAI 2015. LNCS, vol. 9351, pp. 234–241. Springer, Cham (2015). https://doi.org/10.1007/978-3-319-24574-4_28
16. Brosch, T., Tang, L.Y.W., Yoo, Y., Li, D.K.B., Traboulsee, A., Tam, R.: Deep 3D convolutional encoder networks with shortcuts for multiscale feature integration applied to multiple sclerosis lesion segmentation. IEEE Trans. Med. Imaging **35**(5), 1229–1239 (2016)

17. Hashemi, S.R., Salehi, S.S.M., Erdogmus, D., Prabhu, S.P., Warfield, S.K., Gholipour, A.: Asymmetric loss functions and deep densely-connected networks for highly-imbalanced medical image segmentation: application to multiple sclerosis lesion detection. IEEE Access **7**, 1721–1735 (2019)

18. Milletari, F., Navab, N., Ahmadi, S.-A.: V-net: fully convolutional neural networks for volumetric medical image segmentation. In: 2016 Fourth International Conference on 3D Vision (3DV), pp. 565–571. IEEE (2016)

19. Sudre, C.H., Li, W., Vercauteren, T., Ourselin, S., Jorge Cardoso, M.: Generalised dice overlap as a deep learning loss function for highly unbalanced segmentations. In: Cardoso, M.J., et al. (eds.) DLMIA/ML-CDS -2017. LNCS, vol. 10553, pp. 240–248. Springer, Cham (2017). https://doi.org/10.1007/978-3-319-67558-9_28

20. Fidon, L., Li, W., Garcia-Peraza-Herrera, L.C., Ekanayake, J., Kitchen, N., Ourselin, S., Vercauteren, T.: Generalised Wasserstein dice score for imbalanced multi-class segmentation using holistic convolutional networks. In: Crimi, A., Bakas, S., Kuijf, H., Menze, B., Reyes, M. (eds.) BrainLes 2017. LNCS, vol. 10670, pp. 64–76. Springer, Cham (2018). https://doi.org/10.1007/978-3-319-75238-9_6

21. Lin, T.-Y., Goyal, P., Girshick, R., He, K., Dollár, P.: Focal loss for dense object detection. In: Proceedings of the IEEE International Conference on Computer Vision, pp. 2980–2988 (2017)

22. Chen, X., Liew, J.H., Xiong, W., Chui, C.-K., Ong, S.-H.: Focus, segment and erase: an efficient network for multi-label brain tumor segmentation. In: Proceedings of the European Conference on Computer Vision (ECCV), pp. 654–669 (2018)

23. Simonyan, K., Zisserman, A.: Very deep convolutional networks for large-scale image recognition. arXiv preprint arXiv:1409.1556 (2014)

24. Szegedy, C., et al.: Going deeper with convolutions. In: Proceedings of the IEEE Conference on Computer Vision and Pattern Recognition, pp. 1–9 (2015)

25. Pereira, S., Oliveira, A., Alves, V., Silva, C.A.: On hierarchical brain tumor segmentation in MRI using fully convolutional neural networks: a preliminary study. In: 2017 IEEE 5th Portuguese Meeting on Bioengineering (ENBENG), pp. 1–4. IEEE (2017)

Classification with Rejection Option Using the Fuzzy ARTMAP Neural Network

Francisco Felipe M. Sousa[1]([✉]), Alan Lucas Silva Matias[2]([✉]),
and Ajalmar Rego da Rocha Neto[1]([✉])

[1] Federal Institute of Ceará, Fortaleza, CE, Brazil
felipe.moreira@ppgcc.ifce.edu.br, ajalmar@gmail.com
[2] Federal University of Ceará, Fortaleza, CE, Brazil
matiasalsm@gmail.com

Abstract. The ARTMAP networks are machine learning techniques focused on supervised learning, being known mainly for their ability to learn fast, stable, incremental and online. Despite these advantages, the Fuzzy ARTMAP (FAM) suffers from the categories proliferation problem, leading to a reduction in its performance for unknown samples. Such disadvantage is mainly caused by the overlapping region (noise) between classes. The vast majority of work on this issue has been concerned with alleviating the problem. A technique used to improve the performance of a classifier is the rejection option, which is used to retain the classification of a sample if the decision is not considered to be reliable. Therefore, in this paper, we introduce a variant of the Fuzzy ARTMAP to behave as a classifier with the rejection option. The main idea is to create a region of rejection by looking at the place where the categories proliferate since it occurs precisely in the overlapping region. The proposal was validated by conducting experiments with real datasets, as well as by comparing them with other models (MLP, SVM, and SOM) applied with the same rejection option technique.

Keywords: Rejection option · Adaptive Resonance Theory · Fuzzy ARTMAP

1 Introduction

ARTMAP neural networks are neural models for supervised learning based on the Adaptive Resonance Theory [1]. The theory in question seeks to describe how the brain, in a scale of fast and stable learning, and in a totally autonomous way, is able to encode, recognize and predict objects and events in an environment that constantly reorganizes through time. It has, therefore, that the ARTMAP neural networks are based on a theory that proposes an explanation for biological learning.

© Springer Nature Switzerland AG 2019
I. Rojas et al. (Eds.): IWANN 2019, LNCS 11507, pp. 568–578, 2019.
https://doi.org/10.1007/978-3-030-20518-8_47

From this viewpoint, the ARTMAP networks inherit an over-adjustment nature – such property is known as overfitting in the machine learning literature. As it is known, in general, overfitting leads the learning system to an inefficient generalization for unknown samples, giving rise to many incorrect predictions. In face with such a scenario, one can argue that decision-making should be delayed, giving the system the opportunity to identify critical items rather than trying to sort. In machine learning, it is known as rejection choice classification [3,4], or soft decision-making [5], and it is mainly based on the construction of a binary classifier with a third class output, which is usually called the class of rejection.

The rejection option, roughly, consists of a set of techniques that aim to improve the reliability when a learning system is making a decision, being originally formalized in the context of statistical pattern recognition in [4], supported by the minimum risk theory. Patterns that are rejected can be manipulated by other classifiers or manually by a human expert. This paradigm has been adapted to obtain efficient learning models [6].

Bearing this in mind, in this paper, we developed a new variant of the Fuzzy ARTMAP network [2] to act as a classifier with rejection option. In our formulation, we construct a Fuzzy ARTMAP network focused on binary classification problems able to decide, through the training process, whether a category is a rejection category or not. As one can expect, the rejection categories are not categories that map a class, but categories that retain the critical samples and reject them. We must highlight that, as far as we know, this is the first time that such an approach has been developed for an ARTMAP network.

Our paper is organized as follows. In the next section, classification with rejection option is briefly described. In Sect. 3, we describe the Fuzzy ARTMAP architecture. After that, in Sect. 4 we introduce our proposal. The simulation achievements carried out for ROFAM-1C as well as for its counterparts is presented in Sect. 5 – in this section, are presented results for real-world classification problems. At last, Sect. 6 presents the conclusions and future works.

2   Classification with Rejection Option

Given a "complex" dataset, each classifier is bound to misclassify some data samples. Assuming that the input information is represented by an n-dimensional real vector $x = [x_1, x_2, ..., x_n] \in \mathbb{R}^n$, and that the problem data involves only two classes $\{C_{+1}, C_{-1}\}$, the classifier must be able to predict the third class, that is, the class of rejection $\{C_{+1}, C_{-1}, C_{Reject}\}$. There are three different approaches to designing a classifier with rejection options

Method 1: It involves a single binary classifier. When the classifier gives some approximation to the probabilities of the posterior class, $P(C_k|\mathbf{x}), k = 1, 2, ..., K$, the default is rejected if the highest value among the posterior probabilities of K is less than a given threshold T_r. According to Chow [4], the decision is made if:

$$\max_k = [P(C_k|x)] \leq T_r \tag{1}$$

or, equivalently,

$$\max_{k} = [P(\boldsymbol{x}_i|C_k) P(C_k)] \leq T_r \tag{2}$$

where $P(C_k)$ is the a prior probability distribution of the K-th class and $P(\boldsymbol{x}_i|C_k)$ is the conditional probability density function for the input pattern $\boldsymbol{x}_i$ given the k-th class. When the classifier does not provide probabilistic outputs, a specific rejection limit must be used [5]. For this method, the classifier is trained as usual, that is, without initially creating a class of rejection. The rejection region is determined after the training phase, heuristically or based on optimization following some post-training criteria.

Method 2: In this approach, two independent classifiers are used. The former is trained to produce a high probability value for class C_{+1}, while the latter is trained to produce a high probability value for class C_{-1}. When the two classifiers agree with the decision, the corresponding class is generated. Otherwise, the aggregate decision is prone to be unreliable and therefore rejection would be preferable [11].

Method 3: This approach uses a single classifier with a built-in rejection option, that is, the classifier is trained following optimality criteria that automatically take into account the cost of miss-predictions and rejections in its cost function [11–13].

In this paper, we introduce a strategy for Fuzzy ARTMAP with the rejection option paradigm described in method 1.

3   Fuzzy ARTMAP (FAM)

The FAM network uses the Fuzzy ART module and a map-field which maps each category in the Fuzzy ART module to a class. Thus, we must introduce first the Fuzzy ART network followed by the FAM architecture.

3.1   Fuzzy ART (FA)

The FA model works as an unsupervised learning system which creates adaptive hyper-rectangular prototypes to categorize the input data – these prototypes are conventionally called categories. The input patterns are represented as input vectors $\boldsymbol{x}_i \in \mathbb{R}^n$ *MinMax* normalized between 0 and 1, and an alternative formulation for such a input vector, represented by $\overline{\boldsymbol{x}}_i \in \mathbb{R}^{2n}$, is given by the complement coding, where $\overline{\boldsymbol{x}}_i = [\boldsymbol{x}_i, \boldsymbol{x}_i^c]$ and $\boldsymbol{x}_i^c = 1 - \boldsymbol{x}_i$.

The FA categories contains a subset of the input patterns and are described by $\boldsymbol{w}_j = [w_1, ..., w_n]$. For input vectors with complement coding, the categories vectors $\boldsymbol{w}_j = [\boldsymbol{u}_j, \boldsymbol{v}_j^c]$ are $2n$-dimensional, where $\boldsymbol{u}_j, \boldsymbol{v}_j \in \mathcal{R}^n$ are the inferior and superior points of the hyper-rectangle R_j^w from $\boldsymbol{w}_j$. Also, we must highlight that the FA model is a growing model: it starts with 1 category, and as the training is performed new categories are added. In order to maintain such a dynamic, the FA model always keeps one uncoded category $\boldsymbol{w}_{m+1} = \boldsymbol{1} = [1_1, ..., 1_{2n}]$, so that when the model has $m = 0$ coded categories (i.e., when the training process is about to begin), it yet have 1 uncoded category.

The FA activation is based on the winner-take-all algorithm, so that for each input pattern $\overline{x}_i$ we have an activation measure for each category w_j. The category with the maximal activation value becomes active. Defines

$$\phi_j = [\wedge(\overline{x}_1, w_{j1}), \dots, \wedge(\overline{x}_{2n}, w_{j2n})], \tag{3}$$

where $\wedge(\overline{x}_i, w_{ji}) = \min\{\overline{x}_i, w_{ji}\}$ is the fuzzy AND operation. The activation function computed for each category is then given by

$$T_j = \frac{\sum_{i=1}^{2n} \phi_{ji}}{\alpha + \sum_{i=1}^{2n} w_{ji}}, \quad \forall j = 1, \dots, m, \quad \alpha > 0, \tag{4}$$

and the index of the active category is given by

$$J = \arg\min\{T_j : j = 1, \dots, m\}. \tag{5}$$

Each category w_j has its size bounded by the vigilance parameter $\rho \in [0, 1]$. In order to verify if the active category reached the maximum allowed size, the vigilance criterion is performed by

$$\sum_{i=1}^{2n} \phi_{Ji} \geq \rho n, \tag{6}$$

where ϕ_J is related to the active category w_J.

Once satisfied the condition in the Eq. (6), the chosen category w_J matches and the updating rule must be performed as an convex combination between ϕ_J and w_J according to

$$w_J^{new} = \beta\phi_J + (1 - \beta)w_J, \tag{7}$$

where $\beta \in [0, 1]$ is the learning rate parameter.

However, if Eq. (6) is not satisfied, the category J activation value is attenuated by $T_J = -1$. In this situation, a new category with the maximal activation value is (i) obtained and (ii) evaluated in terms of the vigilance criterion by Eq. (6). The steps (i) and (ii) are performed until the vigilance criterion is satisfied or a new category is created.

3.2 FAM Architecture

Define $x_i \in \mathcal{R}^n$ and $y_i \in \mathcal{B}^p$, the input pattern and label, respectively, where p is the number of classes of the problem, $\mathcal{B}$ is the binary set, and y is codified as 1-hot. Also, defines the map-field $\Delta \in \mathcal{B}^{m \times p}$, which associates to each category w_j a class $\delta_j = y_i$, where δ_j is a line vector in the matrix Δ and y_i is an arbitrary expected output class. Also, as the FA maintains a uncoded category w_{m+1}, the map-field maintains a uncoded mapping $\delta_{m+1} = [1_1, \dots, 1_p]$.

In order to train the FAM network we first have to present the input pattern x_i to the Fuzzy ART network. The first operation is to compute the complement coding for x_i as $\overline{x}_i$, followed by the compute of the activation provided by each

$\boldsymbol{w}_j$ (see Eq. (4)). After that, the index J of the active category is obtained through the Eq. (5) and the vigilance criterion is performed (Eq. (6)). When the active category $\boldsymbol{w}_J$ satisfies the vigilance criterion, we must perform the map-field criterion verifying if

$$\sum_{k=1}^{p} \wedge(\delta_{Jk}, y_{ik}) = 1. \tag{8}$$

Notice that the map-field uses the current expected output $\boldsymbol{y}_i$. The operation above, roughly speaking, checks if the current expected output $\boldsymbol{y}_i$ is equal to the class mapped in $\boldsymbol{\delta}_J$ (notice that for the uncoded mapping $\boldsymbol{\delta}_{m+1}$, the Eq. (8) will always be true). If it is not satisfied, we must trigger what we call in the FAM literature the *match tracking* mechanism, which will raises the value of ρ temporarily to $\rho = \sum_{i=1}^{2n} \phi_{Ji}/n + \varepsilon$, where $\varepsilon > 0$, and attenuate the current activation $(T_J = -1)$ in order to activate a new category.

We can see that the FAM network is composed by two constraints (see Eqs. (6) and (8)) which must be satisfied by the active indexation J. The category updating (Eq. (7)) only occurs, if, and only if, both the constraints are satisfied. After updating the active category, we must update the active mapping, which is given by

$$\boldsymbol{\delta}_J^{new} = [\wedge(\delta_{J1}, y_{i1}), \ldots, \wedge(\delta_{Jp}, y_{ip})]. \tag{9}$$

4   Fuzzy ARTMAP with Rejection Option

This section details our proposal, the Fuzzy ARTMAP with Rejection Option (ROFAM-1C). We must highlight that the ROFAM-1C map-field is not the conventional presented in Sect. 3. Instead, it uses the PROBART map-field, which stores the frequency of association between the categories $\boldsymbol{w}_j$ and the expected outputs $\boldsymbol{y}_i$ during the training process. In summary, the ROFAM-1C is composed by two stages: the first consists of training the network and storing the frequencies of association between categories and expected outputs in the PROBART map-field; second, the best threshold value for rejection is found. The details of the modification are described below.

4.1   PROBART Map-Field

The PROBART map-field stores the frequency of association between categories $\boldsymbol{w}_j$ and classes $\boldsymbol{y}_i$ by changing the map-field updating rule to

$$\boldsymbol{\delta}_J^{new} = \boldsymbol{\delta}_J + \boldsymbol{y}_i. \tag{10}$$

Also, in contrast with the conventional FAM map-field, the PROBART map-field do not implements any constraints, such as the criterion described in Eq. (8), and also no *match tracking*. The PROBART map-field only stores the frequency of association between categories and outputs.

With such a modification in the conventional FAM, we can say that if $\delta_{jk} = l$, then l represents the number of times that inputs patterns $\boldsymbol{x}_i$ related to the k-th class were associated with the j-th category.

4.2   FAM with Rejection Option Using One Classifier (ROFAM-1C)

Initially, the ROFAM-1C network is trained with the PROBART map-field instead of the conventional one (see Sect. 4.1). When the training step is finished, the model will use the frequency information stored in Δ to change the label of some categories to the rejection class label. The main idea behind the ROFAM-1C approach is based, precisely, on the development of formal techniques for assigning the rejection class label to certain categories. For more detail, ROFAM-1C requires the following steps.

STEP 1: For an arbitrary dataset, the training step is performed using the FAM network with the PROBART map-field in order to find the categories w_j and frequencies of associations Δ.

STEP 2: With the frequency of association stored in Δ, the probability of association for each category w_j is estimated from the information contained in δ_j through the following equation:

$$P(C_k|w_j) \approx \frac{\delta_{jk}}{\sum_{i=1}^{p} \delta_{ji}}, \tag{11}$$

where C_k is the k-th class.

STEP 3: Change the category label to the rejection class label if

$$IG_j = 1 - \sum_{k=1}^{p} P(C_k|w_j)^2 \geq T_r, \tag{12}$$

where IG_j is the Gini Index and $0 \leq T_r \leq 1$ is a threshold which must be specified.

Concern the Gini index, we can see that when $IG_j \to 0$, then the frequency of association between the j-th category and the classes is homogeneous, indicating that the current category maps successfully only one class. However, when $IG_j \to 1$, it indicates that the j-th category is heterogeneous concern the mapping and the decision-making during classification provided by this category can degenerate the ROFAM-1C performance. Said that, it is noticeable that categories with large IG_j will tend to become rejection categories (see Eq. (12)).

One must also consider in this step the optimization process of the rejection threshold T_r described in Eq. (12). Finding an optimal value for the rejection threshold is described as empirical risk minimization problem as proposed in [4]:

$$\hat{R} = W_r R + E \tag{13}$$

where R is the ratio of rejection, E is the ratio of misclassified patterns, and W_r is the rejection cost whose value must be specified in advance by the user. In summary, given a W_r, one must find a value for the rejection threshold T_r that minimizes $\hat{R}$.

The Fig. 1 shows a example of the ROFAM-1C application to an artificial binary classification problem. The Fig. 1(a, b, and c), shows the ROFAM-1C categories obtained for this problem. In the Fig. 1(d, e, and f), one can see the

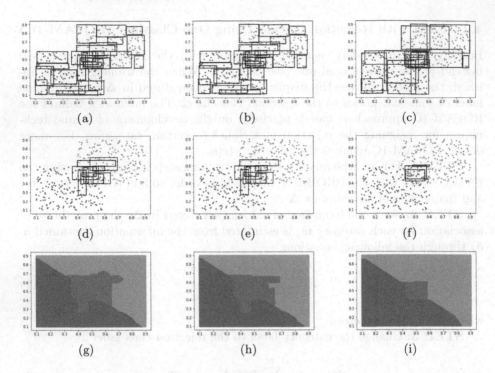

Fig. 1. (a, b, and c) ROFAM-1C categories; (d, e, and f) rejection categories with rejection cost W_r equal to (a, d) 0.04, (b, e) 0.24, and (c, f) 0.48; (g, h, and i) decision boundary with the rejection region in green. (Color figure online)

categories which became rejection categories, so that, the input patterns that activates such categories will be rejected – notice that we used $W_r = 0.04, 0.24$, and 0.48, respectively, and the rejection threshold T_r was optimized for each value of W_r. At last, in the Fig. 1(g, h, and i), we show the decision boundary, with the rejection region in green. One can consider, looking at the green region, that as the rejection cost W_r grows, the number of rejected patterns decreases.

5   Experimental Study and Discussion

In the current section, we present the assessment results for ROFAM-1C when applied to binary real-world based classification problems. We compared the ROFAM-1C with well known machine learning models: Multi-layer Perceptron (MLP), Support Vector Machines (SVM), and Self-organizing Maps (SOM). We must highlight that the same rejection option strategy used in ROFAM-1C was adopted to MLP, SVM and SOM, giving rise to MLP-1C, SVM-1C, and SOM-1C.

In these experiments, we used 7 benchmarking datasets from UCI Machine Learning Repository [9]: Hepatitis (HEP), Vertebral Column Pathology (VCP), Statlog Heart (SLH), Ionosphere (ION), Breast Cancer Wisconsin (BCW),

Table 1. Description of the datasets.

Datasets	#Features	#TrS	#TeS
HEP	19	63	17
VCP	6	247	63
SLH	13	215	55
ION	34	280	71
BCW	9	549	139
PID	8	613	155
SON	60	165	43

Pima Indian Diabetes (PID), and Sonar (SON). Some information about the dimension of the input vector (#Features), the number of training samples (#TrS) and the number of test samples (#TeS) for each dataset can be seen in Table 1.

The performance evaluation involves the classification accuracy and the rejection rate, which were computed for 30 independent realizations, where the training and test samples were randomly chosen for each one. Concerning the model's parameters, they were optimized via k-fold cross-validation with $k = 5$. It is important to highlight that we cannot select only one value for W_r since it is intrinsically dependent on the application. In this way, the function described in Eq. (13) was optimized with three different values for W_r: 0.04, 0.24, and 0.48. As mentioned, the value of W_r is directly related to how many input patterns an expert is willing to reject.

The simulations results can be visualized in the Tables 2, 3, and 4. In this table, we can see the datasets (Dataset), the accuracy and rejection rate metrics (Metric), where we can see the mean accuracy with standard deviation and the mean rejection rate. Besides, statistical tests based on the Friedman test were carried out [10]. Such a test is non-parametric and relies on the null hypothesis that all classifiers are equivalent so that they belong to the same population. Thus, by considering an alternative hypothesis, where all the classifiers are not equivalent, we applied this test with 0.05 significance level in order to identify which results in terms of accuracy were significantly different concerning the proposed and counterpart methods. The symbol ✓ is used to inform the equivalence between classifiers; otherwise, we use the symbol ✗.

Looking at the results, it is possible to notice that in some datasets the proposed model performance is equivalent or better in comparison to the SOM network (see the results for the HEP, VCP, SLH, ION, and BCW datasets in all the tables), and in some cases equivalent to MLP (see the results for HEP, ION, BCW, and SON datasets). However, its performance is, in general, worse than the results obtained by the SVM model. Based on the hypothesis test, ROFAM-1C obtains several statistically different performance from the other methods, particularly for the PID dataset using a rejection cost $W_r = 0.48$. For VCP and HEP sets, ROFAM-1C provides equivalent results compared with its counterparts when using a W_r of 0.04 and 0.24, respectively.

Table 2. Results of the simulations with $W_r = 0.04$

Dataset	Metric	ROFAM-1C	MLP-1C		SVM-1C		SOM-1C	
HEP	ACC	71.39 ± 19.81	75.37 ± 13.03	✓	**86.03 ± 15.74**	✗	78.26 ± 12.25	✓
	REJ	58.82	31.96		69.57		45.10	
VCP	ACC	96.84 ± 03.79	**99.16 ± 02.57**	✓	96.89 ± 03.42	✓	95.05 ± 05.59	✓
	REJ	77.39	70.85		59.31		75.50	
SLH	ACC	87.05 ± 13.98	**97.76 ± 04.75**	✗	96.52 ± 06.59	✗	88.11 ± 08.35	✓
	REJ	74.30	85.96		82.97		69.72	
ION	ACC	94.42 ± 06.85	94.15 ± 03.65	✓	**97.36 ± 02.39**	✗	96.68 ± 04.78	✗
	REJ	70.19	23.05		15.59		73.57	
BCW	ACC	98.63 ± 01.25	98.96 ± 00.89	✗	**99.21 ± 00.77**	✗	98.51 ± 01.01	✓
	REJ	23.19	09.50		10.41		12.69	
PID	ACC	86.30 ± 10.72	93.73 ± 11.73	✗	93.23 ± 06.85	✓	**96.61 ± 03.75**	✗
	REJ	83.23	92.30		79.71		82.56	
SON	ACC	87.03 ± 11.73	86.17 ± 06.99	✓	89.35 ± 04.52	✓	**92.21 ± 06.49**	✗
	REJ	66.88	27.60		07.36		68.22	

Table 3. Results of the simulations with $W_r = 0.24$

Dataset	Metric	ROFAM-1C	MLP-1C		SVM-1C		SOM-1C	
HEP	ACC	72.00 ± 15.87	73.44 ± 12.60	✓	**78.41 ± 13.01**	✓	69.52 ± 13.70	✓
	REJ	38.04	23.92		33.53		31.96	
VCP	ACC	88.88 ± 07.82	**96.61 ± 04.56**	✗	90.96 ± 04.54	✓	87.87 ± 07.15	✓
	REJ	48.89	53.76		21.75		51.43	
SLH	ACC	83.45 ± 05.82	**93.35 ± 06.48**	✗	87.59 ± 03.37	✗	81.60 ± 05.35	✓
	REJ	21.27	55.10		18.48		20.24	
ION	ACC	91.28 ± 04.36	92.31 ± 04.07	✓	**96.60 ± 02.88**	✗	90.18 ± 03.61	✓
	REJ	05.59	13.03		09.01		02.63	
BCW	ACC	97.06 ± 01.40	**98.55 ± 00.98**	✗	98.12 ± 01.18	✗	97.73 ± 01.41	✓
	REJ	04.48	06.52		03.84		03.26	
PID	ACC	80.16 ± 10.63	91.44 ± 08.82	✗	86.42 ± 04.15	✓	**96.42 ± 04.00**	✗
	REJ	53.18	73.35		33.16		82.95	
SON	ACC	83.62 ± 05.77	84.75 ± 06.55	✓	88.73 ± 04.97	✗	**89.07 ± 05.58**	✗
	REJ	04.26	22.02		05.12		19.38	

Table 4. Results of the simulations with $W_r = 0.48$

Dataset	Metric	ROFAM-1C	MLP-1C		SVM-1C		SOM-1C	
HEP	ACC	70.23 ± 11.84	72.78 ± 11.92	✓	$\mathbf{76.31 \pm 12.01}$	✗	73.69 ± 12.59	✓
	REJ	15.88	18.04		16.08		17.06	
VCP	ACC	78.56 ± 04.89	$\mathbf{93.72 \pm 05.88}$	✗	85.70 ± 04.42	✗	80.38 ± 06.20	✓
	REJ	10.21	38.96		06.51		21.75	
SLH	ACC	79.46 ± 05.62	$\mathbf{90.50 \pm 06.99}$	✗	84.03 ± 03.15	✗	81.09 ± 04.78	✓
	REJ	05.15	38.12		05.21		11.39	
ION	ACC	91.28 ± 03.80	91.49 ± 03.98	✓	$\mathbf{95.27 \pm 03.02}$	✗	90.34 ± 003.09	✓
	REJ	01.22	09.19		04.69		02.96	
BCW	ACC	96.43 ± 01.20	$\mathbf{98.21 \pm 01.13}$	✗	97.02 ± 00.94	✓	97.56 ± 01.22	✗
	REJ	01.44	04.79		01.10		02.09	
PID	ACC	70.95 ± 03.78	$\mathbf{86.69 \pm 10.06}$	✗	79.59 ± 02.82	✗	77.06 ± 03.85	✗
	REJ	00.90	52.22		06.82		20.95	
SON	ACC	84.32 ± 05.86	83.73 ± 06.53	✓	$\mathbf{88.03 \pm 05.44}$	✗	86.64 ± 06.17	✗
	REJ	00.78	15.79		04.03		14.96	

6 Conclusions

In this work, we proposed a new variant of the Fuzzy ARTMAP network that incorporates the rejection option technique. Our proposal, called ROFAM-1C requires a single modified and trained FAM in order to assign to some categories the rejection class label. The results obtained by the ROFAM-1C in some datasets is superior or comparable with the results obtained by the SOM-1C model, and equivalent with MLP-1C in some datasets. On the other hand, the ROFAM-1C is in general inferior to SVM. However, when looking at the results obtained by the ROFAM-1C, one must consider, first, that the ROFAM-1C is based on an incremental and online classifier – in fact, FAM-based models can achieve stability in 1 epoch. Sometimes, because of the incremental feature of the FAM network, the order of the input presentation interferes in the model performance. Also, the hyper-rectangular geometrical form of the categories provided by the FAM network may not be suitable for such problems. With this in mind, in future works, we aim to implement the ROFAM-1C framework in ARTMAP models that provide more smooth forms for its categories.

References

1. Grossberg, S.: Adaptive resonance theory: how a brain learns to consciously attend, learn, and recognize a changing world. Neural Netw. **37**, 1–47 (2013)
2. Carpenter, G., et al.: Fuzzy ARTMAP: a neural network architecture for incremental supervised learning of analog multidimesional maps. IEEE Trans. Neural Netw. **3**, 698–713 (1992)

3. El-Yaniv, R., Wiener, Y.: On the foundations of noise-free selective classification. J. Mach. Learn. Res. **11**, 1605–1641 (2010)
4. Chow, C.: On optimum recognition error and reject tradeoff. IEEE Trans. Inf. Theory **16**(1), 41–46 (1970)
5. Ishibuchi, H., Nii, M.: Neural networks for soft decision making. Fuzzy Sets Syst. **34**(115), 121–140 (2000)
6. Sousa, R., da Rocha Neto, A.R., Barreto, G.A., Cardoso, J.S., Coimbra, M.T.: Reject option paradigm for the reduction of support vectors. In: ESANN (2014)
7. Marriott, S., Harrison, R.F.: A modified Fuzzy ARTMAP architecture for the approximation of noisy mappings. Neural Netw. **8**(4), 619–641 (1995)
8. Xu, K.: How has the literature on Gini's index evolved in the past 80 years? Working paper, Department of Economics, Dalhousie University (2003). https://doi.org/10.2139/ssrn.423200
9. Lichman, M.: UCI machine learning repository (2013)
10. Demšar, J.: Statistical comparisons of classifiers over multiple data sets. J. Mach. Learn. Res. **7**, 1–30 (2006)
11. Sousa, R., Mora, B., Cardoso, J.S.: An ordinal data method for the classification with reject option. In: Proceedings of the International Conference on Machine Learning and Applications (ICMLA 2009), pp. 746–750 (2009)
12. Bounsiar, A., Beauseroy, P., Grall-Maes, E.: General solution and learning method for binary classification with performance constraints. Pattern Recogn. Lett. **29**(10), 1455–1465 (2008)
13. Fumera, G., Roli, F.: Support vector machines with embedded reject option. In: Lee, S.-W., Verri, A. (eds.) SVM 2002. LNCS, vol. 2388, pp. 68–82. Springer, Heidelberg (2002). https://doi.org/10.1007/3-540-45665-1_6

About Filter Criteria for Feature Selection in Regression

Alexandra Degeest[1,2]($\boxtimes$), Michel Verleysen[2], and Benoît Frénay[3]

[1] Haute-Ecole Bruxelles Brabant - ISIB, 150 rue Royale, 1000 Brussels, Belgium
adegeest@he2b.be
[2] UCLouvain Machine Learning Group - ICTEAM,
Place du Levant 3, 1348 Louvain-La-Neuve, Belgium
michel.verleysen@uclouvain.be
[3] Faculty of Computer Science, NADI Institute - PReCISE Research Center,
Université de Namur, Rue Grandgagnage 21, 5000 Namur, Belgium
benoit.frenay@unamur.be

Abstract. Selecting the best group of features from high-dimensional datasets is an important challenge in machine learning. Indeed problems with hundreds of features have now become usual. In the context of filter methods, the selected relevance criterion used for filtering is the key factor of a feature selection method. To select an appropriate criterion among the numerous existing ones, this paper proposes a list of six necessary properties. This paper describes then three relevance criteria, the mutual information, the noise variance and the adjusted R-squared, and compares them in the view of the aforementioned properties. Any new, or popular, criterion could be analysed in the light of these properties.

Keywords: Feature selection · Relevance criteria · Regression

1 Introduction

High-dimensional datasets appear now frequently in various domains such as healthcare, marketing or social media, especially in regression. Selecting the most relevant subset of features in high-dimensional datasets has therefore become essential for many purposes: to increase the interpretability of features, to facilitate the learning process, to visualise data, to alleviate the curse of dimensionality, etc [14,16].

Many works, in a variety of domains, focus on methods to reduce the number of features in datasets [2,10,12,17,19,22,23,27]. These methods can be roughly categorised into filters, wrappers and embedded methods. This paper focuses on filter methods, which have the advantage to be fast because they do not require to train any model during the feature selection process, contrarily to wrappers [18,19] and embedded methods [7].

Filter methods rely on a relevance criterion to reduce the set of features to only the most relevant ones. This relevance criterion is therefore the key factor

© Springer Nature Switzerland AG 2019
I. Rojas et al. (Eds.): IWANN 2019, LNCS 11507, pp. 579–590, 2019.
https://doi.org/10.1007/978-3-030-20518-8_48

of a successful filter-based feature selection process. Several relevance criteria exist and are used in feature selection methods on various datasets.

This paper focuses on the necessary properties of a criterion used in filter methods for feature selection in regression. What is needed in a filter criterion in order to obtain the best subset of features with respect to the target or prediction goal (in classification or regression tasks)? Existing criteria are often designed to fulfill a unique purpose: for instance to measure a nonlinear relation or to estimate the noise variance of the distribution. An efficient criterion should probably combine various goals.

This paper does not intend to propose a new filter criterion but, instead, focuses on the diverse properties that make a good relevance criterion, in order to be able to select one among the numerous existing ones. These important properties are listed and discussed in Sect. 3, after an introduction to feature selection in regression with filters in Sect. 2.

It is essential to analyse relevance criteria in view of these properties and these goals in order to understand the strengths and the weaknesses of each of them, and to understand their behaviour according to the type of dataset at hand. Existing criteria are described in Sect. 4 and compared with respect to these properties in Sect. 5. Finally, conclusions are given in Sect. 6.

2   Feature Selection with Filter Methods

Feature selection is an important task in machine learning. It helps to reduce the dimension of the dataset by eliminating redundant and less useful features.

In the context of filter methods for feature selection in regression, a good relevance criterion is necessary to select the most relevant features among all the available ones. The relevance criterion aims at measuring the existing relationship between a feature, or a set of features, and the variable to predict. There exist several relevance criteria based on different measures such as entropy or noise variance.

Filter methods also need a search procedure to find the best feature subset among an exponential number (exponential to the dimensionality of the dataset) of all possible ones that could be extracted from the complete dataset [16]. During the search procedure, the filter criterion is again a strategic factor because it is used to evaluate the relevance of each subset with respect to the target. Implicitly, the search procedure is also used to measure the redundancy between different features or groups of features.

The properties of a filter criterion are therefore essential because they determine the success of a good feature selection process. Understanding why a filter criterion is better for a specific dataset or less good for another one is also important in order to choose the best criterion for every situation.

The next section details some essential properties of a relevance criterion.

3   Properties of a Relevance Criterion for Feature Selection

This section introduces the important properties of a good relevance criterion for feature selection in regression. It also justifies why these properties are important. An analysis of these properties with respect to current filter criteria is realised in Sect. 5.

3.1   Property 1: Ability to Detect Nonlinear Relationships

A good relevance criterion should be able to detect nonlinear relationships between variables (features and target variables) [11,15]. This ability allows the criterion to detect the relevance between a group of features and the target, but also to detect the redundancy between features, even when the relationship is nonlinear, which is most generally the case with real datasets.

3.2   Property 2: Ability to Detect Multivariate Relationships

An efficient relevance criterion must be able to detect any relationship between two variables or, more importantly, between two groups of variables. Indeed, measuring the univariate relation between a single input feature and the target is not sufficient, as some features only contributing to the output when they are combined would not be detected (an obvious example of that phenomenon is a problem where the target is determined by the product of two features).

The necessity for a multivariate criterion is also a direct consequence of the use of greedy search procedures to find the most effective subset of features, such as forward and backward search, genetic algorithms, etc [6,16,24,27].

3.3   Property 3: Estimator Behaviour

Machine learning methods are always used on finite datasets. However, relevance criteria are generally defined in terms of integrals over the data space. In order to use them in practice, an estimator of the criteria, defined on a finite set of data, is needed. The computational complexity and the statistical properties of the estimators are important characteristics that should be taken into account when one needs to choose a criterion for selecting a reduced set of features [3].

3.4   Property 4: Estimator Parameters

In addition to the statistical properties of the estimators, the latter usually require to adjust a parameter whose influence on the quality of the estimation might be important. For example, nearest-neighbours based estimators require to choose the number of neighbours used in the estimation.

The choice of the parameters is sometimes underestimated in the literature, while in practice this choice may be crucial. Criteria whose estimators that do not rely on any parameter, or that rely on parameters having only low influence on the estimation, are therefore more appropriate.

3.5   Property 5: Estimator Behaviour in Small Sample Datasets

The ratio "number of instances/dimensionality" is a very important concept in all machine learning methods. A small sample dataset is a dataset with few instances with respect to the number of features. Many estimators do not work well with these datasets and need many instances to estimate correctly the relevance of features [3,5]. Unfortunately it is not always possible to collect more instances. Therefore, this property of behaving well in small sample scenarios is essential as well for the estimator. Section 5.5 analyses how the different relevance criteria behave in small sample situations.

3.6   Property 6: Invariant Estimator

Among the estimators of relevance criteria, some are not completely invariant to the gradient of the relation between the features and the target, especially in small sample scenarios. Depending on the scaling method or the normalisation method used during the process, the gradient of the relation may vary. However the importance of a feature, or a group of features, with respect to the target should not depend on this gradient. The consequence in feature selection could make a relevance criterion prefer a feature over another one only because of the gradient of their relation with the target, which should not happen.

Section 5.6 shows practically that some relevance criteria are influenced by this gradient of the function in small sample, and some are not.

4   Description of Three Popular Criteria

This section reviews three filter criteria, and their most frequently used estimators, in order to illustrate the strategic properties of a good relevance criterion as listed in Sect. 3.

4.1   Mutual Information

Mutual Information (MI) is a popular criterion for feature selection with filter methods [1,3,4,14,17,26]. It is a symmetric measure of the dependence between random variables (or sets of variables), based on entropy, introduced by Shannon in 1948 [25].

Let X and Y be two random variables, where X represents the set of features and Y the target. MI measures the reduction in the uncertainty on Y when X is known

$$I(X;Y) = H(Y) - H(Y|X) \tag{1}$$

where

$$H(Y) = -\int_Y p_Y(y) \log p_Y(y) dy \tag{2}$$

is the entropy of Y and

$$H(Y|X) = \int_X p_X(x)H(Y|X=x)dx \tag{3}$$

is the conditional entropy of Y given X. The mutual information between X and Y is equal to zero if and only if they are independent. If Y can be perfectly predicted as a function of X, then $I(X;Y) = H(Y)$.

In practice, MI cannot be directly computed because it is defined in terms of probability density functions. These probability density functions are unknown when only a finite sample of data is available. Therefore, MI has to be estimated from the dataset [13]. The estimator introduced by Kraskov et al. [21] is based on a k-nearest neighbour method and results from the Kozachenko-Leonenko entropy estimator [20] $\hat{H}(X) = -\psi(k) + \psi(N) + \log c_d + \frac{d}{N}\sum_{i=1}^{N}\log \epsilon_k(i)$, where k is the number of neighbours, N is the number of instances in the dataset, d is the dimensionality, $c_d = (2\pi^{\frac{d}{2}})/\Gamma(\frac{d}{2})$ is the volume of the unitary ball of dimension d, $\epsilon_k(i)$ is twice the distance from the ith instance to its kth nearest neighbour and ψ is the digamma function. Kraskov estimator of the mutual information is then

$$\hat{I}(X;Y) = \psi(N) + \psi(K) - \frac{1}{k} - \frac{1}{N}\sum_{i=1}^{N}(\psi(\tau_x(i)) + \psi(\tau_y(i))) \tag{4}$$

where $\tau_x(i)$ is the number of points located no further than the distance $\epsilon_X(i,k)/2$ from the ith observation in the X space, $\tau_y(i)$ is the number of points located no further than the distance $\epsilon_Y(i,k)/2$ from the ith observation in the Y space and where $\epsilon_X(i,k)/2$ and $\epsilon_Y(i,k)/2$ are the projections into the X and Y subspaces of the distance between the ith observation and its kth neighbour.

When using MI for feature selection, the relationships between several subsets of features and the target Y are computed with a search procedure. Among these subsets, the one maximising the value of $\hat{I}(X;Y)$ (4) is selected.

4.2 Noise Variance

Noise variance is another popular relevance criterion, whose aim is to estimate the level of noise in a finite dataset. In the context of regression, the noise may be considered as the error in estimating the target as a function of the input features, under the hypothesis that a model could be built.

Let us consider a dataset with N instances, d features X_j, a target Y and N input-output pairs $(\mathbf{x}_i, y_i)$. The relationship between these input-output pairs is

$$y_i = f(\mathbf{x}_i) + \epsilon_i \text{ where } i = 1, ..., N \tag{5}$$

where f is the unknown function between $\mathbf{x}_i$ and y_i, and ϵ_i is the noise, or prediction error, when estimating f. The principle is to select the subsets of features which lead to the lowest prediction error, or lowest noise variance [15].

In practice the noise variance has also to be estimated. One widely used estimator is the Delta Test [8,9,28]. The definition of the Delta Test δ is

$$\delta = \frac{1}{2N} \sum_{i=1}^{N} [y_{NN(i)} - y_i]^2 \tag{6}$$

where $y_{NN(i)}$ is the output associated to $x_{NN(i)}$, $x_{NN(i)}$ being the nearest neighbour of the point x_i.

For selecting features with the noise variance, the same procedure as for the mutual information can be used. But instead of selecting the group of features with the highest mutual information estimation $\hat{I}(X;Y)$, the search procedure selects the group of features with the lowest value of the noise variance estimator δ (6).

4.3 R^2 and Adjusted R^2

R^2, also called the coefficient of determination, is the proportion of the variance in the output variable that can be explained from the input variables; it ranges between 0% (unpredictable) and 100% (totally predictable). The definition of R^2 is

$$R^2 = 1 - \frac{SS_{res}}{SS_{tot}} \tag{7}$$

where $SS_{res} = \sum_i (y_i - f(\mathbf{x}_i))^2$ and $SS_{tot} = \sum_i (y_i - \bar{y})^2$ with i=1,...,n. This coefficient is a statistical measure of how well the regression approximates the target. The R^2 measure automatically increases when features are added to the model. This is the reason why we use its alternative, Adjusted R^2, or R^2_{adj}, for feature selection in regression, more suitable for small sample sizes. Its definition is

$$R^2_{adj} = 1 - \frac{SS_{res}/(n - d - 1)}{SS_{tot}/(n - 1)} \tag{8}$$

where d is the number of selected features in the model and n the sample size. A low R^2_{adj} indicates that the data are not close to the fitted regression line. A high R^2_{adj} indicates the opposite.

The R^2_{adj} criterion used with a linear regression model cannot capture the nonlinear relationships between the features and the target. In order to use the R^2_{adj} in a nonlinear context, local linear approximations are considered. In practice, for each feature of the dataset, for each point of the function f, a linear regression is computed with a number of neighbours k from 4 to $(n-1)$. The R^2_{adj} is computed for every regression. For each value of k, an average of the R^2_{adj} on every point of f is computed. The best mean R^2_{adj} is then selected; it corresponds to a specific number of neighbours k. The feature with the highest value of mean R^2_{adj} is then selected. This is the univariate feature selection strategy, the first step of a search method. The multidimensional feature selection strategy can be implemented similarly.

5   Analysis and Comparison

This section analyses the six strategic properties of a relevance criterion for feature selection, given in Sect. 3. In the view of these properties, the mutual information, the noise variance and the Adjusted R^2 are compared.

5.1   Comparison with Property 1: Non-linearity

As explained in Sect. 3, a good relevance criterion must be able to detect non-linear relationships between variables. Mutual information and noise variance are both able to measure nonlinear relationships between variables. Intrinsically, R^2_{adj} only estimates the quality of a linear regression. However, the method used with R^2_{adj}, described in Sect. 4.3, uses local approximations of the regression and is thus suitable for nonlinear relations between the features and the target.

This property is therefore non-discriminant for the three relevance criteria compared in this section. They can all be used with nonlinear relations between variables (features and target).

5.2   Comparison with Property 2: Multivariate Criterion

As shown in their respective equation, mutual information (1), noise variance (5) and R^2_{adj} (8) can all be used to measure the relation between groups of features.

This property is therefore also non-discriminant for the three relevance criteria compared in this section.

5.3   Comparison with Property 3: Estimator Behaviour

The estimators of the three relevance criteria compared in this paper are all based on a k-nearest neighbour method. Therefore, this property is non-discriminant for them, because the time-complexity is approximately the same.

On the other hand, these estimators behave differently in small sample. This is discussed in Sect. 5.5.

5.4   Comparison with Property 4: Estimator Parameters

As explained in Sect. 4, the estimators of the three relevance criteria are all based on a k-nearest neighbour method. Nonetheless, this method is applied differently for each estimator.

The Kraskov estimator has only this k parameter to adjust. Usually it is set to a number between 6 and 8 for good results [5, 21]. The Delta Test sets by definition its k to 1 [8, 28]. Therefore, this estimator does not have any parameter to adjust. With Adjusted R^2, the range of k is much larger, depending on the size of the dataset and the variables.

In view of this property, Adjusted R^2 has the most complex k parameter to adjust and the Delta Test is the easiest to adjust.

5.5 Comparison with Property 5: Estimator in Small Sample

As discussed in [5], Kraskov estimator and the Delta Test suffer from a bias when comparing smooth and non-smooth features, especially in small sample. An overestimation of the noise variance and an underestimation of the mutual information can occur in small datasets when the function to estimate is not smooth. The biases in the estimations are much more severe when using mutual information than when using the noise variance [5,8].

Adjusted R^2 also underestimates non-smooth functions in small datasets and behaves approximately as the Delta Test, in the sense that the minimal size needed to estimate correctly the same nonlinear relation is approximately the same for Delta Test than for Adjusted R^2.

To illustrate this behaviour, experiments have been performed on simple synthetic datasets. Four different periodic functions have been generated with two different frequencies and two levels of noise :

$$y_1 = f_1(\mathbf{x}) = sin(\mathbf{x}) + \epsilon \quad \text{where } \epsilon \sim \text{N}(0,0.05)$$
$$y_2 = f_2(\mathbf{x}) = sin(3\mathbf{x}) + \epsilon \quad \text{where } \epsilon \sim \text{N}(0,0.05)$$
$$y_3 = f_3(\mathbf{x}) = sin(\mathbf{x}) + \epsilon \quad \text{where } \epsilon \sim \text{N}(0,0.3) \tag{9}$$
$$y_4 = f_4(\mathbf{x}) = sin(3\mathbf{x}) + \epsilon \quad \text{where } \epsilon \sim \text{N}(0,0.3)$$

Figures 1(a), (b), (c), (d) represent the four functions f_1, f_2, f_3, f_4, respectively.

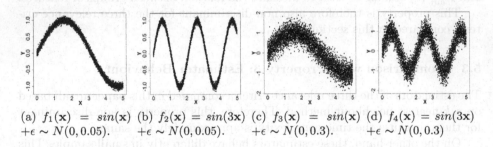

(a) $f_1(\mathbf{x}) = sin(\mathbf{x})$ $+\epsilon \sim N(0, 0.05)$. (b) $f_2(\mathbf{x}) = sin(3\mathbf{x})$ $+\epsilon \sim N(0, 0.05)$. (c) $f_3(\mathbf{x}) = sin(\mathbf{x})$ $+\epsilon \sim N(0, 0.3)$. (d) $f_4(\mathbf{x}) = sin(3\mathbf{x})$ $+\epsilon \sim N(0, 0.3)$

Fig. 1. Experimental data generated with two different frequencies and two levels of noise.

Results are presented in Fig. 2. Figure 2(a) shows that the mutual informa-tion underestimates the non-smooth function f_2 (lower level of noise) over the smooth function f_3 (higher level of noise). Figures 2(b) and (c) show, respec-tively, that the Delta Test and the Adjusted R^2 overestimate the non-smooth function f_2 over the smooth function f_3. Figure 2 also shows that the Delta Test and Adjusted R^2 converge quickly than the mutual information.

5.6 Comparison with Property 6: Estimator Stability

In order to study the estimator stability with respect to the gradient of the relation between the features and the target, illustrative experiments performed

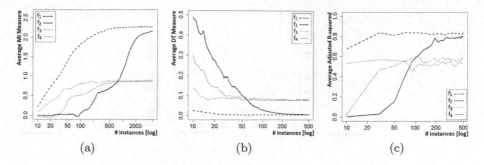

Fig. 2. Average values of (a) MI measures, (b) Delta Test, (c) Adjusted R^2, for two functions with a low level of noise and for two functions with a higher level of noise.

in this paper consider three linear functions with various slopes (Fig. 3). These illustrative experiments are conducted to show the importance of the estimator stability with respect to this gradient and to compare the three criteria described in Sect. 4.

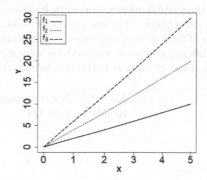

Fig. 3. Experimental data generated with three different slopes.

They have been performed with various sizes of samples, from extremely small to large ones. For each size of the sample, an estimator of the three decision criteria has been used. Results are shown in Fig. 4. The mutual information shows the same results for the three different slopes (Fig. 4(a)), the three measures are superposed, which means that there is no influence of the function slope on its result. The Delta Test shows an influence of the slope of the results in small datasets (Fig. 4(b)). This influence tends to disappear when the size of the datasets sufficiently increases. Adjusted R^2 shows (Fig. 4(c)) no influence of the function slope, the three functions are also superposed, even in small sample scenarios.

For this property, Adjusted R^2 and the mutual information behave better than the Delta Test, in the sense that they offer the same value for the three different functions f_1, f_2 and f_3.

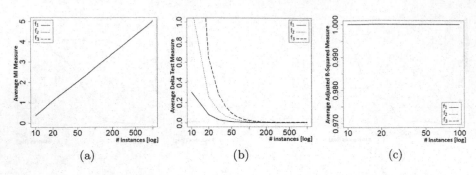

Fig. 4. Average values of MI measures 4(a), Delta Test measures 4(b) and Adjusted R^2 measures 4(c) for three functions f_1, f_2 and f_3, with three different slopes (see Fig. 3).

5.7 Discussion

Table 1 shows a summary of the comparison realised for the three relevance criteria with respect to the six properties presented in this paper.

Table 1. Comparison of the mutual information with Kraskov, the Delta Test and the adjusted R^2. A '+' indicates a good behaviour of the criterion towards this property. A '-' indicates a weakness of the criterion towards this property. The signs '++' or '- -' are only there to show a difference between two criteria with a good (or bad) behaviour towards the property, when one of them is better (or worse) than the other one.

Properties	MI with Kraskov	Noise variance with DT	Adjusted R^2
P1: Non-linearity	+	+	+
P2: Multivariate	+	+	+
P3: Estimator Behaviour	+	+	+
P4: Estimator Parameters	+	+ +	-
P5: Estimator in Small Sample	- -	-	-
P6: Estimator Stability	+	-	+

The three filter criteria proposed in Sect. 4 all respect the two first properties, which make them good candidates for feature selection. The four last properties help to decide between the three criteria, depending on the dataset and the problem at hand. Indeed when comparing the three criteria with the fourth property (P4), the Delta Test does not have any parameter to adjust, which makes it easier to use with respect to the mutual information and the adjusted R^2. In a small sample scenario (P5), the adjusted R^2 seems to behave as the Delta Test for non-smooth functions, which is better than the mutual information. Finally when comparing the criteria with the sixth property (P6), the adjusted R^2 and the mutual information are more stable with respect to the gradient of the function between the features and the target than the Delta Test.

6 Conclusions

This paper proposes six strategic properties of a good relevance criterion for feature selection in regression: two properties for the relevance criterion itself and four properties for the estimator of the relevance criterion. To illustrate the importance of these properties, this paper describes three interesting relevance criteria and compares them with the aforementioned properties. Any relevance criterion used for filters in feature selection could be analysed in the light of these properties.

References

1. Battiti, R.: Using mutual information for selecting features in supervised neural net learning. IEEE Trans. Neural Netw. **5**, 537–550 (1994)
2. Bing, X., Mengjie, Z., Will, N., B., Xin, Y.: A survey on evolutionary computation approaches to feature selection. IEEE Trans. Evol. Comput. 20(4), 606–626 (2016)
3. Brown, G., Pocock, A., Zhao, M., Lujan, M.: Conditional likelihood maximisation: a unifying framework for mutual information feature selection. J. Mach. Learn. Res. **13**, 27–66 (2012)
4. Degeest, A., Verleysen, M., Frénay, B.: Feature ranking in changing environments where new features are introduced. In: 2015 International Joint Conference on Neural Networks (IJCNN), pp. 1–8, July 2015
5. Degeest, A., Verleysen, M., Frénay, B.: Smoothness bias in relevance estimators for feature selection in regression. In: 2018 International Conference on Artificial Intelligence Applications and Innovations (IJCNN), pp. 285–294 (2018)
6. Doquire, G., Verleysen, M.: A comparison of multivariate mutual information estimators for feature selection. In: Proceeding of ICPRAM 2012 (2012)
7. Efron, B., Hastie, T., Johnstone, I., Tibshirani, R.: Least angle regression. Ann. Stat. **32**, 407–499 (2004)
8. Eirola, E., Lendasse, A., Corona, F., Verleysen, M.: The delta test: the 1-nn estimator as a feature selection criterion. In: Proceedings of the 2014 International Joint Conference on Neural Networks (IJCNN), pp. 4214–4222, July 2014
9. Eirola, E., Liitiäinen, E., Lendasse, A., Corona, F., Verleysen, M.: Using the delta test for variable selection. In: Proceedings of ESANN 2008 (2008)
10. François, D., Rossi, F., Wertz, V., Verleysen, M.: Resampling methods for parameter-free and robust feature selection with mutual information. Neurocomputing 70(7–9), 1276–1288 (2007)
11. Frénay, B., Doquire, G., Verleysen, M.: Theoretical and empirical study on the potential inadequacy of mutual information for feature selection in classification. Neurocomputing **112**, 64–78 (2013)
12. Frénay, B., van Heeswijk, M., Miche, Y., Verleysen, M., Lendasse, A.: Feature selection for nonlinear models with extreme learning machines. Neurocomputing **102**, 111–124 (2013)
13. Gao, W., Kannan, S., Oh, S., Viswanath, P.: Estimating mutual information for discrete-continuous mixtures. In: Guyon, I., et al. (eds.) Advances in Neural Information Processing Systems, vol. 30, pp. 5986–5997. Curran Associates Inc, Red Hook (2017)

14. Gómez-Verdejo, V., Verleysen, M., Fleury, J.: Information-theoretic feature selection for functional data classification. Neurocomputing **72**(16–18), 3580–3589 (2009)
15. Guillén, A., Sovilj, D., Mateo, F., Rojas, I., Lendasse, A.: New methodologies based on delta test for variable selection in regression problems. In: Workshop on Parallel Architectures and Bioinspired Algorithms, Toronto, Canada (2008)
16. Guyon, I., Elisseeff, A.: An introduction to variable and feature selection. J. Mach. Learn. Res. **3**, 1157–1182 (2003)
17. Hancer, E., Xue, B., Zhang, M.: Differential evolution for filter feature selection based on information theory and feature ranking. Knowl.-Based Syst. **140**, 103–119 (2018)
18. Karegowda, A.G., Jayaram, M.A., Manjunath, A.S.: Feature subset selection problem using wrapper approach in supervised learning. Int. J. Comput. Appl. **1**(7), 13–17 (2010)
19. Kohavi, R., John, G.H.: Wrappers for feature subset selection. Artif. Intell. **97**, 273–324 (1997)
20. Kozachenko, L.F., Leonenko, N.: Sample estimate of the entropy of a random vector. Probl. Inform. Transm. **23**, 95–101 (1987)
21. Kraskov, A., Stögbauer, H., Grassberger, P.: Estimating mutual information. Phys. Rev. E **69**, 066138 (2004)
22. Li, J., et al.: Feature selection: a data perspective. ACM Comput. Surv. **50**(6), 94:1–94:45 (2017). https://doi.org/10.1145/3136625
23. Paul, J., D'Ambrosio, R., Dupont, P.: Kernel methods for heterogeneous feature selection. Neurocomputing **169**, 187–195 (2015)
24. Schaffernicht, E., Kaltenhaeuser, R., Verma, S.S., Gross, H.-M.: On estimating mutual information for feature selection. In: Diamantaras, K., Duch, W., Iliadis, L.S. (eds.) ICANN 2010. LNCS, vol. 6352, pp. 362–367. Springer, Heidelberg (2010). https://doi.org/10.1007/978-3-642-15819-3_48
25. Shannon, C.E.: A mathematical theory of communication. Bell Syst. Tech. J. **27**(379–423), 623–656 (1948)
26. Vergara, J.R., Estévez, P.A.: A review of feature selection methods based on mutual information. Neural Comput. Appl. **24**, 175–186 (2014)
27. Verleysen, M., Rossi, F., François, D.: Advances in feature selection with mutual information. In: Biehl, M., Hammer, B., Verleysen, M., Villmann, T. (eds.) Similarity-Based Clustering. LNCS (LNAI), vol. 5400, pp. 52–69. Springer, Heidelberg (2009). https://doi.org/10.1007/978-3-642-01805-3_4
28. Yu, Q., Séverin, E., Lendasse, A.: Variable selection for financial modeling. In: Proceedings of the CEF 2007, 13th International Conference on Computing in Economics and Finance, Montréal, Quebec, Canada, pp. 237–241 (2007)

Bistable Sigmoid Networks

Stanislav Uschakow, Jörn Fischer$^{(\boxtimes)}$ (iD), and Thomas Ihme$^{(\boxtimes)}$ (iD)

Mannheim University of Applied Sciences,
Paul-Wittsack-Str. 10, 68163 Mannheim, Germany
{j.fischer,t.ihme}@hs-mannheim.de,
http://services.informatik.hs-mannheim.de/~fischer/index.html,
http://services.informatik.hs-mannheim.de/~ihme/index.html

Abstract. It is commonly known that Hopfield Networks suffer from spurious states and from low storage capacity. To eliminate the spurious states Bistable Gradient Networks (BGN) introduce neurons with bistable behavior. The weights in BGN are calculated in analogy to those of Hopfield Networks, associated with Hebbian learning. Unfortunately, those networks still suffer from small storage capacity, resulting in high reconstruction errors when used to reconstruct noisy patterns. This paper proposes a new type of neural network consisting of neurons with a sigmoid hyperbolic tangent transfer function and a direct feedback. The feedback renders the neuron bistable. Furthermore, instead of using Hebbian learning which has some drawbacks when applied to overlapped patterns, we use the first order Contrastive Divergence (CD^1) learning rule. We call these Networks Bistable Sigmoid Networks (BSN). When recalling patterns from the MNIST database the reconstruction error is zero even for high load providing no noise is applied. For an increasing noise level or an increasing amount of patterns the error rises only moderate.

Keywords: Recurrent neural network · Associative memory · Contrastive divergence · Hopfield network

1 Introduction

Restoring known patterns from noisy or incomplete information is a common task in signal processing. Especially in image classification a slightly different angle on the same object can be thought of as incomplete or noisy data regarding the trained image.

In 1982 John Hopfield introduced the so-called Hopfield Networks [4], a neural model developed from analogies in physical spin glasses. In these networks neurons are coupled symmetrically without self recurrent connection allowing the network to restore known patterns from a sufficient similar input. Using Hebbian learning to adjust the symmetric weight matrix, these networks achieve a storage capacity of $0.14N$, where N is the number of neurons. Trying to store

© Springer Nature Switzerland AG 2019
I. Rojas et al. (Eds.): IWANN 2019, LNCS 11507, pp. 591–600, 2019.
https://doi.org/10.1007/978-3-030-20518-8_49

more patterns results in sporadic spurious states. The output of the network does not correspond to any stored pattern.

In [2,5] Chinarov and Menzinger introduce a Hopfield like network called Bistable Gradient Network. Most of the characteristics like the neighbor coupling, the symmetric weight matrix and Hebbian learning remain as in Hopfield Networks. The only difference is the type of neurons to be used. Whereas Hopfield networks use perceptrons, the Bistable Gradient Network uses a novel artificial neuron, characterized by a bistable behavior of each element. To derive the differential equation we start with the energy function of such a neuron which may be defined as:

$$V(x_i) = V_{bistable} + V_{couple}$$
$$= \left(-\frac{x_i^2}{2} + \frac{x_i^4}{4} \right) + \left(-\gamma \sum_j w_{ij} x_i x_j + b_i \right) \tag{1}$$

The partial derivative to x_i then leads to:

$$\frac{dx_i}{dt} = -\frac{\partial V(x_i)}{\partial x_i} = x_i - x_i^3 + \gamma \sum_{j=1}^{N} w_{ij} x_j + b_i \tag{2}$$

where:

$V_{bistable}$ − bistable part of the neuron
$V_{coupling}$ − coupling part among the coupled neurons
x_i − output of neuron i
γ − coupling strength
w_{ij} − connection weight between neuron i and j
b_i − bias or external input of the neuron i

This equation consists of two parts. The first part $V_{bistable}$ describes the bistable behavior. The second part $V_{coupling}$ especially the weights and the parameter for the coupling strength γ determine the influence of the neighbored neurons on the dynamic behavior. In this context bistable behavior means the output of the neuron remains stable unless a sufficiently high external input is applied. Given such an external input some neurons output are flipped into the other stable state and remain there. This feature widely eliminates the spurious states known from Hopfield Networks and increases the storage capacity for orthogonal patterns. The downside of Bistable Gradient Networks is the novel type of neuron which is far from biologically plausible and the Hebbian Learning rule which has limitations when learning strongly overlapped patterns like the digits from the MNIST database [3].

The rest of the article is organized as follows: The next section introduces Bistable Sigmoid Networks, Sect. 3 shows experimental results, followed by conclusion and further work.

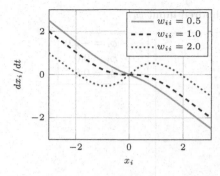

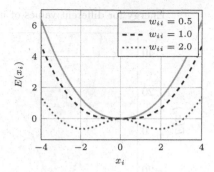

(a) A phase diagram of a neuron with a hyperbolic tangent activation function and a direct feedback of its output to its input. The figure shows that for lower values of w_{ii} only one zero crossing exists, while for $w_{ii} = 2$ three zero crossings appear, two of them marking stable fixed points.

(b) For $w_{ii} = 2$ the energy function $V(x_i)$ has two minima, where the activity of the neuron converges to. The minima correspond to the zero crossings of the left figure and explain the bistability of this neuron with direct feedback.

Fig. 1. Behavior of a single sigmoid neuron

2   Bistable Sigmoid Networks

We propose a novel neural network called Bistable Sigmoid Network. To achieve bistable behavior each neuron with a sigmoid activation function (hyperbolic tangent) uses a direct feedback. The Contrastive Divergence learning rule (CD^1) [1] is used to further increase the storage capacity. Although Hinton applies this learning rule mainly to Restricted Bolzmann Machines it is shown in [3] that a recurrent neural network with bistable elements, in that case the Bistable Gradient Network, can be trained successfully using the CD^1 learning rule:

$$\Delta w_{ij} = \eta(<x_i x_j>_0 - <x_i x_j>_1) \tag{3}$$

with

Δw_{ij} − weight change
η − learning rate
x_i − is the output of neuron i
x_j − is the output of neuron j
$<x_i x_j>_0$ − is $x_i x_j$ before the last activation
$<x_i x_j>_1$ − is $x_i x_j$ after the last activation

As in Hopfield Networks and Bistable Gradient Networks the weight matrix of a Bistable Sigmoid Network is symmetric, but has a non zero diagonal. The dynamic behavior of the output of a single neuron can be described with Eq. (4). In analogy to the dynamics of the Bistable Gradient Networks this equation has two parts. The first part $w_{ii} \tanh(x_i) - x_i$ is the bistable term and the second

Energy for different values of neighbored input ϕ_i.

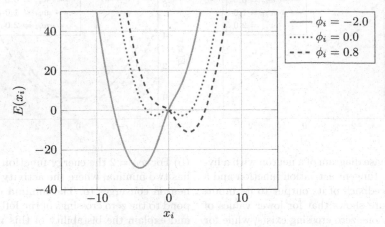

Fig. 2. The energy $E(x_i)$ shown for different values of ϕ_i and $w_{ii} = 2$. We can observe that for values of $\phi_i < -0.5325$ the energy is minimal at $x_{min} < 0$, for values of $\phi_i > 0.5325$ the energy is minimal at $x_{min} > 0$. Only if $0.5325 > \phi_i > -0.5325$, there are two energy minima, one where $x_{min}^1 > 0$, one where $x_{min}^2 < 0$.

one $\phi_i = \sum_{j=1, j \neq i}^{N} w_{ij} x_j$ is the coupling term. The time constant τ determines the size of the time step. A large τ means a small time step, as a small τ means a large time step and the risk to end up in unwanted oscillation.

The differential equation for the change of the neurons activity is:

$$\frac{dx_i}{dt} = \tau \cdot (w_{ii} \tanh(x_i) - x_i + \phi_i) \tag{4}$$

$$\text{with} \quad \phi_i = \sum_{j=1, j \neq i}^{N} w_{ij} x_j + b_i$$

where:

x_i – output of the neuron i
τ – time constant
w_{ii} – feedback weight
γ – coupling strength
ϕ_i – input of neighbor neurons
b_i – bias or external input of the neuron i

Equation (5) shows the energy as the negative integral:

$$E(x_i) = -\int \frac{dx_i}{dt} dx_i = \tau \left[\frac{x_i^2}{2} - w_{ii} \log(\cosh(x_i)) + \phi_i x_i \right] \tag{5}$$

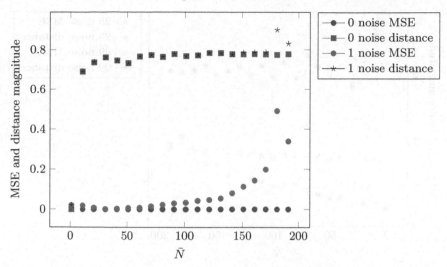

Fig. 3. Showing the mean square error and mean distance for $\bar{N} \in [1, 196]$ and zero respectively one noise pixel applied. Without any noise the error remains zero even for many stored patterns. With one random noise pixel the error becomes more significant with higher load ($\tau = 3.05$, $w_{ii} = 4.15$, $\epsilon = 0.05$).

If the self coupling w_{ii} of neuron i exceeds a specific value, the neuron gets bistable. Figures 1a and 1b show the dynamics and the energy of the neuron for different values of the feedback weight $w_{ii} \in \{0.5, 1.0, 2.0\}$. We set $\tau = 1$ and $\phi_i = 0$.

The neuron reaches a stable state, if its change in time is zero. In this case the energy of the neuron is minimal. For bistability we need two distinct points or energy wells where the output of the neuron can converge to. In Fig. 1a we see those points as zero crossings. We can observe the influence of the feedback weight on the dynamics and of the energy figure. For $w_{ii} > 1$ three distinct zero crossings in Fig. 1a and the corresponding energy extrema in Fig. 1b emerge. The points of minimal energy correspond to the stable points, the local maximum is referred to as the instable point. For the output to flip from one stable state to the other, a sufficiently high external stimulus ϕ_i must be provided, otherwise the neuron remains in its current state. Is the stimulus sufficiently high for a transit, the output of the neuron reaches its opposite state and remains there, even if no further stimulus is provided. Geometrically the stimulus shifts the curve in Fig. 1b along the y axis. If the stimulus is high enough for a transit, only one zero crossing exists thus only one state of minimal energy remains.

The parameter τ has to be chosen high enough, because a small τ is equivalent to a bad discretization with too big time steps. Too small τ then leads to an

Error and distance plot for variable $\bar{N}$

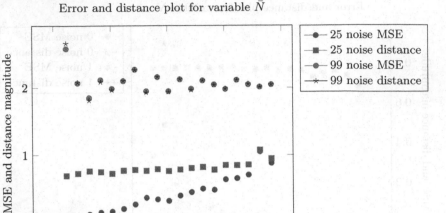

Fig. 4. Showing the mean square error and mean distance for $\bar{N} \in [1, 196]$ and 25 respectively 99 noise pixel applied. For 25 noise pixel with increasing amount of stored pattern the MSE slowly approaches the distance whereas with 99 noise pixel applied the output immediately becomes equal to the distance and hence useless for further processing like classification($\tau = 3.05$, $w_{ii} = 4.15$, $\epsilon = 0.05$).

oscillatory behavior. The feedback weight w_{ii} is responsible for the bistability of the neurons. Chosen too small, the neuron remains monostable.

Figure 2 shows the energy for different values of ϕ_i with fixed $w_{ii} = 2$ and $\tau = 4.4$. w_{ii} and τ are chosen experimentally to demonstrate the desired behavior for these parameters. Other configurations are also feasible. It is shown that increasing/decreasing the stimulus causes one potential well to weaken whereas the other extends until only one well remains. With $\phi_i = -2.0$ the overall minimum of x_i is located at a negative value, whereas with $\phi_i = +0.8$ the overall minimum of x_i is located at a positive value. When ϕ_i turns back to zero, the neuron remains in this state due to the two local minima. The neuron will switch into the opposite state, if a stimulus with the opposite sign is applied. Which well is weakened depends on the sign of the stimulus.

3   Experiments and Results

In this chapter we discuss a network build from bistable sigmoid neurons, the Bistable Sigmoid Network. We test its capacity and its capability to recall noisy patterns. Since all neurons are interconnected it is sufficient to initialize its output with the pattern to learn and activate the layer. Lastly the first order Contrastive Divergence learning rule (CD^1) is applied to adapt the weights. Hereby

Error and distance plot for variable noise with 21 stored pattern

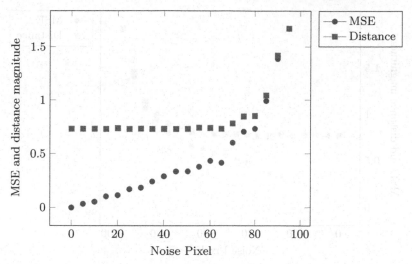

Fig. 5. This figure demonstrates the impact of an increasing noise level from zero noise pixel up to 99 noise pixel for 21 stored patterns. The MSE rises moderate with the number of noise pixel ($\tau = 3.05$, $w_{ii} = 4.15$, $\epsilon = 0.05$).

the feedback weight w_{ii} is not changed. When recalling a pattern the learning rule is not applied, while the activation remains the same.

To evaluate the performance of the network we calculate the sum of the mean square error of each pixel. It is calculated from the difference between the produced output and the desired output which is the original MNIST pattern. To show uniqueness of the output the mean square of the difference between the original and all other patterns, hereinafter referred to as distance, is also calculated. If this distance is of the same size as the mean square error, the output has lost all of its information and does not correlate to the desired output.

For testing we use the MNIST database for handwritten numbers with a reduced resolution of 14×14 pixel To calculate the output of the Eq. (4) we run 10 recursive iterations before proceeding. For the remaining parameters we choose $\tau = 3.05$, $w_{ii} = 4.15$ and for the learning rate $\epsilon = 0.05$. The amount of noise and of the stored patterns are varying. In addition we discovered that learning a pattern several times before continuing with the next pattern reduces the recall error. Therefore we repeat the learning step for 5 times for a single pattern. The learning process for all patterns is repeated for $100 \cdot \bar{N}$ where $\bar{N}$ equals to the number of patterns stored to further improve the properties of our network.

Figure 3 shows the MSE and distance for a network where up to 196 patterns are stored. For each stored pattern zero or one noise pixel is applied to each image. We can observe that recalling the original data with zero noise, the network does not produce any recall errors. Only if we apply one error pixel a larger error

Error and distance plot for variable noise with 51 stored pattern

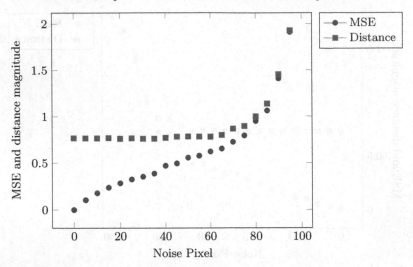

Fig. 6. This figure demonstrates the impact of an increasing noise level from zero noise pixel up to 99 noise pixel for 91 stored patterns. In contrast to Fig. 5 the error is a little bit higher for the same amount of noise ($\tau = 3.05$, $w_{ii} = 4.15$, $\epsilon = 0.05$).

appears for higher loads. Comparing this with Fig. 4 the difference is even more significant. Here the network produces errors even with low loads for 25 error pixel eventually reaching the distance. When the difference of the MSE and distance of the patterns is small the output of the network cannot be differentiated. For 99 error pixel the output immediately becomes indistinguishable.

The next three Figs. 5, 6 and 7 show the impact of the increasing amount of noise for a fixed amount of stored patterns. The network in Fig. 5 stores 21 patterns. Adding noise causes the output error to grow slowly and linearly eventually reaching the distance magnitude. As mentioned before, when the output error reaches the distance magnitude the matching of the output to the corresponding original data becomes next to impossible. Figure 6 shows the error and distance for 51 stored patterns. When comparing this figure with Fig. 5 we can observe that the error increases faster compared to the figure with 21 patterns stored. Still for a small amount of noise the reconstruction error is low enough for a potential classification.

Finally, Fig. 7 shows an extreme case where 191 patterns are stored. Here the MSE becomes almost as high as the distance even with little noise applied. More noise results in a completely indistinguishable output.

Error and distance plot for variable noise with 191 stored pattern

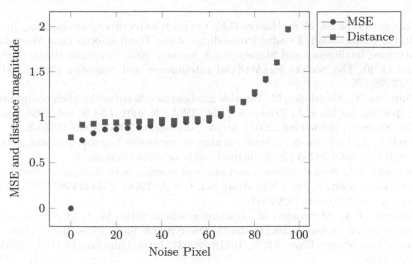

Fig. 7. This figure shows the impact of an increasing noise level from zero noise pixel up to 99 noise pixel for 191 stored patterns. Even a small amount of noise causes the error to be almost the same as the distance to all other patterns thus rendering the output of the network indistinguishable.

4   Conclusion

In this paper a new kind of Hopfield like Neural Network called Bistable Sigmoid Network is presented. It has neurons with sigmoid transfer function and recurrent self connection, which transforms each neuron into a bistable element. We show that Bistable Sigmoid Networks have a storage capacity of up to N patters in a network of N neurons. This means that up to N minima are learned where only noise or an external state change may bring the state out of these minima. The overall capacity of the BSN mainly depends on the noise level applied. Moreover with increasing number of patterns to be stored the reconstruction error rises while a constant amount of noise is applied. So for each application one needs to estimate the expected noise level to choose the network size according to the storage capacity required.

5   Further Work

To further improve the performance of the BSN the feedback weights w_{ii} can also be trained dynamically. The feedback weight determines how much external input must be provided for the neurons output to change its state. This external input is highly dependent on the type of data to be learned. A dynamic learning of the feedback weight eliminates the manual adjustment of this parameter and can hopefully find an even better solution.

References

1. Carreira-Perpiñán, M.A., Hinton, G.E.: On contrastive divergence learning. In: Cowell, R., Ghahramani, Z. (eds.) Proceedings of the Tenth International Workshop on Artificial Intelligence and Statistics, 6-8 January 2005, Savannah Hotel, Barbados, pp. 33–40. The Society for Artificial Intelligence and Statistics (2005). ISBN 0-9727358-1-X
2. Chinarov, V., Menzinger, M.: Bistable gradient neural networks: their computational properties. In: Mira, J., Prieto, A. (eds.) IWANN 2001. LNCS, vol. 2084, pp. 333–338. Springer, Heidelberg (2001). https://doi.org/10.1007/3-540-45720-8_38
3. Fischer, J., Lackner, S.: About learning in recurrent bistable gradient networks. CoRR abs/1608.08265 (2016), https://arxiv.org/abs/1608.08265
4. Hopfield, J.J.: Neural networks and physical systems with emergent collective computational abilities. Proc. Nat. Acad. Sci. U.S.A. **79**(8), 2554–2558 (1982). https://doi.org/10.1073/pnas.79.8.2554
5. McGraw, P.N., Menzinger, M.: Bistable gradient networks. I. Attractors and pattern retrieval at low loading in the thermodynamic limit. Phys. Rev. E Stat. Nonlinear Soft Matter Phys. **67**(2), 16118 (2003). https://doi.org/10.1103/PhysRevE.67.016118

Validation of Unimodal Non-Gaussian Clusters

Luis F. Lago-Fernández[✉], Jesús Aragón, and Manuel Sánchez-Montañés

Escuela Politécnica Superior, Universidad Autónoma de Madrid, 28049 Madrid, Spain
{luis.lago,manuel.smontanes}@uam.es, jesus.aragon@estudiante.uam.es

Abstract. We analyze the influence of the cluster shape on the performance of four cluster validation criteria: AIC, BIC, ICL and NI. First we introduce a method to generate unimodal and radially symmetric clusters whose shape can be interpolated between peaky long-tailed and flat distributions using a single parameter. Normally distributed clusters are obtained as a special case. Then we systematically study the performance of AIC, BIC, ICL and NI when validating clusters of arbitrary shapes. Using problems with two clusters, different inter-cluster distances and different dimensions, we show that, while BIC provides the best results for normally distributed clusters, in a general context with high dimensional data and unknown cluster distributions the use of ICL or NI may be a better choice.

Keywords: Clustering · Cluster validation · Gaussian mixture model

1 Introduction

A common approach to data clustering is based on the assumption of a probabilistic model, usually a mixture of densities, underlying the data. The mathematical structure of these component distributions is assumed to be known (the Gaussian is a common choice) but the distribution parameters must be found, for example by maximizing the log-likelihood using the Expectation-Maximization (EM) algorithm [1]. Once the model parameters have been fitted, clustering may be performed by associating each data point x to the mixture component c that maximizes the posterior probability $p(c|x)$.

A problem with this approach is how to determine the correct number n_c of mixture components. The log-likelihood can not be used for this purpose as it monotonically increases with n_c. In fact, assessing the number of components in a mixture model turns out to be a very difficult problem which, in spite of having received much attention in the literature [2–6], has not been completely resolved. Popular strategies make use of validation criteria that correct the likelihood by adding a term that penalizes the model complexity. Two well known criteria are the Akaike's Information Criterion (AIC) [7] and the Bayesian Information Criterion (BIC) [8,9]. A number of works have explored these and similar strategies

© Springer Nature Switzerland AG 2019
I. Rojas et al. (Eds.): IWANN 2019, LNCS 11507, pp. 601–611, 2019.
https://doi.org/10.1007/978-3-030-20518-8_50

from the point of view of cluster validation [10–13], but in general the evaluation of validation metrics is performed with datasets where the (real) clusters are
generated using Gaussian distributions [14,15]. This introduces a bias since the
same kind of distribution is being assumed for the mixture components. Hence
the observed results might not extrapolate to a real situation where the data
clusters are not necessarily normally distributed.

A few studies tackle the cluster validation problem from this perspective.
The Integrated Classification Likelihood (ICL) [16,17], which is essentially the
BIC criterion penalized by subtraction of the estimated partition mean entropy,
has been shown to outperform AIC and BIC when the focus is clustering rather
than density estimation. It provides a better estimation of the number of mixture
components when the distribution followed by the real clusters does not match
that assumed for the mixture model [16]. Another example is the Negentropy
Increment (NI) [18,19], which is aimed at finding well separated and approximately Gaussian clusters. The results reported in [20] show that the NI is able
to assess the number of clusters more accurately than AIC, BIC or ICL when
the clusters are unimodal but not exactly Gaussian. Although these results are
interesting, the methods were tested on very specific problems, which makes
it difficult to extract general conclusions on how the cluster shape affects the
different validation metrics.

In this article we perform a systematic analysis of the influence of the cluster shape on the outcomes of different validation criteria. We first introduce a
method to generate unimodal and radially symmetric clusters whose degree of
Gaussianity depends on a single shape parameter k. Then we analyze the results
obtained by AIC, BIC, ICL and NI on problems with two clusters as we vary
the cluster shape and the inter-cluster distance, in dimensions ranging between
2 and 5. Our results show that, while BIC provides the best results for Gaussian
clusters, its performance degrades as the clusters move away from Gaussianity.
This effect is more noticeable as the dimension increases. In a general case with
high dimensional data and where the cluster distributions are unknown, the use
of ICL or NI may be a better choice.

2   Cluster Generation

In this section we describe the procedure to generate a cluster of data points. We
assume that the dimension of the space is d, that the cluster is centered around
the origin, and that the number of points to be generated is n. Any cluster
distribution will be unimodal, isotropic and with identity covariance, but its
radial coordinate will depend on a specific shape parameter, k, that determines
its degree of normality. We have Gaussian clusters for $k = 2$. When $k < 2$ we
obtain supergaussian clusters (with a sharper peak and longer tails than the
normal distribution), whilst for $k > 2$ we have subgaussian clusters (flatter than
the normal distribution). Figure 1 shows some examples for $d = 2$.

To generate a cluster, we first obtain n points uniformly distributed on the
surface of a sphere of unit radius by drawing each component from a normal

distribution $N(0,1)$ and normalizing each vector to unit length. Then we multiply each vector by the radial component r, which is drawn from a generalized gamma distribution:

$$f(r) = \frac{k}{\Gamma(d/k)} \frac{r^{d-1}}{a^d} \exp(-(r/a)^k) \tag{1}$$

where $\Gamma(\cdot)$ is the gamma function and the parameter a is chosen so that the covariance matrix is the identity:

$$a = \sqrt{\frac{d \cdot \Gamma(\frac{d}{k})}{\Gamma(\frac{d+2}{k})}} \tag{2}$$

It is not difficult to show that the resulting set of points is distributed according to:

$$\rho(\mathbf{x}) = \frac{\Gamma(d/2)}{\Gamma(d/k)} \frac{k}{2(\pi a^2)^{d/2}} \exp(-(r/a)^k) \tag{3}$$

where $r(\mathbf{x}) = \sqrt{x_1^2 + ... + x_d^2}$. We can recognize the following special cases:

Gaussian cluster (k = 2): Substituting $k = 2$ in Eqs. 2 and 3 we obtain:

$$\rho(\mathbf{x}) = \frac{1}{(2\pi)^{d/2}} \exp(-r^2/2) \tag{4}$$

This is the pdf of a d-variate normal distribution centered at the origin and with covariance matrix $\Sigma = I$. An example is shown in Fig. 1 (third column).

Uniformly distributed cluster (k → ∞): In the limit $k \to \infty$ we have $a = \sqrt{d+2}$ and

$$\rho(\mathbf{x}) = \Gamma(d/2)\frac{d}{2(\pi(d+2))^{d/2}} \tag{5}$$

when $r < a$, and $\rho(\mathbf{x}) = 0$ otherwise. That is, the points are uniformly distributed in the volume limited by a $d - 1$ sphere of radius a centered at the origin. It is easy to show that the expression in Eq. 5 is the inverse of the volume of such a sphere, which ensures normalization of the pdf. Figure 1 (last column) shows an approximation to this distribution using $k = 200$.

Gamma cluster (k = 1): As a last example, when $k = 1$ we have:

$$f(r) = \frac{r^{d-1} \exp(-r/a)}{a^d \Gamma(d)} \tag{6}$$

In this case the radial coordinate follows a gamma distribution with shape parameter given by the dimension d and scale parameter equal to a (see Fig. 1, second column).

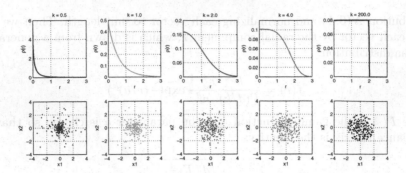

Fig. 1. Top row: $\rho(r)$ versus r for dimension $d = 2$ and different values of the parameter k. Bottom row: random samples of 200 points drawn from each of the distributions in the top row.

3   Experiments

3.1   Problem Definition

For our experiments we generate a set of problems with two clusters distributed according to the previous distributions. The first cluster is centered at the origin, and the second one is shifted a distance b along the first coordinate axis. The inter-cluster distance b takes values between 0 and 5 at intervals of 0.2. The exponent k that characterizes the clusters' shape takes values between 0.2 and 4.0 at intervals of 0.2. We consider dimensions from 2 to 5, and for each configuration ($b \times k \times d$) we generate 40 different random problems. This makes a total of 20800 different problems for each d. The number of points in each cluster depends on the dimension as $n = 10^d$.

3.2   Clustering Approach

We follow a model-based clustering approach, where we use the EM algorithm to fit a mixture of Gaussian components to the problem data and assume that each component is related to a single cluster. For each problem we try different number of components n_c, between 2 and 5. Given the problem and the number of components, 20 different runs of the algorithm are performed, each starting from a different random initial condition. This makes a total of 81 solutions (80 plus the trivial one with $n_c = 1$) for each problem. The optimal solution is then selected according to different validation criteria as explained in the next section.

3.3   Cluster Validation

The first step is to find the optimal solution for each number of components, $\hat{\beta}(n_c) = \min \beta(n_c, i)$, where β is one of the validation criteria described below and the index $i = 1, ..., 20$ represents one of the executions of the EM algorithm for a particular n_c[1]. Then, starting from $n_c = 1$, we compare $\hat{\beta}(n_c)$ to $\hat{\beta}(n_c + 1)$. Only if $\hat{\beta}(n_c + 1) < \hat{\beta}(n_c)$ we proceed to test $\hat{\beta}(n_c + 1)$ against $\hat{\beta}(n_c + 2)$, otherwise the solution with n_c components is selected. If we get to a point where $\hat{\beta}(5) < \hat{\beta}(4)$, we select the solution with $n_c = 5$ as no run of the EM algorithm with $n_c = 6$ has been performed.

The following validation criteria are tested:

Akaike's Information Criterion (AIC). Akaike's criterion [7] measures the goodness of fit of a mixture model by its log-likelihood ($\log L$) corrected with a term that measures the model complexity using the number of parameters in the model (m):

$$\beta_{AIC} = -2 \log L + 2m \tag{7}$$

Bayesian Information Criterion (BIC). The Bayesian Information Criterion [8] is a different correction to the log-likelihood of the model:

$$\beta_{BIC} = -2 \log L + m \log n \tag{8}$$

where m is, as before, the number of model parameters, and n is the number of data points to be fitted.

Integrated Classification Likelihood (ICL). The Integrated Classification Likelihood [16,17] is a modification of the BIC criterion that takes into account the overlap amongst the mixture components. It has been shown to outperform BIC and AIC when the focus is clustering rather than density estimation. In its simplest form, ICL corrects BIC by subtracting a term that measures the mean entropy of the clustering partition:

$$\beta_{ICL} = -2 \log L + m \log n + 2h \tag{9}$$

where the entropy term h is given by:

$$h = -\sum_{i=1}^{n} \sum_{j=1}^{n_c} p_{ij} \log p_{ij} \tag{10}$$

and p_{ij} are posterior probabilities.

Negentropy Increment (NI). The Negentropy Increment of a crisp clustering partition is given by [18]:

$$\beta_{NI} = \frac{1}{2} \sum_{j=1}^{n_c} p_j \log |\Sigma_j| - \sum_{j=1}^{n_c} p_j \log p_j \tag{11}$$

---

[1] Note that $\hat{\beta}(1) = \beta(1)$ as there is only one trivial solution for $n_c = 1$.

where p_j is the prior probability and Σ_j is the covariance matrix of cluster j[2]. Note that this criterion is developed for crisp clustering, so in order to apply it in the present context it is necessary to generate a crisp partition from the outcome of the EM algorithm. We do this by assigning each data point x_i to a single cluster associated to the mixture component j that maximizes the posterior probability p_{ij}.

In [19] a correction to the negentropy increment was introduced to take into account the bias in the evaluation of $\log|\Sigma_j|$ when the number of points in cluster j is too small. Here we follow the same approach.

4   Results

We compute the number of components in the optimal solution provided by each validation criterion for each of the problems. This number is averaged over the 40 problems solved for a given configuration ($b \times k \times d$). Ideally, the average number of components should range between 1 (for problems with small inter-cluster distance where the two clusters are highly overlapped) and 2 (for problems where

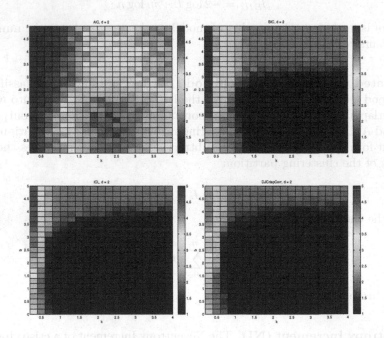

Fig. 2. Average number of components selected by AIC (top-left), BIC (top-right), ICL (bottom-left) and NI (bottom-right) over 40 problems with given cluster distribution (k parameter, x-axis) and inter-cluster distance (b parameter, y-axis). The dimension is $d = 2$.

---

[2] In the original formulation in [18] there is an additional constant term that has been omitted here for the sake of simplicity. This term depends only on the covariance matrix of the entire data set, which is constant for any particular problem.

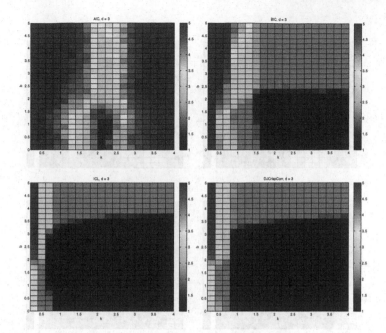

Fig. 3. Average number of components selected by AIC (top-left), BIC (top-right), ICL (bottom-left) and NI (bottom-right) over 40 problems with given cluster distribution (k parameter, x-axis) and inter-cluster distance (b parameter, y-axis). The dimension is $d = 3$.

the two clusters are well separated). We expect that this is true at least for approximately Gaussian clusters ($k \approx 2$). The results of our experiments are shown in Figs. 2, 3, 4 and 5 (one for each dimension d). They show surface plots where k is represented in the x-axis, b is represented in the y-axis, and the average number of components is shown as a color code. In view of these plots, the following conclusions may be extracted.

First, the performance of AIC is very poor for all the problems. It over-estimates the number of components for almost all the tested configurations, providing an acceptable solution only for a narrow interval around the Gaussian case ($k = 2$) for dimensions 4 and 5.

Second, BIC produces good results for lower dimensions ($d = 2, 3$), providing valid solutions (with one cluster for low b and two clusters for high b) for almost the full range of ks. It is only for very small k values that BIC clearly deviates from this ideal behavior, overestimating the number of clusters. For higher dimensions, however, a noticeable reduction of the interval with valid solutions is observed. This interval reduces to a narrow window around $k = 2$ for $d = 5$. On the other hand, BIC is the criterion that allows the highest overlap between the clusters, providing solutions with 2 components at smaller b values than the other methods.

Finally, the ICL and NI criteria show a similar behavior. Both of them present a range of valid solutions that is wider than that of BIC and does not degenerate

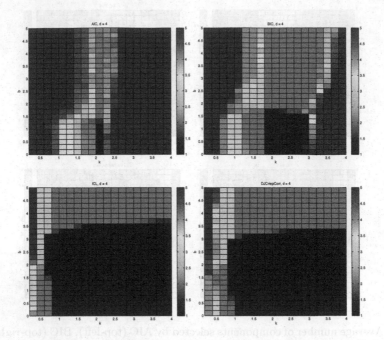

Fig. 4. Average number of components selected by AIC (top-left), BIC (top-right), ICL (bottom-left) and NI (bottom-right) over 40 problems with given cluster distribution (k parameter, x-axis) and inter-cluster distance (b parameter, y-axis). The dimension is $d = 4$.

with the dimension. This range covers all k values except the smallest ones. However, ICL and NI seem to be less tolerant than BIC to the clusters' overlap, with NI behaving slightly better than ICL in this respect.

To quantify these observations we have measured, for each dimension and each validation criterion, both the k-range of valid solutions and the maximum allowed overlap, which we formally define as follows. The set of solutions for a given k is said to be *valid* if: (i) the average n_c is 1 for $b = 0$; (ii) the average n_c is 2 for $b = 5$; and (iii) the average n_c monotonically increases with b. The k-range of valid solutions is the subset of k values for which the obtained solutions are valid. Figure 6 (left) plots this range versus the dimension for each validation method. Note that, as previously observed, ICL and NI provide valid solutions for almost the full range of ks independently of the dimension. The range of valid solutions for BIC covers a wide range of ks for $d = 2$ but it quickly reduces as the dimension increases, while it is practically nonexistent for AIC. The maximum allowed overlap is related to the minimum inter-cluster distance that permits the recognition of two clusters. For a given dimension and validation criterion, this minimum distance is averaged over all ks in the range of valid solutions. Figure 6 (right) shows a plot of this average versus the dimension for each of the 4 validation methods. BIC and AIC are the methods that allow the highest overlap.

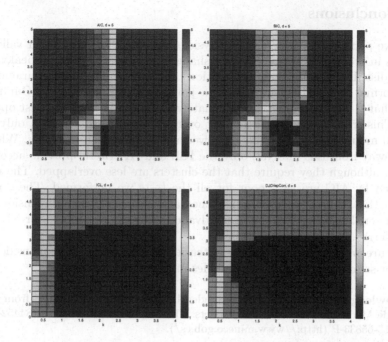

Fig. 5. Average number of components selected by AIC (top-left), BIC (top-right), ICL (bottom-left) and NI (bottom-right) over 40 problems with given cluster distribution (k parameter, x-axis) and inter-cluster distance (b parameter, y-axis). The dimension is $d = 5$.

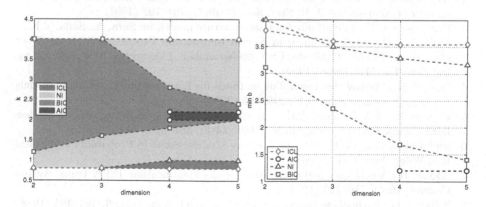

Fig. 6. Left. Range of valid solutions versus dimension for each validation criterion. Right. Minimum inter-cluster distance for which each validation method provides a solution with 2 clusters.

5   Conclusions

We have performed a systematic analysis of the behavior of 4 cluster validation criteria in problems where the cluster shape is interpolated between peaky, long-tailed distributions and flat distributions as we vary the shape parameter k. The normal distribution appears as a particular case for $k = 2$. Our results show that if the clusters are approximately Gaussian ($k \approx 2$) the best option is BIC. This criterion allows the highest overlap between clusters, but only in an interval around $k = 2$ that gets narrower as the dimension increases. When we move away from this interval, only the ICL and NI criteria provide acceptable results, although they require that the clusters are less overlapped. The results provided by AIC are very poor for all the tests we performed. These results suggest that, in real problems where the data are usually high dimensional and the cluster shapes are not known in advance, validation using ICL or NI should be preferred.

Future work contemplates the extension of this research by considering a broader set of experiments that includes more realistic problems.

Acknowledgments. This work was funded by grant S2017/BMD-3688 from Comunidad de Madrid, and by Spanish projects MINECO/FEDER TIN2017-84452-R and DPI2015-65833-P (http://www.mineco.gob.es/).

References

1. Dempster, A., Laird, N., Rubin, D.: Maximum likelihood from incomplete data via the EM algorithm. J. R. Stat. Soc. B **39**(1), 1–38 (1977)
2. Richardson, S., Green, P.J.: On Bayesian analysis of mixtures with an unknown number of components. J. R. Stat. Soc. B **59**(4), 731–792 (1997)
3. Rasmussen, C.E.: The infinite Gaussian mixture model. In: Sara, A., Solla, T.K.L., Müller, K.-R. (eds.) Advances in Neural Information Processing Systems [NIPS Conference, Denver, Colorado, USA, 29 November 4 December 1999], vol. 12, pp. 554–560. The MIT Press (1999)
4. Neal, R.M.: Markov chain sampling methods for Dirichlet process mixture models. J. Comput. Graph. Stat. **9**(2), 249–265 (2000)
5. Figueiredo, M.A.T., Jain, A.K.: Unsupervised learning of finite mixture models. IEEE Trans. Pattern Anal. Mach. Intell. **24**(3), 381–396 (2002)
6. McLachlan, G.J., Peel, D.: Finite Mixture Models. Series in Probability and Statistics. Wiley, New York (2000)
7. Akaike, H.: A new look at the statistical model identification. IEEE Trans. Autom. Control. **19**(6), 716–723 (1974)
8. Schwarz, G.: Estimating the dimension of a model. Ann. Stat. **6**, 461–464 (1978)
9. Fraley, C., Raftery, A.E.: How many clusters? Which clustering method? Answers via model-based cluster analysis. Comput. J. **41**(8), 578–588 (1998)
10. Gordon, A.D.: Cluster validation. In: Hayashi, C., Yajima, K., Bock, H.H., Ohsumi, N., Tanaka, Y., Baba, Y. (eds.) Data Science, Classification and Related Methods, pp. 22–39. Springer, Heidelberg (1998)

11. Bozdogan, H.: Choosing the number of component clusters in the mixture-model using a new information complexity criterion of the inverse-Fisher information matrix. In: Opitz, O., Lausen, B., Klar, R. (eds.) Data Analysis and Knowledge Organization, pp. 40–54. Springer, Heidelberg (1993). https://doi.org/10.1007/978-3-642-50974-2_5

12. Biernacki, C., Celeux, G., Govaert, G.: An improvement of the NEC criterion for assessing the number of clusters in a mixture model. Pattern Recognit. Lett. **20**(3), 267–272 (1999)

13. Bezdek, J.C., Li, W., Attikiouzel, Y., Windham, M.P.: A geometric approach to cluster validity for normal mixtures. Soft Comput. **1**(4), 166–179 (1997)

14. Arbelaitz, O., Gurrutxaga, I., Muguerza, J., Pérez, J.M., Perona, I.: An extensive comparative study of cluster validity indices. Pattern Recognit. **46**(1), 243–256 (2013)

15. Rodriguez, M.Z., et al.: Clustering algorithms: a comparative approach. PLoS ONE **14**, e0210236 (2019)

16. Biernacki, C., Celeux, G., Govaert, G.: Assessing a mixture model for clustering with the integrated completed likelihood. IEEE Trans. Pattern Anal. Mach. Intell. **22**(7), 719–725 (2000)

17. Samé, A., Ambroise, C., Govaert, G.: An online classification EM algorithm based on the mixture model. Stat. Comput. **17**(3), 209–218 (2007)

18. Lago-Fernández, L.F., Corbacho, F.J.: Normality-based validation for crisp clustering. Pattern Recognit. **43**(3), 782–795 (2010)

19. Lago-Fernández, L.F., Sánchez-Montañés, M.A., Corbacho, F.J.: The effect of low number of points in clustering validation via the negentropy increment. Neurocomputing **74**(16), 2657–2664 (2011)

20. Lago-Fernández, L.F., Sánchez-Montañés, M., Corbacho, F.: Fuzzy cluster validation using the partition negentropy criterion. In: Alippi, C., Polycarpou, M., Panayiotou, C., Ellinas, G. (eds.) ICANN 2009. LNCS, vol. 5769, pp. 235–244. Springer, Heidelberg (2009). https://doi.org/10.1007/978-3-642-04277-5_24

Internet Modeling, Communication and Networking

From Iterative Threshold Decoding
to a Low-Power High-Speed Analog VLSI
Decoder Implementation

Werner G. Teich[1(✉)], Heiko Teich[1], and Giuseppe Oliveri[2]

[1] Institute of Communications Engineering, Ulm University, 89069 Ulm, Germany
werner.teich@uni-ulm.de
[2] Institute of Electronic Devices and Circuits, Ulm University, 89069 Ulm, Germany

Abstract. Key capabilities of the fifth generation (5G) of cellular mobile communication systems are increased peak and network data rates and an energy efficient operation. Signal processing plays an important role to meet these goals. Recently, it has been shown that, for the problem of vector equalization, signal processing with analog electronic circuits has a large potential for a high-speed and low-power operation. In this paper we consider the problem of decoding for convolutional self-orthogonal codes. We report on a student project where we used standard off-the-shelf electronic components to realize an analog decoder circuit. The starting point is an iterative threshold decoder. Its structure corresponds to the one of a high-order recurrent neural network (HORNN). Structure as well as weights of the HORNN are given directly by the problem. The dynamics of the HORNN can be implemented in discrete-time, this corresponds to the iterative threshold decoder, or in continuous-time. Both implementations lead to the same asymptotic state, which represents the desired decoder output. The dynamical evolution of the continuous-time HORNN is governed by a system of coupled first-order nonlinear differential equations. Based on that, we design an analog electronic circuit, which solves this set of differential equations. Thus the analog circuit shows a similar dynamical behavior as the continuous-time HORNN, and especially also the same asymptotic state.

Keywords: Iterative threshold decoding ·
Convolutional self-orthogonal codes ·
Continuous-time high-order recurrent neural network ·
Analog VLSI implementation

1 Introduction

The fifth generation of cellular mobile communication systems, also known as "5G", draws large interest in the scientific community as well as in the general public. It provides an improvement of key performance targets. Among others,

© Springer Nature Switzerland AG 2019
I. Rojas et al. (Eds.): IWANN 2019, LNCS 11507, pp. 615–628, 2019.
https://doi.org/10.1007/978-3-030-20518-8_51

an improvement by a factor 100 of the network energy efficiency is required [1]. 5G will shift mobile communications from a human-centric communication to a massive machine-type communication (internet of things). With this machine-to-machine communication such as in sensor networks, autonomeous vehicle, etc. energy efficient operations are crucial.

Signal processing plays an important role to meet this goal. Recently, it has been shown that signal processing with analog electronic circuits has a large potential for a high-speed and low-power energy efficient operation. For a vector equalizer, Oliveri et al. [2] presented an analog electronic circuit which improves energy efficiency by four orders of magnitude. Their point of departure is an iterative vector equalization algorithm which can be represented by a discrete-time recurrent neural network (RNN). The discrete-time RNN is used here as a computational model, i.e., structure and parameters of the RNN are not obtained in a training phase, but are defined directly by the problem statement. The fix point of the RNN corresponds to the desired output, in this case the equalized received vector. Furthermore, it has been shown that a continuous-time RNN can be designed, which exhibits the same asymptotic state as the discrete-time RNN [3]. The continuous-time RNN is described by a system of coupled first-order nonlinear differential equations, which is the basis for the analog[1] electronic implementation.

Recently, analog computing has found an increasing interest [4,5]. In the broader sense, in analog computing a problem is solved by finding a physical system which matches the original problem, i.e., is *analog* to the original problem [4]. That is the structure of an analog computer is adopted to each problem to be solved. On the other hand, in a digital computer, the structure is fixed. A problem is solved by executing a sequence of instructions that are stored in the memory and constitute the algorithm. Therefore, stored-program digital computer are very flexible. They can solve large classes of problems. Disadvantages of digital computer are the increased energy consumption when scaling up the computational speed. Analog computer, on the other hand, allow a large throughput due to their inherent parallel computing capability. Furthermore, in an analog computer inherent physical properties are used to realize specific mathematical functions. This leads to a reduced energy consumption.

In this paper we apply analog computing to the problem of decoding in digital communications. Forward errror correction coding [6,7] is an important part in any modern communication system. Specifically, we consider convolutional self-orthogonal codes (CSOCs) [8]. CSOCs have been introduced by Jim Massey in his PhD thesis as codes suitable for majority logic decoding or threshold decoding (TD) [9]. In the early days of coding theory, the property of self-orthogonality allowed an efficient hardware implementation of the decoder. This made them interesting for applications, despite of the fact that they provide only a moderate coding gain. Much later it was found that this property of self-orthogonality

---

[1] Note, the term *analog* is used here in a twofold meaning. First, it indicates an electronic circuit which matches with the original mathematical problem, and second, it characterizes a circuit for continuous-time and continuous-range signals.

makes these codes also very suitable for iterative decoding [10–12] since their Tanner graph [7] is sparse and has a large girth. Due to their efficient hardware implementation CSOCs have also been included in a standard for fiber-optic transmissions [13].

The starting point for our analog decoder is an iterative threshold decoder for CSOCs. The algorithmic structure of an iterative threshold decoder matches well with the structure of a discrete-time high-order RNN (HORNN) [3,14,15]. The structure as well as the weights of this HORNN are given by the mathematical algorithm. A training phase is not required. Similar as for the RNN, the HORNN can be implemented either as a discrete-time HORNN, i.e., a computational model of iterative threshold decoding (ITD), or as a continuous-time HORNN [3,14,15]. Both implementations have the same asymptotic state (fix point or equilibrium point), which corresponds to the desired decoder output. The dynamical evolution of the continuous-time HORNN is governed by a system of coupled first-order nonlinear differential equations. Based on that we seek an analog electronic circuit which shows a similar dynamical evolution as the continuous-time HORNN, and especially also the same equilibrium state. We report on a student project where we used a simple CSOC and standard off-the-shelf electronic components to realize the analog decoder circuit. However, the same steps can also be performed for more pratical CSOCs or also convolutional self-doubly orthogonal codes (CSO^{2}Cs) [12,16]. Based on our analysis it is also possible to design and build a very large-scale integrated (VLSI) analog electronic circuit which will combine high computational speed and low power consumption (see also [2]).

Decoding with analog nonlinear networks has already been proposed by other authors, see, e.g., Hagenauer et al. [17]. However, their approach was to implement the sum-product algorithm [7] of iterative decoding in analog VLSI. The analog implementation of ITD has the advantage of a much lower complexity, as it requires only one neuron for each information symbol, in contrast to one neuron for each non-zero entry in the parity check matrix for the sum-product algorithm [3]. Generally, iterative sum-product decoding shows a better bit error rate (BER) performance compared to ITD. However, for CSO^{2}Cs He et al. [18] show that the performance of ITD is comparable to the sum-product algorithm.

2 Convolutional Self-orthogonal Codes

2.1 Transmission Model

Figure 1(a) shows the discrete-time transmission model. A sequence of binary source (or information) symbols $q[k] \in \mathbb{F}_2$ is encoded by a CSOC encoder (COD). The resulting sequence of binary code symbols $c[k] \in \mathbb{F}_2$ is mapped (MAP) to the sequence of transmit symbols $x[k] \in \{\pm 1\} \subset \mathbb{Z}$ (cf. Fig. 1(b), antipodal transmission). The transmit symbols are sent over a discrete-time additive white Gaussian noise (AWGN) channel, i.e., $y[k] = x[k] + n[k]$. Here, $n[k] \in \mathbb{R}$ is an i.i.d. Gaussian noise sequence with variance σ^2. The sequence of received symbols $y[k] \in \mathbb{R}$ is the input to the channel decoder (DEC), which gives out the sequence of detected source symbols $\hat{q}[k] \in \mathbb{F}_2$. For simplicity, we used here the standard channel model of coding theory [6,7]. However, the results can easily be extended to other channels, such as the Rayleigh fading channel.

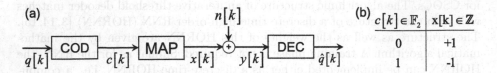

Fig. 1. (a) Discrete-time transmission model. (b) Mapping from code to transmit symbols.

2.2   Convolutional Self-orthogonal Codes: Definition

We restrict ourselfs to CSOCs or CSO^{2}Cs as these codes are most suitable for ITD. CSOCs have been introduced by Jim Massey [9] as a class of convolutional codes (CCs) which are threshold decodable. TD relies on orthogonal parity checks, i.e., considering the parity checks on a specific code symbol $c[m]$, all other code symbols $c[n]$ with $n \neq m$ participate in at most one parity check equation. Later it was found, that these codes can be represented by a Tanner graph with a girth of at least four. This implies that in an iterative decoding scheme we do not see correlations within one iteration step. The concept has been extended to CSO^{2}C with a statistical independence of at least two iterations [16]. To simplify the introduction of iterative decoding, we consider CCs with finite block length. To avoid a loss of data rate, we employ tail-biting convolutional codes [17]. It also avoids a different protection of source symbols at the edges of the block. This results in a cyclic CC $C(N, K, d)$, where N is the length of the code word c, K the length of the information word q, and d the minimum distance of the code. The code rate is given by $r_c = \frac{K}{N}$.

In this paper we consider the simplest[2] CSOC with rate $r_c = \frac{1}{2}$. The code word c of this convolutional block code is obtained as the row vector $c = q \cdot G$ with the $K \times N$ generator matrix G [6]. The parity check matrix H generally is a $(N - K) \times N$ matrix. For the $C(8, 4, 3)$ code we obtain, e.g.,

$$G = \begin{bmatrix} 1\,1\,0\,1\,0\,0\,0\,0 \\ 0\,0\,1\,1\,0\,1\,0\,0 \\ 0\,0\,0\,0\,1\,1\,0\,1 \\ 0\,1\,0\,0\,0\,0\,1\,1 \end{bmatrix} \quad \text{and} \quad H = \begin{bmatrix} 1\,1\,0\,0\,0\,0\,1\,0 \\ 1\,0\,1\,1\,0\,0\,0\,0 \\ 0\,0\,1\,0\,1\,1\,0\,0 \\ 0\,0\,0\,0\,1\,0\,1\,1 \end{bmatrix}. \tag{1}$$

The generator matrix has not been given in standard form in order to show the convolutional character of this block code. It can be clearly seen in the regular

---

[2] Actually it turns out, that this code is also a CSO^{2}C.

structure of the generator matrix. Tail-biting leads to a cyclic structure of the code. The code is comprised of 16 code words with a minimum Hamming weight of $d = 3$. We have adopted a systematic encoding. Code symbols with odd indices (systematic code symbols) corrrespond directly to information symbols and code symbols with even indices correspond to parity symbols. The parity check matrix is also a cyclic matrix. Any valid code word c fulfills the parity checks $\boldsymbol{H} \cdot \boldsymbol{c}^{\mathrm{T}} = \boldsymbol{0}$. It can also be seen that the two parity check equations for the systematic code symbols are orthogonal to each other [9], e.g., for $n = 1$ we have:

$$c[1] \oplus c[2] \oplus c[7] = 0 \tag{2}$$
$$c[1] \oplus c[3] \oplus c[4] = 0. \tag{3}$$

3   Iterative Threshold Decoding

3.1   Majority Logic Decoding

Considering the generator matrix given in (1) we observe that every information symbol $q[k]$ $(k = 1, \ldots, K)$ influences three code symbols. The systematic code symbol $c[2k - 1] = q[k]$, and the two parity symbols

$$c[2k] = q[k] \oplus q[k - 1] \tag{4}$$
$$c[2k + 2] = q[k] \oplus q[k + 1]. \tag{5}$$

Note, that at the edges we have to take care of the cyclic boundary conditions, i.e., $q[0] = q[K]$, $q[K + 1] = q[1]$, and $c[2K + 2] = c[2]$. We assume hard decision decoding for the time being, i.e., the detected transmit and code symbols are given by $(n = 1, \ldots, N)$

$$\hat{x}[n] = \mathrm{sgn}\{y[n]\} \quad \text{and} \quad \hat{c}[n] = \begin{cases} 0 \text{ for } y[n] > 0 \\ 1 \text{ for } y[n] < 0 \end{cases}. \tag{6}$$

Based on the detected symbols, we obtain the following three (hard) replicas for each information symbol $q[k]$:

$$\hat{r}_1[k] = \hat{c}[2k - 1] \tag{7}$$
$$\hat{r}_2[k] = \hat{c}[2k] \oplus q[k - 1] \tag{8}$$
$$\hat{r}_3[k] = \hat{c}[2k + 2] \oplus q[k + 1]. \tag{9}$$

Again the cyclic conditions have to be applied at the block boundaries. The estimate $\hat{r}_1[k]$ is obtained from the direct transmission of the information symbol (systematic code symbol). The other two estimates $\hat{r}_2[k]$ and $\hat{r}_3[k]$ are based on the received parity symbols and the constraints of the code. Note, that $\hat{c}[n]$ and $\hat{r}_l[k]$ $(l = 1, 2, 3)$ are from the finite field $\mathbb{F}_2$, whereus the detected transmit symbols $\hat{x}[k] \in \{\pm 1\}$ are integer numbers.

In reality, the information symbols $q[k \pm 1]$ are not directly available in the decoder. In a first step they can be replaced by the estimates $\hat{c}[2k - 3]$ and

$\hat{c}[2k + 1]$ obtained from the transmission of the systematic code symbols. This leads to

$$\hat{r}_2[k] = \hat{c}[2k] \oplus \hat{c}[2k - 3] \tag{10}$$

$$\hat{r}_3[k] = \hat{c}[2k + 2] \oplus \hat{c}[2k + 1]. \tag{11}$$

In majority logic decoding the decision for an information symbol is based on the majority vote of the available estimates. For an odd number of estimates, a unique decision is always possible. The independence of the estimates is ensured by the orthogonality of the underlying parity check equations. A detected code symbol is never involved in the calculation of more than one replica.

Majority logic decoding is based on three steps. In the first step we calculate preliminary decisions for all information symbols, based on the systematic code symbols only. In the second step, based on the previous preliminary decision of the information symbols, the two (hard) replicas $\hat{r}_l[k], l = 2, 3$ are calculated, cf., (10 and 11). In the last step, all replicas are combined to obtain the final decision for the information symbols. Therefore we can correct one error for the example given. In general, d being the number of independent replicas, we can correct $\lfloor \frac{d-1}{2} \rfloor$ errors.

Majority logic decoding can also be realized by a threshold decision. To do so, we have to replace the detected code symbols $\hat{c}[n]$ in (10 and 11) by the detected transmit symbols $\hat{x}[n]$. The modulo-2 operation in the finite field $\mathbb{F}_2$ becomes a normal multiplication in the field of integer numbers and the replicas $\hat{r}_l[k] \in \mathbb{F}_2$ are mapped to the integer variables $\hat{x}_{r_l}[k] \in \{\pm 1\}$ (cf. also Fig. 1(b)). This leads to the following decision rule for TD ($k = 1, \ldots, K$):

$$\hat{x}_q[k] = \text{sgn} \left\{ \sum_{l=1}^{3} \hat{x}_{r_l}[k] \right\}. \tag{12}$$

Here, $\hat{x}_q[k] \in \{\pm 1\}$ is the integer representation of the decided information symbol $\hat{q}[k] \in \mathbb{F}_2$.

3.2   Iterative Threshold Decoding: Hard Decision

Majority logic decoding or TD can naturally be extended to an iterative algorithm, improving the overall performance of the decoder [10–12]. To do so, we feed back the improved decisions of the third step of majority logic decoding to the second step of calculating the replicas. In integer representation (10 and 11) become ($k = 1, \ldots, K$):

$$\hat{x}_{r_2}^{(i+1)}[k] = \hat{x}[2k] \cdot \hat{x}_q^{(i)}[k - 1] \tag{13}$$

$$\hat{x}_{r_3}^{(i+1)}[k] = \hat{x}[2k + 2] \cdot \hat{x}_q^{(i)}[k + 1]. \tag{14}$$

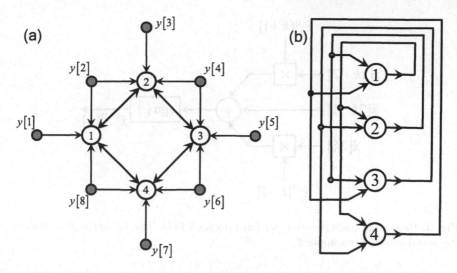

Fig. 2. (a) Graphical representation of ITD. Open circles represent (iterated) informa-tion symbols $\hat{x}_q^{(i)}[k]$, $k = 1, \ldots, 4$ and thus the processing nodes (neurons). Solid circles depict the (fixed) received symbols $y[n]$, $n = 1, \ldots, 8$. Depending on whether we use hard or soft decision ITD, the solid circles calculate the hard decided transmit symbols $\hat{x}[n]$ or the soft L-values $L_c[n]$, respectively the soft bit $\breve{x}[n]$ as input to the processing nodes. (b) The coupling of the processing nodes represented as a single-layer RNN.

Here $(\cdot)^{(i)}$ indicates the $i - th$ iteration of the respective variable. The replica $\hat{x}_{r_1}[k]$ is fixed and obtained directly from the channel output. Thus the hard decision for information symbol $\hat{x}_q[k]$ in iteration step $(i + 1)$ is given by ($k = 1, \ldots, K$):

$$\hat{x}_q^{(i+1)}[k] = \mathrm{sgn}\{\hat{x}_{r_1}[k] + \hat{x}[2k] \cdot \hat{x}_q^{(i)}[k - 1] + \hat{x}[2k + 2] \cdot \hat{x}_q^{(i)}[k + 1]\}. \quad (15)$$

The iteration is initialized by $\hat{x}_q^{(0)}[k] = \hat{x}[2k - 1]$.

A graphical representation of this iterative algorithm for the example given is depicted in Fig. 2(a). Because of the cyclic convolutional code, the network topol-ogy is a ring. The parity symbols are fixed. Information symbols are improved iterativly. Only neighbouring information symbols are coupled. The function of a single processing node (open circles in Fig. 2(a)) calculating hard decisions for the information symbols is depicted in Fig. 3. Figure 2(b) shows the coupling of the processing nodes (neurons) represented as a single-layer recurrent network.

3.3 Iterative Threshold Decoding: Soft Decision

So far the disussion has been based on a hard decision decoding. As is well known, large coding gains can be obtained by soft decision decoding [6]. Insted of performing a hard decision $\hat{x}[n]$ ($\hat{c}[n]$) on the transmit (code) symbols (cf. (6)),

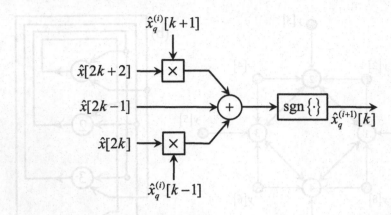

Fig. 3. Processing node (neuron) for hard decision ITD. The integer representation of the variables has been applied.

we introduce the log-likelihood ratios (LLRs) or L-values $L_c[n]$ and the expected values $\breve{x}[n]$ [7,17]:

$$L_c[n] = \ln \frac{\Pr\{c[n] = 0\}}{\Pr\{c[n] = 1\}} = 2\text{atanh}\left(\breve{x}[n]\right) \in \mathbb{R} \tag{16}$$

$$\breve{x}[n] = \tanh \frac{L_c[n]}{2} \in [-1, +1]. \tag{17}$$

For the AWGN channel model used in this paper we obtain for the intrinsic LLRs from the channel: $L_c[n] = \frac{2}{\sigma^2} y[n]$ [7]. Based on that we can calculate three soft replicas for the information symbols using standard log-likelihood algebra [7,17]. For $k = 1, \ldots, K$ we get from the systematic code symbol the intrinsic L-values $L_{r_1}[k] = L_c[2k-1]$. From the parity code symbols we obtain the extrinsic L-values

$$L_{r_2}[k] = 2\text{atanh}\left\{ \tanh\left(\frac{L_c[2k]}{2}\right) \tanh\left(\frac{L_c[2k-3]}{2}\right) \right\} \tag{18}$$

$$L_{r_3}[k] = 2\text{atanh}\left\{ \tanh\left(\frac{L_c[2k+2]}{2}\right) \tanh\left(\frac{L_c[2k+1]}{2}\right) \right\}. \tag{19}$$

L-values and expected values for the information symbols are defined in a similar way as in (16 and 17). They are obtained by $L_q[k] = \sum_{l=1}^{3} L_{r_l}[k]$. Hard decisions for the information symbols are given by $\hat{q}[k] = \begin{cases} 0 \text{ for } L_q[k] > 0 \\ 1 \text{ for } L_q[k] < 0 \end{cases}$.

Similar as for the case of hard decision TD, we can now also improve the decoder output by feeding back the improved LLRs of the information symbols to the calculation of the two extrinsic replicas (18 and 19). This leads to the following iterative method ($k = 1, \ldots, K$):

$$L_{r_2}^{(i+1)}[k] = 2\text{atanh}\left\{ \tanh\left(\frac{L_c[2k]}{2} \right) \tanh\left(\frac{L_q^{(i)}[k-1]}{2} \right) \right\}$$

$$L_{r_3}^{(i+1)}[k] = 2\text{atanh}\left\{ \tanh\left(\frac{L_c[2k+2]}{2} \right) \tanh\left(\frac{L_q^{(i)}[k+1]}{2} \right) \right\}. \quad (20)$$

The first intrinsic replica $L_{r_1}[k]$ remains unchanged. The L-value for information symbol $q[k]$ in iteration step $(i+1)$ is obtained as ($k = 1, \ldots, K$)

$$L_q^{(i+1)}[k] = L_{r_1}[k] + \sum_{l=2}^{3} L_{r_l}^{(i+1)}[k]. \quad (21)$$

Finally, the hard decision for the information symbols in iteration step (i) is given by $\hat{q}^{(i)}[k] = \begin{cases} 0 \text{ for } L_q^{(i)}[k] > 0 \\ 1 \text{ for } L_q^{(i)}[k] < 0 \end{cases}$.

The topology of soft decision ITD is the same as that for hard decision ITD, cf. Fig. 2. Only the hard decision inputs $\hat{x}[n]$ must be replaced by the respective L-values $L_c[n]$ or expected values $\check{x}[n]$. The structure of the processing node is given in Fig. 4.

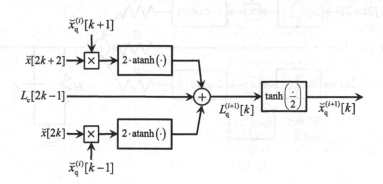

Fig. 4. Processing node (neuron) for soft decision ITD.

4 Analog Decoder Implementation

4.1 High-Order Recurrent Neural Network

The topology of the iterative algorithm as given in Fig. 2(b) resembles the network structure of a single layer RNN. However, the interaction between the neurons is not only a weighted sum of the respective output of the neurons (linear operation). Therefore we have to generalize the concept of a RNN to allow also nonlinear interactions between the neurons. This leads us to the concept of HORNNs [3,14,15]. Thus soft decision ITD as given in (20 and 21) can be considered as a HORNN with a discrete-time dynamics [14]. As shown by Mostafa et al. [15] a continuous-time HORNN with the same topology as the discrete-time HORNN leads to the same asymptotic states.

We adopted the additive model of a neuron [19] to arrive at the continuous-time HORNN. The state variables of the HORNN are represented as potentials. The addition of the L-values is achieved by a current-summing junction, characterized by a low input resistance, unity current gain, and high output resistance. The weights of the neuron are represented by conductances $\frac{1}{R}$. In contrast to the problem of vector equalization [2], here the weights are constant and all of the same value. The structure of the neuron is shown in Fig. 5. Note, that we have moved the activation function $\tanh\left(\frac{\cdot}{2}\right)$ of the neuron from the output to the input. The dynamical variables are now the LLRs $L_k(t), k = 1,\ldots,4$. Also, all outputs from the channel are represented by L-values. The reason for these changes become clear when we discuss the electronic components in Sect. 4.2.

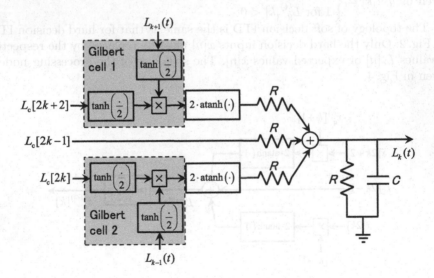

Fig. 5. Processing node (neuron) for the continuous-time HORNN.

The dynamics of the continuous-time HORNN is governed by the following system of coupled first-order nonlinear differential equations ($k = 1,\ldots,4$):

$$RC\frac{dL_k(t)}{dt} = -L_k(t) + L_c[2k-1] + 2\mathrm{atanh}\left\{\breve{x}[2k]\tanh\left(\frac{L_{k-1}(t)}{2}\right)\right\}$$

$$+ 2\mathrm{atanh}\left\{\breve{x}[2k+2]\tanh\left(\frac{L_{k+1}(t)}{2}\right)\right\}. \quad (22)$$

Here, we have used $\check{x}[2k] = \tanh\left(\frac{L_c[2k]}{2}\right)$ and $\check{x}[2k+2] = \tanh\left(\frac{L_c[2k+2]}{2}\right)$, see also (17). Note also, that cyclic boundary conditions apply (cf. Fig. 2). The time-scale for the transient response of the continuous-time HORNN is determined by a time constant $\tau = RC$, assumed to be the same for all neurons. The lower bound is given by the parasitic capacitance and resistance of the circuit.

The equilibrium points of the continuous-time HORNN are given by $\frac{dL_k(t)}{dt} = 0$. On the other hand, the fix points of the discrete-time HORNN (soft decision ITD) are given by $L_q^{(i+1)}[k] = L_q^{(i)}[k]$. Both conditions lead to the same system of non-linear equations. Therefore the equilibrium points of the continuous-time HORNN and the fix points of the discrete-time HORNN coincide. Simulation results showing the BER performance are given in [14].

With the exception of some special codes, analytic results for the convergence of HORNN is still an open problem [3]. It has been shown by simulations, that discrete-time HORNN converge to fix points or limit cycles of length two (for parallel update). Continuous-time HORNN converge to the respective equilibrium points [15].

4.2 Analog Electronic Circuit

The goal is now to build an analog electronic circuit which resembles the dynamical behaviour of the continuous-time HORNN as given in Figs. 2 and 5. To do so, we have to match the mathematical operations as given in the system of coupled first-order nonlinear differential equations (22). Besides the fundamental mathematical operations *addition* and *multiplication*, we have to realize the nonlinear functions *tanh* and *atanh*.

Addition is achieved by a simple current summing junction. Similar as in the case of an analog realization of a vector equalizer [2], we use a four-quadrant analog multiplier (Gilbert cell) [20] based on bipolar junction transistors (BJTs) for the multiplication operation. Matching of the BJTs is achieved by taking the transistors for a given function from the same integrated circuit (IC). BJTs have a natural exponential characteristics. Therefore the differential output current of the Gilbert cell is directly proportional to the product of two tanh functions [17]. Operating at a proper working point, the Gilbert cell not only serves as a multiplier, but also realizes the tanh function, cf. Fig. 5. As in [2], the proper bias is provided by a current mirror. The L-values have an unlimited range. In the hardware implementation they have to be realized as potentials with a finite range. We therefore limit the L-values to $L[k] \in [-5, +5]$. The 2·atanh-function is realized by two serially connected diodes. By cutting short the output of a neuron, a metal-oxid-semiconductor field effect transisor is used as a sequencer (see also [2]).

To realize the continuous-time HORNN, we have used standard off-the-shelf components. The functionality of one neuron as given in Fig. 5 has been implemented on a printed circuit board by eight ICs. Four neurons have been assembled. They are mounted on the main printed circuit board realizing the connectivity between the neurons (cf. Fig. 2). The analog circuit is controlled by an ARDUINO processor and digital-to-analog and analog-to digital converters. Figure 6 shows the actual implemented hardware with the ARDUINO processor above on the left, and two printed circuit boards for two of the four neurons in the foreground.

Fig. 6. Hardware realization of ITD based on a continuous-time HORNN

5   Summary

We have revisited the problem of decoding with analog electronic circuits. Starting from ITD for CSOCs we have derived an analog electronic circuit with the same equilibrium states as the ITD. As this was a proof of concept study, low-cost standard off-the-shelf components have been used for the design and construction of the circuit. However, based on our analysis we expect that it is also possible to design and build a customized analog VLSI circuit, which combines high computational speed and low power consumption. The regular structure of the parity check matrix of the CSOCs (CSO^2Cs) eases the VLSI hardware realization. As the coupling of the neurons is only local, an extension to larger code word length is straightforward. The analysis and design can also be generalized to more complex CSOCs (CSO^2Cs) with larger constraint length and/or larger code rate. An extension to channels with multipath propagation is also possible. This leads to the concept of combined equalization and decoding [17]. If the equalizer is realized by a continuous-time RNN [2], the overall receiver is described by two coupled systems of coupled first-order nonlinear differential equations [21].

Acknowledgment. The authors would like to thank Mohamad Mostafa and Jürgen Lindner for many valuable discussions on the topic of signal processing with RNN. We thank Hermann Schumacher and Andreas Trasser for sharing their expertise in the design of analog circuits.

References

1. IMT Vision - Framework and Overall Objectives of the future development of IMT for 2020 and beyond, Recommendation ITU-R M.2083-0, September 2015
2. Oliveri, G., Mostafa, M., Teich, W.G., Lindner, J., Schumacher, H.: Advanced low power high speed nonlinear signal processing: an analog VLSI example. J. Signal Process. Syst. **89** (2016). https://doi.org/10.1007/s11265-016-1171-0
3. Mostafa, M.: Equalization and Decoding: A Continuous-Time Dynamical Approach. Der Andere Verlag, Uelvesbüll (2014). https://doi.org/10.18725/OPARU-3229
4. Ulmann, B.: Analog Computing, Oldenburg Wissenschaftsverlag GmbH (2013)
5. Tsividis, Y.: Not your Father's analog computer. IEEE Spectr. **55**, December 2017. https://doi.org/10.1109/MSPEC.2018.8278135
6. Bossert, M.: Channel Coding for Telecommunications. Wiley, Chichester (1999)
7. Johnson, S.: Iterative Error Correction: Turbo, Low-Density Parity-Check and Repeat-Accumulate Codes. Cambridge University Press (2010)
8. Wu, W.: New convolutional codes - part i. IEEE Trans. Commun. **23**, 942–956 (1975)
9. Massey, J.L.: Threshold decoding, Massachusetts Institute of Technology, Technical report 410, April 1963
10. Blizard, R.B., Korgel, C.C.: An iterative probabilistic threshold decoding technique. In: IEEE National Telecommunications Conference Record, pp. 13D1–13D-5, December 1972
11. Teich, W.G.: Iterative decoding of one-dimensional convolutional codes. In: 4th International Symposium on Communication Theory and Applications, Ambleside, pp. 52–53, July 1997
12. Cardinal, C., Haccoun, D., Gagnon, F., Batani, N.: Turbo coding using convolutional self-doubly orthogonal codes. In: Proceedings of International Conference on Communications (ICC 1999), pp. 113–117, June 1999
13. Forward Error Correction for High Bit-rate DWDM Submarine Systems, Recommendation ITU-T G.975.1, February 2004
14. Mostafa, M., Teich, W.G., Lindner, J.: Analog realization of iterative threshold decoding based on high-order recurrent neural networks. In: International Conference on Signal Processing and Communication Systems (ICSPCS), Gold Coast, Australia, December 2010
15. Mostafa, M., Teich, W.G., Lindner, J.: Analysis of high order recurrent neural networks for analog decoding. In: 7th International Symposium on Turbo Codes and Iterative Information Processssing (ISTC), Goeteborg, Sweden, September 2012
16. Cardinal, C., Haccoun, D., Gagnon, F.: Iterative threshold decoding without interleaving for convolutional self-doubly orthogonal codes. IEEE Trans. Commun. **51**(8), 1274–1282 (2003)
17. Hagenauer, J., Offer, E., Méasson, C., Mörz, M.: Decoding and equalization with analog non-linear networks. Eur. Trans. Telecommun. **10**, 659–680 (1999)

18. He, Y.-C., Haccoun, D., Cardinal, C.: Performance comparison of iterative BP and threshold decoding for convolutional self-doubly-orthogonal codes. In: IEEE 65th Vehicular Technology Conference (VTC2007-Spring), pp. 2000–2004 (2007)

19. Haykin, S.: Neural Networks: A Comprehensive Foundation. Prentice-Hall, New Jersey (1999)

20. Gilbert, B.: A precise four-quadrant multiplier with subnanosecond response. IEEE J. Solid-State Circuits **SC–3**(4), 365–373 (1968)

21. Mostafa, M., Oliveri, G., Teich, W.G., Lindner, J.: A continuous-time recurrent neural network for joint equalization and decoding - analog hardware implementation aspects. Artificial Neural Networks, Joao Luis G. Rosa, IntechOpen (2016). https://doi.org/10.5772/63387

Machine Learning as a Means to Adapt Requirement Changes for SDN Deployment Process in SDN Migration

Siew Hong Wei[1], Tan Saw Chin[1(✉)], Jason Ng Binlun[1],
Lee Ching Kwang[2], Rizaluddin Kapsin[3], and Zulfadzli Yusoff[2]

[1] Faculty of Computing and Informatics,
Multimedia University, Cyberjaya, Malaysia
hw.siew@actiweb.co, sctanl@mmu.edu.my,
1161302638@student.mmu.edu.my
[2] Faculty of Engineering, Multimedia University, Cyberjaya, Malaysia
{cklee,zulfadzli.yusoff}@mmu.edu.my
[3] Telekom Malaysia Research and Development, Cyberjaya, Malaysia
rizaludin@tmrnd.com.my

Abstract. The deployment of SDN in legacy network has gained popularity across network operators as next generation network architecture. However, full deployment of SDN faces challenges in economical, organizational, and technical aspects. Hence, the deployment of SDN should be incremental over months or even years, and limited numbers of the nodes are upgraded to SDN-enabled one in each period. This forms a hybrid SDN (H-SDN) network which legacy and SDN nodes co-exist in the same network. Importantly, which and when a node should be replaced to SDN node are the common question which impacts the performance of a hybrid SDN network. Efforts to date primarily focus on determining sequence for migration which maximize the performance of traffic engineering (TE) in H-SDN network. However, most works do not take into consideration of the changes that may happen over the periods of SDN deployment. The possibility of these changes requires adaptation techniques to ensure effective migration sequence to cater present and future needs. In this article, we aim to identify the gap and propose the opportunity in which techniques originated from machine learning (ML) may play an important role in solving problem of incremental SDN deployment by alleviating the issues the occur during SDN migration as well as to improve the H-SDN deployment.

Keywords: Machine learning · Software-defined networks ·
Incremental deployment · SDN migration · H-SDN

1 Introduction

Communication networking is the crucial enabling element for many growing Internet applications. The emerging global trends of Internet usage, particularly in Internet of Things (IoT), virtual reality, augmented reality and cloud computing, are urging a transformation of communication networks for high-speed connectivity and dynamic

© Springer Nature Switzerland AG 2019
I. Rojas et al. (Eds.): IWANN 2019, LNCS 11507, pp. 629–639, 2019.
https://doi.org/10.1007/978-3-030-20518-8_52

configurability across various applications. Often, horizontal scaling on network infrastructures is the immediate option of legacy network operator to handle the ever rising demand for dynamic traffic requirements. Consequently, network increases complexity in both architecture and operation. Moreover, a complex network is prone to outages caused by human factors [1]. To address the aforementioned challenges in legacy network, Software-Defined Networking (SDN) has been introduced to offer network programmability over centralized controller [2]. Neither of programmability nor centralized control is totally new in networking. For instance, SwitchWare [3], is proposed to allow dynamically modification of network operations. In some other works, network devices are made programmable by implementing extensible software routers [4, 5]. The distinctness of SDN, as compared to prior works, lies on the fact that it offers simple programmable network devices via the separation of control and data plane in devices. By doing so, SDN offers a simpler environment for programmability over centralized controller. As a result, behaviors of a network can be externally defined though various network applications. This brings three major benefits for SDN: improved performance, enhanced configuration, and greater encouragement for innovation in the field of network operation and architecture [6]. Furthermore, centralized controller has the capability of seeing the global view of network status which allows SDN to optimize and improve network performance as a whole.

While SDN offers promising features for future network operation and management, the deployment of SDN is not without challenges. In general, challenges of SDN deployment comes from three major aspects, economical, organizational and technical [7]. First, the initial deployment cost of SDN is non-negligible, in terms of upgrading or replacing an existing legacy node to SDN-enabled node. This is mainly due to SDN is considered relatively new technology and the cost of SDN-enabled node is expected to reduce as technology getting mature in years to come. For this reason, operators are generally not willing to abolish the existing legacy nodes to enable full SDN deployment. Second, since the operations of SDN network are radically different from the existing legacy network, operators likely need trainings in various areas of operating SDN network, such as network design and management. This implies not only monetary investment but also intangible resources, such as time, to be devoted to this transition from an organization. Third, logic centralization of SDN raises questions in term of resilience, robustness and scalability issues in the network. For instance, how a simple management task such as upgrading software version for centralized controller may impact the network? It is likely that during these events, services may not be available temporarily. However, this temporary unavailability will incur a huge impact on the highly dynamic SDN network and requires the controller to take quick decisions to accommodate for the changes. SDN networks controllers may not be able to handle these tasks efficiently at present. All in all, SDN still requires much optimization and testing before being able to fully replace legacy networks.

Due to various concerns mentioned above, deployment of SDN in legacy network is unlikely to happen within a short time period. One of the feasible solutions for this will be partially deploying SDN devices on selected nodes in legacy network. The co-existence of legacy devices and SDN devices is commonly known as hybrid SDN network. In [7], authors propose four possible models of hybrid SDN to be a transition from the legacy network, and it also discusses the use cases of each model for long and short term transition. These four hybrid SDN models can be briefly categorized in two; SDN nodes deployment in legacy network, and dual stacks hybrid SDN [8]. Most of the literature focuses on first category of hybrid SDN, as it offers several attractive advantages as compared to dual stacks hybrid SDN.

Hybrid SDN realizes the benefits of SDN without full deployment in legacy network. Selected nodes are migrated for SDN functionalities in phases, and thus, huge financial commitment for full SDN deployment is relieved in hybrid SDN. Occasionally, selected SDN node can be deployed in a legacy network for specific fine-grained control requirement without full SDN deployment. Moreover, hybrid SDN offloads tasks from the centralized controller in which those tasks are more robust and effective by traditional distributed routing protocol.

Nevertheless, hybrid SDN poses several challenges together with its benefits. The co-existence of the legacy node and SDN node in a network requires SDN node to be able to interoperate with the legacy node to serve the network as a whole. However, [9–11] have presented viable interoperability between SDN node and non-SDN node. The second challenge to answer for hybrid SDN is that which node and when to replace. The migration of SDN is likely a staged process which spans over months or even years for a large network. Network operators desire to understand the migration sequence of SDN node in the network in order to realize the biggest potential benefits.

In this paper, we aim to identify the gap of hybrid SDN node deployment in literature and propose the possible future direction for works in this aspect. The rest of this paper is organized as follows. Section 2 discusses related works. Section 3 we further elaborate on the uses for Machine Learning (ML) in SDN. In Sect. 4, we identify future opportunities and research directions of ML in SDN migration, and finally we conclude this paper on Sect. 5.

2 Related Works

By selectively placing SDN devices together with switches, hybrid SDN can achieve SDN-like control and management in a legacy network. In this section, we will be discussing works related to hybrid SDN deployment and various options for deploying SDN-enabled devices in legacy network to form hybrid SDN.

The Panopticon approach proposes in [12, 13] constructs virtual SDN network on top of interconnecting legacy and SDN devices in a network. By enforcing waypoint enforcement technique, every packet is guaranteed to traverse through at least one SDN switch, and thus, all packets can be processed by SDN controller. Furthermore, they report that a traditional network can be realized as a SDN network by upgrading 30-40% of legacy nodes. However, the Panopticon approach requires one step architectural change in the network to realize full SDN capability in a hybrid SDN network.

In contrast to the Panopticon approach, [14] identifies that migration towards SDN network should be gradually progressing across multiple time periods, i.e. years, especially for large network like an Internet Service Provider (ISP). They presented SDN nodes migration model to maximize degree of freedom for traffic engineering (TE) and to maximize the programmable traffic in legacy network. The evaluation of [14] shows 54% increase in programmable traffic which can be employed to gain TE benefit.

Other efforts apply optimization techniques, such as Integer Linear Programming (ILP) and heuristic approach to the same problem in looking for SDN migration sequence. Limited budget constraint is often the concern of network operator, especially ISP, during the migration of SDN. With this concern in mind, [15] presented a complete migration trajectory by a greedy heuristic algorithm to maximize the gain of TE over multi-period of SDN deployment. Nevertheless, the greedy algorithm may give only a suboptimal solution in some cases. Likewise, [16] proposes an algorithm to seek for an optimal SDN deployment schedule in multi-period. The proposed algorithm aims to maximize the degree of freedom for TE in which adds more alternative paths by deploying SDN devices. The authors claim that the algorithm proposed outperforms other works especially for large network topologies with increased number of time steps.

Similarly, [17] develops a model to incrementally deploy SDN into the existing network to prevent operational issues. By applying the methodology introduced, the model selectively replaces a minimum subset of the legacy node with a SDN node to optimize the cost-effectiveness. Evaluation done in [17] with real enterprise network suggests that the proposed method reduces 32% of the maximum link usage as compared to legacy distributed routing protocol. TE is used to optimize network performance such as load balancing in hybrid SDN. In [18], TE in OSPF/SDN network is studied further. Authors propose a genetic algorithm to obtain optimized migration sequence which minimizes the maximum link utilization of a network, and the result shows a lower maximum link utilization with 40% of SDN nodes deployed in an ISP network.

Table 1. Comparison of various SDN deployment techniques

Study	Optimization objective	Constraint(s)	Multi-periods	Performance
[14]	(1) Maximize volume of programmable traffic (2) Maximize number of alternative routing paths	(1) Budget	Yes	54% increased gains in programmable traffic, as compared to state-of-the-art techniques
[15]	(1) Maximize number of alternative routing paths	(1) Budget (2) Number of nodes	Yes	
[16]	(1) Maximize number of alternative routing paths	(1) Budget	Yes	
[17]	(1) Minimize the maximum link utilization	(1) Link capacity	Yes	Maximum link utilization decreased by 32% (on average) after migration of 20% of active routers
[18]	(1) Minimize the maximum link utilization	(1) Link capacity	No	Highest benefit achieved at a migration of 40% of routers
[19]	(1) Maximize volume of programmable traffic	(1) Budget	No	Simulation reports that 95% of NCA requires only 10% of upgrading cost
[20]	(1) Maximize number of alternative routing paths & number of flows that can use the paths	(2) Number of nodes	Yes	

In [19], authors study the network control ability (NCA) gain by SDN deployment, and they suggest deployment of SDN should balance between financial investment and NCA. Simulation reports that only 10% of upgrading cost can achieve 95% of NCA in a network. Other than considering number of alternative paths or maximum link utilization, [20] proposes an Optimal Migration Schedule based on Customer's Benefit (OMSB). They argue that prior works overemphasize the controllability of TE while ignoring benefit of customers. In contrast, OMSB considers also the number of customers who can enjoy the alternative paths availed by SDN nodes deployment. A comparison of various deployment techniques discussed above is tabulated in Table 1.

Instead of seeking exact deployment sequence of SDN node in the legacy network, [21, 22] develop a hybrid OSPF/SDN network by dividing a network into multiple small-size domains, and each subdomain of these small networks are interconnected with SDN devices. Traffic traverses among subdomains have no other links but to go through interconnected SDN devices between each subdomain. With this approach, the performance of routing can be improved by tuning the Link State Advertisement (LSA) inside subdomains. Simulation reports that the performance depends on the number of subdomains partition in the network.

In contrast to the others approaches above, [23] propose a hardware device, SDN shim, which can be installed within legacy switches. The SDN shim enables legacy devices for communication routing and the forwarding of control information utilizing the SDN controller. And also, the SDN shim is pre-configured and connected to legacy device via VLAN trunks. With the configuration as such, traffic on switch can propagate to shim and control by the shim in accordance to the SDN control. However, each specific legacy switch requires the complementation of its own specific type of SDN-shim device respectively. The limitation for this solution thus demands for high financial expenditures in order to account for a wide variety of legacy devices, despite the claims that hardware shims are inexpensive.

From Table 1, several other related studies on SDN migration mainly focuses on combination of constraints such as financial budget, limited number of nodes, or link capacity during optimization. However, changes in dynamic network are inevitable. The proposed migration techniques are insufficient to adapt future changes in network. These will hinder the optimization process for deployment of SDN nodes, thus stifling efforts to make improvements on SDN migration.

3 Machine Learning in SDN

Machine Learning (ML), refers to the scientific study of algorithms and statistical models used by computer systems to effectively perform a specific tasks based on patterns and inference [36]. The perks of ML techniques application in SDN is as follows: Firstly, as most ML algorithms are data-driven, a great emphasis is put on the data controlled by the algorithms. One of the characteristics of a SDN centralized controller is its global network view which allows for efficient collection of the required data from the network. This allows for the facilitation of ML algorithm application in the network [37]. Secondly, ML techniques bring about more intelligent decision-making to the SDN controller, by deriving knowledge about the network via various data analysis and network optimization techniques. The "Knowledge Plane" of the network can even recommend certain network adjustments on behalf of the network operator. The network controller could then regulate the traffic flow in the data planes based on the decisions made [37, 38]. Finally, the optimal network solutions based on resource allocation and traffic configuration conducted by ML algorithms can be executed by the SDN networks in real time due to the ease of programmability of the SDN model [39].

The controller of the SDN is able to manipulate the flow of traffic by making various modifications to the switches and flow tables in order to route the flow of traffic more efficiently. The usage of ML algorithms in this case allows for more accurate calculation of the near-optimal routing solution [37]. At the same time, the decision-making with regards to the routing optimization problem in ML algorithms does not require an exact mathematical model of the underlying network [37]. Nevertheless, routing optimization with the aid of ML techniques allows for the saving of financial as well as human resources. The SDN model facilitates network resources management in order to maximize resource utilization. ML techniques and approaches can be used to estimate the number of active SDN nodes by utilization of only the data available at each singular node. This prevents errors in predictions regarding this parameter, thus allowing for better planning of the traffic scheduling among SDN nodes for the time being, while SDN nodes are incrementally deployed in the system during migration [40]. Lastly, traffic classification, which is known as a network function that is crucial for the identification of traffic flow types, allows for network management that is more fine-grained thus assuring more efficient allocation of network resources. ML techniques in this realm allows for more accurate recognition of encrypted traffic as compared to the more commonly known Deep Packet Inspection (DPI) approach [41]. The procedure involves the accumulation of traffic flow, followed by the application of ML techniques to extract information regarding traffic flow. The information extract is then archived for further analysis. Effective traffic classification enhances the TE in SDN networks, thus indirectly contributes to more effective link capacity estimation [37]. A better prediction of link capacity could be crucial to the SDN migration process as a whole.

4 Future Opportunities and Research Directions

The literature strongly suggests that deployment of SDN is likely progressive planning in multiple time steps, especially for a large network like ISP. Often, the performance of TE, such as load balancing, is co-related to the programmable traffic by SDN node deployed. However, the exact future demand of the network cannot be easily predicted. Multi-periods network planning with uncertain demand can be addressed in three different strategies [24], namely incremental network planning, all-periods network planning and adapting all-periods network planning. At the beginning of each time period in incremental planning, the network plans on the basis of recent knowledge and former time periods. However, the solution is optimal for individual time period, but there is no guarantee optimality as a whole for all periods network planning. On the contrary, complete forecast knowledge used in all-periods network planning guarantee optimal solution in overall, provided that forecast knowledge is correct. As such, all-periods planning can offer a lower bound solution as compared to other approaches. The limitation of all-periods planning, however, is often mismatched between the actual situation and forecast knowledge. In adapting all-periods planning, if forecast requirement is not met, planning will need to rework for subsequent time periods and current situation will be adapted into consideration.

Internet traffic is expected to grow at 27% rate per year from 2017–2022 [25], and the cost of deploying a SDN node will decrease roughly 20% over years as technology getting mature [26]. For a large network, there are more factors, topology change and as such, would impact the decision made today for hybrid SDN deployment. Obviously, it is impossible to predict every factor precisely to decide an optimal solution for hybrid SDN deployment. Furthermore, each network is unique and no standards to attain uniformity across networks [27]. Development of efficient algorithms in order to determine solutions to complex problems has proven to be a challenging task, because of the network complexity and diversity. Consequently, the solution of one network is often not applicable for another network of same type without tweaking the solution to fit in the other network. Therefore, adapting All-Period strategy offers a better insight for hybrid SDN deployment, in term of adapting future changes and different scenarios of the network to provide a revised solution. Stochastic Programming [24] and machine learning [28] are two commonly used techniques to incorporate future uncertainty in prediction.

Machine learning (ML), in particular, has been widely applied and reported breakthrough in variety of problems solving from diverse domains [29]. However, successful applications of ML in networking, such as network planning, operation, and management, have been limited [29] for two main obstacles. First, the success of ML technique relies heavily on data [30]. In a legacy network, data can be collected from limited sources due to monitoring overhead [31]. Second, automation is commonly the result of ML. However, heterogeneity of legacy network discourages automation in a network in which control actions that can be applied on legacy network devices are limited [32]. Nevertheless, the inherent natures of SDN with centralized control and programmability, promotes the applicability of ML in networking. [32] Suggests that ML for networking should be considered for a few reasons. It is challenging to build an exact model to represent complex system behaviors. As such, an estimated model of system can be trained by ML using historical data with acceptable accuracy. Often, the same problem in different networks may require to be solved independently due to a different characteristic in networks. ML raises new possibilities to construct a generalized model and apply the trained model in the respective network. Finally, ML technique like Reinforcement Learning can also help decision making task in networking, such as network planning and scheduling [33].

In conclusion, the deployment of SDN nodes in legacy network should consider applying ML techniques for two reasons. Firstly, SDN deployment is likely to span over several months or even years for large network migration. During the multi-periods deployment process, changes in a network, such as traffic growth and topology updates, are inevitable. Efforts attempt to forecast factors for SDN deployment are not resilience to changes. Second, the migration of SDN nodes deployment in legacy network is an offline process in which it does not require real-time responses similar to routing. Thus, ML can be applied to facilitate the training offline with historical traffic data. To date, there are only minimal efforts contributed in similar fashion. In [34], multiple historical traffic matrices (TM) are considered for TE optimization and SDN node deployment. Authors claim that multiple weighted TM is sufficient to depict traffic fluctuations in network and provide a better performance for TE. On the other, [35] proposes a linear regression method to monitor real-time links load in hybrid SDN by selectively placing SDN switches to cover important links in a network. Table 2 suggests several future research directions in the realm of applying ML techniques to improve the incremental deployment process in SDN migration.

Table 2. Future research directions of ML techniques to improve the incremental deployment process in SDN migration

Research direction	Reason	Benefits
Traffic growth	Changes in ISP/backbone networks [42], or recent introduction of augmented reality IoT [43] will cause traffic growth	ML techniques can project potential traffic growth using historical data to schedule the SDN deployment process accordingly
Changes in network topology	Network topology will required to change to accommodate for increase in number of users or network nodes [44]	ML techniques can adapt to the changes of network topology to aid the SDN migration process
Traffic engineering (TE) requirements	The traffic can be split into multiple paths during migration to model different traffic demands [2]	ML could provide adaptation to the changes of traffic splitting and support for better TE performance in SDN migration

5 Conclusion

In this article, we offer an general view regarding the benefits of deployment of the SDN model in legacy networks, and how the distinctive characteristics, centralized control, and programmability, of SDN shines to cater the future demands from internet applications. However, the deployment of SDN in legacy network brings its own set of challenges in technical, economical and organizational aspects. Literature supports that the deployment towards SDN is likely a staged process which spans over several years with multiple time steps. In each time step, only selected number of legacy nodes in network are replaced or upgraded to SDN-enabled nodes. Thus, a network co-exists with both SDN and legacy nodes, more commonly known as hybrid-SDN (H-SDN) network. Several efforts were made to propose architectural design of hybrid SDN.

Meanwhile, some others propose heuristic algorithms in determining the migration sequence to hybrid SDN network. However, SDN migration processes based on the current state of network requirements without considering any changes in network traffic, topology or TE requirements in its solution. As such, the deployment process is unable to adapt to future needs or changes in the network. Owing to the distinct characteristics and unforeseeable changes in the individual networks, such as traffic pattern and topology change, solutions proposed to solve one network problem may not be feasible in the other. Hence, there is a gap between existing literatures in H-SDN deployment, in which the algorithms proposed does not include change adaptation. This article sheds light on the new opportunities ML techniques can bring with regards to solving the SDN migration problem. It has always been challenging to define an exact model to represent the respective problem in it. ML offers the possibility to mold a generalized model to alleviate SDN migration problems and issues, mainly in projecting the future direction of the H-SDN deployment process. In conclusion, we suggest the future extensibility for H-SDN deployment is then possible in the direction of ML.

Acknowledgment. This research work is fully supported by Telekom Malaysia (TM) R&D and Multimedia University (MMU), Cyberjaya, Malaysia. We are very thankful to the team of TM R&D for providing the support to our research studies.

References

1. Juniper Networks Inc.: What's Behind Network Downtime? (2008)
2. Akyildiz, I.F., Lee, A., Wang, P., Luo, M., Chou, W.: A roadmap for traffic engineering in software defined networks. Comput. Netw. **71**, 1–30 (2014)
3. Alexander, D.S., et al.: The switchware active network architecture. IEEE Netw. **12**(3), 29–36 (1998)
4. Morris, R., Kohler, E., Jannotti, J., Kaashoek, M.F.: The click modular router. Most, no, Section 2, pp. 217–231 (2008)
5. Handley, M., Hodson, O., Kohler, E.: XORP: an open platform for network research. ACM SIGCOMM Comput. Commun. Rev. **33**(1), 53–57 (2003)
6. Chen, J., Zheng, X., Rong, C.: Survey on software-defined networking. In: Qiang, W., Zheng, X., Hsu, C.-H. (eds.) CloudCom-Asia 2015. LNCS, vol. 9106, pp. 115–124. Springer, Cham (2015). https://doi.org/10.1007/978-3-319-28430-9_9
7. Vissicchio, S., Vanbever, L., Bonaventure, O.: Opportunities and research challenges of hybrid software defined networks. ACM SIGCOMM Comput. Commun. Rev. **44**(2), 70–75 (2014)
8. Amin, R., Reisslein, M., Shah, N.: Hybrid SDN networks: a survey of existing approaches. IEEE Commun. Surv. Tutor. **20**(4), 3259–3306 (2018)
9. Vissicchio, S., Cittadini, L., Bonaventure, O., Xie, G.G., Vanbever, L.: On the co-existence of distributed and centralized routing control-planes. Proc. - IEEE INFOCOM **26**, 469–477 (2015)
10. Farias, F., Carvalho, I., Cerqueira, E., Abelém, A., Rothenberg, C.E., Stanton, M.: LegacyFlow: bringing openflow to legacy network environments (2011)

11. Stringer, J.P., Fu, Q., Lorier, C., Nelson, R., Rothenberg, C.E.: Cardigan: deploying a distributed routing fabric. In: Proceedings of Second ACM SIGCOMM Workshop on Hot Topics in Software Defined Networking (HotSDN 2013), pp. 169–170 (2013)
12. Levin, D., Canini, M., Schmid, S., Feldmann, A.: Incremental SDN deployment in enterprise networks. In: Proceedings of ACM SIGCOMM 2013 Conference on SIGCOMM - SIGCOMM 2013, p. 473 (2013)
13. Levin, D., Canini, M., Schmid, S., Feldmann, A.: Panopticon: reaping the benefits of partial SDN deployment in enterprise networks. In: USENIX Annual Technical Conference, pp. 333–345 (2014)
14. Poularakis, K., Iosifidis, G., Smaragdakis, G., Tassiulas, L.: One step at a time: optimizing SDN upgrades in ISP networks. In: Proceedings - IEEE INFOCOM, pp. 1–9 (2017)
15. Das, T., Caria, M., Jukan, A., Hoffmann, M.: Insights on SDN migration trajectory. IEEE International Conference on Communications, pp. 5348–5353 (2015)
16. Tanha, M., Sajjadi, D., Ruby, R., Pan, J.: Traffic engineering enhancement by progressive migration to SDN. IEEE Commun. Lett. 22(3), 438–441 (2018)
17. Hong, D.K., Ma, Y., Banerjee, S., Mao, Z.M.: Incremental deployment of SDN in hybrid enterprise and ISP networks. In: Proceedings of the ACM Symposium on SDN Research, SOSR 2016, pp. 1–7 (2016)
18. Guo, Y., Wang, Z., Yin, X., Shi, X., Wu, J., Zhang, H.: Incremental deployment for traffic engineering in hybrid SDN network. In: 2015 IEEE 34th International Performance Computing and Communications Conference on IPCCC 2015 (2016)
19. Jia, X., Jiang, Y., Guo, Z.: Incremental switch deployment for hybrid software-defined networks. In: Proceedings of Conference on Local Computer Networks, LCN, pp. 571–574 (2016)
20. Yuan, T., Huang, X., Ma, M., Zhang, P.: Migration to software-defined networks: the customers' view. China Commun. 14(10), 1–11 (2017)
21. Caria, M., Das, T., Jukan, A., Hoffmann, M.: Divide and conquer: partitioning OSPF networks with SDN. In: Proceedings of 2015 IFIP/IEEE Symposium on Integrated Network and Service Management, IM 2015, pp. 467–474 (2015)
22. Caria, M., Jukan, A., Hoffmann, M.: SDN partitioning: a centralized control plane for distributed routing protocols. IEEE Trans. Netw. Serv. Manag. 13(3), 381–393 (2016)
23. Casey, D.J., Mullins, B.E.: SDN shim: controlling legacy devices. In: Proceedings of Conference on Local Computer Networks, LCN, 26–29 October 2015, pp. 169–172 (2015)
24. Kronberger, C., Schondienst, T., Schupke, D.A.: Impact and handling of demand uncertainty in multiperiod planned networks. In: IEEE International Conference on Communications (2011)
25. Cisco: Cisco Visual Networking Index: Forecast and Trends, 2017–2022 (2018). https://www.cisco.com/c/en/us/solutions/collateral/service-provider/visual-networking-index-vni/white-paper-c11-741490.html#_Toc529314176. Accessed 11 Feb 2019
26. Türk, S., Lehnert, R., Radeke, R.: Network migration using ant colony optimization. In: 2010 9th Conference Telecommunication Media Internet, CTTE 2010 (2010)
27. Aibin, M.: Traffic prediction based on machine learning for elastic optical networks. Opt. Switch. Netw. 30, 33–39 (2018)
28. Vinayakumar, R., Soman, K.P., Poornachandran, P.: Applying deep learning approaches for network traffic prediction. In: 2017 International Conference on Advances in Computing, Communications and Informatics, ICACCI 2017, pp. 2353–2358 (2017)
29. Boutaba, R., et al.: A comprehensive survey on machine learning for networking: evolution, applications and research opportunities. J. Internet Serv. Appl. 9(1), 147–169 (2018)

30. Boser, B.E., Guyon, I.M., Vapnik, V.N.: A training algorithm for optimal margin classifiers. In: Proceeding COLT 1992 Proceedings of Fifth Annual Workshop on Computer Learning Theory, pp. 144–152 (1992)
31. Chen, Z., Wen, J., Geng, Y.: Predicting future traffic using hidden Markov models. In: Proceedings of International Conference on Network Protocols ICNP, vol. 2016–December, NetworkML, pp. 1–6 (2016)
32. Wang, M., Cui, Y., Wang, X., Xiao, S., Jiang, J.: Machine learning for networking: workflow, advances and opportunities. IEEE Netw. **32**(2), 92–99 (2018)
33. Dietterich, T.G.: Reinforcement learning in autonomic computing: a manifesto and case studies. IEEE Internet Comput. **11**, 22–30 (2007)
34. Guo, Y., Wang, Z., Yin, X., Shi, X., Wu, J.: Traffic engineering in hybrid SDN networks with multiple traffic matrices. Comput. Netw. **126**, 187–199 (2017)
35. Cheng, T.Y., Jia, X.: Compressive traffic monitoring in hybrid SDN. IEEE J. Sel. Areas Commun. **36**(12), 2731–2743 (2018)
36. Bishop, C.M.: Pattern Recognition and Machine Learning. Springer, Heidelberg (2006)
37. Xie, J., et al.: A survey of machine learning techniques applied to software defined networking (SDN): research issues and challenges. IEEE Commun. Surv. Tutor. **21**(1), 393–430 (2018)
38. Jose, A.S., Nair, L.R., Paul, V.: Data mining in software defined networking-a survey. In: 2017 International Conference on Computing Methodologies and Communication (ICCMC). IEEE (2017)
39. Xu, G., Mu, Y., Liu: Inclusion of artificial intelligence in communication networks and services. ITU J. ICT Discoveries, Special 1, 1–6 (2017)
40. Del Testa, D., et al.: Estimating the number of receiving nodes in 802.11 networks via machine learning techniques. In: 2016 IEEE Global Communications Conference (GLOBECOM). IEEE (2016)
41. Amaral, P., et al.: Machine learning in software defined networks: data collection and traffic classification. In: 2016 IEEE 24th International Conference on Network Protocols (ICNP). IEEE (2016)
42. Gerber, A., Doverspike, R.: Traffic types and growth in backbone networks. In: Optical Fiber Communication Conference. Optical Society of America (2011)
43. Klymash, M., et al.: Method for optimal use of 4G/5G heterogeneous network resources under M2 M/IoT traffic growth conditions. In: 2017 International Conference on Information and Telecommunication Technologies and Radio Electronics (UkrMiCo). IEEE (2017)
44. Rodrigues, H., et al.: Traffic optimization in multi-layered WANs using SDN. In: 2014 IEEE 22nd Annual Symposium on High-Performance Interconnects. IEEE (2014)

Searching the Shortest Pair
of Edge-Disjoint Paths
in a Communication Network.
A Fuzzy Approach

Lissette Valdés[1]([✉]), Alfonso Ariza[1], Sira María Allende[2], and Gonzalo Joya[1]

[1] Malaga University, Málaga, Spain
luthien44@gmail.com, {aarizaq,gjoya}@uma.es
[2] Havana University, Havana, Cuba
sira@matcom.uh.cu

Abstract. In this paper, we address the problem of finding the shortest pair of edge-disjoint paths between two nodes in a communication network. We use a new cost function named modified fuzzy normalized used bandwidth, which is described as a fuzzy triangular number, thus incorporating the uncertainty generated in calculating this magnitude in a real network. The proposed algorithm uses as a sub-algorithm an adaptation of a Modified Fuzzy Dijkstra algorithm applied in a type V mixed graph with arcs whose costs are negative triangular fuzzy numbers, which has been described in previous work. We prove its effectivity by simulating traffic close to overload with two types of communication sources: regular and priority sending of information. The addressed problem presents a considerable interest in contexts such as finance entities or government services, where privacy and security against external attacks have to be considered.

Keywords: Pair of edge-disjoint paths · Fuzzy number · Dijkstra algorithm · Shortest path

1 Introduction

Communication networks survivability can be defined as the ability of the network to support the committed Quality of Services (QoS) continuously in the presence of various failure scenarios. Basically, the system must remain operational at all times regardless of whether a failure occurs (both in a node and a link). A concept related to network survivability is the capacity of *Self-healing* that can be associated with the redistribution of traffic in the system by using alternative paths between two nodes when it is appropriate. A third crucial concept in communication networks is the guarantee of security in communications against the intrusion of third parties.

© Springer Nature Switzerland AG 2019
I. Rojas et al. (Eds.): IWANN 2019, LNCS 11507, pp. 640–652, 2019.
https://doi.org/10.1007/978-3-030-20518-8_53

In this paper, we address the issues above stated by analysing the problem of searching the shortest pair of edge-disjoint paths between two source and destination nodes in a communication network. The pair of paths can have different purposes: (i) Deviate the information by one of the paths (replacement) when the other suffers a failure. Thus, the network avoids the loss of information that would occur while repairing the issue or looking for an alternative path. (ii) Distribute the data sent between the two paths simultaneously when the network is close to saturation. In this case, it is possible that a package with a certain length cannot completely arrive if a single path sends it. However, this package could reach its destiny if, under the same bandwidth conditions, we divide it into two smaller packages that are transmitted through two paths with no common links. (iii) Distribute the information in a not predictable way between both paths, thus making it impossible to capture a complete message by an intruder.

These tasks have a considerable interest in contexts such as finance entities or government services, where privacy and security against external attacks have to be considered; as well as in networks with a continuous increase in their transmission speed such as the optical ones. Thus, one or several of them have been addressed by different works along the last two decades: [3–6,10,12]. The following characteristics are common to all or a big part of these works: the cost of edges and paths is represented by a real number, in some cases invariant in time; the goal is not to find the shortest pair of paths but only a disjoint pair of them; real traffic in a credible network is not simulated; only a mathematical and/or computational analysis of the proposed methods are presented.

It is not our only interest that both paths are link-disjoint, but also the sum of their costs is minimal. In this way, when both paths are used at the same time, we would be optimizing the cost of sending the information and contributing to the reduction of the saturation of the network. On the other hand, in communication networks, the variables values used to obtain the state of the system over a particular time interval (among these, cost functions in links), have been measured over the previous time interval. This implies a high probability of discrepancy between the calculated value and the real value at each moment. We intend to take this circumstance into account by assuming the function defining the weight (or cost) of links be represented by a fuzzy number. Thus, we model a communication network as a type V fuzzy graph with a set of nodes and another set of links between them.

From the above considerations, as the main contribution, this paper proposes the Fuzzy Shortest Pair of Edge-Disjoint Paths algorithm (FSPPA) to guarantee the reliability of a communication network based on the analysis of its connectivity and the occupied bandwidth of its links. Our aim consists of finding the shortest pair of paths, disjoint in links, between two fixed nodes, in an online manner. We do not consider the possibility of pre-designing the network to ensure the existence of paths duplicity, but we start from a network already in use and look for a pair of edge-disjoint paths if possible.

2   Problem Statement

From a mathematical-computational perspective, we model the network by defining a non-directed *type V fuzzy graph* $\tilde{G} = (V, E, \mathfrak{C})$ that is associated to the system. The uncertainty generated by operating in the time interval t with the values of the variables obtained in the $t-1$ interval, is modeled using a fuzzy logic vision, by means of the use of cost functions defined by triangular fuzzy numbers. For each edge $e \in E$ its cost function $\tilde{C}_e = (a_e, b_e, c_e) \in \mathfrak{C}$ is a *positive triangular fuzzy number*, which is close to the value b_e and is measured with an uncertainty bounded by the values a_e and c_e. In the same way, the cost of a path P, $\tilde{C}_P = (a_P, b_P, c_P)$, is also a positive triangular fuzzy number. We consider the cost of a link as an additive metric, therefore the cost of the path P is defined as $\tilde{C}_P = \sum_{\forall e_i \in P} \tilde{C}_{e_i}$. Graph $\tilde{G}$ is connected such that each vertex can be reached from any other vertex by a sequence of edges. Finally, more than one path may exist between vertices r and t.

Our goal is to find the shortest pair of edge-disjoint paths between the source and destination vertices r and t (pair of paths with no common edges where the sum of the costs of both paths is minimum).

The solution of the search for the shortest pair of edge-disjoint paths in a graph with crisp and positive costs has already been proposed in [2]. In essence, this proposal consists of the following steps: (1) Finding the shortest path of the graph, S, using the Dijkstra algorithm. (2) Modifying the original graph by replacing the edges of S by arcs oppositely oriented, and whose costs are defined as the opposite (negative) number of the costs of said edges (this new graph is called modified graph). (3) Finding the shortest path in the modified graph, P_{aux}, by a modification of the Dijkstra algorithm that allows its convergence in a graph with negative costs. (4) Finally, the shortest pair of edge-disjoint paths is either (S, P_{aux}) or another pair of paths resultant from the combination of S and P_{aux}.

Our algorithm FSPPA constitutes an adaptation of the previous procedure to a type V fuzzy graph. This new approach requires the change of the operations that are involved in the procedure. Thus, we must have a criterion of comparison of fuzzy numbers; we have to find a way to express the negative character of a triangular fuzzy number necessary in the construction of the modified graph $\tilde{G}'$; and we must make an adaptation of the Dijkstra algorithm that can be applied to the graph $\tilde{G}$ and also modify this such that it converges in $\tilde{G}'$ that contains arcs with negative fuzzy costs. These new operations are an essential part of our contribution to this work and are briefly explained in some detail in Sect. 3.

3   Remarkable Concepts Involved in the FSPPA

3.1   Necessary Conditions Under Which a Path Is the Shortest in $\tilde{G}$

Proposition 1 describes an operational condition for the comparison of the costs of two paths based on the ranking criterium proposed in [11].

Proposition 1. *Let P_1 and P_2, be two paths from a source vertex r to a destination vertex t on a type V fuzzy graph $\tilde{G} = (V, E, \mathfrak{C})$ whose costs are triangular fuzzy numbers $\tilde{C}_{P_1} = [a_{P_1}, b_{P_1}, c_{P_1}]$ and $\tilde{C}_{P_2} = [a_{P_2}, b_{P_2}, c_{P_2}]$, respectively. The comparison among both paths is performed based on their total integrals [11]. For a fixed $\alpha \in [0, 1]$, P_1 is shorter (equal) than P_2 if and only if,*

$$(1 - \alpha)a_{P_1} + b_{P_1} + \alpha c_{P_1} < (=)(1 - \alpha)a_{P_2} + b_{P_2} + \alpha c_{P_2} \qquad (1)$$

where the index α represents the degree of optimism of the decision maker.

From Proposition 1, for a fixed $\alpha \in [0, 1]$, we can establish the necessary condition for being a path S a shortest path between r and t in $\tilde{G}$.

$$(1 - \alpha)a_S + b_S + \alpha d_S \leq (1 - \alpha)a_P + b_P + \alpha d_P, \qquad \forall S, P \in \tilde{G} : P \neq S \quad (2)$$

where $\tilde{C}_P = [a_P, b_P, d_P]$ is the cost of path P.

3.2 Fuzzy Dijkstra Algorithm for the Search of the Shortest Path S

For the search of S, we have applied a Fuzzy Dijkstra Algorithm in a type V fuzzy graph. The detailed description of this algorithm appears in a previous work of ours [8].

3.3 General Structure for the Shortest Pair of Edge-Disjoint Paths

Let us denote by P a path with some segments overlapping with S, and by P' a path without any segment overlapping with S. We define as *segment* of a path one edge or several consecutive edges that belong to this path. Let $\gamma_1 \times \gamma_2$ denote the pair of paths in $\tilde{G}$, where "$\times$" symbolizes the combination of the individual paths γ_1 and γ_2. The shortest pair of edge-disjoint paths is then constituted by one of the two configurations in Eq. 3,

$$(\gamma_1 \times \gamma_2)_{\substack{\text{edge-disjoint} \\ \text{shortest pair}}} \in \{S \times P', \ P_1 \times P_2\} \qquad (3)$$

where both paths in the $P_1 \times P_2$ configuration contains a single (but different) segment of path S, in particular each of these segments is at an endpoint vertex of S.

A *break* is an edge of path S that belong to neither of the paths that form the pair of paths, but it is adjacent at each of its endpoint vertices to an edge of each path. The pair $P_1 \times P_2$ contains breaks while $S \times P'$ does not contain any break.

3.4 Definition of a Modified Graph Fuzzy $\tilde{G}'$

The path P_{aux} is found in a new *mixed type V fuzzy graph*, denoted as $\tilde{G}' = (V, A', \mathfrak{C}')$, which is a modification of the original $\tilde{G}$ where: the set V of vertices is the same than in the original graph; the edges and arcs in A' are defined as:

each edge (v_i, v_j) belonging to S in $\tilde{G}$ is replaced by the arc $\overrightarrow{(v_j, v_i)}$. The rest of the edges in $\tilde{G}$ remain the same; and the costs in $\mathfrak{C}'$ are defined as: the cost of each arc is the complementary of the cost of the corresponding edge in $\tilde{G}$ (*complementary of a triangular fuzzy number* [1]. The costs of the remain edges are the same than in $\tilde{G}$.

Path P_{aux} has segments overlapping with S under the following constraints:

(i) On the overlapped segments, the direction of path P_{aux} in $\tilde{G}'$ is opposite to that of the shortest path S in $\tilde{G}$.

(ii) The cost of each arc of P_{aux} is defined as the *complementary of the original triangular fuzzy cost* of its corresponding edge in $\tilde{G}$.

3.5   Conditions for Nonexistence of Negative Cycles

The costs of the introduced arcs in $\tilde{G}'$ are negative triangular fuzzy numbers. Thus, it could be possible the presence of cycles whose total cost is a negative triangular fuzzy number (*negative cycles*), which could lead to the non-convergence of the algorithm. Thus, we have to find under what conditions (α values) the graph $\tilde{G}'$ has no negative cycles.

Any cycle in $\tilde{G}'$ is either:

Mixed cycle: Cycle formed by arcs with costs defined as negative triangular fuzzy numbers replacing the edges of the shortest path S and edges with costs defined as non-negative triangular fuzzy numbers that belong to the rest of $\tilde{G}'$. A mixed cycle can be either positive or negative, according to the sum of the costs of its edges and arcs.

Simple cycle: Cycle composed by only non-negative edges.

Sub-mixed cycle: Cycle with a set of consecutive non-negative arcs and another set of consecutive negative arcs, directed towards the source vertex.

Proposition 2 gives a condition that guarantees the convergence of a shortest path search algorithm in $\tilde{G}'$.

Proposition 2. *Given the type V fuzzy mixed graph $\tilde{G}' = (V, A', \mathfrak{C}')$ defined in Sect. 3.4 under the following conditions:*

– *The costs in $\mathfrak{C}'$ are assumed as triangular fuzzy numbers.*
– *For the comparison of costs, the ranking method proposed by [11] is applied.*
– *Any Mixed cycle in $\tilde{G}'$ is composed by Sub-mixed cycles.*

The following properties hold:

(i) *A Sub-mixed cycle has a non-negative total cost if the parameter α, used for the ranking method, takes any value in the interval $[\frac{1}{2}, 1]$.*

(ii) *Any Mixed cycle is non-negative if every Sub-mixed cycle included in it has a non-negative total cost.*

3.6   Modified Fuzzy Dijkstra Algorithm (MFDA)

We propose a Modified Fuzzy Dijkstra Algorithm (MFDA) that finds the path P_{aux} in the graph $\tilde{G}'$. The term *modified* comes from the fact that this algorithm can find the shortest path in the mixed graph $\tilde{G}'$ that contains arcs with negative fuzzy costs.

Let $\Psi(r, v)$ denote the set of all paths between vertices r and v ($v, r \in V$ with $v \neq r$), and $P^* \in \Psi(r, v)$ be the shortest path between r and v. The cost of P^* is defined in Eq. 4,

$$\tilde{\delta}_{(r,v)} = \tilde{C}_{P^*(r,v)} = \begin{cases} \min\limits_{\forall P \in \Psi(r,v)} \{\tilde{C}_{(P(r,v))}\} & \text{if } \Psi(r, v) \neq \emptyset \\ \infty & \text{if } \Psi(r, v) = \emptyset \end{cases} \tag{4}$$

Then, for each vertex $v \in V$, the MFDA defines an attribute $\tilde{d}(v)$, which is an upper bound on the cost of P^*, we have $\tilde{d}(v) \geq \tilde{\delta}_{(r,v)}$.

Given a vertex $v \in V$, we will denote by $\Gamma(v)$ the set of neighbors of v, that is, adjacent vertices to v. To each vertex is assigned a *label* that is updated during the algorithm execution. At each iteration, the label of vertex $v \in V$ contains its predecessor to v in the (known so far) shortest path from r to v, denoted by $w(v)$, and its corresponding cost $\tilde{d}(v)$. Only when the algorithm ends, we can say that $w(v)$ is the definitive predecessor of v in the shortest path from r to v.

An important roll in MFDA has a set H containing the vertices whose label is potential to be updated. At the end of the algorithm, for each vertex v, $\tilde{d}(v)$ will coincide with $\tilde{\delta}_{(r,v)}$ and $w(v)$ will be its predecessor in $P^*(r, v)$. The pseudo-code of MFDA is described in Algorithm 1, The label of each vertex is initialized in the sub-algorithm Initialize-single-source $(\tilde{G}', r)$. In the sub-algorithm Relaxation $2(u, v, \tilde{C}_{(u,v)}, \alpha)$ the label of a specific vertex can be updated by the replacement of a lower distance from the source and the predecessor of u in this new path. In this case, the vertex is added to H regardless this has already been in this set. Thus, the MFDA guarantees the reentry to H of vertices previously labeled. These modifications in MFDA are significant for graphs like $\tilde{G}'$ with negative costs in some of their edges. Once the destination vertex t is first labeled then $\tilde{d}(t)$ is the distance of the shortest path from r to this vertex $(\tilde{d}(t) = \tilde{\delta}(r, t))$.

3.7   Total Cost of the Shortest Pair of Edge-Disjoint Paths

Once the paths S and P_{aux} are found, the pair of paths described in Eq. 3 is obtained, which can have any of the configurations $S \times P'$ or $P_1 \times P_2$.

In the configuration $P_1 \times P_2$, the common edges/arcs between S and P_{aux} also constitute the breaks between P_1 and P_2. Lets assume there are M breaks between paths P_1 and P_2 where for each break $m = \overline{1, M}$ $(a_m, b_m, c_m) + (-c_m, -b_m, -a_m) = (a - c, 0, c - a)_m$. Then, the cost of $S \times P_{\text{aux}}$ is equal to the sum of the total cost of $P_1 \times P_2$ plus a N_2-*zero triangular fuzzy number* [1] which correspond to the sum of the costs of the overlapping arcs associated to each break. When there are no breaks ($M = 0$), the shortest pair of edge-disjoint paths is of type $S \times P'$ and the cost of the pair $S \times P_{\text{aux}}$ is exactly equal to the cost of $S \times P'$.

Algorithm 1. Mod-FuzzyDijkstra($\tilde{G}', r, t, \alpha$)

```
1: Initialize-single-source($\tilde{G}', r$)                          ▷ Initialization
2: H ← V
3: while H ≠ ∅ do
4:     u ← z|d̃(z) = min {d̃(x)}
                    ∀x∈H
5:     Update H ← H − {u}
6:     if u = t then -end while-
7:     end if
8:     if Γ(u) ≠ ∅ then
9:         for each vertex v ∈ Γ(u) do
10:            if v ≠ w(u) then
11:                Relaxation 2(u, v, C̃_{(u,v)}, α)               ▷ Relaxation 2 of v
12:            end if
13:        end for
14:     end if
15: end while
```

Once we find the paths S and P_{aux} we can compute the total cost of $S \times P_{\text{aux}}$ and detect the existence of breaks, as we describe in Eq. 5,

$$\tilde{C}_{\gamma_1 \times \gamma_2} = \begin{cases} \tilde{C}_{S \times P_{\text{aux}}} - (A_M - C_M, 0, C_M - A_M) & \begin{array}{l} \text{if there are } M \text{ breaks} \\ (\gamma_1 \times \gamma_2 = P_1 \times P_2) \end{array} \\[2ex] \tilde{C}_{S \times P_{\text{aux}}} & \begin{array}{l} \text{if there are no breaks} \\ (\gamma_1 \times \gamma_2 = S \times P') \end{array} \end{cases} \quad (5)$$

where $(A_M - C_M, 0, C_M - A_M) = \sum_{m=1}^{M} (a - c, 0, c - a)_m$.

4 Algorithm to Find the Fuzzy Shortest Pair of Edge-Disjoint Paths (FSPPA)

The FSPPA searches for the shortest pair of edge-disjoint paths in a type V fuzzy graph. In the algorithm, we look first for the shortest path of $\tilde{G}$, denoted as S, by the application of the MFDA. Then, we create a new graph $\tilde{G}'$ through the transformations explained in Sect. 3.4 (Algorithm 2). On the new graph $\tilde{G}'$ we apply the MFDA (Algorithm 1) to find the shortest path P_{aux}. Once the paths S and P_{aux} have been obtained, the common edges between them, called breaks, are eliminated, to create a pair of paths, disjoint in edges, whose total sum of costs is the minimum and has the structure described in Sect. 3.3. The Algorithm 3 computes the total cost of the shortest pair of edge-disjoint paths.

Algorithm 4 shows the pseudocode of the Algorithm for Fuzzy Shortest Pair of edge-disjoint Paths (FSPPA).

Algorithm 2. Build-ModifiedGraph($\tilde{G}, r, t, S$)

    ▷ Creating graph $\tilde{G}'$
1: **for** each $e_i = (u, v)_i \in S$ **do**
2:     Update $e_i \leftarrow \overrightarrow{(v, u)}_i$ ▷ Each edge of S is replaced by an unique arc directed to r
3:     Update $\tilde{C}_{(v,u)_i} = -\tilde{C}_{(u,v)_i}$ ▷ The cost of each arc is defined as the complementary of the original edge cost
4: **end for**

Algorithm 3. Calc-TotalCost($\tilde{G}, M, S, P_{\text{aux}}, \tilde{C}_S, \tilde{C}_{P_{\text{aux}}}$)

    ▷ Computing the total cost of pair (γ_1, γ_2)
1: $\tilde{C}_{(S,P_{\text{aux}})} = \tilde{C}_S + \tilde{C}_{P_{\text{aux}}}$
2: $A_M = \text{sum}(a_{e_i} \,|\, e_i \in M)$, $D_M = \text{sum}(d_{e_i} \,|\, e_i \in M)$
3: $\tilde{C}_{(\gamma_1,\gamma_2)} = \tilde{C}_{(S,P_{\text{aux}})} - (A_M - D_M, 0, D_M - A_M)$

Algorithm 4. FSPPA($\tilde{G}, r, t$)

1: $\alpha = \alpha_0$ ▷ Setting α according to Proposition 2
2: $(S, \tilde{C}_S) = $ Fuzzy-Dijkstra($\tilde{G}', r, t, \alpha$)   ▷ To apply FDA to $\tilde{G}$ to obtain S
3: $\tilde{G}' = $ Build-ModifiedGraph($\tilde{G}, r, t, S$)   ▷ Creating graph $\tilde{G}'$
4: $(P_{\text{aux}}, \tilde{C}_{P_{\text{aux}}}) = $ Mod-FuzzyDijkstra($\tilde{G}', r, t, \alpha$)   ▷ To apply MFDA to $\tilde{G}'$ to obtain P_{aux}
5: $M := \{e_i \,|\, e_i \in S \cap P_{\text{aux}}\}$ ▷ Breaks between S and P_{aux}
6: **if** $M = \emptyset$ **then**
7:     $\gamma_1 := S$ and $\gamma_2 := P_{\text{aux}}$ ▷ If there are no breaks
8: **else**
9:     $(\gamma_1, \gamma_2) := (S \cup P_{\text{aux}}) - \{M\}$ ▷ If there are breaks
10: **end if**
11: $(\tilde{C}_{(\gamma_1,\gamma_2)}) = $ Calc-TotalCost($\tilde{G}, M, S, P_{\text{aux}}, \tilde{C}_S, \tilde{C}_{P_{\text{aux}}}$) ▷ To calculate the total cost of (γ_1, γ_2)

The maximum complexity is O(2 (E + V log V) + D log D), where E is the number of links, V the number of nodes and D is the diameter of the network.

5 Experimentation and Results

We apply the FSPPA in a network with high traffic load. We illustrate the effectiveness of the algorithm when we want to find the shortest pair of edge-disjoint paths where both paths can be used at the same time to distribute the information delivery and contribute to a decrease in network saturation.[1,2]

In experiment 1 we use a set of fixed costs and in experiment 2 we use as the variable indicator of the cost of a link the *modified fuzzy normalized used bandwidth*, defined in Eq. 6,

$$[\widetilde{bw}]_{ij}^{A,B} = A * B^{-[bw]'_{ij}} [\widetilde{bw}]'_{ij}, \qquad \text{for each } e = (i, j) \tag{6}$$

---

[1] The source code of the model used can be downloaded in https://github.com/ aarizaq/flowsimulator.

[2] The configuration for the running of the experiments can be downloaded in https:// github.com/aarizaq/configurationFuzzy/blob/master/omnetpp2.ini.

where $[bw]'_{ij} = \frac{[BW]'_{ij}}{C_{ij}} \frac{C_{\max}}{C_{ij}}$ ($[bw]'_{ij}$ and $[BW]'_{ij}$ are the mean used bandwidth and the normalized mean used bandwidth in link (i,j) at a measurements time interval, respectively; C_{ij} is the capacity of link (i,j)) and $\widetilde{[bw]}'_{ij} = \left([bw]_{ij}^{\min}, [bw]'_{ij}, [bw]_{ij}^{\max}\right)$ ($[bw]_{ij}^{\min}$ and $[bw]_{ij}^{\max}$ are the minimum and maximum value of the Normalized Instantaneous Used Bandwidths measured in the (n-1)-th time interval, respectively). We consider this variable as a triangular fuzzy number and weighted with the values A and B to achieve a displacement to the right and a widening of the fuzzy number. We use the values $A = 10$ and $B = 20$. The update of the links is done every 300 s. This interval is realistic in a network with size as the one we use, and justifies the use of fuzzy numbers in the measure of the link costs since the uncertainty in the value of the variable is high.

5.1   Experiment 1

Experiment 1 consists in, given the network, to apply the FSPPA for all the possible pairs of server and client node. Our focus is to check that our algorithm always find the shortest pair of edge-disjoint paths. Each pair of nodes in the network has at least two alternative paths between them. Thus, when we randomly search for communication between any two nodes, we can apply our algorithm.

We use a network of 12 US cities, whose graphical representation can be found in [7], considering the costs of its links as triangular fuzzy numbers. The costs of links are randomly fixed in the simulation. We perform a single but exhaustive execution due to its high computational cost.

For each node, we apply an algorithm to find the K-shortest paths between this and any other node (K large enough). The costs of the paths found are calculated and organized in ascending order. After a simple but very computationally expensive procedure, we can obtain the shortest disjoint-edge pairs of paths.

On the other hand, we apply the FSPPA algorithm to find the shortest pair of edge-disjoint paths and compare its total cost with that obtained by the exhaustive search.

We have verified that in all cases our algorithm finds the shortest pair of edge-disjoint paths.

5.2   Experiment 2

Once we have shown that our algorithm finds the shortest pair of edge-disjoint paths in all cases, experiment 2 is aimed at checking whether the use of edge-disjoint paths can help guarantee the quality of a given priority traffic in conditions of network overload.

Priority traffic is understood as that generated by a set of nodes to which, for security reasons, a certain level of communication privilege is granted.

Our goal is to find a pair of paths, disjoint in edges, to use them simultaneously to send the fractionated information through both paths. This strategy improves the communication security of priority sources since it is more diffi-cult to capture a complete message by external elements. Also, it increases the quality of the transmission since it decreases the bandwidth necessary in each connection of the edge-disjoint paths, reducing the probability of losses in condi-tions close to saturation. On the other hand, we could create redundant traffic, that is, always sending the same message through two paths at once. This action increases the traffic on the network, but in the case where one of the paths fails, the information arrives anyway by the other path.

In this experiment we use a 57-nodes network inspired by the Nippon Tele-graph and Telephone Corporation (NTT) [9]. We adjust the original network so that each node can be accessed by at least two paths.

The simulation is flow oriented, that is, only two events are simulated: the start and end of a burst, thus guaranteeing the simulation in time of all the shipment by using small information packages. Therefore, a single call is estab-lished to send all the information, and it is not finished until the operation is completely finished. Also, storage is not simulated by queues at the nodes. If a node does not have sufficient capacity to transmit the burst, the data is lost until there is free space or the burst ends, in which case it will be lost entirely. Thus, we are simulating a system without delays, except propagation, which could be assimilated to an Optical Network without delay elements. Consequently, the variable that will determine the quality of the network will be the byte delivery ratio. Thus, we will know whether the data sent is lost or not.

The simulation time has been 10^4 s, and all the links have the same capacity (1 Gb/s).

The traffic is generated through calls with a connection. That is, once a call is established, the chosen path does not change during the entire duration of the call. These calls do not make reservations of resources, so the establishment of calls will never be rejected, but there may be loss of data in it. This restriction facilitates the visualization of the loss of data due to saturation of the links.

Ten replications per experiment have been carried out with different seeds, and the sending of information between nodes follows a probabilistic distribution.

Each node can have two types of communication sources (or independent traffic generators), F1 and F2. Source F1 corresponds to the regular sending of the information. It sends the information only by a single path: the shortest one. Source F2 is a prioritary message source. In nodes with source F2 we can make the information travel either by the shortest pair of edge disjoint path found by our FSPPA or by only one path like sources F1 do. To facilitate the observability of the experiment, sources F2 only communicate with other sources F2.

Description of the Simulation. We measure the quality of the transmission in the network through the Bytes Delivery Ratio (BDR) that is the ratio between delivered and sent bytes.

Experimentation was performed in a network with conditions very close to sat-uration. The source F1 is implemented in all nodes, while the source F2 is only active in a limited number of randomly selected nodes in each experiment. Exper-iments were carried out for sets of 5, 10, 15, 20 and 25 nodes with source F2.

For each set of nodes with source F2, we perform ten simulations. The mean Bytes Delivery Ratio (MBDR) is calculated in each case. The MBDR for all F1 sources is also calculated. We compare the performance of both generated traffics.

Simulation Results. Figure 1 shows the results of experiment 2. On the X-axis the numbers represent the size of the set of F2 source nodes that is used in the experiments, and the Y-axis represents values of MBDR. Each line shows the MBDR of the network for each repetition of the traffic simulations under different conditions.

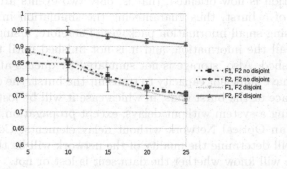

Fig. 1. Simulation results in Experiment 2. The x-axis represents the number of sources F2 in the network and Y-axis the values of MBDR. The Blue line shows the MBDR corresponding to the traffic by sources F1 when sources F2 do not apply FSPPA; Yellow line corresponds to the traffic by sources F1 when sources F2 apply FSPPA; Orange line represents MBDR of sources F2 when they do not apply the FSPPA; Green line shows the MBDR of sources F2 when they apply the FSPPA. (Color figure online)

Blue and Yellow Lines: We compare the behavior of the traffic by source F1 according to the application of the FSPPA by nodes with source F2. When there are few nodes with source F2, both lines are very similar. However, as the number of nodes with source F2 increases, the yellow line separates below the blue one, that is, the nodes with source F1 are harmed even more due to the increase in traffic (they reach the impermissible value of 0.75 approximately). Logically, the MBDR of sources F1 decreases when increases the number of sources F2 applying FSPPA.

Green and Orange Lines: We compare the behavior of traffic by source F2, whether when it applies FSPPA or not. When sources F2 use FSPPA, their MBDR (green line) remains between 0.9 and 0.95 in all cases. However, when these nodes do not apply FSPPA (orange line), their MBDR starts from 0.87 and reaches the impermissible value of approximately 0.72. It is evident that FSPPA effectively privileges the traffic in these nodes.

Blue and Orange Lines: We compare the behavior of both traffics F1 and F2 when the FSPPA is not applied. We can observe that traffic in F1 and F2 sources are more similar when the number of sources F2 increase. That is a reasonable consequence from the fact that the proportion of nodes with both types increase too. For a small number of both sources, the difference is significant, but we can observe that the confidence intervals are very large, so indicating a high variability in this case.

In conclusion, we can say that when we have a network with a small number of "privileged" nodes with source F2, having a strategy that distributes the information sent between two nodes by a pair of links-disjoint paths of minimum cost is very interesting and effective. The algorithm proposed by us provides a solution to make this strategy works.

6   Conclusions

In this paper, we described the applications of a fuzzy shortest pair of edge-disjoint paths algorithm to guarantee the security in a communication network. We associate the network to a type V fuzzy graph, and propose an algorithm (FSPPA) that finds the shortest pair of edge-disjoint paths in the graph. With the goal of illustrating the effectiveness of the algorithm, we apply the FSPPA to a well known network with high traffic load, using the new fuzzy cost function.

Currently, in many communication networks the nodes that are selected as "privileged" form a kind of separate network where only they are part of, which means a higher cost of resources. Therefore, another advantage of our strategy is that it is not necessary to create a separate network to achieve high values of the BDR.

Acknowledgment. This work has been partially supported by the University of Málaga-Andalucía Tech., through its "Plan de Investigación y Transferencia", as well as by its Dpt. of Electronics Technology.

References

1. Bansal, A.: Trapezoidal fuzzy numbers (a, b, c, d): arithmetic behavior. Int. J. Phys. Math. Sci. **2**, 39–44 (2011). ISSN:2010-1791
2. Bhandari, R.: Survivable Networks. Algorithms for Diverse Routing. Kluwer Academic Publishers, Boston (1999). ISBN 0792383818
3. Gottschau, M., Kaiser, M., Waldmann, C.: The Undirected Two Disjoint Shortest Paths Problem (2018). arXiv:1809.03820 [math.CO]
4. He, J., Rexford, J.: Toward internet-wide multipath routing. IEEE Netw. **22**, 16–21 (2008)
5. Lou, W., Fang, Y.: A multipath routing approach for secure data delivery. In: Proceedings Communications for Network-Centric Operations: Creating the Information (2001). https://doi.org/10.1109/MILCOM.2001.986098
6. Maaloul, R., Taktak, R., Chaari, L., Cousin, B.: Two node-disjoint paths routing for energy-efficiency and network reliability. In: 25th International Conference on Telecommunication (ICT 2018), pp. 554–560 (2018)

7. http://hansolav.net/sql/graphs.html. Accessed 10 Mar 2019
8. Valdés, L., Alonso, S.M., Ariza, A., Joya, G.: An implementation of the Dijkstra algorithm for fuzzy costs. Technical report (2018). https://hdl.handle.net/10630/17478
9. OMNeT++ Discrete Event Simulator (2001). https://omnetpp.org/. Accessed 10 Mar 2019
10. Yang, J., Papavassiliou, S.: Improving network security by multipath traffic dispersion. In: Proceedings Communications for Network-Centric Operations: Creating the Information Force (2001). https://doi.org/10.1109/MILCOM.2001.985760
11. Yu, V.F., Dat, L.Q.: An improved ranking method for fuzzy numbers with integral values. Appl. Soft Comput. **14**, 603–608 (2014)
12. Zhu, D., Gritter, M., Cheriton, D.R.: Feedback based routing. ACM SIGCOMM Comput. Commun. Rev. (2002). https://doi.org/10.1145/774763.774774

Expert Systems

Toward Robust Mispronunciation Detection via Audio-Visual Speech Recognition

Mahdie Karbasi[(✉)], Steffen Zeiler, Jan Freiwald, and Dorothea Kolossa

Institute of Communication Acoustics,
Faculty of Electrical Engineering and Information Technology,
Ruhr University Bochum, Bochum, Germany
mahdie.karbasi@rub.de

Abstract. A recent trend in language learning is gamification, i.e. the application of game-design elements and game principles in non-game contexts. A key component therein is the detection of mispronunciations by means of automatic speech recognition. Constraints like quiet environments and the use of close-talking microphones hinder the applicability for language learning games.

In this work, we propose to use multi-modal—specifically audio-visual—speech recognition as an alternative for detecting mispronunciations in acoustically noisy or otherwise challenging environments. We examine a hybrid speech recognizer structure, using either feed-forward or bidirectional long-short term memory (BiLSTM) networks. There are several options to integrate both modalities. Here, we compare early fusion, i.e. the use of one joint audio-visual network, with a turbo-decoding approach that combines contributions from acoustic and visual models. We evaluate the performance of these topologies in detecting some common phoneme mispronunciations, namely the errors in manner (MoA) and in place of articulation (PoA). It is shown that our novel architecture, using deep neural network acoustic and visual submodels in conjunction with turbo-decoding, is very well suited for the task of mispronunciation detection, and that the visual modality contributes strongly to achieving noise-robust performance.

Keywords: Mispronunciation detection · CAPT · Audio-visual speech recognizer · Turbo-decoder · BiLSTM

1 Introduction

Recent technologies, like mobile devices, play an important role in our lives. They are being employed to assist us in our daily work and to ease many small burdens in our day-to-day activities. Increasingly, they are also being used as

This project has received funding from the European Regional Development Fund (ERDF).

© Springer Nature Switzerland AG 2019
I. Rojas et al. (Eds.): IWANN 2019, LNCS 11507, pp. 655–666, 2019.
https://doi.org/10.1007/978-3-030-20518-8_54

teaching tools, for which they are appreciated not only by adults but even by young children. Learning a new language or correcting pronunciation errors can also be aided by apps developed for smart devices. Computer assisted pronunciation training (CAPT) is very useful for different purposes. It can help adults in second-language-learning tasks. Also, in speech therapy, it can be used to help children with speech disabilities to enhance their pronunciation skills and their quality of life. CAPT technologies target two main goals: detection of correct vs. incorrect articulation and assessing the pronunciation to provide feedback to the user [11].

To detect mispronunciations, different approaches have been proposed in the literature. Comparing the user's speech with a reference is a common approach. For instance, the dynamic time warping algorithm is used in [9,10] to find the distance between the student's speech and a reference sample based on different representations of speech such as Mel frequency cepstral coefficients (MFCCs), Gaussian or deep neural network (DNN) posteriograms. Investigating more discriminative features [16,18], specifically designed for the mispronunciation detection task, and modeling the acoustic representation of mispronunciations using classifiers like linear discriminant analysis (LDA) [19] or more recently DNNs [7,11,12] are other approaches.

Employing automatic speech recognizers (ASRs) as a core part of automatic mispronunciation detectors based on posterior probability scores has been reported in many studies dating back from 1997 [17], to more recent approaches in [20,21], where extended recognition networks are proposed to include the likely pronunciation errors in addition to the canonical pronunciations. ASRs have been also used in automatic scoring of elicited speech [6]. It has been shown that the ASR outcome is well correlated with human scorings.

All aforementioned studies are using only audio signals to evaluate the speech of users. However, visual signals are often helpful to speech recognition in noisy surroundings, compensating the informational masking through acoustic noise [1]. The lip movements of a speaker, recorded in visual signals, can even be used alone to recognize speech. There have been many systems developed for automatic lip reading and it has been shown that visual data without any audio can reach a relatively high accuracy in speech recognition [4]. Therefore, it is interesting to employ both of these modalities in CAPT technologies. To the extent of our knowledge, there have not been many works which have used audio-visual speech recognition (AVSR) for such applications. In some earlier works [8,14], different feature extraction and selection methods have been applied to both video and audio with the goal to classify and recognize certain phoneme ambiguities using binary classifiers like support vector machines (SVMs), but no actual recognition engines were employed.

Here, we propose a hybrid AVSR architecture for detecting mispronunciations. Hidden Markov models (HMMs) in combination with DNNs are utilized to build the speech recognizer. To implement DNNs, two different structures, namely feed-forward and BiLSTM, are used and their performance is compared.

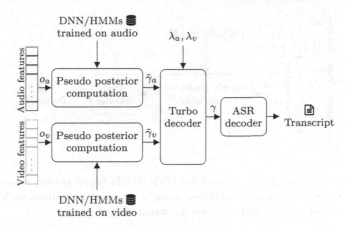

Fig. 1. Block diagram of the proposed audio-visual DNN/HMM-based speech recognizer using turbo-decoding (TD-AVSR). The observation vectors, o_a and o_v have a context size of 15.

To integrate the contributions of both modalities, we compare early fusion, i.e. the use of one joint audio-visual network, with a turbo-decoding approach that combines acoustic and visual model contributions. Several studies have shown that turbo-decoders (TDs) perform very well in combining two streams of probabilities in speech recognition [5,15]. In the following, we show that visual information in a TD structure along with audio helps to enhance the performance in mispronunciation detection.

To detect mispronunciations, we utilize a full speech recognition system with a forced-alignment grammar that includes a parallel connection of the canonical pronunciation and the corresponding possible errors. This setup is evaluated, here, in evaluating the pronunciations of some consonant-vowel (CV) and vowel-consonant (VC) syllables that are prone to mispronunciation. Those phonemes, produced similarly but articulated in a different manner or place in the vocal tract, are likely to be interchanged with one another. We are investigating if an AVSR system can be used to detect such mispronunciations more effectively. Due to its flexible structure, the proposed system can be extended to asses the pronunciation of continuous speech and not just of single embedded items, like specific words or syllables.

2   Audio-Visual Speech Recognition

Combining the information of multiple modalities to achieve highly robust speech recognition is a matter of great recent interest. As the two extreme ends of the spectrum, the modalities can be fused at the signal level, corresponding to so-called *early integration*, or at the decision level, a.k.a. *late integration*. However, best results in multimodal integration have quite consistently been achieved by combining the data at an intermediate level instead. Examples of such integration

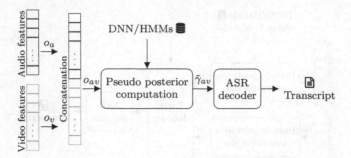

Fig. 2. Block diagram of the audio-visual DNN/HMM-based speech recognizer using an early integration of input modalities, namely concatenation (concat-AVSR). The observation vectors, o_a, o_v and o_{av} have a context size of 15.

schemes are to be found in, e.g., coupled HMMs [13] and TDs [15], which have both proven highly adept at fusing information in such a way, that the combined model consistently outperforms each single modality across all tested conditions.

For this purpose, the introduction of a so-called stream weight has been shown to be vital to dynamically control the contributions of both modalities [2,5]. When comparing audio-visual speech recognition based on TD with AVSR based on coupled HMMs, the TD has consistently won out in terms of clearly superior recognition performance [22]. In this work, we therefore focus on the use of TD as an architecture for audio-visual mispronunciation detection. We additionally introduce its application in a hybrid architecture, where BiLSTM networks contribute acoustic and visual posteriors that are fused by means of TD.

Turbo-decoding uses two decoders, one for the acoustic and one for the visual modality, which iteratively exchange information. In this strategy, the information gained by the decoder of one modality is used as prior knowledge in the decoder of the second modality. This process is being repeated several times, until a maximum agreement is reached between the output of both decoders. This process is explained in detail in [5].

In our proposed TD-AVSR architecture illustrated in Fig. 1, the observation vectors o_a and o_v, extracted from audio and video respectively, along with the separately trained DNN/HMM sets for each modality are used to compute pseudo posteriors $\tilde{\gamma}_a$ and $\tilde{\gamma}_v$ for all HMM states. Although the actual model parameters might be different, all applied state-space models in this paper are equal in the sense, that they are in a bijective relation.

The $\tilde{\gamma}_a$ and $\tilde{\gamma}_v$ as well as the stream weights, λ_a and λ_v, are fed into the turbo-decoder to compute the final posteriors γ. These are used in an ASR decoder (modified Viterbi decoder) to find the best state sequence, and hence, the recognized sequence of words or phonemes.

To have a comparison for the TD-AVSR system, an early integration system of two modalities has also been implemented by concatenating the feature vectors of audio (o_a) and video (o_v). Figure 2 depicts this so-called concat-AVSR setup. The concatenated audio-visual observation, o_{av}, is evaluated by a DNN to get the final pseudo posterior probabilities $\tilde{\gamma}_{av}$. Similar to TD-AVSR, pre-trained DNNs are utilized to compute $\tilde{\gamma}_{av}$. In the last step, an ASR decoder is applied to $\tilde{\gamma}_{av}$, giving the recognized transcription.

In addition to comparison of the two modality integration schemes, two different structures of DNNs, namely feed-forward networks and BiLSTMs, are also implemented and compared in evaluating the mispronunciation detection performance of both concat-AVSR and TD-AVSR.

3 Dataset

In this study we use the audio-visual Grid database [3]. The Grid corpus includes clean audio-visual recordings of 18 male and 16 female speakers (for speaker 21, only audio is available). As visual data, videos from the faces of the speakers have been recorded in parallel to the audio recordings. The sampling rate of the video is 25 Hz and it is available in two resolutions, which are 360×288 and 720×576 pixels. For the presented experiments, the low-resolution video was used.

The database contains 34,000 utterances of British English comprised of 1,000 sentences per speaker. All sentences are semantically unpredictable and follow the same 6 word pattern, arranged as *verb-color-preposition-letter-digit-adverb*.

In order to perform audio-visual mispronunciation detection, in principle, we require an audio-visual database with correct and mispronounced speech. Although the Grid database contains only undisturbed speech and does not include any mispronounced data, it can nevertheless be used to examine the performance of AVSR systems distinguishing between likely phoneme ambiguities in actual mispronounced data. Let us examine the letter section of Grid sentences. For instance, the letter *b* consists of only one consonant-vowel (CV) syllable. Also, a probable mispronunciation, which is the CV syllable *p*, is available in the letter block.

Grid sentences contain 25 different letters, of which 3 pairs are of relevance for this study, because they consist of one syllable and are simultaneously prone to confusion and mispronunciation of their respective consonant part: /f/ – /s/, /t/ – /d/, /b/ – /p/.

We have selected those sentences that include one of these 6 syllables. In addition to the original clean signals from the Grid corpus, 3 types of acoustic noise have been used to generate noisy audio; babble, office and white noise, each mixed with the clean signals at 4 different SNRs (0, 5, 10 and 15 dB). The sampling rate of audio signals was 8000 Hz.

4 Framework Setup

To evaluate the performance of a speech recognizer in detecting mispronunciations, experiments were conducted with audio, video, and audio-visual data. The data was divided randomly into training (90%), development (5%) and test sets (5%). For the two-modality experiments, a synchronization step was performed after feature extraction in order to obtain the same frame rate in both modalities.

Prior to the speech recognizer, there is a feature extraction step that was applied to the signals of both modalities. The extracted feature vectors were then used as observations for the models of the hybrid DNN/HMM speech recognizer. MFCCs with 39 dimensions were extracted as audio features. For video, as introduced in [4], the 190-dimensional slow feature analysis (SFA) feature vectors were used. The SFA feature extraction process is composed of two parts; first the mouth region is extracted from the video files and then the actual visual feature extraction algorithm, here the slow feature analysis, is applied to the mouth region data.

The speech recognizer used in this work has been implemented based on hybrid DNN/HMM models. The ASR models were trained in a speaker-dependent manner with the data from all noise and SNR categories available in the database. Every word of the Grid corpus was modeled with one left-to-right HMM, resulting in 51 whole-word HMMs plus one silence model for the entire Grid corpus. The number of states was chosen as three times the number of phonemes of the word. To discover the optimum weights for audio and video in the turbo-decoding process, an offline grid search was performed on a set of predefined weights using the development data for each SNR.

We implemented two alternative structures of DNNs, a simple feed-forward and a BiLSTM-based network, using Keras with a Tensorflow back-end. Both networks use a ReLU activation function and a softmax output layer, which produces the pseudo state posteriors of the HMM that are then used to generate the recognized word sequences. In all experiments, the DNNs are trained to minimize the categorical cross entropy in a supervised manner. For the optimizer, the Adam algorithm was selected with a learning rate of 0.001. We generated a centered context of length 15 around the input feature vector. We used one whole sentence as a batch. The input dimension depends on the input feature type; 39 for the MFCCs and 190 for the SFA features.

The feed-forward network, in total, consisted of 8 hidden layers plus the input and output layers. The output layer had a dimension of 381, which corresponds to the number of states and represents the state posteriors. The overall topology is shown in Table 1.

For training the BiLSTM network, the weight decay was set to 10^{-6} and the norm clipping length was defined as 1. A batch normalization layer was added after the input tensor. Table 2 shows the implemented BiLSTM topology.

For concat-AVSR, we used larger networks with more parameters. For the feed-forward network, a structure similar to Table 1, but with the following sizes was implemented: 384, 182, 96, 15 × 96, 384, 182, 182, 182. The BiLSTM network used

Table 1. Feed-forward network topology.

Layer type	Output dimensions
Input layer	batch size, 15, input dimension
Dense	batch size, 15, 256
Dense	batch size, 15, 128
Dense	batch size, 15, 64
Flatten	batch size, 15×64
Dense	batch size, 256
Dense	batch size, 128
Dense	batch size, 128
Dense	batch size, 128
Dense + Softmax	batch size, 381

Table 2. The BiLSTM network consists of two BiLSTM layers and one reduction LSTM layer.

Layer type	Output dimensions
Input layer	batch size, 15, input dimension
Batch normalization	batch size, 15, input dimension
BiLSTM	batch size, 15, 256
BiLSTM	batch size, 15, 256
LSTM	batch size, 381
Dense + Softmax	batch size, 381

twice as many units in each BiLSTM layer in comparison to the ones used for the single modality experiments, shown in Table 2. We chose these sizes to keep the learnable parameter count of concat-AVSR roughly equal to that of TD-AVSR.

5 Mispronunciation Detection Using AVSR

In this paper we show how an audio-visual DNN/HMM-based speech recognizer can be used to distinguish the correct pronunciations of CV or VC syllables from their most likely mispronunciations.

First, for each target syllable, we compile an HMM according to a predefined task-specific grammar including possible mispronunciations. Later, we use the AVSR system to detect whether the canonical syllable has been uttered correctly or incorrectly within its continuously spoken sentence. Through task-specific grammars, we force the recognizer to discriminate between certain models including the one correspondent to the target syllable and the remaining models that are treated as error models. These error models are representative of the mispronunciations that can occur for a target syllable.

Table 3. Mispronunciation detection accuracy (in %) with standard deviation for MoA (b vs. p and t vs. d).

Noise type	SNR [dB]	Feed-forward			
		Video	Audio	A-V concat	A-V turbo
Babble	0	-	85.7 (11.0)	82.8 (17.0)	91.4 (9.5)
	5	-	95.1 (7.9)	90.2 (11.7)	96.7 (6.2)
	10	-	96.7 (5.6)	92.2 (8.9)	97.1 (5.4)
	15	-	99.1 (3.0)	93.4 (9.5)	99.1 (3.0)
Office	0	-	68.5 (19.7)	74.2 (19.5)	77.1 (19.2)
	5	-	86.1 (14.2)	82.0 (14.4)	89.7 (11.2)
	10	-	92.6 (10.2)	89.3 (13.5)	93.4 (8.8)
	15	-	95.9 (8.2)	91.0 (13.6)	96.3 (7.6)
White	0	-	93.4 (10.0)	87.3 (15.2)	96.3 (6.1)
	5	-	94.6 (9.0)	90.6 (11.6)	95.5 (8.2)
	10	-	98.3 (4.7)	93.0 (11.1)	98.3 (4.7)
	15	-	98.7 (3.8)	93.4 (10.6)	98.3 (4.4)
Clean	-	56.7 (18.9)	97.9 (4.6)	93.4 (11.9)	97.9 (4.6)
Average	-	56.7 (18.9)	92.6 (5.1)	88.8 (10.0)	94.3 (4.7)
Noise type	SNR [dB]	BiLSTM			
		Video	Audio	A-V concat	A-V turbo
Babble	0	-	87.7 (12.5)	77.1 (17.8)	93.0 (7.9)
	5	-	96.7 (6.6)	83.6 (15.1)	97.1 (5.7)
	10	-	97.9 (4.8)	87.3 (14.5)	98.7 (3.7)
	15	-	98.3 (4.2)	88.5 (14.1)	98.7 (3.8)
Office	0	-	66.1 (18.3)	63.6 (18.1)	79.1 (13.4)
	5	-	82.8 (16.0)	77.1 (17.6)	87.7 (11.4)
	10	-	94.2 (9.1)	78.7 (16.2)	94.6 (8.0)
	15	-	96.3 (7.0)	84.8 (14.9)	97.5 (5.1)
White	0	-	92.6 (9.9)	79.1 (17.8)	95.5 (7.8)
	5	-	97.1 (6.4)	83.2 (15.2)	97.5 (5.1)
	10	-	97.9 (4.7)	87.3 (14.8)	99.1 (3.4)
	15	-	98.7 (3.4)	87.7 (14.8)	99.1 (2.8)
Clean	-	64.8 (17.8)	96.7 (6.4)	90.2 (14.5)	96.7 (6.4)
Average	-	64.8 (17.8)	92.6 (4.1)	81.4 (13.9)	**94.9 (3.2)**

Since a full speech recognition system has been used here, the proposed framework is not limited to pronunciation evaluation of only single words or phonemes. It is easily possible to extend the current framework to detect the mispronunciations of continuous speech.

5.1 Manner of Articulation (MoA)

In the first set of experiments, the letter pairs /b/ – /p/ and /t/ – /d/ from the Grid corpus were utilized as the test speech material. Within each of these pairs, the articulation occurs at the same location. However, the voicing differs within each pair, as /b/ is the voiced version of /p/ and /d/ is the voiced version of /t/. For MoA errors, each of these letters is therefore likely to be mispronounced in such a way as to become too similar to its respective peer.

As described earlier, the decoder grammar only allows for the recognition of a few admissible variants of a letter. For instance, when the target letter is /p/ or /b/, the grammar forces the decoder to distinguish only between /p/ and /b/, and an equivalent setup is used for /t/ and /d/. Table 3 shows the resulting accuracy in detecting manner-of-articulation errors. As can be seen, in general, turbo-decoding AVSR with both DNN structures outperforms all other systems, comprising audio-only, video-only and concatenated bimodal (A-V con-cat) recognition. The video-only column shows that visual information alone is insufficient to reliably detect MoA errors, which is unsurprising. Concatenat-ing the acoustic features with visual data reduces the accuracy from 92.6% to 88.8% and from 92.6% to 81.4% for the feed-forward and BiLSTM networks, respectively. In contrast, TD-AVSR is capable of discovering the most informative cross-modal integration over time and modality. As a result, TD improves the audio-visual accuracy in both networks, reliably performing better than or as well as acoustic-only recognition and improving significantly in noisy conditions.

Comparing the results of both network structures shows that the BiLSTM has a slightly better performance than the feed-forward network in the TD experi-ment. However, in general BiLSTMs did not improve the accuracy in comparison to the feed-forward network. Only in the video-only experiment, there is a clear improvement with BiLTMs.

5.2 Place of Articulation (PoA)

The letters /s/ and /f/ are composed of only one vowel-consonant (VC) syllable. Their voiceless consonant parts are articulated in different places, once as a labio-dental and once as an alveolar fricative. Thus, this pair is representative of errors in place of articulation (PoA), and is used in the second set of experiments to investigate the effect of PoA on the performance of the proposed system.

Again, the grammar was set up to allow only the possible pronunciation variants, /s/ and /f/. Table 4 contains the results of the experiments. As can be seen, TD-AVSR has the best performance in almost all conditions in comparison to the other recognizers. Statistically, there is no significant difference between both TD-AVSR results and those of the concatenated version.

For PoA-error detection, the video-only accuracies, always above 93.5%, are higher than the audio-only results. This indicates that the place of articulation is, on average, more clearly perceivable in the video stream than in the audio signals.

Table 4. Mispronunciation detection accuracy (in %) with standard deviation for PoA (s vs. f).

Noise type	SNR [dB]	Feed-forward			
		Video	Audio	A-V concat	A-V turbo
Babble	0	-	62.6 (28.7)	93.9 (12.7)	97.3 (10.4)
	5	-	68.6 (30.5)	93.9 (11.0)	93.9 (13.6)
	10	-	75.6 (29.4)	93.0 (13.3)	93.0 (15.5)
	15	-	78.2 (26.6)	93.0 (13.3)	95.6 (12.1)
Office	0	-	60.0 (32.4)	96.5 (8.5)	93.9 (21.5)
	5	-	65.2 (31.3)	95.6 (11.2)	96.5 (11.8)
	10	-	69.5 (26.1)	93.9 (12.4)	94.7 (20.4)
	15	-	71.3 (28.4)	93.0 (14.4)	93.9 (14.0)
White	0	-	60.8 (30.9)	92.1 (13.5)	92.1 (17.3)
	5	-	65.2 (26.0)	92.1 (13.5)	95.6 (20.1)
	10	-	73.9 (25.0)	93.0 (19.3)	94.7 (19.8)
	15	-	71.3 (28.8)	93.9 (11.0)	93.9 (20.8)
Clean	-	97.1 (8.1)	76.5 (26.6)	93.9 (12.9)	94.7 (12.6)
Average	-	97.1 (8.1)	70.6 (22.3)	94.8 (11.0)	94.0 (12.4)
Noise type	SNR [dB]	BiLSTM			
		Video	Audio	A-V concat	A-V turbo
Babble	0	-	59.1 (31.3)	94.7 (22.4)	96.5 (14.0)
	5	-	71.3 (30.6)	95.6 (21.5)	97.3 (10.7)
	10	-	81.7 (24.6)	94.7 (10.5)	98.2 (10.3)
	15	-	81.7 (22.7)	94.7 (10.8)	97.3 (10.7)
Office	0	-	55.6 (27.1)	92.1 (22.5)	98.2 (10.3)
	5	-	54.7 (30.6)	94.7 (11.5)	93.0 (16.1)
	10	-	74.7 (23.3)	95.6 (11.2)	94.7 (14.5)
	15	-	73.9 (26.2)	93.9 (11.8)	95.6 (13.5)
White	0	-	60.8 (31.5)	93.9 (21.9)	91.3 (22.4)
	5	-	71.3 (31.5)	94.7 (11.3)	93.0 (22.5)
	10	-	78.2 (31.6)	93.9 (11.8)	97.3 (12.9)
	15	-	80.8 (23.8)	94.7 (10.8)	97.3 (19.2)
Clean	-	93.5 (14.8)	80.0 (24.3)	93.9 (12.3)	96.5 (11.1)
Average	-	93.5 (14.8)	70.7 (19.2)	94.3 (11.5)	**94.9 (11.9)**

Also, the improvements due to the integration of visual features are much greater for PoA than for MoA errors, and—vice versa—the acoustic features in isolation are performing poorly for PoA-error detection in comparison to the easier MoA assessment.

6 Conclusions and Outlook

We have suggested the use of audio-visual speech recognition for the automatic detection of pronunciation errors in speech and language training. Both feed-forward and BiLSTM-based speech recognizers were utilized for this task, and within a novel hybrid architecture, they were used to yield the pseudo-posteriors for a turbo-decoder. Under a broad range of acoustic conditions, it was shown that such a bimodal system can benefit from both modalities and reach a robust performance in adverse acoustic environments. Furthermore, it was shown that intermediate integration of modalities using the turbo-decoding strategy generally provides greater improvements than the early concatenation of acoustic and visual features.

In future work, the proposed AVSR setup should be employed to detect more varieties of mispronunciations, based on recorded data for children's and adult speech. We are especially aiming to develop a mispronunciation assessment for speech-impaired children in this way, and to make it robust enough to allow for its use in diverse environments, e.g. at home or even in the less noisy areas of a kindergarten or day-care center. From the technological point of view, estimating the weights for different modalities in the turbo decoding framework requires additional effort. Prior works have shown that AVSR can benefit from dynamic, frame-by-frame, stream weighting, especially when dealing with highly variable acoustic conditions. It stands to reason that this idea can also boost the performance of our system for different noise conditions and hopefully for an even larger variability of mispronunciations.

References

1. Abdelaziz, A.H.: Comparing fusion models for DNN-based audiovisual continuous speech recognition. IEEE/ACM Trans. Audio Speech Lang. Process. (TASLP) **26**(3), 475–484 (2018)
2. Abdelaziz, A.H., Zeiler, S., Kolossa, D.: Learning dynamic stream weights for coupled-HMM-based audio-visual speech recognition. IEEE/ACM Trans. Audio Speech Lang. Process. **23**(5), 863–876 (2015)
3. Cooke, M., Barker, J., Cunningham, S., Shao, X.: An audio-visual corpus for speech perception and automatic speech recognition. J. Acoust. Soc. Am. **120**(5), 2421–2424 (2006). https://doi.org/10.1121/1.2229005
4. Freiwald, J., et al.: Utilizing slow feature analysis for lipreading. In: Proceedings of ITG, November 2018
5. Gergen, S., Zeiler, S., Hussen Abdelaziz, A., Nickel, R., Kolossa, D.: Dynamic stream weighting for turbo-decoding-based audiovisual ASR. In: Proceedings of ITG, pp. 2135–2139, September 2016
6. Graham, C.R., Lonsdale, D., Kennington, C., Johnson, A., McGhee, J.: Elicited imitation as an oral proficiency measure with ASR scoring. In: LREC (2008)
7. Hu, W., Qian, Y., Soong, F.K., Wang, Y.: Improved mispronunciation detection with deep neural network trained acoustic models and transfer learning based logistic regression classifiers. Speech Commun. **67**, 154–166 (2015)

8. Kjellström, H., Engwall, O., Abdou, S.M., Bälter, O.: Audio-visual phoneme classification for pronunciation training applications. In: Proceedings of the Eighth Annual Conference of the International Speech Communication Association (2007)
9. Lee, A., Glass, J.: A comparison-based approach to mispronunciation detection. In: Proceedings of Spoken Language Technology Workshop (SLT), pp. 382–387 (2012)
10. Lee, A., Zhang, Y., Glass, J.: Mispronunciation detection via dynamic time warping on deep belief network-based posteriorgrams. In: Proceedings of ICASSP, pp. 8227–8231 (2013)
11. Li, K., Qian, X., Meng, H.: Mispronunciation detection and diagnosis in L2 English speech using multidistribution deep neural networks. IEEE/ACM Trans. Audio Speech Lang. Process. **25**(1), 193–207 (2017)
12. Li, W., Chen, N., Siniscalchi, M., Lee, C.H.: Improving mispronunciation detection for non-native learners with multisource information and LSTM-based deep models. In: Proceedings of Interspeech, pp. 2759–2763, September 2017
13. Nefian, A.V., Liang, L., Pi, X., Liu, X., Murphy, K.: Dynamic Bayesian networks for audio-visual speech recognition. EURASIP J. Adv. Signal Process. **2002**(11), 783042 (2002)
14. Picard, S., Ananthakrishnan, G., Wik, P., Engwall, O., Abdou, S.: Detection of specific mispronunciations using audiovisual features. In: Proceedings of Auditory-Visual Speech Processing (2010)
15. Receveur, S., Weiss, R., Fingscheidt, T.: Turbo automatic speech recognition. IEEE/ACM Trans. Audio Speech Lang. Process. **24**(5), 846–862 (2016)
16. Richardson, M., Bilmes, J., Diorio, C.: Hidden-articulator Markov models for speech recognition. Speech Commun. **41**(2–3), 511–529 (2003)
17. Ronen, O., Neumeyer, L., Franco, H.: Automatic detection of mispronunciation for language instruction. In: Proceedings of the Fifth Eurospeech (1997)
18. Tepperman, J., Narayanan, S.: Using articulatory representations to detect segmental errors in nonnative pronunciation. IEEE/ACM Trans. Audio Speech Lang. Process. **16**(1), 8–22 (2008)
19. Truong, K., Neri, A., Cucchiarini, C., Strik, H.: Automatic pronunciation error detection: an acoustic-phonetic approach. In: Proceedings of InSTIL/ICALL Symposium (2004)
20. Wang, Y.B., Lee, L.S.: Improved approaches of modeling and detecting error patterns with empirical analysis for computer-aided pronunciation training. In: Proceedings of ICASSP, pp. 5049–5052 (2012)
21. Wang, Y.B., Lee, L.S.: Supervised detection and unsupervised discovery of pronunciation error patterns for computer-assisted language learning. IEEE/ACM Trans. Audio Speech Lang. Process. (TASLP) **23**(3), 564–579 (2015)
22. Zeiler, S., Nickel, R., Ma, N., Brown, G., Kolossa, D.: Robust audiovisual speech recognition using noise-adaptive linear discriminant analysis. In: Proceedings of ICASSP (2016)

Link Prediction Regression for Weighted Co-authorship Networks

Ilya Makarov[1,2(✉)] and Olga Gerasimova[1]

[1] National Research University Higher School of Economics, 3 Kochnovskiy Proezd, 125319 Moscow, Russia
{iamakarov,ogerasimova}@hse.ru
[2] Faculty of Computer and Information Science, University of Ljubljana, Večna pot 113, 1000 Ljubljana, Slovenia

Abstract. In this paper, we study the problem of predicting quantity of collaborations in co-authorship network. We formulated our task in terms of link prediction problem on weighted co-authorship network, formed by authors writing papers in co-authorship represented by edges between authors in the network. Our task is formulated as regression for edge weights, for which we use node2vec network embedding and new family of edge embedding operators. We evaluate our model on AMiner co-authorship network and showed that our model of network edge representation has better performance for stated regression link prediction task.

Keywords: Co-authorship networks · Recommender systems · Network embedding · Link prediction · Machine learning

1 Introduction

Co-authorship networks are powerful instrument to measure research trends, publication activity and study collaboration patterns in research community. There are various applications of co-authorship and citation networks to recommender systems for searching collaborators on research projects, reading relevant research papers, finding experts based on text of competition application. In order to meet the necessity to bring close researchers inside universities in developing countries and facilitate interdisciplinary research projects the concept of "Scientific matchmaking" was introduced in several research papers on co-authorship networks mining [39,49]. Early unsupervised learning

Sections 1, 2 and 3 on "Knowledge representation, discovery, and processing: a logic-based approach" were supported by the Russian Science Foundation under grant 17-11-01294 and performed at National Research University Higher School of Economics, Russia. Sections 4 and 5 on "Knowledge acquisition and representation for recommender systems" were prepared within the framework of the HSE University Basic Research Program and funded by the Russian Academic Excellence Project '5-100'

© Springer Nature Switzerland AG 2019
I. Rojas et al. (Eds.): IWANN 2019, LNCS 11507, pp. 667–677, 2019.
https://doi.org/10.1007/978-3-030-20518-8_55

approaches for community detection and mining research networks were studied in [5,27,45,52,54,55].

The study on predicting individual collaborations was made by several research groups in Russia [35–37], Brazil [33], UK [11] and many other studies based on open datasets from DBLP and Scholar Google. On the state level the problem of improving research communities was studied at China [25].

In this paper, we study a co-authorship recommender system based on *node2vec* network embeddings [19] and edge characteristics obtained from author node embeddings. We compare our approach with state-of-the-art algorithms for link prediction problem using several edge embedding operators suggested in [19,40] and newly defined for the weighted link prediction (edge weight regression) problem state in the paper. Such obtained system could be applied recommending collaborators and estimating the outcome of such collaborations in temporal timeline. In what follows, we describe solution to link prediction problem leading to evaluation of our recommender system based on co-authorship network embeddings and evaluate its quality in regression and classification tasks for weighted network link prediction.

2   Related Works

2.1   Link Prediction

Network science approach to the problem of predicting collaborations results in the link prediction (LP) problem [28] for temporal networks and missing edges reconstruction in noisy network data. Basically, it is a method to apply standard machine learning framework for graph data considering feature space consisting of pairs of nodes and their features.

Link prediction models are applied in web linking [2], social dating services [4], paper recommender system for digital libraries [21]. A reader can found an up-to-date survey in [47]. Survey on link prediction was published in [53]. LP problem was specifically formulated in [28] based on nodes pairwise similarity measures. Approaches for link prediction include similarity based methods [3], maximum likelihood models [12], and probabilistic models [16,22]. In [51], authors suggesting unsupervised approach for LP problem. Authors of [14,15] suggested temporal link prediction based on matrix factorization technique and noise reduction in large networks. Attribute-based link formation in social networks was studied in [41,44], while deep learning approaches were presented in [6,29,56]. Heterogeneous graph link prediction for predicting links of certain semantic type was suggested in [31,32].

Two surveys on link prediction methods describe core approaches for feature engineering, Bayesian approach and dimensionality reduction were presented in [20,34]. Survey on link prediction was published in [53].

2.2   Collaborator Recommender System as Link Prediction Problem

Although link prediction can help detecting missing links in the current networks, or, on the contrary, abnormal links in fraud detection tasks, the most of LP applications for social networks deals with predicting the most probable persons for future collaboration, which we state as a problem of co-authorship recommender system based on link prediction problem [10, 26, 30]. A survey on co-authorship and citation recommender systems may be found in [42].

We study the problem of recommending collaborator depending on researcher's co-authorship relations, the quality and quantity of publications, and structural patterns based in co-authorship network. We aim to measure the strength of such collaborations in terms of quantity of co-authored papers on Microsoft Academic Graph [46]. We also study the problem of estimating quality of such collaborations based on HSE university co-authorship network [23] with link weights aggregating numeric metrics based on impact-factors and quartiles of the journals for published research papers.

Our model is designed to predict whether a pair of nodes in a network would have certain number of connections and whether we could improve such a prediction in two-step process of predicting the collaboration itself [36–38] and further estimating its quantity/quality.

2.3 Graph Embedding

Recently, new methods of automated graph feature engineering and their applications to machine learning problems on graphs get attention of researchers. The network representation by adjusting each node a vector based on its neighborhood proximity is called network embedding. In general, knowledge representation requires task-dependent feature engineering in order to construct a real-value feature vector for nodes and edges representation. The quality of such an approach will be influenced by domain expert quality, particular task and noise in the data, which is hard to measure in large scale networks. Recently, the theory of constructing numeric embeddings has impacted on machine learning and artificial intelligence leading to creating new task-independent optimization tasks and loss functions instead of manual feature engineering.

New family of random walk based graph embeddings was suggested in such articles as [8, 19, 43, 48]. The results of structural (without node attributes) graph embeddings already showed state-of-art performance on such problems as multiclass node classification and link prediction. A list of surveys on graph embedding models and applications can be found in [7, 9, 13, 17].

3 Edge Feature Generating Based on Node Network Embedding

As we previously mentioned, we use node2vec method to construct node embeddings. Node2vec provides a flexible strategy to make a neighborhood sampling based on biased random walk, which smoothly combines two algorithms such as breadth-first and depth-first samplings. The biased random walk takes into account the second-order and even higher-order proximities. Applying a natural language processing methods such as vectorized technique word2vec to given sequences of neighborhood vertices generated by random walks, node2vec receives node representations to optimize the occurrence probability of neighbor nodes based on the representation of a node.

The property of node2vec is ability to construct close embeddings for nodes belonging to the same network community or that are structurally equivalent. Node2vec has two random walk hyperparameters, p and q, that tune the random walk. The hyperparameter p regulates the chance that the walk revisits a node, while q controls situations, when the walk revisits a node's one-hop neighborhood. Changing settings of the model allows to find a reasonable compromise between learning embeddings to focus on community structures or local structural roles.

However, we are interested in receiving edges vector representation to form a feature space for link prediction task. For edge embedding we applied specific component-wise functions representing edge to node embeddings for source and target nodes of a given edge. This model was suggested in node2vec [19], in which four functions for such edge embeddings were presented (see first three rows in Table 1 for $\alpha = 1, 2$). We consider the simplest and fastest model of edge embedding construction not taking into account papers on joint node-edge graph embedding, such as [1,18]. Presented in the paper model suggests simple generalization of the idea of pooling first-order neighborhood of nodes while constructing edge embedding operator, which is much faster than dimensionality reduction or graph neural networks.

In our previous research [40], we suggested to use another type of operators involving not only edge source and target node representations, but also their neighborhood representations as average over all the nodes in first-order proximity called "neighbour-weighted" L_α operator verifying parameter $\alpha = 1, 2$. The evaluation of such "vertex pooling" techniques was made in [38] for binary classification in link prediction problem.

Now for weighted link prediction we introduce new link embedding operators. Firstly, we consider Neighbor Weighted-L_α^0 operator that differs from Neighbor Weighted-L_α just by not taking the vertex into account when pooling its neighborhood. Such an operator represent purely context based link embedding. Secondly, for weighted link prediction we change average pooling for vertex to weighted average sum of neighbour nodes embeddings multiplied by edge weight and divided by the sum of weights corresponding to the node. We call this Neighbor Double Weighted-L_α operator. For this case we do not place the upper index due to rear case of loops in the presented graph representing the papers written without co-authors. So, following the pooling idea, it is important not only to aggregate information from vertices, but also from their neighbourhood, because an edge appearance can be defined by second-order proximity too. The resulting list of link embeddings is presented in Table 1.

The core idea of original [19] edge embedding operators was to aggregate information for edge from incident nodes (average sum, Hadamard product) or choose L_α, $\alpha = 1, 2$ metric for the difference of embedding vectors to include the idea that similar (by adjacency matrix similarity), i.e., connected nodes should have close embeddings. Our idea follows both of these approaches but adds constraints that edge embedding is defined by second order proximity, meaning that neighbor nodes embeddings may be efficiently included in edge representation

Table 1. Binary operators for computing vectorized (u, v)-edge representation based on node attribute embeddings $f(x)$ for ith component for $f(u, v)$. Parameter $\alpha \geqslant 1$

Symmetry operator	Definition
Average (AVG)	$\dfrac{f_i(u)+f_i(v)}{2}$
Hadamard (MULT)	$f_i(u) \cdot f_i(v)$
Weighted-L_α (WL_α)	$\lvert f_i(u) - f_i(v) \rvert^\alpha$
Neighbor Weighted-L_α (NWL_α)	$\left\lvert \dfrac{\sum\limits_{w \in N(u) \cup \{u\}} f_i(w)}{\lvert N(u) \rvert + 1} - \dfrac{\sum\limits_{t \in N(v) \cup \{v\}} f_i(t)}{\lvert N(v) \rvert + 1} \right\rvert^\alpha$
Neighbor Weighted-L_α^0 (NWL_α^0)	$\left\lvert \dfrac{\sum\limits_{w \in N(u)} f_i(w)}{\lvert N(u) \rvert} - \dfrac{\sum\limits_{t \in N(v)} f_i(t)}{\lvert N(v) \rvert} \right\rvert^\alpha$
Neighbor Double Weighted-L_α ($NDWL_\alpha$)	$\left\lvert \dfrac{\sum\limits_{w \in N(u) \cup \{u\}} f_i(w) \cdot \omega(u,w)}{\sum\limits_{p \in N(u) \cup \{u\}} \omega(u,p)} - \dfrac{\sum\limits_{t \in N(v) \cup \{v\}} f_i(t) \cdot \omega(v,t)}{\sum\limits_{s \in N(v) \cup \{v\}} \omega(v,s)} \right\rvert^\alpha$

construction. In what follows, we evaluate our previous and new edge embedding operators on link prediction problem.

4 Experiments

In order to state weighted link prediction problem we considered the link prediction task for the large network called AMiner [50] containing 4,258,615 collaborations among the 1,560,640 authors. We did not visualize it due to density of the data not allowing to make representative graph drawing. We work with weighted variant of Aminer graph, where an edge weight means the number of publications writing in collaboration between two authors corresponding to incident vertices for the given edge. Another existing weights interpretation that we do not consider in this article is related to publication quality, which is based on publication belonging to quartiles in scientific indexing databases (Q1–Q4, other), which are assigned to a journal or conference.

We construct *Linear Regression* model for predicting number of papers in AMiner graph written in co-authorship. We chose only this machine learning model due to existing time/memory limitations for large networks processing, but also due to high efficiency for our setting based on quality metrics.

We studied the impact of train/test split on different edge embeddings operators while fixing the fastest Regression model for weighted link prediction. We considered train set, consisting of 20%, 40%, 60%, 80% of the graph edges while averaging binary classification quality metrics over 5 negative sampling providing negative examples for non-existent edges with zero weights.

We compared *Mean Absolute Error* (MAE) measuring the average of absolute values of the errors, *Mean Squared Error* (MSE) measuring the average of the squares of the errors, and *Coefficient of Determination* (R^2) measuring proportion of the variance in the predicted variable from the features computed for train and test sets using different edge embeddings. The results on train set

(Fig. 1) and test set (Fig. 2) are expressed in terms of MAE, MSE and R^2 metrics. The least values of MAE and MSE, and the highest values of R^2 represent the best models.

It turns out that Hadamard product outperforms all the others while the closest results was obtained by Neighbor Weighted-L_2 link embeddings. While train size increases all the Neighbor Weighted-L_α and Neighbor Double Weighted-L_α link embeddings achieve MAE less than 0.5, while simple Weighted-L_α and Average sum link embeddings showed the worst performance.

For our experiments We used node2vec network embedding [19] with random walks parameters $p, q = (1, 1)$, dimension of the embedding $d = 128$, length of walks $l = 60$, and number of walks per node equaled $n = 3$. The latter parameters are chosen quite small in order to fit to memory requirements. The node2vec embedding parameters were chosen via MSE optimization over embedding size with respect to Hadamard product edge embedding operators, which was stated to be the best for LP task in [19].

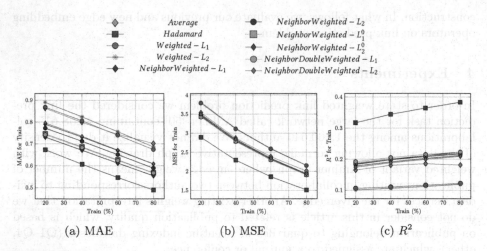

Fig. 1. Linear regression results on train set depending on train size for different link embeddings

5   Discussion

As one could see, well performance of Hadamard product may be due the fact that such link embedding significantly reduce the dimension of link embedding feature space due to sparsity of embedding vector. However, its behavior on link prediction task formulated as binary classification showed bad learning ability of Hadamard product link embedding. In such a case pooling neighborhood must be done after encoding embedding to the lower space and then learning aggregated pooling function similar to [1]. From our experiments we could not deduce

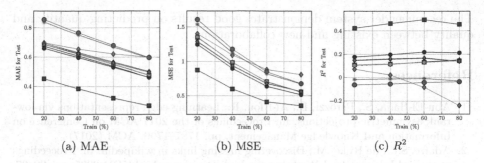

(a) MAE   (b) MSE   (c) R^2

Fig. 2. Linear regression results on test set depending on train size for different link embeddings

whether average or weighted pooling significantly affect the performance to overcome Hadamard link embedding for regression task. On the opposite, classic link prediction task using binary classification based on edge embedding showed the best result on the Neighbor Weighted-L_α operators for HSE and AMiner graphs [38]. We aim to further study the problem of attributed network embeddings based on graph auto-encoders [24] and developed by authors embedding model called JONNEE learning joint node-edge representation of network.

The code for computing all the models with respect to classification evaluation, choosing proper edge embedding operator and tuning hyper-parameters of node embeddings will be uploaded on the project Github http://github.com/makarovia/jcdl2018/.

6   Conclusion

Scientific matchmaking is a new developing area in social network analysis. Its applications to analyzing researcher community and recommending new collaborations may be of interest to individuals, research groups, funding agencies and institutes. We previously showed that formulating the problem of collaborator search in terms of co-authorship network data showed high accuracy in predicting future collaborations for novice researchers [35,39] and analyzing research interaction inside university [36,37].

In this paper, we presented new link embedding operators similar to [38,40] including local proximity of source and target nodes for the edge. We verify the quality of such operators on weighted link prediction problem formulated as weight regression for given co-authorship networks. We considered AMiner graph with weights representing number of co-authored papers. Our approach of solving link prediction with weight regression presented MAE and MSE measures on test results less than 0.5.

Our experiments show that such constructed system may be considered as a recommender system for searching collaborators and simultaneously predictive model for estimating researcher's publishing and collaboration activity.

The recommender system demonstrates good results on predicting quantity and quality between existing and new collaborations.

References

1. Abu-El-Haija, S., Perozzi, B., Al-Rfou, R.: Learning edge representations via low-rank asymmetric projections. In: Proceedings of the 2017 ACM on Conference on Information and Knowledge Management, pp. 1787–1796. ACM (2017)
2. Adafre, S.F., de Rijke, M.: Discovering missing links in wikipedia. In: Proceedings of the 3rd International Workshop on Link Discovery, LinkKDD 2005, pp. 90–97. ACM, New York (2005). http://doi.acm.org/10.1145/1134271.1134284
3. Adamic, L.A., Adar, E.: Friends and neighbors on the web. Soc. Netw. **25**(3), 211–230 (2003)
4. Backstrom, L., Leskovec, J.: Supervised random walks: predicting and recommending links in social networks. In: Proceedings of the Fourth ACM International Conference on Web Search and Data Mining, WSDM 2011, pp. 635–644. ACM, New York (2011). http://doi.acm.org/10.1145/1935826.1935914
5. Barabási, A.L., Pósfai, M.: Network Science. Cambridge University Press, Cambridge (2016)
6. Berg, R.v.d., Kipf, T.N., Welling, M.: Graph convolutional matrix completion. arXiv preprint arXiv:1706.02263 (2017)
7. Cai, H., Zheng, V.W., Chang, K.: A comprehensive survey of graph embedding: problems, techniques and applications. IEEE Trans. Knowl. Data Eng. **30**, 1616–1637 (2018)
8. Chang, S., Han, W., Tang, J., Qi, G.J., Aggarwal, C.C., Huang, T.S.: Heterogeneous network embedding via deep architectures. In: Proceedings of the 21st ACM SIGKDD International Conference on Knowledge Discovery and Data Mining, KDD 2015, pp. 119–128. ACM, New York (2015). http://doi.acm.org/10.1145/2783258.2783296
9. Chen, H., Perozzi, B., Al-Rfou, R., Skiena, S.: A tutorial on network embeddings. arXiv preprint arXiv:1808.02590 (2018)
10. Chen, H., Li, X., Huang, Z.: Link prediction approach to collaborative filtering. In: Proceedings of the 5th ACM/IEEE-CS Joint Conference on Digital Libraries (JCDL 2005), pp. 141–142. IEEE (2005)
11. Cho, H., Yu, Y.: Link prediction for interdisciplinary collaboration via co-authorship network. Soc. Netw. Anal. Min. **8**(1), 25 (2018)
12. Clauset, A., Moore, C., Newman, M.E.: Hierarchical structure and the prediction of missing links in networks. Nature **453**(7191), 98 (2008)
13. Cui, P., Wang, X., Pei, J., Zhu, W.: A survey on network embedding. IEEE Trans. Knowl. Data Eng. **31**(5), 833–852 (2019)
14. Gao, F., Musial, K., Cooper, C., Tsoka, S.: Link prediction methods and their accuracy for different social networks and network metrics. Sci. Program. **2015**, 1 (2015)
15. Gao, S., Denoyer, L., Gallinari, P.: Temporal link prediction by integrating content and structure information. In: Proceedings of the 20th ACM International Conference on Information and Knowledge Management, CIKM 2011, pp. 1169–1174. ACM, New York (2011). http://doi.acm.org/10.1145/2063576.2063744
16. Getoor, L., Taskar, B.: Statistical relational learning (2007)

17. Goyal, P., Ferrara, E.: Graph embedding techniques, applications, and performance: a survey. Knowl.-Based Syst. **151**, 78–94 (2018)
18. Goyal, P., Hosseinmardi, H., Ferrara, E., Galstyan, A.: Capturing edge attributes via network embedding. arXiv preprint arXiv:1805.03280 (2018)
19. Grover, A., Leskovec, J.: Node2vec: scalable feature learning for networks. In: Proceedings of the 22nd ACM SIGKDD International Conference on Knowledge Discovery and Data Mining, KDD 2016, pp. 855–864. ACM, New York (2016). http://doi.acm.org/10.1145/2939672.2939754
20. Hasan, M.A., Zaki, M.J.: A Survey of Link Prediction in Social Networks, pp. 243–275. Springer, Boston (2011). https://doi.org/10.1007/978-1-4419-8462-3_9
21. He, Q., Pei, J., Kifer, D., Mitra, P., Giles, L.: Context-aware citation recommendation. In: Proceedings of the 19th International Conference on World Wide Web, WWW 2010, pp. 421–430. ACM, New York (2010). http://doi.acm.org/10.1145/1772690.1772734
22. Heckerman, D., Meek, C., Koller, D.: Probabilistic entity-relationship models, PRMS, and plate models. Introduction to statistical relational learning, pp. 201–238 (2007)
23. powered by HSE Portal: Publications of HSE (2017). http://publications.hse.ru/en. Accessed 9 May 2017
24. Kipf, T.N., Welling, M.: Variational graph auto-encoders. arXiv preprint arXiv:1611.07308 (2016)
25. Li, J., Xia, F., Wang, W., Chen, Z., Asabere, N.Y., Jiang, H.: ACREC: a co-authorship based random walk model for academic collaboration recommendation. In: Proceedings of the 23rd International Conference on World Wide Web, pp. 1209–1214. ACM (2014)
26. Li, X., Chen, H.: Recommendation as link prediction: a graph kernel-based machine learning approach. In: Proceedings of the 9th ACM/IEEE-CS Joint Conference on Digital Libraries, JCDL 2009, pp. 213–216. ACM, New York (2009). http://doi.acm.org/10.1145/1555400.1555433
27. Liang, Y., Li, Q., Qian, T.: Finding relevant papers based on citation relations. In: Wang, H., Li, S., Oyama, S., Hu, X., Qian, T. (eds.) WAIM 2011. LNCS, vol. 6897, pp. 403–414. Springer, Heidelberg (2011). https://doi.org/10.1007/978-3-642-23535-1_35
28. Liben-Nowell, D., Kleinberg, J.: The link-prediction problem for social networks. J. Assoc. Inf. Sci. Technol. **58**(7), 1019–1031 (2007)
29. Liu, F., Liu, B., Sun, C., Liu, M., Wang, X.: Deep learning approaches for link prediction in social network services. In: Lee, M., Hirose, A., Hou, Z.-G., Kil, R.M. (eds.) ICONIP 2013. LNCS, vol. 8227, pp. 425–432. Springer, Heidelberg (2013). https://doi.org/10.1007/978-3-642-42042-9_53
30. Liu, Y., Kou, Z.: Predicting who rated what in large-scale datasets. SIGKDD Explor. Newsl. **9**(2), 62–65 (2007). https://doi.org/10.1145/1345448.1345462
31. Liu, Z., et al.: Semantic proximity search on heterogeneous graph by proximity embedding. In: AAAI, pp. 154–160 (2017)
32. Liu, Z., et al.: Distance-aware DAG embedding for proximity search on heterogeneous graphs. In: Thirty-Second AAAI Conference on Artificial Intelligence, pp. 2355–2362. AAAI (2018)
33. Lopes, G.R., Moro, M.M., Wives, L.K., de Oliveira, J.P.M.: Collaboration recommendation on academic social networks. In: Trujillo, J., et al. (eds.) ER 2010. LNCS, vol. 6413, pp. 190–199. Springer, Heidelberg (2010). https://doi.org/10.1007/978-3-642-16385-2_24

34. Lü, L., Zhou, T.: Link prediction in complex networks: a survey. Phys. A: Stat. Mech. Its Appl. **390**(6), 1150–1170 (2011)
35. Makarov, I., Bulanov, O., Zhukov, L.: Co-author recommender system. In: Kalyagin, V., Nikolaev, A., Pardalos, P., Prokopyev, O. (eds.) Models, Algorithms, and Technologies for Network Analysis. Springer Proceedings in Mathematics & Statistics, vol. 197, pp. 251–257. Springer, Berlin (2017). https://doi.org/10.1007/978-3-319-56829-4_18
36. Makarov, I., Gerasimova, O., Sulimov, P., Korovina, K., Zhukov, L.E.: Joint node-edge network embedding for link prediction. In: van der Aalst, W.M.P., et al. (eds.) AIST 2018. LNCS, vol. 11179, pp. 20–31. Springer, Cham (2018). https://doi.org/10.1007/978-3-030-11027-7_3
37. Makarov, I., Gerasimova, O., Sulimov, P., Zhukov, L.E.: Co-authorship network embedding and recommending collaborators via network embedding. In: van der Aalst, W.M.P., et al. (eds.) AIST 2018. LNCS, vol. 11179, pp. 32–38. Springer, Cham (2018). https://doi.org/10.1007/978-3-030-11027-7_4
38. Makarov, I., Gerasimova, O., Sulimov, P., Zhukov, L.: Dual network embedding for representing research interests in the link prediction problem on co-authorship networks. PeerJ Comput. Sci. **5**, e172 (2019)
39. Makarov, I., Bulanov, O., Gerasimova, O., Meshcheryakova, N., Karpov, I., Zhukov, L.E.: Scientific matchmaker: collaborator recommender system. In: van der Aalst, W.M.P., et al. (eds.) AIST 2017. LNCS, vol. 10716, pp. 404–410. Springer, Cham (2018). https://doi.org/10.1007/978-3-319-73013-4_37
40. Makarov, I., Gerasimova, O., Sulimov, P., Zhukov, L.E.: Recommending co-authorship via network embeddings and feature engineering: the case of national research university higher school of economics. In: Proceedings of the 18th ACM/IEEE on Joint Conference on Digital Libraries, pp. 365–366. ACM (2018)
41. McPherson, M., Smith-Lovin, L., Cook, J.M.: Birds of a feather: Homophily in social networks. Annu. Rev. Sociol. **27**(1), 415–444 (2001)
42. Ortega, F., Bobadilla, J., Gutiérrez, A., Hurtado, R., Li, X.: Artificial intelligence scientific documentation dataset for recommender systems. IEEE Access **6**, 48543–48555 (2018)
43. Perozzi, B., Al-Rfou, R., Skiena, S.: Deepwalk: online learning of social representations. In: Proceedings of the 20th ACM SIGKDD International Conference on Knowledge Discovery and Data Mining, KDD 2014, pp. 701–710. ACM, New York (2014). http://doi.acm.org/10.1145/2623330.2623732
44. Robins, G., Snijders, T., Wang, P., Handcock, M., Pattison, P.: Recent developments in exponential random graph (p*) models for social networks. Soc. Netw. **29**(2), 192–215 (2007)
45. Scott, J.: Social Network Analysis. Sage, Thousand Oaks (2017)
46. Sinha, A., et al.: An overview of Microsoft Academic Service (MAS) and applications. In: Proceedings of the 24th international conference on world wide web, pp. 243–246. ACM (2015)
47. Srinivas, V., Mitra, P.: Applications of Link Prediction. In: Link Prediction in Social Networks. Springer International Publishing, Cham, pp. 57–61 (2016). https://doi.org/10.1007/978-3-319-28922-9_5
48. Tang, J., Qu, M., Wang, M., Zhang, M., Yan, J., Mei, Q.: Line: large-scale information network embedding. In: Proceedings of the 24th International Conference on World Wide Web, WWW 2015, pp. 1067–1077. International World Wide Web Conferences Steering Committee, Republic and Canton of Geneva, Switzerland (2015). https://doi.org/10.1145/2736277.2741093

49. Tang, J., Zhang, J., Yao, L., Li, J., Zhang, L., Su, Z.: Arnetminer: extraction and mining of academic social networks. In: Proceedings of the 14th ACM SIGKDD international conference on Knowledge discovery and data mining, pp. 990–998. ACM (2008)

50. Tang, J., Zhang, J., Yao, L., Li, J., Zhang, L., Su, Z.: Arnetminer: extraction and mining of academic social networks. In: KDD 2008, pp. 990–998 (2008)

51. Tang, J., Liu, H.: Unsupervised feature selection for linked social media data. In: Proceedings of the 18th ACM SIGKDD International Conference on Knowledge Discovery and Data Mining, KDD 2012, pp. 904–912. ACM, New York (2012). http://doi.acm.org/10.1145/2339530.2339673

52. Velden, T., Lagoze, C.: Patterns of collaboration in co-authorship networks in chemistry-mesoscopic analysis and interpretation. In: 12th International Conference on Scientometrics and Informetrics, pp. 1–12. ISSI Society, Rio de Janeiro (2009)

53. Wang, P., Xu, B., Wu, Y., Zhou, X.: Link prediction in social networks: the state-of-the-art. Sci. China Inf. Sci. **58**(1), 1–38 (2015). https://doi.org/10.1007/s11432-014-5237-y

54. Wasserman, S., Faust, K.: Social Network Analysis: Methods and applications, vol. 8. Cambridge University Press, Cambridge (1994)

55. Yan, E., Ding, Y.: Applying centrality measures to impact analysis: a coauthorship network analysis. J. IST Assoc. **60**(10), 2107–2118 (2009)

56. Zhai, S., Zhang, Z.: Dropout training of matrix factorization and autoencoder for link prediction in sparse graphs. In: Proceedings of the 2015 SIAM International Conference on Data Mining, pp. 451–459. SIAM (2015)

Red-Black Tree Based NeuroEvolution
of Augmenting Topologies

William R. Arellano[1], Paul A. Silva[1], Maria F. Molina[1], Saulo Ronquillo[1],
and Francisco Ortega-Zamorano[2]([✉])

[1] School of Mathematical and Computational Sciences, University of Yachay Tech,
San Miguel de Urcuquí, Ecuador
[2] Department of Computer Science, University of Málaga, Málaga, Spain
fortega@lcc.uma.es

Abstract. In Evolutionary Artificial Neural Networks (EANN), evolutionary algorithms are used to give an additional alternative to adapt
besides learning, specially for connection weights training and architecture
design, among others. A type of EANNs known as *Topology and Weight
Evolving Artificial Neural Networks* (TWEANN) are used to evolve topology and weights. In this work, we introduce a new encoding on an implementation of NeuroEvolution of Augmenting Topologies (NEAT), a type
of TWEANN, by adopting the Red-Black Tree (RBT) as the main data
structure to store the connection genes instead of using a list. This new
version of NEAT efficacy was tested using as case of study some data sets
from the UCI database. The accuracy of networks obtained through the
new version of NEAT were compared with the accuracy obtained from feed-
forward artificial neural networks trained using back-propagation. These
comparisons yielded that the accuracy were similar, and in some cases the
accuracy obtained by the new version were better. Also, as the number of
patterns increases, the average number of generations increases exponentially. Finally, there is no relationship between the number of attributes
and the number of generations.

Keywords: NEAT · Red-black tree · Back-propagation · Classification

1 Introduction

An Artificial Neural Network (ANN) is defined as a computational model that
simulates a biological representation of the interconnections of the units of the
nervous system, the neurons [2]. Nowadays, ANNs are used widely in many different fields, such as natural resources management [7], financial applications [10],
and health [3]. The architecture of an artificial neural network defines the number of layers, the number of neurons per layer and how the interconnections of
the neurons are performed [2].

ANNs have been broadly studied in recent years, mainly considering their
positive results in learning process for different engineering fields. This learning is

© Springer Nature Switzerland AG 2019
I. Rojas et al. (Eds.): IWANN 2019, LNCS 11507, pp. 678–686, 2019.
https://doi.org/10.1007/978-3-030-20518-8_56

achieved through an algorithm experience process that customizes the connection weights to perform tasks. As a consequence, the functionality and feasibility in which the ANN can be implemented for certain fields depends in a great percentage on the complexity patterns that the system implies, and the kind of ANN that is chosen to work with [11].

Deep artificial neural networks help to solve some optimization problems better than shallow networks. Defining which is the most suitable topology for a network can be a very time consuming endeavor as these are complex in nature [5]. Parameters as the number of layers or, the number of neurons per layer are not easy to choose. Other shortcoming is that one topology might work well for a problem but under-perform on others as the topologies are chosen taking in consideration the boundaries of those problems.

Genetic Algorithms (GA) were described by Holland in [4]. For Holland, a GA is a method of Darwinian selection that in conjunction with genetics-inspired operators (crossover, mutation, inversion and selection), moved a population of chromosomes to a new one. A chromosome consists of genes that express a particular trait in the solution. The selection operator is used to choose the chromosomes that will be allowed to reproduce. The crossover operator creates two new offspring by exchanging parts of two chromosomes. The mutation operator randomly changes the value of genes. Finally, the inversion operator reorders the genes structure, by selecting a fragment of the genome and then inverting it.

In EANNs, evolutionary algorithms are used to add another way of adaptation besides learning, where the algorithm is used to perform, and connection weights training and architecture design [11]. EANNs where topology and weights evolve are known as TWEANNs. Many methods for incrementally evolving TWEANNs have been implemented, for example: NEAT [9], *Evolutionary Acquisition of Neural Topologies* (EANT) [6], and *Hypercube-based NeuroEvolution of Augmenting Topologies* (HyperNEAT) [8]. All these methods start from a minimal structure and add any necessary complexity to the topology along the evolutionary process.

In this work we introduce a new encoding format for the structure of NEAT. We define this new encoding as Red-Black Tree based NeuroEvolution of Augmenting Topology (RBTNEAT) because it characterizes the network topology and the connection weights using a typical implementation of NEAT but adopting the RBT as the main data structure to store the connection genes instead of a list. This new approach is inspired by the fact that basic dynamic-set operations such as access and search take $O(m \log (\frac{n}{m} + 1))$ time in the worst case for RBT, whereas those operations take $O(n \cdot m)$ time in the worst case for lists [1], where the values of m and n represent the size of the sets (number of elements in the trees) to be compared, in this case it would be the size of the two genomes to be compared. We compare its performance against a network with back-propagation, using several data sets from the UCI Data Set Problems. Then, its behavior is observed when the number of attributes and patterns vary.

2   Methodology

2.1   NEAT

The process of the algorithm NEAT is described in Algorithm 1. The first population is created by the function *initialPopulation*, and it is composed of networks with connections only between inputs and outputs. To preserve the innovation in NEAT, the population is put through a process of speciation, where the population is divided into species based on a distance function. Then, the population is evaluated and the best fitness is obtained. To measure the genome's fitness Eq. 1 is used, where N is the number of patterns, z is the expected output and y the output of the network. The main part of the algorithm is the while loop, where in each iteration (generation) a new population is obtained by first selecting the parents genomes, then applying the crossover operation over these parents and finally mutating this new population.

$$f = 1 - \frac{1}{N}\sum_{i=1}^{N}(z_i - y_i)^2 \qquad (1)$$

Algorithm 1. NEAT

1: **procedure** NEAT(iterMax, fitnessThreshold)
2:     $population \leftarrow initialPopulation()$
3:     $species \leftarrow speciation(population)$
4:     $iter \leftarrow 0$
5:     $bestFitness \leftarrow evaluate(population)$
6:     **while** $iter < iterMax$ & $bestFitness < fitnessThreshold$ **do**
7:         $parents \leftarrow selection(population, species)$
8:         $population \leftarrow crossover(parents)$
9:         $population \leftarrow mutate(population)$
10:         $bestFitness \leftarrow evaluate(population)$
11:         $species \leftarrow speciation(population)$
12:         $iter \leftarrow iter + 1$
13:     $bestGenome \leftarrow best(population)$
14:     **return** $bestGenome$

In the original NEAT, genomes were encoded as a list of nodes and connection genes. In this work, we introduce a new genome encoding that uses RBT as the data structure to store the connection genes. More details of this new approach are explained in the following section.

2.2   RBTNEAT

In this new approach, a RBT is chosen to store the connection genes as it allows fast crossover, whereas a list is still used to store the nodes. A connection gene indicates the nodes to be connected, the weight, the *innovation number* which is used to track genes that express the same structure, and whether or not the connection is expressed. The connections genes tree is constructed based on each gene's innovation number. For example, Fig. 1 shows how the genome looks like. And, it can be observed that the node genes are stored as a list, whereas the connection genes are stored as a tree. Figure 2 shows the phenotype, the expression of the example genome.

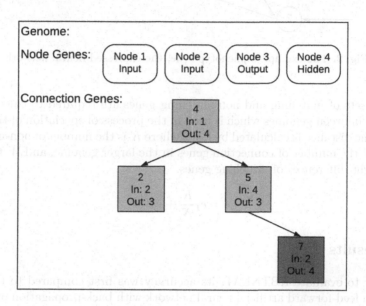

Fig. 1. An example of genome using RBTNEAT encoding.

Crossover is the process whereby two genomes are lined up and exchange their genes to produce offspring. This alignment of the genomes is based on each gene's innovation number. The matching genes are those with matching innovation numbers, whereas the genes without matching innovation numbers are known as disjoint or excess genes. The offspring is obtained by choosing the matching genes from the most fit parent, and choosing randomly the non-matching genes (disjoint or excess).

In NEAT and RBTNEAT[1], the sets of matching and non-matching are calculated by using set operations on lists and red-black trees, respectively. The set of matching genes is obtained by their interception, and the non-matching genes is obtained by their symmetric difference. For lists these set operations have worst time complexity $O(n \cdot m)$, whereas for red-black trees these operations have worst time complexity $O(m \log (\frac{n}{m} + 1))$. Thus, red-black trees are faster for calculating the matching and non-matching genes.

---

[1] Source code available at https://github.com/cptrodolfox/rbtneat.

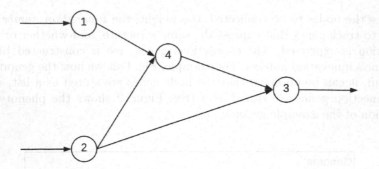

Fig. 2. The example's phenotype (i.e. the expressed neural network).

The sets of matching and non-matching genes are required to calculate the distance between genomes which is used in the process of speciation of the population, the distance is calculated by Eq. 2 where K is the number of non-matching genes, N the number of connection genes in the larger genome, and $\overline{W}$ the average weight differences of matching genes.

$$\delta = c_1 \frac{K}{N} + c_2 \overline{W} \tag{2}$$

3   Results

In order to evaluate RBTNEAT, its accuracy was first compared to the accuracy of a feed-forward artificial neural network with back-propagation on several classification sets of real world medical problems from the UCI database. The classification ability was obtained using a sigmoid activation function.

The network architecture consisted of 10 neurons to obtain the best mean of accuracy for the data sets, and had a learning rate of 0.2 where the synaptic weights were randomly initialize between 1 and -1. These tests were carried out using a 10-fold cross-validation for the training and generalization sets respectively, including a validation set to decide the termination criterion (20% of the training set) with a Hamming window to reduce the oscillations and a maximum of 1000 epochs.

RBTNEAT had a population of 150 networks (genomes) and each started as a full connected network. It ran either 1000 generations or up to it reached a fitness threshold. The probability of adding a new connection was of 0.5 and the probability of adding a new node was of 0.2. A 10-fold cross-validation was also performed on RBTNEAT. The classification ability of both methods can be seen in Table 1.

For evaluating the performance of the approach the time per generation and the number of generations was compared to the number of patterns. Figure 3 shows the evolution of the time execution and the number of generations as a function of the numbers of patterns of the dataset. The card dataset (see

Table 1. Back-propagation and RBTNEAT classification ability in terms of accuracy using several data sets from the UCI Data Sets

Method	# Attributes	# Inputs	BP	RBTNEAT
Blood transfusion	5	748	78.92	79.28
Cancer	9	683	95.56	96.48
Card	51	690	84.10	85.80
Climate	18	540	91.65	94.44
Diabetes	8	768	75.25	76.96
heartc	35	303	79.7	81.15
Ionosphere	34	351	89.23	87.76
Sonar	60	208	75.55	74.48
Statlog	13	270	80.04	83.70
Vertebral column	6	310	85.16	84.52

Table 1) was used to perform this study in which the number of patterns for the learning process was selected from 70 to all patterns (690) in steps of 70. Also, another test was carried out to analyze the impact on the time used per generation and the number of generations if the numbers of attributes (inputs) increase. The card dataset (see Table 1) was also used to perform this study in which the number of attributes for the learning process was selected from 5 to all attributes (50) in steps of 5 (Fig. 4).

4   Discussion

In Table 1 it can be seen that using red-black tree as a main data structure to store the connection genes gives an improvement in 7 of 10 data sets. This means that changing the data structure did not worsen the ability of getting a good classification.

A linear relationship between the number of patterns and the average time per generation is obtained. This may be caused by the increment of operations employed to select the parents genomes done in each species. Additionally, we realize that the average number of generations has exponential growth with respect to the number of patterns. This could be related to speciation. In which more patterns involve more individuals that are protected by competing only within their species. In fact, the speciation step seems to be a critical for the performance of the algorithm but it is also crucial to have a good solution.

In addition, we can appreciate that there is not apparent relationship between the number of attributes and the number of generations. This could be because NEAT make a sort of selection of the attributes that are more relevant to the output in order to get the simplest network possible. The RBTNEAT network on Fig. 5 is an example. Four attributes of the network are unconnected to the rest of the network implying that those attributes are not particularly important

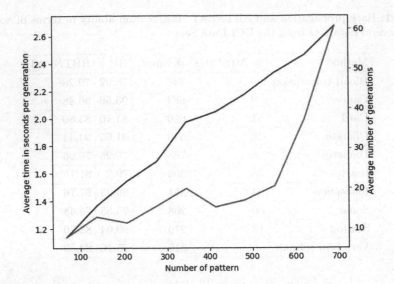

Fig. 3. Evolution of the time execution and the number of generations as a function of the numbers of patterns of the data set

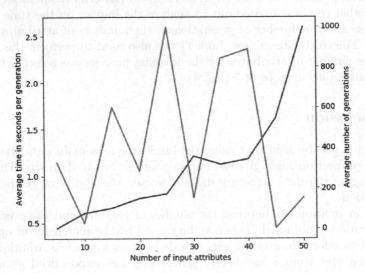

Fig. 4. Time used per generation and the number of generations if the numbers of attributes (inputs) increase.

when predicting output. However, the average time per generation seems to have an exponential relationship with the number of input attributes. The grown of the amount with the number of initial nodes might be the cause.

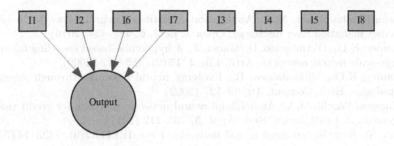

Fig. 5. RBTNEAT example of a winning network for the diabetes data set.

5 Conclusion

RBTNEAT proves to be an effective method to find a well performing topology for an artificial neural network. Results shows that RBTNEAT have similar performance than a feed forward network with back-propagation. Additionally, the number of generations seems to be independent to the number of attributes, but not to the number of patterns. Moreover, the increment in the amount of patterns and attributes cause an increment time per generation. This is due to the growth of the number of operations done to find the parents' genome and evaluating the population.

In future work, RBTNEAT should be analyzed with an initial population of partially connected networks, meaning not all nodes are connected. As, NEAT in general creates a minimal structure, if we start the population with complex structures the resulting search tree increases in size.

Acknowledgements. The authors acknowledge support through grant TIN2017-88728-C2-1-R from MICINN (Spain) that includes FEDER funds and from Plan Propio from Universidad de Málaga (Spain) and the Yachay Tech University (Ecuador).

References

1. Cormen, T.H., Leiserson, C.E., Rivest, R.L., Stein, C.: Introduction to Algorithms. The MIT Press, Cambridge (2009)
2. Floreano, D., Du, P., Mattiussi, C.: Neuroevolution: from architectures to learning. Evol. Intell. **1**(1), 47–62 (2008)
3. Hansapani Rodrigo, C.P.T.: Artificial neural network model for predicting lung cancer survival. J. Data Anal. Inf. Process. **5**, 33–47 (2017)
4. Holland, J.H.: Adaptation in Natural and Artificial Systems, 1st edn. University of Michigan Press, Ann Arbor (1975)
5. Hunter, D., Yu, H., Pukish III, M.S., Kolbusz, J., Wilamowski, B.M.: Selection of proper neural network sizes and architectures-a comparative study. IEEE Trans. Industr. Inf. **8**(2), 228–240 (2012). https://doi.org/10.1109/TII.2012.2187914
6. Kassahun, Y., Sommer, G.: Efficient reinforcement learning through evolutionary adquisition of neural topologies. In: Proceedings of the 13th Annual European Symposium on Artificial Neural Networks, ESANN. d-side Publishing, Bruges, April 2005

7. Saman Mohammadi, M.S.: Application of artificial neural networks in order to predict mahabad river discharge. Open J. Ecol. **6**, 427–434 (2016)
8. Stanley, K.O., D'Ambrosio, D., Gauci, J.: A hypercube-based encoding for evolving large-scale neural networks. Artif. Life J. **15**(2), 185–212 (2009)
9. Stanley, K.O., Miikkulainen, R.: Evolving neural networks through augmenting topologies. Evol. Comput. **10**, 99–127 (2002)
10. Vincenzo Pacelli, M.A.: An artificial neural network approach for credit risk management. J. Intell. Learn. Syst. Appl. **3**, 103–112 (2011)
11. Yao, X.: Evolving artificial neural networks. Proc. IEEE **87**(9), 1423–1447 (1999)

A New Classification Method for Predicting the Output of Dye Process in Textile Industry by Using Artificial Neural Networks

Ahsen Noor Subhopoto[1]($\boxtimes$), Mehmet Akar[1], and Sencer Sultanoglu[2]

[1] Bogazici University, Bebek, 34342 Istanbul, Turkey
{ahsen.subhopoto,mehmet.akar}@boun.edu.tr
[2] Eliar Electronics A.S., Levazim Mh., 34340 Istanbul, Turkey
sencer.sultanoglu@eliarge.com

Abstract. In this paper, a new approach is proposed which predicts the output of the dyeing process in textile industry by using input data consisting of the alarms and/or the interventions during the process. Back-propagation algorithms and radial basis functions are utilized to form the neural networks in predicting whether the dye process is carried out correctly or not before an operator checks it manually. Industrial data are used to test the efficiency of the proposed concept which demonstrates that the success rate is over 85%.

Keywords: Artificial Neural Network (ANN) · Dye process ·
Textile industry · Levenberg-Marquardt (LM) ·
General regression Neural Network (GRNN) ·
Gradient Descent with an Adaptive Learning Rate (GDA)

1 Introduction

Since the 16th century, people have dyed their products by using different available materials [1]. Correct dyeing has always been a difficult task; predicting its outcome (whether it is going to be correct or not) is at least as difficult because of the presence of sensitive parameters that mainly consist of temperature control for different types of textile, raw materials and quality parameters of the products [2]. Conventional methods like spectrophotometric sensors were used to predict the dye processes with the help of different methods including neural network (NN) [3,4]. In [5], the removal of reactive red dye was predicted from the pH of the aqueous solution with the help of a NN model and an adaptive neuro-fuzzy inference system. The Authors in [6] showed that the time required to achieve the desired depth of a shade can be predicted by using ANN. In [7], a neural network which predicts the CIELAB (standardized colour space) values' for cotton fabric dyed with vinyl sulphone reactive dye was developed. The prediction of CIELAB values was also possible for colour changes after the

© Springer Nature Switzerland AG 2019
I. Rojas et al. (Eds.): IWANN 2019, LNCS 11507, pp. 687–698, 2019.
https://doi.org/10.1007/978-3-030-20518-8_57

chemical process for stripped cotton fabrics and for nylon 6,6 using Levenberg-Marquardt (LM) based feed-forward back-propagation neural network algorithm respectively [8,9].

Image processing and fuzzy neural network system approaches were developed in [10] to classify precisely seven kinds of dyeing defects including shade, dye and carrier spots, mist, oil stains, listing, tailing, and uneven dyeing on selvage. In [11], a fuzzy knowledge based expert system was demonstrated which predicts the colour strength of the cotton knitted fabrics. In [12], a feed-forward neural network which predicts the colour alteration values after the spinning process was developed. In [13], a fuzzy-neural network together with an image processing system was established which can identify and analyse printed fabric images and colour pattern editing system by using an automated colour separation algorithm. However, an image processing approach is prone to some limitations, for example light intensity can be a limiting factor. Furthermore, the maximum number of specific colours in an image should be limited in accord with self-organising map (SOM) neural network (NN) [13]. In [14], a neural network system combined with a genetic algorithm which can predict and optimize the dyeing process for nylon and Lyera blend quality was developed. In [15], a NN model which predicts the colour properties of four knitted fabric materials with the laser treatment before the colouring was proposed. The amount of the dye present on the denim fabric which indicates the laser surface engraving treatment technology is found in [16–18]. In [19], another NN model which can predict the colour properties of laser-treated 100% cotton fabric was developed.

In this paper, the object is to predict the output of the dyeing process in advance by using data consisting of alarms and/or interventions during the process which can last up to 2–8 h. By predicting the outcome in real-time, it is possible to correct the process, which in turn reduces the cost and maintenance of the system associated with incorrect dye processes, and preserves the quality of the fabric. In this paper, we utilize artificial neural networks (ANN) to achieve our objective. While different approaches have been used with or without neural networks in the literature as discussed above, none address the problem of predicting the outcome of the dye processes by use of alarms and/or interventions. It is experimentally observed that the proposed technique can work on any type of textile, different colours and properties that are available for the textile industry.

The rest of the paper is organised as follows. Section 2 demonstrates the system implementation carried out for the approach. Section 3 briefly explains the artificial neural networks and the different ANN algorithms which are used in this paper. Later, in Sect. 4, experiments and results are discussed for two different database of the textile industry. Finally, the paper is concluded in Sect. 5.

2   System Implementation

In this section, we explicitly describe the details of our system implementation. Figure 1 shows the flow chart of the methodology carried out in this paper.

The input to the system consists of alarms or interventions or both the alarms and interventions merged together during the process. The output for each dye process of the machine consists of either the successful dye process, i.e., 1 or the unsuccessful dye process, i.e., 0. The detailed descriptions for each input and pre-processing method are explained below.

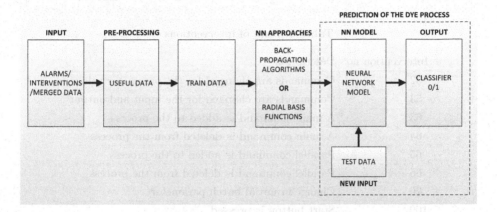

Fig. 1. Methodology of the approach

Alarms. About 1102 different alarms are categorised with the alarm types (0 to 10) are specified in the database. For each machine the same alarm may repeat itself at different times during the process and all alarms are counted until the specific alarm which allows the dye process to be checked manually by an operator. At the end of each process, the number of each alarm is used as the input to the system. Table 1 summarises the alarm types with their descriptions.

Table 1. Data of alarms

Alarm type	Alarm no.	Description
0	100–199	Time exceeded for automatic operations; fabric loading delay
1	100–199	Equipment alarms (e.g., low water, min. speed)
2	100–199	Operator call (e.g., operator call (chemical), get sample)
3	100–199	Manual commands (pH control or fabric loading) time exceeded
4	200–499	Interlock alarms (e.g., volt fault, pressure not provided)
5	500–599	PLC alarms (e.g., no plc communication, connection is available)
6	600–699	Software alarms (e.g., software memory fault)
7	700–799	Calibration alarms (e.g., maximum tank level, high temperature)
8	800–899	Coupling alarms
9	5000	Switch to manual mode alarm (e.g., interventions, mechanical faults)
10	6000	Batch pause alarm (e.g., work order is stopped)

Interventions. The interventions are referred as the interference by an operator corresponding to that specific alarm. These interventions are mostly performed to correct the operation of the machine. There are only 10 specific interventions which are specified in the database that may occur for each machine which are summarised in Table 2. Similar to the extraction of useful alarms, the total available number of interventions before the dye process is checked manually in each process of the machine are counted for training and testing purposes.

Table 2. Data of interventions

Intervention no.	Description
52	Commands are changed for the parameter values
53	Commands are changed for the input and output
63	A main command is added to the process
64	A main command is deleted from the process
65	Parallel command is added to the process
66	Parallel command is deleted from the process
70	Change in initial batch parameter
100	Start button is pressed
105	Start of an additional process
107	Completion of an additional process

Merged Data. The alarms and the interventions as explained above are combined together respectively (that is, the alarms followed by the interventions are added together) to form the "merged" input dataset which gives the correlation between the alarms and the interventions for each process of the machine in real-time. The merged data provide information regarding how the specific intervention which is supposed to correct the operation of a dye process, has an impact on the specific alarm in real-time environment.

2.1   Pre-processing

Initially, each useful dye process of the machine is extracted by considering those processes that should have procedural time of at least 2–4 h depending on the process, and the end time is present in the database for the machine. After the extraction of useful dye processes, input like alarms and interventions for each process can be extracted and pre-processed accordingly. The dataset consisting of alarms or interventions is of size N × M, where N is the total number of alarms (1102) or interventions (10) and M is the total number of dye processes of the machine which is then reduced to P × M, where P = N − U, U is the total number of the alarms or the interventions that do not occur in the machine. Finally, this fixed matrix is fed to the neural network approaches for training and prediction purposes.

3   Artificial Neural Network Models

This section briefly discusses the neural network models used for training and testing the approach. More specifically, two types of feed-forward network; namely multi-layer feed-forward back-propagation neural networks and radial basis functions are reviewed.

3.1   Feed-Forward Back-Propagation Algorithms

There are many algorithms that can be used in the multi-layer feed-forward back-propagation methods and their sole task is to readjust weights in the network. This section briefly discusses several back-propagation (BP) algorithms; Gradient Descent with an Adaptive Learning Rate (GDA), Levenberg-Marquardt (LM) and Scaled Conjugate Gradient (SCG).

Gradient Descent with an Adaptive Learning Rate (GDA). GDA is a class of a gradient descent methods, where the learning parameter α is updated during training which makes this as an improved algorithm of a gradient descent method. In GDA, first initial output and corresponding error are calculated. Another output and corresponding error are calculated and compared with the previous error. Then, at each epoch, if the new error is larger than the previous one by more than a specified value $\tau > 1$, then new biases and weights are rejected, and the learning rate is reduced by a factor $0 < l_d < 1$. Otherwise, the new weights are accepted. Furthermore, if the new error is smaller than the previous one, then α is increased by a factor $l_i > 1$. The following parameters are used in the simulations; $\tau = 1.04$, $l_d = 0.7$ and $l_i = 1.05$. The updating of weights is expressed by (1) where w is the weight matrix, α is the learning rate, E is the error function and g is the gradient expressed by (2):

$$\mathbf{w}_{i+1} = \mathbf{w}_i - \alpha_i \mathbf{g}_i \tag{1}$$

$$\mathbf{g} = \frac{\partial \mathbf{E}}{\partial \mathbf{w}} \tag{2}$$

Scaled Conjugate Gradient (SCG). SCG method is a class of conjugate gradient methods, and it is a supervised learning algorithm for the feed-forward NN. The common BP algorithm uses negative of a gradient to find the minimum of the cost function. SCG works in a similar fashion as that of other gradient back-propagation algorithms except that it searches its minimum cost in a conjugate direction. Refer to [20] for a detailed explanation of the SCG algorithm.

Levenberg-Marquardt (LM). Levenberg-Marquardt algorithm is a combination of two minimization methods: the gradient descent (or steepest descent) and the Gauss-Newton method [21]. The update weight vector which consists of weights and biases is given by (3) where H is the Hessian matrix, I is the identity matrix, α is the learning parameter, g is the gradient expressed by (5) and J is the Jacobian matrix expressed by (6):

$$\Delta \mathbf{w} = [\mathbf{J}'(\mathbf{w})\mathbf{J}(\mathbf{w}) + \alpha \mathbf{I}]^{-1} \mathbf{J}'(\mathbf{w})\mathbf{E} \tag{3}$$

$$\mathbf{H} = \mathbf{J}'(\mathbf{w})\mathbf{J}(\mathbf{w}) \tag{4}$$

$$\mathbf{g} = \mathbf{J}'(\mathbf{w})\mathbf{E} \tag{5}$$

$$\mathbf{J} = \begin{bmatrix} \frac{\partial \mathbf{E}_1(\mathbf{w})}{\partial \mathbf{w}_1} & \frac{\partial \mathbf{E}_1(\mathbf{w})}{\partial \mathbf{w}_2} & \cdots & \frac{\partial \mathbf{E}_1(\mathbf{w})}{\partial \mathbf{w}_n} \\ \frac{\partial \mathbf{E}_2(\mathbf{w})}{\partial \mathbf{w}_1} & \frac{\partial \mathbf{E}_2(\mathbf{w})}{\partial \mathbf{w}_2} & \cdots & \frac{\partial \mathbf{E}_2(\mathbf{w})}{\partial \mathbf{w}_n} \\ \vdots & \vdots & \ddots & \vdots \\ \frac{\partial \mathbf{E}_n(\mathbf{w})}{\partial \mathbf{w}_1} & \frac{\partial \mathbf{E}_n(\mathbf{w})}{\partial \mathbf{w}_2} & \cdots & \frac{\partial \mathbf{E}_n(\mathbf{w})}{\partial \mathbf{w}_n} \end{bmatrix} \tag{6}$$

The learning parameter is automatically adjusted as follows. When α is small or zero, the algorithm becomes Gauss-Newton method. When α parameter is large, LM algorithm becomes steepest descent (or gradient descent). This type of algorithm is mostly used for solving non-linear least squares problems [22, 23] as it avoids vanishing problem that occurs in the gradient descent method and is faster in convergence than any other algorithm.

3.2 Radial Basis Functions (RBF)

Radial basis functions (RBF) architecture is similar to the MLP feed-forward models except that RBF has only a single hidden layer in its architecture. Radial basis function finds the distance between the two points in a multidimensional space. The output of the RBF network is expressed by (7). Unknown N weights "w" in N equations in its simplest form is given by (8), when the activation function ϕ is assumed to be invertible.

$$\mathbf{y}_i(\mathbf{x}) = \sum_{j=1}^{N} \mathbf{w}_{ji}\phi\left(\| \mathbf{x} - \mu_j \|\right) + \mathbf{b}_i \tag{7}$$

$$\mathbf{w} = \phi^{-1}\mathbf{y} \tag{8}$$

$$\phi(\mathbf{r}) = \mathrm{e}^{\frac{-r^2}{2\sigma^2}} \tag{9}$$

In this paper, Gaussian is used as the activation function in a hidden layer of RBF network expressed by (9), where $r > 0$ gives the distance $\| x - \mu \|$ from x (data point) to a centre μ, and σ is a smoothing parameter which is also expressed as "spread" in RBF. As RBF determines the distance between the two points when the distance between the two points is far away, then ϕ gives an output 0, or else the output increases when the distance decreases.

General Regression Neural Network (GRNN). GRNN is a probabilistic-based neural network which performs regression by using Euclidean Norm approach. The design of the GRNN can be achieved in far less time then required for the design and training of an MLP which uses an iterative approach [24]. Usually the number of neurons in the hidden layer is equal to the number of the input data because the clustering of the input training data is performed in the radial layer (hidden layer). The spread value for GRNN is 0.001 in this paper.

Exact Radial Basis Network (RBE). This type of radial basis function can approximate any network with an error very close to zero on the training vectors. In GRNN, the number of neurons is equal to the number of input dataset. However, in RBE, one neuron is added iteratively if the current error is not matched with the goal error (0 in our case). The spread value for RBE is chosen to be 1.

4   Experiments and Results

In this section, two different textile industry's database named as "X" and "Y" respectively, are tested for the prediction. Several results are obtained including different dye processes for various machines, with individual inputs including alarms, interventions and merged data which consist of both inputs (the alarms and the interventions). However, in this section, only alarms data and merged data results are shown. The output for each input corresponding to each process of the machine is either, the correct dye process – 1, or the wrong dye process – 0.

Data Collection. Two different sets of data from textile industry are collected from the companies based in Istanbul, Turkey through MySQL database software. These databases are then used to train and test the neural network models.

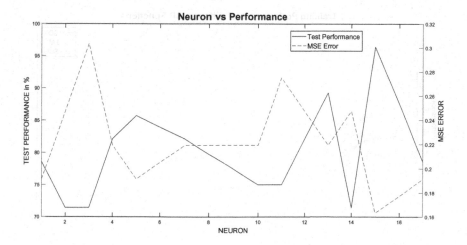

Fig. 2. Performance Vs Neuron

Number of Neurons Used. There is no general rule or formula explicitly to find the optimal number of neurons that could be used for an application [25]. However, in this paper, with trial-and-error method (iteration method), it is noted that 15 neurons in the hidden layer of the BP algorithms gives a good prediction rate compared to any other number of neurons, as seen in Fig. 2. Figure 2 shows that by using 15 neurons in the hidden layer of the SCG algorithm and dividing the merged data of the machine (alarms and interventions) having 137 dye processes into two parts; training (80%) and testing (20%), 96% test accuracy (prediction) with an MSE error of around 0.16 was obtained. By following the same methodology which is carried out in this paper it is experimentally observed that 15 neurons in the hidden layer perform good with another dataset of the industry.

4.1 Database X Results

80% data of specific machine of the database named as "X" is trained by using different BP approaches of the ANN; GDA algorithm, SCG algorithm and LM algorithm. These approaches are compared with each other to get the best approach. All these algorithms give good training rate that is above 90%, however, SCG outperformed training rate of about 99% with a performance very close to zero (i.e., 0.01), as seen in Fig. 3.

Table 3 summarizes the performance of the NN algorithms on the machine having total number of dye processes divided in different parts. Here, data is divided into x/y, as seen in the table, where x is the training data and y is the test data. Note that 100:86 notation means that 100% success is obtained for training and 86% with test data. Algorithm column indicates that different BP algorithms and RBF functions are used for the comparison purpose with different proportions of the alarms and the merged input data. LM-TT is the LM based

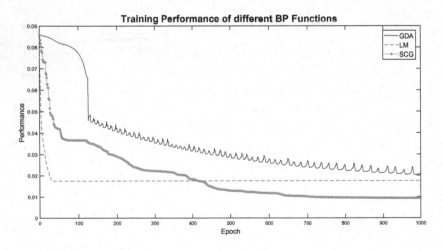

Fig. 3. Performance of different BP functions during training

Table 3. Performance of the machine using different BP algorithms and RBFs with 137 samples

Algorithm	Alarm data			Merged data		
	%Train/Test data			%Train/Test data		
	80/20	85/15	90/10	80/20	85/15	90/10
LM_TT	100:86	100:82	100:67	90:89	90:86	90:87
LM_TL	90:89	90:86	90:87	90:89	90:86	90:87
SCG_TT	99:82	99:77	100:73	100:96	100:86	99:80
SCG_TL	90:89	90:86	90:87	90:89	90:86	90:87
GDA_TT	99:82	99:82	100:80	100:93	99:77	99:80
RBE	100:89	100:86	100:88	100:89	100:86	100:86
GRNN	100:86	100:81	100:93	100:82	100:82	100:87

BP algorithm with tansig as a transfer function used in the hidden and output layer of the network. LM-TL is the LM based BP algorithm with tansig and logsig as a transfer function used in the hidden and the output layer respectively. SCG is the SCG based BP algorithm. From Table 3, it is experimentally noted that both SCG and RBE give good test results for every divided portion of the data comparatively to other algorithms, however, SCG results with merged data having 100% training performance rate and 96% test rate outperformed other mentioned algorithms.

4.2 Database Y Results

Another database is tested to evaluate the efficiency of the proposed approach in different dye-houses. It is experimentally noticed that GDA algorithm tested on

Table 4. Performance of the machine using different BP algorithms and RBF with 327 samples

Algorithm	Alarm data			Merged data		
	%Train/Test data			%Train/Test data		
	80/20	85/15	90/10	80/20	85/15	90/10
LM_TT	97:77	96:74	97:74	98:76	100:70	100:71
LM_TL	84:82	84:84	84:79	86:83	86:84	85:82
SCG_TT	97:79	98:72	97:74	98:79	98:74	98:74
SCG_TL	85:83	85:82	84:82	84:83	84:82	84:82
GDA_TT	90:80	90:76	90:76	94:86	92:86	93:88
RBE	100:83	99:26	100:82	100:83	100:82	100:82
GRNN	99:76	99:66	99:79	100:74	100:68	100:68

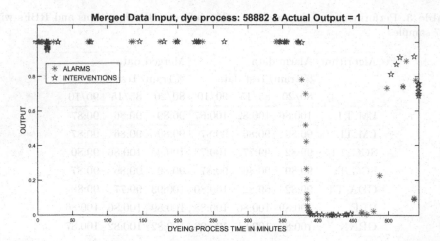

Fig. 4. Real-time Performance of dye process 58882 of the machine 31 using GDA algorithm

the merged data input having 80% of the data for training and 20% for testing the data achieved 94% training performance and 86% test accuracy rate which outperformed any other algorithms as shown in Table 4.

It is also noted that GRNN has almost 100% training accuracy rate when the spread value is close to zero that is 0.001 or 0.0001. However, in GRNN, if the spread value is altered that is when it is 1 or 2 then there is a trade-off between the training performance with the test performance. It is noted that when GRNN has a spread value greater than 1 the training performance is reduced and at the same time test performance is increased.

Real-Time Prediction. Each dye process of the machine consists of different types of alarms or interventions and they are fixed as a NN model which is later used for predicting the output of dye process in online or offline fashion. In offline method, each fixed input contains existence of alarms or interventions before the specific command which is available in the database and is different for each machine. New input whose dimension is always same as that of fixed NN model, is used for the prediction purpose. The specific command allows an operator to check the dye process and if an operator observes that it is not correctly dyed then it is sent back for re-processing. However, in real-time environment, each alarm or intervention are tested with fixed NN model one by one until the presence of specific command.

Gradient Descent with an Adaptive Learning Rate (GDA) is used in real-time application. Figure 4 shows how the interventions change the dye process of the batch-key to correct dye process at the very last minute. It is noted from Fig. 4 that continuous alarms leads to the wrong dye process: 0, however, with the continuous successful interventions led the system to the stable condition towards the end of the process. If this had been a wrong dye process after the unsuccessful interventions then this dye process would be been sent for reprocessing again, subsequently waste of time, resources and many other disadvantages associated with the textile industry.

5   Conclusion

From the experiments, it is noted that the SCG and GDA algorithms give the best prediction accuracy comparatively to other algorithms discussed in this paper. In one of the industries, RBE and GDA have outperformed the prediction accuracy rate of LM and SCG based neural network model. The training result of the RBE convergence of MSE can be low as 0.001 with the number of neurons equal to the input dataset. It is also noted that in another industry dataset the SCG based back-propagation algorithm has outperformed prediction accuracy rate against other mentioned algorithms. The training result of SCG convergence of an MSE can be as low as from 0.03 to 0.01 depending on the machine with 15 neurons in the hidden layer of the network.

Acknowledgments. This work was supported in part by TUBITAK Project 5180025.

References

1. Hana, K.: Natural dyes: their past, present, future and sustainability (2015)
2. Yildirim, P., Birant, D., Alpyildiz, T.: Data mining and machine learning in textile industry. J. WIREs. **8**, e1228 (2018)
3. Jasper, W.J., Kovacs, E.T., Berkstresser, G.A.: Using neural networks to predict dye concentrations in multiple-dye-mixtures. J. Textile Res. **63**, 545–551 (1993)
4. Beck, K.R., Maddera, T.A., Smith, C.B.: Real-time data acquisition in batch dyeing. Textile Chem. and Col. **23**, 23–27 (1991)
5. Kahkha, M.R.R., Piri, J.: Comparison of artificial neural network and neural-fuzzy inference system for photocatalytic removal of reactive red dye. Tech. J. Eng. Appl. Sci. **6**, 39–44 (2016)
6. Senthilkumar, M., Selvakumar, N.: Achieving expected depth of shade in reactive dye application using artificial neural network technique. Dye. Pigment. **68**, 89–94 (2006)
7. Senthilkumar, M.: Modelling of CIELAB values in vinyl sulphone dye application using feed-forward neural networks. Dye. Pigment. **75**, 356–361 (2007)
8. Balci, O., Ogulata, S.N., Sahin, C., Ogulata, R.T.: An artificial neural network approach to prediction of the colorimetric values of the stripped cotton fabrics. Fibers Polymers **9**, 604–614 (2008)
9. Balci, O., Ogulata, S.N., Sahin, C., Ogulata, R.T.: Prediction of CIElab data and wash fastness of nylon 6,6 using artificial neural network and linear regression model. Fibers Polymers **9**, 217–224 (2008)
10. Huang, C.C., Yu, W.H.: Fuzzy neural network approach to classifying dyeing defects. Tex. Res. J. **71**, 100–104 (2001)
11. Hossain, I., Hossain, A., Choudhury, I., Mamun, A.: Fuzzy knowledge based expert system for prediction of color strength of cotton knitted fabrics. J. Eng. Fibers Fabr. **11**, 33–44 (2016)

12. Thevenet, L., Dupont, D., Jolly-Desodt, A.M.: Modeling color change after spinning process using feed-forward neural networks. Color. Res. Appl. **28**, 50–58 (2002)
13. Xu, B., Lin, S.: Automatic color identification in printed fabric images by a fuzzy-neural network. AATCC Rev. **2**, 42–45 (2002)
14. Kuo, C.F.J., Fang, C.C.: Optimization of the processing conditions and prediction of the quality for dyeing nylon and lycra blended fabrics. Fibers Polymers **7**, 344–351 (2006)
15. Kan, C.W., Song, L.J.: An artificial neural network model for prediction of colour properties of knitted fabrics induced by laser engraving. Neural Process. Lett. **44**, 639–650 (2016)
16. Kan, C.W., Yuen, C.W.M., Cheng, C.W.: Technical study of the effect of CO_2 laser surface engraving on the colour properties of denim fabric. Color. Technol. **126**, 365–371 (2010)
17. Kan, C.W.: CO_2 laser treatment as a clean process for treating denim fabric. J. Clean. Prod. **66**, 624–631 (2014)
18. Hung, O.N., Chan, C.K., Kan, C.W., Yuen, C.W.M., Song, L.J.: Artificial neural network approach for predicting colour properties of laser-treated denim fabrics. Fibers Polymers **15**, 1330–1336 (2014)
19. Hung, O.N., Chan, C.K., Kan, C.W., Yuen, C.W.M., Song, L.J.: Using artificial neural network to predict colour properties of laser-treated 100% cotton fabric. Fibers Polymers **12**, 1069–1076 (2011)
20. Moller, M.F.: A scaled conjugate gradient algorithm for fast supervised learning. Neural Netw. **6**, 525–533 (1990)
21. Marquardt, D.W.: An algorithm for least-squares estimation of nonlinear parameters. J. Soc. Ind. Appl. Math. **11**, 431–441 (1963)
22. Hagan, M.T., Menhaj, M.: Training feedforward networks with the Marquardt algorithm. IEEE Trans. Neural Netw. **5**, 989–993 (1994)
23. Tiwari, S., Naresh, R., Jha, R.: Comparative study of backpropagation algorithms in neural network based identification of power system. Int. J. Comput. Sci. Inf. Technol. **5**, 93–107 (2013)
24. Bendu, H., Deepak, B., Murugan, S.: Application of GRNN for the prediction of performance and exhaust emissions in HCCI engine using ethanol. Energy Convers. Manag. **122**, 165–173 (2016)
25. Xu, S., Chen, L.: A novel approach for determining the optimal number of hidden layer neurons for FNN's and its application in data mining. In: Proceedings of 5th ICITA, pp. 683–686 (2008)

An Efficient Framework to Detect and Avoid Driver Sleepiness Based on YOLO with Haar Cascades and an Intelligent Agent

Belmekki Ghizlene[1], Mekkakia Zoulikha[1] ⓘ, and Hector Pomares[2](✉) ⓘ

[1] Université des Sciences et de la Technologie d'Oran Mohamed Boudiaf (USTO-MB), Oran, Algeria
[2] Department of Computer Architecture and Technology, CITIC-UGR Research Center, University of Granada, Granada, Spain
hector@ugr.es

Abstract. In this paper we present a new approach to discern and handle driver's drowsiness. This task is usually based only on its detection, without providing any intelligent feedback appropriated to the situation of the driver, and focusing only on the eyes. The response is usually a simple beep alarm which is not enough to wake up or keep the driver awake all along the road. The innovation in our method resides first in the use of a combination of Haar cascades and deep convolutional neural networks for fast detection of the state of the driver and second, the use of an intelligent assistant agent who will follow up the driver by the front camera of his phone, and tries to take care of his security, and the security of the others.

Keywords: Driver drowsiness detection · Deep neural networks · CNN · Assistant agent · YOLO real-time object detection · Haar cascade

1 Introduction

Nowadays a prominent cause of mortality on the road is due to "sleeping on the wheel". With the advent of new technologies and interfaces, there are a growing number of new solutions to this problem, such as wearable interfaces like vibrant stop sleep rings or ringing earpieces, and also graphic interfaces based usually on the face and eye tracking like Mercedes drive assist and INNOV-EYE technology for stop sleeping. In [1] a system is proposed that uses both template matching and features matching to attain drivers' eyes' localization. In [2] a very fast parallel system is used for eye tracking and in [3] the authors propose to make out the state of the eye by calculating the distance between the eyelids. The authors in [4] use eye tracking glasses and propose an effective method to measure fatigue from electro-oculogram signals. In [5] a Haar cascade method is used to detect the face and the eye and a circular Hough transform to track eye gaze, so that the algorithm can be used in real time using a Raspberry Pi. In another interesting approach, the authors in [6] propose to use the Gabor Walvet features extraction for yawning detection, and finally the approach [7] is based on deep learning and uses multi-task cascaded convolutional networks

© Springer Nature Switzerland AG 2019
I. Rojas et al. (Eds.): IWANN 2019, LNCS 11507, pp. 699–708, 2019.
https://doi.org/10.1007/978-3-030-20518-8_58

(MTCNN) to detect the face and as second step uses DDDN driver drowsiness detection network a fusion of 4 CNN for the face landmarks to detect the drowsiness.

However these techniques are not fast enough or do not cover the whole reach of the addressed problematic. In this paper, we propose an approach that tries to take advantage of the presented previous methods and provides a whole solution to the problem. In our case we need to do more than detection, so we add a virtual assistant agent who will be able to communicate with the driver in case of tiredness and save him/her in case of sleepiness, and make it easy to get in the car so it should be on the smartphone.

2   Methodology

The first phase of our framework (see Fig. 1) is image pre-processing, starting with the detection of the head, eyes and mouth of the driver. Current state-of-the-art multi-detection methods can be classified into these three main categories:

- Detection by sliding boxes: they start by decomposing the image in a lot of bounding boxes and try to detect what is in each bounding box and tell if there is a recognized object, then minimalize the number of the bounding boxes depending on the detection and repeat it once again until all the recognized objects are detected. Examples of this method are VGGNET [8], the inception model [9] and MTCNN [10].
- Detection by proposal regions detection: in this case, the method starts by previously segmenting the image into proposal regions and then create a bounding box for each region and detect the objects in the boxes. Examples of this method are RCNN [11], Mask-RCNN [12] and Fast RCNN [13].
- One shot multi detector: the advantage of these methods is a fast detection plus localization of the different objects by just looking once at the image. They first divide the image in a grid, then predict bounding boxes and extract multiple feature maps from them. Examples of this method are SSD (single shot detection) [14] and YOLO (you only look once) [15].

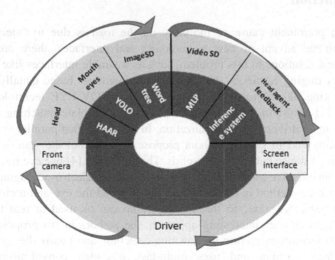

Fig. 1. Flowchart and structure of the framework for driver drowsiness detection and assistance

In our approach we need to make a very fast detection since the method should work in real time. Therefore, the latter methods seem to be the more promising ones. YOLO, which has been inspired from the GOOGLEnet model for image classification, has inception modules that are based on very small convolutions in order to reduce the number of parameters of the network (see Fig. 2). There are several faster versions of YOLO, such as YOLOv2 also known as YOLO9000 [16] where batch normalization has been added to regularize the model and the k-means algorithm is used to get the right anchor box's dimensions. Although there exists a new version of YOLO, called YOLOv3 [17], which improves the previous version for small object detection, in our case, we will make use of YOLO-LITE [18], which is a simplified version of YOLO9000 for computer systems without GPU.

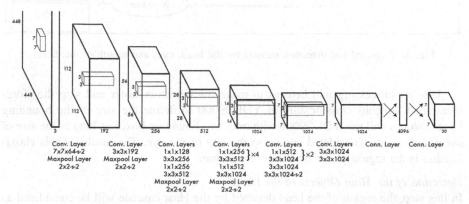

Fig. 2. Image taken from [15] that shows the internal architecture of the deep neural network used in the first version of YOLO

Since in our case the objects to be detected are very few (head, mouth and both eyes of the driver), and are located in very deterministic zones (the driver is not supposed to change very much from his/her normal position) some optimizations can be made to the system (see Fig. 3 for an overview of the proposed detection method):

Detection of the Head by Haar Cascade
In our approach, the initialization step of the YOLO9000 method is replaced by a Haar cascade for the detection of the head. As we have commented before, the reason behind this lies in the fact that we know what we are looking for in the image so we can limit the zone of search and additionally we know that the both eyes and mouth have approximately the same small size compared to the head zone, so the clusters of the bounding boxes will be small too.

Inspired by the concept of wavelets, the Haar cascade uses a convolutive mask consisting of two rectangles, one in black and the other in white, of small size that grows with each iteration and delimits two adjacent regions whose sums of their pixels is used to characterize them. This is a very quick method thanks to this binary nature (black or white pixels). In our case, we will use the Viola and Jones algorithm with the addition of rotated Haar-like features as proposed in [19].

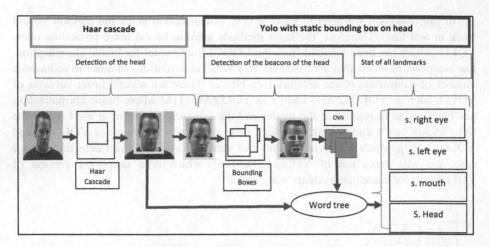

Fig. 3. Proposed fast detection method for the head, eyes and mouth of the driver.

This pre-processing will help us to make the detection faster and skip the afore-mentioned k-means classification of YOLO9000 to define the size of the bounding boxes. So, we will run YOLO9000 with only two bounding boxes as long as the size of the mouth can be considered similar to the one of the eye. The results of this classification is the region of the head of every frame.

Detection of the Main Objects of the Head
In this step, the region of the head detected by the Haar cascade will be considered as an input image for the deep neural network of YOLO9000. Every input image is divided into an $S \times S$ grid, where each grid cell is responsible of predicting B bounding boxes where an object can be (in our case, left eye, right eye or mouth) and we obtain a confidence score, which is calculated as:

$$C = \Pr(Object) \times IOU_{pred}^{truth}$$

where $\Pr(Object)$ is the probability of having an object, and IOU_{pred}^{truth} represents the intersection over union between the predicted box and the real box, i.e. the result of dividing the area of overlap between those boxes by the union of both boxes. Only the highest confidence boxes will be kept to run YOLO9000 on them. In Fig. 4 it is shown the essentials of this part of the detection method is pseudo-code. The width and height of the bounding boxes are predicted as offsets from cluster centroids and since the classes (left eye, right eye, mouth) have approximately the same size we will use only one bounding box height and width.

Begin
Divide the image in S×S grids
For (i=0, i<=S, i++) {
Predict B bounding boxes confidence for each grid
Calculate the probability of having an object in each bounding box
Delete all the bounding boxes with the lowest probability
Run k-means clustering on the training bounding boxes to find good priors
}
Run the CNN on clustering boxes to get the prediction of those boxes
Output: {Left eye box, Right eye box, Mouth box}
End

Fig. 4. Pseudo-code of the detection method for the eyes and mouth of the driver.

Detection of the Status of the Mouth and Eyes of the Driver
In this stage of the methodology, we need to get information about the mouth and eyes of the driver in order to ascertain whether the driver has been yawning or whether he is blinking too much and so on. In order to do so, the classification layers of the CNN are replaced by a semantic graph just like in YOLO9000. However, since the input image is the class of the head, we only need to provide an ontology of the objects that are inside the head, together with the synonyms. The ontology generated is shown in Fig. 5.

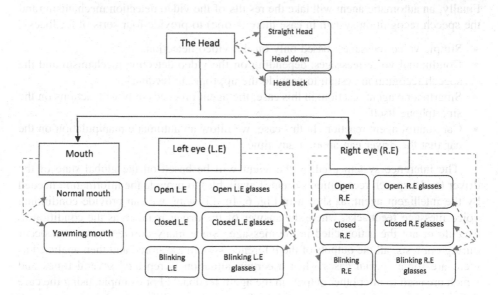

Fig. 5. Ontology of the head used to replace the last layers of classification of the CNN used in YOLO where the continuous arrow represent the relation of part of and the discontinuous ones represent the synonym-code of the detection method for the eyes and mouth of the driver.

Detection of the Driver's Status

Every class of the ontology (word tree) is assigned a risk level from where we can eventually obtain the driver's status as a measure of how tired this driver seems to be. More specifically, this risk level will range between 0, which is considered as the safe state, and -1, the most dangerous one (see Table 1 for the complete list of assigned risk levels). Therefore, the global driver's status of the instance image (θ) can be considered as a probability $Pg(\theta)$ that consists in the summation of the different risk levels of each detected region:

$$Pg(\theta) = \sum_{i=1}^{i=4} r(i) \rightarrow [-2.5, 0]$$

Table 1. Risk levels

Word	Classes and risk levels
Head	Straight 0, turned down -1, turned left -0.3, turned right 0.3.
Left eye	Open 0, closed -1, blinking -0.3, open with glasses 0, closed with glasses-1, blinking with glasses -0.3
Right eye	Open 0, closed -1, blinking -0.3, open with glasses 0, closed with glasses-1, blinking with glasses -0.3
Mouth	Open 0, closed 0, yawning -0.5

The Feedback Agent

Finally, an automatic agent will take the results of the video detection mechanism (and the speech recognition system in case there is one) to provide four sorts of feedbacks:

- Simple voice messages: based only on the video detection.
- Conditional voice messages: depending on the video detection mechanism and the speech recognition system to provide the appropriate feedback.
- Smartphone agent reaction: in this case, the agent proceeds only with actions on the smartphone itself.
- Car control agent reaction: In this case, we allow an automatic manipulation on the car that the driver can abort at any time.

The inference system used by the agent will be based on the global state of the driver (and a speech recognition system if available) to provide the information needed by the intelligent agent, as shown in Fig. 6. In this way, we can provide conditional voice message feedbacks, where the answer of the driver is taken as the condition to execute or not the action cited in the message. Since many different cases can occur during our tests, all the outputs of the inference system are stored and then analyzed as there are some special cases when a certain situation is repeated several times and which are used to infer other outputs in the agent feedback. For example, using the case definition given in Table 2, the repetition of case 5 during different video sequences are used to infer case 6, even if the case 6 is not detected directly.

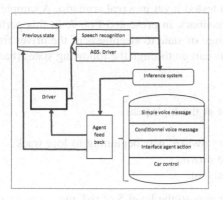

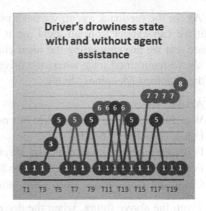

Fig. 6. Agent & driver interaction process.   **Fig. 7.** Ideal agent performance in case of starting drowsiness

3 Preliminary Results and Discussion

Although the main goal of this paper is to present the key ideas of the framework to detect and avoid driver sleepiness, in this section we are going to show some preliminary results that we have obtained. For that purpose, the implementation of the first prototype has been made using the Java NetBEans IDE adding TensorFlow with Bazel for YOLO and OPENCV for the Haar cascade. Protégé 4.2 is used to make the ontology to discern the driver's status. The dataset we will use in these preliminary experiments is the Head Pose Database [20] which is composed of head pose images of 15 people with and without transparent glasses. Each frame of images of video sequences contains 93 images of different poses of the same person.

Once all the specified landmarks of the head are detected and represented by words in the word tree (our ontology), the diagnostic of the state of the driver can be made by the relations between those words, the predicate set used to determine each relationship between the words. These relationships are expressed in the form of (if {action (1) &&/or (action 2) &&/or (action 3)....} then (action 9)}. In our case, it generated a set of classes that represent the state of the driver from the state of the head, the eyes and the mouth as shown in the following table:

Table 2. The set of predicates found and its driver diagnostic classes

Classes	Set of predicates
Conscious driver	If {(head straight) && (open E.L) && (open E.R) (normal mouth)}
Yawning driver	If {(head straight) && (blinking E.L) && (blinking E.R) (yawning mouth)}
Tired driver	If (head back) && (blinking E.L) && (blinking E.R) (normal mouth)}
Sleeping driver	If {(head straight) && (close E.L) && (closed E.R) && (normal mouth)}
Unconscious driver	If {(head down) && (undetected E.L) && (undetected R.E) && (undetected mouth)}

As for the feedback agent, we have not tested it yet in a real scenario. A sample of what we can expect from an ideal agent feedback in presented Fig. 7. In this performance test we are simulating the ideal change of state to sleepiness of the driver from the above-mentioned scale where the drivers can go through the following states (based on the Stanford Sleepiness Scale):

- Active, wide awake.
- Was functioning at a high level but not at the high peak.
- Felt a little foggy headed.
- Felt foggy headed, had difficulty staying awake, was beginning to lose track.
- Felt sleepy, woozy, and preferred to lie down.
- Could not stay awake, sleep onset imminent.

From the above figure, when the driver gets to the level 5 of risk the agent tries to influence the state of the driver. The first message is to tell the distance and time left to get to destination. The attention and motivation of the driver changes by this message in the second case 5. The suggestion to turn on the radio or play a song in the repetition of the state 5 is not successful and leads to the inference of the case 6. Now the agent tries to make the driver aware about his/her state and suggests him the nearest coffee shop or station. It tries to make the driver more awake by talking to him/her. This will decrease his/her sleepiness. In the next case 5 of our simulation the feedbacks change to talk the driver about the weather. The success of this feedback depends not only on the driver's state but also on the environment surrounding him/her (Table 3).

Table 3. Agent feedback depending on the driver's state

Case	Classes	Agent feedback		
Case 1	Correct posture	Standby		
Case 2	Side distraction	Voice message — Please slow down and try to focus on the road		
Case 3	Simple pivot	Standby		
Case 4	Undetected driver	Voice message — the driver has just forgotten his phone in the car		
Case 5	Tiredness detected	Conditional voice message		
		Tell to the driver the time and distance left to go to destination	Turn on the radio or suggest to play the most heard playlist music motivation	Talk about the weather
Case 6	Drowsiness detected	Make the driver aware of his alarming stat by advising for a pause	Suggest the nearest cafe shop or relay	
Case 7	Sleepiness detected	ring the alarm		
Case 8	Driving with sleepiness	Slow down the car	Turn on the hazard lights of the car	Call the local police and communicate to them their critical situation and geographical
Case 9	Asleep driver	Stop the car	Close the windows and put the security	Send a code to the police to unlock the car

4 Conclusions

In this paper we have presented a new framework to avoid driver's drowsiness, not only by alarming him in his sleepiness, but by trying to assist him by giving some solutions using an inference system that can handle the most frequent situations of the driver on the steering wheel. In summary, it is based on these three major steps: (a) Eyes and mouth identification and characterization: by starting with a Haar cascade for the Head detection to restrict the landmark detection only in this region; then a YOLO9000 algorithm is used for the eyes and mouth detection, by using two kind of bounding boxes since the size of the eyes is approximated to size the mouth. (b) Sleepiness state identification: By using a word tree to define the exact state of every detected landmark, then a deep neural network will do the video diagnostic and try to estimate the global state of the driver. (c) Intelligent assistance: Based on an inference system, an agent will provide feedback to the driver. As future research we plan to test other one single shot detection methods and make a comparison among them. We will also be adding some image preprocessing as well as edge detection to detect the eyes behind the sunglasses which can have an effect of tinted transparency. Finally, we hope to be able to treat more situations, like drivers in crisis or having a sudden body anomaly such as epileptic seizures.

Acknowledgements. This work was done as part of the Erasmus Plus program, from the USTO University of Sciences and Technology of Oran to the University of Granada Spain (December 2017/May 2018 in Granada).

Funding. This research was partially funded by the Spanish Ministry of Economy and Competitiveness (MINECO) Project TIN2015-71873-R together with the European Fund for Regional Development (FEDER).

Conflicts of Interest. The authors declare no conflict of interest. The funding sponsors had no role in the design of the study; in the collection, analyses, or interpretation of data; in the writing of the manuscript, and in the decision to publish the results.

References

1. Eriksson, M., Papanikolopoulos, N.: Eye-tracking for detection of driver fatigue. In: Proceedings of Conference on Intelligent Transportation Systems (1997)
2. Kao, W., Tsai, J., Chui, Y.C.: Parallel computing architecture for eye tracking systems. In: IEEE International Conference on Consumer Electronics (2017)
3. Fuhl, W., Santini, T., Kasnci, E.: Fast and robust eyelid outline and aperture detection in real-world scenarios. In: IEEE Winter Conference on Applications of Computer Vision (2017)
4. Gao, X., Zhang, Y., Zheng, W., Lu, B.: Evaluating driving fatigue detection algorithms using eye tracking glasses. In: 7th Annual International IEEE EMBS Conference on Neural Engineering (2015)

5. Stan, O., Miclea, L., Centea, A.: Eye-gaze tracking method driven by Raspberry Pi applicable in automotive traffic safety. In: Second International Conference on Artificial Intelligence, Modelling and Simulation (2014)
6. Xiao, F., Bao, C.-Y., Yan, F.-S.: Yawning detection based on Gabor wavelets and LDA. J. Beijing Univ. Technol. **35**, 409–413 (2009)
7. Gonçalves, M., et al.: Sleepiness at the wheel across Europe, a survey of 19 countries. J. Sleep Res. **24**(3), 242–253 (2015)
8. Simonyan, K., Zisserman, A.: Very deep convolutional networks for large scale image recognition. arXiv:1409.1556v6 (2015)
9. Szegedy, C., et al.: Going deeper with convolutions. In: IEEE Conference on Computer Vision and Pattern Recognition (2014)
10. Zhang, K., Zhang, Z., Li, Z., Qiao, Y.: Joint face detection and alignment using multi-task cascaded convolutional networks. IEEE Signal Process. Lett. **23**(10), 1499–1503 (2016)
11. Girshick, R., Donahue, J., Darrell, T., Malik, J.: Rich feature hierarchies for accurate object detection and semantic segmentation. In: Proceedings of the IEEE Computer Society Conference on Computer Vision and Pattern Recognition (2013)
12. He, K., Gkioxari, G., Dollar, P., Girshick, R.: Mask R-CNN. In: IEEE International Conference on Computer Vision (2017)
13. Girshick, R.: Fast R-CNN. In: IEEE International Conference on Computer Vision (2015)
14. Liu, W., et al.: SSD: single shot MultiBox detector. In: Leibe, B., Matas, J., Sebe, N., Welling, M. (eds.) ECCV 2016. LNCS, vol. 9905, pp. 21–37. Springer, Cham (2016). https://doi.org/10.1007/978-3-319-46448-0_2
15. Redmon, J., Divvala, S., Girshik, R., Farhadi, A.: You only look once: unified, real-time object detection. In: IEEE Conference on Computer Vision and Pattern Recognition Workshops (2017)
16. Redmon, J., Farhali, A.: YOLO9000 better faster stronger. In: IEEE Conference on Computer Vision and Pattern Recognition (2017)
17. Redmon, J., Farhadi, A.: YOLOv3: an incremental improvement. arXiv:1804.02767v1 (2018)
18. Pedoeem, J., Huang, R.: YOLO-LITE: a real-time object detection algorithm optimized for non-GPU computers. arXiv:1811.05588 (2018)
19. Lienhart, R., Maydt, J.: An extended set of Haar-like features for rapid object detection. In: Proceedings International Conference on Image Processing (2002)
20. Gourier, N., Hall, D., Crowley, J.L.: Estimating face orientation from robust detection of salient facial features. In: International Workshop on Visual Observation of Deictic Gestures (2004)

Fingerprint Retrieval Using a Specialized Ensemble of Attractor Networks

Mario González[1][(✉)], Carlos Dávila[1], David Dominguez[2], Ángel Sánchez[3], and Francisco B. Rodriguez[2]

[1] Intelligent and Interactive Systems Lab (SI2 Lab), Universidad de Las Américas, Quito, Ecuador
{mario.gonzalez.rodriguez,carlos.davila.armijos}@udla.edu.ec
[2] Grupo de Neurocomputación Biológica, Dpto. de Ingeniería Informática. Escuela Politécnica Superior, Universidad Autónoma de Madrid, 28049 Madrid, Spain
{david.dominguez,f.rodriguez}@uam.es
[3] ETSII, Universidad Rey Juan Carlos, 28933 Móstoles, Madrid, Spain
angel.sanchez@urjc.es

Abstract. We tested the performance of the Ensemble of Attractor Neural Networks (EANN) model for fingerprint learning and retrieval. The EANN model has proved to increase the random patterns storage capacity, when compared to a single attractor of equal connectivity. In this work, we tested the EANN with real patterns, i.e. fingerprints dataset. The EANN improved the retrieval performance for real patterns more than tripling the capacity of the single attractor with the same number of connections. The EANN modules can also be specialized for different patterns sets according to their characteristics, i.e. pattern/network sparseness (activity). Three EANN modules were assigned with skeletonized fingerprints (low activity), binarized (original) fingerprints (medium activity), and dilated/thickened fingerprint (high activity), and their retrieval was checked. The more sparse the code the larger the storage capacity of the module. The EANN demonstrated to improve the retrieval capacity of the single network, and it can be very helpful for module specialization for different types of real patterns.

Keywords: Pattern sparseness · Module specialization · Hopfield network · Synaptic dilution · Storage capacity

1 Introduction

Attractor networks have many desirable capabilities for pattern processing, such as performing pattern denoising and completion. Extremely diluted connectivity is preferred for low wiring cost, and low computational cost of the extensive recursive dynamics update [1]. However, the storage capacity of the diluted network is moderate, and for any application, storage of a large number of patterns is required [2–5]. In this sense, an Ensemble of Attractor Neural Networks (EANN) model have been proposed in [6,7]. The EANN model have proved

© Springer Nature Switzerland AG 2019
I. Rojas et al. (Eds.): IWANN 2019, LNCS 11507, pp. 709–719, 2019.
https://doi.org/10.1007/978-3-030-20518-8_59

to increase the network storage capacity using a divide-and-conquer strategy, where each diluted module of the ensemble stores and retrieves a subset of patterns, improving the capacity respect to a single densely connected module of equal connectivity [6,7]. These works have dealt with random patterns to quantify the model storage capacity limits. The ensemble modules dilution helps in making the transition from retrieval to non-retrieval be smooth. Also, dividing the pattern load in modules helps in the cross talk noise induced in each step of learning/retrieving, allowing a faster convergence of the ensemble. In the present paper we propose to deal with real patterns, namely, fingerprints dataset that have been explored before for the single attractor [4,8], with moderate retrieval of fingerprint patterns. In this work, we want to investigate if the ensemble improves the retrieval capacity (number of stored patterns) using the same connectivity of the single network. Each of the attractor modules of the proposed EANN model performs a holistic processing of the fingerprints, not needing to extract the fingerprint minutiae [9,10].

The EANN improved the retrieval capacity of the single attractor, with ten modules more than tripling the number of retrieved patterns. Also, a specialization of the EANN model is proposed for learning fingerprints datasets with different levels of pattern activities, namely, skeletonized, normal binarized, and thickened fingerprints. Each module specialized in a type of fingerprint uses a particular threshold, which is determined by the specific activity of involved patterns, to optimize the retrieval. We showed that the sparseness of the code influenced the retrieval capacity of the modules, with the more sparse (lower activity) module retrieving a larger number of patterns.

The rest of the paper is organized as follows. Section 2 describes the learning and retrieval dynamics of the individual modules, as well as EANN model and their performance measures. Section 3 presents the results in terms of retrieval performance comparing the storage capacity of the EANN with a single attractor. Also, a proof of concept of EANN modules specialization is presented for fingerprint sets with different levels of activity. Section 4 concludes the paper and discusses the main findings of the work.

2   EANN Model

In this section the base module of the EANN model, the neural coding, the network topology, learning and retrieval dynamics, are described. Finally, in this section, a schematic representation of the modularized EANN system is illustrated for the fingerprint retrieval and specialization.

2.1   Modules Coding, Topology and Dynamics

The state of a neuron in a network of N units is defined, at any discrete time t, by a set of N binary variables $\boldsymbol{\tau}^t = \{\tau_i^t \in 0, 1; i = 1, \ldots, N\}$, where 1 and 0 represent, respectively, active and inactive states. The target of network will be to recover a set of patterns, i.e. the fingerprints $\{\boldsymbol{\eta}^\mu, \mu = 1, \ldots, P\}$ that have been stored by a learning process. Each pattern corresponds to a stable fixed point attractor and the network retrieval state satisfies $\tau^t = \eta^\mu$, for large enough time t. The patterns are encoded as a set of binary variables $\eta^\mu = \{\eta_i^\mu \in 0, 1; i = 1, \ldots, N\}$, with mean activity $a^\mu = \langle \eta^\mu \rangle \equiv \sum_i \eta_i^\mu / N$ [8,11].

The synaptic couplings between the neurons i and j are given by the adjacency matrix

$$J_{ij} \equiv C_{ij} W_{ij}, \tag{1}$$

where the topology matrix $C = \{C_{ij}\}$ describes the connectivity structure of the neural network and $W = \{W_{ij}\}$ is the matrix with the learning weights. The topology matrix corresponds to a symmetric random network [12], with a network degree of K, that is each node is connected to K other nodes. The network is then characterized by the dilution parameter $\gamma = K/N$, which represents the connectivity ratio of the ANN.

For convenience in the sequel, we use the normalized variables, for the pattern encoding and network states, respectively, $\xi \equiv \frac{\eta - a}{\sqrt{A}}$, $\sigma \equiv \frac{\tau - q}{\sqrt{Q}}$, where $a \equiv \langle \eta \rangle$, $A \equiv Var(\eta) = a(1 - a)$, and $q \equiv \langle \tau \rangle$, $Q \equiv Var(\tau) = q(1 - q)$. Here a and q are the pattern and neuronal activities, respectively.

In terms of these normalized variables, the neuron dynamics [11,13] can be written as

$$\sigma_i^{t+1} = g(h_i^t - \theta(a), q_i^t), \tag{2}$$

$$h_i^t \equiv \frac{1}{K} \sum_j J_{ij} \sigma_j^t, \ i = 1, \ldots, N, \tag{3}$$

where h_i^t represents the neuron's input field, and the gain function is given by

$$g(x, y) \equiv [\Theta(x) - y]/\sqrt{y(1 - y)}. \tag{4}$$

Here the step function is used: $\Theta(x) = 1, x \geq 0$, $\Theta(x) = 0, x < 0$.

The weight matrix W is updated according to the Hebb's learning rule,

$$W_{ij}^\mu = W_{ij}^{\mu-1} + \xi_i^\mu \xi_j^\mu. \tag{5}$$

Weights start at $W_{ij}^0 = 0$ and after P learning steps, they reach the value $W_{ij} = \sum_\mu^P \xi_i^\mu \xi_j^\mu$. A field threshold is necessary to keep the neural activity close to that of the learned patterns,

$$\theta(a) = \frac{1 - 2a}{2\sqrt{A}} \tag{6}$$

is used, where a corresponds to the mean activity of the learning pattern set, and $A = a(1 - a)$ to the activity variance [8,11,14].

In order to characterize the retrieval ability of the network modules, the overlap is used,

$$m \equiv \frac{1}{N} \sum_i^N \xi_i \sigma_i, \tag{7}$$

which is the statistical correlation between the learned pattern ξ_i and the neural state σ_i. For $m = 1$ and $q \sim a$ one has a perfect retrieval of the pattern by the network, for $m = 0$ no retrieval is achieved, and for intermediate values the pattern is retrieved with noise. Hence, the value of m is a measure of the retrieval quality of the pattern performed by the network [8, 13]. Together with the overlap m, the load ratio $\alpha = P/K$ is useful to evaluate the network retrieval permanence.

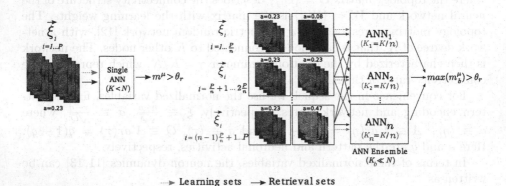

Fig. 1. Schematic representation of the model. Left: single attractor network with entire fingerprint dataset. Right: ensemble of attractor networks with pattern subsets. (Color figure online)

2.2 Ensemble of Attractor Neural Networks (EANN)

A schematic representation of the single ANN is presented in Fig. 1-left. The connectivity ratio γ is diluted with $K < N$. A set of P fingerprints ξ is presented to the network in a learning phase, represented with the red dashed arrow. Then, this set of fingerprints is presented in a retrieval phase in order to test the recall abilities of the network in terms of the retrieved patterns load α, and the quality of the retrieval m. This is represented with the solid black arrow.

In Fig. 1-right, a schematic representation of an ensemble of ANN modules with a number of n components is presented. The connectivity in each ANN_b module b is highly diluted with $K_b \ll N$, $b \in \{1, ..., n\}$. The set of patterns is divided into disjoint subsets of uniform size $P_b = P/n$, and each pattern subset is learned by its corresponding ANN_b module as represented with the red dashed arrows. E.g. $\{\xi^\mu, \ \mu = 1, ..., P/n\}$ for the first module $ANN_b, b = 1$, as shown in Fig. 1-right. The solid black arrows in Fig. 1-right, represent the retrieval stage, in which all the pattern subsets are presented to all ANN modules in order to test the discrimination among them. The target patterns are considered as retrieved by the ANN module with the higher overlap value over the retrieval threshold λ, i.e. $max(m_b^\mu) > \lambda$. For comparison purpose, the retrieval threshold is assumed to take the same value $\lambda = 0.5$ for each component in the ensemble, as well as, for the single ANN system.

Two use cases are depicted in Fig. 1 according to the experiments carried out in this work. In the first use case, the fingerprint dataset with $a \sim 0.23$ is used to measure the retrieval capacity of the single network with $n = 1, K = 300$ schematically represented in Fig. 1-left. This is compared with retrieval of the EANN with $n = 2, K_b = 150$ and $n = 10, K_b = 30$ modules, using the same fingerprint dataset as shown schematically in Fig. 1-right for the fingerprints with $a \sim 0.23$ in the left panels. The computational cost of the single ANN and the ANN ensemble are the same, as we showed in [6], one uses $K = K_b \times n$. The original dataset can be divided in subsets in order to increase the storage capacity when compared with single attractor.

In the other use case, the EANN can perform specialization, where the fingerprints can be divided in subsets with similar characteristics such as pattern activity. In this case an EANN with $n = 3, K_b = 100$ modules is used. Each module specializes its retrieval dynamics using a threshold adapted for the following pattern subsets activities: skeletonized fingerprints (activity $a \sim 0.0844$, and field threshold $\theta(a) = 1.4951$), normal binarized ($a \sim 0.2258$, $\theta(a) = 0.656$), and thickened fingerprints ($a \sim 0.4660$, $\theta(a) = 0.0681$), as depicted in Fig. 1-right for the right panels datasets, each with the corresponding activity marked in each panel. Note that the same dataset have been used differing the morphological operation over the binarized images to achieve the respective activity as described in Fig. 2.

2.3   EANN Information Measures

In order to evaluate the EANN performance, the retrieval efficiency R is defined as the number of learned patterns that are successfully retrieved $R = \frac{P_r}{P_l}$, where P_r is the overall number of retrieved patterns that satisfy $m^\mu > \lambda$, and P_l is the overall number of patterns presented to the network during the learning phase. One has that $P_l \geq P_r$. When the super-index b is used, P_r^b, P_l^b refer to the ANN_b module b in the ensemble. Here $\lambda = 0.5$ is used as the retrieval threshold. The mean retrieval overlap M is calculated as the mean retrieval overlap over all patterns subset $\mu \in 1, 2, \ldots, P_l$, $M = \langle m^\mu \rangle = 1/P_l \sum_{\mu=1}^{P_l} m^\mu$. It is worth noting that in the case of the ANN ensemble, the retrieval pattern load is calculated as $\alpha_R = \frac{P_r}{K_b \times n}$, where n is the number of subnetworks. Thus, we use $K_b \times n = K$ constant for all network ensembles studied, where K is the connectivity of the single "dense" network. Also, it is of worth to define the pattern gain G of the ANN ensemble by taking the single ANN system retrieval performance in terms of recovered patterns (P_r^s) as baseline, and it is given by $G = P_r^e/P_r^s$. Here P_r^e stands for the number of total recovered patterns by the ANN ensemble and P_r^s stands for the patterns recovered by the single network at the maximum retrieval pattern load $max(\alpha_R)$.

Fig. 2. Fingerprint dataset. From left to right: gray-scale fingerprint, skeletonized fingerprint (activity $a \sim 0.08$), binarized fingerprint ($a \sim 0.23$), dilated (thickened) fingerprint ($a \sim 0.47$).

3 Fingerprint Retrieval Results

3.1 EANN Retrieval Performance

Figure 2 depicts an example of processed fingerprint from the dataset collections from the International Fingerprint Verification Competition [15]. The fingerprint patterns were preprocessed in the following steps: image enhancement, binarization, morphological operations and cropping [4]. Figure 2 shows a grayscale fingerprint image obtained from the sensor (left) and the pre-processing outputs: skeletonized, binarized, and thickened fingerprints respectively. The image size is 263×340 pixels, giving a pattern/network size of $N = 89420$ sites. The total connectivity is $K = 300$ for the single and the EANN systems. The binarized dataset (second from the right) is used to test the EANN retrieval performance, in terms of the storage capacity, comparing the pattern gain over a single network. The skeletonized, binarized and thickened datasets are used as a proof of concept of the EANN modules specialization for retrieving patterns with different levels of activity.

Figure 3 shows the retrieval performance of the EANN model for $n = 1$, $n = 2$ and $n = 10$ modules. The mean retrieval overlap of the patterns M, the retrieval efficiency R and the pattern load α_R of the systems are plotted. The maximum value of α_R can be used to determine a critical value for the capacity of the systems, the point where the system storage is not increasing. The value $max(\alpha_R)$ occurs at the vertical red dashed line. For the single attractor $n = 1, K = 300$, in Fig. 3-left, $max(\alpha_R) = 0.0567$ occurs at $P_l = 26$, that is, the network have been presented with $P_l = 26$ patterns in the learning phase. At this point the mean retrieval overlap of the patterns is $M = 0.7350$ and the retrieval efficiency is $R = 0.6539$, that is, the networks retrieves $P_r = 17$ patterns with an overlap $m^{\mu} > 0.5$. This value $P_r = 17$ is comparatively lower, for the single system in relation with the following EANNs systems with $n = 2$ and $n = 10$ modules. Figure 3-middle depicts an EANN system with $n = 2, K_b = 150$ modules. The value of $max(\alpha_R) = 0.09$ corresponds to $P_l = 34$. The systems mean retrieval overlap is $M = 0.8172$ with a retrieval efficiency $R = 0.7941$, that is the systems retrieves $P_r = 27$ patterns with an overlap over the aforementioned

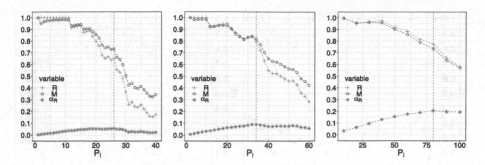

Fig. 3. EANN retrieval performance. Set of binarized fingerprints activity and field threshold of $a \sim 0.23, \theta = 0.656$. Single ANN with $n = 1, K_b = 300$ (left); EANNs with $n = 2, K_b = 150$ (middle); and $n = 10, K_b = 30$ (right) modules n and connectivity per module K_b respectively. (Color figure online)

threshold. The EANN model with $n = 2$ increases the retrieval in 10 patterns, with a gain $G = 27/17 = 1.5882$, that is around 59% more patterns are stored by the system with $n = 2$ modules. Finally, Fig. 3-right shows an EANN with $n = 10, K_b = 30$ modules, with $max(\alpha_R) = 0.09$ at $P_l = 80$, and $R = 0.775$ corresponding to $P_r = 62$ patterns retrieved, with a mean overlap $M = 0.7359$. In this case, the pattern gain $G = 62/17 = 3.6471$, that is, the EANN with $n = 10$ modules manages to more than triple the storage number of fingerprints compared with the single network. The EANN systems improves significantly the capacity retrieval when the number of modules increases. Surprisingly, this pattern gain is larger for real patterns than for random ones [6,7]. The EANN subset division of the dataset manages to keep the cross-talk noise lower for correlated patterns than the single network.

Figure 4, shows a fingerprint retrieval example for the single $n = 1, K = 300$ network (left) and the EANN system with $n = 10$ modules (right). For each pattern μ learned by the system is depicted their corresponding retrieval overlap m^μ. The examples correspond to observing the retrieval of the patterns at one point of Fig. 3. Figure 4-left shows the single network presented with the task to learn $P_l = 30$ patterns. This number of patterns is slightly over the saturation point of the single attractor, to demonstrated how stressed it is with a moderate pattern load. The single network retrieves $P_r = 11$ with $m^\mu > 0.5$. The mean retrieval overlap $M = 0.4953$ is depicted as the solid blue line. The single network retrieval abilities degrades considerably for a relative low number of patterns. The ensemble system increases the number of stored/retrieved patterns. Figure 4-right shows the retrieval for the $n = 10, K_b = 30$ modules EANN. $P_l = 80$ patterns are presented to the EANN systems, which succeeds in retrieving $P_r = 65$ patterns with a mean retrieval overlap of $M = 0.756$ (blue line). The EANN systems with $n = 10, K_b = 30$ increases significantly the number of retrieved patterns when compared with the single network $n = 1, K = 300$ with equal connectivity.

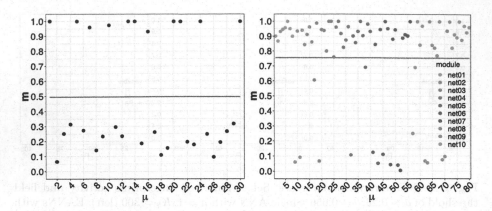

Fig. 4. Fingerprint dataset (activity $a \sim 0.23$) retrieval examples. Left: Single network $n = 1, K = 300$ presented with learned patterns $P_l = 30$ fingerprints. Right: EANN with $n = 10, K_b = 30$ with $P_l = 80$. (Color figure online)

3.2   EANN Modules Specialization

An interesting application of the ensemble, besides increasing the retrieval performance of the single network, is the EANN systems specialization. The EANN modules can be used for storing different subsets with different characteristics. As a proof of concept, the EANN system is tested for fingerprint patterns with different levels of activity (Fig. 2). Each of the EANN system modules with $n = 3, K_b = 100$ specializes as follows. Module 1: skeletonized fingerprints ($a \sim 0.0844,\ \theta(a) = 1.4951$), module 2: normal binarized ($a \sim 0.2258,\ \theta(a) = 0.656$), and module 3: thickened fingerprints ($a \sim 0.4660,\ \theta(a) = 0.0681$). Figure 5 top-left panel presents the retrieval results for $P_l = 30$ patterns, that is, 10 patterns are assigned to each module. The modules 1 and 2 retrieved all patterns in the skeletonized and binarized datasets with a high overlap m. The module 3 only retrieves 4 thickened patterns above $m^\mu > 0.5$. Increasing the number of patterns that are presented to the EANN system to $P_l = 45$ patterns, the module 1 retrieves all 15 skeletonized patterns, module 2 drops one of the binarized patterns, and the module 3 is unable to retrieve a single thickened fingerprint above $m^\mu > 0.5$. Increasing further to $P_l = 60$, the module 2 degrades its retrieval performance dropping 5 of their 20 patterns. Finally, for $P_l = 75$, 25 patterns per module, module 1 still recovers all skeletonized fingerprints, while module 2 only retrieves 6 patterns. The performance of each specialized module can also be observed by their corresponding mean retrieval overlaps depicted with dashed lines. The more sparse is the code better is the retrieval performance of the module. This is expected as the sparse coding is favorable to increase the network capacity, because the cross-talk term between stored patterns decreases [16]. The capacity of the specialized EANN system could be improved if the connections, instead of uniform distributed as presented in this work, i.e., the modules could be rearranged to the modules with lower retrieval, according to the particularities of the datasets, in this case, form the skeletonized module to the larger activities modules.

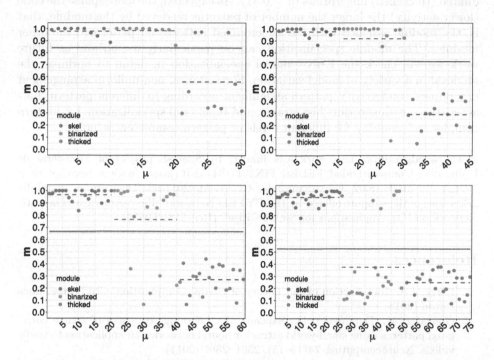

Fig. 5. Specialization of EANN modules. Module 1: skeletonized fingerprints (activity and field threshold $a \sim 0.0844$, $\theta = 1.4951$), module 2: normal binarized ($a \sim 0.2258$, $\theta = 0.656$), and module 3: thickened fingerprints ($a \sim 0.4660$, $\theta = 0.0681$). Retrieval results for learned patterns $P_l = \{30, 45, 60, 75\}$ are depicted from left to right and top to bottom respectively. The continuous blue line means retrieval overlap of the whole system (M). The dashed lines represent mean retrieval overlap of each module. (Color figure online)

4   Conclusions

We evaluated the Ensemble of Attractor Neural Networks (EANN) model performance for learning and retrieval of fingerprint patterns. With the same connectivity cost the EANN system with $n = 10, K_b = 30$ modules more than triple the storage capacity of the single network $n = 1, K = 300$, that is a pattern gain of $G = 3.65$. This gain is higher that the one found for random patterns [6,7], indicating that the modularization helps to reduce the cross-talk noise for highly correlated patterns. This is an interesting result, given that real world applications, implies dealing with structured patterns, where the EANN system will be more robust retrieving a larger number of such correlated patterns.

Also, the EANN modules were specialized for patterns with different levels of activity: skeletonized fingerprints ($a \sim 0.08$), binarized fingerprints ($a \sim 0.23$), dilated (thickened) fingerprints ($a \sim 0.47$). As expected, the more sparse the code (lower activity) the larger the number of patterns retrieved by the module, that is, the module assigned with the skeletonized patterns outperformed the other modules. The module specialization can be particularly useful, and as future work we will check the EANN model specialization in detail to optimize the retrieval in a variety of model variation. For instance, non-uniform arrangement of modules' connectivity, pattern assignment according to different pattern characteristics, i.e. fingerprints histograms, and functional specialization, i.e. pattern completion/denoising for different levels of pattern completeness/noise.

Acknowledgments. This work was funded by Spanish project of Ministerio de Economía y Competitividad/FEDER TIN2017-84452-R (http://www.mineco.gob.es/), UDLA SIS MGR.18.02, UAM-Santander CEAL-AL/2017-08. The authors gratefully acknowledge the support offered by the CYTED Network: "Ibero-American Thematic Network on ICT Applications for Smart Cities" (Ref: 518RT0559).

References

1. Hertz, J.A.: Introduction to the Theory of Neural Computation. CRC Press, Boca Raton (2018)
2. González, M., Dominguez, D., Sánchez, Á.: Learning sequences of sparse correlated patterns using small-world attractor neural networks: an application to traffic videos. Neurocomputing **74**(14–15), 2361–2367 (2011)
3. Barra, A., Bernacchia, A., Santucci, E., Contucci, P.: On the equivalence of hopfield networks and Boltzmann machines. Neural Networks **34**, 1–9 (2012)
4. González, M., Dominguez, D., Rodríguez, F.B., Sanchez, A.: Retrieval of noisy fingerprint patterns using metric attractor networks. Int. J. Neural Syst. **24**(07), 1450025 (2014)
5. Fachechi, A., Agliari, E., Barra, A.: Dreaming neural networks: forgetting spurious memories and reinforcing pure ones. Neural Networks **112**, 24–40 (2019)
6. Gonzalez, M., Dominguez, D., Sanchez, A., Rodriguez, F.B.: Increase attractor capacity using an ensembled neural network. Expert Syst. Appl. **71**, 206–215 (2017)
7. González, M., Dominguez, D., Sánchez, Á., Rodríguez, F.B.: Capacity and retrieval of a modular set of diluted attractor networks with respect to the global number of neurons. In: Rojas, I., Joya, G., Catala, A. (eds.) IWANN 2017. LNCS, vol. 10305, pp. 497–506. Springer, Cham (2017). https://doi.org/10.1007/978-3-319-59153-7_43
8. Doria, F., Erichsen Jr., R., González, M., Rodríguez, F.B., Sánchez, Á., Dominguez, D.: Structured patterns retrieval using a metric attractor network: application to fingerprint recognition. Physica A **457**, 424–436 (2016)
9. Jiang, L., Zhao, T., Bai, C., Yong, A., Wu, M.: A direct fingerprint minutiae extraction approach based on convolutional neural networks. In: 2016 International Joint Conference on Neural Networks (IJCNN), pp. 571–578. IEEE (2016)
10. Tang, Y., Gao, F., Feng, J., Liu, Y.: Fingernet: an unified deep network for fingerprint minutiae extraction. In: 2017 IEEE International Joint Conference on Biometrics (IJCB), pp. 108–116. IEEE (2017)

11. Dominguez, D., González, M., Rodríguez, F.B., Serrano, E., Erichsen Jr., R., Theumann, W.: Structured information in sparse-code metric neural networks. Physica A **391**(3), 799–808 (2012)
12. Albert, R., Barabási, A.L.: Statistical mechanics of complex networks. Rev. Mod. Phys. **74**(1), 47 (2002)
13. Dominguez, D., González, M., Serrano, E., Rodríguez, F.B.: Structured information in small-world neural networks. Phys. Rev. E **79**(2), 021909 (2009)
14. González, M., del Mar Alonso-Almeida, M., Avila, C., Dominguez, D.: Modeling sustainability report scoring sequences using an attractor network. Neurocomputing **168**, 1181–1187 (2015)
15. Dorizzi, B., et al.: Fingerprint and on-line signature verification competitions at ICB 2009. In: Tistarelli, M., Nixon, M.S. (eds.) ICB 2009. LNCS, vol. 5558, pp. 725–732. Springer, Heidelberg (2009). https://doi.org/10.1007/978-3-642-01793-3_74
16. Dominguez, D., Bollé, D.: Self-control in sparsely coded networks. Phys. Rev. Lett. **80**(13), 2961 (1998)

Evolutionary and Genetic Algorithms

A Fixed-Size Pruning Approach
for Optimum-Path Forest

Leonardo da Silva Costa[1], Gabriel Santos Barbosa[2],
and Ajalmar Rêgo da Rocha Neto[1(✉)]

[1] Federal Institute of Ceará, IFCE, Fortaleza, Ceará, Brazil
leonardoscifce@gmail.com, ajalmar@gmail.com
[2] Federal Institute of Ceará, IFCE, Maracanaú, Ceará, Brazil
gabrielsantos.ifce@gmail.com

Abstract. Optimum-Path Forest (OPF) is a graph-based classifier that
has achieved remarkable results in various applications. OPF has many
advantages when compared to other supervised classifiers, since it is free
of parameters, achieves zero classification errors on the training set without overfitting, handles multiple classes without modifications or extensions, and does not make assumptions about the shape and separability
of the classes. Despite these advantages, it still suffers with a high computational cost required to execute its classification process, which grows
proportionally to the size of the training set. In order to overcome this
drawback, we propose a new approach based on genetic algorithms to
prune irrelevant training samples and still preserve accuracy in OPF
classification. In our proposal, named FSGAP-OPF, the standard reproduction and mutation operators are modified so as to maintain the number of pruned patterns with a fixed-size. To evaluate the performance of
our method, we tested its generalization capabilities on datasets obtained
from the UCI repository. On the basis of our experiments, we can say
that FSGAP-OPF is a good alternative for classification tasks and can
also be used in problems where the memory consuming is crucial.

Keywords: Optimum-path forest · Genetic algorithms ·
Pattern recognition · Pruning

1 Introduction

Optimum-Path Forest (OPF) [13] is a graph-based classifier that has achieved
remarkable results in various applications [1,3,6], including electrocardiogram
heartbeat signal classification [8], intrusion detection in computer networks [16]
and epilepsy diagnosis [11].

OPF has many advantages when compared to other supervised classifiers,
since it is free of parameters, achieves zero classification errors on the training
set without overfitting, handles multiple classes without modifications or extensions, and does not make assumptions about the shape and separability of the

I. Rojas et al. (Eds.): IWANN 2019, LNCS 11507, pp. 723–734, 2019.
https://doi.org/10.1007/978-3-030-20518-8_60

classes [14]. Despite these advantages, it still suffers with a high computational cost required to execute its classification process, which grows proportionally to the size of the training set. Therefore, a pattern recognition task would benefit from the adoption of pruning techniques in the training samples of OPF, aiming to reduce the computational time required to execute its classification phase and also to build models that can be applied in problems where the memory consuming is crucial.

In order to overcome these issues, we propose a method based on genetic algorithms (GAs) to prune irrelevant or similar samples in the training graph of OPF, while keeping the classifier's accuracy in a similar level. In the proposed approach, the percentage of pruned samples is determined before the pruning process. Moreover, by using GAs, we can solve the graph pruning problem, which is combinatorial with non-polinomial complexity, in an efficient way.

To evaluate the performance achieved by the proposed strategy, the method will be aplied in real world datasets obtained from UCI Machine Learning Repository[1]. Besides that, a performance analysis against two methods used for pruning OPF [12,15] will be conducted.

The remaining of this paper is organized as follows: Sect. 2 presents the fundamentals of the OPF classifier. In Sect. 3 we present the pruning algorithms (OPF-TPPA and ROB-OPF) we use for comparison. In Sect. 4, we introduce genetic algorithms which are necessary to understand our proposal presented in Sect. 5 and then, in Sect. 6, we present and discuss our experimental results. Finally, the paper is concluded in Sect. 7.

2   Optimum-Path Forest Classifier

In this section, we explain the fundamentals of the optimum-path forest classifier, by presenting its process of training and classification.

2.1   Training

Consider Z a λ-labeled dataset and let Z_1 and Z_3 be training and test sets, respectively, with $|Z_1|$ and $|Z_3|$ samples, such that $Z = Z_1 \cup Z_3$. Additionally, consider $s \in Z$ a n-dimensional sample and $d(s, v)$ a function that calculates the distance between two samples s and v, in which $v \in Z_1$. Let $G^{tr} = (Z_1, A)$ be a complete graph (Fig. 1a) derived from the training set so that the arcs are weighted by the distance function d. A path π_s specifies a sequence of adjacent and distinct nodes in G^{tr} ending at the node s. A cost $f(\pi_s)$ is assigned to each path π_s given by a connectivity function f.

The training phase basically consists of finding a set $S \subseteq Z_1$ of prototypes, that is, the most representative elements of each class, and generating an optimum-path forest. One way of estimating the prototypes is to choose them from regions with overlapping classes, since they are regions very susceptible to

---

[1] https://archive.ics.uci.edu/ml/.

classification errors. In fact, the prototypes are the nodes of different classes that have the lowest distance d between any another (Fig. 1c), which can be found by computing a minimum-spanning tree (MST) [2] in G^{tr} (Fig. 1b). Note that each class can be represented by multiple prototypes and there must be at least one prototype per class. Optimum paths are computed from the prototypes to each training sample, such that each prototype becomes root of an optimum-path tree composed by its most strongly connected samples (Fig. 1d).

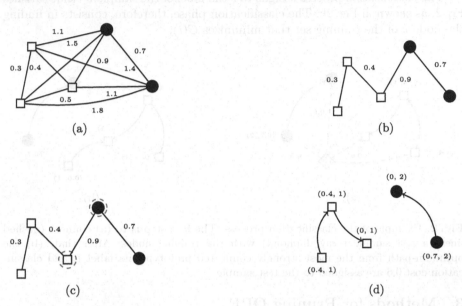

Fig. 1. (a) Complete weighted graph derived from a certain training set. (b) MST of the complete graph. (c) Prototypes selected as being the adjacent elements of different classes in the MST (circled nodes). (d) Optimum-path forest resulting from applying the path cost function $fmax$ with the two prototypes. The arrow indicates the predecessor node in the optimum-path. The entries (x, y) next to the nodes stand, respectively, for the cost and the label of the samples.

The OPF algorithm, which is used to compute the optimum-path forest (OPF model), may be used with any smooth path cost function that can group samples with similar properties [5]. This work used the path cost function $fmax$ [10], which is calculated as follows:

$$fmax(\langle s \rangle) = \begin{cases} 0 & \text{if } s \in S, \\ +\infty & \text{otherwise} \end{cases} \quad (1)$$

$$fmax(\pi.\langle s, t \rangle) = \max\{fmax(\pi), d(s, t)\}$$

The function $fmax(\langle \pi \rangle)$ computes the maximum distance between adjacent samples in π, when π is not composed by a single sample (trivial path). The concatenation of a path π and an arc (s, t) is represented by $\pi.\langle s, t \rangle$.

The implementation used here for the OPF algorithm is the one presented in [13].

2.2 Classification

During the classification phase, each sample $t \in Z_3$ is connected to all samples of Z_1, as though t were part of the training graph (Fig. 2a). The node t will receive an optimum cost $C(t)$ computed as

$$C(t) = \min\{\max\{C(v), d(v, t)\}\}, \forall v \in Z_1. \tag{2}$$

The classification process assigns to t the label of the sample v^* that satisfies Eq. 2, as shown in Fig. 2b. The classification phase, therefore, consists in finding the node v of the training set that minimizes $C(t)$.

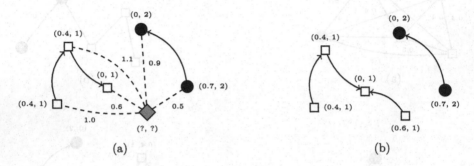

(a)   (b)

Fig. 2. Example of the classification process. The first step is to (a) connect (dashed lines) a test sample (gray diamond) with the training nodes. After that, (b) the optimum-path from the most strongly connected prototype, its label 1, and classification cost 0.6 are assigned to the test sample.

3   Methods for Pruning OPF

In this section, we present OPF-TPPA and ROB-OPF methods used to obtain pruned OPF classifiers. Both of them reduce the number of training samples by removing the ones that did not participate on the process of classification of new nodes. These methods are compared with our proposal in Sect. 6.

3.1   OPF Training Patterns Pruning Algorithm

The OPF Training Patterns Pruning Algorithm (OPF-TPPA [12]) uses a third evaluation set Z_2 to verify which samples from the training set Z_1 are less relevant in order to reduce its size. To do so, the algorithm projects an instance I of the OPF classifier from Z_1 and evaluates it on Z_2.

As previously mentioned in Sect. 2, the node $v^* \in Z_1$ can be identified as the one that satisfies Eq. 2. In this regard, each node $v^* \in Z_1$ that participated in the classification process of a node $t \in Z_2$ is marked as relevant and so is its predecessors (the parents in the path until reaching the root). After finishing the evaluation of Z_2, the unmarked (irrelevant) samples are removed from Z_1 and the classifier is re-trained using the new training set.

After that, the samples of the test set Z_3 can be classified using the pruned model.

3.2 Robust Pruning of Training Patterns for OPF

Robust Pruning of Training Patterns for OPF (ROB-OPF [15]) is an improvement of the OPF-TPPA method that is able to obtain higher pruning rates in the training samples of the OPF classifier. In order to do this, ROB-OPF makes use of three algorithms: (i) Algorithm 1 (*OPF Algorithm*), which assigns one optimum path $P^*(t)$ from S to each training sample t in a nondecreasing order of minimum cost, such that the graph is partitioned into an optimum-path forest, (ii) Algorithm 2 (*Learning Algorithm*), which projects a classifier using the training set Z_1 and evaluates its performance on Z_2 during T iterations, in order to select the instance of classifier (Z_1) with highest accuracy on Z_2. It essentially assumes that the most informative samples in Z_2 are the misclassified ones. It replaces these samples by nonprototype samples in Z_1, chosen randomly. Lastly, (iii) Algorithm 3 (*Learning-with-Pruning*), which combines Algorithm 2 with the identification and elimination of irrelevant samples from Z_1 to reduce its size.

The idea, basically, is: divide the dataset into Z_1, Z_2 and Z_3. Use Z_1 and Z_2 as input for Algorithm 2, which is trained on Z_1 and classified on Z_2 in order to create a representative set Z_1 with relevant samples (misclassified on Z_2). Subsequently, use these new sets Z_1 and Z_2 in the pruning process. Train the classifier in Z_1 and classify the samples of Z_2. When classifying each sample of Z_2, mark all the nodes of the path of the sample in Z_1 that was used to classify the other sample in Z_2. After classifying all samples in Z_2, remove the unmarked samples from Z_1, and start the process again, that is, execute Algorithm 2 another time. This process (learning and pruning) is repeated until no irrelevant sample exists in Z_1.

4   Genetic Algorithms

According to Srinivas and Patnaik [17], GAs are a branch of evolutionary algorithms created by Holland [7] and, as such, can be defined as a search technique inspired by the theory of evolution through natural selection, proposed by Darwin. This meta-heuristic can also be used to solve optimization problems with large search spaces containing many solutions. Its characteristics enable problems to be solved without the assumption of linearity, continuity, differentiability or convexity of the objective function, therefore having certain advantages over many classical optimization and search methods.

In GAs, a set of individuals (population) is evolved in a similar way as the evolution of the species, by natural selection. The aptitude of an individual (solution) is measured by a function, named fitness function, that quantifies how good each solution is to the problem at hand. Each individual is composed of one or more chromosomes, typically coded as a sequence of bits, i.e., strings of 0s and 1s. The individuals can be modified by mutation or combined in new evolutionary cycles (generations) with other individuals in order to generate new solutions by reproduction processes which use crossover. One can calculate the fitness value after decoding the chromosome and the best solution of the problem to be solved has the best fitness value. To prevent the current generation from being worse than the previous one, a strategy known as elitism can be adopted, in which the most apt individuals will remain for subsequent generations. This process is repeated until a determined stop criterion is satisfied, for example after a certain number of generations has been reached or when the fitness of the individuals converge to a particular value. When the stop criterion is satisfied, the genetic algorithm must return the best found solution [9].

5   Proposal: FSGAP-OPF

The method proposed in this work, called Fixed-Size Genetic Algorithm for Pruned OPF (FSGAP-OPF), emerges from the application of GAs to the OPF classifier with the objective of selecting and pruning less representative samples that would be used in the classification phase, therefore generating a compact training set, without affecting the classifier's accuracy. To do so, we model an individual that contains only one chromosome, which is represented by a vector of genes that can assume only the values "one" (true) or "zero" (false), to indicate, respectively, the presence or absence of a sample in the training set Z_1. Hence, each individual has as many samples in Z_1 as the number of genes set up to "one". After that, genetic operators will be chosen to assist the process of reproduction, mutation and elitism.

To achieve a fixed-size training set, we project the reproduction operator in such a way that the number of bits set up to "one" is kept unchanged. In this way, it is possible to determine the pruning rate we want to achieve before the pruning process occurs. It will also be necessary to elaborate a fitness function, which will take into consideration the classification error rate of a certain individual. This function will have its return value minimized by the GA. When the GA reaches its stop criterion, the best found individual will be selected.

5.1   Individuals or Chromosomes

Considering the proposed method, our individual, which is composed of only one chromosome, can be described as $\mathbf{c} = [g_1 \ g_2 \ldots g_i \ldots g_{N-1} \ g_N]$, where $g_i \in \{0, 1\}$ refers to the i-th training sample in the vector $\mathbf{c}$ and N is equal to $|Z_1|$. In this codification, the genes of $\mathbf{c}$ with value equal to 1 represent the samples that compose the newest training set, whereas the genes equal to 0 stand for the pruned samples. For example if the chromosome is set-up as $\mathbf{c} = [1 \ 1 \ 0 \ 1 \ 0 \ 1]$, we have pruned two samples.

5.2   Reproduction

In order to keep the pruning rate constant, that is, with a fixed number of samples in the training set, we codified the restrictions of our problem so as to keep the number of bits set up to "one" unchanged in each individual in the reproduction process. We would not reach this desired property if we allowed a direct

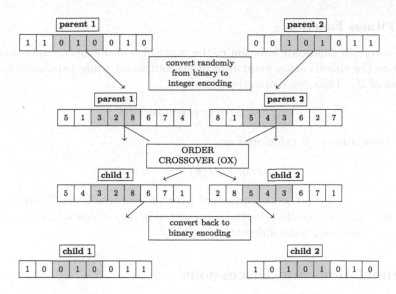

Fig. 3. Transformation process between binary-integer encoding for application of OX crossover operator followed by the transformation between integer-binary encoding.

swap between individuals, since invalid individuals would be produced. Therefore, we propose a codification based on a transformation of binary-encoding individuals into integer-encoding ones to which we are able to apply permutation operators, such as partially mapped crossover (PMX), cycle crossover (CX) and order crossover (OX) operators [4].

Figure 3 depicts the way the transformation between binary and integer encodings happens and then how the OX operator[2] works over the integer-encoding parents. After such process, each offspring is turned back to the binary-encoding so that each child will be suitable to belong to the next population.

5.3   Mutation

In order to escape from local minima and produce stronger chromosomes, we used the bit flip operator to perform the mutation mechanism. This operator works by selecting one or more random bits and flipping them. More specifically, we take the genes with values 0's and 1's from an individual and create two new subsets containing the same values of bits from the individual (one subset with only "ones" and the other with only "zeros"), mapping their original positions (index) on the original individual. Then, we choose random bits from each new subset and flip their values in the corresponding positions in the original individual. This procedure guarantees that the number of 0's and 1's is kept unchanged after the mutation mechanism is completed.

---

[2] The OX operator was the one used in this work.

5.4   Fitness Function

Given a certain candidate solution $\mathbf{c}$, the main idea of our fitness function is to minimize the classification error rate of an individual while pruning irrelevant samples of Z_1. Thus, we define the fitness as

$$fitness(\mathbf{c}) = error\_rate(\mathbf{c}) \qquad (3)$$

where $error\_rate(\mathbf{c})$ is calculated as

$$error\_rate(\mathbf{c}) = 1 - acc \qquad (4)$$

in which acc is the accuracy computed over an evaluation set Z_2 and can be measured using the method (formula) presented in [13], which takes into account that the classes may have different sizes in Z_2[3].

6   Simulations and Discussions

We evaluate here the FSGAP-OPF capacity of generating compact training sets while keeping the classifier's accuracy at a similar level. We compare its performance with those obtained by OPF-TPPA, ROB-OPF and OPF before pruning, by applying them on the datasets presented in Table 1.

Table 1. List of datasets evaluated in this work.

Data set	Abbreviation	# Patterns	# Attributes	# Classes
Balance Scale	BS	625	4	3
Australian Credit Approval	ACA	690	14	2
Haberman's Survival	HS	306	3	2
Leaf	LE	340	15	15
Pima Indians Diabetes	PID	768	8	2
Vertebral Column	VC	310	6	3

Initially, we selected 60% of the dataset samples for composing Z_1, 20% for Z_2 and the remaining 20% for Z_3. We then trained the OPF classifier using Z_1 and passed its trained model as input for FSGAP-OPF, OPF-TPPA, and ROB-OPF so that we could work with the same model in all algorithms. The number of iterations (T) defined in ROB-OPF is 50. The GA used in FSGAP-OPF randomly creates a population of feasible solutions. Each solution is a string of binary values in which each bit represents the presence (1) or absence (0) of a sample in Z_1. For subsequent generations, we take into account the fact that

---

[3] Similar definition will be applied for measuring the accuracy in Z_3 in all methods presented in this work.

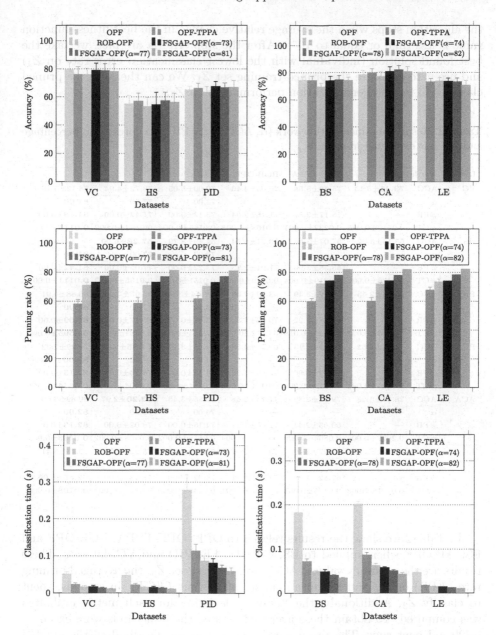

Fig. 4. Accuracy, pruning rate and classification time obtained by the methods employed in this work applied in the datasets presented in Table 1.

80% of new individuals are generated from the reproduction process shown in Fig. 3, 10% of the best individuals are selected due to an elitist selection scheme, and then the remaining individuals are obtained by mutation. Our population has 100 individuals and the maximum number of generations is 150. However, the algorithm stops when the average relative change in the best fitness function value is less than or equal to 10^{-5}. After the stop criteria are met, we take the best found solution (individual with the lowest classification error rate on Z_2) and use it to compute the pruned training set Z_1. We can then use the pruned classifier to test its generalization performances on Z_3.

Table 2. Results obtained in this work (α, ACC and #PR are expressed in percentage and #CT is expressed in seconds.)

Dataset	Metric	OPF	OPF-TPPA	ROB-OPF	FSGAP-OPF		
VC	ACC	**79.45±5.13**	75.98±5.58	76.05±4.43	79.00±5.05	78.72±4.94	78.12±5.46
	α	—	—	—	73.00	77.00	**81.00**
	#PR	—	58.17±2.87	71.16±2.04	73.12±0.00	77.42±0.00	**81.18±0.00**
	#CT	0.0510±0.006	0.0232±0.004	0.0168±0.003	0.0165±0.004	0.0134±0.002	**0.0113±0.001**
HS	ACC	55.15±5.92	57.07±5.66	53.21±6.34	54.49±8.79	**57.28±6.00**	56.29±6.31
	α	—	—	—	73.00	77.00	**81.00**
	#PR	—	58.58±3.96	71.01±2.13	73.22±0.00	77.05±0.00	**81.42±0.00**
	#CT	0.0476±0.006	0.0219±0.004	0.0159±0.002	0.0151±0.003	0.0135±0.002	**0.0111±0.001**
PID	ACC	65.13±2.43	66.19±3.49	63.39±4.06	**67.44±3.48**	66.65±3.29	66.88±4.91
	α	—	—	—	73.00	77.00	**81.00**
	#PR	—	61.78±2.22	70.63±1.31	73.04±0.00	77.17±0.00	**81.09±0.00**
	#CT	0.2788±0.041	0.1140±0.018	0.0872±0.013	0.0813±0.012	0.0681±0.008	**0.0595±0.009**
BS	ACC	74.41±2.62	74.29±3.15	69.28±4.33	74.08±3.19	**74.85±3.08**	74.29±2.32
	α	—	—	—	74.00	78.00	**82.00**
	#PR	—	59.89±1.96	72.75±1.33	74.13±0.00	78.13±0.00	**82.13±0.00**
	#CT	0.1825±0.081	0.0717±0.006	0.0491±0.002	0.0484±0.005	0.0404±0.001	**0.0344±0.001**
ACA	ACC	78.51±3.23	79.97±2.89	77.19±3.68	80.85±3.45	**82.20±2.97**	80.89±3.60
	α	—	—	—	74.00	78.00	**82.00**
	#PR	—	60.05±2.44	72.05±1.11	74.15±0.00	78.02±0.00	**82.13±0.00**
	#CT	0.2017±0.006	0.0864±0.006	0.0628±0.004	0.0578±0.002	0.0508±0.002	**0.0433±0.003**
LE	ACC	**80.35±2.70**	73.22±2.49	73.89±2.51	73.65±2.27	73.22±2.71	70.75±3.41
	α	—	—	—	74.00	78.00	**82.00**
	#PR	—	67.82±1.99	73.48±1.91	74.02±0.00	78.43±0.00	**82.35±0.00**
	#CT	0.0471±0.003	0.0175±0.001	0.0152±0.001	0.0145±0.000	0.0125±0.000	**0.0108±0.000**

In Table 2 we show the results related to OPF, OPF-TPPA, ROB-OPF and FSGAP-OPF when applied to BS, CA, HS, LE, PID and VC datasets. These results are the accuracy values (ACC) over the test set Z_3, the averaged pruning rate (#PR) on Z_1 and the mean classification time (#CT), in seconds, spent to classify Z_3. Additionaly, the standard deviation for each metric evaluated was computed. To obtain these averaged results, these methods were executed in 20 indepent runs. The pruning rate was calculated by dividing the number of pruned nodes by the total number of training samples. Besides that, the α symbol in FSGAP-OPF algorithm denotes the desired pruning rate one wants to achieve before the pruning process. In Fig. 4, we show a graphic representation of the results obtained in this study.

As can be seen in more details in Table 2, our proposed method, FSGAP-OPF, was able to obtain good results in terms of #PR and, consequently, #CT values, when compared to the other pruning methods. The reduction of classification time and the size of Z_1 was very expressive, reaching the best results against all algorithms. Our method was also able to obtain best accuracy values in most datasets, even though the OPF without pruning obtained higher accuracies in some cases. This shows that our method can be safely applied in classification problems in an efficient way. Another interesting point, as expected, is that the α value is equivalent to the pruning rate we obtained in the results, as shown in Table 2. One can also notice that the standard deviation for the pruning rate (#PR) was zero for each α value, since α was constant for the 20 independent runs, thus generating training sets with the same size. Besides that, it is worth noting that even when the value of α increased the accuracy values obtained by our method were kept at a similar level, or even improved.

7 Conclusions

In this work, we propose a fixed-size pruning algorithm based on genetic algorithms able to identify and eliminate irrelevant samples from the training graph of the OPF classifier. The method is called FSGAP-OPF and is capable of keeping the accuracy values at a similar level as the original OPF algorithm without pruning. In this method, the desired pruning rate is established before the pruning process takes place by codifying the reproduction and mutation mechanisms so as to keep the number of pruned nodes unchanged for each individual. The fitness function is guided by the accuracy values obtained by the pruned classifier.

To evaluate the performance of FSGAP-OPF, we tested its generalization capabilities on public datasets from the UCI repository and compared its results with those obtained by OPF-TPPA, ROB-OPF and OPF without pruning. On the basis of our experiments, we can state that FSGAP-OPF is a promising alternative for classification tasks, since it outperformed OPF, OPF-TPPA and ROB-OPF in terms of pruning rate (#PR) and classification time (#CT) without affecting, or even improving, the accuracy values (ACC). We highlight again that as our proposal achieves fixed-size models, it can help to solve problems where the memory consuming is crucial, such as embedded systems, and helps to control the maximum size that the models must reach after the training process.

References

1. Chen, S., Sun, T., Yang, F., Sun, H., Guan, Y.: An improved optimum-path forest clustering algorithm for remote sensing image segmentation. Comput. Geosci. **112**, 38–46 (2018)
2. Cormen, T.H., Leiserson, C.E., Rivest, R.L., Stein, C.: Introduction to Algorithms. MIT Press, Cambridge (1990)
3. Costa, K.A., Pereira, L.A., Nakamura, R.Y., Pereira, C.R., Papa, J.P., Falcão, A.X.: A nature-inspired approach to speed up optimum-path forest clustering and its application to intrusion detection in computer networks. Inf. Sci. **294**, 95–108 (2015)
4. Eiben, A.E., Smith, J.E., et al.: Introduction to Evolutionary Computing, vol. 53. Springer, Berlin (2003). https://doi.org/10.1007/978-3-662-05094-1

5. Falcão, A.X., Stolfi, J., de Alencar Lotufo, R.: The image foresting transform: theory, algorithms, and applications. IEEE Trans. Pattern Anal. Mach. Intell. **26**(1), 19–29 (2004)
6. Fernandes, S.E., Pereira, D.R., Ramos, C.C., Souza, A.N., Gastaldello, D.S., Papa, J.P.: A probabilistic optimum-path forest classifier for non-technical losses detection. IEEE Trans. Smart Grid **10**, 3226–3235 (2018). https://ieeexplore.ieee.org/abstract/document/8329170
7. Holland, J.H.: Adaptation in Natural and Artificial Systems: An Introductory Analysis with Applications to Biology, Control, and Artificial Intelligence. MIT press, Cambridge (1992)
8. da S. Luz, E.J., Nunes, T.M., De Albuquerque, V.H.C., Papa, J.P., Menotti, D.: ECG arrhythmia classification based on optimum-path forest. Expert Syst. Appl. **40**(9), 3561–3573 (2013)
9. Man, K.F., Tang, K.S., Kwong, S.: Genetic Algorithms: Concepts and Designs. Springer Science & Business Media, New York (2012)
10. Miranda, P.M., Falcão, A.X., Rocha, A., Bergo, F.P.: Object delineation by-connected components. EURASIP J. Adv. Signal Process. **2008**(1), 467928 (2008)
11. Nunes, T.M., Coelho, A.L., Lima, C.A., Papa, J.P., de Albuquerque, V.H.C.: Eeg signal classification for epilepsy diagnosis via optimum path forest-a systematic assessment. Neurocomputing **136**, 103–123 (2014)
12. Papa, J.P., Falcão, A.X.: On the training patterns pruning for optimum-path forest. In: Foggia, P., Sansone, C., Vento, M. (eds.) ICIAP 2009. LNCS, vol. 5716, pp. 259–268. Springer, Heidelberg (2009). https://doi.org/10.1007/978-3-642-04146-4_29
13. Papa, J.P., Falcao, A.X., Suzuki, C.T.: Supervised pattern classification based on optimum-path forest. Int. J. Imaging Syst. Technol. **19**(2), 120–131 (2009)
14. Papa, J.P., Falcão, A.X., Suzuki, C.T.N., Mascarenhas, N.D.A.: A discrete approach for supervised pattern recognition. In: Brimkov, V.E., Barneva, R.P., Hauptman, H.A. (eds.) IWCIA 2008. LNCS, vol. 4958, pp. 136–147. Springer, Heidelberg (2008). https://doi.org/10.1007/978-3-540-78275-9_12
15. Papa, J.P., Falcao, A.X., de Freitas, G.M., de Avila, A.M.H.: Robust pruning of training patterns for optimum-path forest classification applied to satellite-based rainfall occurrence estimation. IEEE Geosci. Remote Sens. Lett. **7**(2), 396–400 (2010)
16. Pereira, C.R., Nakamura, R.Y., Costa, K.A., Papa, J.P.: An optimum-path forest framework for intrusion detection in computer networks. Eng. Appl. Artif. Intell. **25**(6), 1226–1234 (2012)
17. Srinivas, M., Patnaik, L.M.: Genetic algorithms: a survey. Computer **27**(6), 17–26 (1994)

Constraint Exploration of Convolutional Network Architectures with Neuroevolution

Jonas Dominik Homburg[✉][iD], Michael Adams[✉][iD], Michael Thies[✉],
Timo Korthals[✉][iD], Marc Hesse[✉][iD], and Ulrich Rückert[✉]

Cognitronics and Sensor Systems Group, CITEC, Bielefeld University,
Bielefeld, Germany
{jhomburg,madams,mthies,tkorthals,mhesse,rueckert}
@techfak.uni-bielefeld.de

Abstract. The effort spent on adapting existing networks to new applications has motivated the automated architecture search. Network structures discovered with evolutionary or other search algorithms have surpassed hand-crafted image classifiers in terms of accuracy. However, these approaches do not constrain certain characteristics like network size, which leads to unnecessary computational effort. Thus, this work shows that generational evolutionary algorithms can be used for a constrained exploration of convolutional network architectures to create a selection of networks for a specific application or target architecture.

Keywords: Neuroevolution · Genetic programming ·
Convolutional neural networks · Architecture search

1 Introduction

Automating the design process of neural network structures using evolutionary algorithms is an established technique [6]. Today's compute capacities allows to use these methods to produce more complex networks and to discover new network structures. Network architecture search algorithms commonly try to maximize a single quantitative measurement like accuracy which often leads to ever-growing networks with the drawback of longer inference time and higher energy consumption. With the increasing amount of mobile and embedded devices, neural networks not only need to have high quality but also need to satisfy specific hardware limits in terms of size and efficiency.

This work uses neuroevolution to create network architectures which represent a balanced compromise between network size in terms of trainable parameters and quality in terms of accuracy. At first, most significant parameters of a generational evolutionary method are evaluated on a subset of the search space of network structures. After a suitable configuration has been found, the evolutionary algorithm is used to generate diverse network structure balancing applied constraints. The presented results are created with a single GPU setup using the MNIST [12] data set.

© Springer Nature Switzerland AG 2019
I. Rojas et al. (Eds.): IWANN 2019, LNCS 11507, pp. 735–746, 2019.
https://doi.org/10.1007/978-3-030-20518-8_61

In the following, Sect. 2 introduces work related to automated architecture search for convolutional neural networks followed by a description of the performed analysis in Sect. 3. The search space (Sect. 3.1) and the operations (Sects. 3.2 and 3.3) on the search space are explained in detail. Finally, the results are discussed in Sect. 4 followed by Sect. 5 commenting on the selection of most significant parameters of the evolutionary algorithm as well as assumptions and restrictions for the search space.

2 Related Work

Neuroevolution has already been used to generate network structures in many different ways. For example evolutionary programming has been used to generate modular neural networks [14], multilayer perceptrons have been developed with G-Prop [2] and the evolutionary system EPNet has been proposed [25]. Moreover structures similar to convolutions have been generated with compositional pattern producing networks [24], convolutional networks have been evolved using an indirect encoding [20] and the computation process has been accelerated by asynchronous algorithms [22]. Other methods like reinforcement learning [13], hill-climbing [4], random search [1] and grid search [26] have also been used to search for network structures and are available in frameworks like Auto-Keras [9], TPOT [15] or auto-sklearn [5].

Recent literature applies neuroevolutionary algorithms to automated architecture search for image recognition tasks [13,16,18,27]. As stated out by Real et al. [17], network structures generated by neuroevolution can surpass hand-crafted state-of-the-art networks in terms of accuracy. Instead of minimizing a known architecture using techniques like Deep Compression [7], this work aims to generate smaller architectures already by design. In comparison to previous work, the used method will be based on a generational evolutionary algorithm and will use operations to mutate network structures as well as operations for recombination. The network itself will be encoded with a tree structure which excludes bypassing connections to generate multiple distinct network structures.

3 Methods

This work uses a generational evolutionary algorithm which can be parameterized in terms of selection strategy, replacement scheme and generation size instead of a pool with asynchronous updates like previous work [23]. Before a particular combination of parameters is used to evolve complex artificial neural network structures, the effects of multiple selection strategies, replacement schemes and generation sizes are evaluated. At first, the generation size is kept constant at a value of six to select combinations of selection strategies and replacement schemes.

In order to select neural networks for the creation process, five different selection strategies S (proportional, linear ranking, exponential ranking, tournament, truncation) are analyzed on the number of generations they need to reach the maximum in a predefined search space described in Sect. 3.1.

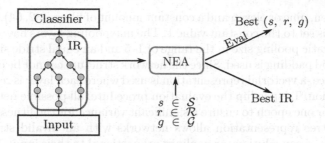

Fig. 1. A neuroevolutionary algorithm (NEA) generates network structures which are a trade-off between the optimization parameters based on an inner representation (IR), with the use of a selection strategy (s), a replacement scheme (r) and a generation size (g). This process returns the best IR with respect to the search space and is evaluated (Eval) to obtain the best parameter triplet.

Neural networks for new generations are created by mating the selected networks and mutating their descendants. Afterwards, the generation replacement is done by using one of the schemes $\mathcal{R}$ (generational replacement, n-elitism, n-weak-elitism, delete-n and delete-n-last).

After testing all combinations of selection strategies and replacement schemes, a set of five promising combinations are varied in their generation size $g \in \mathcal{G}$ where $\mathcal{G} = \{x \in \mathbb{N} \mid 6 \leq x \leq 42\}$. The most promising combination of selection strategy, replacement scheme and population size, which makes a compromise between the computational cost for each generation and the number of generations, is used for evolving convolutional network structures (see Fig. 1).

3.1 Search Space

The neural network structure is separated into convolutional feature extraction and classification. While the neuroevolutionary algorithm has been able to evolve a convolutional feature extraction, the configuration for the classification realized with fully connected layers has not been altered. The feature extraction is represented by a tree structure with the layer types *flattening, convolutional, maxpooling* and *input*. The flattening layer is always used as the root node. Internally the flattening is applied on each inner node before all results are concatenated and passed to the classifier. For the inner nodes, convolutional and maxpooling layers can be used. Each inner node can have either several inner nodes as children or a single terminal node (input layer). The convolutional layer can be varied in terms of filter count, kernel size, stride size and padding. Similar to the convolutional layer, the maxpooling layer can be varied in terms of pooling size, stride size and padding. Every convolutional layer contains a rectified linear activation to ensure non-linearity.

For the first evaluations, a subset of the described search space is used. Here, the feature extraction consists of two convolutional layers each followed by a maxpooling layer. The convolutional layers are limited to quadratic kernels with a size of three, five, seven, nine or eleven and a constant amount of filters (32, 64). Likewise, the stride size is set to the constant value 1. The maxpooling layers have been limited to a quadratic pooling size in the range of 1–5 and an equal stride size. For all four layers, valid padding is used. Since the network structure cannot be changed in this search space, a vectorial representation is used where each layer is represented by one dimension. To speedup the evaluation procedure, all possible networks are trained once for one epoch to ensure a significant variance for the fitness values.

Since the tree representation allows networks with an invalid structure in terms of oversized convolutions or poolings proportional to their input, the methods for mutation and crossover are restricted to produce only valid network structures. These methods are described in the following.

3.2 Mutation

To mutate the feature extraction layers of a given neural network four basic methods are used: *modifying an inner node, adding an inner node, duplicating a subtree* and *deleting a subtree* which can be divided in the groups *adding nodes, modifying nodes* and *removing nodes*. The root node is excluded from the mutation process while only some of the operations can be applied on leaf nodes. The inner nodes can be mutated by all operations as long as specific constraints are met. The operations can be uniformly drawn as well as hierarchically where the selection is separated in uniformly selecting a group and in uniformly selecting an operation within the group.

The methods duplicating or deleting a subtree cannot be applied on leaf nodes. Duplicating a leaf node would violate the constraint that an inner node can only have one leaf node or an arbitrary number of inner nodes. Deleting a leaf node would result in a subtree without input with the effect of being unused. Besides, a subtree given by an inner node can be deleted only if the parent node in the original tree has more than one child node. A subtree given by an inner node can be duplicated at any time. These restrictions result in an invalid tree structure only for the very rare case that all child nodes of inner nodes are effected by the delete operation.

If a new inner node is added or the given node is modified, the possible configurations are limited by the parent node of the given inner node on the one hand and by the parent node of the newly generated tree on the other hand. This has the effect that later mutations can be restricted by earlier mutations but minimizes the amount of invalid tree structures generated.

A new inner node is always added between the given node and its parent node. Since input dimensions of a layer can be varied in a certain range without changing the output dimensions, this range is used to find a set of possible configurations for the new node. The configuration for the new node is uniformly drawn.

Modifying a node is a hierarchical process. The first decision is keeping the node type or using a different one. For a different node type, a new possible configuration is found randomly. If the node type is not modified, keeping or changing the padding is equally likely but is kept the same if no suitable configuration could be found for the changed padding. Afterwards, the remaining parameters are randomly chosen from a set of possible configurations.

3.3 Crossover

The standard crossover operator on tree structures [10] would allow interchanges between all nodes. Using this method, the amount of invalid network structures would outnumber the valid structures. Therefore, a node can only be replaced by nodes with the same or larger output dimensions. If it is not possible to find a suitable node for replacement in the other network structure, a node which should be replaced, but cannot, is not changed at all. If a node gets exchanged, the replacing node is adapted to meet the dimensional constraints if necessary. The new node can be adapted by changing its parameters or by adding a new node to reduce the output dimensions. The used adjustment is chosen by the difference between the replaced and the replacing node output dimensions. It is worth mentioning that by design it is more likely that a new node is inserted for a larger difference than for similar output dimensions. If the replacing node is modified, its parameters are selected likewise in the modifying process for mutations.

4   Results

For the following results, the network structures are trained on a single NVIDIA GTX 1080 Ti with the well known MNIST data set. To prevent training on the test data set, the train data set is split into three subsets preserving the class balance. Two subsets of 5,000 images each are used for validation during training and generating a fitness score respectively. The remaining 50,000 images are used for training. Afterwards, the best network is tested on the 10,000 images of the MNIST test data set using [8].

To find suitable combinations of selection strategy and replacement scheme, the subset of the search space is used to analyze all possible combinations of the implemented methods. The combinations are then compared based on the amount of generations needed to find the optimal solution in the subset of the search space and the amount of different, reviewed network structures. Since non-deterministic methods are involved in the generation of new network configurations, every analysis is performed 10,000 times. An example of the results is shown as whiskerplots in Fig. 2. In comparison, combinations with a replacement scheme based on elitism performed better than similar combinations with a replacement scheme based on deletion.

After testing all combinations of selection strategies and replacement schemes, a set of five promising combinations are chosen for varying the generation size. To achieve the required amount of new networks, the selection size is increased as well. Considering that the previously found configurations may rely on the generation size, their parameters are adopted without changing as well as adjusted proportionally to the population size. Since the amount of new network structures per generation follows the population size, the product of the amount of generations and the population size is used for comparison.

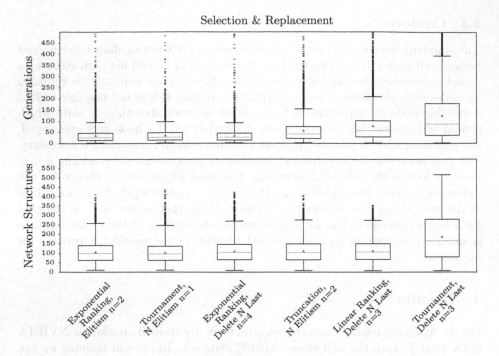

Fig. 2. Combinations of selection strategies and replacement schemes evaluated based on the amount of network structures and the amount of generations. The mean in these whiskerplots is represented by a green triangle whereas the median is visualized by a orange line. (Color figure online)

Figure 3 shows the best performing configuration consisting of an exponential ranking with elitism based replacement where the elite group size is preserved constantly. 1,000 values for each population size are visualized in a whiskerplot like fashion. In comparison, a linear ranking selection with a deletion based replacement scheme and an adapted parameter for the amount of removed network structures has an overall worse performance. In particular, using the number of generations times generation size as a measurement, it increases along with the generation size even faster than it increases for best performing configuration.

This configuration is then used to evolve convolutional feature extractors represented as a tree structure using the operators described in Sects. 3.2 and 3.3. The mutation operations can be picked uniformly or in a hierarchical fashion where the modifications are grouped in adding, modifying and removing nodes. Each group is assigned with the same probability to be chosen to provide a modification operation. The fitness function is altered to penalize network structures with a high amount of trainable parameters. Therefore, an interval of trainable parameters is normalized and values outside are clipped to the boundaries. The lower parameter boundary is motivated by the fact that a network structure with very few parameters cannot be competitive whereas the upper parameter boundary is given by the application for the network structures.

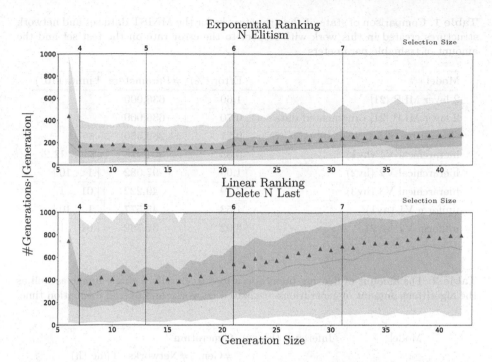

Fig. 3. The correlation between generation size and number of generations visualized by a line plot. The mean is visualized in orange whereas the median is represented by green triangles. The upper and lower whiskers correspond to the bright curves whereas the interquartile range is visualized by the dark curve. (Color figure online)

One of the best state-of-the-art algorithms on the MNIST data set is using convolutional filters and an ensemble of five networks [19]. Since each of these networks has about 2,200,000 trainable parameters, the proposed approach was used to evolve a network architecture within a range of 2,200,000 and 11,200,000 trainable parameters.

The best generated networks with respect to the search space are shown in Table 1 (hv1, hv2, hv3, uv1, uv2). Especially the network hv1 achieves an error of 0.83% with only 805,000 parameters and thus outperforms the well known 2 layer MLP [21] trained without the augmented data set. To further decrease the number of parameters, the boundaries have been lowered to 100,000–10,000,000 and 10,000–1,000,000 for the hierarchical and the uniform selection of mutation operations. Both methods found network structures with a higher error rate but fewer parameters than the previously evolved network.

In addition, the generated neural networks are analyzed based on the amount of floating point operations to classify a single image (see Table 2). A visualization of their inner representation reveals that the evolved networks are linear models as well as wide and complex structures (see Fig. 4). These major differences can also be observed in the misclassified images. Figure 5 highlights that the analyzed networks do not necessarily share images assigned with a wrong label.

Table 1. Comparison of state-of-the-art networks for the MNIST data set and network structures created in this work with respect to the error rate on the test set and the amount of trainable parameters.

Model	Error (%)	#Parameters	Limits (10^6)
2 layer MLP [21]	1.60	636,000	
2 layer MLP [21] (augmented data set)	0.70	636,000	
Committee of 5 CNNs [19] (C5CNN)	0.21	11,289,550	
hierarchical V1 (hv1)	0.83	804,978	[2.2 .. 11.2]
hierarchical V2 (hv2)	1.05	97,082	[.1 .. 10.]
hierarchical V3 (hv3)	3.19	40,222	[.01 .. 1.]
uniform V1 (uv1)	1.53	45,577	[.1 .. 10.]
uniform V2 (uv2)	3.02	32,762	[.01 .. 1.]

Table 2. The amount of floating point operations to classify a single image as well as the algorithms amount of generations, evaluated networks and overall evaluation time.

Model	Inference (Flops)	Generation		
		#Gen.	#Networks	Time (h)
hierarchical V1	451,597,615	90	900	≈126
hierarchical V2	60,878,149	52	520	≈28
hierarchical V3	184,411	30	300	≈4
uniform V1	6,530,539	50	500	≈27
uniform V2	4,777,225	49	490	≈24

Furthermore the evolutionary process is evaluated. The network structures are generated within 30 to 90 generations training 10 new networks per generation (see Table 2). The diversity in terms of degree and depth of the trained networks is visualized in a multidimensional whiskerplot with whiskers including the minimum and maximum values (see Fig. 4).

5   Discussion

The preceding analysis covered a choice of selection strategies along with replacement schemes and the population size. Since Fig. 3 shows that it should be considered to use a population size larger than 6 and smaller than 21, the prior choice of selection strategies and replacement schemes could be reproduced with suitable population size. Furthermore, the mutation rate of 40% was kept constant for all analyses. As one of the main parameters of an evolutionary algorithm, a different mutation rate could improve the performance for certain analyzed configurations which performed less well in this evaluation.

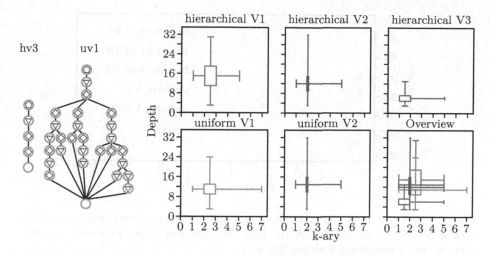

Fig. 4. Evolved network structures (left) hierarchical V3 (hv3) and the uniform V1 model (uv1). Style: rhombus i.e. convolution, triangle i.e. maxpooling, pentagon i.e. classifier, empty i.e. input. The distribution of maximum depth and degree over all evaluated network structures visualizing the diversity of observed architectures for each final outcome (right).

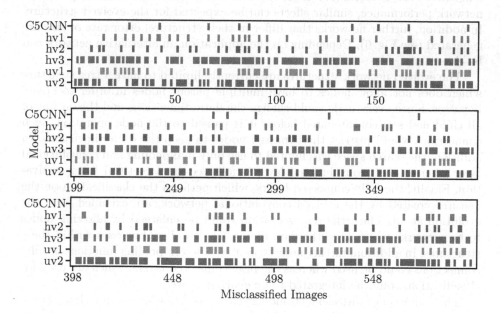

Fig. 5. Comparison of misclassified images within the MNIST test data set of 10,000 images. The 598 misclassified images are renumbered in the order they occur in the test set.

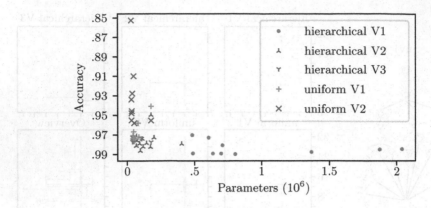

Fig. 6. The fittest networks in terms of size and accuracy of each generation. Since each generation is extended with the best two networks of the previous generation some networks contribute multiple times.

In comparison, the evolved network structures can compete with state-of-the-art networks considering their size and outperform non-ensemble architectures. Without augmenting the data set, an evolved structure with lower error rate and smaller size is found (see Table 1). Since data augmentation can improve the network performance, similar effects can be expected for the evolved structure. In addition, further networks that differ in their structure, error rate or size are generated (see Fig. 6). Depending on the application an alternative network can be chosen.

The evolved network structures are currently limited to have a tree structure which does not allow nodes to have multiple parent nodes in order to create bypasses. Since every node could change the data dimensions and the output of all child nodes is concatenated before it is passed to the node operation, such connections would restrict the bypassed nodes to keep the data dimensions. In addition, the search space was limited to use only maxpooling and convolutional layer whereas each convolutional layer is followed by a rectified linear activation. Finally, the fully connected layers, which perform the classification on the features created by the evolved convolutional network, are excluded from the evolution at all. Thus, the representation could be enlarged by allowing nodes to be networks which keep the data dimensions to allow bypasses within these structures. In addition, further node types like different pooling functions or fully connections could be used whereas the fully connected layers, which are used for classification, could be integrated to be evolved as well.

The search space with elements modeled as rooted, non-isomorphic, k-ary trees, is rich in solutions. Simplifying possible networks as binary trees with a maximum depth of ten, the final network structures are generated by evaluating less than one per mille of the search space. Since the proposed method has been tested on the MNIST data set, additional analyses on other data sets like CIFAR-100 [11] or ImageNet [3] data set is planned to be investigated in future work.

6   Conclusion

This work has shown that neuroevolution is capable of generating narrow and linear as well as wide and complex models. Furthermore, an evolutionary algorithm has been used to maximize accuracy and minimize size in terms of amount of parameters within applied interval. In this manner, network structures have been created that outperform hand crafted non-ensemble neuronal network structures for classification on the MNIST data set. Furthermore, it is possible to constrain the trade-off between accuracy and size. This enables the demand-oriented generation of networks for specific applications or target architectures.

Acknowledgement. The research leading to these results has received funding from the European Union Seventh Framework Programme (FP7) under grant agreement no 604102 and the EU's Horizon 2020 research and innovation programme under grant agreements No 720270 (Human Brian Project, HBP). It has been further supported by the European Fund for Regional Development under Grant IT-1-2-001 and the Cluster of Excellence Cognitive Interaction Technology "CITEC" (EXC 277) at Bielefeld University, which is funded by the German Research Foundation (DFG).

References

1. Bergstra, J., Bengio, Y.: Random search for hyper-parameter optimization. J. Mach. Learn. Res. **13**, 281–305 (2012)
2. Castillo, P.A., Merelo, J., Prieto, A., Rivas, V., Romero, G.: G-Prop: global optimization of multilayer perceptrons using gas. Neurocomputing **35**(1–4), 149–163 (2000)
3. Deng, J., Dong, W., Socher, R., Li, L.J., Li, K., Li, F.-F.: ImageNet: a large-scale hierarchical image database (2009)
4. Elsken, T., Metzen, J.H., Hutter, F.: Simple and efficient architecture search for convolutional neural networks. arXiv preprint arXiv:1711.04528 (2017)
5. Feurer, M., Klein, A., Eggensperger, K., Springenberg, J., Blum, M., Hutter, F.: Efficient and robust automated machine learning. In: Cortes, C., Lawrence, N.D., Lee, D.D., Sugiyama, M., Garnett, R. (eds.) Advances in Neural Information Processing Systems, vol. 28, pp. 2962–2970. Curran Associates Inc., New York (2015)
6. Fogel, D.B., Fogel, L.J., Porto, V.: Evolving neural networks. Biol. Cybern. **63**(6), 487–493 (1990)
7. Han, S., Mao, H., Dally, W.J.: Deep compression: compressing deep neural networks with pruning, trained quantization and Huffman coding. arXiv preprint arXiv:1510.00149 (2015)
8. Homburg, J.D.: JonasDHomburg/CECNAN: Release v1.0, February 2019. https://doi.org/10.5281/zenodo.2580089
9. Jin, H., Song, Q., Hu, X.: Auto-Keras: efficient neural architecture search with network morphism (2018)
10. Koza, J.R.: Genetic programming as a means for programming computers by natural selection. Stat. Comput. **4**(2), 87–112 (1994)

11. Krizhevsky, A., Hinton, G.: Learning multiple layers of features from tiny images. Technical report, Citeseer (2009)
12. LeCun, Y., Cortes, C., Burges, C.: The MNIST database of handwritten digits. The Courant Institute of Mathematical Sciences (1998)
13. Liu, C., et al.: Progressive neural architecture search. In: Ferrari, V., Hebert, M., Sminchisescu, C., Weiss, Y. (eds.) ECCV 2018. LNCS, vol. 11205, pp. 19–35. Springer, Cham (2018). https://doi.org/10.1007/978-3-030-01246-5_2
14. Liu, Y., Yao, X.: Evolving modular neural networks which generalise well. In: Proceedings of 1997 IEEE International Conference on Evolutionary Computation (ICEC 1997), pp. 605–610. IEEE (1997)
15. Olson, R.S., Bartley, N., Urbanowicz, R.J., Moore, J.H.: Evaluation of a tree-based pipeline optimization tool for automating data science. In: Proceedings of the Genetic and Evolutionary Computation Conference 2016, GECCO 2016, pp. 485–492. ACM, New York (2016). https://doi.org/10.1145/2908812.2908918
16. Pham, H., Guan, M.Y., Zoph, B., Le, Q.V., Dean, J.: Efficient neural architecture search via parameter sharing. arXiv preprint arXiv:1802.03268 (2018)
17. Real, E., Aggarwal, A., Huang, Y., Le, Q.V.: Regularized evolution for image classifier architecture search. arXiv preprint arXiv:1802.01548 (2018)
18. Real, E., et al.: Large-scale evolution of image classifiers. arXiv preprint arXiv:1703.01041 (2017)
19. Romanuke, V.V.: Training data expansion and boosting of convolutional neural networks for reducing the MNIST dataset error rate (2016)
20. Schrum, J.: Evolving indirectly encoded convolutional neural networks to play tetris with low-level features (2018)
21. Simard, P.Y., Steinkraus, D., Platt, J.C.: Best practices for convolutional neural networks applied to visual document analysis. In: Proceedings of the Seventh International Conference on Document Analysis and Recognition, vol. 2, p. 958. IEEE (2003)
22. Stanley, K.O., Bryant, B.D., Miikkulainen, R.: Real-time neuroevolution in the NERO video game. IEEE Trans. Evol. Comput. 9(6), 653–668 (2005)
23. Stanley, K.O., Miikkulainen, R.: Evolving neural networks through augmenting topologies. Evol. Comput. 10(2), 99–127 (2002)
24. Verbancsics, P., Harguess, J.: Generative neuroevolution for deep learning. arXiv preprint arXiv:1312.5355 (2013)
25. Yao, X., Liu, Y.: Towards designing artificial neural networks by evolution. Appl. Math. Comput. 91(1), 83–90 (1998)
26. Zagoruyko, S., Komodakis, N.: Wide residual networks. arXiv preprint arXiv:1605.07146 (2016)
27. Zoph, B., Le, Q.V.: Neural architecture search with reinforcement learning. arXiv preprint arXiv:1611.01578 (2016)

Impact of Genetic Algorithms Operators on Association Rules Extraction

Leila Hamdad[1(✉)], Karima Benatchba[2], Ahcene Bendjoudi[4], and Zakaria Ournani[3]

[1] LCSI, Ecole nationale Supérieure en Informatique ESI, Oued ESmar, El Harrach, Algeria
l_hamdad@esi.dz
[2] LMCS, Ecole nationale Supérieure en Informatique ESI, Oued ESmar, El Harrach, Algeria
k_Benatchba@esi.dz
[3] Ecole nationale Supérieure en Informatique ESI, Oued ESmar, El Harrach, Algeria
oz_ournani@esi.dz
[4] DTISI, CERIST Research Center, Algiers, Algeria
abendjoudi@cerist.dz

Abstract. The first goal of this paper is to study the impact of Genetic Algorithms (GA's) components such as encoding, different crossover and replacement strategies on the number and quality of extracted association rules. Moreover, we propose a strategy to manage the population. The later is organized in sub-populations where each one encloses same size rules. Each sub-population can be seen as a population on which a GA is applied. Hence, we propose two GAs, a sequential one and a parallel one. All tests are conducted on two types of benchmarks: synthetic and real ones of different sizes.

Keywords: Association rules · Genetic Algorithm · Population · Parallel · CPU · Apriori

1 Introduction

Association Rule Mining (ARM) is one of the most important task of data mining. It consists of extracting useful correlations among items over transactional databases. This task is used in many fields including marketing, medical, bioinformatic, telecommunication, etc. ARM can be defined as follows: Let $I = \{i_1, i_2, \ldots, i_n\}$ be a set of items and $T = \{t_1, t_2, \ldots, t_N\}$ a set of transactions, each transaction contains a number of items. And let A and B two item-sets of I, an association rule is defined by:

$$R : A \rightarrow B,$$

where A is called antecedent of the rule, B the consequence and $B \cap A = \emptyset$ ([1]).

© Springer Nature Switzerland AG 2019
I. Rojas et al. (Eds.): IWANN 2019, LNCS 11507, pp. 747–759, 2019.
https://doi.org/10.1007/978-3-030-20518-8_62

Two measures are used to evaluate the quality of the rules: *Support* and *Confidence*. *Support* measures the rule generality and is defined as follows:

$$Support(A \cup B) = Supp(A, B) = \frac{|A \cup B|}{N}.$$

Where N is the number of transactions in a database. An association rule (AR) $R : A \rightarrow B$ has a *Support* S if $S\%$ of transactions contain both A and B. *Confidence* measures rule validity and is defined as follows:

$$Confidence(R) = Conf(R) = \frac{|A \cup B|}{|A|}.$$

$A \rightarrow B$ has a *Confidence* C if $C\%$ of transactions that contain A, contain also B. The interesting rules extracted are those with *Support* greater than *Minsup* and *Confidence* greater than *Minconf*, where *Minsup* and *Minconf* are two thresholds given by the user.

Nowadays, the most used algorithm to deal with ARMs is APRIORI [1]. This algorithm is composed of two phases. The first one generates frequent itemsets. The second one extracts association rules from those itemsets. The first phase is the most time consuming as it is an NP-hard problem [12]. Execution time grows exponentially as the size of the database grows (2^n itemsets to read through for n item database). In the literature, one can find many works on the use of metaheuristics to extract ARs. They are approximate methods based on an intelligent browse of the search space, providing a solution which may not be optimal, but of good quality, in a shorter time [6]. Although the main purpose of using metaheuristics has been to speed up ARM, the continuous growth of data keeps this execution time prohibitive. This led to the proposition of parallel rule mining algorithms. Indeed, to handle very large datasets, parallel computing techniques have been used to accelerate the extraction by carrying out several tasks simultaneously. In this paper, we are interested in Genetic Algorithms (GA) for association rules extraction. This work has two objectives: First, to study the impact of GAs' components such as encoding and different crossover, mutation and replacement strategies on the number and quality of extracted association rules. Second, we propose a strategy to manage the population. The latter is divided in sub-populations where each one encloses same size rules. On each sub-population a GA is applied. Hence, we propose two GAs, a sequential one and a parallel one. All tests are carried on a set of real and synthetic benchmarks.

The rest of the paper is organized as follows: in Sect. 2, we start by presenting some works that have used GA for rule extraction. Then we present and introduce our population's organization. In Sects. 3 and 4, the impact of those strategies on the number and quality of extracted rules and performance are discussed. In Sect. 5, we present our conclusion.

2   Genetic Algorithm for Rule Mining

The most used metaheuristic in rule mining is Genetic Algorithm (GA). One can find in the literature several works that differ according to chromosome encoding or strategies used for the different GA components. One can cite GENAR

(GENetic Association Rules) proposed by [9]. Each chromosome represents a rule and each gene defines an item to which an interval of value is associated. Only numerical items are considered. It uses a usual crossover (uniform mask) and a simple mutation that consists on changing genes [9]. One year later, the same authors proposed GAR (Genetic Association Rules) which is similar to GENAR. The difference is that an individual represents an itemset. As a result, the second phase of APRIORI algorithm has to be used to generate rules. ASGARD (Adaptative Steedy state Genetic Algorithm for Rules Discovery) introduced by [6] is a GA where chromosomes represent rules of different sizes. The author uses an elitist insertion and proposes a generalization operator which role is to delete randomly an item from the chromosome. ARMGA (Association Rules Mining Genetic Algorithm) [13] proposes to represent several rules by a chromosome, distinguishing the antecedent part and the consequent one. It uses a two points crossover, a mutation by value and a fitness based on a support measure. An improved algorithm was proposed two years later in (2011) by [11], called AGA (Adaptative Genetic Algorithm). This improvement focuses on the crossover and mutation. The two proposed operators are more adapted to the problem. It uses crossover and mutation matrices. MBAREA (Mining Boolean Association Rules Algorithm) was proposed in [7]. It is based on ARMGA, using the same representation of chromosomes, the same crossover and fitness. The main difference is that, the values of items are not saved since they are binary ones. Other metaheuristics to solve the problem of association rules have been proposed: BSO-ARM (Bees Swarm Optimisation for Association Rules Mining) based on bees swarm [4], PSOARM (Particules Swarm Optimisation for Association Rules Mining) based on Particle swarm [8] and ACOR (Ant colony Optimisation for Rules) based on ants [10].

2.1 Chromosomes Encoding

We will be using the two types of encoding found in the literature. For the first encoding, a chromosome represents a rule as shown in Fig. 1 and contains different items. The last one represents the conclusion part and the others the antecedent [13]. For the second encoding, a chromosome consists of several rules. Many items are considered in the conclusion part. Hence, an index is used to separate the two parts of the rule.

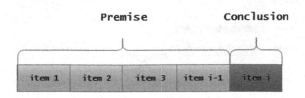

Fig. 1. One rule encoding.

2.2   Fitness

Usually support and confidence are used to evaluate rules ([5], [4]). The fitness function can be defined as follows:

$$Fitness = \frac{\alpha \times Support + \beta \times Confidence}{\alpha + \beta}$$

Where α and β are two parameters in $[0, 1]$.

2.3   Genetic Operators

Selection, crossover, mutation and replacement are applied on a population of chromosomes for a number of iterations. At each iteration, chromosomes are selected randomly in pairs on which crossover is applied. The two selected chromosomes are crossed with a probability PC to give two new solutions. As we wish to study the impact of crossover on the quality of the result, we propose to study the three types of crossover shown in Fig. 2 (one point, two points and uniform). Each time, a chromosome is generated, its fitness is computed. This value helps to decide whether the new chromosome will be added to the new population. We propose to test two replacement techniques after a crossover. In the first one, a parent is replaced by its son if the cost of the latter is better. The second is to replace the worst chromosome in the population by the new one if its cost is better. Each chromosome is mutated with a probability PM. The mutation consists of changing a random item in a chromosome with another item selected randomly. This mutation might generate a new association leading to a diversification.

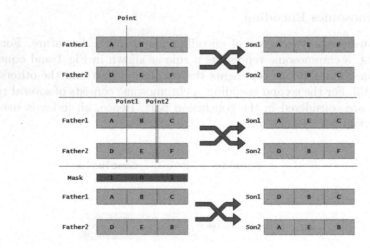

Fig. 2. Crossover types.

2.4   GA's Population

We propose to test two types of population's organization. The first one is the usual. It consists of individuals of different sizes generated randomly during initialisation. For the second, we propose to organize the population into classes where each class contains same size rules. Each class can be seen as a sub-population. Distribution of the population in classes has two goals: The first one is proposed to be sure to have several rules of different sizes. The second one is to give preference to smallest rules which usually are the relevant ones. The strategy of distribution is made according to a decreasing arithmetic progression. This sequence represents parts to be allocated to each class in order to give more chances to classes containing chromosomes of smaller sizes. Thus, for a population of size N, we have $N-1$ classes. The first class contains chromosomes of size 2, the last one contains chromosomes of size N. Each class will have an assigned size: Class $N - 1$, will be assigned a part, Class $N - 2$ will assigned two parts and so one. Class 1 will be assigned $N - 1$ parts. The number of parts is equal to $K = \frac{N(N-1)}{2}$. At the end, we have N classes containing a different number of individuals. For each class, a number of chromosomes are generated randomly while respecting the size of individuals in a class.

3   Tests and Results

In order to test the impact of encoding, operators and different parameters of GA on quality of extracted rules, several tests have been conducted. Recall that quality is evaluated according to the number of obtained rules and support and confidence of each rule. Moreover, the impact of parallelism on CPU on execution time is also tested. For both quality and execution time, the impact of the proposed population organization is evaluated. For this purpose, several datasets of different sizes are used. Each reported result is an average of 10 executions.

3.1   Datasets Description

Real ([3]) and synthetic ([2]) datasets are used; their characteristics are given in Table 1. The size of a dataset is essentially represented by the number of transactions. The used datasets are classified (small or medium) according to their size (Table 1). For this, we used two parameters: total number of items in transactions and number of transactions. A dataset is considered small if its total number of items is less than 100 000. It is considered medium if the total number of items is in [100 000, 1 000 000] and has more than 100 000 transactions.

3.2   Impact of Chromosome Encoding

In the literature two types of encoding are used: In the first, one an individual represents one rule while in the second it represents several. The goal of

Table 1. Datasets caracteristics.

Size	Dataset	Nbr of transactions	Nbr of items	Total Nbr of items in transactions	Type
Small	SPEC-heart	187	86	8 415	Real
	Forests	246	206	15 332	Real
	C20d10	2 000	386	40 000	Synt
Medium	Chess	3 176	75	121448	Real
	Mushroom	8 416	128	193568	Real
	T25i10d10	9 976	1 000	247148	Synt
	C73d10	10 000	2 178	750000	Synt
	T10i4d100	98395	1 000	995244	Synt
	Retail	88 162	16 469	996754	Real
	T20i6d100	99922	1 000	1988427	Synt
	Pumsb_star	49 046	7 116	2524993	Real
	Connect	67 557	129	2972508	Real
	Pumsb	49 046	7 116	3679219	Real

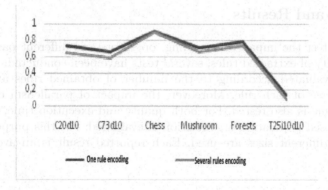

Fig. 3. Average fitness according to the type of encoding.

our tests is to evaluate the impact of encoding on the number and quality of extracted rules. The results of the execution of GA with the following parameters (two points exchange crossover with $PC = 0.9$, item mutation with $PM = 0.1$, replacement of parents, population size (50), population type: classes, number of iteration = 100) while using the two encodings are shown in Fig. 3. It is clear that the first encoding (one rule) gives best results with the small dataset C20d10 and medium one C73d10 since it gives highest fitness. Otherwise for the rest, the two encodings are almost similar with a slight advantage for the first one. This is explained by the fact that the second codification encodes several rules in a chromosome. However, grouping two rules in a chromosome means that the

Table 2. Quality according to type of population organization.

Dataset	Size	Population organized in classes Nbr rules (nbitems)	MS[a]	MC[b]	One population Nbr rules (nbitems)	MS	MC
C20d10	100	22 (16)	0.41	0.97	41 (20)	0.47	0.94
	200	89 (26)	0.48	0.98	107 (33)	0.42	0.96
Chess	100	95 (26)	0.87	0.98	96 (22)	0.90	0.98
	200	187 (30)	0.90	0.98	190 (26)	0.88	0.97
Mushroom	100	83 (33)	0.41	0.98	89 (29)	0.44	0.96
	200	160 (46)	0.38	0.98	177 (47)	0.37	0.98
T25i10d10	100	2 (3)	TS	0.38	3 (4)	TS	0.22
	200	3 (4)	TS	0.35	6 (6)	TS	0.21
C73d10	100	10 (10)	0.42	0.80	24 (13)	0.48	0.87
	200	30 (20)	0.54	0.94	52 (28)	0.47	0.92
T10i4d100	100	3 (5)	TS	0.23	2 (3)	TS	0.11
	200	4 (5)	TS	0.2	3 (3)	TS	0.07
Retail	100	2 (3)	TS	0.32	2 (3)	TS	0.06
	200	2 (3)	TS	0.30	2 (3)	TS	0.03
T20i6d100	100	1 (2)	TS	0.25	2 (3)	TS	0.28
	200	3 (4)	TS	0.31	3 (4)	TS	0.27
Pumsb_ star	100	2 (3)	0.26	0.56	4 (5)	0.28	0.62
	200	3 (4)	0.37	0.82	6 (6)	0.33	0.78
Connect	100	88 (28)	0.89	0.98	99 (25	0.90	0.97
	200	184 (38)	0.91	0.99	200 (39)	0.88	0.95
Pumsb	100	3 (4)	0.33	0.86	17 (12)	0.40	0.90
	200	6 (6)	0.60	0.74	10 (10)	0.52	0.87

[a] Mean Support
[b] Mean Confidence

support and confidence of the new rule does not exceed the minimum of support and confidence of those two rules. This can decrease the quality of the rule. In the following tests, we use one rule encoding as it gives association rules with best quality.

3.3 Impact of Population Type and Size

Our objective in the following tests is to evaluate the organization strategy for the population. Recall that we propose to organize the population into classes where each one contains rules of same sizes. To measure this impact, we have conducted tests with the following GA parameters (two points exchange crossover with

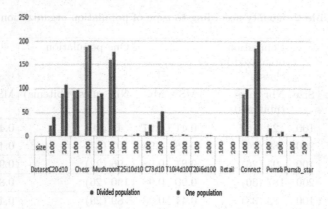

Fig. 4. Number of rules according to type of organization and size of population.

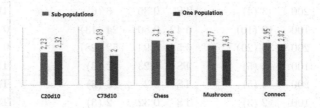

Fig. 5. Average size of extracted rules.

$PC = 90\%$, Item mutation with $PM = 0.1$, replacement of parents, number of iteration $= 100$). For each test, we varied the population size (100 and 200) as reported in the first column of Table 2.

Table 2 shows that organizing population in classes gives usually a better confidence. One can notice that increasing the size of population gives more rules and increases the confidence. Figure 4 gives an overview of the evolution of number of rules according to the type and size of the population. We notice that for all datasets, we obtain more rules with one population than with divided population.

Figure 5 shows the average number of rules returned when using the two types of population. One can notice that the average is higher for the population organized in classes than the other one. This is due to the convergence toward rules of smallest size. In the following Table 3, we present the different results obtained while varying the number of classes in the population. The number of classes has an effect on the number of extracted rules and not on the average fitness. Indeed increasing the number of classes generates a higher rule's size. In the following tests, organized population in classes is used.

Table 3. Impact of number of classes on quality of extracted rules.

Dataset	Number of classes	1	2	3	4
Forests	Number of rules	95	95	93	90
	Average Fit	0.78	0.78	0.78	0.77
C20d10	Number of rules	54	40	32	26
	Average Fit	0.67	0.73	0.68	0.70
Chess	Number of rules	98	97	95	92
	Average Fit	0.93	0.94	0.93	0.94
Mushroom	Number of rules	80	76	73	66
	Average Fit	0.68	0.69	0.69	0.70

3.4   Impact on the Type of Crossover and Replacement

The impact of the type of crossover used is not trivial or predictable. Hence, we have tested three types of crossover: One point, two points and uniform mask. The parameters of the GA for those tests are ($PC = 90\%$, Item mutation with $PM = 0.1$, replacement of parents, population size (50)) while using different number of iterations. Figures 6 and 7 show that the two points crossover gives more often the best result in terms of number of obtained rules and quality of rules (fitness). The uniform mask crossover gives more rules than one point crossover. Two types of replacement have been used: Children replacing parents and replacement of worst individuals (elitist strategy). Table 4 gives a summary of obtained results.

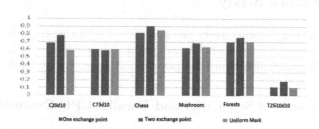

Fig. 6. Average fitness of extracted rules according to type of crossover.

Average fitness is best when using elitist replacement as worst individuals are replaced by better ones. Furthermore this type of replacement can lead to convergence in number of items contained in rules (a small set of items leads to same appearing items in all the rules), because the strongest individuals are more likely to be involved in crossover and spread in the population. In contrast, the replacement of parents can cause the loss of a good individual as it can be replaced by a chromosome of less quality.

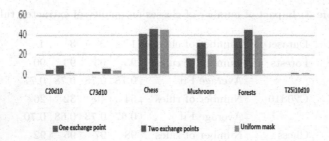

Fig. 7. Number of extracted rules according to the type of crossover.

Table 4. Evolution of quality of solution according to replacement strategy.

| Dataset | Child rep parents | | | Elitist rep | | |
	Nbr rls[a]	MF[b]	Mnbr items	Nbr rls	MF	Mnbr items[c]
C20d10	9	0.68	10	13	0.70	10
C73d10	5	0.62	8	2	0.68	3
Chess	47	0.88	24	49	0.92	21
Mushroom	27	0.63	21	29	0.68	19
Forests	44	0.73	33	40	0.76	28
T25i10d10	2	0.13	3	2	0.14	3

[a]Number of rules.
[b]Mean Fitness.
[c]Mean number of items.

4   Performance Study

In this section, performance tests are conducted to measure and compare the execution time of GA in sequential and parallel cases. The aim is to show the impact of CPU parallelism and type population on execution time.

4.1   Performance of Sequential and Parallel CPU Executions

In what follows, we present the results of three GA execution types: Sequential Execution (S-E), computation of parallel fitness on CPU (P-E CPU(fitness)), the parallel GA algorithm on sub-populations (P-E CPU(classes)). The size of the population is fixed to 200 chromosomes and the number of iterations to 100. Table 5 shows the results of the different executions on several datasets. Table 5 shows that P-E CPU(fitness) is more advantageous for large instances. Indeed for small instances, the execution time is 20 times slower than the sequential one. With the increase of the size of the database, especially the number of transactions, the fitness computation becomes time consuming and the synchronization time becomes negligible. This helps to achieve an acceleration[1] of 3 on

---

[1] Acceleration = Sequential time/Parallel time.

the Kosarak dataset and 3.5 on the Accidents dataset compared to the sequential version. Hence, P-E CPU(fitness) offers an acceleration which increases with the size of the database to reach a maximum factor equal to the number of execution units in the CPU. It will therefore offer better acceleration on a rather large dataset. P-E CPU(classes) GA always gives good results, whether on small or large datasets. The obtained acceleration is in the neighborhood of 3.5. Figure 8 compares the slopes of the evolutions of the two methods (sequential and parallel). On a one population, we compare the two modes of execution (S-E, P-E CPU (fitness)). These comparisons between these execution modes give similar results and the same remarks as in the case of a population divided into sub-populations (P-D-S) (Fig. 8(a)). The difference is that the implementation of the GA on a one population is faster than execution on sub-populations.

Table 5. Obtained execution time according to sequential and parallel versions on the two types of populations.

| Dataset | S-E[a] | Pop: Sub-populations | | | Pop: One population | |
		P-E[b] CPU (fitness)	P-E CPU (classes)	S-E	P-E CPU (fitness)	
SPEC-heart	3	55	1	2	54	
Forests	4	53	1	3	8	
C20d10	13	1 m 3 s	4	9	58	
Chess	43	1 m 18 s	11	35	1 m 18 s	
Mushroom	1 m 14	1 m 44	21	1 m 12	1 m 21	
T25i10d10	2 m 31 s	2 m 16 s	37	1 m 30 s	1 m 41 s	
C73d10	5 m 9 s	3 m 36 s	1 m 26 s	3 m	2 m 48 s	
T10i4d100	9 m 23 s	5 m 33 s	2 m 32 s	5 m 6 s	3 m 55 s	
Retail	7 m 32 s	4 m 37 s	2 m 7 s	5 m 28 s	3 m 35 s	
T20i6d100	19 m 51 s	8 m 45 s	5 m 15 s	14 m 49 s	7 m 15 s	
Pumsb_ star	17 m 43 s	8 m 2 s	5 m 5 s	11 m 28 s	6 m 22 s	
Connect	19 m 33 s	8 m 40 s	5 m 45 s	17 m 7 s	7 m 57 s	
Pumsb	26 m 50 s	10 m 38 s	7 m 14 s	17 m 24 s	8 m 28 s	
Kosarak	1 h 3 m 17 s	21 m 35 s	17 m 52 s	45 m 5 s	17 m 20 s	
Accidents	1 h 25 m 1 s	25 m 21 s	24 m 44 s	53 m 8 s	26 m 8 s	

[a]Sequential Execution.
[b]Parallel Execution.

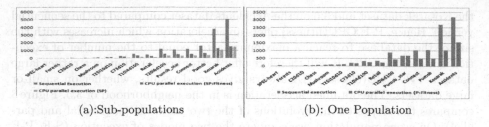

(a):Sub-populations   (b): One Population

Fig. 8. Evolution of the time execution according to the size of the datasets

5   Conclusion

In this work, we deal with GA for extraction of association rules. They are widely used in literature. Our contribution is to study the impact of different strategies of GA components such as encoding, crossover, mutation on their behavior, since different strategies are used in the literature but in our knowledge, there is no previous work for comparing these strategies. This comparison is done according to the number of rules obtained and the fitness (support and confidence). Besides this, we proposed to organize the GA population into sub-populations where each one contains same size rules. The conducted tests on this strategy show that this organization gives interesting results on quality and execution time. Parallel Execution on CPU are also studied and the tests on different datasets show the interesting gain in time execution.

References

1. Agrawal, R., Srikant, R.: Fast algorithms for mining association rules. In: Proceedings of the 20th International Conference Very Large Data Bases, VLDB, vol. 1215, pp. 487–499 (1994)
2. Bacha, K., Lichman, M.: UCI machine learning repository. https://archive.ics.uci.edu/ml/datasets.html. Accessed 20 Apr 2016
3. Bruno, E.: Building multi-core ready java applications (2008). https://dzone.com/articles/building-multi-core-ready-java. Accessed 13 Apr 2016
4. Djenouri, Y., Bendjoudi, A., Mahdi, M., et al.: GPU-based bees swarm optimization for association rules mining. J. Supercomput. **71**(4), 1318–1344 (2015)
5. Indira, K., Kanmani, S.: Performance analysis of genetic algorithm for mining association rules. Int. J. Comput. Sci. Issues **9**(2) (2012)
6. Jourdan, L.: Métaheuristiques pour l'extraction de connaissances: Application à la génomique. Université des Sciences et Technologie de Lille I, Ph.D. (2003)
7. Kabir, M.M.J., Xu, S., Kang, B.H., Zhao, Z.: A new evolutionary algorithm for extracting a reduced set of interesting association rules. In: Arik, S., Huang, T., Lai, W.K., Liu, Q. (eds.) ICONIP 2015. LNCS, vol. 9490, pp. 133–142. Springer, Cham (2015). https://doi.org/10.1007/978-3-319-26535-3_16
8. Kuo, R.J., Chao, C.M., Chiu, Y.T.: Application of particle swarm optimization to association rule mining. Appl. Soft Comput. **11**(1), 326–336 (2011)

9. Mata, J., Alvarez, J.L., Riquelme, J.C.: Mining numeric association rules with genetic algorithms. In: Kurková, V., Neruda, R., Kárný, M., Steele, N.C. (eds.) Artificial Neural Nets and Genetic Algorithms, pp. 264–267. Springer, Vienna (2001). https://doi.org/10.1007/978-3-7091-6230-9_65

10. Moslehi, P., Bidgoli, B.M., Nasiri, M., Salajegheh, A.: Multi-objective numeric association rules mining via ant colony optimization for continuous domains without specifying minimum support and minimum confidence. Int. J. Comput. Sci. Issues (IJCSI) 8 (2011)

11. Wang, M., Zou, Q., Liu, C: Multi-dimension association rule mining based on adaptive genetic algorithm. In: International Conference on Uncertainty Reasoning and Knowledge Engineering (URKE). IEEE (2011)

12. Wijsen, J., Meersman, R.: On the complexity of mining quantitative association rules. Data Min. Knowl. Discovery 2(3), 263–281 (1998)

13. Yan, X., Zhang, C., Zhang, S.: Genetic algorithm-based strategy for identifying association rules without specifying actual minimum support. Expert Syst. Appl. 36(2), 3066–3076 (2009)

The Problems of Selecting Problems

Alberto de la Encina, Natalia López, Ismael Rodríguez,
and Fernando Rubio[✉]

Dpto. Sistemas Informáticos y Computación, Facultad Informática,
Universidad Complutense de Madrid, 28040 Madrid, Spain
{albertoe,natalia,isrodrig,fernando}@sip.ucm.es

Abstract. We face several teaching problems where a set of exercises has to be selected based on their capability to make students discover typical misconceptions or their capability to evaluate the knowledge of the students. We consider four different optimization problems, developed from two basic decision problems. The first two optimization problems consist in selecting a set of exercises reaching some required levels of coverage for each topic. In the first problem we minimize the total time required to present the selected exercises, whereas the surplus coverage of topics is maximized in the second problem. The other two optimization problems consist in composing an exam in such a way that each student misconception reduces the overall mark of the exam to some specific required extent. In particular, we consider the problem of minimizing the size of the exam fulfilling these mark reduction constraints, and the problem of minimizing the differences between the required marks losses due to each misconception and the actual ones in the composed exam. For each optimization problem, we formally identify its approximation hardness and we heuristically solve it by using a genetic algorithm. We report experimental results for a case study based on a set of real exercises of Discrete Mathematics, a Computer Science degree subject.

Keywords: Computational complexity · Optimization · Education · Genetic algorithms

1 Introduction

Every time a new topic has to be taught to the students of any subject from any academic level, several examples are usually presented to help understanding the general description. However, selecting an appropriate set of examples is not a trivial task (see e.g. [9,10]). Notice that the teacher has to take into account the most common misconceptions the students may have. For instance, 6-year old students may think the number that is carried over in additions is always 1, and novice

This work has been partially supported by Spanish project TIN2015-67522-C3-3-R, the UCM project PIMCD 2016_176, and by Comunidad de Madrid as part of the programs S2013/ICE-2731 (NGREENS-CM), and S2018/TCS-4339 (BLOQUES-CM) co-funded by EIE Funds of the European Union.

© Springer Nature Switzerland AG 2019
I. Rojas et al. (Eds.): IWANN 2019, LNCS 11507, pp. 760–772, 2019.
https://doi.org/10.1007/978-3-030-20518-8_63

programming students may think the variable controlling a while loop is always incremented by one unit every iteration. Thus, in order to cover all possible misconceptions, a good set of examples has to be selected: For each possible misconception, there should be at least one example that cannot be solved successfully in case the misconception is present. By doing so, students having any misconceptions will notice them, as examples will make any contradiction between their view and the correct view arise.

Given a set of examples covering misconceptions on a given topic, we tackle the teaching problem of finding a subset of examples that covers all the possible misconceptions a given number of times. For instance, let us assume we have 5 possible misconceptions (A, B, C, D, and E) and 5 possible examples where Example 1 covers errors A, B, C, and E; Example 2 covers C and D; Example 3 covers B and C; Example 4 covers A, B, and D; and Example 5 covers C, D, and E. Besides, suppose we want to cover each misconception at least *twice*. Then, it is not necessary to present all the examples (although they would obviously cover each misconception at least twice), it is enough to present examples 1, 4, and 5.

Going one step forward, we will also apply those notions to a related teaching activity. When we want to evaluate the knowledge of the students with an exam, we should be able to detect and assess the misconceptions of the students. Thus, the exam should include exercises whose elaboration is sensitive to the main misconceptions (i.e. exercises which will be failed if some misconceptions are present), and each misconception should minor the score of the exam appropriately, in some desired extent. Let us suppose each misconception is assigned a given desired grade penalty. Then, designing a suitable exam does not consist just in covering all misconceptions a minimum number of times (as we can do when teaching the topic in the classroom), as this would wrongly increase the relative penalty of all misconceptions which are covered by a proportion of exercises higher than desired. For instance, in the example presented before, in case an exam includes the five examples 1–5 (in this case, exam exercises), a student with a misconception of type C will fail 4 out of 5 exercises, while a student with a misconception of type A will only fail 2 out of 5 exercises. Given how many points should ideally be penalized by each misconception, the second teaching problem tackled in this paper will be selecting the appropriate set of exercises to cover each misconception the specified number of times, or to minimize the *distance* to such optimal distribution in case it is not possible to obtain the exact solution required.

These two teaching tasks motivate the definition of the four optimization problems faced in this paper. On one hand, for the task of selecting a set of exercises reaching some target coverage levels for each misconception, we will consider the problem of minimizing the time required to teach them and the problem of maximizing the surplus coverage of misconceptions beyond their minimum required levels. On the other hand, for the task of composing exams where each misconception yields its corresponding required mark reduction, we will consider the problem of minimizing the time required to complete the exam and the problem of minimizing the differences between the required mark losses due to the misconceptions and the actual losses in the composed exam. As we will see, all four optimization problems are NP-hard, meaning that no algorithm can find optimal solutions for them in reasonable time when the size of the problem is not trivial. Moreover, the *approximability* (see e.g. [1,4,18]) of these four problems is not good either, as all of them are Log-APX-hard (some even worse). Being Log-APX-hard means that it is not possible to find good approximations to their optimal solutions: If P $\neq$ NP, then it is not possible to find polynomial-time algorithms guaranteeing that their solutions will be a given fix percentage worse than the optimal ones in the worst case.

In order to heuristically solve these problems, we will use genetic algorithms [6,15,23], as they have proven to be successful solving many different types of optimization problems (see e.g. [3,5,14,16,17,21,22]). Given a set of misconceptions, a set of examples (exercises), the time required to present (and solve) each example in the classroom, the list of misconceptions covered by each example, the maximum total time available to present all examples, and the number of times each misconception has to be presented by means of examples (or the expected mark loss each misconception should produce in an exam), the algorithms will face the aforementioned optimization problems.

In order to test our algorithms, we consider a concrete subject: Discrete Mathematics. This subject is taught to first-year students of the Computing Engineering Degree, and it covers many different concepts, including probability, graph theory, recursion, induction, number theory, etc. Although most of the concepts are independent, a single exercise usually combines several of them. The teachers of the subject have created a list of 261 exercises covering 55 possible misconceptions. For each exercise, we know the time required to present it, the time required to solve it in an exam, and the misconceptions covered by it. Given this information, our algorithms are used to both design exams and select a list of exercises to be presented in the classroom to introduce a given topic (or to summarize the main topics of the subject during the last lessons). Experimental results are reported.

The rest of the paper is structured as follows. In the next section we formally define the problems considered in our work. Then, in Sect. 3 we present algorithms solving the problems and we present the results obtained when using them to deal with the set of exercises of a concrete subject. Finally, in Sect. 4 we present our conclusions and lines for future work.

2   Formal Description of the Problems

In this section we formally present our problems. The first problem under consideration, that is, the problem of selecting the exercises to be taught in the classroom so that they cover all required misconceptions some numbers of times, is introduced next. The most basic version of the problem will be its *decision* version: given the coverage level of each misconception provided by each exercise (i.e. the *capability* of the solution of the exercise to make students with the corresponding misconception realize their view is wrong), the required levels of coverage of all misconceptions, and the maximum time to be spent presenting exercises in the classroom, it asks whether there exists a set of exercises meeting all these requirements. We also consider two associated optimization problems. In the first one, we select the set of exercises meeting the required minimum misconception coverage conditions which also *minimizes* the total time required to introduce the chosen set of exercises. This is the goal of a teacher interested in minimizing the time needed to present a set of problems covering all target errors, in order to maximize the productivity of the teaching time. Alternatively, it could be the case that we have a fix amount of time available to present exercises. In such situation, we do not need to minimize the total required time, but to select a set of problems fitting into the available time. In this case, we may be interested in taking profit of all the available time to cover the misconceptions as much as possible (as long as all misconceptions reach their corresponding minimum required coverage levels). The corresponding problem is the following: for all solutions fitting into the maximum time and reaching all minimum coverage levels for each error, we want to maximize the total coverage *surplus* (i.e. the excess of coverage over the required minimums) of all misconceptions. In this case, the teacher can define the relative contribution of the surplus coverage of each misconception, in such a way that exceeding the required coverage of some misconceptions is more valuable than exceeding the required coverage of others.

Hereafter we consider that, given a vector v, v^T denotes the transpose of v.

Definition 1. Let us consider that we have n questions or problems available $(P_1, \ldots, P_n)$ and m misconceptions to be covered $(M_1, \ldots, M_m)$. Let $c \in \mathbb{N}^n$ be a vector of naturals where each value $c_i \in \mathbb{N}$ denotes the time needed to present and solve problem P_i, $A \in \mathbb{R}^{m \cdot n}$ be a matrix where each value $a_{ij} \in \mathbb{R}$ denotes the *coverage level* of the misconception M_i provided by problem P_j, and $b \in \mathbb{R}^m$ be a vector where each value $b_j \in \mathbb{R}$ denotes the minimum coverage we want to reach for each misconception M_j. Finally, let $w \in \mathbb{R}^m$ be a vector where each $w_j \in \mathbb{R}$ denotes the contribution of the surplus coverage of misconception M_j. Then,

- Given $k \in \mathbb{N}$, the *Misconception Coverage* problem, denoted by MC, is the decision problem of finding out whether there exists $x \in \{0,1\}^n$ such that $c^T \cdot x \leq k$ and $A \cdot x \geq b$.
- The *Misconception Coverage Time Optimization* problem, denoted by MCTO, is the problem of minimizing $c^T \cdot x$ subject to $x \in \{0,1\}^n$ and $A \cdot x \geq b$.
- Given $k \in \mathbb{N}$, the *Misconception Coverage Surplus Optimization* problem, denoted by MCSO, is the problem of maximizing $\sum_i w_i \cdot ((\sum_j a_{ij} \cdot x_j) - b_i)$ subject to $c^T \cdot x \leq k$, $x \in \{0,1\}^n$, and $A \cdot x \geq b$.

$\square$

We have the following properties.

Theorem 1. We have

(a) MC $\in$ NP-complete.
(b) MCTO $\in$ Log-APX-complete.
(c) MCSO $\in$ NPO-complete.

Proof. Let us start proving (b). In order to do it, we have to provide a Log-APX-hardness preserving polynomial reduction from a **Log-APX-hard** problem into MCTO. We consider an S-reduction (see e.g. [4]) from *Minimum Set Cover*, MSC [8]. Given a collection $\mathscr{C}$ of sets $\mathscr{C} = \{S_1, \ldots, S_k\}$ with $\bigcup_{S \in \mathscr{C}} S = \{e_1, \ldots, e_p\}$, MSC consists in picking a subset $\mathscr{C}'$ of $\mathscr{C}$ such that $\bigcup_{S \in \mathscr{C}'} S = \{e_1, \ldots, e_p\}$ in such a way that $|\mathscr{C}'|$ (i.e., the number of sets of $\mathscr{C}'$) is minimized. The construction of an S-reduction from MSC into MCTO can be done as follows. From a MSC instance $\mathscr{C}$, we define an MCTO instance in the same terms as in Definition 1 where

- $n = k$, $m = p$,
- $c = 1^n$, $b = 1^m$,
- For all $1 \leq i \leq m$ and $1 \leq j \leq n$,

$$a_{ij} = \begin{cases} 1 \text{ if } e_i \in S_j \\ 0 \text{ otherwise} \end{cases}$$

We can see that there exists a solution to this MCTO instance with cost $q \in \mathbb{N}$ iff there exists a solution to the original MSC instance with the same cost q. Note that each linear inequality $a_{i1} \cdot x_1 + \ldots + a_{in} \cdot x_m \geq 1$ imposed by the matrix constraint $A \cdot x \geq 1^m$ requires that at least one of the sets containing element e_i is picked (in particular, picking $x_j = 1$ in MCTO represents picking the set S_j in MSC). Since all costs in c are set to 1, the cost of any solution for the MCTO instance is the number of variables x_j which are set to 1, which equals the cost of picking the corresponding sets as a solution for the corresponding MSC instance. Therefore, the optimal solution for the MSC solution has cost $o \in \mathbb{N}$ iff the optimal solution for the MCTO solution has cost o. We conclude that there is an S-reduction from MSC into MCTO, which proves MCTO $\in$ **Log-APX-hard**.

Besides, it is easy to see that problem CIP [11] generalizes MCTO. Since CIP admits a logarithmic approximation, MCTO does too, so MCTO $\in$ **Log-APX**. We conclude that MCTO $\in$ **Log-APX-complete**.

In order to prove (a), we can trivially adapt the previous S-reduction to polynomially reduce *Set Cover* (the decision version of MSC) into MC, which implies the NP-hardness of MC. Since MC $\in$ NP (as we can check $A \cdot x \geq b$ and $c^T \cdot x \leq k$ in polynomial time), we conclude MC $\in$ NP-complete.

In order to prove (c), we S-reduce the NPO-complete problem *Max Ones* [12] (that is, the problem of satisfying all the clauses of a 3-CNF formula while maximizing the number of propositional symbols that are set to $True$) to MCSO as follows. For each disjunctive clause $C_i = y_1^i \vee y_2^i \vee y_3^i$ of the original Max Ones instance (where each y_j^i is a literal, i.e. it is p or $\neg p$ for some propositional symbol p), we create a misconception M_i which represents the clause C_i to be covered.

Moreover, for each propositional symbol p, we also create two misconceptions M_p and M_p', whose coverage will represent that p has been assigned some value ($True$ or $False$) and that p has been assigned a $True$ value in particular, respectively. This completes the set of misconceptions, and problems (questions) are defined as follows. For each propositional symbol p, we create two problems PT_p and PF_p representing that symbol p is set to $True$ or $False$, respectively. Each problem PT_p covers misconceptions M_p, M_p', and all misconceptions M_i such that some literal y_1^i, y_2^i or y_3^i is of the form p, whereas problem PF_p covers the misconception M_p and all misconceptions M_i such that some literal y_1^i, y_2^i or y_3^i is of the form $\neg p$.

The idea of the transformation is that if we choose problem PT_p then the propositional symbol p will be $True$, whereas if we choose problem PF_p then the propositional symbol p will be $False$. Notice that if a problem of the form PT_p or PF_p is chosen and that problem covers the misconception M_j, then clause C_j will be satisfied.

Regarding the required coverage of each misconception in the constructed instance of MCSO (vector b), we require to cover each misconception M_i and M_p at least once (as clauses must be satisfied and proposition symbols must be given some value, respectively), and the extra contribution for being covered more than needed (vector w) will be 0 for both kinds of misconceptions. As we want to maximize the number of propositional symbols set to $True$, for all propositional symbols p we need to reward the problems PT_p. We do it by giving some contribution to covering each misconception M_p', but we do not force to cover them because they are not needed to satisfy the original 3-CNF instance. So, the required coverage for misconceptions M_p' in vector b will be 0, whereas the weight of their contribution for covering them more than needed, defined in vector w, will be 1. Notice that, as M_p' is only covered by problem PT_p (not by PF_p), valuations where propositional symbols p are set to $True$ in the original $Max\ Ones$ instance are preferred.

The cost of all problems will be 1, and the maximum total cost allowed for the whole MCSO instance (i.e. k) will be the number of propositional symbols. Recall that, for each propositional symbol p, we need to cover M_p at least once. Thus, we need to choose at least one of the problems PT_p or PF_p. However, as the maximum total cost allowed is the number of propositional symbols, it is not possible to take at the same time PT_q and PF_q for some propositional symbol q (which would mean that the symbol q is $True$ and $False$ at the same time in the $Max\ Ones$ solution), because in that case the total cost would be greater than the number of propositional symbols. Since every valid solution of MCSO requires that the total cost due to the cost of problems is lower than or equal to the number of propositional symbols, all MCSO solutions denote a correct valuation of the propositional symbols.

Any valid solution of MCSO covers all misconceptions M_i, which means it satisfies all clauses C_i of the original $Max\ Ones$ problem. Moreover, it also provides a correct valuation of all the propositional symbols (that is, a symbol and the negation of the same symbol cannot hold at the same time). Regarding the total value to be maximized in the MCSO problem, it corresponds to the contribution due to the coverage of the M'_p misconceptions. That is, it equals the number of problems PT_p that have been selected, which equals the number of propositional symbols p that are set to $True$ in $Max\ Ones$. Thus, in case we have a solution to MCSO with value x, the corresponding solution to $Max\ Ones$ has exactly the same value x. This means that we have not only a PTAS-reduction from $Max\ Ones$ to MCSO, but also an S, strict, and AP-reduction. So, we conclude that MCSO $\in$ NPO-hard. Moreover, as we can trivially prove MCSO $\in$ NPO (because its constraints can be checked in polynomial time), we conclude that MCSO $\in$ NPO-complete.

The Log-APX and NPO completeness of problems MCTO and MCSO implies, in particular, that both are Log-APX-hard. This means that if P $\neq$ NP then these problems cannot belong to APX, which in turn implies that no polynomial-time algorithm can guarantee that the ratio between the optimal solutions and the solutions found by the algorithm will be bound by some constant in the worst case. Thus, rather than seeking for specific-purpose approximation algorithms for these problems, using generic-purpose algorithms like Genetic Algorithms (which do not guarantee any performance ratio in the worst case but are known to provide good results on average) is a suitable choice in this case.

Next we introduce our second kind of problems, where we want to design an exam (set of exercises) where the total weight of each misconception is exactly that required by the teacher. Again, a decision problem and two optimization problems are defined. In the decision problem, we check whether there exists an exam with the desired coverage of misconceptions which also takes, to be completed, less time than some maximum allowed time. Two related optimization problems are considered too. In the first one, we minimize the time needed to complete the exam. In the second problem, the exact coverage is relaxed, though we wish to minimize the cumulated differences between the expected coverage of each misconception and the coverage reached by the composed exam.

Definition 2. Let us consider that we have n questions or problems available $(P_1, \ldots, P_n)$ and m misconceptions to be covered $(M_1, \ldots, M_m)$. Let $c \in \mathbb{N}^n$ be a vector of naturals where each value $c_i \in \mathbb{N}$ denotes the time needed to deal with problem P_i, $A \in \mathbb{R}^{m \cdot n}$ be a matrix where each value $a_{ij} \in \mathbb{R}$ denotes the coverage of the misconception M_i provided by problem P_j, and $b \in \mathbb{R}^m$ be a vector where each value $b_j \in \mathbb{R}$ denotes the exact coverage we want to reach for each misconception M_j. Then,

- Given $k \in \mathbb{N}$, the *Exact Exam Coverage* problem, denoted by EEC, is the problem of finding out whether there exists $x \in \{0,1\}^n$ such that $c^T \cdot x \leq k$ and $A \cdot x = b$.
- The *Exact Exam Coverage Optimization* problem, denoted by EECO, is the problem of minimizing $c^T \cdot x$ subject to $x \in \{0,1\}^n$ and $A \cdot x = b$.
- Given $k \in \mathbb{N}$, the *Approximate Exam Coverage Optimization* problem, denoted by AECO, is the problem of minimizing $\sum_i |(\sum_j a_{ij} \cdot x_j) - b_i|$ subject to $c^T \cdot x \leq k$ and $x \in \{0,1\}^n$.

$\square$

We have the following properties.

Theorem 2. We have

(a) EEC $\in$ NP-complete.
(b) EECO $\in$ NPO-complete.
(c) AECO $\in$ Log-APX-hard.

Proof. Let us start proving (c). We can AP-reduce *Minimum Set Cover* (defined in the proof of Theorem 1) to AECO as follows.

Given a collection $\mathscr{C}$ of sets $\mathscr{C} = \{S_1, \ldots, S_k\}$ with $\bigcup_{S \in \mathscr{C}} S = \{e_1, \ldots, e_p\}$, we construct an instance of AECO as follows. We consider the elements $\{e_1, \ldots, e_p\}$ as misconceptions and we add an extra misconception e_0 that we do not want to cover (i.e. we want to cover it 0 times). We create a problem for each set S_i that we call $Problem_{S_i}$, and we consider that this problem covers $k^2 + 1$ times each misconception $e_j \in S_i$, whereas it covers only once the misconception e_0. Besides, for each $i \in \{1, \ldots, p\}$ and $j \in \{1, \ldots, k-1\}$ we define an additional problem $Problem_{ij}$ that covers the misconception e_i exactly $k^2 + 1$ times (and it does not cover e_0). Our coverage goals will be covering each misconception e_i with $i \in \{1, \ldots, p\}$ exactly $k \cdot (k^2 + 1)$ times, and covering each misconception e_0 exactly 0 times. The cost of each problem will be 0, so we will not need to define any total cost limit k in the constructed AECO instance (it could be assigned any arbitrary value). There exist two different solutions to the AECO instance:

1. Solutions where the accumulated distance to the desired misconception coverage is at least $k^2 + 1$. In this case, the approximation ratio of this solution in AECO (i.e. the ratio between the value of the solution and the optimal solution) is $(k^2 + 1)/k$ or worse (i.e. bigger). In order to prove it, let us reason about the optimal solution. Note that, if we construct an exam using all problems of type $Problem_{S_i}$, these problems alone will make each misconception be covered at least $k^2 + 1$ times and at most $k(k^2 + 1)$ times (depending on the number of problems covering the misconception). The remaining coverage until reaching the desired $k(k^2 + 1)$ coverage for each misconception can be obtained by selecting as many problems of type $Problem_{ij}$ as needed (notice that, for each misconception, there are $k - 1$ of them, so if problems of the form $Problem_{S_i}$ cover it at least once, then the $k(k^2 + 1)$ desired coverage can be reached by taking the appropriate number of $Problem_{ij}$ problems). Hence, the distance from the optimal exam to the exact desired coverage can only be due to the number of times misconception e_0 is covered, which is a value in $1, \ldots, k$. That is, the cost of the optimal exam is lower than or equal to k. Hence, for a solution whose cost is greater than or equal to $k^2 + 1$, its approximation ratio is greater than or equal to $(k^2 + 1)/k$.
2. Solutions whose accumulated distance to the desired coverage is in the $1, \ldots, k$ range. This situation can only happen if, for each misconception, we have chosen k problems covering it. Since there are $k - 1$ problems of the form $Problem_{ij}$ covering it, at least one of them is of the form $Problem_{S_i}$, so the MSC element represented by this misconception is actually covered.

We conclude that all elements of the original MSC instances are covered by the corresponding AECO solution. Note that the cost of the AECO solution is the number of times e_0 is covered, which equals the number of taken problems of the form $Problem_{S_i}$. In turn, this is the number of sets taken in the MSC instance. Thus, a solution to AECO has value x iff the corresponding solution to MSC has x value. This applies to optimal solutions too. Thus, in this case the approximation ratios of solutions is exactly the same in both problems, MSC and AECO.

The AP-reduction between both problems will map approximation ratios as follows. AP-reductions require that, if the approximation ratio for the second problem is $\leq r$, then the approximation ratio for the first problem is $\leq 1 + \alpha \cdot (r-1)$ for some constant α. Let us use $\alpha = 1$. Then, if the approximation ratio of AECO is $\leq r$, the approximation ratio of MSC must also be $\leq r$. Let us describe the function mapping AECO solutions back into MSC solutions. If we have an AECO solution of type (1) (that is, the deviation from the desired coverage is at least $k^2 + 1$, so the approximation ratio for AECO is greater than or equal to $(k^2 + 1)/k$), then we return the MSC solution which selects *all* subsets S_i (whose approximation ratio in MSC is $\leq k$, as any set cover needs to select at least one subset). Thus, in this case the approximation ratio of the returned MSC solutions is at least as good as the ratio of the original AECO solution, as required by the AP-reduction. On the contrary, when we have a solution of type (2), the approximation ratios are equal in both problems. Thus, we have an AP-reduction in both cases. Thus, AECO $\in$ Log-APX-hard.

In order to prove (a), we can trivially adapt the previous AP-reduction to polynomially reduce *Set Cover* (the decision version of MSC) into EEC, which implies the NP-hardness of EEC. Since EEC $\in$ NP (as we can check $A \cdot x = b$ and $c^T \cdot x \leq k$ in polynomial time), we conclude EEC $\in$ NP-complete.

In order to prove (b), we S-reduce (see e.g. [4]) the NPO-complete problem *Min Ones* [13] (that is, the problem of satisfying all clauses of a 3-CNF formula while *minimizing* the number of propositional symbols which are set to *True*) to EECO as follows. For each clause C_i of the original *Min Ones* instance, we define a misconception M_i. Besides, for each propositional symbol p, we define two problems PT_p and PF_p, representing setting the propositional symbol p to *True* or *False*, respectively. Each problem PT_p covers all misconceptions M_i such that setting $p = True$ satisfies clause C_i, whereas problem PF_p covers all misconceptions M_i such that $p = False$ satisfies C_i.

In addition to that, for each propositional symbol p we create a new misconception MP_p such that both PT_p and PF_p cover misconception MP_p. Finally, for each clause C_i we create two additional problems $PD1_i$ and $PD2_i$ such that they cover misconception M_i. Problems of the form PT_p have cost 1, whereas the rest of the problems have cost 0.

We require to cover each misconception M_i exactly three times, and each misconception MP_p exactly once (so that p and $\neg p$ cannot be true at the same time). If the selected problems of the form PT_p and PF_p satisfy some clause C_i, then the corresponding misconception M_i will be covered once, two times, or three times by these problems (depending on the number of literals of C_i satisfied by the valuation represented by the selected problems of the form PT_p and PF_p). In this case, covering misconception M_i exactly three times (as required in the EECO instance) just requires taking zero, one, or two problems of the form $PD1_i$ and $PD2_i$, whose cost is 0. We conclude that any solution for the EECO instance represents a valuation of the propositional symbols such that all clauses C_i are satisfied, and the goal of EECO is minimizing the number of symbols set to 1 (because only PT_p problems have a cost greater than zero, in particular 1). Thus, the cost of a solution for the EECO instance is exactly the same as the cost of the corresponding solution of the *Min Ones* instance, and this applies to optimal solutions too. Hence, we have an S-reduction between both problems, so EECO ∈ NPO-hard.

Moreover, as we can trivially prove that EECO ∈ NPO (because its constraints can be checked in polynomial time), we conclude that EECO ∈ NPO-complete.

Again, since both optimization problems are Log-APX-hard (note that the hardness in NPO implies it), using Genetic Algorithms rather than tailored specific-purpose algorithms for these problems is a suitable choice.

3 Algorithms

For each of the four optimization problems described in the previous section (MCTO, MCSO, EECO, and AECO) we have implemented a genetic algorithm. The basic representation in all cases is the same. Each individual in the population (i.e., each chromosome) is the bit vector $x = (x_1, \ldots, x_n)$ to be set, where n is the number of questions available in the system. Each bit denotes whether the corresponding question is to be included or not in the solution. That is, a bit vector $(1, 0, 0, 1)$ represents a solution where only the first and the last question are included in the solution. In addition to the representation of chromosomes, we need to deal with the representation of all the information of the problem. In particular, we need to handle a vector representing the time needed to present and to solve each problem (i.e. its cost), a matrix indicating the coverage level of each question for each misconception, and a vector denoting the minimum coverage required for each misconception (in the case of MCTO and MCSO) or the exact desired coverage for each misconception (in the case of EECO and AECO).

The difference among the four problems is the fitness function to be optimized. In the case of MCTO, we minimize the cost of the time vector (i.e., the sum of costs due to all values c_k such that x_k is set to 1) if all constraints hold (i.e., $A \cdot x \geq b$). If the constraints do not hold, then the fitness function returns a value bigger than the sum of costs of all questions. Moreover, in this case the returned value is also proportional to the *distance* to fulfilling all constraints.

In the case of MCSO, we maximize the extra coverage of the questions, that is, the sum of additional coverage for each of the misconceptions

$$\sum_i w_i \cdot ((\sum_j a_{ij} \cdot x_j) - b_i)$$

if all constraints hold (i.e, the minimum coverage is satisfied for each misconception, that is, $A \cdot x \geq b$, and the total time required to deal with all the questions is small enough, i.e. $c^T \cdot x \leq k$). When the constraints do not hold, the fitness function returns a value smaller than the minimum possible value, and proportional to the distance to fulfilling all constraints.

In the case of EECO we minimize the same time cost as in MCTO, but subject to a stronger constraint ($A \cdot x = b$). If the constraints do not hold we proceed in the same way as in MCTO. In the case of AECO the constraints are the same as in MCSO, but we have to *minimize* a function, not to maximize. In particular, the fitness function computes the *distance* to the optimal coverage distribution, that is, we want to minimize $|A \cdot x - b|$. If the constraints do not hold, the fitness function returns a value bigger than the maximum possible distance, and this value is proportional to the distance to fulfilling the constraints.

Note that, given any three variables x_i, x_j, x_k represented at different positions of the chromosome, there is no (a priori) reason to believe that the suitability of the value given to x_i will depend more on the value given to x_j than on the value given to x_k. Consequently, the crossover function just randomly picks which genes (i.e., bits x_i) are inherited from each parent, that is, no specific crossover point is considered.

Regarding other aspects of the algorithm, we use elitist selection, so that the quality of the best solution of each generation increases monotonically. In order to fix the parameters configuration of the algorithm, we have performed a wide range of experiments over a set of randomly generated instances of the problems. After analyzing the results, we have set the population size to 100, the mutation probability to 0.01, and the number of iterations to 1000. In order to avoid over-tunning [2,19,20], we clearly separate this training phase from the evaluation phase. Thus, we have provided such fix Genetic Algorithm configuration to the teachers of the subject Discrete Mathematics, and they have used it to generate both exams and sets of exercises to be presented during the course. Their satisfaction with the tool was very high. First, they acknowledged that the selection of exercises to be presented in the last sessions of the course improved a little bit, as they could maximize the coverage of misconceptions. However, the main improvement they acknowledged appears in the selection of exercises to create exams. Adjusting an exam so that each misconception minors the marks of the students the desired amount of points is a really difficult task. In fact, in many situations it is impossible to do that, and we can only approximate the desired weights for each error. By using our tool, the task of creating exams is simplified. This is particularly useful when several *similar* exams have to be created to evaluate different students of the same subject in different days.

4   Conclusions and Future Work

In this work we have considered the problem of selecting a set of exercises fulfilling different conditions. By doing so, we can use it to select exercises covering a list of misconceptions a given amount of times and minimizing the time needed to present them; or to cover these misconceptions with a given amount of time while maximizing the extra coverage; or to select a set of exercises to create an exam where each misconception minors the marks of the student an exact amount of points (while minimizing the time needed to solve the exam); or to design an exam which can be solved in a given amount of time, while minimizing the distance between the desired mark reductions due to misconceptions and the actual ones.

We have proved the approximability complexity of these problems. Moreover, we have provided genetic algorithms solving them. The usefulness of such algorithms has been proved by applying them in the context of a specific subject of our university. The teachers of this subject (Discrete Mathematics) have used these algorithms for generating both exams and lists of exercises to be presented to students, obtaining satisfactory results.

As future work we plan to perform two independent extensions. First, our aim is to apply our algorithms with other repertories of exercises from other subjects. Second, in order to speedup the computations, we plan to provide a parallel version of our implementations by using the parallel library described in [7].

References

1. Ausiello, G., Paschos, V.Th.: Reductions that preserve approximability. In: Handbook of Approximation Algorithms and Metaheuristics: Methologies and Traditional Applications, vol. 1 (2018)
2. Birattari, M.: Tuning Metaheuristics. A Machine Learning Perspective, vol. 197. Springer, Heidelberg (2009). https://doi.org/10.1007/978-3-642-00483-4
3. Chen, S.-H.: Genetic Algorithms and Genetic Programming in Computational Finance. Springer, Heidelberg (2012)
4. Crescenzi, P.: A short guide to approximation preserving reductions. In: Proceedings of the 12th IEEE Conference on Computational Complexity, pp. 262–273. IEEE (1997)
5. Dasgupta, D., Michalewicz, Z.: Evolutionary Algorithms in Engineering Applications. Springer, Heidelberg (2013)
6. de Jong, K.: Evolutionary Computation: A Unified Approach. MIT Press, Cambridge (2006)
7. de la Encina, A., Hidalgo-Herrero, M., Rabanal, P., Rubio, F.: A parallel skeleton for genetic algorithms. In: Cabestany, J., Rojas, I., Joya, G. (eds.) IWANN 2011. LNCS, vol. 6692, pp. 388–395. Springer, Heidelberg (2011). https://doi.org/10.1007/978-3-642-21498-1_49
8. Feige, U.: A threshold of ln n for approximating set cover. J. ACM (JACM) 45(4), 634–652 (1998)
9. Hidalgo-Herrero, M., Rodríguez, I., Rubio, F.: Testing learning strategies. In: Fourth IEEE Conference on Cognitive Informatics 2005, (ICCI 2005), pp. 212–221. IEEE (2005)
10. Hidalgo-Herrero, M., Rodríguez, I., Rubio, F.: Comparing learning methods. Int. J. Cogn. Inf. Nat. Intell. (IJCINI) 3(3), 12–26 (2009)
11. Kolliopoulos, S.G., Young, N.E.: Approximation algorithms for covering/packing integer programs. J. Comput. Syst. Sci. 71, 495–505 (2005)

12. Kratsch, S., Marx, D., Wahlström, M.: Parameterized complexity and kernelizability of max ones and exact ones problems. In: Hliněný, P., Kučera, A. (eds.) MFCS 2010. LNCS, vol. 6281, pp. 489–500. Springer, Heidelberg (2010). https://doi.org/10.1007/978-3-642-15155-2_43

13. Kratsch, S., Wahlström, M.: Preprocessing of min ones problems: a dichotomy. In: Abramsky, S., Gavoille, C., Kirchner, C., Meyer auf der Heide, F., Spirakis, P.G. (eds.) ICALP 2010. LNCS, vol. 6198, pp. 653–665. Springer, Heidelberg (2010). https://doi.org/10.1007/978-3-642-14165-2_55

14. Man, K.F., Tang, K.S., Kwong, S.: Genetic Algorithms for Control and Signal Processing. Springer, Heidelberg (2012)

15. Mitchell, M.: An Introduction to Genetic Algorithms. MIT Press, Cambridge (1998)

16. Oreski, S., Oreski, G.: Genetic algorithm-based heuristic for feature selection in credit risk assessment. Expert Syst. Appl. 41(4), 2052–2064 (2014)

17. Pal, S.K., Wang, P.P.: Genetic Algorithms for Pattern Recognition. CRC Press, Boca Raton (2017)

18. Paschos, V.Th.: An overview on polynomial approximation of NP-hard problems. Yugoslav J. Oper. Res. 19(1), 3–40 (2009)

19. Rabanal, P., Rodríguez, I., Rubio, F.: On the uselessness of finite benchmarks to assess evolutionary and Swarm methods. In: Proceedings of the Companion Genetic and Evolutionary Computation Conference, GECCO 2015, pp. 1461–1462. ACM (2015)

20. Rabanal, P., Rodríguez, I., Rubio, F.: Assessing metaheuristics by means of random benchmarks. Procedia Comput. Sci. 80, 289–300 (2016)

21. Rodríguez, I., Rabanal, P., Rubio, F.: How to make a best-seller: optimal product design problems. Appl. Soft Comput. 55(C), 178–196 (2017)

22. Rodríguez, I., Rubio, F., Rabanal, P.: Automatic media planning: optimal advertisement placement problems. In: 2016 IEEE Congress on Evolutionary Computation (CEC), pp. 5170–5177. IEEE (2016)

23. Sastry, K., Goldberg, D.E., Kendall, G.: Genetic algorithms. In: Burke, E., Kendall, G. (eds.) Search Methodologies, pp. 93–117. Springer, Heidelberg (2014). https://doi.org/10.1007/978-1-4614-6940-7_4

Unsupervised Learning Bee Swarm Optimization Metaheuristic

Souhila Sadeg[1]([⊠]) [iD], Leila Hamdad[2], Mouloud Haouas[1],
Kouider Abderrahmane[1], Karima Benatchba[1], and Zineb Habbas[3]

[1] Ecole Nationale Supérieure d'Informatique, LMCS, Oued Smar, Algiers, Algeria
{s_sadeg,bm_haouas,bk_abderrahmane,k_benatchba}@esi.dz
[2] Ecole Nationale Supérieure d'Informatique, LCSI, Oued Smar, Algiers, Algeria
l_hamdad@esi.dz
[3] Université de Lorraine - LORIA, Metz, France
zineb.Habbas@univ-lorraine.fr

Abstract. In this work, we investigate the use of unsupervised data mining techniques to speed up Bee Swarm Optimization metaheuristic (*BSO*). Knowledge is extracted dynamically during the search process in order to reduce the number of candidate solutions to be evaluated. One approach uses *clustering* (for grouping similar solutions) and evaluates only clusters centers considered as representatives. The second uses *Frequent itemset mining* for guiding the search process to promising solutions. The proposed hybrid algorithms are tested on MaxSAT instances and results show that a significant reduction in time execution can be obtained for large instances while maintaining equivalent quality compared to the original *BSO*.

Keywords: Combinatorial optimization · Metaheuristics ·
Unsupervised learning · BSO · MaxSAT

1 Introduction

In the last decade, increasing interest in the use of data mining and machine learning techniques to enhance metaheuristics' performances has been observed. These combinations have provided very efficient hybrid algorithms that outperformed original ones for many optimization problems methods. Interesting surveys have been published in the literature and propose taxonomies following certain criteria such as the kind of knowledge extraction (online or offline), the aim of hybridization (improving the quality or speeding up the search process), and the part of the metaheuristic that will benefit from the acquired knowledge (initialization, evaluation, population management...etc.) [2,3,15].

There are several approaches to speed up metaheuristics by using data mining techniques. The first one consists in reducing the objective function computation cost by building approximate models of the function. Another way is to divide the set of candidate solutions using clustering techniques and determine for each

© Springer Nature Switzerland AG 2019
I. Rojas et al. (Eds.): IWANN 2019, LNCS 11507, pp. 773–784, 2019.
https://doi.org/10.1007/978-3-030-20518-8_64

cluster a reference individual called representative. Only representatives are evaluated. Fitness of other solutions is approximated according to their associated representative. A good survey on fitness approximation has been proposed in [6]. Metaheuristics can also be sped up through the reduction of the search space size by exploiting only candidate solutions that has interesting properties. A good example of this approach is DM-GRASP [11,13] which is an hybridization of GRASP metaheuristic and data mining techniques. Patterns are extracted periodically from an elite set of solutions in order to guide the search, during the following iterations, and lead to a more effective exploration of the search space. The resulting hybrid method gave promising results both in terms of solution quality and execution time. It has been applied to several optimization problems such as set packing problem [11], maximum diversity problem [13] and p-median problem [10]. A survey of the applications of DM-GRASP can be found in [14].

In this work we investigate the use of data mining techniques to speed up BSO metaheuristic [4] which is a nature inspired local search based metaheuristic that offers a good balance between intensification and diversification. However, the analysis of the numerical results in previous works [4,12] showed that more than 70% of execution time is spent on evaluating candidate solutions. In order to speed up the search process, we propose two different approaches that reduce the number of solutions to be evaluated by extracting knowledge during the search process [1,5,8,9]. The first one consists of grouping candidate solutions into clusters and evaluating only the centers considered as their representatives while the second approach extracts patterns from the best solutions found after some iterations of *BSO* in order to guide the search in the following iterations. In this hybridization, we use *clustering* and *frequent itemset mining* which are unsupervised data mining techniques since they do not need labeled data to extract knowledge.

Hybrid algorithms provided by these combinations, named *CLS-BSO* and *FIM-BSO*, are used for solving MaxSAT which is an optimization variant of the well-known satisfiability (SAT) problem.

In the rest of the paper, we first describe in Sect. 2 the original version of BSO applied to MaxSAT, then we present in Sect. 3 our proposed hybrid algorithms. Section 4 discusses the experimental results and presents a statistical study used to check the experimental observations.

2   BSO for MaxSAT Problem

A SAT instance is defined by a set of n boolean variables $X = \{x_1, x_2, ..., x_n\}$, a Conjunctive Normal Formula (CNF) F of a set of m clauses $C = \{c_1, c_2, ..., c_m\}$ where each clause is a disjunction of literals, each literal being a variable from X or its negation. The objective of MaxSAT is to find an assignment that maximizes the number of satisfied clauses. BSO is an iterative local search process based on a population of artificial bees cooperating to solve an instance of an optimization problem [4] following three key steps inspired by foraging behavior of natural bees: after determining a reference solution *RefSol*, the first step consists in generating a set of candidate solutions called SearchRegion and assigning them to the bees in order to perform a local search. At the end of this second step, a new *RefSol* is selected among the best solutions stored by the bees in the *Dance* table. To ensure a good balance between diversification and intensification, a maximum number of chances is granted to each SearchRegion and reference solutions are stored in a *Tabu* list to avoid cycles. General BSO algorithm is described bellow.

Algorithm 1. BSO General algorithm

1: Generate the first reference solution *RefSol* randomly or via a heuristic
2: **While** non stopping criterion **Do**
3: - Determine *SearchRegion* from *RefSol*
4: - Assign a solution from *SearchRegion* to each bee
5: **For** each bee k
6: - Perform a local search
7: - return the best found solution
8: **EndFor**
9: Choose the new reference solution
10: **EndWhile**
11: Return the best solution found

Applying BSO to MaxSAT requires adapting its different steps to the specificities of the problem.

- *SearchRegion.* It is represented by a set of N solutions (N being the number of bees in the swarm) obtained by inverting $1/flip$ variables in *RefSol*. The obtained solutions will be equidistant from *RefSol*. To ensure that they are as distinct as possible, two strategies are used to generate them. In the first one, the kth solution is generated by flipping the variables separated by $flip$ bits starting at the kth variable. Let $n = 20$ and $flip = 5$. If the variables are indexed from 0 to 19 then as illustrated in Fig. 2, solutions s_0, s_1, s_2, s_3 and s_4 are obtained respectively by flipping the following bits: (0, 5, 10, 15), (1, 6, 11, 16), (2, 7, 12, 17), (3, 8, 13, 18) and (4, 9, 14, 19). In the second strategy, the kth solution is obtained by flipping $n/flip$ contiguous bits starting at the kth bit. Following the previous example the solutions s_0, s_1, s_2, s_3 are obtained respectively by flipping the following bits: (0, 1, 2, 3), (4, 5, 6, 7), (8, 9, 10, 11), (12, 13, 14, 15) and (16, 17, 18, 19)
- *Bee search process*: It is a simple local search that starts from the assigned solution and returns the best neighbor.
- *Selection of the reference solution*: At the end of each iteration, a new reference solution is selected among those returned by bees and stored in *Dance*. For a good balance between intensification and diversification, the best returned solution called *BestSol* is firstly compared to *BestGlobalSol* which is the best solution obtained since the begining of the search process. If *BestGlobalSol* is improved, intensification is performed and *BestSol* is

selected to be the next *RefSol*. If *BestSol* is worse than *BestGlobalSol* then the number of chances initialized to *MaxChances* is decremented. After consuming all granted chances to the current search region, the diversification process is launched. It consists in choosing from *Dance* table the solution that is the furthest from solutions stored in *Tabu*. In case all solutions in *Dance* are in *Tabu*, *RefSol* is generated randomly.

3   Unsupervised Learning BSO for MaxSAT

In this section, we propose two hybrid versions of BSO that integrate unsupervised learning techniques in order to reduce the number of solutions to be evaluated: *CLS-BSO* that uses clustering and *FIM-BSO* that uses frequent item set mining.

3.1   CLS-BSO Algorithm

clustering is a descriptive task of data mining that consists in partitioning the data into homogeneous clusters using similarity measures. We integrate it here in the local search performed by the bees in order to group candidate solutions in clusters and evaluate only the centers which are considered as representative. In our context, similarity between solutions is measured by the similarity of boolean assignments to variables. Several clustering algorithms exist following different approaches. In our proposed hybrid algorithm *CLS-BSO*, we chose to experiment *k-means* [7], the most popular one.

Algorithm 2. CLS-BSO

Require: *nbClusters: number of clusters* ≥ 1
Begin
  for Each iteration of BSO **do**
    -The clustering algorithm k-means is applied to the set of candidate solutions of
    each bee, groups them into clusters and returns their clusters's centers.
    - Each bee evaluates the obtained centers and stores the best one in *Dance*.
  end for
End

3.2   FIM-BSO Algorithm

The frequent item set mining problem can be defined as follows: If D is a set of transactions where each transaction is a subset from the set of items $I = \{i_1, i_2, ..., i_n\}$, a frequent itemset f, with support s, is a subset of I which occurs in at least $s\%$ of the transactions in D.

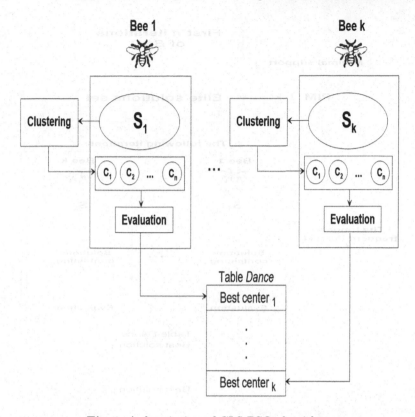

Fig. 1. A description of CLS-BSO algorithm

In the second hybrid algorithm named *FIM-BSO*, the search process is guided in order to evaluate only the most promising solutions. The basic idea consists in analyzing the best solutions found after some iterations of BSO in order to extract their common characteristics (patterns). During the next iterations only solutions presenting these patterns are evaluated.

Patterns are extracted using FIM technique on an elite set of solutions containing the best solutions evaluated in the previous n iterations (n being an empirical parameter). In the context of MaxSAT, the patterns are sets of variables assignments that frequently appear in the best solutions evaluated previously. Among the frequent itemsets, the longest is selected as the pattern to look for in the future candidate solutions, and only solutions containing this itemset are evaluated.

4   Experimental Results

In this section, we present the numerical results obtained by *BSO* and the hybrid versions *CLS-BSO* and *FIM-BSO* on 79 MaxSAT instances from random, crafted

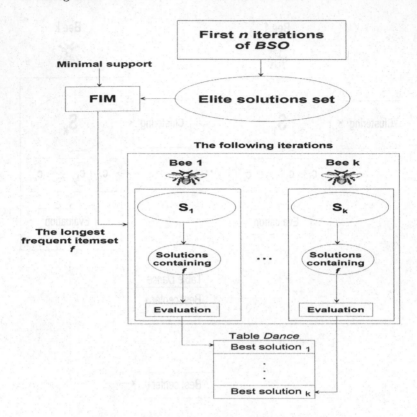

Fig. 2. A description of FIM-BSO algorithm

and industrial categories of the MaxSAT evaluation 2016[1]. Algorithms have been implemented in Java, Weka[2] has been used to perform data mining tasks and experiments have been performed on Intel(R) Xeon(R) CPU E5-2650 2.00 GHz.

The values of algorithms' parameters used in our experiments presented in Table 1. They have been adjusted, for random and crafted instances, using a tool for optimizing algorithms parameters named *SMAC* (Sequential Model-based Algorithm Configuration)[3]. Concerning the industrial instances, a manual tuning has been performed to find the values that allow to obtain the best compromise between solution quality and execution time.

For the sake of clarity, in the remainder of this section, instances are split into two groups following their size. The first group contains 49 small and medium size instances and the second one contains 30 large instances.

---

[1] www.maxsat.udl.cat.

[2] http://www.cs.waikato.ac.nz/ml/weka/.

[3] http://www.cs.ubc.ca/labs/beta/Projects/SMAC/.

Algorithm 3. FIM-BSO

Require: $1 \leq n < maxIterations$, $minimalSupport$ (%)

Begin

Execute n iterations of BSO and save all the solutions stored in *Dance*

Extract the set of frequent itemsets

Select the longest itemset I

for the remaining iterations **do**

Each bee evaluates only solutions containing I

end for

End

Table 1. Algorithms' parameters values

Algorithm	Parameter	Random	Crafted	Industrial
BSO	flip	10	16	2
	maxChances	5	3	2
	maxIter	34	36	2
	nbees	20	15	2
	maxRL	10	2	2
CLS-BSO	nbClusters	5	5	5
FIM-BSO	nbIterations	1	1	1
	minimalSuppport	0.9	0.9	0.9

4.1   Results in Terms of Solution Quality

The obtained results in terms of solution quality are presented in Fig. 3 for small and medium instances while Fig. 4 shows the results for large instances. We can see that CLS-BSO is outperformed by both BSO and FIM-BSO for all instances and more specifically for large instances. Indeed, we see in Fig. 4 that BSO and FIM-BSO give very good results with almost the same average at around 0.05% whereas CLS-BSO is less performing with an average error between 0.2 and 0.25. This can be explained by the fact that 5 clusters is probably a too small number for the large instances. Indeed, choosing the appropriate number of clusters is sensitive and crucial. If too few clusters are considered, the cluster centers are less representative because of the large number of solutions contained in each cluster. In this case, evaluating only the centers narrows the search space exploration which might result in missing promising solutions.

4.2   Results in Terms of Execution Time

In terms of execution time, the average results illustrated in Figs. 5 and 6 show clearly that hybrid algorithms that CLS-BSO is generally the fastest algorithm with an overall gain of time equal to 79% for small and medium instances, and equal to 62.5% for large instances. On the other hand, we can notice that FIM-BSO is more interesting for large instances for which execution time is

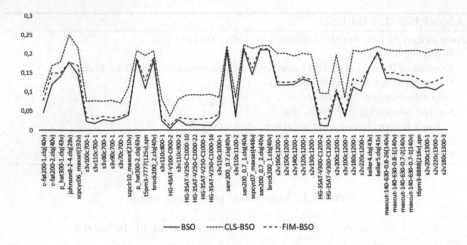

Fig. 3. Average solution quality for small and medium instances

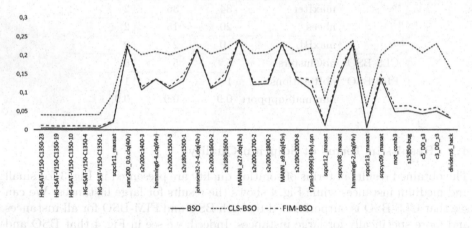

Fig. 4. Average solution quality for large instances

reduced by 78% instead of 13% for small and medium size instances. Besides, it outperforms CLS-BSO for largest instances as shown in Fig. 6 which shows that using clustering increases the computational costs for large instances even with a small number of clusters.

5   Statistical Study

In order to check if the observations concluded from experimental results are statistically significant, we perform Kruskall Wallis and Wilcoxon statistical comparison tests on both solution quality and execution time results. This is done separately on small and medium instances and on large ones.

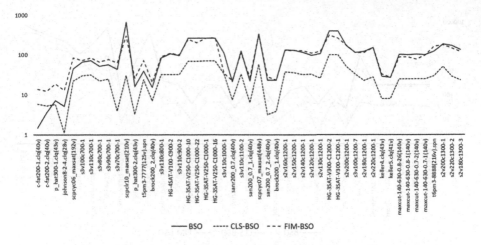

Fig. 5. Average execution time for small and medium instances

5.1 Comparison on Small and Medium Instances

The comparison test on quality gives a $p - value = 3.308e - 05$ which means that differences between the three algorithms on solution quality are significant. Therefore, a pairwise comparison is performed that shows that $CLS\text{-}BSO$ results are statistically different from those of BSO and $FIM\text{-}BSO$ with $p - value = 0.00020$ and $p - value = 0.00037$ respectively, while BSO and $FIM\text{-}BSO$ are statistically equivalent with a $p - value = 0.10221$.

Concerning execution time, the test shows also a significant difference between algorithms with $p - value = 4.314e - 09$. Pairwise test concludes that $FIM\text{-}BSO$ and BSO are equivalent with $p - value = 0.94$. On the other hand, a significant difference is shown between $CLS\text{-}BSO$ and $FIM\text{-}BSO$ with $p-value = 5.8e-09$ and between $CLS\text{-}BSO$ and BSO with $p-value = 3.3e-07$. This comparisons on time and quality are confirmed by the boxplots in Fig. 7.

5.2 Comparison on Large Instances

The comparison test on quality gives a $p - value = 0.00053$ which means that differences on solution quality are significant. Hence, a pairwise comparison is performed that shows that with $p - value = 0.00097$, $CLS\text{-}BSO$ results are statistically different from those of BSO and $FIM\text{-}BSO$. On the other hand, the latter two are statistically equivalent with a $p - value = 0.39305$ such as for small and medium size instances.

Concerning execution time, the test shows also a significant difference between algorithms with $p - value = 0.02423$. Pairwise test shows a significant difference is shown between BSO and $FIM\text{-}BSO$ with $p - value = 0.034$ which validates that considerable time reduction is achieved on large instances by using *frequent itemset mining* as illustrated in Fig. 8.

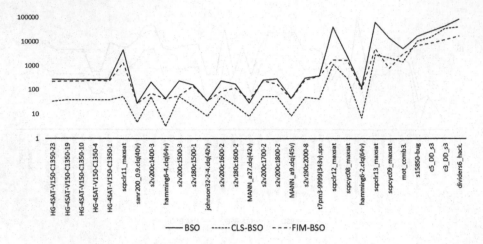

Fig. 6. Average execution time for large instances

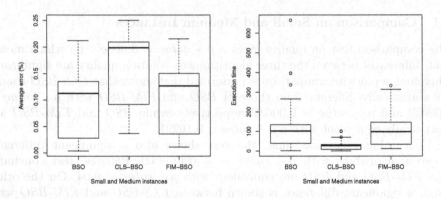

Fig. 7. Results in terms of quality and execution times for small and medium instances

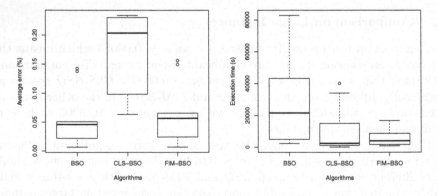

Fig. 8. Results in terms of quality and execution times for large instances

6 Conclusion

In this work, we proposed to speed up the *BSO* metaheuristic for solving the MaxSAT problem by reducing the number of solutions to be evaluated. To this purpose, two unsupervised learning techniques are integrated to the local search performed by the bees which provided two hybrid algorithms. The first one named *CLS-BSO* integrates clustering to local search in order to group similar solutions. Assuming that solutions in the same cluster have approximately same quality, only the centers of the clusters are evaluated which reduces considerably the number os solutions to evaluate. The second proposed hybrid algorithm named *FIM-BSO*, uses frequent itemset mining to extract the longest frequent itemset from the best solutions found in the first iterations of BSO. In the following iterations, only solutions containing this itemset are considered as promising ones and are consequently evaluated. Experimental results showed that *CLS-BSO* reduces considerably the execution time when it is applied to solve small instances but becomes more time consuming for large instances. Furthermore, it is difficult to determine a priori the right number of clusters that gives the good balance between solution quality and execution time. Indeed, a too small number narrows search space exploration by loosing promising candidate solutions while a too big number increases considerably the execution time. The results showed also good performances for *FIM-BSO* that gave very good results statistically equivalent to those of the original BSO. Moreover, it significantly reduces execution time especially for large instances. In our future works, we will perform additional tests to tune *CLS-BSO* and *FIM-BSO* and study more precisely the effect of the parameters' values on their performances. We also aim to investigate the use of other data mining and machine learning techniques to enhance *BSO* performances.

References

1. Adibi, M.A., Shahrabi, J.: A clustering-based modified variable neighborhood search algorithm for a dynamic job shop scheduling problem. Int. J. Adv. Manuf. Technol. **70**(9–12), 1955–1961 (2014)
2. Battiti, R., Brunato, M.: Reactive search optimization: learning while optimizing. In: Gendreau, M., Potvin, J.Y. (eds.) Handbook of Metaheuristics, vol. 146, pp. 543–571. Springer, Boston (2010). https://doi.org/10.1007/978-1-4419-1665-5_18
3. Calvet, L., de Armas, J., Masip, D., Juan, A.A.: Learnheuristics: hybridizing metaheuristics with machine learning for optimization with dynamic inputs. Open Math. **15**(1), 261–280 (2017)
4. Drias, H., Sadeg, S., Yahi, S.: Cooperative bees swarm for solving the maximum weighted satisfiability problem. In: Cabestany, J., Prieto, A., Sandoval, F. (eds.) IWANN 2005. LNCS, vol. 3512, pp. 318–325. Springer, Heidelberg (2005). https://doi.org/10.1007/11494669_39
5. Farag, M.A., El-Shorbagy, M., El-Desoky, I., El-Sawy, A., Mousa, A.: Genetic algorithm based on k-means-clustering technique for multi-objective resource allocation problems. Br. J. Appl. Sci. Technol. **8**(1), 80–96 (2015)

6. Jin, Y.: A comprehensive survey of fitness approximation in evolutionary computation. Soft Comput. **9**(1), 3–12 (2005)
7. MacQueen, J., et al.: Some methods for classification and analysis of multivariate observations. In: Proceedings of the Fifth Berkeley Symposium on Mathematical Statistics and Probability, Oakland, CA, USA, vol. 1, pp. 281–297 (1967)
8. Martins, S.D.L., Rosseti, I., Plastino, A.: Data mining in stochastic local search. In: Marti, R., Panos, P., Resende, M. (eds.) Handbook of Heuristics. Springer, Cham (2016)
9. Plastino, A., Barbalho, H., Santos, L.F.M., Fuchshuber, R., Martins, S.L.: Adaptive and multi-mining versions of the DM-GRASP hybrid metaheuristic. J. Heuristics **20**(1), 39–74 (2014)
10. Plastino, A., Fuchshuber, R., Martins, S.D.L., Freitas, A.A., Salhi, S.: A hybrid data mining metaheuristic for the p-median problem. Stat. Anal. Data Min. ASA Data Sci. J. **4**(3), 313–335 (2011)
11. Ribeiro, M.H., Plastino, A., Martins, S.L.: Hybridization of grasp metaheuristic with data mining techniques. J. Math. Model. Algorithms **5**(1), 23–41 (2006)
12. Sadeg, S., Hamdad, L., Benatchba, K., Habbas, Z.: BSO-FS: bee swarm optimization for feature selection in classification. In: Rojas, I., Joya, G., Catala, A. (eds.) IWANN 2015. LNCS, vol. 9094, pp. 387–399. Springer, Cham (2015). https://doi.org/10.1007/978-3-319-19258-1_33
13. Santos, L.F., Ribeiro, M.H., Plastino, A., Martins, S.L.: A hybrid GRASP with data mining for the maximum diversity problem. In: Blesa, M.J., Blum, C., Roli, A., Sampels, M. (eds.) HM 2005. LNCS, vol. 3636, pp. 116–127. Springer, Heidelberg (2005). https://doi.org/10.1007/11546245_11
14. Santos, L.F., Martins, S.L., Plastino, A.: Applications of the DM-GRASP heuristic: a survey. Int. Trans. Oper. Res. **15**(4), 387–416 (2008)
15. Talbi, E.G.: Combining metaheuristics with mathematical programming, constraint programming and machine learning. Ann. Oper. Res. **240**(1), 171–215 (2016)

QBSO-FS: A Reinforcement Learning Based Bee Swarm Optimization Metaheuristic for Feature Selection

Souhila Sadeg[1]([envelope]) [ORCID], Leila Hamdad[2], Amine Riad Remache[1],
Mehdi Nedjmeddine Karech[1], Karima Benatchba[1], and Zineb Habbas[3]

[1] Ecole nationale Supérieure d'Informatique, LMCS, Oued Smar, Algiers, Algeria
{s_sadeg,em_karech,ea_remache,k_benatchba}@esi.dz
[2] Ecole nationale Supérieure d'Informatique, LCSI, Oued Smar, Algiers, Algeria
l_hamdad@esi.dz
[3] Université de Lorraine - LORIA, Metz, France

Abstract. Feature selection is often used before a data mining or a machine learning task in order to build more accurate models. It is considered as a hard optimization problem and metaheuristics give very satisfactory results for such problems. In this work, we propose a hybrid metaheuristic that integrates a reinforcement learning algorithm with Bee Swarm Optimization metaheuristic (BSO) for solving feature selection problem. QBSO-FS follows the wrapper approach. It uses a hybrid version of BSO with Q-learning for generating feature subsets and a classifier to evaluate them. The goal of using Q-learning is to benefit from the advantage of reinforcement learning to make the search process more adaptive and more efficient. The performances of QBSO-FS are evaluated on 20 well-known datasets and the results are compared with those of original BSO and other recently published methods. The results show that QBO-FS outperforms BSO-FS for large instances and gives very satisfactory results compared to recently published algorithms.

Keywords: Feature selection · Hybrid metaheuristic ·
Bee swarm optimization · Reinforcement learning · Q-learning

1 Introduction

Feature selection (FS) is a widely used pre-processing task that improves the performances of data mining and machine learning techniques such as classification. It consists in selecting the most relevant attributes among a set in order to reduce the dimensionality of the datasets while retaining sufficient information to perform classification. Indeed, irrelevant or redundant features can lead to accuracy decrease and to an unnecessary increase of computational cost [3].

According to [6], a feature selection process mainly consists of four steps: subset generation, subset evaluation, stopping criterion and result validation. The generation procedure is a search procedure that generates subsets of features.

© Springer Nature Switzerland AG 2019
I. Rojas et al. (Eds.): IWANN 2019, LNCS 11507, pp. 785–796, 2019.
https://doi.org/10.1007/978-3-030-20518-8_65

Each of them is evaluated according to an evaluation function in order to save the best one. This procedure is repeated iteratively until a stopping criterion is reached. It can be a predefined number of features, a predefined number of iterations, or some evaluation function's value.

According to how and when the relevance of the selected features is evaluated, a feature selection algorithm generally falls in one of the two main following approaches: filter and wrapper [15]. In the filter approach, the algorithm evaluates the features on a criterion without involving any learning making them fast and more appropriate for big datasets. On the other hand, wrapper approach searches through the feature subset space and uses a learning algorithm to evaluate a feature subset. Algorithms based on wrapper approach are computationally more expensive because measuring classification accuracy requires more resources than simply measuring an information gain for example. However, wrappers generally achieve better results compared to filters and are widely used to find feature subsets that fit a given classifier [16].

Feature selection problem is challenging because of the size of the search space which increases exponentially with respect to the cardinality of the original set N. Indeed, an exhaustive search requires to find the best subset among the 2^N candidates which is computationally intractable [6]. Therefore, methods based on heuristics or random search are used to find good solutions in reasonable time following an evaluation function and a stopping criterion.

Various works are proposed for solving FS problem using metaheuristics. Most of them use the genetic algorithm [11,17,18,26]. Metaheuristics based on Swarm Intelligence have also been applied to feature selection such as Ant Colony Optimization (ACO) [12,14], and Particle Swarm Optimization (PSO) [5,25].

Even though metaheuristics are very efficient in solving hard optimization problems, the rising amount of data makes this task more and more difficult. Hence, we observe an increasing interest on helping metaheuristics by combining them with methods and techniques from different fields such as machine learning in order to enhance their performances. These combinations generally provide more efficient hybrid algorithms that outperform original ones [1,4,22]. Reinforcement learning (RL) is among machine learning techniques that can be combined to heuristics or metaheuristics for solving combinatorial optimization problems. The main goal is to make the original algorithms adaptive. In fact, reinforcement learning has many advantages [24]: (1) it does not require a complete model of the underlying problem since learning is performed by gathering experience referred to as trial-error. (2) It is also very suitable to fully decentralized problems because it allows applying independent learning agents. (3) Moreover, it is not computationally expensive because it often uses a single update formula.

In this paper we propose a hybrid metaheuristic that integrates a reinforcement learning algorithm to Bee Swarm Optimization metaheuristic (BSO) for solving feature selection. BSO [10] is a swarm intelligence algorithm inspired from the foraging behavior of natural bees that has been successfully applied to various optimization problems [2,7–9,19]. Its satisfactory performances can be explained by a good balance between intensification and diversification, which leads respectively to a good exploitation and exploration of the search space. In [20], BSO was applied to feature selection and gave very interesting results. Here, we propose to integrate Q-learning algorithm to BSO as a substitute to the simple local search performed by the bees.

The rest of the paper is organized as follows: Sect. 2 presents the general algorithm of BSO and describes its application to feature selection problem. In Sect. 3, we introduce the reinforcement learning and present the Q-learning algorithm. Section 4, describes the proposed hybrid algorithm QBSO-FS. Section 5 is dedicated to the experimental results, numerical results of QBSO-FS are given and its performances are compared to those of BSO-FS and other algorithms in the literature. Finally, we conclude by giving perspectives and future works.

2   BSO-FS: Bee Swarm Optimization for Feature Selection

2.1   BSO Metaheuristic

Bee Swarm Optimization (BSO) metaheuristic is inspired from the interesting foraging behavior of real bees which is characterized by collective intelligence that allows self-organization, self-adaptation to the environment and dynamic tasks assignment. It is an iterative search process based on a population of artificial bees cooperating for solving an instance of an optimization problem by imitating the foraging behaviour described above. It is based on 3 key steps:

- Determination of the search region;
- Bee search process;
- Selection of the reference solution.

At the first iteration, an initial solution is generated randomly or via a heuristic. It is considered as the first reference solution *RefSol* and used to determine a set of N solutions called *SearchRegion*. Each of these solutions are assigned to a bee as a starting point for a local search. The best found solution by each bee is passed on to its congeners via a table named *Dance* from which a new *RefSol* is selected. In order to avoid cycles, reference solutions are stored in a *Tabu* list.

Two parameters play an important role in BSO. The first one named flip is used to determine the set of solutions that defines the search region. These solutions are equidistant from *RefSol* by a distance inversely proportional to flip. Hence, the value of this parameter must be chosen carefully in order to ensure a good coverage of the search space. The second important parameter called *MaxChances* indicates the number of chances granted to a search region before escaping to another one. Its role is to avoid getting stock in local optima. *MaxChances* value is considered while selecting the reference solution and allows to guarantee a good balance between exploitation and exploration by applying judiciously intensification and diversification principles. Every time the current solution is improved, intensification is performed. However, after *MaxChances* iterations, if no improvement is noticed, a diversification is launched. It consists in selecting from *Dance* table the most distant solution from all previous reference solutions stored in the *Tabu* list. The algorithm stops when the optimal solution is found or the maximum number of iterations is reached. The BSO general algorithm is given below.

Algorithm 1. BSO General algorithm

Input: An instance of a combinatorial optimization problem
Ouput: The best solution found
1: *RefSol* an initial solution found randomly or via a heuristic
2: **while** not condition of stop **do**
3:     Insert *RefSol* in *Tabu* list
4:     Determine *SearchRegion* from *RefSol*
5:     Assign a solution from *SearchRegion* to each bee
6:     **for** each bee k **do**
7:         Perform a local search
8:         Store the result in the table *Dance*
9:     **end for**
10:    Choose the new reference solution folowing
11: **end while**
12: **Return** *bestGlobalSol*

2.2   BSO-FS: BSO for Feature Selection

BSO was applied to feature selection by adapting the general algorithm to the specificities of the problem [20].

- **Solution encoding**: A solution is represented by a binary vector of length n, where n is the original number of features. A position of the vector is set to 1 if the corresponding feature is selected and to 0 otherwise.
- **Fitness**: It represents the quality of the solution and is noted $f(s)$. It is represented by the classification accuracy returned by the used classifier and is computed as follows:

$$accuracy = \frac{number\,of\,true\,positive + number\,of\,true\,negative}{total\,population} \quad (1)$$

  Note that if two solutions have the same fitness, the one that uses less features is considered as the best one.
- **Search region**: It is a set of solutions generated from the reference solution by flipping in refSol a number of bits equal to n/flip, flip being an empirical parameter. The size of this set equals the number of bees since each solution will be assigned to one bee as a starting point of its local search. The value of flip has an impact on the performance of the research process because it

determines the distance between refSol and the solutions which define the search region. Indeed, a too small value will favor the exploration instead of exploitation of the search space, while a too high value might lead the algorithm to converge to a local optimum.

3   Q-Learning Algorithm

Q-learning falls in the class of reinforcement learning algorithms which deals with learning by interacting with the environment. it is defined by "a way of programming agents by reward and punishment without needing to specify how the task is to be achieved" [13]. Indeed, the learner (intelligent agent) performs an action based on its state and receives from the environment a reward or a punishment. To maximize the expected sum of rewards, the agent prefers actions performed in the past. More formally, let an environment described by a set of states $S = \{s_1, s_2, ..., s_n\}$ and $A = \{a_1, a_2, ..., a_n\}$ be a set of actions that the agent can select in each state s_i in S. Each moment an action a_t is performed for a state S_t, the agent receives a reward r_t. The role of the agent is to learn a policy $\pi : S \rightarrow A$, which maximizes the value:

$$V^*(s_t) = r_t + \gamma r_{t+1} + \gamma^2 r_{t+2}, ... \tag{2}$$

where $0 \leq \gamma \leq 1$ is a discount parameter. If γ is close to 0, the agent tends to chose the immediate rewards. If it's close to 1, the agent tends to consider the long-term reward.

When learning, the agent faces an important choice: exploitation or exploration?. The question is: when to promote the exploration of the unknown states and actions relatively to the exploitation of early visited states and actions to accumulate more Rewards.

To deal with RL, three approaches exist, MDP (Markov Decision Process) approach, MC (Monte Carlo) approach and TD (Temporel Difference) one [21]. This latter is a fusion between MDP and MC. It computes the value of the current state based on estimates from other states. The most used algorithm in this approach is the Q-learning. It consists to learn the value of Q which is the reward received immediately upon executing action a_t from state s_t, plus the value (discounted by γ) computed using Eq. 2. Q is estimated recursively based on this equation:

$$Q(s, a) = r(s, a) + \gamma \max Q(\delta(s, a), a\prime)$$

$\delta(s, a)$ denotes the state resulting from applying action a to state s.

The one-step Q-learning, is defined by

$$Q(s, a) = (1 - \alpha)Q(s, a) + \alpha(r(s, a) + \gamma \max Q(\delta(s, a), a\prime))$$

where $\alpha \in [0, 1]$ represent a learning rate. The Q-learning algorithm is given bellow:

Algorithm 2. Q-learning

1: **for** each $(s\ ,\ a)$ **do**
2:     initialize the table entry $\widehat{Q}(s,a)$ to zero.
3: **end for**
4: Observe the current state s
5: Do forever
6: Select an action a and execute it
7: Receive immediate reward r
8: Observe the new state $s\prime$
9: Update the table entry for $\widehat{Q}(s,a)$ as follows:

$$\widehat{Q}(s,a) \leftarrow r + \gamma \max \widehat{Q}(s\prime, a').$$

10: $s \leftarrow s\prime$

---

4   QBSO-FS: The Proposed Hybrid Algorithm

In this paper, we propose to integrate Q-learning to BSO in order to allow bees learning during the search process from their own experience. Indeed, in BSO, after a bee is assigned a solution, it performs a classical local search where it evaluates all the solutions in its neighborhood and returns the best one. Thus, the bee never uses its experience during this process. We propose in this work to replace the local search performed by bees by a Q-learning algorithm where each bee is considered as an intelligent agent which gathers experience during its search process and benefits from the experiences of other bees. The states constituting the environment are all the possible solutions feature subsets in the neighborhood of the bee. As mentioned in Sect. 2.2, a solution is a boolean vector that indicates which features belong to the features subset. An action in our context consists in adding or removing a feature from the current subset which consists in flipping a bit in the current solution. The reward r_t associated to a couple (s_t, a_t) is computed by taking into account classification accuracy as the main criterion and the size of the features subset as a second criterion.

Let $A_t = \{a_{t_1}, a_{t_2}, ..., a_{t_n}\}$ be the set of possible actions of current state s_t and s_{t+1} be the next state achieved after selecting an action from A_t. If we note by $Acc(s)$ the classification accuracy obtained using the feature subset represented by s and by $nbFeatures(s)$ the size of this set then, the reward r_t is given by:

$$\begin{cases} r_t \leftarrow Acc(s_{t+1}) \text{ if } Acc(s_t) < Acc(s_{t+1}) \\ r_t \leftarrow Acc(s_{t+1}) - Acc(s_t) \text{ if } Acc(s_t) > Acc(s_{t+1}) \\ r_t \leftarrow \frac{1}{2} * Acc(s_{t+1}) \text{ if } nbFeatures(s_t) > nbFeatures(s_{t+1}) \\ r_t \leftarrow -\frac{1}{2} * Acc(s_{t+1}) \text{ if } nbFeatures(s_t) < nbFeatures(s_{t+1}) \end{cases}$$

In order to reduce the size of the search space, we propose to consider among the actions in A_t only those that maintain the similarities between the current solution s_t and $bestGlobalSol$. As an example, let consider a dataset with four features an let the current $bestGlobalSol$ = (0011). If the current solution s_t= (0101), then its neighborhood is composed of the following solutions: (1101), (0001), (0111) and (0100). Instead of considering all of them, we perform a XOR operation between s_t and $bestGlobalSol$ to obtain the position of values to flip in s_t which gives the subset of neighbors to keep. In our example (0011) XOR (0101) = (0110) which means that only the solutions obtained by flipping the second and third values in s_t will be considered, that is, (0001) and (0111).

5   Experimental Results

In order to evaluate the proposed hybrid algorithm QBSO-FS and compare it to the original one BSO-FS, both have been implemented in Python and 30 runs have been performed on Colaboratory[1] which is a Jupyter[2] notebook environment that offers an Intel(R) Xeon(R) 2.30 GHz and 12 GB of RAM.

5.1   Datasets and Parameters' Values

We used 20 datasets available in UCI Machine Learning Repository[3] that are frequently studied in machine learning. We considered datasets with considerable diversity in characteristics which are the number of features, classes and instances. According to the number of features, we considered small-sized datasets with less than 19 features, medium-sized datasets having between 20 and 40 features, and large sized datasets having more than 50 features. A summary of these datasets is shown in Table 1.

Table 1. Datasets characteristics

Dataset	Features	Instances	Classes	Dataset	Features	Instances	Classes
Iris	4	150	2	Lymphography	18	148	8
Diabets	8	768	2	Vehicule	18	846	4
Glass	9	214	7	Hepatitis	19	155	2
Breastcancer	10	699	2	Spect	22	267	2
Vowel	10	901	15	German	24	1000	2
Heart-C	13	303	5	WDBC	30	569	2
Heart-StatLog	13	270	2	Ionosphere	34	351	2
Wine	13	178	3	Lung-Cancer	56	32	3
Congress	16	435	2	Sonar	60	208	2
Zoo	16	101	7	Movementlibras	90	360	15

Parameters' values have a great impact on algorithms' performances. In our work, a manual tuning was performed to find the optimal values that give the best compromise between classification accuracy and running time. Table 2 shows the values used in our experiments.

---

[1] https://research.google.com/colaboratory/faq.html.
[2] https://jupyter.org/.
[3] http://archive.ics.uci.edu/ml/.

Table 2. Parameters' values

	Parameter	Value
BSO	Flip	5
	maxChance	3
	nBees	10
	maxIterations	10
	LsIterations	10
Q-Learning	α	0.9
	γ	0.1
	ϵ	0.1

5.2   Comparison Between QBSO-FS and BSO-FS

Table 3 presents the results obtained using QBSO-FS and BSO-FS The numerical results given in table show that for all datasets, reduced number of features allow to achieve better classification accuracy than the initial set of features. They also show that BSO-FS and QBSO-FS give generally the same best accuracy except for Ionosphere dataset for which QBSO-FS outprforms BSO-FS. We can also note that QBSO-FS achieves the best accuracies with less features in many datasets. As an example an accuracy equal to (100%) is achieved for Lung-Cancer dataset with only 16 features among 56 against 18 features for BSO-FS. The results show also that both BSO-FS and QBSO-FS are very stable since their average accuracy is equal to the best obtained for almost all datasets.

In terms of time execution, Fig. 1 shows that QBSO-FS and BSO are equivalent in terms of execution time for small and medium size instances while QBSO-FS is faster for larger datasets. Figure 2, illustrates the number of evaluated solutions by both algorithms for each instances. We can easily see that for the large size instances QBSO-FS evaluates less solutions which can be explained by the fact that the increase in the number of previously evaluated solutions allows the search process to benefit from the experience gathered reach good solutions more efficiently.

5.3   Comparison with Other Algorithms

In order to further evaluate the performance of QBSO-FS, we compare our results with 3 recently published works that also use KNN classifier. Table 4 presents the results of QBSO-FS, Particle Swarm Optimization (PSO) [27], Hybrid Binary Bat Enhanced Particle Swarm Optimization (HBBEPSO) algorithm [23] and Genetic Algorithm (GA). We can see that QBSO-FS outperforms the other algorithms for 8 among 10 datasets used in the experiments. Indeed PSO achieved a better performance for Breastcancer (97%) and for Spect dataset (78%), GA achieved the best result for this dataset with (77%) against 76.67% for QBSO-FS. For the remaining datasets (Wine, Sonar, Heart-StatLog, Ionosphere, ZOO, WDBC, Congress, Lymphography) QBSO-FS obtained the best results.

Table 3. Comparison between QBSO-FS and BSO-FS

Dataset	QBSO-FS			BSO-FS		
	Avg.Features	Avg.Acc (%)	Best.Acc	Avg.Features	Avg.Acc (%)	Best.Acc
Iris	2,27	97,33	97,33	3,87	97,33	97,33
Diabets	4	70,65	70,65	4	70,65	70,65
Glass	6	75	75	6,47	75	75
Breastcancer	3,43	96,29	96,29	3,73	96,29	96,29
Vowel	8,4	99,34	99,45	9	99,45	99,45
Heart-C	6	58,71	58,71	4,53	58,59	58,71
Heart-StatLog	5	82,59	82,59	5	82,59	82,59
Wine	6,83	95,56	95,56	6,73	95,56	95,56
Congress	6	97,95	97,95	6	97,95	97,95
Zoo	8,17	100	100	9,47	100	100
Lymphography	9,87	43,31	43,33	10	43,33	43,33
Vehicule	8	74,47	74,47	8	74,47	74,47
Hepatitis	7	72,67	72,67	6,87	71,96	72,67
Spect	16	76,67	76,67	16	76,67	76,67
German	9	74,2	74,2	9	74,2	74,2
WDBC	12,3	94,54	95,61	13	94,55	94,56
Ionosphere	11,27	95,81	96,94	11,5	95,59	96,11
Lung-Cancer	20,97	98,25	100	25,13	99,67	100
Sonar	28	98,22	99,05	26,37	99,04	99,52
Movementlibras	31,87	92,49	93,33	31,53	93,13	93,61

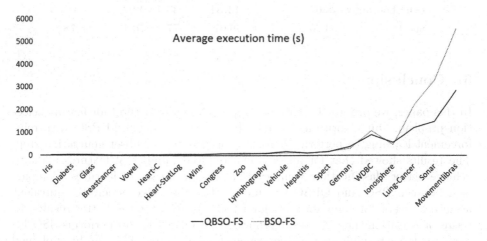

Fig. 1. Comparison in terms of execution time

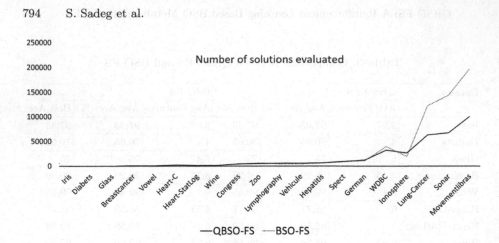

Fig. 2. Comparison in terms of number of evaluated solutions

Table 4. Comparison between QBSO-FS and other algorithms in terms of accuracy

Dataset	Acc. without FS	QBSO-FS	PSO	HBBEPSO	GA
Wine	81.67	**95.56**	95	90.1	93
Sonar	80.95	**98.22**	74	70.4	73
Breastcancer	61.86	96.29	**97**	95.8	96
Heart-StatLog	60.74	**82.59**	78	81.3	82
Ionosphere	86.39	**95.81**	84	83.1	83
Zoo	99.09	**100**	83	87.3	88
WDBC	90.18	**94.54**	94	94.5	94
Congress	94.55	**97.95**	94	94.4	94
Lymphography	26.67	**43.31**	35.4	42.2	-
Spect	63.70	76.67	77	76.4	**78**

6   Conclusion

In this paper, we proposed a wrapper approach based method for feature selection problem. Our algorithm, named QBSO-FS, uses a hybrid BSO with reinforcement learning to perform the search process in its subset generation step. The results showed that for all datasets, reduced number of features allow to achieve better classification accuracy than the initial set of features. Comparison between BSO and QBSO-FS showed that they give generally equivalent accuracy except for some datasets for which QBSO-FS gives better results. In terms of execution time, the results showed that QBSO-FS outperforms BSO-FS for large datasets while they are equivalent for small ones. This can be explained by the adaptive search used by Q-learning algorithm that exploits previous solutions evaluation. Finally, the comparison of the results of QBSO-FS with those of recently published metaheuristics showed that our algorithm gives very satisfactory results and outperforms other algorithms for 8 datasets among 10. In our future works, we plan to apply QBSO-FS for larger datasets and perform a deeper parameters tuning in order to study their impact on the algorithms' performances.

References

1. Battiti, R., Brunato, M.: Reactive search optimization: learning while optimizing. In: Gendreau, M., Potvin, J.Y. (eds.) Handbook of Metaheuristics, pp. 543–571. Springer, Boston (2010). https://doi.org/10.1007/978-1-4419-1665-5_18
2. Belkebir, R., Guessoum, A.: A hybrid BSO-CHI2-SVM approach to Arabic text categorization. In: 2013 ACS International Conference on Computer Systems and Applications (AICCSA), pp. 1–7. IEEE (2013)
3. Blum, A.L., Langley, P.: Selection of relevant features and examples in machine learning. Artif. Intell. **97**(1–2), 245–271 (1997)
4. Calvet, L., de Armas, J., Masip, D., Juan, A.A.: Learnheuristics: hybridizing metaheuristics with machine learning for optimization with dynamic inputs. Open Math. **15**(1), 261–280 (2017)
5. Chuang, L.Y., Tsai, S.W., Yang, C.H.: Improved binary particle swarm optimization using catfish effect for feature selection. Expert Syst. Appl. **38**(10), 12699–12707 (2011)
6. Dash, M., Liu, H.: Feature selection for classification. Intell. Data Anal. **1**(1–4), 131–156 (1997)
7. Djeffal, M., Drias, H.: Multilevel bee swarm optimization for large satisfiability problem instances. In: Yin, H., Tang, K., Gao, Y., Klawonn, F., Lee, M., Weise, T., Li, B., Yao, X. (eds.) IDEAL 2013. LNCS, vol. 8206, pp. 594–602. Springer, Heidelberg (2013). https://doi.org/10.1007/978-3-642-41278-3_72
8. Djenouri, Y., Drias, H., Chemchem, A.: A hybrid bees swarm optimization and tabu search algorithm for association rule mining. In: 2013 World Congress on Nature and Biologically Inspired Computing, pp. 120–125. IEEE (2013)
9. Drias, H., Mosteghanemi, H.: Bees swarm optimization based approach for web information retrieval. In: 2010 IEEE/WIC/ACM International Conference on Web Intelligence and Intelligent Agent Technology, vol. 1, pp. 6–13. IEEE (2010)
10. Drias, H., Sadeg, S., Yahi, S.: Cooperative bees swarm for solving the maximum weighted satisfiability problem. In: Cabestany, J., Prieto, A., Sandoval, F. (eds.) IWANN 2005. LNCS, vol. 3512, pp. 318–325. Springer, Heidelberg (2005). https://doi.org/10.1007/11494669_39
11. Huang, J., Cai, Y., Xu, X.: A hybrid genetic algorithm for feature selection wrapper based on mutual information. Pattern Recogn. Lett. **28**(13), 1825–1844 (2007)
12. Kabir, M.M., Shahjahan, M., Murase, K.: A new hybrid ant colony optimization algorithm for feature selection. Expert Syst. Appl. **39**(3), 3747–3763 (2012)
13. Kaelbling, L.P., Littman, M.L., Moore, A.W.: Reinforcement learning: a survey. J. Artif. Intell. Res. **4**, 237–285 (1996)
14. Ke, L., Feng, Z., Ren, Z.: An efficient ant colony optimization approach to attribute reduction in rough set theory. Pattern Recogn. Lett. **29**(9), 1351–1357 (2008)
15. Liu, H., Motoda, H.: Less is more. In: Computational Methods of Feature Selection, pp. 16–31. Chapman and Hall/CRC (2007)

16. Liu, H., Yu, L.: Toward integrating feature selection algorithms for classification and clustering. IEEE Trans. Knowl. Data Eng. **4**, 491–502 (2005)
17. Oliveira, L.S., Sabourin, R., Bortolozzi, F., Suen, C.Y.: A methodology for feature selection using multiobjective genetic algorithms for handwritten digit string recognition. Int. J. Pattern Recogn. Artif. Intell. **17**(06), 903–929 (2003)
18. Rostami, M., Moradi, P.: A clustering based genetic algorithm for feature selection. In: 2014 6th Conference on Information and Knowledge Technology (IKT), pp. 112–116. IEEE (2014)
19. Sadeg, S., Drias, H.: A selective approach to parallelise bees swarm optimisation metaheuristic: application to max-w-sat. Int. J. Innovative Comput. Appl. **1**(2), 146–158 (2007)
20. Sadeg, S., Hamdad, L., Benatchba, K., Habbas, Z.: BSO-FS: bee swarm optimization for feature selection in classification. In: Rojas, I., Joya, G., Catala, A. (eds.) IWANN 2015. LNCS, vol. 9094, pp. 387–399. Springer, Cham (2015). https://doi.org/10.1007/978-3-319-19258-1_33
21. Sutton, R.S., Barto, A.G., et al.: Introduction to Reinforcement Learning, vol. 135. MIT press, Cambridge (1998)
22. Talbi, E.G.: Combining metaheuristics with mathematical programming, constraint programming and machine learning. Ann. Oper. Res. **240**(1), 171–215 (2016)
23. Tawhid, M.A., Dsouza, K.B.: Hybrid binary bat enhanced particle swarm optimization algorithm for solving feature selection problems. Appl. Comput. Inf. (2018)
24. Wauters, T., Verbeeck, K., De Causmaecker, P., Berghe, G.V.: Boosting metaheuristic search using reinforcement learning. In: Talbi, E.G. (ed.) Hybrid Metaheuristics. SCI, pp. 433–452. Springer, Heidelberg (2013). https://doi.org/10.1007/978-3-642-30671-6_17
25. Xue, B., Zhang, M., Browne, W.N.: Particle swarm optimisation for feature selection in classification: novel initialisation and updating mechanisms. Appl. Soft Comput. **18**, 261–276 (2014)
26. Yang, J., Honavar, V.: Feature subset selection using a genetic algorithm. In: Liu, H., Motoda, H. (eds.) Feature Extraction, Construction and Selection, pp. 117–136. Springer, Boston (1998). https://doi.org/10.1007/978-1-4615-5725-8_8
27. Zawbaa, H.M., Emary, E., Parv, B.: Feature selection based on antlion optimization algorithm. In: 2015 Third World Conference on Complex Systems (WCCS), pp. 1–7. IEEE (2015)

Advances in Computational Intelligence

Device-Free Passive Human Counting with Bluetooth Low Energy Beacons

Maximilian Münch[1](✉) and Frank-Michael Schleif[2]

[1] Department of Computer Science, University of Applied Sciences Würzburg-Schweinfurt, 97074 Würzburg, Germany
`maximilian.muench@fhws.de`
[2] School of Computer Science, University of Birmingham, Edgbaston, B15 2TT Birmingham, UK

Abstract. The increasing availability of wireless networks inside buildings has opened up numerous opportunities for new innovative smart systems. For a lot of these systems, acquisition of context-sensitive information about attendant people has evolved to a key challenge. Especially the position and distribution of attendants significantly influence the system's service quality. To meet this challenge, several types of sensor systems have been presented over the last two decades. Most of these systems rely on an active mobile device that has to be carried by the tracked entity. Contrary to the so-called device-based active systems, device-free passive sensing systems are grounded on the idea of detecting, tracking, and identifying attendant people without carrying any active device or to actively taking part in a localization process. In order to obtain information about the position or the distribution of present people, these systems quantify the impact of the physical attendants on radio-frequency signals. Most of device-free systems rely on the existing WiFi infrastructure and device-based active concepts, but here we want to focus on a different approach. In line with our previous research on presence detection with Bluetooth Low Energy beacons, in this paper, we introduce a strategy of using those beacons for a device-free passive human counting system.

Keywords: Device-free passive sensing · Bluetooth Low Energy · Passive human counting · Regression · Applied machine learning

1 Introduction

Progressively advancements of smart systems, e.g. smart home systems, smart factories, or intelligent energy systems, rely on a continuously increasing amount of context-sensitive information. In the past few years, the expansion of existing technology by intelligent sensors has resulted in a variety of innovative businesses, products, and services that integrate context-sensitive information [27,30]. For many of these systems, the number and position of attendant people are valuable information that can significantly improve the quality of its

© Springer Nature Switzerland AG 2019
I. Rojas et al. (Eds.): IWANN 2019, LNCS 11507, pp. 799–810, 2019.
https://doi.org/10.1007/978-3-030-20518-8_66

services [24]. Especially for intelligent safety systems, getting information about human presence in particular indoor areas can be of decisive importance. In case of a building fire, for example, rescue forces may be led directly into areas with a high number of attendants. Also within evacuation scenarios of larger buildings, such as football arenas, theatres, or shopping malls, these systems may support the rescue. As a result of this, these systems might avoid highly crowded areas and choose a longer but safer route instead.

Various approaches using a wide variety of technologies have been proposed in order to tackle these challenges of detecting and tracking entities in a specific building area [9]. Thereby, the majority of these approaches is based on Wi-Fi as it is already integrated into the infrastructure of most buildings. However, many of these attempts struggled with the fluctuations of Wi-Fi signal strength, which are occasionally unpredictable with a high load last in the wireless network. Therefore, deploying an independent Bluetooth Low Energy (BLE) network became an adequate alternative to Wi-Fi technology [5]. Currently, both Bluetooth and Wi-Fi enjoy widespread applications in indoor positioning systems [9,21,28]. Depending on whether the tracked entity has to carry an active device or not, these systems can be classified into *device-based active* and *device-free passive* sensing systems. Since device-based active (dba) sensing systems require tracked entities to carry a signal transmitting tracking device, device-free passive (DfP) sensing systems have no need for a tracking device or the active participation within the localization or detection process.

In line with our previous research on *device-free passive presence detection* with BLE beacons [12], an extension of this system to a *device-free passive human counting system* is introduced in this article. Although BLE beacons have proven to be an adequate alternative to Wi-Fi, there are only a few systems that integrate BLE [5,13] and almost no systems that combine BLE beacon technology and DfP concept [22].

An overview of previous work in the field of DfP sensing is given in Sect. 2. Section 3 comprises a summary of our sensing system, the study design and some preliminary tests on BLE with various numbers of attendants. Section 4 covers the data evaluation models used in this approach. The results of the experiments are shown in Sect. 5, followed by a summary and an outlook on further research in Sect. 6.

2   Related Work

Location determination has attracted a great deal of attention in the research community since there is no adequate equivalent to GPS for indoor [10]. This resulted in many location determination systems based on other technologies such as ultrasonics [15], infrared [25] and radio frequency [1]. As all of these systems require an active tracking device, they are referred to as device-based active systems. In the early years, dba sensor systems with Wi-Fi technology gained considerable popularity in research. However, nearly all of them had to manage the vast impact of human presence on received signal strength (RSS) [11].

This multipath fading and shadowing effects were regarded exclusively as negative phenomena until they became the basis of the *sensorless sensing* concept in [26]. Thereby, signal strength fluctuations were recognised between transmitters and receivers. These fluctuations were used for detecting the presence of humans. In [29], a similar concept was introduced with device-free passive localization.

Due to the shadowing and absorbing effect of human bodies, the presence of any human obstacle close to the experimental area caused changes in the signature of wireless signals. With this characteristic footprint on RSS using various monitoring points, the position of the attendant person was determined. Overviews of other Wi-Fi based device-free passive sensing systems provide [4,14,28]. The majority of these DfP sensing systems is based on Wi-Fi as it is already part of entirely each building.

Despite the inexpensive acquisition, the high flexibility, and the independence to network traffic, there are hardly any combinations of the dfp concept and BLE beacons. Best to the author's knowledge, the only combination of BLE and dfp is used in a motion detector for a remote elder care support system [22]. In this approach, the authors constructed a human motion detector consisting of B as transmitters and tablets as signal receivers. A fluctuation in RSS pattern enabled caregivers to monitor the behavior of care recipients within their residential rooms.

In our previous research on BLE-based device-free sensing systems, a system prototype of a device-free passive presence detection using BLE beacons was provided [12]. We presented a system for deciding whether a room is occupied or empty with an accuracy of ≈98%. Contrary to this, the focus of this article is not only on determining if, but also to estimate how many people are in a room. Such a system, being able to estimate the number of attendants, is called a *device-free passive human counting* system. Best to the author's knowledge, the presented system prototype is the first approach in estimating the number of attendants with BLE beacons in the context of dfp sensing.

3   BLE-Based DfP Human Counting

In line with our previous research, this system prototype represents the next stage in DfP sensing system taxonomy described in [23]. In general, such systems are grounded on the fact that the signal pattern of RF signals is affected by the presence of people. The physical attendance of a human body close to the supervised area results in fluctuations of the signal strengths. Figure 1 illustrates the main difference between device-based active (a) and device-free passive (b) systems: In (a), the transmitter device is attached closely to the tracked entity. This device permanently transmits signals that are acquired by the receivers. With the signal's reception, the *received signal strength* (RSS) is monitored by the receivers. What we call the RSS is actually a measure of the received signal's power level in decibels. In a multipath environment, a wireless signal takes many different paths on the way from the transmitter to the receiver. Finally, RSS represents all path-specific power levels of a time window as one single value.

Based on these RSS values, the distance to the transmitter can be determined. In (b), the transmitter is fixed stationary such that the signals are always transmitted from the same location. As a result, the signal strengths of an empty room are quite similar and often nearly identical. However, the presence of people in this area absorbs, reflects, or shadows the signal. Therefore, the receiver obtains a signal with reduced signal strength.

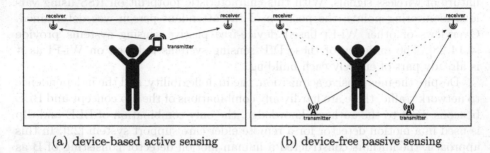

(a) device-based active sensing   (b) device-free passive sensing

Fig. 1. Difference between a device-based active (a) and a device-free passive (b) sensing system. In (a), signal strengths mainly depend on the distance between transmitter (smartphone) and the receiver (antenna). In (b), the signal strengths depend on whether a human is in the signal's line-of-sight.

Knowing the location of the different transmitters, the reduced RSSs can be used, for example, to determine the location or the number of attendants.

Since BLE beacons are used in this system prototype for transmitting the signals, a Bluetooth Low Energy network must be set up. Here, *Bluetooth Low Energy network* refers to a composition of transmitters which are consistently emitting a low energy signal using the Bluetooth protocol. Thereby, the transmitted signal has a lower information content than traditional Bluetooth signals. Such signals may be considered as radio waves from radio stations where all receivers within the range of the transmitters can detect the signal. Typically, a data frame consists of only a few pieces of information including the transmitter's ID, a small information package and a timestamp. This data is enriched by the measured signal intensity, called the Received Signal Strength Indicator (RSSI). Subsequently, such a transmitter is referred to as *BLE beacon* or *beacon*.

In an application scenario, the RSSI is recorded as time series for a given time window. Subsequently, the time series has to be analysed in further steps. Potential applications could be the tracking of people, environment monitoring, localization of entities, presence detection and the counting of crowds within a building. In the following, our system considers the task of estimating the number of attendants referred to as *human counting* system.

3.1   System Composition and Study Design

Common dfp sensing systems consist of three components: signal transmitters, signal receivers, and an application for either storing the data or start further data processing [29].

1. *signal transmittance* is accomplished by beacons. The allocation of transmitters is essential to maximise the impact of attendants on RSS. Preferably, the beacons are arranged in such a way that attendants interrupt or disturb the direct signal line between transmitters and receivers.
2. *signal receivers* are accomplished by USB antennas with built-in Texas Instruments CC2540 chips.
3. *storage application:* after signal reception at the receiver, two scenarios are available: either an application stores the incoming data in a database or the packages are directly processed in the context of streaming [3].[1]

As the following study design has been proven successful in our previous research, we follow the same line in this approach. Using the same environment of our presence detection prototype, this BLE-based dfp human counting is applied in the environment of a lecture room with different room occupancies. The supervised area of this lecture room comprises a total of 120 stationary seats in 8 rows. In order to get a better understanding of the advantages and disadvantages within different transmitter and receiver allocations, ten different test setups were investigated during the test signal recording. For this purpose, we tracked the impact of the different arrangements of transmitters and receivers on the variance of RSS. The best test results occurred in the arrangement with transmitters placed beneath the attendant's seats and the receivers close to the ceiling and laterally to the seating rows. In this context, 'best test results' means the biggest difference between the RSS variances of an empty room and those of an occupied room. Following this, $b = 24$ beacons (3 transmitters per row) were located beneath the attendant's seats in order to ensure a high absorption in an occupied and less absorption in an empty room. In our system, we used $r = 4$ receivers. During the preliminary tests, a reduction of the number of beacons and receivers was also considered. However, due to its complexity, a reduction of beacons and receivers needs to be investigated separately in further research.

3.2   Data Acquisition and Preprocessing

Data Acquisition: To obtain a broad overview of the RSS characteristics at different degrees of occupancy, about $L = 80$ lectures were monitored, whereby each lecture has a length of 90 min. In this time frame, 5 *measurement sessions*

---

[1] In this paper, we followed the line of a batch scenario. When using this approach in the context of streaming, it should be considered that the beacons used here transmitted only once a second. In order to reduce the observed time window while receiving a sufficient amount of RSS values, the transmission frequency of the beacons has to be adjusted.

were recorded in small slots with a length of $T = 300$ s. At first, a window size of $T = 30$ min was taken for this system prototype. As the fluctuations remained the same as long as the external circumstances did not change, the window size was shortened to $T = 300$ s. Using an even smaller time window may result in not enough RSS values arriving in a window to ensure a successful evaluation. Especially in high occupancy scenarios, only a fractional part of the originally transmitted signals arrive at the receiver since attendants absorb those signals with low signal strength. But exactly these signals are essential to estimate areas with high attendance. Loss of these signals leads to a contradictory result of a low attendance estimation but a high number of attendants.

In total, $N = 80 \cdot 5 = 400$ measurement sessions have been recorded, denoted as measurement set $\mathbf{X}$. If attendants have left or entered the supervised area during a measurement session, the respective 300-s recording was removed from further analysis. New measurement sessions were recorded for each lecture until 5 successful measurement sessions were obtained. For each of the 5 measurement sessions, we stored the respective number of attendants as y_i. Each $\mathbf{x}_i \in \mathbf{X}$ is represented in a summarized form by the $r \times b \times T = 28.800$ RSS values, as explained in the preprocessing (see below). Therefore, each $\mathbf{x}_i$ is labelled with the monitored number of attendants $\mathbf{y}_i$.

$\mathbf{X}$ will later be used as the basic data set in a case study for regression to predict the number of attendants within the measurement area. This *raw* data is subsequently denoted by DS1. After the first step of preprocessing to summarize each measurement session (see below), we obtain dataset DS2 or X, in which each $x_i \in X$ is a summary of one measurement session $\mathbf{x}_i \in \mathbf{X}$. Step 2 of the preprocessing integrates spatial relationship between receivers and transmitters. The result of this preprocessing step is denoted by DS3 or $\hat{X}$.

Preprocessing: After recording of the time series within the lectures, DS1 needs to be examined with respect to the signal strength of a beacon over a period of time with various occupancies. For this purpose, the progressions with different degrees of occupancy are depicted in Fig. 2. As illustrated in this figure, both the average RSS and the RSS variances increase significantly along with a higher count of attendants. With a high number of attendants in the measurement area represented by *60 attendants*, the average RSSI of this scenario exhibits a significantly lower value and a substantially higher degree of variance compared to the other two scenarios. In medium occupied rooms (*30 attendants*), the average RSSI is slightly higher and features a smaller variance. The RSSI progression without any attendants (*0 attendants*) occurred almost constant, except in a few cases. In this context, the RSSI mean value and the RSSI fluctuation give some indications whether a room is occupied or empty. Also the range from the maximum and the minimum of the RSS series for the different room occupancies as shown in Fig. 2 indicates a relationship to variance as a relevant feature.

In former work on sensorless sensing [26,29], preprocessing was widely limited to standard signal processing approaches like filtering or smoothing. This still

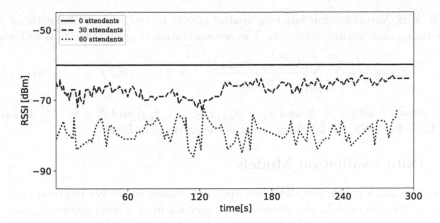

Fig. 2. Time series of the received signal strengths with various room occupancies

keeps rather high dimensional datasets, not well-suited for real applications, e.g., in embedded systems. To reduce the dimensionality of the RSS time series, statistical moments have been calculated to summarise the encoded information. Here, we follow this line and use the 4 features m: mean, variance, trimmed mean and trimmed variance on each measurement session along dimension T. We obtain a four-dimensional vector for each r and b pair leading to a matrix $\mathbf{X}'$ with $\mathbf{X}' \in \mathbb{R}^{N \times r \times b \times 4}$, with $\mathbf{x}' \in \mathbf{X}'$. Trimmed mean and trimmed variance have been calculated using an α-trimmed filter (with $\alpha = 0.05$) following [16], a kind of noise reduction.

The $\mathbf{x}'$ is further summarised by the following equation:

$$x_{i,f} = \frac{1}{r \cdot b} \sum_{k=1}^{b} \sum_{j=1}^{r} (\mathbf{x}')_{k,j}^{i,f} \qquad i = 1, \ldots, N, \ f = 1, \ldots, m \qquad (1)$$

We obtain X with $x_i \in X$ and $x_i = [x_{i,1}, x_{i,2}, x_{i,3}, x_{i,4}]$ and $X \in \mathbb{R}^{N \times 4}$. Further, for each $x_i \in X$ we have a label y_i, identical with $\mathbf{y}_i$. This data subsequently referred to as DS2. The spatial allocation of transmitters and receivers is not taken into account in [16].

As transmitters differ in distances to the receivers, implementing a weighting factor for each beacon according to its beacon-receiver-distance (brd) is useful when aggregating the whole RSS series to one single value per slot. Therefore, we introduce a normalization matrix $W \in \mathbb{R}^{b \times r \times 2}$ with $w_{kjs} \in W$. The normalization terms w_{kjs} are determined as follows: (1) we measure the signal pattern in an empty room with respect to all beacons and receivers. (2) we calculate the minimum and maximum of the respective averaged RSS. (3) minimum and maximum are stored as subsequent scaling values for a Min/Max-Normalization. In the final preprocessing of the real measurements, the slots are normalised using the normalization matrix W such that the brds are taken into account. In the experiments, we also provide results for data with standard

Min/Max-Normalization ignoring spatial effects in brd and results without α-trimming and feature reduction. The normalization is applied to X as follows:

$$\hat{x}_{i,f} = \frac{1}{r \cdot b} \sum_{k=1}^{b} \sum_{j=1}^{r} \frac{(\mathbf{x}'^{i,f}_{k,j} - w_{k,j,1})}{w_{k,j,2}} \qquad i = 1, \ldots, N, \ f = 1, \ldots, m \qquad (2)$$

We obtain $\hat{x}$ with $\hat{x}_i \in X$ and $\hat{x}_i = [\hat{x}_{i,1}, \hat{x}_{i,2}, \hat{x}_{i,3}, \hat{x}_{i,4}]$ and $\hat{X} \in \mathbb{R}^{N \times 4}$, denoted as DS3. For DS3, the labelling remains the same as in DS2.

4   Data Evaluation Models

The relationship between data and response values is always interesting as it can be used to explain the structure or mechanism of a complicated system or to construct a predictive model. This analysis is typically done via a regression analysis, where least squares regression (LSR) is the most popular one.

With kernelization of LSR, a major restriction of regression, where data points and estimated values have to be in linear relation, are relaxed by selecting a more expressive feature transformation or kernel functions. Due to the flexibility and simplicity of the regression model, it is widely applied in various domains such as image processing, bioinformatics and other [17].

The kernel function can be very generic. Most prominent are the linear kernel with $k(x, x') = \langle \phi(x), \phi(x') \rangle$ where $\langle \phi(x), \phi(x') \rangle$ is the Euclidean inner product and ϕ is the identity mapping, or the RBF kernel $k(x, x') = \exp\left(-\frac{||x-x'||^2}{2\sigma^2}\right)$, with $\sigma > 0$ as a free scale parameter [18].

4.1   Regression Models

For this kind of supervised learning problems, we have a set of training data $(x_1, y_1) \ldots (x_N, y_N)$ from which to estimate the parameters β to fit an underlying function. Each x_i is a vector of feature measurements for the ith case, restricted to an individual column (*mean, trimmed mean, variance, trimmed variance* as described above) of X. A variety of regression models have been proposed in the literature [2,6,7]. A recent review of regression methods and applications is given in [20]. The simplest one is a lazy learning approach named of k nearest neighbor strategy. For kNN regression we average the dependent variable outcomes of all k nearest neighbors for prediction. Typically, the value of k is chosen between 1, ... 10. More complicated parametric approaches are given by least squares regression and support vector regression (SVR) as detailed in the following.

Least Squares Regression: The most popular estimation method is least squares, in which we pick the coefficients $\beta = (\beta_0, \beta_1, \ldots, \beta_p)^\top$, where p is the order of the underlying polynomial function, to minimize the residual sum of squares, see Eq. (3):

$$E = \sum_{i=1}^{N} (y_i - f(x_i))^2 \qquad (3)$$

where $f(x_i) = \beta_p \cdot x_i^p + \ldots + \beta_0$ in case of a polynomial function of degree p. After a few basic mathematical reformulations we end up with:

$$\beta = (X^T \cdot X)^{-1} X^T y \tag{4}$$

Depending on the structure of the input data X and the parameter vector beta, the former derivation can be used to express linear and polynomial regression. For further details, we refer to [8].

Support Vector Regression: Again we have training data with a continuous output label $\{(x_1, y_1), \ldots, (x_N, y_N)\} \subset X \times \mathbb{R}$. We define $Y = [y_1, \ldots, y_N]^\top$ as the values of the output function. y_i is the respective label of the input as defined above. In ϵ-SV regression [19], our goal is to find a function $f(x)$ that has at most ϵ deviation from the actually obtained targets y_i for all the training data, and at the same time is as flat as possible. In its simplest from $f(x)$ may be given as $f(x) = <w, x>$ where $w \in \mathbb{R}$. We skip the explicit use of the bias term $b \in \mathbb{R}$ in f. The aim is to fit the data to the output function such that $f(x_i) - \hat{f}(x_i) \leq \pm\epsilon$, where $\hat{f}(x_i)$ is the prediction for a given x_i and ϵ is a user-defined parameter controlling the permitted error of the prediction. Using the so-called epsilon-loss, this can be formalized as:

$$\min \frac{1}{2}||w||^2 \quad s.t. \begin{cases} y_i - <w, x_i> -b \leq \epsilon \\ <w, x_i> +b - y_i \leq \epsilon \end{cases} \tag{5}$$

As shown, e.g. in [19], the parameter vector w can be completely described as a linear combination of the training patterns x_i. In Eq. 5 the data and parameters occur only by means of inner products which permits the use of the kernel trick. Accordingly, the problem can be formulated using the kernel function k only (for detailed deviations see e.g. [19]). The kernelized SVR problem can be solved using a quadratic problem solver, providing a unique solution given the kernel function k is positive semi-definite.

5 Experiments and Results

We evaluate our proposed system and preprocessing concept on the aforementioned datasets using least squares regression, k-nearest neighbor- and support vector regression [8]. All parameters were determined individually for DS1, DS2, and DS3 in a grid search on an independent data set in a pre-study. To evaluate the model's performance, we used a ten-fold cross-validation. Here we made sure that measurements of the same lecture (measurement sessions) have not caused any bias. Only one of 5 measurements of a measurement session was randomly selected for the training set and none of the others in the test set. For comparison of the methods, root mean square error (RMSE) was chosen as measure since its value directly represents the misestimation of the number of attendants. The results are reported in Table 1.

Table 1. Comparison of regression models for the several data sets: without any preprocessing (DS1), after aggregation and preprocessing (DS2), and after normalization with distance-depending weighting factors (DS3).

Model	DS1	DS2	DS3
kNN regression	18.81(3.24)	6.64(1.28)	5.66(0.84)
Least Squares Regression	18.79(3.23)	8.49(0.94)	7.64(0.33)
Polynomial Regression	**18.78(3.22)**	6.41(1.02)	5.44(0.32)
Support Vector Regression	19.72(2.49)	**6.40(1.08)**	**5.42(0.40)**

The use of *raw* data as in DS1 for the models not only results in poor performance (computational complexity of SVR is $\mathcal{O}(n^2)$) but also provides false estimations of $\approx$18 people. With the preprocessing from DS1 to DS2, we were able to reduce the high dimensional RSS series for each beacon-receiver pair to four statistical moments. Each transmitter-receiver pair in DS2 is now represented by *mean, variance, trimmed mean* and *trimmed variance*. The aggregation to those four features results in a successful regression by SVR of $\approx$6.40 shown in Table 1. By weighting DS2 using a transmitter-receiver-depending matrix, we were able to further reduce the false estimated number of attendants to $\approx$5.42 shown as DS3. As Table 1 illustrates, the different regression models performed nearly same (except LSR in DS2 and DS3). Thus, for an application in a real-time intelligent security system, other factors such as runtime and complexity are more decisive.

6   Conclusions and Future Work

In this paper, a system prototype for a BLE-based dfp human counting was presented. As supervised measurement area, an auditorium was equipped with transmitters and receivers to determine the number of attendants. In total, the number of attendant people was recorded in 80 lectures. With the preprocessing from our previous research comprising several steps of aggregation, we reduced the dimensionality of the dataset. As a result, not only complexity was reduced, but also runtime and accuracy were improved. Using the raw data without any preprocessing, the estimation of the number of attendants ended up with an accuracy of $\approx \pm 20$ people. In a real hazardous situation, e.g. a building fire, an incorrect estimation of 20 people would result in fatal consequences. For example, assuming the number of attendants was underestimated at 20 instead of the real 40, not enough rescue workers might be sent to the corresponding building area. Conversely, in case of overestimation, too many rescue forces might be sent to areas in which are hardly any people. At first, we reduced the false estimations to 6.40 by our preprocessing steps. This false estimation rate was further decreased to 5.42 by using the distance-depending weighting matrix. In a system that is about saving people's lives, a miscalculation of ± 5.42 is not the optimum yet. Nevertheless, the system can give a valuable indication of the distribution of people in buildings, so that a large number of task forces can still be sent to the place with a high estimation despite 5.42 miscalculation. Finally, we provided a system prototype in which we used BLE for a device-free human counting system.

Future work in this area will be twofold: on the one hand side, the enhancement of the presented counting system is an interesting area of research. This comprises both reducing the technical equipment and improving data and methods for a more accurate estimation. In this context, the use of other regression models such as lasso and ridge regression or the use of neural networks would be interesting to increase the results. On the other hand, following the taxonomy described in [23], the extension of dfp human counting to further stages *localization*, *tracking*, and *identification* are interesting fields of research.

References

1. Bahl, P., Padmanabhan, V.N.: Radar: an in-building rf-based user location and tracking system. In: Proceedings IEEE INFOCOM 2000. Conference on Computer Communications. Nineteenth Annual Joint Conference of the IEEE Computer and Communications Societies, vol. 2, pp. 775–784 (2000)
2. Bertuletti, S., Cereatti, A., Della, U., Caldara, M., Galizzi, M.: Indoor distance estimated from bluetooth low energy signal strength: comparison of regression models. In: 2016 IEEE Sensors Applications Symposium (SAS), pp. 1–5 (2016)
3. Bifet, A., Gavaldà, R., Holmes, G., Pfahringer, B.: Machine Learning for Data Streams with Practical Examples in MOA. MIT Press, Cambridge (2018)
4. Deak, G., Curran, K., Condell, J.: A survey of active and passive indoor localisation systems. Comput. Commun. **35**(16), 1939–1954 (2012)
5. Faragher, R., Harle, R.: Location fingerprinting with bluetooth low energy beacons. IEEE J. Sel. Areas Commun. **33**(11), 2418–2428 (2015)
6. Faraway, J.: Extending the Linear Model with R: Generalized Linear, Mixed Effects and Nonparametric Regression Models. CRC Press, Boca Raton (2006)
7. Harrell, F.: Regression Modeling Strategies: With Applications to Linear Models, Logistic and Ordinal Regression, and Survival Analysis. Springer Series in Statistics. Springer, Cham (2015). https://doi.org/10.1007/978-3-319-19425-7
8. Hastie, T., Tibshirani, R., Friedman, J.: The Elements of Statistical Learning: Data Mining, Inference and Prediction, 2nd edn. Springer, New York (2009)
9. Mainetti, L., Patrono, L., Sergi, I.: A survey on indoor positioning systems. In: 2014 22nd International Conference on Software, Telecommunications and Computer Networks (SoftCOM), pp. 111–120 (2014)
10. Mautz, R.: Indoor positioning technologies. Ph.D. thesis, ETH Zurich (2012)
11. Mistry, H.P., Mistry, N.H.: Rssi based localization scheme in wireless sensor networks: A survey. In: 2015 Fifth International Conference on Advanced Computing Communication Technologies, pp. 647–652 (2015)
12. Münch, M., Huffstadt, K., Schleif, F.: Towards a device-free passive presence detection system with bluetooth low energy beacons. In: 27th European Symposium on Artificial Neural Networks (ESANN) (2019)
13. Oosterlinck, D., Benoit, D.F., Baecke, P., de Weghe, N.V.: Bluetooth tracking of humans in an indoor environment: an application to shopping mall visits. Appl. Geogr. **78**, 55–65 (2017)

14. Pirzada, N., Nayan, M.Y., Hassan, F.S.M.F., Khan, M.A.: Device-free localization technique for indoor detection and tracking of human body: a survey. Procedia Soc. Behav. Sci. **129**, 422–429 (2014). 2nd International Conference on Innovation, Management and Technology Research
15. Priyantha, N.B., Chakraborty, A., Balakrishnan, H.: The cricket location-support system. In: Proceedings of the 6th Annual Intern. Conference on Mobile Computing and Networking. MobiCom 2000, pp. 32–43. ACM, New York (2000)
16. Sabek, I., Youssef, M.: Multi-entity device-free wlan localization. In: 2012 IEEE Global Communications Conference (GLOBECOM), pp. 2018–2023 (2012)
17. Schleif, F., Tiño, P.: Indefinite proximity learning: a review. Neural Comput. **27**(10), 2039–2096 (2015)
18. Scholkopf, B., Smola, A.J.: Learning with Kernels: Support Vector Machines, Regularization, Optimization, and Beyond. MIT Press, Cambridge (2001)
19. Smola, A.J., Schölkopf, B.: A tutorial on support vector regression. Stat. Comput. **14**(3), 199–222 (2004)
20. Stulp, F., Sigaud, O.: Many regression algorithms, one unified model: a review. Neural Networks **69**, 60–79 (2015)
21. Subhan, F., Hasbullah, H., Rozyyev, A., Bakhsh, S.T.: Indoor positioning in bluetooth networks using fingerprinting and lateration approach. In: 2011 International Conference on Information Science and Applications, pp. 1–9 (2011)
22. Sugino, K., Katayama, S., Niwa, Y., Shiramatsu, S., Ozono, T., Shintani, T.: A bluetooth-based device-free motion detector for a remote elder care support system. In: 2015 IIAI 4th International Congress on Advanced Application Informatics, pp. 91–96 (2015)
23. Teixeira, T., Dublon, G.: A survey of human-sensing: methods for detecting presence, count, location, track, and identity. ACM Comput. Surv. **5**, 59–69 (2010)
24. Turgut, Z., Aydin, G.Z.G., Sertbas, A.: Indoor localization techniques for smart building environment. Procedia CS **83**, 1176–1181 (2016)
25. Want, R., Hopper, A., Falcão, V., Gibbons, J.: The active badge location system. ACM Trans. Inf. Syst. (TOIS) **10**(1), 91–102 (1992)
26. Woyach, K., Puccinelli, D., Haenggi, M.: Sensorless sensing in wireless networks: implementation and measurements. In: 2006 4th International Symposium on Modeling and Optimization in Mobile, Ad Hoc and Wireless Networks, pp. 1–8 (2006)
27. Wu, S., et al.: Survey on prediction algorithms in smart homes. IEEE Internet Things J. **4**(3), 636–644 (2017)
28. Xiao, J., Zhou, Z., Yi, Y., Ni, L.: A survey on wireless indoor localization from the device perspective. ACM Comput. Surv. **49**(2), 1–31 (2016)
29. Youssef, M., Mah, M., Agrawala, A.: Challenges: device-free passive localization for wireless environments. In: Proceedings of the 13th Annual ACM International Conference on Mobile Computing and Networking. MobiCom 2007, pp. 222–229. ACM, New York (2007)
30. Zanella, A., Bui, N., Castellani, A., Vangelista, L., Zorzi, M.: Internet of things for smart cities. IEEE Internet Things J. **1**(1), 22–32 (2014)

Combining Very Deep Convolutional Neural Networks and Recurrent Neural Networks for Video Classification

Rukiye Savran Kızıltepe[1(✉)], John Q. Gan[1], and Juan José Escobar[2]

[1] School of Computer Science and Electronic Engineering,
University of Essex, Colchester, UK
{rs16419,jqgan}@essex.ac.uk
[2] Department of Computer Architecture and Technology, CITIC,
University of Granada, Granada, Spain
jjescobar@ugr.es

Abstract. Convolutional Neural Networks (CNNs) have been demonstrated to be able to produce the best performance in image classification problems. Recurrent Neural Networks (RNNs) have been utilized to make use of temporal information for time series classification. The main goal of this paper is to examine how temporal information between frame sequences can be used to improve the performance of video classification using RNNs. Using transfer learning, this paper presents a comparative study of seven video classification network architectures, which utilize either global or local features extracted by VGG-16, a very deep CNN pre-trained for image classification. Hold-out validation has been used to optimize the ratio of dropout and the number of units in the fully-connected layers in the proposed architectures. Each network architecture for video classification has been executed a number of times using different data splits, with the best architecture identified using the independent T-test. Experimental results show that the network architecture using local features extracted by the pre-trained CNN and ConvLSTM for making use of temporal information can achieve the best accuracy in video classification.

Keywords: Deep learning · Video classification · Action recognition · Convolutional Neural Networks · Recurrent Neural Networks

1 Introduction

In recent years, the number of published videos has exponentially increased due to wider access to internet and more accessible hardware. Analyzing the content of videos is essential for both users and suppliers. Users want to know what type of movies they watch and to reach the videos which are related to their interests. Similarly, suppliers are required to manage advertisements, contents and to suggest interest-based videos to users. For these purposes, video classification has become an important research area in recent years.

© Springer Nature Switzerland AG 2019
I. Rojas et al. (Eds.): IWANN 2019, LNCS 11507, pp. 811–822, 2019.
https://doi.org/10.1007/978-3-030-20518-8_67

However, classification of high-dimensional data is a challenging task. Action recognition from videos is a highly active research area with the state-of-the-art systems still being far from human performance.

Convolutional Neural Network (CNN) has been demonstrated as a powerful architecture for image classification achieving the state-of-the-art results in recent years. The performance of CNN in action recognition from videos is not as satisfied as those achieved in image-based tasks such as detection [14], pattern recognition [22], segmentation [8] and classification [13,16]. While deep networks have been shown to be very effective in the mentioned image processing tasks, it is still unclear how to properly extend these methods to the video domain.

Both the extension of CNNs to video classification and the use of Recurrent Neural Networks (RNN) to understand video content have achieved relatively outstanding results [1,6]. However, video classification remains a challenging problem because of the temporal information which can be used to achieve better performance. Thus, motivated by these results, in this paper seven different network architectures are proposed to classify human actions from videos using RNNs on top of either local or global features extracted by the CNN architecture called VGG-16 that was pre-trained on ImageNet dataset. Experiments are conducted to explore appropriate approaches to using temporal information by RNNs and find out whether local or global features extracted by the pre-trained CNN can achieve better video classification performance.

The rest of the paper is organized as follows: Related work is reviewed in Sect. 2 and the proposed methods are described in Sect. 3. Experiments are explained in Sect. 4, and results are reported in Sect. 5, followed by conclusion in Sect. 6.

2   Related Work

There are several approaches to video classification, ranging from text-based [4], audio-based [18] to visual based. In this study, the visual-based approach is followed due to the rich information in visual context.

Traditional feature extraction methods are usually time-consuming and require feature engineering and domain knowledge. The recent trend to extract robust feature representations is applying deep learning to raw image or video data. CNNs involve two main parts: a sequence of convolution and pooling layers for feature extraction and densely connected layers for classification. Pre-trained networks have been used to extract new features from different instances using their first part. In the case of convolutional networks, the convolutional part of the pre-trained network is included in the place of feature extractor to obtain interesting features from new samples. Therefore, representations learned by the convolutional part of previously trained model on large datasets can be used without domain knowledge [13,16].

The great success in learning robust feature representations with CNN from raw images has encouraged the use of the deep features for video classification. The common idea is to handle a video as an accumulation of consecutive frames.

The feature representations could be extracted using a feed-forward pass up to a specific fully-connected layer with pre-trained deep models on ImageNet [5] such as AlexNet [13], GoogleNet [17], VGGNet [16] and ResNet [10]. Then, these features can be fused as representations in video level as the input of a classifier for video classification.

In a CNN architecture for video classification, the main issue is the way of combining appearance and motion information. The existing architectures can be categorized into two approaches: single-stream and two-stream networks.

Single-stream networks handle spatiotemporal features together in contrast to two-stream networks. Karpathy et al. conducted an empirical evaluation of CNNs considering the methods to fuse local temporal information from frame sequences [12]. Their network, which benefits from integrating spatiotemporal information, made significant improvements in classification accuracy contrast to traditional feature-based baseline methods (55.3% to 63.9%). They showed the capability of CNNs in learning features from video data. They suggested that local motion information might not be very essential especially for dynamic dataset. However, Castro et al. [3] highlighted the importance of motion information in video classification challenge by showing that the use of visual information only is inadequate in motion-based action recognition.

Zha et al. utilized an image-trained CNN to extract features from different layers of deep models [24]. They showed that CNN features can gain satisfactory recognition performance in video classification by using the motion information added via late fusion on the UCF-101 dataset. However, the spatiotemporal features learnt could not capture movement information properly in single-stream networks.

Two-stream networks tackle with spatial and temporal information in separate networks. Simonyan and Zisserman proposed a two-stream CNN which involves spatial and temporal streams merged by fusion [15]. They used the temporal stream on multi-frame dense optical flow to recognize actions from movements while the spatial stream to recognize actions on raw video frames. Their two-stream model achieved better results on the UCF-101 and HMDB-51 datasets than the models consisting of either spatial or temporal stream only [15]. Following this work, the two-stream architecture has been extended in other studies.

Tran et al. proposed a deep 3D CNN as a feature extractor by using sequence of RGB frames. Deconvolutional layer is used to interpret model decision in their study. They also showed that 3D CNNs are more convenient than 2D CNNs to learn spatiotemporal features [19].

Yao et al. proposed a novel 3D CNN-RNN encoder-decoder which can capture both local spatial and temporal information. With an attention mechanism within this framework, they capture global context as well [21].

Feichtenhofer et al. introduced a novel two-stream network fusion architecture. Two streams are fused after convolution layer, and the output is fused for spatiotemporal loss evaluation [7].

Carreira and Zisserman compared two-stream architectures consisting of a combination of 3D models. Two 3D networks are employed for two streams. Spatial stream includes frames stacked in time dimension rather than single frames. The pre-trained two-stream network on ImageNet and Kinetics datasets outperformed the existing models [2].

3   Methods

In this section, we describe the proposed architectures of the networks for action recognition in detail. The methods for pre-processing and feature extraction are also described.

3.1   Preprocessing

To capture the temporal information between the consecutive frames, we used the combination of multiple techniques. The primary preprocessing was removing the very similar consecutive frames by extracting one frame per second. Secondly, we resized the remaining frames using an interpolation technique from the original image frame size of 320×240 to 240×240.

3.2   Feature Extraction

The VGG-16 was trained on 1000 images per each of 1000 categories in 2014 ImageNet Large Scale Visual Recognition Competition (ILSVRC) using a subset of ImageNet dataset by Simonyan and Zisserman [16]. It consists of 16 convolutional layers with quite small convolution filters (3×3). It has been showed that the learnt representations can be generalized to other datasets; therefore, we decided to use the pre-trained VGG-16 as a feature extractor for video classification. VGG-16 can be used to extract both local features employing only the convolutional base and global features including the fully-connected layers, as well.

3.3   Different VGG-16 Based Architectures for Video Classification

Using transfer learning, in this paper seven video classification architectures are proposed as shown in Fig. 1 for action recognition using the UCF-101 dataset. Hold-out validation is used to determine the dropout parameter, the number of units in both the fully-connected layers and the LSTM, number of filters and kernel size in the ConvLSTM. The layers in the pre-trained VGG-16 are fixed and used for feature extraction, and the newly added fully-connected layers are trained using the UCF-101 training data for video classification.

As baseline models, VGG-16-VOTE (a) and VGG-16-VOTE (b) are built based on the general approach which treats frames as single images. For each frame of a video, softmax produces a score for each possible video class. VGG-16-VOTE determines the class of the video by voting in terms of the maximum score among all the frames. The features extracted by the pre-trained VGG-16 convolutional base are fed into the fully connected layer (FC-512) in VGG-16-VOTE (a). To explore the effect of global features, VGG-16-VOTE (b) includes the fully-connected layers of VGG-16.

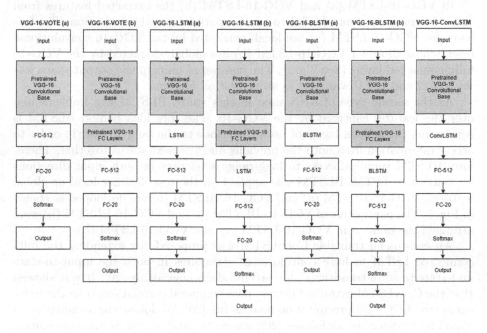

Fig. 1. The architectures of networks used in the experiment VGG-16-VOTE (a) and VGG-16-VOTE (b) are designed as baseline methods using the pre-trained VGG-16 by [16].

LSTM is one of the most common approaches for sequence modelling, which has been demonstrated as a robust method for representing long-range dependencies in the previous studies [11,23]. The main advantage of LSTM is having its memory cell c_t which accumulates the state information. The cell is modified by controlling the input gate i_t and forget gate f_t. Once the cell is fed with new input, it accumulates input information provided that the input gate is on. If the forget gate is activated, the previous cell information c_{t-1} could be forgotten. The output gate o_t checks the current cell output c_t to decide whether it is propagated to the final state h_t or not. In this study, we follow the hidden layer function of LSTM described as follows [9], where '·' denotes the Hadamard product:

$$i_t = \sigma(W_{xi}x_t + W_{hi}h_{t-1} + W_{ci}c_{t-1} + b_i)$$
$$f_t = \sigma(W_{xf}x_t + W_{hf}h_{t-1} + W_{cf}c_{t-1} + b_f)$$
$$c_t = f_t \cdot c_{t-1} + i_t \cdot \tanh(W_{xc}x_t + W_{hc}h_{t-1} + b_c) \tag{1}$$
$$o_t = \sigma(W_{xo}x_t + W_{ho}h_{t-1} + W_{co} \cdot c_t + b_o)$$
$$h_t = o_t \cdot \tanh(c_t)$$

In VGG-16-LSTM (a) and VGG-16-LSTM (b) the extracted features from video frames are fed into LSTM to access spatiotemporal information. The features for VGG-16-LSTM (a) are local, extracted by the VGG-16 convolutional base, whilst those for VGG-16-LSTM (b) are global, extracted by the VGG-16 fully-connected layers (See Fig. 1). The number of units in the output space was set to 512 and ReLU was used as an activation function.

As the second type of RNN used in this study, BLSTM was implemented over the features captured by the VGG-16. The core idea behind BLSTM is connecting two hidden layers of both directions to the same output in order to access information from both the previous and the next status together. Therefore, BLSTM supplies access to long-range sequences in both input directions in contrast to LSTM. BLSTM was applied on the top of either local or global features. VGG-16-BLSTM (a) and VGG-16-BLSTM (b) are developed as shown in Fig. 1. The parameters of VGG-16-BLSTM (a) and VGG-16-BLSTM (b) were arranged the same as in VGG-16-LSTM (a) and VGG-16-LSTM (b).

An end-to-end trainable ConvLSTM was proposed "by extending the fully connected LSTM to have convolutional structures in both the input-to-state and state-to-state transitions" for precipitation nowcasting [20]. It was showed that the ConvLSTM extracted better spatiotemporal correlations than the fully-connected LSTM for precipitation nowcasting [20]. We follow the formulation of ConvLSTM described as follows [20], where '⊛' and '·' denote the convolution operator and Hadamard product, respectively:

$$i_t = \sigma(W_{xi} \circledast \mathcal{X}_t + W_{hi} \circledast \mathcal{H}_{t-1} + W_{ci} \cdot C_{t-1} + b_i)$$
$$f_t = \sigma(W_{xf} \circledast \mathcal{X}_t + W_{hf} \circledast \mathcal{H}_{t-1} + W_{cf} \cdot C_{t-1} + b_f)$$
$$C_t = f_t \cdot C_{t-1} + i_t \cdot \tanh(W_{xc} \circledast \mathcal{X}_t + W_{hc} \circledast \mathcal{H}_{t-1} + b_c) \tag{2}$$
$$o_t = \sigma(W_{xo} \circledast \mathcal{X}_t + W_{ho} \circledast \mathcal{H}_{t-1} + W_{co} \cdot C_t + b_o)$$
$$\mathcal{H}_t = o_t \cdot \tanh(C_t)$$

Inspired by the mentioned study, the ConvLSTM is used to construct a new architecture for video classification, VGG-16-ConvLSTM as shown in Fig. 1, by taking the advantage of its capacity in capturing spatiotemporal information throughout time series. We add one ConvLSTM layer on top of the spatial feature maps extracted by VGG-16 and used the hidden states for video classification. The ConvLSTM layer has 64 hidden states and 7×7 kernels, and the stride of convolution is set to 2 in our experiment.

4   Experiments

In this section, the dataset, experiment design and experiment setup and evaluation are described.

4.1   Dataset

In this study, the UCF-101 dataset is used to evaluate the neural network architectures utilized to classify human actions from video clips. The dataset contains 13320 clips from 101 non-overlapping classes, with frame size of 240 to 360 pixels. UCF-101 has defined 3 training-testing splits, which facilitates bench-marking algorithms. In this experiment, only the first twenty categories of the data set were used (due to limited time and computing facility), and the first training-testing split was followed to generate train and testing data. Figure 2 presents an overview of the distribution of categories in both training and testing data. In addition, Fig. 3 shows some training and testing examples from the UCF-101 data set. In this experiment, hold-out method was used to split training data into the following two groups: 70% of training data for training and 30% of training data for validation. Testing dataset was never used during training and validation, but only used for producing the testing accuracy of each network architecture.

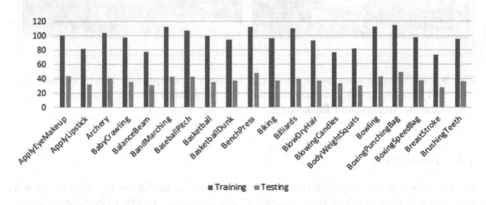

Fig. 2. The distribution of categories in training and test splits.

4.2   Experiment Design

During the training process, we apply parameter tuning with the hold-out validation technique. Specifically, if the training is finished with the first possible set of parameters, the next training begins with the second set of possible parameters, and so on. When the validated training is finished, the scores of both training and validation are averaged. The best parameters are identified based on the validation scores. Then, the model with the best parameters is evaluated on the testing data by predicting unseen test videos' classes.

4.3   Experiment Setup

The proposed network architectures are implemented by using Tensorflow-gpu v1.12 on an NVDIA GTX Titan X GPU using the CUDA v9.0 toolkit. The batch size is set to 128, and the cost is minimized by using the Stochastic Gradient Descent (SGD) optimizer. Dropout is used as a regularization method. We keep the neurons active with the probability $p = 0.5$ after the first fully-connected layer.

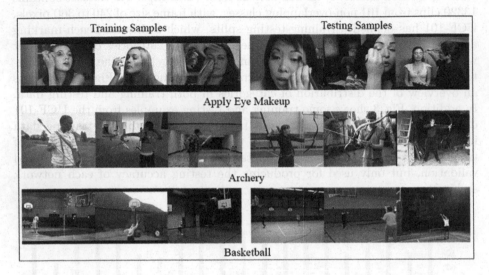

Training Samples

Testing Samples

Apply Eye Makeup

Archery

Basketball

Fig. 3. (a) Training samples on the left; (b) Testing samples on the right from the UCF-101 Human Action Dataset. Samples are from three different categories: Apply Eye Makeup, Archery, and Basketball.

4.4   Evaluation

In our experiment classification accuracy is used as performance metric, which is the percentage of correctly classified video instances. In the experiment, seven different video classification architectures were compared based on their accuracy scores through statistical significance test. We collected 14 training, validation and test accuracy scores for each architecture under the same conditions but different data splits for training, validation and testing. In this context, the dependent variable is testing accuracy while the independent variable is classification architecture. A significance level of 0.05 was considered during the analysis. The distribution of performances was evaluated using Shapiro-Wilk test, which shows that the testing accuracy scores are normally distributed at the 0.05 level of significance. Moreover, we applied T-test to compare pairs of network architectures in terms of testing accuracy and reported t-scores and p-values in Sect. 5.

5   Results and Discussion

In this study, the first twenty categories of the UCF-101 data set have been used to obtain results. Training and testing data were separated using the first part of the train/test splits enabling bench-marking algorithms. Hold-out validation technique was applied to identify the optimal values for dropout parameter, the number of units in LSTM, the number of filters, strides and kernel size in ConvLSTM, and the number of units in dense layers.

Table 1. Average accuracy scores achieved by different architectures

Model	Training accuracy	Validation accuracy	Testing accuracy
VGG-16-ConvLSTM	99.24%	97.45%	82.04%
VGG-16-LSTM (a)	98.86%	97.91%	81.27%
VGG-16-BLSTM (a)	97.76%	95.19%	76.20%
VGG-16-VOTE (a)	78.71%	76.55%	73.20%
VGG-16-LSTM (b)	78.41%	79.47%	67.62%
VGG-16-VOTE (b)	73.70%	67.05%	64.04%
VGG-16-BLSTM (b)	89.17%	75.37%	61.18%

Table 1 demonstrates the average of 14 results for each architecture. It is clear that VGG-16-ConvLSTM outperformed the remaining architectures (82.04%). It is followed by VGG-16-LSTM (a) (81.27%). VGG-16-BLSTM (a) was the third best architecture (76.20%). Interestingly, the baseline method with local features, VGG- 16-VOTE (a), achieved higher accuracy than the architectures extracting global features.

Regarding the results on the testing dataset, we did the independent T-test for possible pairs of the network architectures to identify whether the difference in their testing accuracy is statistically different. Table 2 provides the results of the T-Test, where t, df and p represent t-score, degree of freedom, and probabilistic significance, respectively. It is clear that there is a significant difference between all architectures at the 0.05 level of significance.

The best model is VGG-16-ConvLSTM, and the second-best model is VGG-16-LSTM (a) without the fully-connected layers of the pre-trained VGG-16. The third-best architecture is VGG-16-BLSTM (a) which performs well using local features captured by the VGG-16. These results show that ConvLSTM plays an important role in improving the video classification performance by making better use of temporal information between frames. They also show that local features extracted by the pre-trained CNN are more effective than its global features for video classification. Because the pre-trained CNN was constructed on a large amount of training images, it can overcome overfitting problems to some extent, especially when the local features extracted by the pre-trained CNN are used for video classification, as indicated by the results in Table 2.

Table 2. Independent-Samples T-Test results

Architecture	Mean	Std.	t	df	p
VGG-16-ConvLSTM	82.04%	0.01151	2.109	22.473	.046
VGG-16-LSTM (a)	81.27%	0.00757			
VGG-16-ConvLSTM	82.04%	0.01151	7.070	17.076	.000
VGG-16-BLSTM (a)	76.20%	0.02871			
VGG-16-ConvLSTM	82.04%	0.01151	9.196	15.929	.000
VGG-16-VOTE (a)	73.20%	0.03407			
VGG-16-ConvLSTM	82.04%	0.01151	29.357	24.821	.000
VGG-16-LSTM (b)	67.60%	0.01436			
VGG-16-ConvLSTM	82.04%	0.01151	50.232	21.222	.000
VGG-16-VOTE (b)	64.04%	0.00687			
VGG-16-ConvLSTM	82.04%	0.01151	35.477	21.579	.000
VGG-16-BLSTM (b)	61.18%	0.01875			
VGG-16-LSTM (a)	81.27%	0.00757	6.387	14.79	.000
VGG-16-BLSTM (a)	76.20%	0.02871			
VGG-16-LSTM (a)	81.27%	0.00757	8.643	14.28	.000
VGG-16-VOTE (a)	73.20%	0.03407			
VGG-16-LSTM (a)	81.27%	0.00757	31.492	19.704	.000
VGG-16-LSTM (b)	67.60%	0.01436			
VGG-16-LSTM (a)	81.27%	0.00757	63.025	25.761	.000
VGG-16-VOTE (b)	64.04%	0.00687			
VGG-16-LSTM (a)	81.27%	0.00757	37.164	17.127	.000
VGG-16-BLSTM (b)	61.18%	0.01875			

6   Conclusion

This paper proposes seven network architectures for video classification by incorporating spatial and temporal features extracted from the VGG-16, LSTM, BLSTM and ConvLSTM. First, we optimized the baseline models, which are VGG-16-VOTE (a) and (b) with local and global features, respectively. Then, we optimized ConvLSTM and LSTM to extract higher-level features from frame sequences, based on the features extracted by the pre-trained CNN (VGG-16). Experimental results show that combining pre-trained CNN with ConvLSTM achieved the highest performance among the seven network architectures for video classification, demonstrating the importance of effective integration of spatiotemporal information in video classification.

Acknowledgment. This research was partially supported by the Republic of Turkey Ministry of National Education. The authors would also like to acknowledge the help of Martin Balla in conducting the experiment for this paper.

References

1. Ballas, N., Yao, L., Pal, C., Courville, A.: Delving deeper into convolutional networks for learning video representations. arXiv preprint arXiv:1511.06432 (2015)
2. Carreira, J., Zisserman, A.: Quo vadis, action recognition? A new model and the kinetics dataset. In: Proceedings of the IEEE Conference on Computer Vision and Pattern Recognition, pp. 4724–4733 (2017)
3. Castro, D., et al.: Let's dance: learning from online dance videos. arXiv preprint arXiv:1801.07388 (2018)
4. Darji, M.C., Patel, D.N., Shah, Z.H.: A review on audio features based extraction of songs from movies. Int. J. Adv. Eng. Res. Dev. 2348–4470 (2015)
5. Deng, J., Dong, W., Socher, R., Li, L.J., Li, K., Fei-Fei, L.: ImageNet: a large-scale hierarchical image database. In: Proceeding of the IEEE Conference on Computer Vision and Pattern Recognition, pp. 248–255 (2009)
6. Donahue, J., et al.: Long-term recurrent convolutional networks for visual recognition and description. In: Proceedings of the IEEE Conference on Computer Vision and Pattern Recognition, pp. 2625–2634 (2015)
7. Feichtenhofer, C., Pinz, A., Zisserman, A.: Convolutional two-stream network fusion for video action recognition. In: Proceedings of the IEEE Conference on Computer Vision and Pattern Recognition, pp. 1933–1941 (2016)
8. Girshick, R., Donahue, J., Darrell, T., Malik, J.: Rich feature hierarchies for accurate object detection and semantic segmentation. In: Proceedings of the IEEE Conference on Computer Vision and Pattern Recognition, pp. 580–587 (2014)
9. Graves, A.: Generating sequences with recurrent neural networks. arXiv preprint arXiv:1308.0850 (2013)
10. He, K., Zhang, X., Ren, S., Sun, J.: Deep residual learning for image recognition. In: Proceedings of the IEEE Conference on Computer Vision and Pattern Recognition, pp. 770–778 (2016)
11. Hochreiter, S., Schmidhuber, J.: Long short-term memory. Neural Comput. **9**(8), 1735–1780 (1997)
12. Karpathy, A., Toderici, G., Shetty, S., Leung, T., Sukthankar, R., Fei-Fei, L.: Large-scale video classification with convolutional neural networks. In: Proceedings of the IEEE Conference on Computer Vision and Pattern Recognition, pp. 1725–1732 (2014)
13. Krizhevsky, A., Sutskever, I., Hinton, G.E.: ImageNet classification with deep convolutional neural networks. In: Advances in Neural Information Processing Systems, pp. 1097–1105 (2012)
14. Ren, S., He, K., Girshick, R., Sun, J.: Faster R-NN: towards real-time object detection with region proposal networks. In: Advances in Neural Information Processing Systems, pp. 91–99 (2015)
15. Simonyan, K., Zisserman, A.: Two-stream convolutional networks for action recognition in videos. In: Advances in Neural Information Processing Systems, pp. 568–576 (2014)
16. Simonyan, K., Zisserman, A.: Very deep convolutional networks for large-scale image recognition. arXiv preprint arXiv:1409.1556 (2014)
17. Szegedy, C., et al.: Going deeper with convolutions. In: Proceedings of the IEEE Conference on Computer Vision and Pattern Recognition, pp. 1–9 (2015)
18. Takahashi, N., Gygli, M., Van Gool, L.: AENet: learning deep audio features for video analysis. IEEE Trans. Multimed. **20**(3), 513–524 (2018)

19. Tran, D., Bourdev, L., Fergus, R., Torresani, L., Paluri, M.: Learning spatiotemporal features with 3d convolutional networks. In: Proceedings of the IEEE International Conference on Computer Vision, pp. 4489–4497 (2015)
20. Xingjian, S., Chen, Z., Wang, H., Yeung, D.Y., Wong, W.K., Woo, W.C.: Convolutional LSTM network: a machine learning approach for precipitation nowcasting. In: Advances in Neural Information Processing Systems, pp. 802–810 (2015)
21. Yao, L., et al.: Describing videos by exploiting temporal structure. In: Proceedings of the IEEE International Conference on Computer Vision, pp. 4507–4515 (2015)
22. Yin, X., Liu, X.: Multi-task convolutional neural network for pose-invariant face recognition. IEEE Trans. Image Process. **27**(2), 964–975 (2018)
23. Yue-Hei Ng, J., Hausknecht, M., Vijayanarasimhan, S., Vinyals, O., Monga, R., Toderici, G.: Beyond short snippets: Deep networks for video classification. In: Proceedings of the IEEE Conference on Computer Vision and Pattern Recognition, pp. 4694–4702 (2015)
24. Zha, S., Luisier, F., Andrews, W., Srivastava, N., Salakhutdinov, R.: Exploiting image-trained CNN architectures for unconstrained video classification. arXiv preprint arXiv:1503.04144 (2015)

Towards Applying River Formation Dynamics in Continuous Optimization Problems

Pablo Rabanal, Ismael Rodríguez, and Fernando Rubio[⊠]

Dpto. Sistemas Informáticos y Programación, Facultad de Informática,
Universidad Complutense de Madrid, C/ Prof José García Santesmases,
28040 Madrid, Spain
prabanal@fdi.ucm.es, {isrodrig,fernando}@sip.ucm.es

Abstract. River Formation Dynamics (RFD) is a metaheuristic that
has been successfully used by different research groups to deal with a wide
variety of discrete combinatorial optimization problems. However, no
attempt has been done to adapt it to continuous optimization domains.
In this paper we propose a first approach to obtain such objective, and we
evaluate its usefulness by comparing RFD results against those obtained
by other more mature metaheuristics for continuous domains. In particular, we compare with the results obtained by Particle Swarm Optimization, Artificial Bee Colony, Firefly Algorithm, and Social Spider Optimization.

Keywords: Swarm intelligence · Metaheuristics · Optimization ·
River Formation Dynamics

1 Introduction

Swarm intelligence methods [9] are heuristic problem-solving methods where a
set of simple entities interact with each other according to their local information. The goal of these interactions is to collaboratively obtain a good solution
to a given problem. Many swarm intelligence metaheuristics have been proposed
in the literature, both for discrete combinatorial optimization problems (see e.g.
ACO: Ant Colony Optimization [7,8] or RFD: River Formation Dynamics [19,21])
and for continuous domain optimization problems (see e.g. PSO: Particle Swarm
Optimization [14] or ABC: Artificial Bee Colony [13])

Briefly, River Formation Dynamics (RFD) is a water-based metaheuristic [27]
that consists on copying the geological forces that form rivers. When rivers fall
through steep slopes, they erode some soil from the ground and transport it
within the water. Later, this sediment is deposited in flatter areas of the river,

This work has been partially supported by Spanish project TIN2015-67522-C3-3-R, and
by Comunidad de Madrid as part of the program S2018/TCS-4339 (BLOQUES-CM)
co-funded by EIE Funds of the European Union.

I. Rojas et al. (Eds.): IWANN 2019, LNCS 11507, pp. 823–832, 2019.
https://doi.org/10.1007/978-3-030-20518-8_68

where the water moves more slowly. In this way, the altitude of points traversed by the river iteratively changes, and the whole path tends to form a ever decreasing slope. The river and tributaries courses change along time, and eventually the formed river (together with its tributaries) represents an efficient way to gather all the rain water in some geographical area and send it to the sea. In fact, the final form of the river constitutes an efficient tradeoff between finding short paths from all raining points towards the sea (i.e. finding shortest paths) and forming a small tree of river and tributaries (i.e. finding a small spanning tree): the first goal improves if more tributaries are used, whereas the latter goal encourages collecting water by using *meanders* instead.

RFD fits particularly well in NP-hard problems consisting in creating a kind of tree, as the two tendencies commented before (i.e. finding short paths or small spanning trees) can easily be leant towards either way by means of parameters (see [22]). RFD has been applied to deal with several classical NP-hard optimization problems (see e.g. [20,21,23]) and has also been applied to solve industrial problems such as network routing [2,10], robot navigation [25], VLSI design [6], or optimization in electrical power systems [1]. The interested reader is referred to [24] for a detailed survey covering the main applications of RFD. It is worth noting that, roughly speaking, RFD can be thought as a derivative-oriented version of ACO: In ACO, entities (ants) tend to move to those nodes where some value (pheromone trail) is higher, whereas in RFD, drops tend to move next to those nodes where the *difference* between the values (altitudes) at the origin and the destination nodes is higher (the flow is larger in steeper slopes).

It is worth pointing out that, so far, RFD has been applied only to problems where the solution space is discrete. In this paper we face the problem of developing a continuous version of RFD. Note that adapting a swarm method to the continuous domain may be relatively straightforward if the entities are the solutions themselves, because most of times the operators defining how each entity affects other entities can be easily generalized from the discrete to the continuous domain. However, it might not be so straightforward if solutions are defined by a structure drawn by the entities over some environment (e.g. ACO or our target method here, RFD), particularly if this structure consists of sequences of steps between consecutive neighbor solutions (points): If a continuous domain is adopted, then these paths contain an infinite amount of points, so an alternative representation would be required to denote them. More importantly, the typical expected outputs of a continuous problem are not naturally represented as a structure over an environment (e.g. a sequence of steps, a path, a round trip, or a tree over a graph). On the contrary, they are typically (and naturally) viewed as a point in a continuous-dimensional space.

Thus, in order to create a continuous version of RFD, it is a sensible choice abandoning our previous RFD view, where the output of the algorithm is a given structure (e.g. path, tree, etc.), and adopting the more natural view that the output is a point in the continuous space. Then, each drop represents a possible solution that moves around the search space. However, rather than guiding the movement of entities in terms of the fitness at each possible destination (as in PSO), each drop will consider the slopes towards other known positions (other drops). The higher the slope, the higher the probability of moving in such direction. Thus, we consider a gradient-oriented version of continuous optimization metaheuristics.

The rest of the paper is structured as follows. In the next section we present our proposal to adapt RFD to continuous domains. Then, in Sect. 3 we analyze the usefulness of the approach by comparing its results against those of other metaheuristics. Finally, our conclusions and future work plans are shown in Sect. 4.

2   Continuous RFD

In the continuous version of RFD we had to decide what the drops would represent, and how they would move through the solution space. The most natural approach is to consider each drop as a solution to the problem, that is, a position in the search space. The evaluation of the position of a drop will be the height at which it is (the value of the solution). To move the drops we use the slopes between them as in the discrete version of the algorithm. In this way, the drops will move with higher probability approaching those drops that are at a lower height (following the slopes of maximum gradient) and, therefore, closer to the optimum.

Next, we describe in detail the RFD scheme for solving minimization continuous problems.

In Fig. 1, the main steps of the presented algorithm can be seen.

```
initializeVariables()
initializeDrops()
while (not endingCondition())
    for each drop d
        if numEvals > 0 then
            if noImprove(d) then
                createDrop(d,computeRangeLimit())
            elseif not isBestDrop(d) then
                moveDrop(d)
            else
                moveBestDrop(d)
            end if
        end if
    end for
    computeRangeFactor()
end while
```

Fig. 1. Continuous RFD scheme

2.1 Initialization

In the `initializeVariables()` phase the following variables are initialized:

- `numDrops` represents the number of drops (entities) used.
- `numEvals` represents the maximum number of function evaluations allowed.
- `numStepsNoImprove`: for each drop it indicates the number of steps it can be moved without improving its previous solution. When the drop is moved `numStepsNoImprove` steps without improving, it is created again.
- `accuracy` indicates the precision, i.e. the number of decimal numbers, of the solution.
- `moveLimit` represents the percentage that limits the movement of a drop (it is used in the `moveDrop(d)` phase).
- `rangeFactor` variable allows us to focus (or unfocus) the search in a more concrete (or general) area. It will be explained later in the context of the `computeRangeFactor()` method.
- `time` represents the maximum time of execution measured in seconds.

After that, in `initializeDrops()` each drop is randomly created in the search space. For each dimension i, a random value between the minimum - $min(i)$- and maximum value -$max(i)$- of dimension i is generated defining the drop position $position_d$. In this process the best drop is stored. The best drop will be the drop with the best solution found so far, that is, the drop whose value is the minimum supposing we are minimizing a function f, where this value is the evaluation of function f in the drop position $position_d$, that is, $value = f(position_d)$.

2.2 Main Loop of the Algorithm

After the initialization has taken place, the body of the loop is executed until the `endingCondition()` is satisfied. The execution of the loop finishes when the required accuracy is achieved, when the maximum number of evaluations of the function is reached, or when the time has expired.

In the main loop, three different strategies are used to move each drop if `numEvals>0`:

- If the drop has not improved in the last `numStepsNoImprove` steps (that is, `noImprove(d)`), then the drop is randomly created in the search space (`createDrop(d,computeRangeLimit())`) depending on the `rangeFactor` variable. This variable *narrows* the dimensions where the drop can be created. Each dimension i is *reduced* in the `computeRangeLimit()` function as follows:

$$range(i) = (max(i) - min(i)) * \texttt{rangeFactor}$$

These ranges will limit where the drop can be created. To create the new position of the drop, first we choose a drop (cd) in a random manner depending on its values. Those drops with lower values will be selected with higher probability. Second, and having into account the position of the chosen drop ($position_{cd}$) and the dimension limit $range$, we compute the position of the drop d. For each dimension i we randomly choose a value for $position_d(i)$ in range

$$[position_{cd}(i) - (range(i)/2),\ position_{cd}(i) + (range(i)/2)]$$

without exceeding the limits of the dimensions.

- If the drop is not the best drop at this moment (`not isBestDrop(d)`) then it is moved (`moveDrop(d)`) depending on the slopes between the drop d and the rest of the drops. First, these slopes are computed as follows:

$$slope(d, ad) = (position_d - position_{ad})/distance(d, ad)$$

where ad is another drop and $distance(d, ad)$ is the Euclidean distance between both drops. There exist two exceptional cases to deal with cases with non-decreasing slopes: (a) When $slope(d, ad) = 0$ we assign an epsilon slope: $slope(d, ad) = \epsilon$; (b) When $slope(d, ad) < 0$ we assign

$$slope(d, ad) = 0.1/(Abs(slope) + 1)$$

in order to allow drops climbing ascendent slopes (with a very low probability). Second, one drop is chosen as destination (dd) in a random way depending on the slopes: The higher the slope, the higher the probability of choosing that drop as destination. $slope_{dd}$ will represent the slope between drop d and dd. Third, once the destination is selected, we compute the direction in which the drop will be moved. For each dimension i,

$$direction(i) = position_{dd}(i) - position_d(i)$$

Fourth, the new position of d is calculated. In particular, for each dimension i we have:

$$newPosition_d(i) = position_d(i) + step(i)$$

where $step(i) = direction(i) * slope_{dd}$. However, there is a limit given by the expression

$$limit(i) = (max(i) - min(i)) * (\text{moveLimit}/100)$$

If $|step(i)| > limit(i)$ then $step(i) = limit(i)$ if $step(i) >= 0$, and $step(i) = -limit(i)$ in other case. Of course, the movement cannot exceed the limit values $min(i)$ and $max(i)$. Fifth, once the new position is known, the new value is computed: $newValue = f(position_d)$. Sixth and finally, if the new position improves the previous one, the drop is moved to the new position: $position_d = newPosition_d$. In other case, the drop remains in its previous position.

- In `moveBestDrop(d)` phase, the best drop is moved using the golden spiral. In this case, we forget about the river analogy, and we take inspiration from [18] to analyze the surroundings of the current position. By using the golden spiral we analyze with more probability nearby points, but trying to find a good tradeoff with positions located farther. More precisely, we modify two randomly chosen dimensions i and $j = (i + 1)$ *mod dimensions* according to the following expression:

If $exp \bmod 4 = 0$ or $exp \bmod 4 = 1$ then:

$$newPosition_d(i) = position_d(i) + addend$$

Else:

$$newPosition_d(i) = position_d(i) - addend$$

If $exp \bmod 4 = 0$ or $exp \bmod 4 = 3$ then:

$$newPosition_d(j) = position_d(j) + addend$$

Else:

$$newPosition_d(j) = position_d(j) - addend$$

where $addend = 1/\varphi^{exp}$, $\varphi = (1 + \sqrt{5})/2$ is the golden ratio, and exp takes values from 1 to `accuracy` $* 10$ (increasing it one by one) for each pair of dimensions i and j. The rest of dimensions remain unchanged. Again, the drop is moved only if the new position improves the previous position.

In the last step of the loop, the `computeRangeFactor()` method modifies the variable `range_factor`. If the solution has not been improved in the last $numEvalsOneLoop$ evaluations of function f, then:

$$\texttt{rangeFactor} = \texttt{rangeFactor} * 2$$

that is, the range where a drop can be created is duplicated. In other case, the range is halved:

$$\texttt{rangeFactor} = \texttt{rangeFactor}/2$$

$numEvalsOneLoop = numDrops + accuracy * 10$ is the number of evaluations of function f in one loop of the algorithm, because we have $numDrops$ evaluations for every movement or creation of a drop, and $accuracy * 10$ evaluations when moving the best drop. Let us remark that the values of `range_factor` are limited to $10^{-accuracy}$ as minimum, and 1 as maximum.

After creating (`createDrop(d,computeRangeLimit())`) or moving a drop (`moveDrop(d)` or `moveBestDrop(d)`) we compare if the best solution has been improved, and if it is the case then the new best solution is stored.

Table 1. Results for a benchmark of constrained optimization problems

Problem	PSO	ABC	FF	SSO	RFD
f_1	−15	−15	−14.99	−15	−15
f_2	−0.79	−0.79	−0.785	−0.802	−0.716
f_3	−30662.821	−30664.923	−30662.032	−30665.538	−30665.538
f_4	−6958.369	−6958.022	−6950.114	−6961.008	−6961.814
f_5	24.475	26.58	28.54	24.306	24.704
f_6	−0.749	−0.75	−0.749	−0.75	−0.75
f_7	0.05416	0.05398	0.05417	0.05394	0.08615
f_8	963.925	962.642	965.428	961.999	961.719

3 Experiments

In order to assess the usefulness of our approach, we have conducted a set of experiments to compare the performance of RFD against that obtained by other more mature metaheuristics. In particular, we consider two types of case studies. First, we consider a benchmark of eight well-known optimization problems obtained from [16]. Then, we consider three real-world optimization problems dealing with concrete engineering problems. In particular, we deal with the tension/compression spring design problem [4], the welded beam design problem [3], and the speed reducer design problem [12]. All cases can be described as optimization problems where a minimization has to be done subject to fulfill a given set of constraints. For all the case studies, we compare the results obtained by RFD with the results obtained by Particle Swarm Optimization (PSO [14]), Artificial Bee Colony (ABC [13]), Firefly Algorithm (FF [28]), and Social Spider Optimization (SSO [5]).

Table 1 summarizes the results obtained with the first benchmark, while Table 2 summarizes the results obtained with the three real-world engineering optimization problems. In all cases, the results correspond with the average of 30 independent executions of each algorithm. Regarding the parameter tuning, in the case of RFD the number of drops (numDrops) used was 50; numStepsNoImprove was set to 10; the accuracy value varies between 4 and 8, depending on the problem; the moveLimit value used in the experiments was 90; the initial value of rangeFactor is 1 in all cases; while the time was set to values from 10 to 300 s. Regarding the other metaheuristics, we have used the configurations and the results described in [5].

In order to appropriately compare the metaheuristics, we perform a statistical test. In general, a Friedman test can be used to check whether the hypothesis that all methods behave similarly (the null hypothesis) holds or not. However, since the number of metaheuristics under consideration is low, using a Friedman aligned ranks test is more recommended in this case. This test does not rank methods for each problem separately (as Friedman test does), but construct a global ranking where values of all methods and problems are ranked together.

Table 2. Results for a benchmark of real-world optimization engineering problems

Problem	PSO	ABC	FF	SSO	RFD
Tension/compression	0.0148631	0.0128507	0.0129307	0.0127649	0.0127486
Welded-beam	2.01115	2.16736	2.19740	1.74646	1.727833
Speed reducer	3079.262	2998.063	3000.005	2996.113	2994.805

In Friedman aligned ranks test, for each problem the difference of each method with respect to the average value for all methods is considered, and next all values of all problems are ranked together. Table 3 shows the results of applying an Aligned Friedman test, considering five metaheuristics and eleven case studies (that is, putting together all the case studies from both benchmarks). As it can be seen, RFD obtains the highest overall score, with a very small difference over SSO. In fact, a more detailed analysis using Holm's procedure shows that there is not an statistical relevant difference between RFD and SSO. Although the null hypothesis can not be rejected to differentiate RFD and SSO when considering both benchmarks together, we can try to analyze each case study independently. In this case, it is worth to mention that SSO outperforms RFD in the first set of examples, while RFD outperforms SSO in the case of the real-world engineering optimization problems. That is, RFD behaves better when the problems are harder.

Table 3. Ranking aligned friedman results

Metaheuristic	Ranking
RFD	21,5455
SSO	22,7273
ABC	26,7273
PSO	29,9545
FF	39,0455

4 Conclusions

We have provided a first approach to adapt RFD to deal with continuous domain optimization problems. The results we have obtained are promising. In particular, RFD obtains competitive results against ABC, PSO, FF, and SSO. However, there is still plenty of space for improvement. Let us remark that in our current approach we do not take profit from a basic RFD issue in discrete domains: erosion. In fact, our main line of current work is integrating erosion into continuous RFD. The basic idea is to use an alternative fitness function f' recording erosion, where this f' function is computed by using a data structure that records information about all the positions that have been explored so far by the algorithm.

In addition to including erosion, we are also working on improving the performance of the algorithm by providing a parallel implementation of our metaheuristic. In this sense, we are using the parallel functional language Eden (see e.g. [11,15,17]) to extend our library of parallel versions of metaheuristics (see [26]) to deal with RFD.

Acknowledgments. The authors would like to thank Alberto de la Encina for valuable suggestions about the development of a version of RFD to deal with continuous domain optimization problems.

References

1. Abood, H.G., Sreeram, V., Mishra, Y.: Optimal placement of PMUs using river formation dynamics (RFD). In: 2016 IEEE International Conference on Power System Technology (POWERCON), pp. 1–6, September 2016
2. Amin, S.H., Al-Raweshidy, H.S., Abbas, R.S.: Smart data packet ad hoc routing protocol. Comput. Netw. **62**, 162–181 (2014)
3. Cagnina, L.C., Esquivel, S.C., Coello Coello, C.A.: Solving engineering optimization problems with the simple constrained particle swarm optimizer. Informatica **32**(3), 319–326 (2008)
4. Coello Coello, C.A.: Use of a self-adaptive penalty approach for engineering optimization problems. Comput. Ind. **41**(2), 113–127 (2000)
5. Cuevas, E., Cienfuegos, M.: A new algorithm inspired in the behavior of the social-spider for constrained optimization. Expert Syst. Appl. **41**(2), 412–425 (2014)
6. Dash, S., Dey, S., Joshi, D., Trivedi, G.: Minimizing area of VLSI power distribution networks using river formation dynamics. J. Syst. Inf. Technol. **20**(4), 417–429 (2018)
7. Dorigo, M., Birattari, M., Stutzle, T.: Ant colony optimization. IEEE Comput. Intell. Mag. **1**(4), 28–39 (2006)
8. Dorigo, M., Di Caro, G.: Ant colony optimization: a new meta-heuristic. In: Proceedings of the 1999 IEEE Congress on Evolutionary Computation, CEC 1999, vol. 2, pp. 1470–1477. IEEE (1999)
9. Eberhart, R.C., Shi, Y., Kennedy, J.: Swarm Intelligence. Morgan Kaufmann, Burlington (2001)
10. Guravaiah, K., Leela Velusamy, R.: Energy efficient clustering algorithm using RFD based multi-hop communication in wireless sensor networks. Wirel. Pers. Commun. **95**(4), 3557–3584 (2017)
11. Hidalgo-Herrero, M., Ortega-Mallén, Y., Rubio, F.: Analyzing the influence of mixed evaluation on the performance of Eden skeletons. Parallel Comput. **32**(7–8), 523–538 (2006)
12. Jaberipour, M., Khorram, E.: Two improved harmony search algorithms for solving engineering optimization problems. Commun. Nonlinear Sci. Numer. Simul. **15**(11), 3316–3331 (2010)
13. Karaboga, D., Akay, B.: A modified artificial bee colony (ABC) algorithm for constrained optimization problems. Appl. Soft Comput. **11**(3), 3021–3031 (2011)
14. Kennedy, J.: Particle swarm optimization. In: Sammut, C., Webb, G.I. (eds.) Encyclopedia of Machine Learning, pp. 760–766. Springer, Boston (2011). https://doi.org/10.1007/978-0-387-30164-8

15. Klusik, U., Peña, R., Rubio, F.: Replicated workers in Eden. In: Constructive Methods for Parallel Programming (CMPP 2000). Nova Science (2000)
16. Liang, J.J., et al.: Problem definitions and evaluation criteria for the CEC 2006 special session on constrained real-parameter optimization. J. Appl. Mech. **41**(8), 8–31 (2006)
17. Loogen, R.: Eden – parallel functional programming with Haskell. In: Zsók, V., Horváth, Z., Plasmeijer, R. (eds.) CEFP 2011. LNCS, vol. 7241, pp. 142–206. Springer, Heidelberg (2012). https://doi.org/10.1007/978-3-642-32096-5_4
18. López, N., Núñez, M., Rodríguez, I., Rubio, F.: Introducing the golden section to computer science. In: Proceedings First IEEE International Conference on Cognitive Informatics, pp. 203–212. IEEE (2002)
19. Rabanal, P., Rodríguez, I., Rubio, F.: Using river formation dynamics to design heuristic algorithms. In: Akl, S.G., Calude, C.S., Dinneen, M.J., Rozenberg, G., Wareham, H.T. (eds.) UC 2007. LNCS, vol. 4618, pp. 163–177. Springer, Heidelberg (2007). https://doi.org/10.1007/978-3-540-73554-0_16
20. Rabanal, P., Rodríguez, I., Rubio, F.: Solving dynamic TSP by using river formation dynamics. In: Fourth International Conference on Natural Computation (ICNC 2008), pp. 246–250. IEEE (2008)
21. Rabanal, P., Rodríguez, I., Rubio, F.: Applying river formation dynamics to solve NP-complete problems. In: Chiong, R. (ed.) Nature-Inspired Algorithms for Optimisation. SCI, vol. 193, pp. 333–368. Springer, Heidelberg (2009). https://doi.org/10.1007/978-3-642-00267-0_12
22. Rabanal, P., Rodríguez, I., Rubio, F.: Applying RFD to construct optimal quality-investment trees. J. Univers. Comput. Sci. **16**(14), 1882–1901 (2010)
23. Rabanal, P., Rodríguez, I., Rubio, F.: Studying the application of ant colony optimization and river formation dynamics to the steiner tree problem. Evol. Intell. **4**(1), 51–65 (2011)
24. Rabanal, P., Rodríguez, I., Rubio, F.: Applications of river formation dynamics. J. Comput. Sci. **22**, 26–35 (2017)
25. Redlarski, G., Dabkowski, M., Palkowski, A.: Generating optimal paths in dynamic environments using river formation dynamics algorithm. J. Comput. Sci. **20**, 8–16 (2017)
26. Rubio, F., de la Encina, A., Rabanal, P., Rodríguez, I.: A parallel swarm library based on functional programming. In: Rojas, I., Joya, G., Catala, A. (eds.) IWANN 2017. LNCS, vol. 10305, pp. 3–15. Springer, Cham (2017). https://doi.org/10.1007/978-3-319-59153-7_1
27. Rubio, F., Rodríguez, I.: Water-based metaheuristics: how water dynamics can help us to solve NP-hard problems. Complexity (2019)
28. Yang, X.-S.: Firefly algorithm, Lévy flights and global optimization. In: Bramer, M., Ellis, R., Petridis, M. (eds.) Research and Development in Intelligent Systems XXVI, pp. 209–218. Springer, London (2010). https://doi.org/10.1007/978-1-84882-983-1_15

Go for Parallel Neural Networks

David Turner$^{(\boxtimes)}$ and Erich Schikuta$^{(\boxtimes)}$

Faculty of Computer Science, University of Vienna, RG WST, 1090 Vienna, Austria
davidturner@gmx.at, erich.schikuta@univie.ac.at

Abstract. Training artificial neural networks is a computationally intensive task. A common and reasonable approach to reduce the computation time of neural networks is parallelizing the training. Therefore, we present a data parallel neural network implementation written in Go. The chosen programming language offers built-in concurrency support, allowing to focus on the neural network instead of the multi-threading. The multi-threaded performance of various networks was compared to the single-threaded performance in accuracy, execution time and speedup. Additionally, two alternative parallelization approaches were implemented for further comparisons. Summing up, all networks benefited from the parallelization in terms of execution time and speedup. Splitting the mini-batches for parallel gradient computation and merging the updates produced the same accuracy results as the single-threaded network. Averaging the parameters too infrequently in the alternative implementations had a negative impact on accuracy.

Keywords: Neural network simulation · Backpropagation ·
Parallelization · Go programming language

1 Introduction

Parallelization is a classic approach for speeding up execution times and exploiting the full potential of modern processors. Thus, we present in this paper a data parallel implementation of the training phase of artificial neural networks and show the feasibility of the Go language as implementation framework. Go is a fairly new programming language that makes concurrent programming easy. With the optimal number of threads, Go programs can fully exploit modern multi-core processors. Still, not every algorithm has to profit from multi-core execution, as parallel execution might add a non-negligible overhead. This can also be the case for data parallel neural networks, where accuracy problems usually occur, as the results have to be merged. The authors of [5] present a parallel backpropagation algorithm additionally dealing with the accuracy problem by using a MapReduce and Cascading model. In [3], two novel parallel training approaches are presented for face recognizing backpropagation neural networks. The authors use the OpenMP environment for classic CPU multithreading and CUDA for parallelization on GPU architectures. Aside from that, they differentiate between structural data parallelism and topological data parallelism.

© Springer Nature Switzerland AG 2019
I. Rojas et al. (Eds.): IWANN 2019, LNCS 11507, pp. 833–844, 2019.
https://doi.org/10.1007/978-3-030-20518-8_69

Most work will just differentiate between data parallelism and model parallelism. [8] offers a comparison of different parallelization approaches on a cluster computer. The results differ depending on the network size, data set sizes and number of processors. Besides parallelizing the backpropagation algorithm for training speed up, alternative training algorithms like the Resilient Backpropagation described in [10] might lead to faster convergence. One major difference to standard backpropagation is that every weight and bias has a different and variable learning rate. A detailed comparison of both network training algorithms is given in [9] in the case of spam classification. The authors conclude, that resilient backpropagation can be a promising choice for training neural networks for time-sensitive machine learning applications.

The paper is structured as follows: In the next section we introduce the Go programming language and its characteristics. In Sect. 3 we lay out the fundamentals and techniques of the artificial neural network model and our parallelization approach. Accuracy, performance and speedup analysis is presented in Sect. 4. The lessons learned form our approach are listed in Sect. 5. The paper closes with the conclusion of our work done.

2   The Go Programming Language

Go, often referred to as Golang, is a compiled, statically typed, open source programming language developed by a team at Google and released in November 2009. It is distributed under a BSD-style license, meaning that copying, modifying and redistributing is allowed under a few conditions. It is designed to be expressive, concise, clean and efficient [1]. Hence, Go compiles quickly and is as easy to read as it is to write. That being said, built-in support for concurrency is one of the most interesting aspects of Go, offering a great advantage over older languages like C++ or Java. One major component of Go's concurrency model are goroutines, which can be thought of as lightweight threads with a negligible overhead, as the cost of managing them is cheap compared to threads. If a goroutine blocks, the runtime automatically moves any blocking code away from being executed and executes some runnable code, leading to high-performance concurrency [7]. Communication between goroutines takes place over channels, which are derived from "Communicating Sequential Processes" found in [2]. A channel can be used to send and receive messages from the type associated with it. Since receiving can only be done when something is being sent, channels can be used for synchronization, preventing race conditions by design.

3   Fundamentals

Artificial Neural Networks are networks made of artificial neurons. The first artificial neuron was invented in 1943 by Warren McCulloch and Walter Pitts, known as the McCulloch-Pitts-Neuron and described in [6]. It is the simplest form of an artificial neuron, only accepting binary input. If the summed input is larger than a certain threshold value, the neuron's output is 1, else 0. Today's neurons have weighted, real number inputs and use an activation function instead of a threshold value. Although many different kinds of ANNs exist, artificial neurons are typically organized in layers. The first and the last layer are also called input and output layer. If the signal in a network moves in only one direction, that is to say from the input to the output layer, the network is called a *feedforward neural network*, also known as *multilayer perceptron (MLP)*.

Forwardpropagation. To calculate an output in the last layer, the input values need to get propagated through each layer. This process is called forward propagation and is done by applying an activation function on each neuron's corresponding input sum. The input sum z for a neuron k in the layer l is the sum of each neuron's activation a from the last layer multiplied with the weight w:

$$z_k^l = \sum_j (w_{kj}^l a_j^{l-1} + b_k^l) \tag{1}$$

Backpropagation. For proper classification the network obviously has to be trained beforehand. In order to do that, a cost function, telling us how well the network performs, like the cross entropy error with expected outputs e and actual outputs x,

$$C = -\sum_i e_i log(x_i) \tag{2}$$

has to be defined. The aim is to minimize the cost function by finding the optimal weights and biases with the gradient descent optimization algorithm. Therefore, a training instance gets forward propagated through the network to get an output. Subsequently, it is necessary to compute the partial derivatives of the cost function with respect to each weight and bias in the network:

$$\frac{\partial C}{\partial w_{kj}} = \frac{\partial C}{\partial z_k} \frac{\partial z_k}{\partial w_{kj}} \tag{3}$$

$$\frac{\partial C}{\partial b_k} = \frac{\partial C}{\partial z_k} \frac{\partial z_k}{\partial b_{kj}} \tag{4}$$

As a first step, $\frac{\partial C}{\partial z_k}$ needs to be calculated for every neuron k in the last layer L:

$$\delta_k^L = \frac{\partial C}{\partial z_k^L} = \frac{\partial C}{\partial x_k^L} \varphi'(z_k^L) \tag{5}$$

In case of the cross entropy error function, the error signal vector δ of the softmax output layer is simply the actual output vector minus the expected output vector:

$$\delta^L = \frac{\partial C}{\partial z^L} = x^L - e^L \tag{6}$$

To obtain the errors for the remaining layers of the network, the output layer's error signal vector δ^L has to be back propagated through the network, hence the name of the algorithm:

$$\delta^l = (W^{l+1})^T \delta^{l+1} \odot \varphi'(z^l) \tag{7}$$

$(W^{l+1})^T$ is the transposed weight matrix, $\odot$ denotes the Hadamard product or entry-wise product and φ' is the first derivative of the activation function.

Gradient Descent. Knowing the error of each neuron, the changes to the weights and biases can be determined by

$$\Delta w_{kj}^l = -\eta \frac{\partial C}{\partial w_{kj}^l} = -\eta \delta_k^l x_j^{l-1} \tag{8}$$

$$\Delta b_k^l = -\eta \frac{\partial C}{\partial b_k} = -\eta \delta_k^l \tag{9}$$

The constant η is used to regulate the strength of the changes applied to the weights and biases and is also referred to as the learning rate, x_j^{l-1} stands for the output of the jth neuron from layer $l-1$. The changes are applied by adding them to the old weights and biases.

3.1 Parallelism

A fundamental distinction is made between two parallelization techniques. The approach to parallelism used in the implementation of this work is often referred to as data parallelism, since all threads or goroutines train identical networks concurrently on a different subset of the data, see Fig. 1. This stands in contrast to model parallelism, where the model gets split instead of the data. While model parallelism produces better results on very large neural networks, data parallelism profits from large data sets [8] and performs better on networks where all weights fit into memory.

The difficulty with data parallelism is the combination of the parameters computed by each network copy. One way to merge the weights and biases is by averaging them after each thread finished network training on its subset of the training data. However, this is not recommended, as each individually trained network might find a different optimum with diverging weights and biases. In that case, the combined neural network could perform even worse on the whole data set than the individual networks, which were trained only on a subset of the training examples. Better accuracy can be achieved by synchronizing the individual networks from time to time during training. Therefore, the parameters of all neural network copies are averaged periodically, i.e. after a specified number of mini-batches or after a few minutes. The resulting parameters are then broadcasted back to the copies.

Another technique is merging the parameter updates instead of the weights and biases themselves. This can happen asynchronously or synchronously. The asynchronous variant has an increased speedup, as the threads do not wait for each

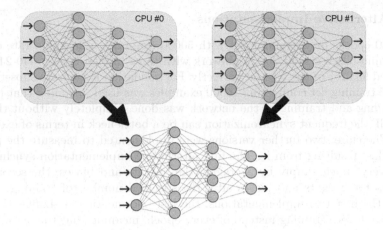

Fig. 1. Data parallelism. Identical networks train on different subsets of the data

other. In each iteration, a network instance requests the current parameters from a separate thread, computes the gradients for a mini-batch and communicates them to the parameter thread, which applies the updates [12]. By the time one thread finishes its mini-batch, the parameters might have already been updated several times. This is called gradient staleness and usually results in accuracy loss, as the computed gradient is no longer parallel to the steepest descent direction at the current weights and biases [11]. The implementation presented in the following section is a synchronized version. Each thread computes the gradients for a subset of the mini-batch. When all gradients of the mini-batch are computed by the network copies and accumulated, one copy updates its weights and biases accordingly and broadcasts its new parameters to the other copies. Splitting the mini-batches and synchronizing after every mini-batch update produces exactly the same accuracy results of the network as the sequential algorithm would do. This comes at the cost of a decreased speedup because of the frequent synchronization. Evidently, the algorithms speedup profits from larger mini-batch sizes.

4 Performance Evaluation

4.1 Data Set

The chosen MNIST database contains a training set of 60000 examples and a test set of 10000 examples of handwritten digits [4]. It is a widely used data set for benchmarking classification algorithms. The training data was further split into 50000 training examples and 10000 validation examples. The network trains on the 50000 examples, while the evaluation and tuning of the parameters use the validation set for better generalization. The final performance evaluation is done on the test set.

4.2   Alternative Implementations

After 30 epochs on a training set with 50000 examples, a learning rate of 0.25 and a mini-batch size of 256, a network with three hidden layers (440-240-120) classified 98.31% of the test set correctly. For validation a separate subset of the MNIST training set comprising 10000 examples was used, meaning that parameter finding and training of the network was done completely without the test set. Still, the frequent synchronization can be a bottleneck in terms of execution time. Therefore, two further versions were implemented to measure the performance loss resulting from it. The first alternative implementation synchronizes after every epoch simply by averaging the weights and biases, the second one averages the weights and biases after the specified number of training epochs. Unlike the first two implementations, the last one mentioned shuffles the data only once before training instead of every epoch, meaning that the network will train in the same order and on the same batches for the configured number of epochs.

4.3   Environment

The tests were carried out on three different systems. The first system has a dual Intel Xeon X5570 quad-core processor setup and 24 GB memory. The processors work with a basic clock rate of 2,93 GHz per physical core and support hyper-threading (thus 16 logical cores in total). Each processor has an 8 MB L3-Cache. The second system utilizes an AMD Ryzen Threadripper 1950X hexadeca-core processor and 32 GB memory. The processor comprises 32 logical cores, a 32 MB L3-Cache and works with a basic clock rate of 3,4 GHz per physical core. The third system has an Intel Core i7-5820K hexa-core processor and 16 GB memory. The processor has a 15 MB L3-Cache and also supports hyper-threading. The basic clock rate is 3,3 GHz. The first system will be further referred to as 8-Core, the second one as 16-Core and the third one as 6-Core.

4.4   Performance

Each test was performed with the default source (=goSeed(1)) of the *math/rand* package, a fixed learning rate of 0.25 and 15 epochs. The networks were trained on 50000 training set examples. As described earlier, *implementation 1* synchronizes after every mini-batch, *implementation 2* after every epoch and *implementation 3* at the end of the training (depending on the configured number of epochs).

Accuracy. Implementation 2 & 3 can not keep up with the first version's accuracy when multithreaded because the parameters get synchronized too infrequently. This happens regardless of the network size. Shuffling the data and averaging the weights and biases after every epoch improves the overall accuracy of implementation 2 a bit, but does not affect the accuracy loss curve when

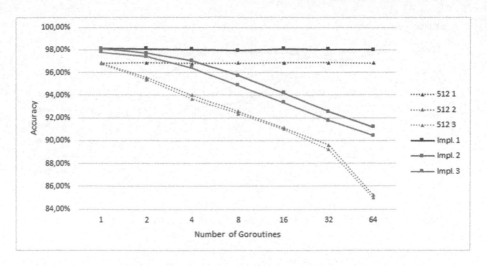

Fig. 2. 784-800-10 network accuracy on the test data set with a mini-batch size of 128 (solid line) and a mini-batch size of 512 (dotted line)

compared to implementation 3. Decreasing the number of neurons from 800 to 450 and 100 increases the error rate in general, with the smallest network performing worst. As seen in Fig. 2, all three implementations profit from a smaller mini-batch size and drop in accuracy with a larger size, although networks with more than one hidden layer generalize better with slightly larger mini-batch sizes. Splitting 800 neurons into three hidden layers (440-240-120) produces a bit more irregular results but generally improves the accuracy. Splitting 450 as well as 100 hidden neurons (250-135-65/55-30-15) affects the curves in a similar way, although in those cases implementation 1 does not profit from the deeper networks in each instance because of the irregularities, which get stronger the smaller the deep network is. Figure 3 shows a comparison between a network with 450 hidden neurons in one hidden layer and a network with 450 hidden neurons split in three layers. Especially implementation 2 produces noticeably better results with the deeper network. The accuracy results do not differ between the systems, since all three computers performed the tests with the same seed.

Execution Time. Shuffling and synchronizing after every epoch in implementation 2 produces marginal additional execution time overhead compared to implementation 3. While the difference can be seen in networks with 100 neurons in the hidden layers, where training time is short in general, it vanishes in larger ones, where the training takes considerably more time. Both versions outperform implementation 1 in every test case, although the results converge with a greater mini-batch size as seen in Fig. 4. Besides impacting the error rate, splitting neurons into three hidden layers also reduces the training time significantly in comparison to networks with the same number of hidden neurons in one hidden layer. Figure 5 shows the execution times of the two 450 hidden

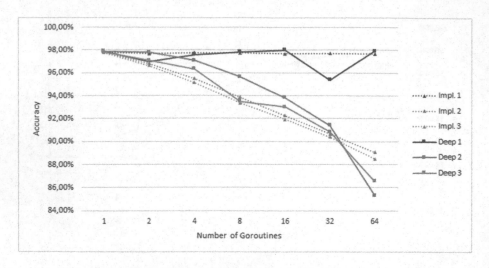

Fig. 3. 784-250-135-65-10 network accuracy (solid line) on the test data compared to a 784-450-10 network (dotted line) with a mini-batch size of 256

neuron networks compared in Fig. 3. Interestingly, the 16-core machine and the 8-core machine generally both perform best when run with eight goroutines. The 6-Core system runs fastest with four goroutines, although it should be stated, that the tests were only carried out with powers of two. Most of the time, the 16-Core and the 6-Core system already outperformed the 8-Core systems peak performance with only two goroutines, showing that the performance per core had a greater impact on the training time than the number of cores.

Speedup. The three test systems achieve different speedup results. While the 16-Core machine performs best in networks with 100 or 450 hidden neurons, the 8-Core machine shows a greater speedup than the others in networks with 800 hidden neurons.

All three systems show better speedup values when training smaller networks, although the 8-Core machine scales equally and sometimes even better in networks with 800 hidden neurons than in networks with 450 hidden neurons. This can be seen in Fig. 6 and compared with the results of the 6-Core and the 16-Core system visualized in Figs. 7 and 8. Figure 6 also shows the exceptional results of the 8-Core system in networks with 800 hidden neurons, where the speedup of implementation 1 and 2 increases even with 64 goroutines. This happens regardless of the mini-batch size. Although having lower peak values, the 8-Core machine also scales better than the other systems in networks with 450 hidden neurons. While the speedup from implementation 1 with up to eight goroutines and a mini-batch size of 512 is comparable with the speedup from implementation 2 and 3, it decreases with a lower mini-batch size as already mentioned earlier and shown in Fig. 4. Networks with three hidden layers achieve a better speedup than networks with the same

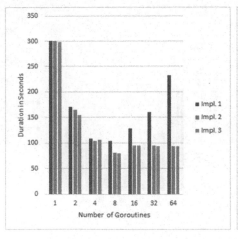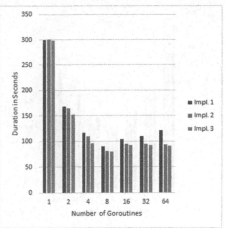

Fig. 4. 8-Core 784-100-10 network execution time with a mini-batch size of 128 on the left and a mini-batch size 512 on the right

number of neurons in one hidden layer. Overall, the 8-Core system scales best but runs slowest with its two quad-core processors.

5   Lessons Learned and Future Work

5.1   Parallelization

Data parallel neural networks can achieve a great speedup compared to single threaded networks. Implementation 1, where the mini-batches get split into equal parts for each thread, probably has limited possible applications, despite the good accuracy rates. If large mini-batches are usable on the chosen data set and not too many threads are used, it can be a valuable choice. Otherwise, particularly with a higher number of threads, parameter averaging with more frequent synchronization than in implementation 2 and 3, or alternative strategies, for example asynchronous ones, should be chosen. In general, data parallelism is a simple and effective parallelization strategy, preferably for large data sets.

5.2   Deep Neural Networks

The used hyper-parameters, which were mostly chosen by trial and error, like the learning rate, are probably not the best ones in each instance. Still, the results were satisfying, especially for networks with three hidden layers. The initial purpose was to see if networks executed faster when the neurons from one hidden layer were split into more hidden layers. Surprisingly, the deep networks, although not being that deep in fact, proved to be not only noticeably faster but also more accurate. 800 hidden neurons were split to 440 neurons in the first

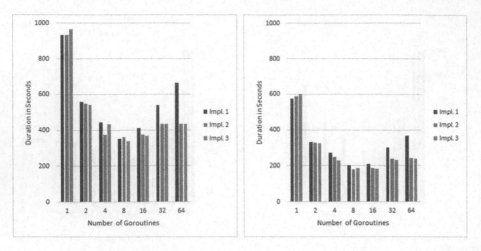

Fig. 5. 16-Core 784-450-10 network (left) execution time compared to a 784-250-135-65-10 network (right) with a mini-batch size of 256

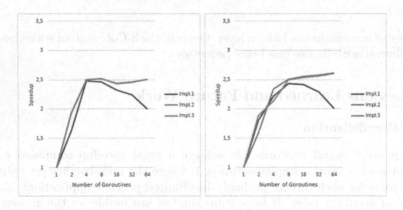

Fig. 6. 8-Core 784-450-10 network (left) speedup compared to a 784-800-10 network (right) with a mini-batch size of 512

hidden layer, 240 in the second and 120 in the third. 450 and 100 hidden neurons were split in a similar way (250-135-65/55-30-15). Distributing the neurons this way seems to be a good choice.

5.3 Matrix Multiplication

Unfortunately, the implementation presented in this paper does not exploit the full potential of mini-batches. The features of each training instance are stored in a vector and multiplied with the weight matrices. The same happens with the error in the backpropagation phase. Instead of doing these series of matrix-vector multiplications in a less efficient for-loop, it is highly recommended to

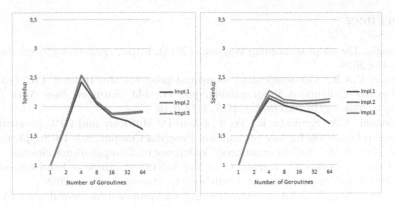

Fig. 7. 6-Core 784-450-10 network (left) speedup compared to a 784-800-10 network (right) with a mini-batch size of 512

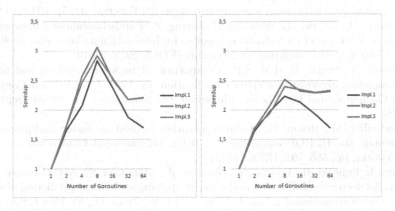

Fig. 8. 16-Core 784-450-10 network (left) speedup compared to a 784-800-10 network (right) with a mini-batch size of 512

save the features of all mini-batch examples in a weight matrix and to perform a more efficient matrix-matrix multiplication for each mini-batch. Particularly GPUs should profit greatly from this variation.

6   Conclusion

Data parallelism proved to be an efficient parallelization strategy. In combination with the programming language Go, a parallel neural network implementation is programmed as fast as a sequential one, as no special efforts are necessary for concurrent programming thanks to Go's concurrency primitives, which offer a simple solution for multithreading. Further work will focus on comparing the parallel performance of neural networks written in Go with implementations written in other languages. Necessary to that end is the improvement of the Go implementation, starting by catching up on the missed opportunities mentioned above.

References

1. Google: The go programming language (2018). https://golang.org/doc/. Accessed 06 Jan 2018
2. Hoare, C.A.R.: Communicating sequential processes. In: Hansen, P.B. (ed.) The Origin of Concurrent Programming, pp. 413–443. Springer, New York (1978). https://doi.org/10.1007/978-1-4757-3472-0_16
3. Huqqani, A.A., Schikuta, E., Ye, S., Chen, P.: Multicore and GPU parallelization of neural networks for face recognition. Procedia Comput. Sci. **18**(Supplement C), 349–358 (2013). 2013 International Conference on Computational Science
4. LeCun, Y., Cortes, C., Burges, C.J.: The MNIST database of handwritten digits (1998). http://yann.lecun.com/exdb/mnist/. Accessed 06 Jan 2018
5. Liu, Y., Jing, W., Xu, L.: Parallelizing backpropagation neural network using mapreduce and cascading model. Comput. Intell. Neurosci. **2016**, 11 (2016)
6. McCulloch, W.S., Pitts, W.: A logical calculus of the ideas immanent in nervous activity. Bull. Math. Biophys. **5**(4), 115–133 (1943)
7. Meyerson, J.: The go programming language. IEEE Softw. **31**(5), 104 (2014)
8. Pethick, M., Liddle, M., Werstein, P., Huang, Z.: Parallelization of a backpropagation neural network on a cluster computer. In: International Conference on Parallel and Distributed Computing and Systems (PDCS 2003) (2013)
9. Prasad, N., Singh, R., Lal, S.P.: Comparison of back propagation and resilient propagation algorithm for spam classification. In: Fifth International Conference on Computational Intelligence, Modelling and Simulation, pp. 29–34, September 2013
10. Riedmiller, M., Braun, H.: A direct adaptive method for faster backpropagation learning: the RPROP algorithm. In: IEEE International Conference on Neural Networks, pp. 586–591. IEEE (1993)
11. Sato, I., Fujisaki, R., Oyama, Y., Nomura, A., Matsuoka, S.: Asynchronous, data-parallel deep convolutional neural network training with linear prediction model for parameter transition. In: Liu, D., Xie, S., Li, Y., Zhao, D., El-Alfy, E.S.M. (eds.) ICONIP 2017. LNCS, vol. 10635, pp. 305–314. Springer, Cham (2017). https://doi.org/10.1007/978-3-319-70096-0_32
12. Zhang, W., Gupta, S., Lian, X., Liu, J.: Staleness-aware async-SGD for distributed deep learning. In: Proceedings of the Twenty-Fifth International Joint Conference on Artificial Intelligence, pp. 2350–2356. AAAI Press (2016)

Using Boolean- and Self-Enforcing-Networks for Mathematical E-Tutorial Systems

Christina Klüver$^{(\boxtimes)}$ and Jürgen Klüver

CoBASC Research Group, Universitätsstr. 12, 45117 Essen, Germany
{christina.kluever,juergen.kluever}@uni-due.de

Abstract. Mathematical thinking as an important instrument in science is a stumbling block for many students in the first years. A lot of investigations occur to help the students understanding the principles of mathematics. The proposed tutorial system for the basics focuses on the analysis and visualizations of the solution algorithms and solution processes with Boolean Networks and Self-Enforcing Networks. The students can check not only the correctness of their results, but also if the solution steps are complete. In addition, in case of wrong results the students check in which step of the solution they made a mistake and what kind of mistake. The goal is to promote the explorative learning and to help understanding the problems through self-recognition.

Keywords: Boolean Networks · Self-Enforcing Networks ·
Visualization of solution algorithms · Tutorial systems

1 Introduction

Many students develop "math anxiety", which often begins as early as the second grade (for an overview [1, 2]). Numerous studies have worldwide taken place over the last decades to understand the reasons for the different anxiety forms, the accordingly learning disorders [3–6], and to try different "feed-back models" [7].

Concepts as "Massive Open Online Courses" (MOOCs), "Learning Management Systems" (LMS) were over years developed, research fields as "Learning Analytics" (LA) and "Educational Data Mining" (EDM) have emerged [8, 9].

AI- and Machine Learning-techniques such as Fuzzy-Logic, Decision Tree, Bayesian Networks, Neural networks (Back Propagation and Self-Organized Maps), Genetic Algorithms and Hidden Markov Models are used for developing E-Learning platforms, and "adaptive educational systems", or to analyze the learning behavior [8, 10]. Accordingly, "Intelligent Tutoring Systems" (e.g. [11–15]) or a "digital pen learning system" (DPLS) [16] are at disposal to improve the knowledge in mathematics, to name only a few.

Most of the developed systems verify only the end results of given math problems, but not the solution process, which is important to enable an individual feedback and to find out possible knowledge gaps. In this article such a system is described, namely the coupled system of Boolean Networks and Self-Enforcing Networks to enable a differentiated analysis and the visualization of the false solutions and correct ones in a tutorial system.

© Springer Nature Switzerland AG 2019
I. Rojas et al. (Eds.): IWANN 2019, LNCS 11507, pp. 845–856, 2019.
https://doi.org/10.1007/978-3-030-20518-8_70

Although the importance of visualization is also analyzed for decades [17, 18], no study was found with visualizations of the solution processes and the solution algorithm. The used techniques in this study allow different dynamical visualizations, as shall be shown. The goal of the project is to promote the explorative learning.

The remaining organization of the paper is as follows: in the next section the conceptual development is presented. In section three the used techniques are shortly described, followed by first results in section four. Finally, further investigations are discussed.

2   Conceptual Development

The proposed method to solve mathematic problems with an E-tutorial system is to give not only the problems and results, but also a complete "solving algorithm" and the solution steps; the advantage is that each step can be checked individually. In addition, the solving algorithms can be visualized for each problem, depending on the errors in respect to the solution steps and/or to the arguments.

The whole system consists of different components: the first system contains different classes of problems, which are defined in Boolean Networks (BN); the second system, a Self-Enforcing Network (SEN) is generated by the BN consisting of the correct solution. The students solve a problem in the third system, e.g. on a web-platform, using LaTeX rendering of math formulas or clicking through mathematical symbols, evaluated by the SEN. The feedback is dynamically delivered by the BN (Fig. 1).

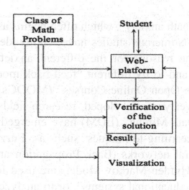

Fig. 1. An overview of the whole system

Depending on the exercise level the student should first solve the problems according to the complete solving algorithm, meaning e.g. that mathematical arguments shall not be summarized. This can be done only on a higher exercise level.

As an example, for such a general solving algorithm the problem of quadratic equations using the pq-formula is demonstrated (Fig. 2):

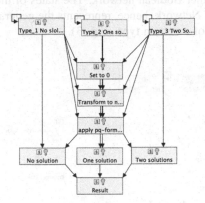

Fig. 2. The solving algorithm

In this case the general solving algorithm requires following steps:

– set the equation to 0
– transform the equation into the normal form
– apply the pq-formula
– specify the result

This example contains three different mathematical "types", each defined according to the general solving algorithm, and enables to give a large number of problems with no-, one-, or two concrete solutions.

The verification of the solutions is done by the network-systems, containing all problems, solving algorithms, and the according solving processes. Each solution of the user will be checked for completeness and correctness.

It is also important to mention that a learner does not know how many steps and solutions are needed; a learner has only the information that a problem must completely be solved (see next section).

As in any leaning system, it is not possible to anticipate every possible mistake or different solutions (which can be correct, but "complicated") by all users. To avoid a simple "false"-output if a solution is not as expected, we have introduced the node "unknown". In this case a teacher or a tutor can in addition analyze the results.

For concretization reasons, first the used techniques are briefly described.

3   Boolean Networks, Self-Enforcing Networks, and the User Platform

The *Boolean Network* (BN) contains nodes, Boolean variables and according functions. The topology is as usual preserved in the adjacent matrix and the dynamics is accordingly determined by the functions. We have developed a tool for BN, which allows an easy construction of the adjacent matrix and the according functions [19]. Additionally the definition of "complex nodes" is also possible, meaning that each node can consist of another Boolean network. The states of the nodes are dynamically visualized. The Boolean Network, hence, represents the general solution algorithm that is needed for problems of a certain type (Fig. 3).

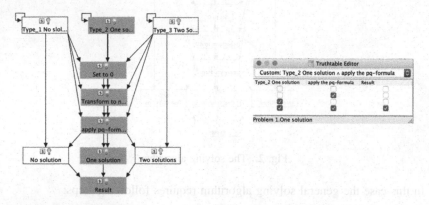

Fig. 3. A visualized solution algorithm for a problem of Type_2 (One solution). On the right side the "Truth table Editor" allows an easy definition of the functions.

For mathematical problems the nodes describe the problems and the different steps that are necessary for the specific solution.

As mentioned, in this example 3 different "problem-types" of quadratic equations are inserted into the BN and also the according solution processes.

To give an example, the following problem should be solved (Type_2 Problem_1):

$$3x^2 - 30x + 70 = -5 \tag{1}$$

This quadratic equation represents a concrete mathematical problem of Type_2 (Fig. 2), meaning that the node is turned-off to "true". According to the solving algorithm, the following nodes, which contain the correct solutions, are only activated if each step is correctly solved (Fig. 3).

The visualization shows dynamically step-by step the respective next activated node. In the presented case the given problem of a quadratic equation ends after 4 steps, i.e. a solution has been found. In the last step the given result is in addition checked. All solving algorithms are passed automatically to the next system, namely to the self-organized learning neural network:

The *Self-Enforcing Network* (SEN) is developed by our research group [19–23]. The task of a SEN is usually the structuring of data sets. The units of a SEN are chiefly (a) specific objects, and (b) attributes that characterize the objects. The affiliation of the different attributes to the specific objects is mainly inserted to the SEN via a semantical matrix. This matrix generates according to certain rules the initial weight matrix of the SEN.

In this case SEN contains as attributes the solving algorithms and the concrete formula for each solving process; the number of the attributes depend on the number of the solving steps.

The objects are the numbers of the math problems, in this case Type_1 to Type_3. A teacher has in addition the possibility to insert typical "mistakes" as objects, e.g. the object "result is false" and to include a feed-back according to the error.

For the same example the attributes in SEN are as follows (Fig. 4):

Attributes
✚ ✖ ⬆ ⬇ ⊢ ⊣
Compact View   Expert View
Name
No solution
One solution
Two solutions
Set to 0
Transform to normal form
apply pq–formula
3x^2-30x+75=0
x^2-10x+25=0
x1=-(-10)/2+√(((-10)/2)^2-25)∧x2=-(-10)/2-√(((-10)/2)^2-25)
X=5

Fig. 4. Excerpt of the attributes for a concrete problem of Type_2

The semantical matrix, which is essential for the SEN contains the attributes **a** and objects **o**, covering the correct and complete solutions for each quadratic equation, which should be solved. In the example the attributes for the concrete mathematical problem have the value 1.0 in the correspondent matrix components; the other attributes belonging to other mathematical problems all have the values 0.0 (Fig. 5).

Semantic Matrix											⬜ ☒
✚ ✖ ⬆ ⬇ ⤜ ⬚ ⬚ ⬜								Filter Rows			✎
Raw Normalized Weighted											
Object Name	No solution	One solution	Two soluti...	Set to 0	Transfor...	apply pq...	3x^2-30...	x^2-10x...	x1=-(-1...	Result	Unkno...
Type_2 Problem_1	0.00	1.00	0.00	1.00	1.00	1.00	1.00	1.00	1.00	5.00	0.00

Fig. 5. Values in the semantical matrix for the given problem

SEN starts as mentioned by analyzing the values of the semantical matrix and by transforming the attribute values **a** of the associated objects **o** into the weight matrix w_{ao} of the network according to the learning rule (2) with the restriction that only values unequal zero are changed:

$$w(t+1) = w(t) + \Delta w, \text{ and}$$
$$\Delta w = c * w_{ao}$$

(2)

c is an equivalent to the learning rate and a constant defined as $0 \leq c \leq 1$.

The weight matrix w_{ao}, hence, is generated from the semantical matrix and not randomly.

Using the linear activation function $a_j = \sum w_{ij} * a_i$ and c = 0.03, SEN needs only one iteration to learn the classification of each problem and the correspondent attributes, namely the solving algorithms and the solving steps.

After the learning process is finished a possible user-solution for a problem is inserted as a new input vector. The results of a SEN system are *visualized* in different ways to allow a fast interpretation [21]: (a) the visualization of the computed similarities according to the highest activated neuron (ranking), and (b) the smallest difference between the vectors (distance). Having both computed values it is possible to ensure unambiguous results. The results of SEN in respect of a new input vector containing a correct solution of the given problem is shown in Fig. 6:

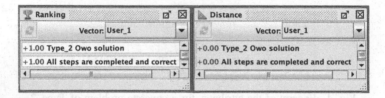

Fig. 6. The computed values of ranking and distance

Both computed results have on the top the correct association with the problem of Type_2 – this object (neuron) is most strongly activated (left side) and the distance to the new input vector is = 0 (right side). A solution can only be unambiguous when the first results are equal referring to the highest activation (ranking) and the smallest difference (distance).

A teacher has this way the possibility to check if the developed solving algorithm, correct solutions, and possible mistakes are well-defined.

To put it into a nutshell: The task of the BN is the representation of the specific solution path for a certain problem. Subsequently the adjacency matrix is transformed into the SEN, which consists of the problems as objects and the according solutions and solution steps as attributes.

This enables the SEN to control if the solving algorithm and the concrete solutions are complete and correct solved by a learner.

For the development of the prototype we have used the server-based system JACK, developed at the University of Duisburg-Essen. Figure 7 shows the user view:

As already mentioned, a learner does not know how many steps are needed to solve the problem. The mathematical expressions can be inserted through the formula editor. The user can decide if an additional step is needed or if the solution should be submitted.

Exercise "Problem 1"

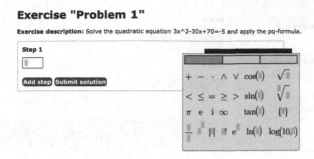

Exercise description: Solve the quadratic equation 3x^2-30x+70=-5 and apply the pq-formula.

Fig. 7. The user view in JACK

4 The Analysis of the Solutions

The potentiality of the whole system is first presented by the given example, assuming that a user has submitted the correct solving algorithm and complete solving steps but the result contains one careless mistake or typing error result.

First, the solution in each step is checked. If a solution is correct, the value in the input vector, representing the actual user, is set to 1; in case of errors the value is 0. Considering that the result is not correct the computed result is shown in Fig. 8:

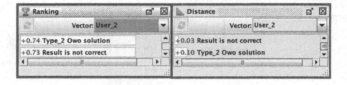

Fig. 8. Result of the SEN-analysis

In contrast to the result shown in Fig. 5 the computed ranking values are lower than 1.0 and the distance is 0.03, which indicates that the result has a mistake.

The result of SEN is transformed into the BN and a user gets the following information (Fig. 9):

On the left side the BN shows that all steps of the solving algorithm are complete and correct; only the result is false. On the right side the submitted solution by the user is also shown on the result page. In this case the user can (hopefully) recognize immediately the error.

The BN shows also if a step is left (Fig. 10). In this case the quadratic equation which should be solved is:

$$2x^2 + 4x = 70 \tag{3}$$

The problem belongs in this case to the "Type_3", namely a problem, which has two solutions:

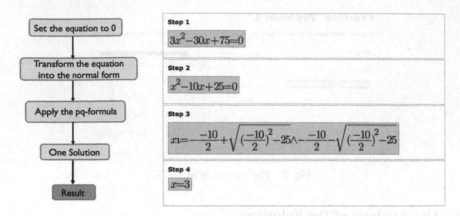

Fig. 9. The (dynamically) visualized results by the BN.

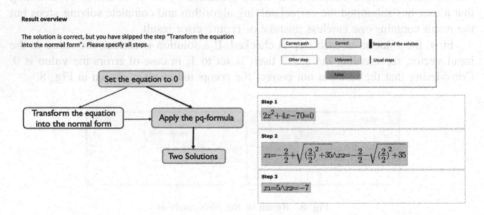

Fig. 10. The valuation and visualization of the submitted solution

Again, first SEN analyzes the solution. For the attribute "Transform the equation into the normal form" no value of 1 is available, meaning that again the end activation of the neuron representing the object type_3 (Problem_1) and the distance are not equal to 1 (ranking), and 0 (distance) respectively. The learner gets immediately the information that the solving algorithms is not complete and has the possibility to insert the missing step.

In the last example a learner has submitted a solution with one (by the system) "unknown" solution step and a false result for the given problem. In this case the analysis of SEN transforms accordingly the results to the BN (Fig. 11):

In this case it seems that a user has not really tried to solve the problem. The system is able to "know" that the solution x = 5 belongs to another problem type (Type_2 Problem 1), but it is false for the actual problem. For the first time the BN shows the whole solving algorithm to help the user to understand the problem class.

Result overview

After the second step your solution shows several weaknesses.
Please have a look to the usage of pq-formula.

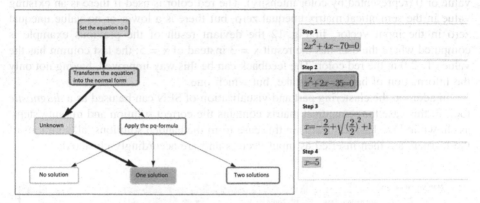

Fig. 11. Submitted solution with several mistakes

5 Conclusion and Further Work

The first promising results show the potential of such a tutorial system for mathematics. Different visualizations can support the students to understand the own possible problems through self-recognition. The prototype uses the web-platform JACK [24], which is a server-based system for the management of computer-aided assessments and exercises at the university of Duisburg-Essen.

In addition, it is planned to analyze how often the "help-buttons" are used to find out if there are knowledge-gasps. Having the processing time for the solutions, one can analyze if some errors occur because the problems are maybe too easy and therefore boring; in this case the level of exercises can be changed. In contrast, if the processing time is long, one can think about starting with simple problems to reach a higher motivation because of the positive feedbacks.

The SEN-Tool has several functionalities, not yet implemented in the prototype, as e.g. a component-wise analysis, checking if there are any differences between the trained data and the input data (Fig. 12).

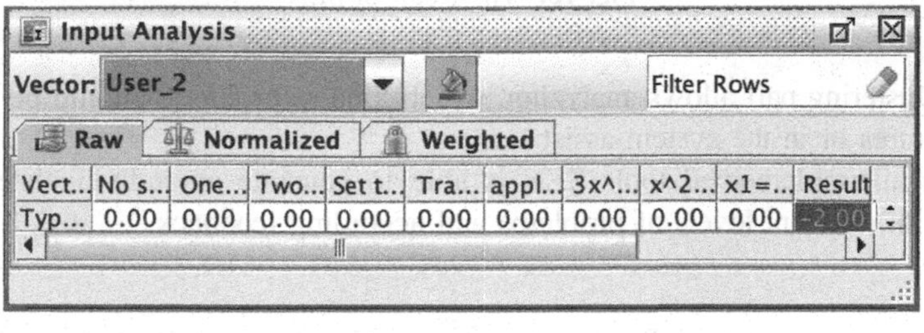

Fig. 12. The result of the component-wise analysis

If there exists no difference, the distance is of course 0.00 and the values are represented without labeling; the green color indicates that the input vector has a positive value for this attribute, the semantical matrix contains an according smaller value or 0 (represented by color intensity). The red color is used if there is an existing value in the semantical matrix unequal zero, but there is a lower or no value unequal zero in the input vector. In Fig. 12 the deviant result of the previous example is computed where the user has the result x = 3 instead of x = 5; the last column has the value of −2.0 in the red color. The feedback can be this way improved, having not only the information of having a mistake, but which one.

In addition the clustering part and visualization of SEN can be used as a *diagnostic* tool. In this case the semantical matrix contains the correct solution and missing steps, as shown in Fig. 13. SEN learns in this case from the handed solutions; all solutions of the learners are then inserted as input vectors and are accordingly clustered:

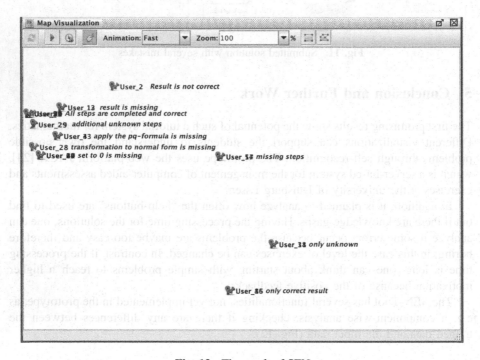

Fig. 13. The result of SEN

The clustering part allows analyzing, which kind of problems should be addressed in the lectures or in the system assistance.

Using all implemented tools SEN is able to generate an individualized tutorial system with different levels of problems and according feedback.

References

1. Sorvo, R., et al.: Math anxiety and its relationship with basic arithmetic skills among primary school children. Br. J. Educ. Psychol. **87**, 309–327 (2017)
2. Marshall, E.M., Staddon, R.V., Wilson, D.A., Mann, E.V.: Addressing maths anxiety and engaging students with maths within the curriculum. MSOR Connections **15**(3), 28–35 (2017)
3. Mercer, C.D., Miller, S.P.: Teaching students with learning problems in math to acquire, understand, and apply basic math facts. Remedial Educ. **13**(3), 19–35 (1992)
4. Ginsburg, H.P.: Mathematics learning disabilities: a view from developmental psychology. J. Learn. Disabil. **30**(1), 20–33 (1997)
5. Duval, R.A.: Cognitive analysis of problems of comprehension in a learning of mathematics. Educ. Stud. Math. **61**(103), 103–131 (2006)
6. Straehler-Pohl, H., Pais, A., Bohlmann, N.: Welcome to the jungle. an orientation guide to the disorder of mathematics education. In: Straehler-Pohl, H., Bohlmann, N., Pais, A. (eds.) The Disorder of Mathematics Education, pp. 1–15. Springer, Cham (2017). https://doi.org/10.1007/978-3-319-34006-7_1
7. Faber, J.M., Luyten, H., Visscher, A.J.: The effects of a digital formative assessment tool on mathematics achievement and student motivation: results of a randomized experiment. Comput. Educ. **106**, 83–96 (2017)
8. Almohammadi, K., Hagras, H., Alghazzawi, D., Aldabbagh, G.: A Survey of artificial intelligence techniques employed for adaptive educational systems within E-learning platforms. JAISCR **7**(1), 47–64 (2017)
9. Dutt, A., Ismail, M.A., Herawan, T.: A systematic review on educational data mining. IEEE Access **5**, 15991–16005 (2017)
10. Montebello, M.: AI Injected e-Learning: The Future of Online Education. Springer, Cham (2017). https://doi.org/10.1007/978-3-319-67928-0
11. AbuEloun, N.N., Abu Naser, S.S.: Mathematics intelligent tutoring system international. J. Adv. Sci. Res. **2**(1), 11–16 (2017)
12. Paiva, R.C., Ferreira, M.S., Frade, M.M.: Intelligent tutorial system based on personalized system of instruction to teach or remind mathematical concepts. J. Comput. Assist. Learn. **33**(4), 370–381 (2017)
13. Xin, Y.P., Tzur, R., Hord, C., Liu, J., Young Park, J., Si, L.: An intelligent tutor-assisted mathematics intervention program for students with learning difficulties learning disability. Quarterly **40**(1), 4–16 (2017)
14. Lin, S., Thomas, D.: Inquiry-based science and mathematics using dynamic modeling. SCIREA J. Math. **2**(2), 28–40 (2017)
15. Chakraborty, U., Konar, D., Roy, S., Choudhury, S.: Intelligent evaluation of short responses for e-learning systems. In: Satapathy, S.C., Prasad, V.K., Rani, B.P., Udgata, S.K., Raju, K. S. (eds.) Proceedings of the First International Conference on Computational Intelligence and Informatics. AISC, vol. 507, pp. 365–372. Springer, Singapore (2017). https://doi.org/10.1007/978-981-10-2471-9_35
16. Huang, C.S.J., Su, A.Y.S., Yang, S.J.H., Liou, H.-H.: A collaborative digital pen learning approach to improving students' learning achievement and motivation in mathematics courses. Comput. Educ. **107**, 31–44 (2017)
17. Duval, R.: Representation, vision and visualization: cognitive functions in mathematical thinking. basic issues for learning. In: Proceedings of the Annual Meeting of the North American Chapter of the International Group for the Psychology of Mathematics Education, pp. 3–26 (1999)

18. Vale, I., Barbos, A.: The importance of seeing in mathematics communication. J. Eur. Teach. Educ. Netw. **12**, 49–63 (2017)
19. Klüver, C., Klüver, J.: Soft computing tools for the analysis of complex problems. In: Impe, J.F.M., Logist, F. (Eds.) Proceedings of the 1st International Simulation Tools Conference & Expo – SIMEX 2013, pp. 23–30 (2013)
20. Klüver, C., Klüver, J.: Self-organized learning by self-enforcing networks. In: Rojas, I., Joya, G., Gabestany, J. (eds.) IWANN 2013. LNCS, vol. 7902, pp. 518–529. Springer, Heidelberg (2013). https://doi.org/10.1007/978-3-642-38679-4_52
21. Klüver, C.: Self-enforcing neworks (SEN) for the development of (medical) diagnosis systems. In: Proceedings of the International Joint Conference on Neural Networks (IJCNN): IEEE World Congress on Computational Intelligence (IEEE WCCI), Vancouver, pp. 503–510 (2016)
22. Klüver, C.: A self-enforcing network as a tool for clustering and analyzing complex data. Procedia Comput. Sci. **108**, 2496–2500 (2017). ICCS, Zürich
23. Klüver, C., Klüver, J., Zinkhan, D.: A self-enforcing neural network as decision support system for air traffic control based on probabilitstic weather forecasts. In: Proceedings of the IEEE International Joint Conference on Neural Networks (IJCNN). Anchorage, AK, pp. 729–736 (2017)
24. Schwinning, N., Kurt-Karaoglu, F., Striewe, M., Zurmaar, B., Goedicke, M.: A framework for generic exercises with mathematical content. In: Proceedings of the International Conference on Learning and Teaching in Computing and Engineering (LaTiCE 2015), pp. 70–75 (2015)

Digital Implementation
of a Biological-Plausible Model
for Astrocyte Ca²⁺ Oscillations

Moslem Heidrapur[iD], Arash Ahmadi[iD], and Majid Ahmadi[✉][iD]

University of Windsor, Windsor, ON N9B 3P4, Canada
{heidarp,arash.ahmadi,m.ahmadi}@uwindsor.ca

Abstract. New findings show that astrocytes are important parts of the information processing in brain and believed to be responsible for some brain diseases such as Alzheimer and Epilepsy. Astrocytes generate Ca²⁺ waves and release neuro-transmitters over a large area. To study astrcoytes, one need to simulate large number of biologically realistic models of these cells alongside neuron models. Software simulation is flexible but slow. On the other hand, hardware simulation has advantage of running parallel, is more energy efficient and much faster. This work presents a digital hardware which can effectively implement nonlinear differential equations of astrocyte. As proof of concept, the design was simulated and implemented on the Field Programmable Gate Array (FPGA) device. As the results indicated, proposed hardware was capable of replicating the astrocyte in cellular level.

Keywords: Neuromorphic · Astrocyte · Digital hardware design · FPGA · Spiking Neural Network (SNN)

1 Introduction

Understanding the information processing algorithms in the brain can help to develop more intelligent, efficient, fault tolerant and much faster computing devices [7,17,30]. Moreover, studying biological reactions in the brain, plays a key role to find the mechanisms underlying neurological and psychiatric diseases [3,9]. Additionally, it could be the initial step toward building a platform to test the drugs made for brain diseases prior to test on live animals [31]. To study brain, first, we need to replicate its components and the way they are interconnected and interact with each other. However, the complexity of the brain, due to the large number of cells [2] and their communication pathways, make this a challenging task. The path to this goal commences by proper modelling and building a simulation platform.

This work focuses on the astrocytes, a type of glial cells which were believed to have only nutritional and supportive roles for neurons. New experimental results, however, shows that astrocytes participate in neuronal plasticity [4,24–27,29], development of neuronal pathologies [3,9], sleep functions [6] and brain

© Springer Nature Switzerland AG 2019
I. Rojas et al. (Eds.): IWANN 2019, LNCS 11507, pp. 857–868, 2019.
https://doi.org/10.1007/978-3-030-20518-8_71

self-repair ability [19]. Although, astrocytes do not generate action potentials the way neurons do but instead, propagate Ca^{2+} waves and release transmitters over a large area and long time span.

Several mathematical models are presented for astrocytes [4,23,26,28,29]. All of these models (except the model in [26] where a high level mathematical specifications are utilized) are biologically-plausible models which describe cellular phenomena and physiological parameters. Comparing these two types of models, the abstract models are simpler to understand and cheaper to simulate whereas they can not investigate dynamics, complexity and emergent nonlinear coherence that arise when large number of neurons and glial cells are coupled [8]. But in general, the choice of the model depends on the scale, purpose and cost of the simulation.

As for simulation platform, it could be computer based or Very Large Scale Integration (VLSI) dedicated hardware. Computers are flexible and available for every one which make them best choice for small networks. But for large networks, VLSI systems are more efficient and affordable in comparison with supercomputers. Field Programmable Gate Arrays (FPGA) provide a viable platform to simulate neural networks [12,13,18,21]. However, limited resources is one of main challenges of large scale neural network FPGA implementations [20].

This work presents an approach to simulate the biologically-plausible astrocyte model presented in the [4] on FPGA. The advantage of this model is incorporation of glutamate regulation of Ca^{2+} waves. The model equations include nonlinear terms as division, non-integer roots, multiplication, quadratic, cubic, quartic etc. which make the FPGA implementation difficult. This work uses linearization techniques to design a hardware capable of replicating the behavior of astrocyte on FPGA. In designing process, search algorithms were utilized to find the most efficient parameters for linearization that both reduce the implementation cost and maintain the accuracy of the model.

Researchers already presented circuits for simplified and abstract models of astrocytes [1,15] while this on-FPGA hardware is capable of simulating the astrocyte down to ions levels. This is very useful to study calcium dynamics and calcium-based learning algorithms that are getting a lot of attentions recently [5, 11,15,25]. The FPGA implementation results is further compared with another work that also implemented a biological realistic model on FPGA [16].

The rest of the paper is organized as follows. The astrocyte model is described in Sect. 2. Section 3 presents linearized models and evaluation of the accuracy of proposed models. Design and hardware implementation are discussed in Sect. 5. The paper concludes in Sect. 6.

2   Background

In general, when an action potential arrives at the pre-synaptic terminal, neurotransmitters release into synaptic cleft as shown in Fig. 1. Among them, glutamate is the major excitatory neurotransmitter in the nervous system and is critically involved in many functions [22]. Glutamate in the synaptic cleft quickly binds to glutamate receptors on the post-synaptic terminal. Activation of these glutamate receptors leads to the depolarization of the post-synaptic terminal and eventually could result in the excitatory post-synaptic potential. The released glutamate may spill over the synaptic cleft and bind to the extra-cellular part of astrocytic Metabotropic GLUtamate Receptors (mGluRs). Binding of glutamate to mGluRs promotes opening of a few Inositol Trisphosphate (IP3) channels. As a consequence, intracellular Ca^{2+} slightly increases. Since the opening probability of IP3 channels nonlinearly increases with Ca^{2+} concentration, such an initial amount boosts the opening probability of neighboring channels which In turn, leads to a further increase of Ca^{2+}.

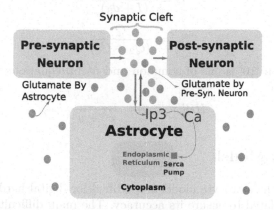

Fig. 1. Mechanism of glial regulation of synaptic transmission. (1-) Release of glutamate (Glu) from pre-synaptic terminal activates astrocyte receptors (2-) evoking an increase in IP3 and consequently Ca^{2+} levels (3-) and release of glutamate from glia.

Action of Ca^{2+} release, reverses at high cytoplasmic Ca^{2+} concentrations, when inactivation of IP3R channels takes place and SERCA pumps, which its activity increases with cytoplasmic Ca^{2+}, quickly pumps back excess cytoplasmic Ca^{2+} into the Endoplasmic Reticulum (ER). The intracellular Ca^{2+} concentration consequently recovers toward basal value which suppresses IP3 channels activation. If the glutamate stimulation continues, intracellular IP3 level remains high enough to repeat the cycle into oscillations of Ca^{2+} and IP3 ions.

Glutamate in astrocyte, acts as modulator of the calcium oscillations which modulates the frequency (FM), amplitude (AM) or combination of both (AFM) [4]. Such oscillations further trigger the release of glutamate by astrocyte which effect the same or other synapses considering it's comparatively larger size. The G-ChI model that describe calcium and IP3 dynamics is shown in Eqs. 1 to 8. The parameters values and descriptions are available in [4].

$$\frac{dI}{dt} = \nu_\beta Hill(0.7, \gamma, K_R(1 + \frac{K_p}{K_r} Hill(1, Ca, K_\pi)))\frac{\nu_\delta}{1 + \frac{I}{k_\delta}} Hill(2, Ca, K_{PLC\delta})$$
$$- \nu_{3K} Hill(4, Ca, K_D) Hill(1, I, K_3) - r_{5P} I \tag{1}$$

$$\frac{dCa}{dt} = (r_C m_\infty^3 n_\infty^3 h(i)^3 + r_L)(C_{ER} - Ca) - \nu_{ER}.Hill(2, Ca, K_{ER}) \tag{2}$$

$$\frac{dh}{dt} = \frac{h_\infty - h(i)}{\tau_h} \tag{3}$$

$$\tau_h = \frac{1}{\alpha_2(Q_2 + C_a)} \tag{4}$$

$$n_\infty = hill(1, C_a, d5) \tag{5}$$

$$m_\infty = hill(1, I, d1) \tag{6}$$

$$Q = \frac{(d_2 I + d_1)}{(I + d_3)} \tag{7}$$

$$h_\infty = \frac{Q}{(Q_2 + C_a)} \tag{8}$$

The Hill equation [10] is one of common and useful equations in bio-chemistry which is given by:

$$Hill(n, x, y) = \frac{x^n}{x^n + y^n} \tag{9}$$

3   Modified Model

In this section, the astrocyte model is modified for digital hardware implementation and simulated to ensure its accuracy. The main difficulty for the circuit implementation of the G-ChI model lies in nonlinearity of the expressions which describe the biochemical reactions. To bypass the problem, these nonlinearities were substituted with equivalent linear terms. The objective is to find a sequence of linear functions, L, so that for domain of nonlinear function N:

$$|L - N| \leq \epsilon \tag{10}$$

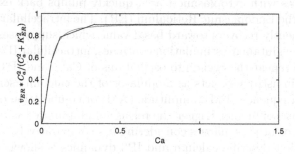

Fig. 2. Computer simulation of the term $\nu_{ER} * C_a^2/(C_a^2 + K_{ER}^2)$ (red line) and it's linear equivalent term (black line) as described in Eq. 11, calculated with $\epsilon = 0.02$. (Color figure online)

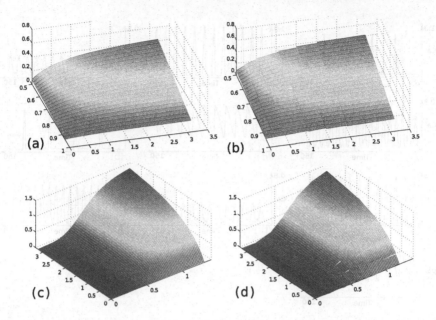

Fig. 3. Computer simulation of the nonlinear and their linear equivalent term as described in Eqs. 14 and 15, obtained with $\epsilon = 0.02$. (a) Simulation of the term $d2*(I+d_1)/(I+d_3)*(1-h_a)$ and (b) it's linear substitute. (c) Simulation of the term $v_{3k}Hill(4, C_a, k_D)Hill(1, I, k_3)$ and (d) it's linear substitute.

where ϵ is maximum acceptable error. For one variable function N, L functions is as follows:

$$L = \alpha x + \beta \tag{11}$$

and for two variable function L, L is:

$$L = \alpha x + \beta y + \delta \tag{12}$$

The algorithm for finding α, β and δ for each segment of two variable L described in follows:

For given numbers of $\{\epsilon,\ f(0,0),\ f(0,1),\ f(0,1), ...\}$ find real numbers $\alpha, \beta, \delta, m, n$ such that:

$$|f(i,j) - (\alpha i + \beta j + \delta)| \leq \epsilon \ \ for \ \ 0 \leq i \leq m \ \ and \ \ 0 \leq j \leq n \tag{13}$$

and $n * m$ is maximal.

The similar algorithm was used for one variable L. The algorithm begins by selecting the largest possible restricted plate (longest possible single line) which satisfies the error restriction and repeats the process by selecting restricted plates

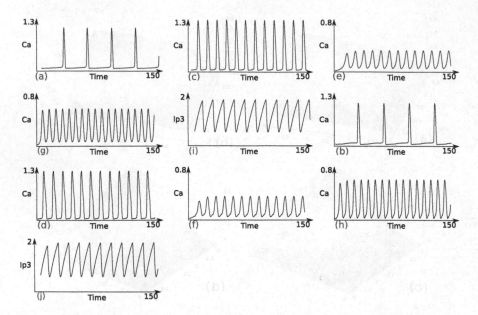

Fig. 4. Computer simulation of the calcium and Ip3 oscillations in original (first row) and linearized (second row) models, correspond to different values of glutamate: (a,b) calcium oscillations for FM mode and glutamate = 0.05, (c,d) calcium oscillations for FM mode and glutamate = 1.5, (e,f) calcium oscillations for AM mode and glutamate = 0.15, (g,h) calcium oscillations for AM mode and glutamate = 1.5, (i,j) Ip3 oscillations for FM mode and glutamate = 0.15.

(lines) of the maximum area (length) from the left-hand sides (side) towards remaining of the function to be approximated. The number of linear segments depends on the value of maximum tolerable error ϵ. For the larger values, there are less number of segments which results in more deviation from the original term and vice versa. To compare the results, the nonlinear equations were linearized using three values of $\epsilon = 0.04$, 0.03 and 0.02. Figure 2 and shows the simulation of one variable term:

$$N1 = \frac{v_{ER}\, C_a^2}{C_a^2 + K_{ER}^2} \tag{14}$$

and Fig. 3 displays simulation of the nonlinear terms:

$$\begin{cases} N2 = \dfrac{d2\,(I + d_1)\,(1 - h)}{(I + d_3)} \\ N3 = \dfrac{v_{3k}\, C_a^4\, I}{(C_a^4 + k_D^4)\,(I + k_3)} \end{cases} \tag{15}$$

and their respected linearized equivalent for $\epsilon = 0.02$. As these figures indicate, despite using much simpler computational units, the linearized equivalent terms follow their respected nonlinear equations closely. For a quantitative comparison, normalized Root Mean Square Deviation (NRMSE) error [14] was calculated for the nonlinear term N and linearized term L as:

$$NRMSE = \frac{\sqrt{\frac{\sum_{i=0}^{n}(L-N)^2}{n}}}{N_{max} - N_{min}} \tag{16}$$

where N_{max} and N_{min} are minimum and maximum of the N and n is number of points that this error is calculated. Table 1 presents the values of this error for nonlinear terms in Eqs. 14 and 15. The small values denote the resemblance of the suggested linear substitute terms. Furthermore, the nonlinear terms in the Eqs. 1 to 8 were substituted by their respective linear terms to develop a linearized model for astrocyte. Figure 4 shows the simulation of the proposed and the original model for different modes and values of glutamate. As it can be seen in the figure, the responses of the models are very similar.

Table 1. NRMSE calculation of nonlinear terms in Eqs. 14, 15 and their corresponding linear term.

	$\epsilon = 0.02$		$\epsilon = 0.03$		$\epsilon = 0.04$	
	No. l.s.	NRMSE	No. of l.s.	NRMSE	No. of l.s.	NRMSE
$N1$	5	0.008	4	0.033	3	0.039
$N2$	9	0.006	6	0.009	4	0.011
$N3$	16	0.0080	11	0.0122	8	0.0213

No. of l.s.: number of the linear segments

4   Hardware Implementation

The linearized model presented in the previous section, make it possible to implement the nonlinear astrocyte differential equations on FPGA. In this section an efficient fixed point hardware for astrocyte is designed and physically implemented. The reason that we use a fixed point instead of a floating point hardware, is that despite lower range and precision, it is in general faster and cheaper.

To design the hardware, in the first step, the Eulers method was used to numerically solve the Eqs. 1 to 8. Small step size $(1/256)$ was selected to ensure minimum error in the method. The scheduling diagram for execution of this operations are shown in Fig. 5. Since nonlinear terms was already linearized, the operations that hardware will perform only includes addition, subtraction, comparison and arithmetic shifts. There are three major blocks correspond to three differential Eqs. 1, 2 and 3. The α, β and δ are parameters of the linear segments in Eqs. 11 and 12 for $\epsilon = 0.02$. Multiplication in these and other constants, was performed with shift and add operations. To do this, first, constants were represented with sequence of $+$, $-$ and 0 symbols $(+ - 0)$ which each position representing and addition or subtraction. For instance, multiplication in 15 was performed by $2^3 * x - x$, which in hardware computed with three arithmetic right

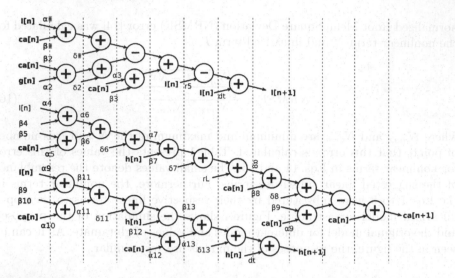

Fig. 5. As Soon As Possible (ASAP) scheduling diagram for the implementation of the linearized model.

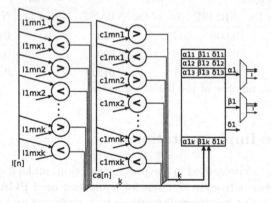

Fig. 6. The logic unit for selecting α, β and δ based on the value of ca, $Ip3$ for each linear segment.

shift x minus x. Since each nonlinear term was replaced by a sequence of linear terms, α, β and δ change with the range of the function. Therefore, a control unit is required to coordinate these parameters with range of ca, $Ip3$ and h. This unit for $\alpha 1$, $\beta 1$ and $\delta 1$ is shown in the Fig. 6. In this figure, based on the current value of ca and $Ip3$, the comparators generate the address for selecting the $\alpha 1$, $\beta 1$ and $\delta 1$ from the memory $M1$. These values will further be decoded into addition/subtraction operations. Further, word size of the functional units was determined to prevent over/under flow and preserving the precision. For this purpose, factors such as range of variables and coefficients as well as number of right/left shift operations were taken into account. With those concerns, 34 word size was selected, 24 bit for the fraction and 10 bit for the integer part.

5 Implementation Results

To prove validation of the proposed architecture, it was first described as a Hardware Description Language (HDL) code (Verilog) using a Finite State Machine (FSM). Moreover, Verilog code was simulated and then implemented on the Spartan-6 XC6SLX75T FPGA. The on-FPGA values of ca, $Ip3$ and h were converted to analog and observed on oscilloscope which for Ca^{2+} oscillation is shown in the Fig. 7. Further, Table 2 compares implementation resources and operation frequency of the proposed hardware with another work that implements a biological realistic astrocyte model. In this work no DSP multiplier is used while [16] reports extensive use of that units. Being multiplier-less gives this work two advantages. First is higher speed and throughput. Multiplication is a slow arithmetic operation. Using such number of multipliers significantly degrade the operation frequency and throughput of the system. However, comparing with this work, the proposed method in [16] will have lower error and deviation from the software floating point simulation.

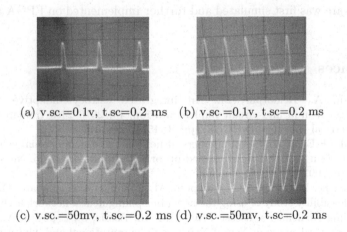

(a) v.sc.=0.1v, t.sc=0.2 ms   (b) v.sc=0.1v, t.sc=0.2 ms

(c) v.sc.=50mv, t.sc.=0.2 ms (d) v.sc.=50mv, t.sc=0.2 ms

Fig. 7. Oscilloscope photos of on-FPGA Ca^{2+} oscillation in astrocyte for different modes and values of glutamate. (a) FM, glutamate = 0.2. (b) FM, glutamate = 2 (c) AM, glutamate = 0.2. (d) AM, glutamate = 2.

Second advantage is area efficiency. Number of DSP multipliers are limited in FPGAs. For instance, the FPGA that we used in this work has 132 DSP fast multipliers. As result, not even one DSP based implementation can be fitted into this FPGA. Nevertheless, in the proposed design, almost %28 of the FPGA LUTs is used and more number could be implemented on the high-end FPGAs.

Table 2. Device utilization of the XILINX Spartan 6 Lx75

	4In.LUT	Frequency	DSPs
[16]	≈1400	NA	≈180
This work (FM)	13667	89.35 MHz	0
This work (AM)	13119	88.62 MHz	0

The benefit of a multiplier-less design relative to the one presented in [16] will show more on ASIC implementation. As FPGA is a already fabricated device with a number of multipliers. Implementing proposed hardware on silicon will have considerably less cost and have better performance.

6   Conclusion

A biological G-ChI astrocyte model for Ca^{2+} oscillation was modified for hardware implementation. Simulation data reveals that this models follow the original model with an acceptable accuracy. The simplicity of the models, which only consist of add/sub and shift operations, made it possible to implement the nonlinear astrocyte equations effectively on hardware. The HDL code describing the hardware was first simulated and further implemented on FPGA as proof of concept.

References

1. Ahmadi, A., Heidarpur, M.: An integrated astrocyte-adaptive exponential (AAdEx) neuron and circuit implementation. In: 2016 24th Iranian Conference on Electrical Engineering (ICEE), pp. 1545–1550 (2016)
2. Azevedo, F.E., et al.: Equal numbers of neuronal and nonneuronal cells make the human brain an isometrically scaled-up primate brain. J. Comp. Neurol. **513**(5), 532–541 (2009)
3. Capecci, E., Morabito, F.C., Campolo, M., Mammone, N., Labate, D., Kasabov, N.: A feasibility study of using the neucube spiking neural network architecture for modelling Alzheimer's disease EEG data. In: Bassis, S., Esposito, A., Morabito, F.C. (eds.) Advances in Neural Networks: Computational and Theoretical Issues. SIST, vol. 37, pp. 159–172. Springer, Cham (2015). https://doi.org/10.1007/978-3-319-18164-6_16
4. De Pittà, M., Goldberg, M., Volman, V., Berry, H., Ben-Jacob, E.: Glutamate regulation of calcium and IP3 oscillating and pulsating dynamics in astrocytes. J. Biol. Phys. **35**(4), 383–411 (2009)
5. Falcke, M.: Reading the patterns in living cells -the physics of Ca^{2+} signaling. Adv. Phys. **53**(3), 255–440 (2004)
6. Fellin, T., Ellenbogen, J., De Pitt, M., Ben-Jacob, E., Halassa, M.: Astrocyte regulation of sleep circuits: experimental and modeling perspectives. Front. Comput. Neurosci. **6**, 65 (2012)

7. Friedl, K.E., Voelker, A.R., Peer, A., Eliasmith, C.: Human-inspired neurorobotic system for classifying surface textures by touch. IEEE Robot. Autom. Lett. **1**(1), 516–523 (2016)
8. Gerstner, W., Kempter, R., van Hemmen, J.L., Wagner, H.: A neuronal learning rule for sub-millisecond temporal coding. Nature **383**(6595), 76–78 (1996)
9. Ghosh-Dastidar, S., Adeli, H.: A new supervised learning algorithm for multiple spiking neural networks with application in epilepsy and seizure detection. Neural Netw. **22**(10), 1419–1431 (2009)
10. Goutelle, S., et al.: The hill equation: a review of its capabilities in pharmacological modelling. Fundam. Clin. Pharmacol. **22**(6), 633–648 (2008)
11. Graupner, M., Brunel, N.: Calcium-based plasticity model explains sensitivity of synaptic changes to spike pattern, rate, and dendritic location. Proc. Natl. Acad. Sci. **109**(10), 3991–3996 (2012)
12. Heidarpour, M., Ahmadi, A., Rashidzadeh, R.: A cordic based digital hardware for adaptive exponential integrate and fire neuron. IEEE Trans. Circuits Syst. I: Regul. Pap. **63**(11), 1986–1996 (2016)
13. Heidarpur, M., Ahmadi, A., Kandalaft, N.: A digital implementation of 2D hindmarsh-rose neuron. Nonlinear Dyn. **89**(3), 2259–2272 (2017)
14. Hyndman, R.J., Koehler, A.B.: Another look at measures of forecast accuracy. Int. J. Forecast. **22**(4), 679–688 (2006)
15. Jokar, E., Soleimani, H.: Digital multiplierless realisation of a calcium based plasticity model. IEEE Trans. Circuits Syst. II: Express Briefs **PP**(99), 1 (2016)
16. Karim, S., et al.: Assessing self-repair on FPGAs with biologically realistic astrocyte-neuron networks. In: 2017 IEEE Computer Society Annual Symposium on VLSI (ISVLSI), pp. 421–426, July 2017
17. Lee, J.H., et al.: Real-time gesture interface based on event-driven processing from stereo silicon retinas. IEEE Trans. Neural Netw. Learn. Syst. **25**(12), 2250–2263 (2014)
18. Liu, J., Harkin, J., Maguire, L., McDaid, L., Wade, J., McElholm, M.: Self-repairing hardware with astrocyte-neuron networks. In: 2016 IEEE International Symposium on Circuits and Systems (ISCAS), pp. 1350–1353 (2016)
19. Liu, J., Harkin, J., McDaid, L., Halliday, D.M., Tyrrell, A.M., Timmis, J.: Self-repairing mobile robotic car using astrocyte-neuron networks. In: 2016 International Joint Conference on Neural Networks (IJCNN), pp. 1379–1386 (2016)
20. Maguire, L., McGinnity, T., Glackin, B., Ghani, A., Belatreche, A., Harkin, J.: Challenges for large-scale implementations of spiking neural networks on FPGAs. Neurocomputing **71**(13), 13–29 (2007). Dedicated Hardware Architectures for Intelligent Systems Advances on Neural Networks for Speech and Audio Processing
21. Matsubara, T., Torikai, H., Hishiki, T.: A generalized rotate-and-fire digital spiking neuron model and its on-fpga learning. IEEE Trans. Circuits Syst. II: Express Briefs **58**(10), 677–681 (2011)
22. McIver, S.R., Faideau, M., Haydon, P.G.: Astrocyte-neuron communications. In: Cui, C., Grandison, L., Noronha, A. (eds.) Neural-Immune Interactions in Brain Function and Alcohol Related Disorders, pp. 31–64. Springer, Boston (2013). https://doi.org/10.1007/978-1-4614-4729-0_2
23. Nadkarni, S., Jung, P.: Modeling synaptic transmission of the tripartite synapse. Phys. Biol. **4**(1), 1 (2007)
24. Nedergaard, M., Ransom, B., Goldman, S.A.: New roles for astrocytes: redefining the functional architecture of the brain. Trends Neurosci. **26**(10), 523–530 (2003)

25. Pitt, M.D., Brunel, N., Volterra, A.: Astrocytes: orchestrating synaptic plasticity? Neuroscience **323**, 43–61 (2016). Dynamic and metabolic astrocyte-neuron interactions in healthy and diseased brain
26. Postnov, D., Ryazanova, L., Sosnovtseva, O.: Functional modeling of neural-glial interaction. Biosystems **89**(13), 84–91 (2007). Selected Papers presented at the 2005 6th International Workshop on Neural Coding
27. Tewari, S.G., Majumdar, K.K.: A mathematical model of the tripartite synapse: astrocyte-induced synaptic plasticity. J. Biol. Phys. **38**(3), 465–496 (2012)
28. Volman, V., Ben-Jacob, E., Levine, H.: The astrocyte as a gatekeeper of synaptic information transfer. Neural Comput. **19**(2), 303–326 (2007)
29. Wade, J.J., McDaid, L.J., Harkin, J., Crunelli, V., Kelso, J.A.S.: Bidirectional coupling between astrocytes and neurons mediates learning and dynamic coordination in the brain: a multiple modeling approach. PLOS ONE **6**(12), 1–24 (2011)
30. Wall, J.A., McDaid, L.J., Maguire, L.P., McGinnity, T.M.: Spiking neural network model of sound localization using the interaural intensity difference. IEEE Trans. Neural Netw. Learn. Syst. **23**(4), 574–586 (2012)
31. Wallach, I., Dzamba, M., Heifets, A.: AtomNet: a deep convolutional neural network for bioactivity prediction in structure-based drug discovery. arXiv preprint arXiv:1510.02855 (2015)

Evolving Balancing Controllers for Biped Characters in Games

Christopher Schinkel Carlsen(✉) and George Palamas(✉)

Aalborg University Copenhagen, Copenhagen, Denmark
ccarls14@student.aau.dk, gpa@create.aau.dk

Abstract. This paper compares two approaches to physics based, balancing systems, for 3D biped characters that can react to dynamic environments. The first approach, based on the concept of proprioception, use a neuro-controller to define the position and orientation of the joints involved in the motion. The second approach use a self-adaptive Proportional Derivative (PD) controller along with a neural network. Both neural networks were trained using a Genetic Algorithm (GA). The study showed that both approaches were capable of achieving balance and the GA proved to work well as a search strategy for both the neuro-controller and the PD-controller. The results also showed that the neuro-controller performed better but the PD-controller was more flexible and capable to recover under external disturbances such as wind drag and momentary collisions with objects.

Keywords: Evolutionary · Neural-network · Neuro-controller ·
Virtual-agent · Procedural animation

1 Introduction

In 1981 the arcade game Donkey Kong was released. The game characters and animations were sprite based and were all experienced in the same way as responses to specific actions. The first set of tools that allowed for more expressive movement of virtual characters was based on forward and inverse kinematics. This approach use a small set of key-frames and interpolation techniques to generate the intermediate frames. The main challenge of inverse kinematics is the selection of the key-frames and the proper interpolation technique. Additionally, the quality of the motion depends on the animator's skills and quite often some knowledge regarding mechanical or physical principles are needed for the characterization of the motion states.

In inverse kinematics the animator sets a desired position of the end-effector of a kinematic chain and the system determines the joint parameters that provide the desired position in space. Inverse kinematics can have zero, one or multiple solutions. There are many approaches to solving the system of equations but usually an inverse Jacobean matrix is used. However, kinematic techniques are often unable to provide a realistic motion and thus physically based simulations

© Springer Nature Switzerland AG 2019
I. Rojas et al. (Eds.): IWANN 2019, LNCS 11507, pp. 869–880, 2019.
https://doi.org/10.1007/978-3-030-20518-8_72

might be required [1]. Procedural animation differs from traditional animation since the animation is being procedurally generated by a system, such as an algorithm, instead of a human animator. Within this context, the animator's role is being redefined. There are multiple approaches to designing a system capable of producing an animation. Sims [2], was one of the first to show that it is possible to evolve animation, such as jumping or forward motion, by evolving the morphology and behavior of virtual creatures, based on a Genetic Algorithm. Procedural animation is an efficient way of exploring a large space of possible solutions. By utilizing a generative system, a large number of unique animations can be produced [4]. Galanter, mentioned that generative systems have been here for quite some time in different practices: music, design and architecture, but it is only recently that they made their way to physics based animation for games. Games such as "Human: Fall Flat" and "Gang Beasts" are based on controlling a physically simulated rag-doll. The user can indirectly control the rag-doll's movement, in an unpredictable and unique way. "Grow Home: A Buds Life" use Center Of mass (COM) to calculate where to place the feet of the characters at each step. However, this application cannot be credited as physically correct, as the main characters always land on their feet. Rag-dolls and rag-doll physics have been used many years in the game industry, mostly for death animations.

1.1   Using a Physics Simulator

Physics based motion can make the animations more interesting and fluent because it resembles real life situations where the same laws of physics apply. Moreover, these animations do not become repetitive and might create interesting movement strategies that animators would never think of. But this can also lead to the aspect of Uncanny Valley. If the animation closely resembles the human gait it might end up in a poor immersive experience. Current research around procedural animation is centered in the realm of physics based motion. There are a few different approaches being used namely, Central Pattern Generators (CPG), Proportional Integral Derivative controllers (PID), motion tracking, Inverted Pendulum Models (IPM) and Finite State Machines (FSM). These techniques are usually combined in order to achieve balance, gait or other type of motion. The biped must learn to balance before learning to walk and this is the main reason to achieve balance for virtual agents. The ability to change which foot a biped balance on, should achieve a standing walk. This paper shows a comparative study between two biped balancing control strategies: (i) Control based on a Proportional Derivative-controller (PD-controller); (ii) Control based on a neuro-controller. In both controllers the optimization strategy based on a Genetic Algorithm (GA).

2   Background

In order to synthesize physics based motion a rag-doll is required. This is a structure that consists of a set of rigid bodies tied together with joints. There are two approaches to apply forces in a physics environment. One is similar to human beings and animals where different muscles, placed on the armature, contract and expand generating torque on the effected joints. The other, less computational and more common approach, is to apply torques directly on the joints, ignoring the muscles. [3] used a Hill-type muscle system for their biped locomotion, while [5] used torque models for their biped control.

The Inverted Pendulum Model. The IPM is a key concept for bipedal balancing strategies and has been proposed multiple times in the robotics and bio-mechanics literature [6]. The dynamics of the IPM can be determined in real time by the Center of Mass (COM), Center of Pressure (COP) and Zero Moment Point (ZMP). For complex structures such as a biped, the COM is calculated as the summation of each individual limb's COM. The formula for a complex joint system is:

$$\sum_{i=1}^{N}(i.m * i.COM)/m(total) \tag{1}$$

Where N is the number of limbs, i.m is the mass of the limb, i.COM is the COM of the limb and m(total) is the total weight of the entire object. In case of a three dimensional space the formula need to be applied to all three dimensions. The concept of Zero Moment Point (ZMP) was introduced in 1969 by Vukobratovic [7]. This concept greatly simplifies the complex dynamics of bipedal robots, for instance, Honda's humanoid robots [8]. Vukobratovic defined ZMP as the point on the ground where the tipping moment acting on the biped, from inertia and gravity forces, equals zero.

Proportional-Integral-Derivative Controller. The PID Controller is one of the most established closed loop control systems used to continuously modulate an output action through a transfer function, by associating error measurements with the action. The PID controller is commonly used in industrial applications and is well known to provide optimized control for linear systems but might not be well suited for non-linear systems. The controller can be described with the following equation.

$$Kp * e(t) + Ki * \int_0^t e(t)d(t) + Kd * de(t)/dt \tag{2}$$

There are many heuristic methods to determine the control parameters Kp, Ki and Kd, such as the Ziegler-Nichols method [9]. This method sets the Ki and Kd to be equal to zero and then adjusts the Kp until a stable oscillation is established. The Kp value, along with the oscillation period are used to tune the Ki and Kd constants. Many recent approaches have been based on Genetic Algorithms (GA) to optimize the parameters. The PID-controller then can be used in conjunction with the inverse pendulum model (IPM). Imagine a biped balancing on one leg. The system knows the position of the COM (mass of the body) and the COP (foot which the biped is balancing on). In order to achieve balance the COM should be above the COP. The error then is the difference between these two positions. This error could be used in the PID-controller to automatically adjust the applied torques to the joints of the biped. The further away the COM is from the COP the more torque will be generated. As the COM approaches the COP the velocities become higher but the differential component (D) ensures that these will be regulated. This means that the COM and COP remain at a constant distance. The integral component (I) calculates the accumulated error over time and has the tendency to minimize the distance between COM and COP. If stability is achieved then very low forces are required to retain the balanced state.

Neuro-Controller. This is a type of adaptive controller that has the form of a multi-layer neural network. A neuro-controller regulates an action based on an adaptable set of weights. The main inspiration draws from the human control actions that are directly regulated by the brain, through proprioception. Moreover, neural networks act as universal approximators which are of primary importance in the development of nonlinear controllers. Among the different methods of training a neuro-controller, Neuro-Evolution of Augmenting Topologies (NEAT) provides a very powerful approach where a Genetic Algorithms is used to optimize the topology of the network (the hyper parameters) instead of optimizing the weights [10].

2.1   Genetic Algorithm Optimization

A Genetic Algorithm (GA) is a search procedure that can be used for optimizing a set of parameters. As Goldberg stated about GAs "They efficiently exploit historical information to speculate on new search points with expected improved performance" [11]. GAs have been proven to work well for game design, especially for procedural content generation, state/action selection, modeling player experiences or strategy selection [12]. Some examples include the optimization of spaceship design [13], data-driven generation of game scenarios and modeling player's experience [14], motion control [15] for biped, for quadruped [16] or driving. A GA can be used to solve, but is not limited to, problems such as: Tuning the PID controller's constant values or optimize a neural network instead of using back-propagation. It can also be used to optimize hyper parameters such as network size, topology or activation function.

2.2   Related Work

Hodgins et al. [17] applied control algorithms to make a biped maintain balance while performing various athletic tasks. [18] used a GA as search strategy for optimal PID control of a 2D biped. [19] used gravity compensation techniques to provide soft landings for a swing-foot. [20] used a PID-controller to minimize energy consumption by finding the optimal position of the COM. [5] developed a simple biped locomotion controller based on a a Finite State Machine (FSM). Their system used a PD-controller and through the FSM they manually set the target pose from motion captured data. [21] used a neuro-controller in conjunction with an FSM to control joint angles and a recurrent neural network to apply the forces and torques. [22] used a Truncated Fourier Series (TFS) to define the angular trajectories of the joints. They also used a GA to optimize their TFS variables and applied gait generators to joints in order to produce walking strategies. In their velocity-driven method [23] used motion capture data, to calculate the desired joint angular velocities and subsequent joint torques, in order to achieve these velocities. Another approach that used IPM for motion generation proposed by [24]. Their method calculated the targeted poses and gravity compensation strategies of each limb's COM so as to create an interactive editor of walking gaits and character's limb morphology. An extended 3D linear inverted pendulum model proposed by [25], where they used a central pattern generator (CGP) to generate the vertical COM trajectory along with a walking pattern. A number of balance strategies proposed, where virtual forces applied to compensate for gravity, control speed, and general balance control [15]. In their approach they formulated a fitness function for forward motion based on the principle of COM distance from origin while keeping COM in a higher position. In almost all the aforementioned approaches, whether it is IPM, PIDs or Central Pattern Generators (CGP), the main factor is the use of virtual forces applied to the pelvis or the COM which were used to stabilize the biped in a balanced pose, since this is the predecessor to a successful walking gait.

3   Experimental Procedure

From this analysis it became clear the need for a comparative analysis, based on robustness and computational complexity, of the two most prominent approaches; the PD-controller and the neuro-controller. The PD-controller is widely used in robotics as a feedback based controller, while the neuro-controller relies on an internal representation of weights and functions to process data instead of straightforward equations like in a PD-controller. The task under test was one legged balance of a simulated biped with nine (9) degrees of freedom (DOF). Additionally, the controllers exposed to random wind drag and random momentary collisions from flying objects. The Unity3D used for the simulation of the biped character and the python libraries Keras and TensorFlow used as the back-end for the neural networks and the genetic algorithms.

3.1   System Architecture

The system consists of two main modules: the physics based simulator implemented in Unity3D and the external controller implemented with the Keras and Tensorflow frameworks. A communication between the two modules was established through Open Sound Control (OSC) protocol. The controller represented the server with a client instantiated in the simulated biped. The rag-doll represented with rigid bodies and the limbs were connected with configurable joints and angular constraints. For each epoch, a population of bipeds was tested until a criteria was met. If the stabilizing oscillation exceeded a threshold, the procedure was terminated. This heuristic approach might be sub-optimal but it ensured that each condition was reaching a balancing state. When the biped hit the ground, with anything other than the feet, a termination signal was triggered and the biped was reset and a new one was instantiated.

Genetic Algorithm. Table 1 shows the parameters used for the Genetic Algorithm. This used random crossover for the 10 fittest chromosomes (truncation selection) and elitism for the top 5 fittest individuals which found to be very important for a fast convergence to a solution. Elitism ensures that good solutions are not lost from one generation to the next. The fitness value of the GA used to train the NN was a function based on the following tasks:

- Reward for time no other limbs than feet touch the ground.
- Punish for every undesired action.
- Punish proportionally to the zError and xError.

Table 1. Parameters used for the Genetic Algorithm

Population	Selection	Crossover	Mutation
100	Truncation selection (top 10)	Uniform crossover	5%
	Elitism (top 5)		

Neural Network. All Neural Networks were feed forward with a single hidden layer. The dimensions were different for each controller. For each individual in the population a single network was assigned. The network inputs, from positions and velocities were normalized, except of the rotations that were represented as quaternions. The architecture of each of this networks can be seen in Table 2.

Table 2. Neural network topologies

Type	Input size	Hidden layer size	Number of outputs
P-controller	5	10	2
PD-controller	5	10	4
Neuro-controller	30	60	2

3.2   Evolvable PD-Controller

During the first phase, an experiment conducted to examine the possibility to use the simpler P-controller, instead of a PD-controller. For this experiment the right thigh and knee of the biped was set, with a regular PD controller, to a fixed angle so the leg was lifted off the ground. This PD-controller was tuned with a variation of the Ziegler-Nichols method. An adaptive P-controller was trained based on these setting. The Kp value was dynamically updated through a feed forward NN. The network had the following structure: 5 inputs, 10 neurons in the hidden layer and 2 outputs. The inputs were the distances between COM and COP (zError and xError respectively) and the velocities of the pelvis. The two outputs, which represented the Kp values, used to apply the appropriate torques at the pelvis. However, the evolvable P-controller failed to maintain balance as this acted like an oscillator with the COM position moving around the COP position. This was not surprising due to overshooting phenomena that follow this type of controllers. During the second phase, an experiment conducted involving an adaptive PD-controller, using the same premises as phase one. The neural network had a structure of: 5 inputs, 10 neurons in the hidden layer and 4 outputs (Fig. 1. Again, the inputs were the zError, xError and the velocities of the pelvis with the outputs being the Kp and Kd variables of the controller. This NN had 4 outputs while the NN controlling the P-controller had only 2 outputs. After a number of unsuccessful epochs, the system reached stability while performing one-legged balance.

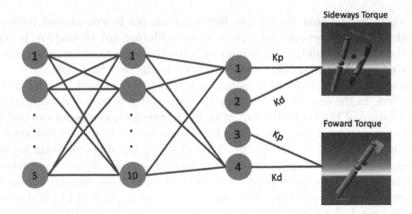

Fig. 1. The neural network for the adaptive PD-controller.

3.3   Evolvable Neuro Controller

The same fitness function and modeled biped used for the neuro-controller. Once again, the right thigh and knee were kept in a lifted position through a regular PD-controller. The NN had the following structure: 30 inputs, 60 neurons in the hidden layer, 2 outputs (Fig. 2). The inputs were mapped to the positions, rotations and velocities of the pelvis the left balancing thigh and the left balancing knee. The outputs were mapped as torques applied to the pelvis. After a number of failed attempts the simulated biped successfully performed one-legged balance.

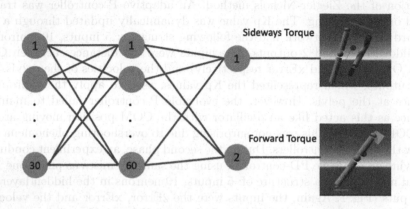

Fig. 2. The neural network for the neuro-controller.

4   Results

It is worth noting that during the first experiments it was almost impossible to achieve balance because all bipeds were oscillating out of control. It wasn't until the energy consumption introduced, into the fitness function, when the first stabilized biped evolved. This was attributed to the fact that optimized energy consumption led to minimized motion, such as jitter, caused by over-reacting controllers. In the case of the P-Controller the system acted as an oscillator due to the nature of the controller but also due to precision errors introduced from the physics engine. In the case of the PD-controller the fastest convergence to a solution happened after 164 epochs. After reaching a stable condition it continued exploring and developing the strategy but without significant improvements. The time for finding a stable solution tested with two similar simulations for both the neuro-controller and the PD-controller. As can be seen in Fig. 3 the neuro-controller reached a better fitness value than both the P and PD-controllers.

The PD-controller did not reach the same fitness as the neuro-controller, but it was close. This might be attributed to the fact that the integral part was missing which would most likely make the controller to end up in a steady state error closer to this of the neuro-controller. The P-controller almost immediately started to oscillate and very early went out of control. The PD-controller performed a dumping oscillation while converging to a specific value. The neuro-controller performed a smoother convergence and reached a better solution.

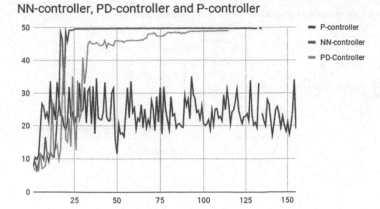

Fig. 3. Comparison of fitness for P, PD and Neuro-controller.

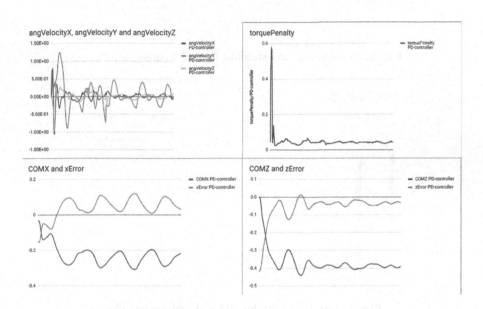

Fig. 4. Performance and stability for the PD-controller.

Figures 4 and 5, depicts the angular velocities, the torque penalty and xError and zError as functions of time. Figure 6 depicts a comparison of the PD and the neuro-controllers, in terms of energy consumption.

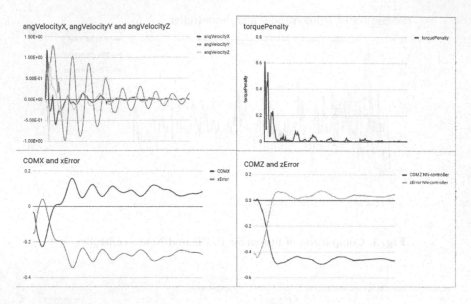

Fig. 5. Performance and stability for the Neuro-controller.

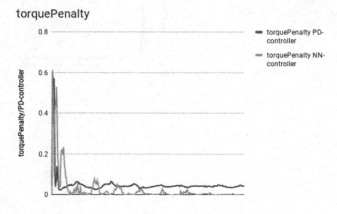

Fig. 6. Torque penalty for PD and neuro-controller.

5   Conclusions

Defining a descriptive fitness function is a challenging task which requires a deep understanding of the problem at hand. It was only after many observations that we became aware of how a biped should learn to balance. The biped was observed performing extreme movements, exhibiting oscillating behavior, fixating its upper limbs downwards and sometimes forces would make it fly upwards.

Although GAs performed well as global optimization strategy, sometimes took a lot of time to converge to a proper solution. Neuro-controllers were more precise, as these depend on the fitness function and not on observed errors. Moreover, for a very specific task the neuro-controller can outperform the PD-controller, in terms of precision, because much more information is required about the current state, which also means that it can become subject to over-fitting. The PD-controller was more adaptive than the neuro-controller. This means that once the parameters have been tuned it is be able to cope with mild physical disturbances and states that haven't experienced before. This also means that PD-controllers can not be subject to over-fitting. The neuro-controller had also the simplest implementation. Once the biped was set up the controller could be trained. The PD-controller took more effort to stabilize, but once the functionality was tuned this could be applied to more joints. In general, both controllers performed well for the given balancing task. The PD-controller was more flexible and the neuro-controller was more precise. Depending on the task, the neuro-controller could be more explorative than the PD-controller.

References

1. Parent, R.: Interpolation and basic techniques. In: Computer Animation (2007). https://doi.org/10.1016/b978-155860579-4/50004-5
2. Sims, K.: Evolving virtual creatures (2005). https://doi.org/10.1145/192161.192167
3. Geijtenbeek, T., van de Panne, M., van der Stappen, A.F.: Flexible muscle-based locomotion for Bipedal creatures. ACM Trans. Graph. https://doi.org/10.1145/2508363.2508399
4. Galanter, P.: Computational aesthetic evaluation: past and future. In: McCormack, J., d'Inverno, M. (eds.) Computers and Creativity, pp. 255–293. Springer, Heidelberg (2012). https://doi.org/10.1007/978-3-642-31727-9
5. Yin, K.K., Loken, K., van de Panne, M.: SIMBICON, simple biped locomotion control. ACM Trans. Graph. **26**, 105 (2007)
6. Yokoi, K., Kanehiro, F., Hirukawa, H., Kaneko, K., Kajita, S.: The 3D linear inverted pendulum mode: a simple modeling for a biped walking pattern generation (2002). https://doi.org/10.1109/iros.2001.973365
7. Vukobratovic, M., Juricic, D.: Contribution to the synthesis of biped gait. IEEE Trans. Bio-med. Eng. (2008). https://doi.org/10.1109/tbme.1969.4502596
8. Hirai, K., Hirose, M., Haikawa, Y., Takenaka, T.: The development of Honda humanoid robot. In: Proceedings - IEEE International Conference on Robotics and Automation (1998). https://doi.org/10.1109/ROBOT.1998.677288
9. Ziegler, J.G., Nichols, N.B.: Optimum settings for automatic controllers. InTech (1995). https://doi.org/10.1115/1.2899060
10. Stanley, K.O., Miikkulainen, R.: Evolving neural networks through augmenting topologies. Evol. Comput. (2002). https://doi.org/10.1162/106365602320169811
11. Goldberg, D.E.: Genetic Algorithms in Search, Optimization, and Machine Learning (1989). https://doi.org/10.1007/BF01920603
12. Risi, S., Togelius, J.: Neuroevolution in games: state of the art and open challenges. IEEE Trans. Comput. Intell. AI Games (2015). https://doi.org/10.1109/TCIAIG.2015.2494596
13. Liapis, A., Yannakakis, G.N., Togelius, J.: Neuroevolutionary constrained optimization for content creation. In: 2011 IEEE Conference on Computational Intelligence and Games, CIG 2011 (2011). https://doi.org/10.1109/CIG.2011

14. Luo, L., Yin, H., Cai, W., Zhong, J., Lees, M.: Design and evaluation of a data-driven scenario generation framework for game-based training. IEEE Trans. Comput. Intell. AI Games (2017). https://doi.org/10.1109/TCIAIG.2016.2541168
15. Geijtenbeek, T., Pronost, N.: Interactive character animation using simulated physics. Comput. Graph. Forum (2012). https://doi.org/10.1111/j.1467-8659.2012.03189.x
16. Qiu, G.Y., Wu, S.H.: Self-adjusting locomotion on a partially broken-down quadrupedal biomorphic robot by evolutionary algorithms. In: 2012 IEEE International Conference on Robotics and Biomimetics, ROBIO 2012 - Conference Digest (2012). https://doi.org/10.1109/ROBIO.2012.6490942
17. Hodgins, J.K., Wooten, W.L., Brogan, D.C., O'Brien, J.F.: Animating human athletics (2005). https://doi.org/10.1145/218380.218414
18. Cheng, M.Y., Lin, C.S.: Genetic algorithm for control design of biped locomotion. J. Robot. Syst. (1997). https://doi.org/10.1002/(SICI)1097-4563(199705)14:5⟨365::AID-ROB3⟩3.0.CO;2-N
19. Ayhan, O., Erbatur, K.: Biped robot walk control via gravity compensation techniques. In: IECON Proceedings (Industrial Electronics Conference) (2004). https://doi.org/10.1109/IECON.2004.1433381
20. Kho, J.W., Lim, D.C., Kuc, T.Y.: Implementation of an intelligent controller for biped walking robot using genetic algorithm. In: 2006 IEEE International Symposium on Industrial Electronics, vol. 1, pp. 49–54 (2006)
21. Heinen, M.R., Osrio, F.S.: Applying genetic algorithms to control gait of simulated robots. In: Electronics, Robotics and Automotive Mechanics Conference, CERMA 2007 - Proceedings (2007). https://doi.org/10.1109/CERMA.2007.4367736
22. Shafii, N., Javadi, M.H.S., Kimiaghalam, B.: A truncated fourier series with genetic algorithm for the control of biped locomotion. In: IEEE/ASME International Conference on Advanced Intelligent Mechatronics, AIM (2009). https://doi.org/10.1109/AIM.2009.5229814
23. Tsai, Y.Y., Lin, W.C., Cheng, K.B., Lee, J., Lee, T.Y.: Real-time physics-based 3D biped character animation using an inverted pendulum model. IEEE Trans. Vis. Comput. Graph. (2010). https://doi.org/10.1109/TVCG.2009.76
24. Coros, S., Beaudoin, P., van de Panne, M.: Generalized biped walking control. ACM Trans. Graph. (2010). https://doi.org/10.1145/1778765.1781156
25. Hong, Y.D., Park, C.S., Kim, J.H.: Stable bipedal walking with a vertical center-of-mass motion by an evolutionary optimized central pattern generator. IEEE Trans. Industr. Electron. (2014). https://doi.org/10.1109/TIE.2013.2267691

Computational Biology and Bioinformatics

Feature Selection and Assessment of Lung Cancer Sub-types by Applying Predictive Models

Sara González[1], Daniel Castillo[2(✉)], Juan Manuel Galvez[2], Ignacio Rojas[2], and Luis Javier Herrera[2]

[1] Instituto de Biomedicina de Sevilla, IBIS/Hospital Universitario Virgen del Rocio/CSIC, University of Sevilla, Sevilla, Spain
[2] Computer Architecture and Technology Department, University of Granada, Granada, Spain
cased@ugr.es

Abstract. The main goal of this study is the identification of a robust set of genes having the capability of discerning among the different sub-types of lung cancer: Small Cell Lung Carcinoma (SCLC), Adenocarcinoma (ACC), Squamous Cell Carcinoma (SCC) and Large Cell Lung Carcinoma (LCLC). To achieve this goal, an overall differentially expressed genes analysis was performed by using data from gene expression microarrays publicly stored at NCBI/GEO platform. Once the analysis was done, a total of 60 Differential Expressed Genes (DEGs) were selected and then used in the development of predictive models combining supervised machine learning and feature selection algorithms. This provided a reduced and specific gene signature that allows identifying the sub-type of lung cancer of new samples. The predictive models designed are assessed in terms of accuracy, f1-score, sensitivity and specificity. Finally, a set of public web platforms having biological information on genes, were used in order to determine the relation that exists between the final subset of genes and the addressed sub-types of lung cancer.

Keywords: Lung cancer · Microarray · Gene expression · Feature selection · Machine learning

1 Introduction

Lung cancer is the most common cancer and the main cause of death by cancer in men, followed by prostate cancer and colorectal. In women, lung cancer has the second and third position in mortality and incidence, respectively [1].

There are two main types of lung cancer: non small cell lung cancer (NSCLC) and small cell lung cancer (SCLC). NSCLC has three differentiated sub-types, namely: adenocarcinoma (ACC), squamous cell cancer (SCC) and large cell cancer (LCLC) [2].

© Springer Nature Switzerland AG 2019
I. Rojas et al. (Eds.): IWANN 2019, LNCS 11507, pp. 883–894, 2019.
https://doi.org/10.1007/978-3-030-20518-8_73

The identification of genetic biomarkers associated with lung cancer allows the early prognostic and the right treatment. This is critical nowadays, as this could be the difference between live and death. For that, it is crucial to know what genes could be promoting disorders in one or more biological process that finally cause, in this case, any of the different sub-types of lung cancer.

For several decades, Microarray technology has allowed studying the alteration at gene expression level with the purpose of finding genes involved in pathologies of genetic source. This technology is highly widespread and known and is based on the capability of the complementary molecules to hibridate among themselves to determine the gene expression values of each studied gene in the analyzed samples [3]. Through this process, the over-expressed or inhibited genes can be identified in tumor samples when comparing to normal samples.

Previous studies performed by Sanchez-Palencia et al. have used this technique in the identification of biomarkers for the sub-types ACC and SCC, for a reduced number of patients [4]. Others like Yanaihara et al. have studied the molecular profiles of lung cancer by using microRNA data [5].

This work is aimed to identify a reduced set of genes that has the ability to discern among the five contemplated states (ACC, SCC, LCLC, SCLC and Control). This set was later used to design and compare a number of predictive models. These models can perform the prediction of the state of samples not seen before in the learning process, with a great reliability.

2   Data Description

The total amount of lung cancer samples have been obtained from the public repository NCBI/GEO [6]. Concretely, 13 series stored and publicly available in this repository have been used. These series have samples from the different sub-types of lung cancer addressed in the study, as well as healthy samples. A total of 851 samples from the different series conform the final dataset. Table 1 shows certain information about each of the 13 series. For each series, the information about the NCBI/GEO ID Microarray platform, the excluded outliers and the number of samples from the different lung cancer states is shown. Finally, the table includes the total sum of samples that conform each lung cancer and control states.

3   Methodology

This section shows the methodology implemented to carry out this study and has two subsections. The first one is focused in the Differentially Expressed Genes analysis for the states (subtypes of lung cancer and control) addressed in the study. The second one is focused in the application and explanation of the Computer Intelligence-based predictive models proposed for this research. Figure 1 shows the whole pipeline where the first three steps correspond to the data pre-processing and the last step corresponds to the predictive models design process.

Table 1. Table with the information about the 13 series used in this study. For each series the information about the GEO ID, the platform used, the removed outliers and the number of samples from different subtypes that each series has shown.

Serie	Platform	Index outliers	SCLC	LCLC	SCC	ACC	Control	Total
GSE7670	Affymetrix HG-U133A	61,46,34	-	2	0	29	2	33
GSE99316	Affymetrix HG-U133B	36	23	-	-	-	-	23
GSE43580	Affymetrix HG-U133_Plus_2	24,30,43,46,58,59, 82,87,124,128,133	-	-	69	70	-	139
GSE73160	Affymetrix GPL11028	1,10,21,32,42,63	62	-	-	-	-	62
GSE3268	Affymetrix HG-U133A	-	-	-	10	-	-	10
GSE40275	Human Exon 1.0 ST Array GPL15974	10,18	15	3	5	14	41	78
GSE18842	Affymetrix HG-U133_Plus_2	60,63,74,84,85	-	-	-	-	42	42
GSE37745	Affymetrix HG-U133_Plus_2	39,52,55,57,58,71, 82,79,95,155	-	21	64	-	-	85
GSE41271	Illumina HumanWG-6 v3.0	83,84,88,138,145, 185,201,202,203, 223,253	-	3	74	180	-	257
GSE12771_1	Illumina GPL6097	5,8,13	-	-	-	-	24	24
GSE12771_2	Illumina GPL6102	112,152,237,238, 239,240,241,242	-	-	-	-	40	40
GSE39345	Illumina GPL6104	21,23,24,25,39	-	-	-	-	20	20
GSE21933	Phalanx Human GPL6254	2,4,22,30	-	-	9	10	19	38
	Total		100	29	231	303	188	851

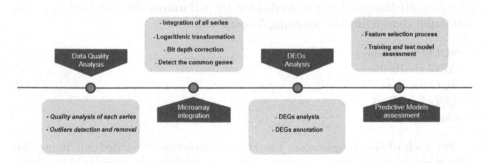

Fig. 1. Pipeline designed for this study to first analyze and integrate the addressed series and to finally evaluate the extracted DEGs by using predictive models.

3.1 Diferential Expression Data Processing Using R

In this study, series that belong to different Microarray technologies were used. For that, the pre-processing step is crucial in order to ensure the right harmony among them in order to achieve the integration without losing biological information. In the analysis of data from biological sources, the programming language R and the Bioconductor repository are the most commonly used tools in the scientific community [7–9]. This pre-processing step, as well as the posterior integration, correspond to the first three steps in Fig. 1, which are next detailed:

- Quality analysis of the samples with the aim of finding remaining outliers.
- Removal low quality samples (outliers).
- Integration of the different series belonging to different Microarray platforms.
- Application of the same logarithmic transformation to reach the right cohesion of the data.
- Correction of the bit depth of the data in order to equalize the series.
- Equalize the number of genes among the series through the search of common genes.
- Differentially Expressed Genes analysis.
- Genes annotations.

The integration step combines the 13 series that belong to different platforms into an unique super-serie, ensuring the equal conditions in the next steps among the integrated samples. To achieve it, as previously mentioned, a strong and rigorous pre-processing step for each series was required.

The Differential Expressed Genes analysis step determine which genes are differentially expressed in tumor samples in comparison with normal samples, and, if this differential expression is significative. To carry out the analysis, limma package was used to statistically compare the expression of the selected samples in pursuit of the DEGs among the compared classes [10]. This process is performed by leaving the 20% of the data out in order to have a test dataset to verify the suitability of the DEGs with data no seen before. The DEGs extraction for this multiclass problem was carried out by performing different bi-class DEGs extraction considering the following 5 comparisons:

- SCLC vs. Rest
- ACC vs. Rest
- SCC vs. Rest
- LCLC vs. Rest
- Control vs. Rest.

For each of the comparisons, the DEGs extraction is carried out fixing the restriction values in limma package. The first restrictive parameter that allows to know if a gen is or not expressed, is the p-value (The value of statistic significance according to the statistical test t-Student). Moreover, the Fold Change is used as restrictive parameter (Existing differences among the mean expression of the analyzed conditions) in conjunction with the p-value to decide the DEGs candidates.

Applying a p-value lower or equal to 0.05 and a Fold Change greater or equal to log2(1.1), a total of 12 genes for each of one the comparisons proposed before were extracted, being joined finally in a set of 60 EGs to assess.

Furthermore, in order to assess the robustness of the whole DEG extraction process, the training-test subdivision was finally repeated in a 5 cross-validation manner and the results discussed.

3.2   Predictive Models Development

The genes identified as relevant in the previous comparisons, determine the features in the dataset used in the development of the predictive models. Due to the global 5-fold process implemented in the preprocessing step, there are 5 training datasets, conformed by the 80% of the total amount of samples and 5 test datasets with the 20% of the samples.

Feature Selection

Feature selection techniques are widely applied in the machine learning scope to reduce the curse of dimensionality, specially in those presenting a large number of features in comparison with the number of samples. Concretely, in the area of gene expression, after a first preprocessing step selecting a group of candidate DEGs, feature selection can be used to identify a more reduced gene signature for a certain disease or group of diseases, types of lung cancer in this case, removing redundant information and retaining the information that allows to differentiate among the classes. In the feature selection process carried out, we distinguish two stages:

- Feature selection: several feature selection algorithms exist in the biomedical literature. These algorithms try to select a group of features taking into account a number of criteria such as mutual information, distance measures, performance of the features in a classification process, etc. In general these algorithms measure the relationship among all the features, generating a ranking of these in function of their relevance with the objective class. The algorithms used and compared in this work are minimum Redundancy Maximum Relevance (mRMR), Relief (Rel) and Random Forest (RF) [11–13].
- Wrapper selection: the Support Vector Machine (SVM) and k-Nearest Neighbor (k-NN) classifiers are used in an incremental manner, in the search for the optimal set of genes [14–16]. This technique consists in creating as many models as number of features in the dataset, following the ranking given by the feature selection algorithm, in order to identify the subset bringing the best performance.

The identification of the optimal number of genes is thus supported by the performance of the classifiers in the training dataset, trying always to achieve a reduced number of genes, nevertheless containing the maximum information to discern among the addresses classes. If the feature selection is carried out right, the DEGs would be optimal for the approached problem and, they are expected to be related to a greater or lesser extent with the pathology.

Training and Validation of the Predictive Models

To perform the assessment of the DEGs, 6 alternative approaches were compared, by combining each classifier with each of the feature selection algorithms proposed. The possible combinations are mRMR+SVM, Rel+SVM, RF+SVM, mRMR+KNN, Rel+KNN y RF+ KNN, measuring for each combination the accuracy, the sensitivity and the specificity, on both the training and test datasets.

Due to the 5-fold global process performed on the available dataset, five different feature rankings can be obtained for each feature selection algorithm, and thus different models with different sets of features will be assessed on different test data. The idea of this process is to assess on the one hand the methodology proposed in this work, and on the hand assess and compare the differences in the gene signature obtained for each of the five executions.

Moreover, through the use of the web platforms *targetValidation* and *Genecards*, which provide information about the biological background of the genes, the relationship between the gene signature identified by the proposed complete approach and the pathology addressed can be inspected [17,18].

4   Results and Discussion

This section shows the results of the simulations proposed for this study. As previously stated in the Methods section, different combinations of feature selection and classification algorithms were considered and compared. First, results using a single training-test subdivision of the complete dataset is used (first of the 5-fold global execution). Figure 2 shows the results obtained by k-NN for the three different feature selection methods on the training dataset. With only 10 genes, the results of mRMR and RF are above 88%, leaving Rel in third position with less than 80% of accuracy using the same number of genes. The same trend is repeated in Fig. 3 but this time using SVM, also on the training dataset. Both algorithms reach the 90% when the classification is performed using the total number of DEGs.

Given this behaviour in the training dataset, 10 genes were selected as candidate size for the gene signature.

Then, for the test dataset, and for up to 10 genes, Fig. 4 represents a matrix of plots in which the first column shows the results for the first fold for k-NN and the second column the same but for SVM. The first row plots the accuracy and it can be seen how the trend in test is similar to training for both classification algorithms, bringing mRMR a globally better performance than the other two feature selection algorithms. The second row plots the sensitivity, presenting RF better results than mRMR. The third row shows the specificity plots, achieving mRMR better results. The figure demonstrate how the models trained and the selected DEGs are useful to discern among the proposed lung cancer classes. Table 2 shows the numerical results of the previous plots for k-NN and SVM and their combination with the feature selection algorithms.

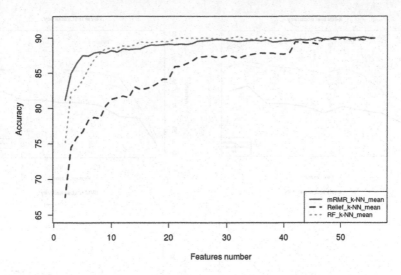

Fig. 2. Validation results of k-NN for the three different feature selection methods for the first fold.

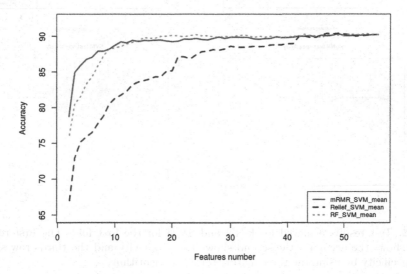

Fig. 3. Validation results of SVM for the three different feature selection methods for the first fold.

Relying on the results of the previous figures, it was decided to take the first 10 genes for the 5-fold global execution, as the difference in terms of accuracy between the classification with 10 genes and with the 60 selected DEGs is minimal. For the 5 fold execution, results were very similar than those observed for the first fold. It was again observed that the best models with few genes are

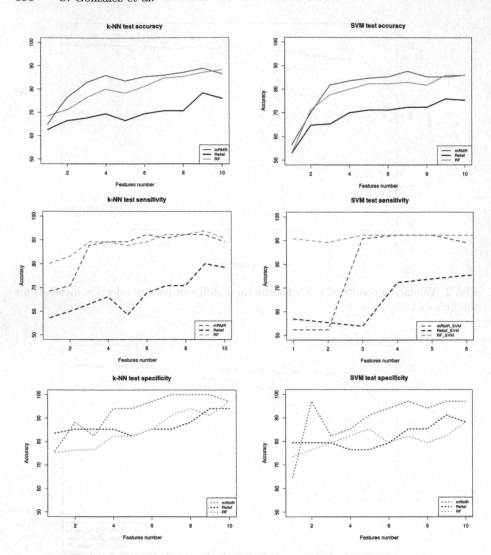

Fig. 4. Test results achieved for k-NN and SVM for the first fold. The first row of plots shows the accuracy, the second shows the sensitivity and the thirds row shows the specificity by using the three feature selection algorithm.

the models in which mRMR are used as feature selection algorithm. In view of this, the most relevant genes for execution of the mRMR in each of the five folds were taken. The idea was to search for the common genes in the top 10 rankings selected by the 5 different mRMR executions. A total of 6 commons genes were found, being those genes independent from the test datasets, and thus can be considered as very robust genes.

Table 2. Classification results for test for the first fold for each combination of the feature selection algorithms with the classifiers by using the top 10 of DEGs. The table shows the accuracy, the mean sensitivity and the mean specificity.

	Accuracy			Sensitivity			Specificity		
	mRMR	Rel	RF	mRMR	Rel	RF	mRMR	Rel	RF
SVM	85.88%	75.29%	85.88%	89.23%	75.38%	90.76%	97.05%	88.23%	88.23%
k-NN	86.47%	75.88%	88.23%	89.23%	78.46%	90.76%	97.05%	94.11%	97.05%

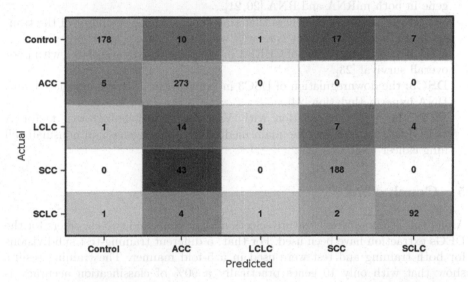

Fig. 5. Confusion matrix that shows the sum of the confusion matrices of the 5 different folds (test datasets) and the information of the global accuracy (86.21%), f1-score (74%), sensitivity (73.1%) and specificity (96.12%) when the 6 mRMR common genes are used for classifying.

Finally, the assessment on all the test datasets was repeated, this time using the 6 common genes. The results showed a mean accuracy of the five execution of 86% with a standard deviation of 2.15%. Moreover, Fig. 5 shows the confusion matrix conformed by the sum of the 5 confusion matrices from the 5 different test datasets used. In the Figure, the LCLC class is confounded with the other classes related with lung cancer. This may be due to the low number of samples available for this class. The rest of the classes are well classified achieving outstanding results for both the accuracy and the specificity (86.21% and 96.12% respectively). The f1-score and the sensitivity are lower than the other two metrics due to the bad classification results for the LCLC class, attaining a 74% the f1-score and a 73.1% the sensitivity. If all the sub-types of lung cancer are treated as the same class and the confusion matrix calculated (cancer class vs health class), the results show an accuracy of 98%, a sensitivity of 94.68% and a specificity of 98.94%. These results corroborate the relationship of those genes with lung cancer. The biological information about those genes was searched and all of them are practically related with cancer or lung cancer according to the web platform mentioned before. The 6 common genes are the following:

- **NONO:** no exist evidences that probe the relationship of this gene with lung cancer. However, the protein codified by this gene interacts with a robust biomarker of NSCLC [19].
- **DSG3:** exists studies that correlate SCC with a differential expression of this gene in both miRNA and RNA [20,21].
- **SH2D3C:** there are evidences that this gene is a gene regulator of the transcription factor ELF5, that is related with lung cancer [22].
- **CHEK1:** high expression of CHEK1 in lung tumors was associated with poor overall survival [23].
- **DSC3:** the downregulation of DSC3 in lung cancer cells was associated with DNA hypermethylation [24].
- **FST:** this gene has relation with Vascular endothelial growth factor A (VEGFA) that is one of the main mediators of angiogenesis in nonsmall cell lung cancer (NSCLC) [25].

5   Conclusions

Along this study different feature selection algorithms and two classifiers for the DEGs extraction have been used. For that, 5 different training-test subdivisions for both training and test were used in a 5-fold manner. The training results show that with only 10 genes practically a 90% of classification accuracy is reached. The test results support the selection of these genes by reaching the similar results than the training assessment but on unseen data. After inspecting the results on a single training-test subdivisions, results were similar in the rest of them. Finally, a total of 6 genes were selected as those being selected by the three feature selection algorithms, ensuring their robustness for the problem. In a posterior search of those 6 genes in biological web platform, a clear relationship with cancer or lung cancer were discovered. Then they were assessed on the complete dataset showing a high accuracy for this complex and critical multiclass problem.

For further researches, it could be very interesting to take into account a possible integration of these data with data from RNA-seq technology. This integration has been performed in previous works on other pathologies such as breast cancer or leukemia, among other diseases, achieving outstanding results [26–28].

Acknowledgements. This research has been possible thanks to the support of project: TIN2015-71873-R (Spanish Ministry of Economy and Competitiveness – MINECO – and the European Regional Development Fund – ERDF).

References

1. Bray, F., Ferlay, J., Soerjomataram, I., Siegel, R.L., Torre, L.A., Jemal, A.: Global cancer statistics 2018: GLOBOCAN estimates of incidence and mortality worldwide for 36 cancers in 185 countries. CA: Cancer J. Clin. **68**(6), 394–424 (2018)
2. Cooper, W.A., et al.: The textbook on Lung Cancer: time for personalized medicine. Ann. Transl. Med. **3**(7), 86 (2015)
3. Schena, M., Shalon, D., Davis, R.W., Brown, P.O.: Quantitative monitoring of gene expression patterns with a complementary DNA Microarray. Science **270**(5235), 467 (1995)
4. Sanchez Palencia, A., et al.: Gene expression profiling reveals novel biomarkers in nonsmall cell lung cancer. Int. J. Cancer **129**(2), 355–364 (2011)
5. Yanaihara, N., et al.: Unique microRNA molecular profiles in lung cancer diagnosis and prognosis. Cancer Cell **9**(3), 189–198 (2006)
6. Barrett, T., Troup, D.B., Wilhite, S.E., Ledoux, P., Rudnev, D., Evangelista, C., et al.: NCBI GEO: mining tens of millions of expression profiles database and tools update. Nucl. Acids Res. **35**(suppl. 1), D760–D765 (2007)
7. R Core Team: R: A language and environment for statistical computing (2013)
8. Gentleman, R.C., Carey, V.J., Bates, D.M., Bolstad, B., Dettling, M., Dudoit, S., et al.: Bioconductor: open software development for computational biology and bioinformatics. Genome Biol. **5**(10), R80 (2004)
9. Galvez, J.M., Castillo, D., Herrera, L.J., Roman, B.S., Valenzuela, O., Ortuno, F.M., et al.: Multiclass classification for skin cancer profiling based on the integration of heterogeneous gene expression series. PLoS ONE **13**(5), 1V (2018). https://doi.org/10.1371/journal.pone.0196836
10. Smyth, G.K.: Limma: linear models for Microarray data. In: Gentleman, R., Carey, V.J., Huber, W., Irizarry, R.A., Dudoit, S. (eds.) Bioinformatics and computational biology solutions using R and Bioconductor. SBH, pp. 397–420. Springer, New York (2005). https://doi.org/10.1007/0-387-29362-0_23
11. Ding, C., Peng, H.: Minimum redundancy feature selection from microarray gene expression data. J. Bioinform. Comput. Biol. **3**(02), 185–205 (2005)
12. Hira, Z.M., Gillies, D.F.: A review of feature selection and feature extraction methods applied on microarrays data. Adv. Bioinform. **2015**, 13 (2015)
13. Diaz Uriarte, R., de Andres, S.A.: Gene Selection and classification of microarray data using Random forest. BMC Bioinform. **7**, 3 (2006)
14. Cortes, C., Vapnik, V.: Support vector networks. Mach. Learn. **20**(3), 273–297 (1995)
15. Noble, W.S.: What is a support vector machine? Nature Biotechnol. **24**, 1565–1567 (2006)
16. Parry, R., Jones, W., Stokes, T., Phan, J., Moffitt, R., Fang, H., et al.: K nearest neighbor models for Microarray gene expression analysis and clinical outcome prediction. Pharmacogenomics J. **10**(4), 292 (2010)
17. Carvalho-Silva, D., et al.: Open Targets Platform: new developments and updates two years on. Nucl. Acids Res. **47**(D1), D1056–D1065 (2019). https://doi.org/10.1093/nar/gky1133
18. Safran, M., et al.: GeneCards Version 3: the human gene integrator. Database **2010**, baq020 (2010)
19. Chen, Z., et al.: cAMP/CREB-regulated LINC00473 marks LKB1-inactivated lung cancer and mediates tumor growth. J. Clin. invest. **126**(6), 2267–2279 (2016)

20. Savci-Heijink, C.D., Kosari, F., Aubry, M.C., Caron, B.L., Sun, Z., Yang, P., Vasmatzis, G.: The role of desmoglein-3 in the diagnosis of squamous cell carcinoma of the lung. Am. J. Pathol. **174**(5), 1629–1637 (2009)
21. Saaber, F., Chen, Y., Cui, T., Yang, L., Mireskandari, M., Petersen, I.: Expression of desmogleins 13 and their clinical impacts on human lung cancer. Pathol.-Res. Pract. **211**(3), 208–213 (2015)
22. Zhang, F., et al.: Identification of key transcription factors associated with lung squamous cell carcinoma. Med. Sci. Monit.: Int. Med. J. Exp. Clin. Res. **23**, 172 (2017)
23. Chen, Z., et al.: MiR-195 suppresses non-small cell lung cancer by targeting CHEK1. Oncotarget **6**(11), 9445 (2016)
24. Cui, T., et al.: The p53 target gene desmocollin 3 acts as a novel tumor suppressor through inhibiting EGFR/ERK pathway in human lung cancer. Carcinogenesis **33**(12), 2326–2333 (2012)
25. Frezzetti, D., et al.: Vascular endothelial growth factor a regulates the secretion of different angiogenic factors in lung cancer cells. J. Cell. Physiol. **231**(7), 1514–1521 (2016)
26. Wang, Z., Gerstein, M., Snyder, M.: RNA-Seq: a revolutionary tool for transcriptomics. Nature Rev. Genet. **10**(1), 57–63 (2009)
27. Castillo, D., Galvez, J.M., Herrera, L.J., Roman, B.S., Rojas, F., Rojas, I.: Integration of RNA-Seq data with heterogeneous Microarray data for breast cancer profiling. BMC Bioinform. **18**(1), 506 (2017). https://doi.org/10.1186/s12859-017-1925-0
28. Castillo, D., et al.: Leukemia multiclass assessment and classification from Microarray and RNA-Seq technologies integration at gene expression level. PLoS ONE (2019). https://doi.org/10.1371/journal.pone.0212127

Energy-Time Analysis of Convolutional Neural Networks Distributed on Heterogeneous Clusters for EEG Classification

Juan José Escobar[1]($\boxtimes$), Julio Ortega[1], Miguel Damas[1],
Rukiye Savran Kızıltepe[2], and John Q. Gan[2]

[1] Department of Computer Architecture and Technology, CITIC,
University of Granada, Granada, Spain
{jjescobar,jortega,mdamas}@ugr.es
[2] School of Computer Science and Electronic Engineering, University of Essex,
Colchester, UK
{rs16419,jqgan}@essex.ac.uk

Abstract. Training a deep neural network usually requires a high computational cost. Nowadays, the most common way to carry out this task is through the use of GPUs due to their efficiency implementing complicated algorithms for this kind of tasks. However, training several neural networks, each with different hyperparameters, is still a very heavy task. Typically, clusters include one or more GPUs that could be used for deep learning. This paper proposes and analyzes a distributed parallel procedure to train multiple Convolutional Neural Networks (CNNs) for EEG classification, in a heterogeneous CPU-GPU cluster and in a Desktop PC. The procedure is implemented in $C++$ and with the MPI library to dynamically distribute the hyperparameters among the nodes, which are responsible for training the corresponding CNN by using Python, Keras, and TensorFlow. The proposed algorithm has been analyzed considering running times and energy measures, showing that when more nodes are used, the procedure scales linearly and the lowest running time is obtained. However, the desktop PC provides the best energy results.

Keywords: CPU-GPU clusters · Energy-time analysis ·
EEG classification · Convolutional Neural Networks ·
Hybrid Master-worker algorithms

1 Introduction

A few decades ago, with the rapid growth of the computing capacity of microprocessors and the reduction of lithography in the manufacturing process, a large number of mathematical problems could be addressed through computational simulation. Later, more complex problems that require much more computing capacity emerged, which resulted in the dawn of the era of General Purpose

© Springer Nature Switzerland AG 2019
I. Rojas et al. (Eds.): IWANN 2019, LNCS 11507, pp. 895–907, 2019.
https://doi.org/10.1007/978-3-030-20518-8_74

computation on GPU (GPGPU) [8]. GPUs, which were initially designed for 3D graphics acceleration, began to be used for specific scientific calculations of great complexity, and later for a wider range of problems in cooperation with CPUs [5,10]. However, for a number of years, the rate of increase in computing power has started to fall due to the Moore's law, with R&D more focused on energy efficiency due to environmental issues, and less on increasing CPU and GPU brute power. Even the introduction of multi-core processors is not enough to reverse this trend since most applications are not optimized to use more than one core, and the efficient use of multiple cores is still a topic of active research in computer science. In addition, there is a problem with the exponential growth of data generated by applications, which is impossible to cope without the development of new processing techniques. Thus, currently, the use of clusters to distribute the workload constitutes the mainstream approach to take advantage of technology improvements and overcome the barrier of hardware limitation.

There are two commonly-used methods for parallelizing neural network training across clusters: (i) model parallelism [13], where different nodes in the distributed system are responsible for the computations in different parts of a single network (e.g. a layer of the network), and (ii) data parallelism [15], in which different nodes have a complete copy of the model and each one handles a different portion of the data. The results from each node are somehow combined. These two approaches are focused on the parallelization of a single neural network. However, the problem considered in this paper is to train multiple CNNs with different hyperparameters, and either of those two methods is not efficient given the large number of synchronization operations required between the nodes. Thus, we aim to provide a different and more efficient approach, where each CNN is evaluated independently in each node of the heterogeneous cluster, which involves multiple CPUs and other accelerators such as GPUs. Although the use of clusters for neural networks has been proposed in some previous studies, the parallelization on a heterogeneous platform of a procedure with the characteristics of our target application is less frequent in the literature.

The paper is organized as follows: Sect. 2 describes issues in EEG classification and the architecture of the implemented CNN to evaluate the proposed master-worker algorithm, which is detailed in Sect. 3. Section 4 describes the experimental setup to carry out the experimental work and analyzes the results obtained. Finally, Sect. 5 summarizes the conclusions and gives an insight into possible future work.

2   Convolutional Neural Networks for EEG Classification

Many classification or modeling tasks in bioinformatics deal with patterns defined by a large number of features, and thus require feature selection techniques in order to remove redundant, noisy-dominated, and irrelevant inputs from the patterns. Moreover, these high-dimensional classification problems have frequently to be addressed with the number of training patterns much lower than the number of features, which is known as the curse of dimensionality problem [2]. A good feature selection can provide an improvement in the accuracy of the classifiers when reducing the dimensionality. EEG classification, applied to Motor Imagery (MI) based Brain Computer Interface (BCI) tasks [4], is one of the most interesting research areas due to its potential application in different fields, such as games [14], or in health care where some patients have suffered amputations and/or paralysis in a limb [3]. BCI applications based on EEG classification, like the one considered in this paper, is the perfect example of the aforementioned curse of dimensionality problem, which can be caused by the following reasons, among others: (i) the presence of noise or outliers since EEG signals have a low signal-to-noise ratio; (ii) the need to represent time information in the features because brain signal patterns are related to changes in time, and (iii) the non-stationary character of EEG signals, which may change quickly over time or between experiments. The problem is that the MI BCI paradigm uses series of amplifications and attenuations of short duration conditioned by limb movement imagination, also known as Event Related Desynchronization (ERD) and Event Related Synchronization (ERS) [9]. The task of ERD/ERS analysis is complex because the signals are weak and noisy, and occur at different locations of the cortex, at different instants within a trial, and in different frequency bands. Moreover, there is no consistency in the patterns among subjects, and the patterns can even change within a session for the same subject. This scenario usually leads to high-dimensional patterns making the number of available patterns to conduct ERD/ERS analysis significantly less than the number of features.

This way, a feature selection method is mandatory to reduce the dimensionality of the input patterns. Nevertheless, as the size of the search space depends exponentially on the number of possible features, an exhaustive search for the best feature set is almost impossible when the dimensionality is too high and the performance evaluation is based on the accuracy of a classifier. Even for a modest number of features, feature selection procedures based on branch-and-bound, simulated annealing, evolutionary algorithms, or combinations of them have been previously proposed [6,7]. Currently, with the recent boom in deep learning due to the availability of large databases, new processing algorithms, and the performance growth of GPUs, as discussed in Sect. 1, some neural networks such as Convolutional Neural Networks (CNNs) are beginning to be used for EEG classification, which demands ofr more computational power. Although its most common application is image and video recognition, and natural language processing, this kind of neural network is widely used for EEG classification due to its good performance, as it has been previously demonstrated [11]. CNNs can automatically extract the most discriminating spatio-temporal features before the classification step, directly from the raw signal. Thus, in this paper, a two-dimensional CNN (2D-CNN) is adopted as the EEG classifier in order to evaluate the performance of our proposed parallel algorithm. We consider $N_C = 128$ different combinations of hyperparameters to build different CNN architectures, which are obtained from combining two possible values for each of the seven hyperparameters, as Table 1 shows. The parameters are arbitrary, although they have been chosen to vary the complexity of the CNNs and to obtain valid and coherent results in their outputs. Figure 1 shows a general scheme of the proposed procedure and the CNN topology, which is as follows:

Table 1. Values for the different hyperparameters used in the CNN. N_F: number of filters for the convolution layers with a kernel size of $K_S \times K_S$; H_L: number of hidden layers; F_C: number of fully-connected layers with H_N hidden neurons

N_F	K_S	H_L	F_C	H_N	Epochs	K-folds
16	3	1	1	100	50	4
32	9	2	2	200	100	6

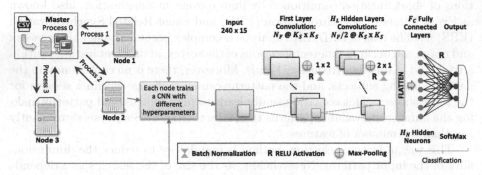

Fig. 1. Scheme of the proposed distributed parallel procedure, and the 2D-CNN topology. The master distributes different combinations of hyperparameters among the computing nodes. Each of them trains a CNN with the received hyperparameters. The CNN is composed by a 2D-convolutional layer, followed by H_L hidden layers (also convolutional), a flattening operation, F_C fully-connected layers, and the output layer. After all convolutional layers, the batch normalization and max-pooling operations are applied. Each fully-connected layer contains half of the neurons of the previous layer

1. The first layer is a 2D-convolutional operation whose input shape is a 240×15 matrix since an EEG signal is composed of 240 samples from 15 electrodes. The convolution applies N_F filters with a kernel size of $K_S \times K_S$.
2. A batch normalization operation is applied to normalize the activations of the first layer at each batch.
3. A max-pooling operation of size 2×1 is applied to reduce the size in the samples's dimension of the previous convolutional layer.
4. H_L hidden layers are applied, which are also convolutional operations. Each convolution applies half of the filters used by the previous layer but keeps the kernel size. After each hidden layer, a batch normalization operation, and a max-pooling of size 1×2 to reduce the electrode's dimension are applied. All convolution layers compose the feature extraction part, which should be able to extract the most relevant spatio-temporal features of the EEGs.

5. A flattening operation is used to reshape the output of the last convolutional layer to vectors. This is mandatory before using any fully-connected layer.

6. F_C fully-connected layers (FC), together with the output layer compose the classification part. The first fully-connected layer contains H_N hidden neurons. The next one contains half of the neurons of the previous layer, and so on. The last FC layer is connected to the output layer, which has three units since the EEG problem considered in this paper has three possible classes.

3   MPI-Based Distributed Master-Worker Algorithm

Algorithm 1 details our distributed parallel algorithm, schematized in Fig. 1. It has been implemented by using OpenMPI [12] as a Message Passing Interface (MPI) library to dynamically distribute combinations of hyperparameters over the different workers used in the cluster. We employ this approach instead of a static distribution to avoid imbalanced workload since the execution time depends mainly on two factors: (i) the combination of hyperparameters to train a neural network and (ii) the computing capabilities of each device because the cluster is a heterogeneous environment. The Scheduler function (line 1) is divided into two sections: one for the master process (lines 2–17), and the other for the workers (lines 18–27), where each process has a unique identifier (rank) that is used to identify them unequivocally. Firstly, the function receives the input parameters necessary to perform correctly the procedure, such as the total number of combinations of hyperparameters to evaluate (N_C), the CSV filename with these combinations ($CSVFile$), the Python file with the CNN implementation ($PyFile$), the number of workers or nodes available to do the job (N_W), and the device to execute the CNN (CPU or GPU).

The master (MPI process with rank number 0), begins to read the CSV file with the combinations of hyperparameters (line 3). Each row of the file contains a different combination and is composed of multiple columns (i.e. the hyperparameters). In line 4, by using an if-else statement, the program checks the total number of workers running the application. If the condition becomes true, no workers are available to help with the task an thus the master is the responsible for doing the entire job, i.e., to perform all N_C combinations of hyperparameters by using the CPU or GPU allocated in the node to which it belongs. In other words, the master also includes the worker role and for this specific case only one MPI process is enough. Apparently, this functionality might not be relevant, since a solution could be to call the program with an MPI process that acts as master and another one that acts as a worker. However, logically this presents a problem with the saturation of CPU and memory resources, since each MPI process is mapped to a thread of the same node and thus less CPU resources are available to train the CNN. In addition, other overheads such as the message passing between MPI processes, or their synchronization requirements have to be taken into account as it affects to the execution time and the energy consumed. All these issues have been considered giving the master the ability to do all the job without the need for more processes.

If the `if-else` statement becomes false, it means that the CNNs are not computed by the master. At this point, master and workers are synchronized and ready to start the MPI section. The master asynchronously starts to attend the requests of each worker, and dynamically distributes the rows of the CSV file until there is no more work to do (lines 11–15). The communication protocol by the master to distribute a row to a worker is simple: a request reception operation with the MPI function `Irecv` (line 12), a send operation with the MPI function `Isend` (line 13), and the decrease of the available work counter (line 14). Once all N_C combinations (rows) have been evaluated, the master sends the signal (END_SIGNAL) to the workers to notify that there are no more rows to be distributed and the MPI section has ended (line 16).

Starting from the case where more than one MPI process is running, while the master process is scheduling the distribution of rows/combinations, the workers (line 18) perform the training of the CNN. At the beginning, each worker W_j requests to the master for a new combination of hyperparameters (line 19), which is stored in C_l (line 20). However, the signal END_SIGNAL could also have been received, and therefore the loop of line 21 would not be executed since that means that there is no work available for this worker. If this is not the case, the worker proceeds in a cyclical way to perform the task (line 23), request the master for more work (line 24), and wait for the answer (line 25) until the signal to finish is received. On the other hand, regardless of whether the training of a CNN is done by a worker or by the master (in case there are no workers), the procedure is as follows: From the application a call is made to the Python interpreter by invoking the command processor through the function `system` (lines 7 and 23). This function requires as argument the command to be executed, which is composed mostly from the hyperparameters, and its use is equivalent to executing it through a terminal of the operating system. The command is a concatenation of, in this order: Name of the Python interpreter, name of the Python file with the implementation of the CNN, the hyperparameters, and the device that will run the CNN (CPU or GPU). Taking into account the possible combinations of seven hyperparameters that could be obtained from the values shown in Table 1, an example of the command to be executed would be:

```
"python3 cnn.py 2 9 4 100 1 100 32 GPU"
```

Algorithm 1. Pseudocode of the master-worker algorithm. The master dynamically attends the requests of each worker, which trains a CNN with the combination of parameters received by calling a Python interpreter. If no worker is detected, the master performs all combinations

---

1 **Function** Scheduler(N_C, $CSVFile$, $PyFile$, N_W, Dev)

    Input: Number of combinations to perform, N_C
    Input: File containing all N_C combinations, $CSVFile$
    Input: Python file with the CNN implementation, $PyFile$
    Input: Number of workers, N_W
    Input: Device to execute the CNN, Dev

2   **if** *Master* **then**

        `// `C_i`;`$\forall i = 1, ..., N_C$` is a string like "<ARG_1>...<ARG_N>"`
3          $C \leftarrow$ readCSV($CSVFile$, N_C)

4   **if** *Only_master* is detected **then**

5   **repeat**

6                  $CmdExec \leftarrow$ append("python3", C_i, Dev)
7                  $StatusCode \leftarrow$ system($CmdExec$)

8   **until** all N_C combinations are been evaluated;

        `// Distribution of combinations. Start MPI section`
9   **else**

10             $RemainingWork \leftarrow N_C$
11   **repeat**

12                 $C_R \leftarrow$ The worker W_j requests a new combination
13                 $C_i \rightarrow$ Send a combination to the worker W_j
14                 $RemainingWork \leftarrow RemainingWork - 1$

15   **until** $RemainingWork$ is 0;

            `// End MPI section`
16             $END\_SIGNAL \rightarrow$ broadcastSignal(N_W)

17   **end**

18   **else**

19         $1 \rightarrow W_j$ requests the master a new combination
20         $C_l \leftarrow W_j$ receives a new combination from the master
21   **repeat**

22             $CmdExec \leftarrow$ append("python3 ", C_l, Dev)
23             $StatusCode \leftarrow$ system($CmdExec$)
24             $1 \rightarrow W_j$ requests the master a new combination
25             $C_l \leftarrow W_j$ receives a new combination from the master

26   **until** the END_SIGNAL is received from the master;

27   **end**

28 **End**

---

4   Experimental Results

In this section, we analyze the performance of our MPI codes, running on a desktop computer (PC) with Ubuntu (18.04), and on a 4-node cluster that executes CentOS (v7.4.1708) with 32 GB of RAM memory and connected by Gigabit Ethernet. The $C++$ source code of the scheduler algorithm has been compiled with the GNU compiler (GCC 4.8.5), and the CNN is executed by a Python interpreter (v3.6.5), and developed with Keras (v2.2.4) and TensorFlow (v1.12) APIs. When multiple nodes are used, the front-end node is dedicated to the master process and the other three are the workers. The characteristics of the CPU and GPU devices of each platform are shown in Table 2. The instantaneous power and the energy consumption for each node of the cluster have been measured by a watt-meter we have developed based on Arduino Mega. It provides, in real-time, four measures per second for the instantaneous power (in Watts) and the accumulated consumed energy (in $W \cdot h$). Moreover, the measures of instantaneous power and energy consumed by the switch, necessary for the communications between nodes, are included too (below 5 W).

Table 2. CPU-GPU characteristics of the platforms used in the experiments

	CPU		GPU		
	Model	Threads/MHz	Model	Cores/MHz	RAM/MHz
Node 1	2x Intel Xeon E5-2620 v2	24/2,100	Tesla K20c	2,496/706	5 GB/5,200
Node 2	1x Intel Xeon E5-2620 v4	16/2,100	Tesla K40m	2,880/745	12 GB/6,008
Node 3	2x Intel Xeon E5-2620 v4	32/2,100			
PC	1x Intel i7 4770K	8/3,500	Titan Xp	3,840/1,582	12 GB/11,408

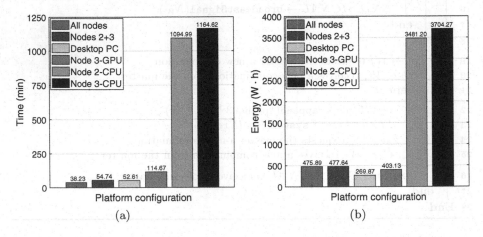

Fig. 2. Performance evaluation for different platform configurations after evaluating the N_C combinations of hyperparameters: (a) Running time; (b) Energy consumed

In our experiments, we have used fourteen datasets from the BCI Laboratory at the University of Essex, as described in [1]. They correspond to human subjects and each one includes 178 EEG patterns composed of 3,600 features, which can belong to three different types of movement, or classes. For simplicity, only the results for one dataset are shown since the time required to compute one of them is the same. In addition, we do not provide an analysis of the quality of the solutions provided by the different CNNs evaluated because the scope of this paper is more related with the energy-time behavior of the proposed algorithm. However, the classification accuracy obtained from the hyperparameters evaluated is between 61.81% and 79.13%, and 72.55% on average.

Figure 2a shows the average execution time resulting from evaluating all possible CNNs. Observing the values corresponding to the use of the cluster (i.e. all except the "*Desktop PC*" configuration), it can be seen that the more nodes are used, the shorter the execution time is obtained. Although this is the expected behavior, what is really interesting is the scalability of the program when more nodes are dedicated to the task. The execution time when nodes 2 and 3 are used, theoretically, should be half the execution time when only node 3 works, but what happens is that the time is even less, which is known as super-linear speedup. Taking into account that TensorFlow also executes some operations on the CPU, the difference is due to the fact that the CPU of node 2 is more efficient than the one of node 3 despite having fewer cores/threads. This can be verified by observing the CPU execution times for both nodes in the figure. The behavior can be explained if we take into account that the number of CPU sockets of node 2 is 1, while node 3 has 2. It has been checked that during the execution of the program the threads constantly migrate between the different cores of the processor, and also, internally the sockets share the necessary information, which introduces a considerable overhead. Therefore, we believe that if two nodes exactly equal to node 3 had been used, the speedups would be very close to 2. Finally, using all the compute nodes provide a time reduction equal to 3. This is pure coincidence, since continuing with the previous logic the speedup should be greater than 3. However, it seems that the lower power of the Tesla K20c of node 1 and the overhead caused by its 2 CPU sockets derive in a reduction of performance, obtaining a speedup equal to 3. In any case, it seems that the algorithm scales in the correct way.

On the other hand, the executions on GPU have been compared with the equivalent ones on CPU, and the result confirms that the CPUs are not currently the best devices for this type of problems, since the time to perform the task is around 10 times slower than the worst possible scenario when using GPUs. Although the CPU of node 3 has 16/32 cores/threads, it has been observed that on average the percentage of load of the processor does not exceed 35% during execution, so the CPU is not being exploited (45% for node 2). This could be due to one or several factors: (i) that neural networks are not highly parallelizable in CPU; (ii) TensorFlow is not optimized for these devices, perhaps derived from point (i); (iii) the model of the neural network used in this paper is not complex enough to take advantage of the full architecture. It has also been checked that the percentage of GPU usage is in the range of 68–85% when the program is executed. Looking at Fig. 3, which indicates the instantaneous power corresponding to the platform configurations shown in Figs. 2a and b (except for CPU, since they are very long), it can be seen that in all configurations the pattern of the power is repeated (going up and down twice), and the general tendency is to increase as time progresses. This is because the hyperparameters that create a more complex neural network model are evaluated at the end of each stretch. If the instantaneous power is higher when the model is more complex, and given that the device usage is associated with energy consumption, it can be concluded that at least point (iii) is affecting performance, and not only the CPU but also the GPUs. Concerning the performance between CPU and GPU, the CPU is negatively affected by its low usage and also because does not have the same data parallelism capacity present in GPU architectures. Summarizing, all this makes the GPUs greatly outperform the CPUs, and therefore the acceleration of neural networks is much more efficient.

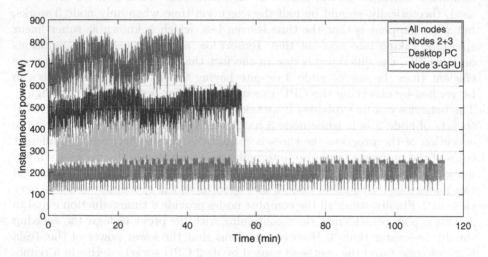

Fig. 3. Evaluation of the instantaneous power for different platform configurations during the execution of the N_C combinations of hyperparameters

From Fig. 3, it can also be observed that the use of more nodes not only entails a reduction in the execution time but also an increase of the instantaneous power, as is logical. However, the instantaneous power of the "*Desktop PC*" configuration can be highlighted, since it is able to perform the task in a slightly less time than needed by the cluster with nodes 2 and 3 simultaneously, but with the instantaneous power to just over half. The reasons are obvious: The GPUs architecture of the cluster is older (Maxwell) compared to the desktop PC (Pascal) besides that the Titan Xp is more powerful than the Teslas, having approximately the same TDP ($\sim$250 W). Moreover, it must be taken into account the energy consumption of the 3 nodes (front node, and nodes 2 and 3), which include RAM, the CPUs, hard disks, etc. From all this it can be seen that the GPU architecture acquires great importance and, based on the energy consumptions obtained, it could be better to acquire more modern architectures and amortize their cost by decreasing the number of nodes and the monetary cost for energy consumption.

Figure 2b shows the energy consumed (or accumulated) after executing the program. As expected, its behavior should be analogous to the obtained with execution times, since the energy consumed depends on the time, the instantaneous power, and also on the number of devices used. The shape of the bars for the CPU executions remains constant. However, the interesting thing is to see how the number of nodes and the type of architecture affects the energy consumed. It can be seen that using all nodes simultaneously produces practically the same energy consumption than using only the most powerful nodes (2 and 3), but with the difference that the execution time is less. Using only node 3 leads to a reduction in energy consumption of approximately 15.6% with respect to the other two configurations with the cluster, but its execution time is 3x slower than when using all nodes, and approximately 2x if the nodes 2 and 3 are used. Again, the Titan Xp allocated in the Desktop PC provides the most remarkable results, since it is the configuration with the lowest energy consumption, while offering good times and is the simplest platform. Analyzing the data, it seems that there is a certain tie between using the desktop PC and all nodes of the cluster. To clarify this, we are going to establish the energy-time product as a measure of performance, $P = time \cdot Energy$, and substituting in Eq. (1):

$$P_{AllNodes} = 38.23 \cdot 475.89 = 1.819327 \cdot 10^4$$
$$P_{DesktopPC} = 52.61 \cdot 269.87 = 1.419786 \cdot 10^4 \tag{1}$$

As P is lower for the *"Desktop PC"* configuration, it can be concluded that this option provides the best results. But depending on the context, it may be preferred the option that consumes less energy or provides faster execution.

5   Conclusions

A distributed parallel master-worker procedure that uses the MPI library has been devised to reduce the execution time necessary to train multiple convolutional neuronal networks for EEG classification. It makes possible to use simultaneously all nodes of the cluster though message-passing communications to benefit from data-level parallelism offered mainly by the GPUs allocated in each node. Also, a scheme where the master is able to perform all work has been developed with the objective of avoiding extra communications between the MPI processes. Different platform configurations and parallel alternatives have been compared by analyzing both speed and energy consumption as performance measures. The experimental results show that when using all nodes, the best running time is obtained, and it seems that the program scales linearly when more nodes are used. On the other hand, the configuration with the desktop PC provides the lowest energy consumed and the best energy-time product.

Only 128 combinations of parameters have been evaluated for each experiment. However, the number of hyperparameters to be optimized in a neural network is usually much higher, and if more possible values for each of them are taken into account, the search space becomes a NP-hard problem. Therefore, observing the high execution times obtained in the experimental results, it is clear that it is necessary to design strategies of parallelization for clusters where several nodes containing GPUs are used, since in the future the complexity of the neural networks and the volume of data to process will increase.

With respect to the EEG classification, in order to evaluate the scalability of the distributed parallel program, the combinations of hyperparameters for the CNNs have been previously fixed. However, the correct way to train a neuronal network is by optimizing these hyperparameters. Thus, as future work we propose a combined method between an optimization algorithm that is able to automatically select the best hyperparameters for the CNN (e.g. a genetic algorithm), and the distributed algorithm proposed in this paper to explore the maximum combinations of hyperparameters as quickly as possible.

Acknowledgements. This research was funded by grants TIN2015-67020-P (Spanish "Ministerio de Economía y Competitividad"), PGC2018-098813-B-C31 (Spanish "Ministerio de Ciencia, Innovación y Universidades"), and ERDF funds. We would also like to thank the BCI Laboratory of the University of Essex for allowing us to use their databases. The Titan Xp used for this research was donated by the NVIDIA Corporation.

References

1. Asensio-Cubero, J., Gan, J., Palaniappan, R.: Multiresolution analysis over simple graphs for brain computer interfaces. J. Neural Eng. **10**(4), 21–26 (2013)
2. Bellman, R.: Adaptive Control Processes: A Guided Tour. Princeton University Press, Princeton (1961)
3. Birbaumer, N., Cohen, L.: Brain-computer interfaces: communication and restoration of movement in paralysis. J. Physiol. **579**(3), 621–636 (2007)
4. Brumberg, J.S., Burnison, J.D., Pitt, K.M.: Using motor imagery to control brain-computer interfaces for communication. In: Schmorrow, D.D.D., Fidopiastis, C.M.M. (eds.) AC 2016. LNCS (LNAI), vol. 9743, pp. 14–25. Springer, Cham (2016). https://doi.org/10.1007/978-3-319-39955-3_2
5. Collet, P.: Why GPGPUs for evolutionary computation? In: Tsutsui, S., Collet, P. (eds.) Massively Parallel Evolutionary Computation on GPGPUs. NCS, pp. 3–14. Springer, Heidelberg (2013). https://doi.org/10.1007/978-3-642-37959-8_1
6. Ortega, J., Asensio-Cubero, J., Gan, J., Ortiz, A.: Classification of motor imagery tasks for BCI with multiresolution analysis and multiobjective feature selection. BioMed. Eng. OnLine **15**(1), 73 (2016)
7. Ortega, J., Ortiz, A., Martín-Smith, P., Gan, J.Q., González-Peñalver, J.: Deep belief networks and multiobjective feature selection for BCI with multiresolution analysis. In: Rojas, I., Joya, G., Catala, A. (eds.) IWANN 2017. LNCS, vol. 10305, pp. 28–39. Springer, Cham (2017). https://doi.org/10.1007/978-3-319-59153-7_3
8. Owens, J., et al.: A survey of general-purpose computation on graphics hardware. Comput. Graph. Forum **26**(1), 80–113 (2007)

 9. Pfurtscheller, G.: EEG event-related desynchronization (ERD) and event-related synchronization (ERS). Electroencephalogr. Clin. Neurophysiol. **103**(1), 26 (1997)
10. Raju, K., Niranjan, N.: A survey on techniques for cooperative cpu-gpu computing. Sustain. Comput.: Inf. Syst. **19**, 72–85 (2018)
11. Tabar, Y., Halici, U.: A novel deep learning approach for classification of EEG motor imagery signals. J. Neural Eng. **14**(1), 016003 (2016)
12. The Open MPI Project: OpenMPI documentation. https://www.open-mpi.org/doc/. Accessed 19 Nov 2018
13. Wawrzynek, J., Asanovic, K., Kingsbury, B., Johnson, D., Beck, J., Morgan, N.: Spert-II: a vector microprocessor system. Computer **29**(3), 79–86 (1996)
14. Wei, R., Zhang, X., Dang, X., Li, G.: Classification for motion game based on EEG sensing. In: ITM Web Conferences, vol. 11, p. 05002 (2017)
15. Zou, Y., Jin, X., Li, Y., Guo, Z., Wang, E., Xiao, B.: Mariana: tencent deep learning platform and its applications. VLDB **7**(13), 1772–1777 (2014)

The Frequent Complete Subgraphs in the Human Connectome

Máté Fellner[1], Bálint Varga[1], and Vince Grolmusz[1,2(✉)]

[1] PIT Bioinformatics Group, Eötvös University, Budapest 1117, Hungary
{fellner,balorkany,grolmusz}@pitgroup.org
[2] Uratim Ltd., Budapest 1118, Hungary
grolmusz@uratim.com
http://pitgroup.org

Abstract. While it is still not possible to describe the neuronal-level connections of the human brain, we can map the human connectome with several hundred vertices, by the application of diffusion-MRI based techniques. In these graphs, the nodes correspond to anatomically identified gray matter areas of the brain, while the edges correspond to the axonal fibers, connecting these areas. In our previous contributions, we have described numerous graph-theoretical phenomena of the human connectomes. Here we map the frequent complete subgraphs of the human brain networks: in these subgraphs, every pair of vertices is connected by an edge. We also examine sex differences in the results. The mapping of the frequent subgraphs gives robust substructures in the graph: if a subgraph is present in the 80% of the graphs, then, most probably, it could not be an artifact of the measurement or the data processing workflow. We list here the frequent complete subgraphs of the female and the male braingraphs of 414 subjects, each with 463 nodes, with a frequency threshold of 80%, and identify complete subgraphs, which are more frequent in male and female connectomes, respectively. We hope that the deep structural analysis of the human connectome will motivate the construction of novel artificial neural networks in the near future.

1 Introduction

Diffusion MRI-based macroscopic mapping of the connections of the human brain is a technology that was developed in the last 15 years [13,22,28,30]. Applying the method, we are able to construct braingraphs, or connectomes, from the diffusion MRI images [5,13,18]: the vertices of the graph are anatomically labeled areas of the gray matter (called "Regions of Interests", ROIs), and two such ROIs are connected by an edge, if a complex workflow, involving either deterministic or probabilistic tractography, finds axonal fibers between the two ROIs. Therefore, one can construct graphs, with up to 1015 nodes and several thousand edges, from the MR image of each subject.

The analysis of these graphs is a fast-developing and an important area today: these connections form the "hardware" of all brain functions on a macroscopic

© Springer Nature Switzerland AG 2019
I. Rojas et al. (Eds.): IWANN 2019, LNCS 11507, pp. 908–920, 2019.
https://doi.org/10.1007/978-3-030-20518-8_75

level [22, 28, 30]. Naturally, it would be exciting to map the neuronal scale human connectome, too: here the nodes were the individual neurons, and two nodes (or neurons), say X and Y, would be connected by a directed edge, say (X, Y), if the axon of X were connected to a dendrite of Y. Unfortunately, to date, the neuronal-level connectome of only one adult organism is described: that of the nematode *Caenorhabditis elegans* with 302 neurons, in the year 1986 [41]. In larval state, two more neuronal level connectomes are published: the larva of the fruitfly *Drosophila melanogaster* [25], and the tadpole larva of the *Ciona intestinalis* [27]. Despite of some exciting, very recent developments [42], the complete connectome of the adult *Drosophila melanogaster* with 100,000 neurons, is not determined yet. Humans have 80 billion neurons in their brains. Therefore, the mapping and the analysis of the neuronal scale human connectome is out of our reach today.

There are numerous results published for the analysis of the diffusion MRI-computed human connectomes, e.g., [13, 22, 28, 30]. Our research group has also contributed some more graph-theoretically oriented analytical methods, like the comparison of the deep graph theoretical parameters of male and female connectomes [34–36], the parameterizable human consensus connectome [31, 33], the description of the individual variability in the connections of the major lobes [19], the discovery of the Consensus Connectome Dynamics [17, 20, 32, 37] the description of the frequent subgraphs of the human brain [7], and the Frequent Neighborhood Mapping of the human hippocampus [8].

The reasons of our graph-theoretical approach are listed as follows:

(i) Graph theory has a long history of great successes, starting with the paper of Leonhard Euler in 1741 on Königsberg's bridges [6].

(ii) "Pure mathematical" graph theory and its applications in computer engineering reached an exceptionally high level of development in the late XXth and in the early XXIst century, just mentioning three famous examples: the Strong Perfect Graph Theorem [4], Szemerédi's Regularity Lemma [38], and the intricate parallel algorithms for multiprocessor routing in [21].

(iii) Graph theoretical definitions and notations are well-developed, clear, and usually catch the deep and most relevant properties of the networks examined.

1.1 Frequent Edges and Subgraphs: A Robust Analysis

The data acquisition and processing workflow, whose results are the braingraphs or structural connectomes, has numerous delicate steps. Naturally, errors may occur in MRI recordings and processing, as well as in segmentation, parcellation, tractography and graph computation steps [13, 16, 23]. When we have hundreds of high-quality MR images, we can analyze the *frequently appearing* graph edges or subgraphs, in order to derive robust, reproducible results, appearing in high fraction of the brains imaged. By analyzing only the frequently appearing structural elements, the great majority of data acquisition and processing errors will be filtered out.

Our first effort for describing frequent edges in human connectome was the construction of the Budapest Reference Connectome Server [31,33], in which the user can select the frequency threshold $k\%$ of the edges, and the resulting consensus connectome contains only those edges, which are present in at least $k\%$ of the subjects. The generated consensus connectome can be both visualized and downloaded at the site https://pitgroup.org/connectome/.

The frequent, connected subgraphs of at most 6 edges are mapped in the human connectome in [7]. The frequencies were compared between female and male connectomes, and strong sex differences were identified: there are connected subgraphs, which are significantly more frequent in males than in females, and there are a higher number of connected graphs that are more frequent in females than in males.

The direct connections of important brain areas are of special interest: correlations between the present or missing connections and psychological tests results or biological parameters may enlighten the fine structure-function relations of our brain. For error-correction reasons, the frequent neighbors of the relevant brain areas form the robust objects of study: small errors in the data processing workflow will most probably have no effects on the frequent connections. In our work [8], we have introduced the method of the Frequent Neighborhood Mapping, which describes the frequent neighbor sets of the given nodes of the braingraph. In [8], we have demonstrated the method by mapping the frequent neighborhoods of the human hippocampus: one of the most deeply studied part of the brain. We have mapped the frequent neighbor sets of the hippocampus, and we have found sex differences in the frequent neighbor sets: males have much more frequent neighbor sets of the hippocampus than the females; therefore, the neighborhoods of the men's hippocampi are more regular, with less variability than those of women. This observation is in line with the results of [34–36], where we have shown that the female connectomes are better expander graphs than the braingraphs of men.

In the present contribution, we are mapping the frequent complete graphs of the human connectome, based on the large dataset of the Human Connectome Project [24]. Our dataset contains the braingraphs of 414 subjects. A recently appeared work [29] deals with complete subgraphs in braingraphs of 8 subjects, each with 83 nodes. Our results are derived from 414 braingraphs, each of 463 nodes. Therefore, we are able to find frequent structures, i.e., frequent complete subgraphs in our dataset of 414 graphs (while it is not feasible to derive frequent structures from 8 graphs).

1.2 Cliques vs. Complete Subgraphs

Here we intend to clarify some graph theoretical terms. A complete graph on v vertices contains (undirected) edges, connecting all the $\binom{v}{2} = v(v-1)/2$ vertex-pairs: that is, in a complete graph, each pair of vertices are connected by an edge.

If we have a graph G on n vertices, we can look for the complete subgraphs H of G: all the vertices and the edges of H need to be vertices and edges of G (i.e., H is a subgraph of G), and, moreover, H needs to be a complete graph.

The complete subgraph of the maximum vertex-number of G is called a clique. The clique number of graph G, $\omega(G)$, is the number of vertices in the largest complete subgraph of G. Computing the clique number $\omega(G)$ is a well-known hard problem: it is NP-hard [10], that is, it is not probable that one could find a fast (i.e., polynomial-time) algorithm for computing $\omega(G)$. Moreover, in general, not only the exact value of $\omega(G)$ is hard to compute, but it is also very difficult to approximate, even roughly [14]. In special cases, however, when the number of the vertices is only several hundred, and the graph is not too dense, that is, it has not too many edges, then all the frequently appearing complete subgraphs can be computed relatively quickly by the *apriori* algorithm [1,2]. The computational details are given in the Materials and Methods section.

Our goal in the present contribution is to map the frequently appearing complete subgraphs in human connectomes. We need to make clear that our analysis is done on 463-vertex braingraphs. Therefore, if a complete subgraph is found, it does not imply the neuronal level existence of complete subgraphs. It implies, however, that the macroscopic ROIs, corresponding to the vertices of the complete graphs discovered, are connected densely to each another, probably even on the neuronal level.

In the literature one may find numerous references to the "rich club property" of some networks, related to the braingraph [3,40]. Here we prefer using classical graph theoretical terms instead of this "rich club property", consequently, we intend to map those densely connected subgraphs of the human connectomes, which form complete graphs, and appear in at least the 80% of the all braingraphs considered.

2   Discussion and Results

First we review the frequent complete subgraphs of the human braingraph, next we analyze the significant differences in their frequencies in males and in females.

2.1   Frequent Complete Subgraphs of the Human Connectome

Supporting Table S1 contains the complete subgraphs of the human connectomes appearing in at least 80% of the graphs of the 414 subjects examined. In each row, the vertices of the complete subgraphs are listed, together with their frequencies of appearance. Note, that the vertices of a complete graph uniquely determine its edges. The list is redundant in the following sense: if a k-vertex complete graph has frequency at least 80%, then all of its complete subgraphs are also listed. We find that this redundancy helps in the analysis of the results, as it will be clear from what follows.

We would like to emphasize the following very simple, but powerful fact: If a given subgraph U has a frequency, say $\ell\%$, then all subgraphs of U has frequency at least $\ell\%$. This is the central point in the apriori algorithm [1,2], and it was noted and applied in [7,8].

The ROIs in Table S1 carry the names of the resolution-250 parcellation labels (where the number 250 refers to the approximate number of vertices in each hemisphere; the graphs of resolution-250 contain 465 vertices, not just 250), based on the Lausanne 2008 brain atlas [12] and computed by using FreeSurfer [9] and CMTK [5,11], given at https://github.com/LTS5/cmp_nipype/blob/master/cmtklib/data/parcellation/lausanne2008/ParcellationLausanne2008.xls. The "lh" and the "rh" prefixes abbreviate the "left-hemisphere" and "right-hemisphere" terms of localizations.

2.2   Complete Subgraphs Appearing in Each Subject

Here we list the maximal complete subgraphs from supporting Table S1, which are present in all of the braingraphs, and contains at least three nodes:

L1: (Left-Caudate)(Left-Pallidum)(Left-Putamen)(Left-Thalamus-Proper),
L2: (Left-Hippocampus)(Left-Putamen)(Left-Thalamus-Proper),
L3: (Left-Putamen)(Left-Thalamus-Proper)(lh.insula_1)
R1: (Right-Caudate)(Right-Pallidum)(Right-Putamen)(Right-Thalamus-Proper)
R2: (Right-Hippocampus)(Right-Putamen)(Right-Thalamus-Proper)
R3: (Right-Putamen)(Right-Thalamus-Proper)(rh.insula_2)
R4: (rh.superiorfrontal_7)(rh.superiorfrontal_8)(rh.superiorfrontal_9)

Note that L1 and L2 correspond to R1 and R2, and L3 almost corresponds to R3. Complete graph R4 has no correspondence in the left hemisphere (which are present in each subject), but in the left hemisphere, the superiorfrontal regions are also connected densely, as one can verify easily from Table S1.

We believe that the connections between the above-listed areas are very strong in each subject: so strong that they are not affected by measurement errors and individual variability.

2.3   The Largest Frequent Complete Subgraphs

The largest complete subgraphs, which are present in at least the 80% of the subjects, have seven vertices, and they are located in the left hemisphere. The first one connects the left putamen with six vertices in the left frontal lobe (B1), the second one connects the left caudate and the left putamen ROIs to five left frontal areas (B2):

B1: (Left-Putamen) (lh.lateralorbitofrontal_4) (lh.lateralorbitofrontal_6) (lh.lateralorbitofrontal_7) (lh.parstriangularis_3) (lh.rostralmiddlefrontal_12) (lh.rostralmiddlefrontal_9)

B2: (Left-Caudate)   (Left-Putamen)   (lh.lateralorbitofrontal_7)   (lh.medialorbitofrontal_2) (lh.rostralanteriorcingulate_1) (lh.rostralmiddlefrontal_12) (lh.rostralmiddlefrontal_9)

There are 48 different 6-vertex complete subgraphs, which are present in at least 80% of the connectomes. Only 6 of these are situated in the right hemisphere, the other 42 are in the left hemisphere.

2.4   Complete Subgraphs Across the Hemispheres

Since the neural fiber tracts, connecting the two hemispheres of the brain, are very dense in the corpus callosum, their tractography in the diffusion MR images is difficult since the fiber-crossings cannot always be tracked reliably [15,26].

We have found only relatively few frequent complete subgraphs of the human connectome, which have nodes from both hemispheres. Here we list those, which are present in more than 80% of the braingraphs studied; therefore, they are most probably not false positives. Again, we are listing only the maximal complete subgraphs for clarity. We note that most ROIs in the list are the parts of the striatum: each complete subgraph contains either a caudate nucleus or a nucleus accumbens of either the right- or the left hemisphere:

A1: (Left-Accumbens-area)(Left-Caudate)(Left-Thalamus-Proper)(Right-Caudate)
A2: (Left-Accumbens-area)(Left-Caudate)(Right-Caudate)(lh.rostralanteriorcingulate_1)
A3: (Left-Accumbens-area)(Left-Thalamus-Proper)(Right-Thalamus-Proper)
A4: (Left-Caudate)(Left-Thalamus-Proper)(Right-Caudate)(Right-Thalamus-Proper)
A5: (Left-Caudate)(Right-Caudate)(lh.caudalanteriorcingulate_1)(lh.caudalanteriorcingulate_2)
A6: (Left-Caudate)(Right-Caudate)(lh.caudalanteriorcingulate_1)(lh.rostralanteriorcingulate_1)
A7: (Left-Caudate)(Right-Caudate)(rh.caudalanteriorcingulate_1)
A8: (Left-Caudate)(Right-Caudate)(rh.rostralanteriorcingulate_2)
A9: (Left-Thalamus-Proper)(Right-Accumbens-area)(Right-Thalamus-Proper)

2.5   Counts of the Hippocampus, Thalamus, Putamen, Pallidum and the Amygdala in the Frequent Complete Subgraphs

In this section we count the appearances of certain ROIs in the frequent complete subgraphs, with a frequency threshold of 80%. Our results show that there are considerable differences between the hemispheres in these numbers: The right hippocampus and the right amygdala are present in much more complete subgraphs than the left ones; the left thalamus-proper, the left putamen and the left pallidum are present in much more complete subgraphs than the right ones (Table 1).

Table 1. The number of appearances of the hippocampus, the amygdala, the thalamus-proper, the putamen and the pallidum in the frequent complete subgraphs, with a frequency threshold of 80%, in each hemisphere. The right hippocampus and the right amygdala are present in much more complete subgraphs than the left ones; the left thalamus-proper, the left putamen and the left pallidum are present in much more complete subgraphs than the right ones.

	Left	Right
Hippocampus	187	247
Amygdala	66	99
Thalamus-proper	265	175
Putamen	1041	673
Pallidum	149	123

2.6   Sex Differences

Mapping sex differences in the human connectome is a hot and fast-developing area of research. In our earlier works we have shown - first in the literature - that in numerous well-defined graph theoretical parameters, women have "better connected" braingraphs than men [34–36]. In the work [7] we have mapped the frequent subgraphs of the human brain of at most 6 vertices, and have found sex differences: there are numerous frequent connected subgraphs, which are more frequent in men than in women, and, similarly, which are more frequent in men than in women. In the study of [8], we have mapped the neighbor-sets of the human hippocampus and found also significant sex differences in these sets.

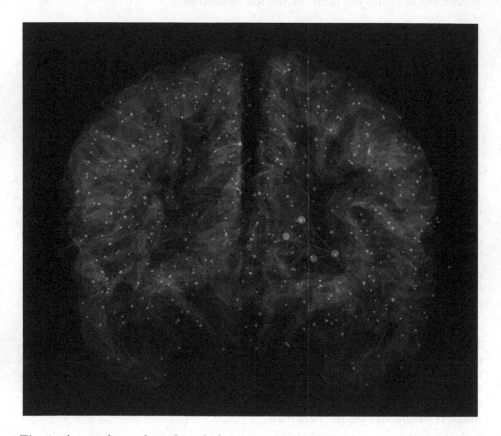

Fig. 1. A complete subgraph with four vertices, which is present in all 414 human connectomes examined. The vertices of the depicted subgraph are right-caudate, right-pallidum, right-putamen, right-thalamus-proper. The supporting Table S1 contains all the complete subgraphs with frequency of at least 80%, Table S2 contains the complete subgraphs, where the frequency of their appearance in females is significantly higher (p = 0.01) than in males; Table S3 contains those, where the frequency is significantly higher in males than in females. The supporting tables are available at http://uratim. com/cliques/tables.zip.

Here we compare the frequencies of the complete subgraphs of the connectomes of men and women. We have found significant differences in the frequencies of some complete subgraphs, with the advantage at men and women, too.

We have found much more complete subgraphs with significantly higher frequency in men than in women. More exactly, Supporting Table S2 lists 224 complete subgraphs, with significantly higher frequency in females than in males, while Table S3 lists 812 complete subgraphs, where their frequencies in males were higher than in females (with $p = 0.01$, and the inclusion threshold was a minimum 80% for the larger frequency).

This observation, in a sense, shows that men's connectomes show less interpersonal variability in complete subgraphs than those of women. This observation is in contrast with our findings in [7], where we have shown that women have much more 6-vertex frequent subgraphs than men: but in [7] we required connectedness, and not completeness.

3   Materials and Methods

3.1   The Data Source and the Graph Computation

The data source of the present study is the website of the Human Connectome Project at the address http://www.humanconnectome.org/documentation/S500 [24]. The dataset contains the HARDI MRI data of healthy human young adults between the ages of 22 and 35 years.

The CMTK toolkit [5], together with the FreeSurfer tool and the MRtrix tractography processing tool [39] were applied in the graph generation. In the MRtrix tool, we have applied randomized seeding and deterministic streamline method, with 1 million streamlines. We have studied here graphs with 463-vertex resolution. The parcellation data is given in the CMTK nypipe GitHub repository https://github.com/LTS5/cmp_nipype/blob/master/cmtklib/data/parcellation/lausanne2008/ParcellationLausanne2008.xls.

Further particularities of the graph processing workflow are described in [18], where the http://braingraph.org repository is also given. The braingraphs, analyzed here, can be accessed at the https://braingraph.org/cms/download-pit-group-connectomes/ site, choosing the "Full set, 413 brains, 1 million streamlines" option.

3.2   The Algorithm

In general, in a graph G, finding the size of the largest complete subgraph, called the *clique-number*, and denoted by $\omega(G)$, is a hard problem: it is NP-hard [10]. Naturally, finding the largest complete subgraph itself cannot be easier than finding its size $\omega(G)$; therefore, it is also hard.

Finding the largest complete subgraphs in sparse graphs (i.e., graphs with relatively few edges, compared to the number of its vertices) is usually not a very difficult task, since in these graphs, regularly, there are not too many large complete subgraphs. Finding only the frequently appearing complete subgraphs further simplifies the computational tasks, and we can apply an algorithm, which resembles in many points to the apriori algorithm for finding frequent item sets [1,2], and this algorithm is very fast in the practice.

Now we describe the algorithm: A frequent complete subgraph is characterized by the list of its vertices and the set of its edges. At the beginning, for an (undirected) edge (v_i, v_j), let these two lists be given as $([v_i, v_j], \{(v_i, v_j)\})$, where $i < j$.

In general, the vertices of the complete subgraphs are listed in the increasing order of their indices, and the vertices of each edge are listed also in the increasing order of its indices; otherwise, the particular order of the edges is indifferent, since they are elements of an unordered set, and they are stored also as a set.

Now we describe the generating, "apriori" step. Let

$$L_1 = ([v_1, v_2, ..., v_k], \{(v_1, v_2), (v_1, v_3), ...(v_{k-1}, v_k)\})$$

and

$$L_2 = ([u_1, u_2, ..., u_k], \{(u_1, u_2), (u_1, u_3), ...(u_{k-1}, u_k)\})$$

be two frequent complete subgraphs of size k. If the first $k - 1$ vertices of L_1 and L_2 are the same, and the last ones differ, we will consider generating a new, $k + 1$-vertex complete graph, as follows: if $v_1 = u_1, v_2 = u2, ...v_{k-1} = u_{k-1}$ and $v_k \neq u_k$, then, by the notation $v_{k+1} = u_k$, we verify the suitable frequency of the complete graph $L = ([v_1, v_2, ...v_k, v_{k+1}], \{(v_1, v_2), (v_1, v_3), ...(v_k, v_{k+1})\})$.

It is easy to see that in the edge list only the last one, (v_k, v_{k+1}) is new, all the others are already the edges of the frequent subgraphs L_1 or L_2.

In generating L one needs to make sure that the vertices in the vertex-list are ordered by their indices, and that the frequency of L is above the inclusion threshold.

The apriori generating step is correct, since if L is frequent, then both L_1 and L_2 were frequent. Additionally, every $k + 1$-vertex complete graph is generated only once, since the vertices are in increasing order: $(v_1, v_2, \ldots, v_{k+1})$ can be generated only from $(v_1, ...v_k)$ and $(v_1, ...v_{k-1}, v_{k+1})$.

4 Conclusions

By an apriori-like algorithm, we have mapped the frequent ($>80\%$) complete subgraphs of 414 subjects, each with 463 vertices. The largest frequent complete subgraph has 7 vertices. Most of the largest frequent subgraphs are located in the left hemisphere. We have also identified the frequent complete subgraphs, containing vertices from both hemispheres, and identified complete subgraphs with significant frequency-differences between the sexes. We have found that men have much more frequent complete subgraphs than women: this result contrasts our earlier finding [34], where we have shown that women have much better connectivity-related parameters in their connectomes than men.

5 Data Availability

The data source of this work is published at the Human Connectome Project's website at http://www.humanconnectome.org/documentation/S500 [24]. The parcellation data, containing the anatomically labeled ROIs, is listed in the CMTK nypipe GitHub repository https://github.com/LTS5/cmp_nipype/blob/master/cmtklib/data/parcellation/lausanne2008/ParcellationLausanne2008.xls.

The braingraphs, computed by us, can be accessed at the https://braingraph.org/cms/download-pit-group-connectomes/ site, by choosing the "Full set, 413 brains, 1 million streamlines" option. Here we have used exclusively the 463-node graphs.

The Supplementary Tables are available on-line at the address http://uratim.com/cliques/tables.zip. Table S1 contains the list of all the complete subgraphs of 414 human connectomes with a minimum frequency of 80%. Table S2 contains the complete subgraphs, where the frequency of their appearance in females is significantly higher ($p = 0.01$) than in males; Table S3 contains those, where the frequency is significantly higher in males than in females. In both Tables S2 and S3 a frequency cut-off 80% is applied to the larger frequency of the appearance in the sexes: only those significant differences are listed, where the larger of the frequencies of males and females are at least 80%.

Acknowledgments. Data were provided in part by the Human Connectome Project, WU-Minn Consortium (Principal Investigators: David Van Essen and Kamil Ugurbil; 1U54MH091657) funded by the 16 NIH Institutes and Centers that support the NIH Blueprint for Neuroscience Research; and by the McDonnell Center for Systems Neuroscience at Washington University. VG and BV were partially supported by the VEKOP-2.3.2-16-2017-00014 program, supported by the European Union and the State of Hungary, co-financed by the European Regional Development Fund, VG and MF by the NKFI-126472 and NKFI-127909 grants of the National Research, Development and Innovation Office of Hungary. BV and MF was supported in part by the EFOP-3.6.3-VEKOP-16-2017-00002 grant, supported by the European Union, co-financed by the European Social Fund.

References

1. Agrawal, R., Imielinski, T., Swami, A.N.: Mining association rules between sets of items in large databases. In: Buneman, P., Jajodia, S. (eds.) Proceedings of the 1993 ACM SIGMOD International Conference on Management of Data, Washington, D.C., 26–28 May 1993, pp. 207–216. ACM Press (1993)
2. Agrawal, R., Srikant, R.: Fast algorithms for mining association rules in large databases. In: Bocca, J.B., Jarke, M., Zaniolo, C. (eds.) Proceedings of the 20th International Conference on Very Large Data Bases, VLDB 1994, vol. 1215, pp. 487–499. Kaufmann Publishers Inc. (1994)
3. Ball, G., et al.: Rich-club organization of the newborn human brain. Proc. Natl. Acad. Sci. USA **111**(20), 7456–7461 (2014). https://doi.org/10.1073/pnas.1324118111
4. Chudnovsky, M., Robertson, N., Seymour, P., Thomas, R.: The strong perfect graph theorem. Ann. Math. **164**(1), 51–229 (2006)

5. Daducci, A., et al.: The connectome mapper: an open-source processing pipeline to map connectomes with MRI. PLoS ONE **7**(12), e48121 (2012). https://doi.org/10.1371/journal.pone.0048121

6. Euler, L.: Solutio problematis ad geometriam situs pertinentis. Commentarii Academiae Scientarum Imperialis Petropolitanae **8**(1), 128–140 (1741). http://eulerarchive.maa.org//docs/originals/E053.pdf

7. Fellner, M., Varga, B., Grolmusz, V.: The frequent subgraphs of the connectome of the human brain. arXiv preprint arXiv:1711.11314 (2017)

8. Fellner, M., Varga, B., Grolmusz, V.: The frequent network neighborhood mapping of the human hippocampus shows much more frequent neighbor sets in males than in females. arXiv preprint arXiv:1811.07423 (2018)

9. Fischl, B.: FreeSurfer. Neuroimage **62**(2), 774–781 (2012)

10. Garey, M.R., Johnson, D.S.: Computers and Intractability: A Guide to the Theory of NP-Completeness. W.H Freeman, Holtzbrinck (1979)

11. Gerhard, S., Daducci, A., Lemkaddem, A., Meuli, R., Thiran, J.P., Hagmann, P.: The connectome viewer toolkit: an open source framework to manage, analyze, and visualize connectomes. Front Neuroinform **5**, 3 (2011). https://doi.org/10.3389/fninf.2011.00003

12. Hagmann, P., et al.: Mapping the structural core of human cerebral cortex. PLoS Biol. **6**(7), e159 (2008). https://doi.org/10.1371/journal.pbio.0060159

13. Hagmann, P., Grant, P.E., Fair, D.A.: MR connectomics: a conceptual framework for studying the developing brain. Front Syst. Neurosci. **6**, 43 (2012). https://doi.org/10.3389/fnsys.2012.00043

14. Håstad, J.: Clique is hard to approximate within $n^{1-epsilon}$. In: 37th Annual Symposium on Foundations of Computer Science, FOCS 1996, Burlington, Vermont, USA, 14–16 October 1996, pp. 627–636. IEEE Computer Society (1996). https://doi.org/10.1109/SFCS.1996.548522

15. Hofer, S., Frahm, J.: Topography of the human corpus callosum revisited-comprehensive fiber tractography using diffusion tensor magnetic resonance imaging. NeuroImage **32**, 989–994 (2006). https://doi.org/10.1016/j.neuroimage.2006.05.044

16. Jbabdi, S., Johansen-Berg, H.: Tractography: where do we go from here? Brain Connect. **1**(3), 169–183 (2011). https://doi.org/10.1089/brain.2011.0033

17. Kerepesi, C., Szalkai, B., Varga, B., Grolmusz, V.: How to direct the edges of the connectomes: dynamics of the consensus connectomes and the development of the connections in the human brain. PLOS One **11**(6), e0158680 (2016). https://doi.org/10.1371/journal.pone.0158680

18. Kerepesi, C., Szalkai, B., Varga, B., Grolmusz, V.: The braingraph. org database of high resolution structural connectomes and the brain graph tools. Cogn. Neurodyn. **11**(5), 483–486 (2017)

19. Kerepesi, C., Szalkai, B., Varga, B., Grolmusz, V.: Comparative connectomics: mapping the inter-individual variability of connections within the regions of the human brain. Neurosci. Lett. **662**(1), 17–21 (2018). https://doi.org/10.1016/j.neulet.2017.10.003

20. Kerepesi, C., Varga, B., Szalkai, B., Grolmusz, V.: The dorsal striatum and the dynamics of the consensus connectomes in the frontal lobe of the human brain. Neurosci. Lett. **673**, 51–55 (2018). https://doi.org/10.1016/j.neulet.2018.02.052

21. Leighton, F.T.: Introduction to parallel algorithms and architectures: Arrays, Trees, Hypercubes. Elsevier, Amsterdam (1992)

22. Lichtman, J.W., Livet, J., Sanes, J.R.: A technicolour approach to the connectome. Nat. Rev. Neurosci. **9**(6), 417–422 (2008). https://doi.org/10.1038/nrn2391

23. Mangin, J.F., Fillard, P., Cointepas, Y., Le Bihan, D., Frouin, V., Poupon, C.: Toward global tractography. Neuroimage **80**, 290–296 (2013). https://doi.org/10.1016/j.neuroimage.2013.04.009

24. McNab, J.A., et al.: The human connectome project and beyond: initial applications of 300 mT/m gradients. Neuroimage **80**, 234–245 (2013). https://doi.org/10.1016/j.neuroimage.2013.05.074

25. Ohyama, T., et al.: A multilevel multimodal circuit enhances action selection in Drosophila. Nature **520**, 633–639 (2015). https://doi.org/10.1038/nature14297

26. Reginold, W., et al.: Tractography at 3T MRI of corpus callosum tracts crossing white matter hyperintensities. AJNR. Am. J. Neuroradiol. **37**, 1617–1622 (2016). https://doi.org/10.3174/ajnr.A4788

27. Ryan, K., Lu, Z., Meinertzhagen, I.A.: The CNS connectome of a tadpole larva of Cona intestinalis (L.) highlights sidedness in the brain of a chordate sibling. ELife **5**, 16962 (2016). https://doi.org/10.7554/eLife

28. Seung, H.S.: Reading the book of memory: sparse sampling versus dense mapping of connectomes. Neuron **62**(1), 17–29 (2009). https://doi.org/10.1016/j.neuron.2009.03.020

29. Sizemore, A.E., Giusti, C., Kahn, A., Vettel, J.M., Betzel, R.F., Bassett, D.S.: Cliques and cavities in the human connectome. J. Comput. Neurosci. **44**(1), 115–145 (2018)

30. Sporns, O., Tononi, G., Kötter, R.: The human connectome: a structural description of the human brain. PLoS Comput. Biol. **1**(4), e42 (2005). https://doi.org/10.1371/journal.pcbi.0010042

31. Szalkai, B., Kerepesi, C., Varga, B., Grolmusz, V.: The budapest reference connectome server v2. 0. Neurosci. Lett. **595**, 60–62 (2015)

32. Szalkai, B., Kerepesi, C., Varga, B., Grolmusz, V.: High-resolution directed human connectomes and the consensus connectome dynamics. arXiv:1609.09036, September 2016

33. Szalkai, B., Kerepesi, C., Varga, B., Grolmusz, V.: Parameterizable consensus connectomes from the human connectome project: the budapest reference connectome server v3.0. Cognit. Neurodyn. **11**(1), 113–116 (2017). https://doi.org/10.1007/s11571-016-9407-z

34. Szalkai, B., Varga, B., Grolmusz, V.: Graph theoretical analysis reveals: women's brains are better connected than men's. PLoS One **10**(7), e0130045 (2015). https://doi.org/10.1371/journal.pone.0130045

35. Szalkai, B., Varga, B., Grolmusz, V.: The graph of our mind. arXiv preprint arXiv:1603.00904 (2016)

36. Szalkai, B., Varga, B., Grolmusz, V.: Brain size bias-compensated graph-theoretical parameters are also better in women's connectomes. Brain Imag. Behav. (2017). https://doi.org/10.1007/s11682-017-9720-0, also in arXiv preprint arXiv:1512.01156

37. Szalkai, B., Varga, B., Grolmusz, V.: The robustness and the doubly-preferential attachment simulation of the consensus connectome dynamics of the human brain. Scientific Reports **7**, 16118 (2017). https://doi.org/10.1038/s41598-017-16326-0

38. Szemeredi, E.: Regular Partitions of Graphs. In: Colloq. Internat. CNRS, Univ. Orsay, Orsay, 1976. vol. 260. CNRS (1975)

39. Tournier, J., Calamante, F., Connelly, A., et al.: Mrtrix: diffusion tractography in crossing fiber regions. Int. J. Imaging Syst. Technol. **22**(1), 53–66 (2012)

40. van den Heuvel, M.P., Sporns, O.: Rich-club organization of the human connectome. J. Neurosci. **31**(44), 15775–15786 (2011). https://doi.org/10.1523/JNEUROSCI.3539-11.2011

920   M. Fellner et al.

41. White, J., Southgate, E., Thomson, J., Brenner, S.: The structure of the nervous system of the nematode caenorhabditis elegans: the mind of a worm. Phil. Trans. R. Soc. Lond. **314**, 1–340 (1986)
42. Zheng, Z., et al.: A complete electron microscopy volume of the brain of adult drosophila melanogaster. Cell **174**, 730–743.e22 (2018). https://doi.org/10.1016/j.cell.2018.06.019

Author Index

Printed in the United States
By Bookmasters